REMEDIATION OF HEAVY METALS IN THE ENVIRONMENT

REMEDIATION OF HEAVY METALS IN THE ENVIRONMENT

EDITED BY

JIAPING PAUL CHEN

LAWRENCE K. WANG

MU-HAO SUNG WANG

YUNG-TSE HUNG

NAZIH K. SHAMMAS

CRC Press
Taylor & Francis Group
Boca Raton London New York

CRC Press is an imprint of the
Taylor & Francis Group, an **informa** business

CRC Press
Taylor & Francis Group
6000 Broken Sound Parkway NW, Suite 300
Boca Raton, FL 33487-2742

© 2017 by Taylor & Francis Group, LLC
CRC Press is an imprint of Taylor & Francis Group, an Informa business

No claim to original U.S. Government works

Printed on acid-free paper
Version Date: 20161021

International Standard Book Number-13: 978-1-4665-1001-2 (Hardback)

Library of Congress Cataloging-in-Publication Data

Names: Chen, Jiaping Paul, editor. | Wang, Lawrence K., editor. | Hung,
Yung-Tse, editor. | Shammas, Nazih K., editor. | Wang, Mu Hao Sung, 1942-
editor.
Title: Remediation of heavy metals in the environment / edited by Jiaping
Paul Chen, Lawrence K. Wang, Yung-Tse Hung, Nazih K. Shammas, and Mu-Hao
S. Wang.
Description: Boca Raton : CRC Press, 2016. | Series: Advances in industrial
and hazardous wastes treatment | Includes bibliographical references and
index.
Identifiers: LCCN 2016042348 | ISBN 9781466510012 (hardcover : alk. paper)
Subjects: LCSH: Heavy metals--Environmental aspects.
Classification: LCC TD196.M4 R46 2016 | DDC 628.5/2--dc23
LC record available at https://lccn.loc.gov/2016042348

Visit the Taylor & Francis Web site at
http://www.taylorandfrancis.com

and the CRC Press Web site at
http://www.crcpress.com

Printed and bound in the United States of America by Sheridan

Contents

Preface

Any element that has molecular weight of above 40 is defined as heavy metal. A heavy metal should have such basic properties as electro- and thermal-conductivity when it is in its elemental form. It is appropriate to use the phrase "double-edged sword" to describe the importance and environmental risk of heavy metals, as they are useful resources and can cause harmful effects to humans.

It is well noted that some heavy metals are essential to living microorganisms, plants, animals, and humans. Their absence may negatively affect the growth and functionality of living creations. On the other hand, almost all heavy metals play key roles in many industrial production processes; some of the products are used every day, such as LCD monitors and smart phones.

However, the presence of any heavy metal in its ionic form in excessive quantities is harmful to human beings. One important characteristic of heavy metals is that they are nonbiodegradable, which makes it more difficult to decontaminate them from wastewater and contaminated sites. Therefore, it is desirable to measure, understand, and control the heavy metal concentrations in the environment.

The environmental consequences of heavy metals usually cannot be immediately recognized and handled as the risks are chronic and not acute. Some historical tragedies include Minamata disease resulting in severe mercury poisoning and Gulf War syndrome mainly due to depleted uranium. Great effort has been made for the clean-up of contaminated sites and treatment of wastewater. However, more work has to be done in order to have a heavy-metal-risk-free society. For example, some water utilities companies still use lead water pipes, which may leach lead into tap water. Another example is that several extremely toxic heavy metals are still used or present in the production of electronic products or recycled chemicals.

Recently, high levels of lead in tap water in several cities in the United States have raised great concern among the public. For example, 6000 to 12,000 children in Flint, Michigan, have been exposed to high levels of lead in the water supply, which is 13,000 times the lead concentration found in nearby areas for months, without any official warning. This so-called Flint water crisis may require more than 200 million USD for medical care, infrastructure upgrades, and replacement of lead pipes.

The key questions facing us are *What is happening? How can we avoid such? Can we do something before accidents happen?* It is rather important for environmental professionals, government officials, educators, and the public to have updated knowledge and experience in heavy metals in the environment for awareness, management, and remediation.

We are happy to work with Taylor & Francis and CRC Press to develop a book series of *Industrial and Hazardous Wastes Treatment*, one of which is *Remediation of Heavy Metals in the Environment*, contributed by a group of environmental scientists, engineers, and educators from several countries in the world who are experts in the relevant subjects of studies. Since the area of heavy metal in the environment is rather broad, collective contributions are selected to better represent the most up-to-date and complete knowledge.

Remediation of Heavy Metals in the Environment covers most recently updated information on heavy metals. Chapter 1 addresses toxicity, sources, and treatment of key heavy metals such as copper, nickel, and zinc. Nanotechnology for bioremediation is described in Chapter 2. Chapter 4 provides a detailed description of technologies for remediation of heavy metal contaminated soils. A series of low-cost adsorbents is presented in Chapter 5. Treatment of metal finishing wastes is given in Chapters 6 and 11. Stabilization of cadmium in waste incineration residues is described in Chapter 7. Treatment technologies of arsenic and chromium are discussed in Chapters 8, 9, and 12. E-waste is an emerging environmental problem; its disposal and recycling are described in Chapter 10. Finally, treatment technologies of photographic processing waste and barium containing wastewater are discussed in Chapters 13 and 14.

This book can be used as a reference book for environmental professionals for learning and practice. Readers in environmental, civil, chemical, and public health engineering and science as well as governmental agencies and non-governmental organizations will find valuable information from this book to trace, follow, duplicate, or improve on a specific industrial hazardous waste treatment practice as well as manage the currently existing systems.

The editorial team and the authors would like to thank many people who have provided encouragement and support during the period when this book was prepared. Also, our family members, colleagues, and students have done a good job in supporting us during the writing of the text. Taylor & Francis senior editor Joseph Clements provided strong support to the team for years. Without all these people, the completion of this book would have been impossible.

Jiaping Paul Chen, Singapore
Lawrence K. Wang, New York
Mu-Hao Sung Wang, New York
Yung-Tse Hung, Ohio
Nazih K. Shammas, California

Editors

Jiaping Paul Chen has been a professor at the National University of Singapore since 1998. His research interests are water/wastewater treatment and modeling. He has published 3 books, more than 100 journal papers, and book chapters with more than 4600 citations and 39 H-index. He has been an inventor of seven technology patents, an awardee of the Sustainable Technology Award (IChemE), guest professor of the Huazhong University of Science and Technology and Shandong University, and Distinguished Overseas Chinese Young Scholar of NNSF. He has been recognized as an author of highly cited papers of ISI Web of Knowledge. Dr. Chen earned an MEng at Tsinghua University, China, and a PhD at the Georgia Institute of Technology, USA.

Lawrence K. Wang has over 35 years of experience in facility design, construction, plant operation, and management. He has expertise in water supply, air pollution control, solid waste disposal, water resources, waste treatment, and hazardous waste management. He is a retired dean/director/vice president of the Lenox Institute of Water Technology, Krofta Engineering Corporation, and Zorex Corporation, respectively, in the United States. Dr. Wang is the author of more than 700 papers and 44 books, and he is credited with 24 U.S. patents and 5 foreign patents. He earned a BSCE at National Cheng-Kung University, Taiwan, ROC, an MSCE at the Missouri University of Science and Technology, USA, an MS at the University of Rhode Island, USA, and a PhD at Rutgers University, USA.

Mu-Hao Sung Wang has been an engineer, an editor, and a professor serving private firms, governments, and universities in the United States and Taiwan for over 36 years. She is a licensed professional engineer and a Diplomate of the American Academy of Environmental Engineers. Her publications are in the areas of water quality, modeling, wastewater management, NPDES, flotation, and analytical methods. Dr. Wang is the author of more than 50 publications and an inventor of 14 US and foreign patents. She earned a BSCE at National Cheng Kung University, Taiwan, ROC, an MSCE at the University of Rhode Island, USA, and a PhD at Rutgers University, USA.

Yung-Tse Hung has been a professor of civil engineering at Cleveland State University since 1981. He is a Fellow of the American Society of Civil Engineers and has taught at 16 universities in 8 countries. His research interests and publications have been involved with biological processes and industrial waste treatment. Dr. Hung is credited with more than 470 publications and presentations on water/wastewater treatment. He earned a BSCE and an MSCE at National Cheng-Kung University, Taiwan, and a PhD at the University of Texas at Austin, USA. He is the editor of *International Journal of Environment and Waste Management*, *International Journal of Environmental Engineering*, and *International Journal of Environmental Engineering Science*.

Nazih K. Shammas is an environmental consultant and professor for over 45 years. He is an ex-dean/director of the Lenox Institute of Water Technology and an advisor to Krofta Engineering Corporation, USA. Dr. Shammas is the author of more than 250 publications and 15 books in the field of environmental engineering. He has experience in environmental planning, curriculum development, teaching, scholarly research, and expertise in water quality control, wastewater reclamation and reuse, physicochemical and biological processes, and water and wastewater systems. He earned a BE at the American University of Beirut, Lebanon, an MS at the University of North Carolina at Chapel Hill, and a PhD at the University of Michigan, USA.

Contributors

A. Olanrewaju Alade
Department of Chemical Engineering
Ladoke Akintola University of Technology
Ogbomoso, Nigeria

O. Sarafadeen Amuda
Department of Pure and Applied
    Chemistry
Ladoke Akintola University of
    Technology
Ogbomoso, Nigeria

Hamidi Abdul Aziz
School of Civil Engineering
Universiti Sains Malaysia
Pulau Pinang, Malaysia

Rajasekhar Balasubramanian
Department of Civil and Environmental
    Engineering
National University of Singapore
Singapore

Jiaping Paul Chen
Division of Environmental Science
    and Engineering
National University of Singapore
Singapore

Irvan Dahlan
School of Chemical Engineering
Universiti Sains Malaysia
Pulau Pinang, Malaysia

Haoran Dong
College of Environmental Science and
    Engineering
Key Laboratory of Environmental Biology
    and Pollution Control
Hunan University
Changsha, China

Miskiah Fadzilah Ghazali
School of Civil Engineering
Universiti Sains Malaysia
Pulau Pinang, Malaysia

R. Jamshidi Gohari
Advanced Membrane Technology Research
    Centre (AMTEC)
University Teknologi Malaysia
Johor, Malaysia
and
Department of Chemical Engineering
Islamic Azad University
Bardsir, Iran

P. Gopinath
Nanobiotechnology Laboratory, Centre for
    Nanotechnology
Indian Institute of Technology Roorkee
Uttarakhand, India

Xiaohong Guan
State Key Laboratory of Pollution Control
    and Resources Reuse
College of Environmental Science and
    Engineering
Tongji University
Shanghai, China

Sie-Tiong Ha
Universiti Tunku Abdul Rahman
Kampar, Perak, Malaysia

Joseph F. Hawumba
Biochemistry Department
Makerere University
Kampala, Uganda

Yung-Tse Hung
Department of Civil and Environmental
    Engineering
Cleveland State University
Cleveland, Ohio

Ahmad F. Ismail
Advanced Membrane Technology Research
    Centre (AMTEC)
University Teknologi Malaysia
Johor, Malaysia

Obulisamy Parthiba Karthikeyan
Department of Civil and Environmental
    Engineering
National University of Singapore
Singapore

Pei-Sin Keng
International Medical University
Kuala Lumpur, Malaysia

Milos Krofta
Lenox Institute of Water Technology
Lenox, Massachusetts

S. Raj Kumar
Nanobiotechnology Laboratory, Centre for
    Nanotechnology
Indian Institute of Technology Roorkee
Uttarakhand, India

Woei Jye Lau
Advanced Membrane Technology Research
    Centre (AMTEC)
University Teknologi Malaysia
Johor, Malaysia

Siew-Ling Lee
Ibnu Sina Institute for Fundamental Science
    Studies
Universiti Teknologi Malaysia
Johor, Malaysia

Takeshi Matsuura
Advanced Membrane Technology Research
    Centre (AMTEC)
University Teknologi Malaysia
Johor, Malaysia
and
Department of Chemical and Biological
    Engineering
University of Ottawa
Ottawa, Ontario, Canada

Siew-Teng Ong
Universiti Tunku Abdul Rahman
Kampar, Perak, Malaysia

Liu Qing
Institute of Urban Environment
Chinese Academy of Sciences
Xiamen, Fujian, China

Nazih K. Shammas
Lenox Institute of Water Technology
Pasadena, California

Kaimin Shih
Department of Civil Engineering
University of Hong Kong
Hong Kong, China

Minhua Su
Department of Civil Engineering
University of Hong Kong
Hong Kong, China

Lawrence K. Wang
Lenox Institute of Water Technology
Newtonville, New York

Mu-Hao Sung Wang
Lenox Institute of Water Technology
Newtonville, New York

Yu-Ming Zheng
Institute of Urban Environment
Chinese Academy of Sciences
Xiamen, Fujian, China

1 Toxicity, Sources, and Control of Copper (Cu), Zinc (Zn), Molybdenum (Mo), Silver (Ag), and Rare Earth Elements in the Environment

*O. Sarafadeen Amuda, A. Olanrewaju Alade, Yung-Tse Hung,
Lawrence K. Wang, and Mu-Hao Sung Wang*

CONTENTS

ABSTRACT

There are more than 20 heavy metal toxins contributing to a variety of adverse health effects in humans. Exposed individuals experience different behavioral, physiological, and cognitive changes depending on the type of the toxin and the degree of exposure by the individual. This chapter presents the sources of exposure, toxicity, and control technologies of Cu, Zn, Mo, Ag, and rare earth elements in the environment.

1.1   INTRODUCTION

Toxic substances are generally poisonous and cause adverse health effects in both man and animals. Some chemical substances can be of use at certain concentrations, beyond which they become toxic. The toxicity of a substance is based on the type of effect it causes and its potency. Exposure to such toxic substance is via inhalation, ingestion, or direct contact. Both long-term exposure (chronic) and short-term exposure (acute) may cause health effects that manifest immediately or later in life. The concentration of trace metals is increasing as a result of releases into the air and water as well as their heavy use in products for human consumption. The impact of these heavy metals at toxic concentration produces behavioral, physiological, and cognitive changes in an exposed individual. These impacts are well documented based on reports of accidental human exposure and animal studies (1–120). Most agencies that specialize in the study of toxicity of substances, which are either consumed or not consumed by human, terrestrial, and aquatic animals as well as plants, have set lower and upper allowable concentrations of the substances. Concentrations above the upper limit and beyond exposure time are toxic and exhibit health effects ranging from intestinal and neurotic to death.

In this chapter, the toxicity, sources, environmental issues (121–134), and specific control technologies (121–141) of selected heavy metals (Cu, Zn, Ag, and Mo) and rare earth elements (REEs) are discussed.

1.2 COPPER

1.2.1 COPPER AND ITS COMPOUNDS

Copper is a malleable light reddish-brown metallic element. It is represented by the symbol "Cu" and assigned with atomic number of 29 and atomic weight of 64 on the periodic table of elements (1,2). Copper occurs naturally in rock as a wide range of mineral deposits either in its pure or compound form. It has also been found in soils, water, and sediments (3,4). It can also be found in areas designated for municipal incineration, metal smelting sites (4), foundries, and power plants as a result human activities (1).

Copper sulfate and copper oxide are the most widely distributed naturally form of copper compound, Table 1.1, however, it combines with other metals like zinc and tin to form alloys such as brass and bronze, respectively (1,2).

1.2.2 CHARACTERISTICS OF COPPER

Copper is a solid metal at room temperature and possesses good electrical and thermal conductivity. It does not react with water but reacts slowly with oxygen present in the air to form a thin film of dark-brown copper oxide. It does not react in sulfide, ammonia, and chloride media. There are 29 identified isotopes of copper ranging from 52 to 80. Only two of these, ^{63}Cu and ^{65}Cu, are stable and occur naturally. ^{63}Cu shows a predominant existence (69%).

1.2.3 INDUSTRIAL PRODUCTION OF COPPER

About 35% of world's copper is produced in Chile, while 11% is produced in the United States and the remaining percentage comes from Indonesia, the former Soviet Union, Peru, Zambia, China, Poland, and the Democratic Republic of Congo (1). Copper ore is often extracted from large open pit mines as copper sulfides.

1.2.4 INDUSTRIAL APPLICATIONS OF COPPER

Copper is widely used in pure form or as an alloy, in the production of electrical conductors and wires, sheet metals, pipe and plumbing fixtures, coins, cooking utensils, and other metal products (3,5). Copper compounds have received wide application in the agricultural sector where they are being used as fungicides. It is also used in the treatment of water, particularly, to eliminate algae. Other applications include production of preservative lumbering, tannery, and textiles. Copper also

TABLE 1.1
Concentration and Distribution of Copper in Environment

Distribution	Concentration (ppm)
Earth's crust	50
Soil	2–250
Copper production	
Facilities	7,000
Plants (dry weight basis)	10

Source: US EPA. *Environmental Technology Verification Report Environmental Bio-detection Products Inc.* Toxi-chromotest. Washington DC: US Environmental Protection Agency, June, 2006. EPA/600/R-06/071 and NTIS PB2006-113524

form an essential component in ceramic, glaze, and glass works. In the form of Fehling's solution, copper compounds have application in chemistry for the determination of reducing sugars.

1.2.5 TOXICITY AND RELATED HAZARDS

Copper is an essential micronutrient that is required by plants and animals for growth and other body metabolisms. Higher concentrations of copper in both animals and plants are toxic and result in adverse health effects and stunted growth, respectively (7).

1.2.5.1 Route of Exposure

Environmental pollution due to copper is mostly anthropogenic ranging from mining, smelting, incineration, and water treatment processes, while pollution through natural origin is as a result of wind and rain erosion and through eruption of volcanoes (2), animals are expose to copper through inhalation of contaminated air, ingestion of contaminated water and food, and through skin contact with soil that is contaminated with copper (5). Exposure to copper can occur in plant through deposition on leaves and stem and through absorption of contaminated water in the soil.

1.2.5.2 Toxicity of Copper

The toxicity of copper can be traced to its ability to accept and donate single electrons while undergoing the change of oxidation state (8). The health effects of the acute ingestion of copper or copper compounds by man and animals include gastrointestinal ulcerations and bleeding, acute hemolysis and hemoglobinuria, hepatic necrosis, nephropathy, cardiotoxicity, tachycardia, and tachypnea. Other effects include dizziness, headache, convulsions, lethargy, stupor, and coma, all of which are central nervous system related effects.

Recent cases of accident and research-based studies of copper toxicity are reported in Table 1.2, based on the effect experienced by the victims.

Copper bioavailability in water is always higher than in other environmental media particularly in diets where it is a function of its solubility as well as the types of complexes in which it is present. These complexes often inhibit copper absorption (12). Chronic toxicity in human often results in liver and liver related diseases (Table 1.3), such as Wilson, hepatic, and renal diseases (13). Wilson disease impacts cases of acute toxicity of copper leading to liver disease (14).

Information on the studies of toxicity of copper in animals is sparsely available and these cover mainly physiological, biochemical, and pathological aspects of copper metabolism or chronic toxicity of copper in comparison to the copper concentration standard in human diets. As such the level of acute copper toxicity demonstrated in these animals cannot be adopted as standard for humans. The maximum contaminant level goal (MCLG) for copper is given as 1.3 mg/L (21,22); this concentration indicates gastro intestinal symptoms. The International Programme on Chemical Safety, (IPCS) (23), stipulated 2–3 mg Cu per day as the upper limit of acceptable range of copper intake and these values have received World Health Organization (WHO) acceptance (24) (Table 1.4).

TABLE 1.2
Health Effects of Chronic Toxicity of Copper in Humans and Animals

Health Effect	Victims	References
Acute hemolytic anemia	Humans; sheep	(13)
Cessation of menstruation and osteoarthritis	Humans	(14)
Neurological abnormalities	Rats	(15,16)
Prevention of embryogenesis	Women	(17,18)
Enhancement of endogenous		
Oxidative reaction leading to DNA damage	Humans	(19,20)

TABLE 1.3
Acute Copper Toxicosis Resulting from Oral Exposure (Ingestion)

Exposure Cases	Health Effect	Copper Exposure Measurement (mg/L)	References
43 individuals in single point source contact in hotel	Acute illness	4.0–70	(9)
5 drank water with over night build up of copper	Abdominal symptoms	>1.3	(10)
60 adult women of low socio-economics status	Gastrointestinal effect	≥3	
	No symptoms	>5	(11)

TABLE 1.4
Maximum Limits of Copper for Environmental Releases and Human Exposure

Medium	Individual	Concentration	Body Responsible
Lakes and streams	Aquatic organisms	1.0 ppm	US EPA
Drinking water	Humans	1.3 ppm	US EPA
Workroom air	Humans (workers)	0.2 mg/m^3 (copper fume)	OSHA
		1.0 mg/m^3 (copper dust)	OSHA
Workplace air	Workers	0.1 mg/m^3 (copper fumes)	NIOSH
		1.0 mg/m^3 (copper mist)	NIOSH
Dietary (RDA)	Adult	0.9 mg/day	NAIM
Dietary (RDA)	Lactating women	1.3 mg/day	NAIM
Dietary (RDA)	Children (0–3 years)	0.34 mg/day	NAIM
Dietary (RDA)	Children (4–8 years)	0.44 mg/day	NAIM

Source: The facts on copper, Dartmouth College, Hanover, NH, http://www.dartmouth.edu/, 2015.

Note: US EPA—US Environmental Protection Agency; OSHA—Occupational Safety and Health Administration; NIOSH—National Institute for Occupational Safety and Health; and NAIM—National Academics Institute of Medicine.

1.2.6 INTERACTION OF COPPER WITH OTHER ELEMENTS

Many divalent cations such as copper, cadmium, cobalt, lead, and zinc influence the synthesis of metallothionein as a result of their binding properties. However, this competition often leads to the physiological regulation of these elements in relation to the presence and concentration of the other (25). The effects of interaction of copper with other elements are as shown in Table 1.5.

1.3 ZINC

1.3.1 ZINC AND ITS COMPOUNDS

Zinc, one of the most common elements in the Earth's crust, is found in the air, soil, and water and is present in all foods. Zinc in its pure elemental (or metallic) form is a bluish-white, shiny metal. Zinc is commonly used in the industry to coat steel and iron as well as other metals to prevent rust and corrosion; this process is called galvanization. Metallic zinc, when mixed with other metals forms alloys such as brass and bronze. A zinc and copper alloy is used to make pennies in the United States. Metallic zinc is also used to make dry cell batteries (33).

Zinc can combine with other elements, such as chlorine, oxygen, and sulfur, to form zinc compounds such as zinc chloride, zinc oxide, zinc sulfate, and zinc sulfide. Most zinc ore found naturally

TABLE 1.5
Effect of Interaction of Copper with Other Essential Elements

Combination	Effects	References
Zinc and copper	Induction of intestinal metallothionein synthesis leading to poor systemic absorption of copper	(26)
	Reductions in erythrocyte superoxide dismutase in women	(27)
Molybdenum and copper	Decrease in copper uptake leading to copper utilization and toxicity	(28,14)
Ferrous iron and copper	Decrease in copper absorption in intestine	(12,29)
Stannous tin and copper	Decrease in copper absorption in the intestine	(12,30)
Selenium and copper	No significant hepatic and histological alterations in rats subjected to study	(31,32)

in the environment is present in the form of zinc sulfide. Zinc sulfide and zinc oxide are used to make white paints, ceramics, and other products. Zinc enters the air, water, and soil as a result of both natural processes and human activities. Zinc, in most cases enters the environment through mining, purifying of zinc, lead, and cadmium ores, steel production, coal burning, and burning of wastes. These activities can increase zinc levels in the environment. Waste streams from zinc and other metal manufacturing and zinc chemical industries, domestic waste water, and run off from soil containing zinc can discharge zinc into waterways. The level of zinc soil increases mainly from disposal of zinc wastes from metal manufacturing industries and coal ash from electric utilities (33). Zinc is present in the air mostly as fine dust particles, which eventually settles over land and water. Zinc in lakes or rivers may settle on the bottom, dissolve in water, or remain as fine suspended particles. Fish can ingest zinc in from the water and from their feeding habits. Depending on the type of soil, some zinc from hazardous waste sites may percolate into the soil and thus cause contamination of groundwater. Zinc may be ingested by animals through feeding or drinking of water containing zinc.

1.3.2 CHARACTERISTICS OF ZINC

Zinc is not found in its free state in nature but can be processed from its ore. This bluish-white element has melting and boiling points of 419.5°C and 908°C, respectively (34). It tends to form a covalent bond with sulfide and oxides (35) and show amphoteric characteristics (36). On exposure to air, it forms a coat of zinc oxide, which covers the underlying metal and gives it anti-corrosion properties. In anaerobic condition, it may form zinc sulfide. Zinc influences membrane stability in humans and plants and plays a role in the metabolism of proteins and nucleic acids (37).

1.3.3 PRODUCTION OF ZINC

Zinc is essentially produced from its ore excavated from both underground and open pits through an electrolytic process involving the leaching of zinc oxide, from calcined ore, with sulfuric acid, leading to the formation of zinc sulfate solution. The solution is then subjected to electrolysis after which zinc deposits are collected on cathode electrodes (38). About 90% of the zinc production comes from zinc sulfide, ZnS (sphalerite) (35). In 2001, world production of zinc was 8,850,000 metric tons and the United States contributed about 799,000 metric tons (39).

1.3.4 APPLICATION OF ZINC

Industrially, zinc is widely used as protective coating on metals such as iron and steel that are highly susceptible to corrosion. It is also used in the production of zinc-based alloys involving other metals such as aluminum, copper, titanium, and magnesium. In 2002, over 50% of the zinc produced in the United States was used for galvanizing, while about 20% was used for zinc-based alloys, and the remaining specifically went for the production of bronze and brass (40). Other applications of zinc include production of campaigned zinc which is used in a wide range of industries as an essential material for production (35,37,38).

1.3.5　Toxicity and Related Hazard

Zinc has been reported to play important roles in plants and humans, particularly in the metabolism of proteins and nucleic acids (37). It, however, affects human and animals when taken in higher concentration and for a prolonged time thus resulting in adverse health effects (33).

1.3.5.1　Route of Exposure

Principal human activities leading to the release of zinc and zinc compounds into the environment are zinc mining, purification, and decomposition. Run off from ore, production, and waste sites distribute zinc into water ways and over soil. Leaching of waste sites and other areas contaminated with zinc eventually contaminate underground water. Fine dust particles from the production sites are often bound to aerosols (41) and are later washed down by rain, snow, or wind onto land, water, and vegetation. In water, zinc is present in suspended form, dissolved form or bound to suspended matter (42). Human and animals are exposed to zinc and its compound through ingestion of food, water, and soil. Use of zinc-plated and zinc-based products such as paints and batteries are other sources of exposure in addition to occupation exposure which involves inhalation, food consumption, and skin contact. Zinc accumulates in aquatic organisms which invariably form human diets (33). Plant species, soil pH, and the composition of the soil greatly influence the accumulation of zinc in plants (43).

1.3.5.2　Toxicity

The exposure of animals and humans to acute concentration of zinc and its compounds often results in adverse health effects. The inhalation of a high concentration of zinc dust for a prolonged periods of time results in flu like symptoms, fever, sweating, headache, and subsequent weakness (44). Oral exposure to zinc often interferes with the essential body metabolism of copper and this may result in hematological and gastrointestinal effects as well as decrease in cholesterol levels in the body. Zinc is often absorbed in the small intestine and its uptake from a normal diet ranges from 26% to 33% when taken with food (45,46). Zinc in animal blood does not undergo metabolism, but interacts with protein or forms soluble chelating complexes. Recent cases of accident and research-based studies of zinc toxicity are reported in Table 1.6.

Generally ingestion of zinc at a high concentration causes decrease in cholesterol levels and copper metalloenzyme activity (51,52) and other health effects such as hematological gastrointestinal and immunotoxicity (53) (Table 1.7).

Inhalation of zinc, in the form of zinc oxide fumes or zinc chloride from the smoke of bombs, shows different adverse health effects including dryness and irritation of the throat, and other effects, which manifest after hours when exposure persists for 1 or 2 days (35) (Table 1.8).

Zinc is essentially needed in human nutrition; the recommended dietary allowance (RDA) is given in Table 1.9.

1.3.6　Interaction of Zinc with Other Elements

Metabolism leading to toxicity is often activated or deactivated by the presence of other elements in both plant and animal. For a particular element under study, many studies have shown the interaction between zinc and other metals to be of significant reaction. Few of these studies are summarized in Table 1.10.

TABLE 1.6

Zinc Toxicity Resulting from Oral Exposure

Exposure Cases	Health Effect	Zinc Exposure (mg/day)	References
21 men and 26 woman fed with zinc for 6 weeks	Abdominal cramps, nausea, and vomiting	2–15	(47,48)
31 men and 38 women fed with zinc for 1 year	Lower mean serum certainties, lower total serum protein, lower serum curare acid and higher mean corpuscular hemoglobin (Hb)	20–150	(49)
9 men and 11 women fed with zinc for 8 weeks	Increase in plasma zinc concentration and decrease in DNA oxidation	45	(50)

TABLE 1.7

Health Effect of Chronic Toxicity of Zinc in Animals

Health Effects	Species	References
Decrease in erythrocytes and Hb levels; total and differential leukocyte levels. Percentage increase in reticulocytes and polychromatophilic erythrocytes	13 males and 16 females and Wistar rats	(62)
Decrease in Hb level and serum capper. Increase in serum and tissue	7–8 male New Zealand white rabbits	(63)
Negative effect on retention of learned behavioral response	A group of 9–12 male and female Swiss mice	(64)
Increase in lavage fluid parameters	Hartley Guinea pigs and 344 Fischer rats	(57)
Distortion of chromosome structure of sperm	10 male Sprague-Dawley rats	(65)

TABLE 1.8

Zinc Toxicity Resulting from Inhalation Exposure

Exposure Case	Health Effect	Concentration of Zinc	References
Shipyard workers who sprayed zinc onto steel surfaces	Aches and pains, dyspnea, dry cough, lethargy, and fever	–	(54)
Workers exposed to zinc oxide fumes	Impaired lung function	–	(55,56)
4 adults exposed to zinc oxide fumes for 2 h	Chills, muscle/joint pain, chest tightness, dry throat, and headache	5 mg/m³	(57)
A group of 13 healthy nonsmoking individuals exposed to zinc oxide fumes for 2 h	Fatigue, muscle ache, and cough	0–5 mg/m³	(58)
20 Chinese workers exposed to zinc oxide over a single 8 h workday	No significant health effect detected or reported	0–36.3 mg/m³	(59)
13 soldiers exposed to zinc chloride smoke during combat exercise	Decrease in lung diffusion capacity, plasma level of fibrinogen elevated at 1–8 weeks postexposure	Unknown	(60)
3 patients exposed to zinc chloride for 1–5 min	Two died of edema, pulmonary sepsis, emphysematic changes, and necrosis. The third revealed severe restrictive pulmonary dysfunction	Unknown	(61)

TABLE 1.9
Recommended Dietary Allowance (RDA) Requirement of Zinc at Various Life Stages and Gender

Life Stage Group	RDA (mg/day)	
	Male	Female
0–12 months	≤3	≤3
1–3 years	3	3
4–8 years	5	5
9–13 years	8	8
14–18 years	11	9
19–50 years	11	8
Above 50 years	11	8
Pregnant women	–	11
Lactating women	–	12

Source: US EPA. *Toxicological Review of Zinc and Compound.* Washington, DC: US Environmental Protection Agency, 2005. EPA/635/R-05/002.

TABLE 1.10
Effects of Interaction of Zinc with Other Essential Elements

Combination	Effect	References
Copper and zinc	Induction of intestinal metallothionein synthesis leading to poor systemic absorption of copper	(67)
Calcium and zinc	No significant interference with absorption of zinc nor changes in hair or serum zinc	(68,69)
Iron and zinc	Significant lower percentage zinc absorption particularly in pregnant women	(70)
	Increased dietary iron intake result in diminished absorption of zinc	(71)
Cadmium and zinc	Likely decrease of toxicity and carcinogenicity of cadmium	(72,73)
	Inhibition of zinc absorption toxic at level of cadmium is possible	(44,68)
Lead and zinc	No significant evidence of possible interference of absorption of zinc by lead and vice versa	(74,75)
Cobalt and zinc	Study animals (rats) demonstrated protection against the testicular toxicity of cobalt in the presence of zinc	(76)

1.4 SILVER

1.4.1 Silver and Its Compounds

Silver is a ductile and white metallic element represented by the symbol Ag and assigned with atomic number 47 and atomic weight 247.8014 on the periodic table of the element. It is found in the environment mostly as silver sulfide (AgS) or in combination with other metals (77). Its primary source is the ore while other sources include new scrap generated in the manufacturing of silver-containing products (Table 1.1). The anthropogenic sources of silver in the environment include smelting operations, coal combustion, production and disposal of silver-based photographic and

TABLE 1.11

Maximum Concentration of Silver Distribution in the United States

Destination	Concentration	Sites	References
Air near smelter	36.5 ng/m^3	Idaho	(78)
Seawater	8.9 µg/L	Galveston	(79)
Soil	31 mg/kg	Idaho	(78)
Liver of marine mammals	1.5 mg/kg	–	(80)
Mushrooms	110 mg/kg	–	(81)

electrical materials, and cloud seeding (78). The larger percentage of the lost silver is immobilized in the form of minerals, metals, or alloys in the terrestrial ecosystem which serve as their destination (Table 1.11). About half of the emitted silver into the environment is precipitated some kilometers away from its point source (77).

1.4.2 CHARACTERISTICS OF SILVER

Silver is a solid metal at room temperature but occurs naturally in several oxidation states, which include Ag^0, Ag^+, Ag^{2+}, and Ag^{3+} (77) and forms compounds with sulfide, bicarbonate, and sulfate (77). Ag^{2+} and Ag^{3+} are more effective oxidizing agents than Ag^0 and Ag^+ but are relatively unstable in an aqueous environment with a temperature close to 100°C. Silver exists as silver sulfhydrate (AgSH) or as a polymer HS—Ag—S—Ag—SH at the lowest concentration in the aqueous phase. However, at higher concentration, it exists as colloidal silver sulfide or polysulfide complexes (82). Only two isotopes of silver, ^{107}Ag and ^{109}Ag are stable and exist naturally, the other 20 isotopes do not exist naturally. Some compounds of silver, like silver oxalate (AgC_2O_4), silver acetylide (Ag_2C_2) and silver azide ($Ag N_3$) are potential explosives.

1.4.3 PRODUCTION OF SILVER

The current world estimate of mine production of silver is given as 15.5 million kg (83) and their distribution is given in Table 1.12.

TABLE 1.12

World Major Mine Producers of Silver

Country	Percentage Production (%)
Mexico	17
USA	14
Peru	12
USSR (former)	10
Canada	9
Others	38

Source: Eisler R. *Silver Hazards to Fish, Wildlife and Invertebrates: A Synoptic Review.* Washington, DC: US Department of the Interior, National Biological Service, 44pp. (Biological Report 32 and Contaminant Hazard Reviews Report 32), 1997.

The open pit or underground mining methods are the predominantly used methods for the mining of silver. The excavated ore is upgraded through floatation, smelting, and a series of other processes; the pure silver is extracted using an electrolytic process (electrolysis) (77).

1.4.4 APPLICATION OF SILVER

The use of silver has been dated back to the historical period of man's civilization where it was used in ornamental materials, utensils, coinage, and even as basis of wealth. It has, however, been used in recent times as raw material for the production of a variety of other products. The industrial use of silver in the United States is summarized in Table 1.13.

Silver is also used in the water purification process because of its bacteriostatic property; it has equally been employed in food and drugs processing (77). Silver is used medically for the treatment of burns and as an antibacterial agent. It is also used as catalyst in the industrial production of some chemicals such as formaldehyde and ethylene oxide.

1.4.5 TOXICITY AND RELATED HAZARDS

Silver has been reportedly used in food and for medical purposes by man (77); however, its release and eventual contact with both plants and animals in the environment in toxic concentrations results with adverse health consequences.

1.4.5.1 Route of Exposures

Silver from anthropogenic sources is often transported over a long range and reaches the soil through wet and dry deposition and eventual sorption to soils and sediments. Silver reaches underground water through leaching which is influenced mostly by the decreasing pH of the soil (77). The presence of silver in marine environments is also influenced by salinity, as a result of its strong affinity for chloride ions (84). The presence of silver in some aquatic organisms varies considerably with the ability of such organisms to bioaccumulate silver (85). Animals and humans can be exposed to varying concentrations of silver as a result of occupation, skin contact, ingestion, and inhalation. Skin contact and inhalation are generally occupational exposure; however, further contact occurs through the use of ornaments and other domestic products made of silver, and the use of silver utensils leads to ingestion (77).

1.4.5.2 Toxicity of Silver

Exposure to a high concentration of silver through different routes resulted in adverse health effects in people. Inhalation of dust containing a high concentration of silver compound, like $AgNO_3$ or

---

TABLE 1.13
Use of 50% of Refined Silver Produced in the United States (1990)

Product	Percentage Used
Photographic and x-ray	50
Electrical and electronic	25
Electroplated ware, sterling ware, and jewelry	10
Brazed alloys	5
Others use (products)	10

Source: ATSDR. *Toxicological Profile for Silver.* Atlanta, GA: US Department of Health and Human Services, Public Health Service, Agency for Toxic Substances and Disease Registry (TP-90-24), 1990.

TABLE 1.14
Acute Silver Toxicosis Resulting from Exposure

Exposure Cases	Health Effect	Exposure Level
112 workers exposure to work place silver nitrate and silver oxide	Rise in blood silver 0.6 µg silver/100 mL blood	0.039–0.378 mg silver/m^3
Workers at photographic facility	Presence of silver in blood, urine, and fecal samples	0.001–0.1 mg/m^3

Source: ATSDR. *Toxicological Profile for Silver.* Atlanta, GA: US Department of Health and Human Services, Public Health Service, Agency for Toxic Substances and Disease Registry (TP-90-24), 1990.

TABLE 1.15
Effect of Toxicity of Silver on Terrestrial Plants

Plant Species	Effect	Exposure Level	References
Lettuce	Adverse effect on germination	0.7 mg silver/L	(88)
Rye grass	Adverse effect on germination	7.5 mg silver/L	(88)
Seeds of corn, oat, turnip, soybean, spinach in silver rich soil	No adverse effect on germination	106 mg silver/kg dry soil	(89,90)
Seed of Chinese cabbage and lettuce	Adverse effect on germination	106 mg silver/kg dry soil	(89,90)

AgO, may cause trachea-related problems like lung and throat irritation as well as stomach pain. Skin contact with silver demonstrates rashes, swelling, and inflammation (77). Silver demonstrates high level of toxicity to aquatic plants and animals particularly in its ionic state (86,87). Recent cases of accident and research-based studies of silver toxicity are reported in Table 1.14.

Generally, the accumulation of silver is higher in aquatic media than in soils and thus, aquatic animals are expected to be more affected by the toxicity of silver than terrestrial animals (87,88). Most aquatic organisms demonstrate a high accumulation of silver at nominal concentration of 0.5–4.5 µg/L and corresponding health effects include stunted growth, muting, and histopathology (78). Studies have shown that silver accumulations in aquatic organisms such as marine algae are due mainly to adsorption rather than uptake (88). Accumulation of silver by terrestrial plants is relatively slow and only affects the plant growth but higher concentration may eventually lead to the plant's death. Tables 1.15 through 1.17 report cases of toxicity of silver in terrestrial plants.

1.4.6 INTERACTIONS OF SILVER WITH OTHER ELEMENTS

Interaction of silver and other metals usually influences absorption, distribution, and excretion of one or more of the metals (77). Though silver demonstrates good dissociation in water media, its interaction with other elements is sparsely documented in the literature. However, studies show interaction between silver and selenium increases deposition of insoluble silver salt in body tissue (77).

1.5 MOLYBDENUM

1.5.1 MOLYBDENUM AND ITS COMPOUND

Molybdenum is a transition metallic element existing in five oxidation states (II–VI). It has a silvery white color in its pure metal form and is more ductile than tungsten (101). It has a melting point of 2623°C (102) but boils at a temperature above 600°C (103). Molybdenum largely exists in association with other elements and molybdate anion (MoO_4^{2-}) is its predominant form found in soil and

TABLE 1.16
Toxicity of Silver Nitrate (AgNO₃) in Aquatic Animals

Organism	End Point	Silver Concentration (μg/L)	Reference
Protozoan	24-h LC$_{50}$	8.8	(91)
Chlamydomonas	96-h LC$_{50}$	200	(92)
Chlamydomonas	250-h LC$_{50}$	100	(92)
Mussel	110-h LC$_{50}$	1000	
Asiatic Clam	21-day NOEC	7.8	(93)
Flatworm	96-h LC$_{50}$	30	(88)
Snail	96-h LC$_{50}$	300	(88)
Copepods	48-h LC$_{50}$	43	(94)
Amphipod	10-h LC$_{50}$	20	(95)
Amphipod	96-h LC$_{50}$	1.9 (1.4–2.3)	(93)
Daphnia magna	96-h LC$_{50}$	5	(88)
Mayfly	96-h LC$_{50}$	6.8	(93)
Rainbow trout	96-h LC$_{50}$ at 25% salinity seawater acclimatized	401	(96)
Tide pool sculpin	96-h LC$_{50}$	331 (25% salinity)	(97)
Mosquito fish (juvenile)	96-h LC$_{50}$	23.5 (17.2–27.0)	(93)
Bluegill	96-h LC$_{50}$	31.7 (24.3–48.4)	(93)
Coho salmon	96-h LC$_{50}$	11.1 (7.9–15.7)	(86)
Rainbow trout (juvenile)	144-h LC$_{50}$	48	(93)
	96-h LC$_{50}$	11.8	(98)
	96-h LC$_{50}$	19.2 (16–23.1)	(86)
Arctic graying (juvenile)	96-h LC$_{50}$	11.1 (9.2–13.4)	(99)
Arctic grayling alevin	96-h LC$_{50}$	6.7 (5.5–8.0)	(100)
European eel	96-h LC$_{50}$ in soft, low-chloride (10 μmol/L)	34.4	(100)
Leopard frog	EC$_{10}$ based on mortality or abnormal development of embryos and larvae	0.7–0.8	(100)
	EC$_{50}$ based on mortality or gross ferata of embryos and larvae	10	

Note: EC$_{50}$—Median effective concentration; LC$_{50}$—Median lethal concentration; and NOEC—No observed effect concentration.

TABLE 1.17
Toxicity of Silver (AgNO₃) to Terrestrial Animals

Animal	Effect	Route	Concentration Silver	Reference
Mice	Lethal	Intraperitoneal injection	13.9 mg/kg body weight	(77)
Rabbits	Lethal	Intraperitoneal injection	20.0 mg/kg body weight	
Dogs	Lethal	Intravenous injection	50.0 mg/kg	
Rats	Lethal	Drinking water	1586 mg/L for 37 weeks	
Mice	Sluggishness	Drinking water	95 mg/L for 125 days	
Guinea pigs	Reduced growth	Skin contact	81 mg/cm^2 for 8 weeks	

natural waters. It ranks the 42nd and 25th most abundant element in the universe and the earth's oceans, respectively (102). It is represented with the symbol "Mo" and assigned with atomic number of 42 and atomic weight of 95.94 on the periodic table of the elements.

1.5.2 CHARACTERISTICS OF MOLYBDENUM

Molybdenum reacts with oxygen at high temperature to form molybdenum trioxide but cannot do so at room temperature. Its oxidation states of +2, +3, +4, +5, and +6 and these are well illustrated in the range of compounds it forms with chlorides, which are molybdenum (II) chloride ($MoCl_2$), molybdenum (III) chloride ($MoCl_3$), molybdenum (V) chloride ($MoCl_5$), and molybdenum (VI) chloride ($MoCl_6$). Molybdenum also forms quadruple bonds with other transition metals. Of the 35 known isotopes of molybdenum, only seven occur naturally and five of these are stable. The unstable isotopes usually decay to form niobium, technetium, and ruthenium which are equally isotopes (104). Molybdenum-98, comprising 24.14% of all the molybdenum isotopes, is the most common isotope (104).

1.5.3 PRODUCTION OF MOLYBDENUM

Countries such as the United States, Canada, Chile, Russia, and China are the world's largest producers of molybdenum materials (104). Molybdenum mines are located in Colorado, British Columbia, northern Chile, and southern Norway with each having different compounds and grade of molybdenum. Molybdenum is mined from its ore but can equally be recovered as a byproduct of copper and tungsten mining (101).

1.5.4 APPLICATION OF MOLYBDENUM

Molybdenum is widely used in the manufacturing of heat-resistant parts used in the aircraft, automobile, and electric industries (103). Due to its high corrosion resistance and weldability, molybdenum is used in the production of high-strength steels such as stainless steel, tool steel, cast irons, and other high-temperature alloys (101,102). Another important application of molybdenum is in the oil industry where it is used as an additive in engine oil, due to its high resistance, extreme temperatures, and high pressure. Molybdenum is also used in its pure or compound form as raw material in the adhesive and fertilizer industries (102). Isotopes of molybdenum, particularly, technetium 99, is used in many medical procedures. About 0.25 mg maximum daily dose is used to treat patients with malabsorption and hypoproteinemia and in preoperative nutritional support. Molybdenum has been medically implicated as an agent contributing to the decrease of dental caries, and incidence of cancer (105).

1.5.5 TOXICITY AND RELATED HAZARDS

Molybdenum is an essential trace element and has received medical acceptance for the treatment of patients, however, at high concentration, molybdenum is toxic, leading to adverse health effects (106).

1.5.5.1 Route of Exposure

Molybdenum, unlike other trace metals, has a higher record of exposure through direct or indirect ingestion than through inhalation and occupational exposures. Some significant dietary sources of molybdenum include green beans, eggs, sunflower seeds, wheat flour, lentils, cereal grain, canned vegetables, nuts, and some animal parts such as kidney and liver (103); most natural water contains a low level of molybdenum. Furthermore, reports are available on the molybdenum content of individual plant species though molybdenum uptake by plants is influenced positively by increasing soil pH.

TABLE 1.18
Acute Molybdenum Toxicity Resulting from Exposure

Exposure Case	Health Effect	Exposure Level	Reference
An individual on self-exposure to improve health	Hallucination extended to mal seizures and finally psychosis (lucor molybdenum)	Cumulative 13.5 mg on the 18th day	(105)

Humans can also be exposed to molybdenum in areas near mining sites (107). Molybdenum is an essential constituent of xanthine oxidase and aldehyde oxidase which are two important enzymes responsible for the formation of uric acid and chemical oxidation of aldehydes, respectively.

1.5.5.2 Toxicity of Molybdenum

Health effects of acute exposure to molybdenum or molybdenum compounds by human and animals include diarrhea, depressed growth rate, anemia, and gout-like symptoms. Others are anorexia, headache, arthralgia, myalgia chest pain, nonproductive cough, and testicular atrophy at chronic level (108). Toxicity of molybdenum also affects the activities of alkaline phosphate which eventually result in bone abnormalities. Recent cases, based on the effect experienced by the victims, of accident and research-based studies of molybdenum toxicity are reported in Table 1.18.

Molybdenum toxicity in humans following ingestion and its subsequent absorption from the stomach into the bloodstream which is favored by increased acidity is referred to as molybdenosis which has symptoms similar to the disease of copper deficiency (109,110). Study on uptake of trace elements in neuron culture indicated a high affinity of neurons for molybdenum (111). The recommended average daily intake of molybdenum is approximately 0.3 mg, while the WHO recommended a maximum level 0.07 mg/L of molybdenum in drinking water (107).

1.5.6 Interactions of Molybdenum with Other Elements

The presence of molybdenum alongside some other trace elements influences the rate and amount of absorption of molybdenum in human systems. The presence of copper and molybdenum in the liver often leads to the prevention of accumulation of molybdenum.

1.6 RARE EARTH ELEMENTS

REEs are group of metallic elements with unique physical and similar chemical properties (112). They are 17 in number with relative atomic masses ranging from 139 to 175; they are relatively soft and malleable metals with bright silver luster. They are principally found together in various combinations in many areas (113). The common among them is cerium which is relatively more abundant than lead or copper. Other members of REEs include lanthanum, praseodymium, promethium, neodymium, samarium, europium, gadolinium, terbium, dysprosium, holmium, erbium, thulium, ytterbium, lutetium, yttrium, and scandium.

1.6.1 Characteristics of REEs

REEs are malleable, ductile, and soft metals. Some of them oxidize readily in air, and some are attacked by cold water, slowly or rapidly, while others are attacked only by hot water. Some REEs react directly with elemental carbon, nitrogen, boron, selenium, silicon, phosphorus, sulfur, and the halogens. When heated, REEs exhibit cubic structural changes within hexagonal face-centered and body-centered structures. Because of their similar molecular structure, they are difficult to separate however, their unique chemical structure as a group of elements is often explored for scientific advantages (114) and most of them exist as isotopes.

1.6.2 PRODUCTION OF REEs

The REEs are often found together in various combinations in ores such as monatize, allanite, bastnasite, xenotime, gadolinite, samarskite, fergusonite, apatite, euxenite, and others (113). Some of the countries where they are located include Norway, Sweden, the United States, Australia, India, Canada, and Brazil.

1.6.3 APPLICATION OF REEs

Due to their unique properties, REEs form important elements for research purposes and for modern industrial applications (115). REEs have equally received a wide application in a variety of nonnuclear industries and agriculture (116); some of them are also used as tracers for the presence of other elements (117). Some of the applications of the REEs are summarized in Table 1.19.

The REEs are essential components in the production of the world's strongest permanent magnets that are widely used in the automobile industry. Some are of them are heavily used in medical and nuclear research activities.

1.6.4 TOXICITY AND RELATED HAZARDS CAUSED BY REEs

Some REEs and their compound have a low to moderate toxicity rating and as such must be handled with care.

1.6.4.1 Exposure Route

Since the REEs are mostly used in nuclear, nonnuclear, and agricultural applications there is a higher probability for occupational exposure than for ingestion and inhalation (116). Since standard

TABLE 1.19
Applications of REEs

REE	Industrial	Research
Lanthanum	Carbon lighting, production of alkali resistance and optical glasses, production of nodular cast iron	Hydrogen sponge alloys
Cerium	Production of optical glass, electrodes, ceramics, fireworks, metallurgical alloys also in printing, dyeing, and textile processes	Tracer bullet catalytic converter in automobile
Praseodymium	Production of alloys and arc cores for lights. Also in glass coloring	
Neodymium	Production of purple glass and carbon-arc rods	Dopant for glass lasers
Promethium		Luminescent
Samarium	Production of infrared absorbing glass, and constituent television phosphor	Neutron absorber in nuclear reactors, dopant for glass lasers
Terbium		Solid state and laser dopant
Erbium	Production of metallurgical products coloring of glass and porcelain	Nuclear research
Thulium		Isotope used in radiation units
Ytterbium	Production of alloys	Lasers source for irradiation devices and radiography
Yttrium	Production of optical glasses, ceramics color television tubes, and alloys	
Europium		Detection of chrome in contaminated water diagnostic and therapeutic tools

safety measures are involved during nuclear application of some of the rare earth metals, humans and animals are less susceptible to contamination, at least to toxic levels. Skin absorption of REEs has not been well documented while the gastrointestinal tract and lungs show poor absorption.

1.6.4.2   Toxicity REEs

Toxic levels of REEs have shown different health abnormalities in humans and animals. Acute exposure results in irritation, pneumonitis, bronchitis, and edema, while chronic exposure results in irritation to skin and pneumoconiosis. In the study carried out by Zhang et al. (113), REEs were discovered to have effects on the human central nervous system, cardiovascular, and immune systems. Lanthanum affects the uptake of glutamic acid at the nerve ends. Concentration of 6–6.7 mg of REEs in food consumed per day leads to subchronic toxicity (113). The concentrations of REEs in plants are influenced by the difference in REEs concentration in the soil and in plant species (118). The concentrations of lanthanum in ferns and the needles of the Norway spruce were reported to be about 700 ng/g and <10 ng/g, respectively (117,119), however, for a plant grown in REEs contaminated soils, concentrations of REEs are more in the roots than the leaves, while the least concentration is detected in the seeds (120).

1.6.4.3   Interaction of REEs with Other Elements

Most REEs are rarely released into the environment, and where needed for medical use, they are handled with care. It is however been recorded that some REEs affect the activities of some enzymes as a result of the presence of some elements. Lanthanum inhibits the activity of calcium and magnesium adenosine-triphosphate in enzymes (119).

1.7   CONTROL OF SELECTED HEAVY METALS AND REEs

1.7.1   CONTROL OF SOLID METAL CONTAINING WASTES IN THE ENVIRONMENT

Precious metal solid wastes, such as silver-containing and gold-containing parts are usually separated by skilled workers for recycle and reuse. The metal smelter or refinery for both precious metal and nonprecious metal are similar to each other. However, the recycling of precious metals from electronic parts frequently takes place in developing countries using informal processes due to their low labor costs. These informal operations may include manual dismantling, open burning to recover precious metals, desoldering of printed wiring boards over coal fires, and acid leaching in open vessels, which all have high potential for significant adverse human health and environmental impacts.

Much of the solid nonprecious metal containing wastes, such as copper pipes/sheets, brass products (copper/zinc), bronze products (copper/tin), beryllium-copper, monel (copper/nickel), gunmetal (copper/tin), automobile parts (die-cast zinc alloys), foundry dusts (50% zinc), electric arc furnace dust (zinc and lead), etc. can be recycled by scrap metal merchants in industrial as well as developing countries. The merchants usually sort and accumulate the nonprecious solid metallic wastes until a sufficient quantity has been accumulated to provide sufficient quantities for a smelter to refine and recover the metals for reuse (121–123). For 1992, the world production of refined copper and zinc was estimated to be about 11.1 million tones and 7.2 million tones, respectively. Of the refined copper and zinc, some 4.25 million tonnes were derived from the copper containing solid waste materials, and 1.4 million tonnes were obtained from the processing of zinc containing solid waste materials.

Example: Electric arc Furnace Dust from Secondary Steel Production for Zinc and Lead Recovery.

This is a general description of processes involved and environmental considerations for the recovery of particular materials (121). The zinc-lead containing solid waste arises as a fine powder collected in bag filters from air filtration of gases from electric arc furnaces.

The following paragraph describes the recovery processes: (a) Fine powder is pelletized and kept damped down to make it less dispersible and thereby reduce its release to the environment. The principal hazard is as a nuisance although the lead in dust is very toxic to humans. (b) The waste material is stored in open bays having concrete walls on hard stands and is drained to a wastewater treatment facility nearby. (c) The solid waste material is loaded into watertight trailers by front-end loaders and sheeted down. Vehicles pass through a wheel-wash and are hosed down before leaving the site of refined metal production. Wash waster is drained to the site wastewater treatment facility. (d) Solid waste material is off-loaded at a processor facility by tipping into underground hoppers. Empty trucks are washed down before leaving site. Washings are collected and transferred to the site wastewater treatment facility. (e) Solid waste material is processed using the Waelz process by being fed by conveyor, together with coke and silica, into a rotary kiln. The zinc and lead content of solid waste is removed as fume and dust which is collected by electrostatic precipitators and bag filters in series. Filtered material is now in a suitable form to be conveyed to the Imperial Smelting ISF primary production process situated nearby. (f) The byproduct from the Waelz process is slag which is discharged from the rotary kiln and quenched in water. Slag is in suitable form and in nonleachable composition for use as road making or similar material.

Molybdenum is a silver-white solid or gray-black powder without odor. It is insoluble in water. Molybdenum waste is flammable in the form of dust or powder (<9 μm), which may ignite during intensive mechanical treatment. Care must be taken to handle the waste material because the molybdenum dust–air mixtures may be explosive (126). When there is a fire caused by molybdenum containing powders, firefighters must wear full face, self-contained breathing apparatus with full protective clothing to prevent contact with skin and eyes. Only suitable extinguishing media (powder not water) should be used to control the fire hazard. For proper disposal of solid molybdenum powder waste, the only known method is the solidification process using cement or other solidifying agents. More research is needed in this area.

REEs milling and processing is a complex, ore-specific operation that has potential for environmental contamination when not properly controlled and managed. Waste streams with the greatest pollution potential are the tailings and their associated treatment and storage. Heavy metals and radionuclides associated with REE tailings pose the greatest threat to human health and the environment if not controlled (134).

Example: Environmental Damage from Radioactivity of REE

This is an example of issues associated with radionuclides from REE production. A refinery being built by an Australian mining company processing REE minerals in Malaysia may cause a threat of radioactive pollution there. The refinery is one of Asia's largest radioactive waste cleanup sites. The plant is meant to refine slightly radioactive ore from the Mount Weld mine in Western Australia, which is trucked to Fremantle and transported to Malaysia by container ship. The Australian mining company expects to produce nearly a third of the worldwide demand for REE, excluding China. Public concerns have raised the environmental and public health issues. New technologies and management processes are being developed to reduce the risk of environmental contamination (130–133).

1.7.2 Control of Liquid Metal Containing Wastes in the Environment

Silver ions in the fixer of a photographic process can kill or damage microorganisms in a public biological wastewater treatment plant (WWTP) or in a private septic waste treatment facility, and is toxic to humans. Silver is the most common precious metal in photographic processing wastes. Such wastes must be evaluated for hazardous characteristics. There are four hazardous characteristics (ignitability, corrosivity, reactivity, and toxicity), but only the characteristic of toxicity usually applies to photographic wastes. The US EPA recommended test method for toxicity is called the toxicity characteristic leaching procedure (TCLP), which may test for 38 different chemical constituents, and may cost about US$ 3000. An environmental manager may decide whether or not to analyze their liquid waste to determine whether it is hazardous based on his/her knowledge of the wastes. A silver recovery unit is commercially available for recovering silver from a liquid silver-bearing waste (124). If a silver recovery unit is an integral part of the photographic processing equipment (i.e., "a totally enclosed treatment facility"), only silver flake or silver-bearing filter cartridges will be generated as solid wastes. The silver flake generated by electrolytic recovery systems, or cartridges containing silver sludge from metallic replacement systems, routinely exhibit the characteristic of toxicity for silver. The waste generators may either recover silver on-site using a silver recovery unit, or send at least 75% of the annual volume of this silver-bearing wastes (i.e., untreated spent fixer) for off-site reclamation. Commercial silver recyclers in some states of the United States must be registered with the state environmental agencies.

Copper and zinc ions are also toxic to humans and the environment when their concentrations reach toxic levels. Generators of liquid copper-bearing wastes or zinc-bearing wastes are usually industrial production plants. Their environmental managers may use either material safety data sheets information, or other industrial process information to decide whether or not a TCLP analytical test is needed. Industrial effluent is usually produced in large quantities, and therefore cannot use small recovery units for recovering copper and zinc. Large-scale physicochemical treatment facilities, such as ion exchange (135), reverse osmosis (136), chemical precipitation (137), lime and soda-ash softening (138), electrodialysis reversal (139), and chemical coagulation, sedimentation/flotation and filtration (140,141) can be potential industrial pretreatment processes. After copper and zinc are significantly removed, the pretreated industrial effluents can then be discharged into a sewer system leading to a biological WWTP for final treatment.

The most significant environmental impact from contaminant sources associated with hard-rock mining of REE is to surface water and ground water quality. Documented environmental impacts also have occurred to sediments, soil, and air. Increased demand and reduced supply of REE, along with the knowledge of the quantities available in waste products, has resulted in expanded R&D efforts focused on the identification of alternatives to REE, and the recycling of REE (130–133).

1.8 SUMMARY

Exposure of humans to heavy metal toxins has increased over the years. This has been due to the dramatic increase in the overall environmental load of heavy metal toxins by explosive industrialization, which humans have depended on for economic, social, and political reasons. As a result of increased usage of these metals, heavy metal toxins concentrations are increasing at alarming rate in drinking water, soil, air, and vegetation. These metals are present in virtually all human endeavors, from construction materials to cosmetics, medicines to processed foods, fuel sources to agents of destruction, appliances to personal care products. In fact, it is very difficult to avoid exposure to the bulk of harmful heavy metals in our environment. Since the threat of heavy metal toxicity can neither be totally neutralized in our environment nor can the application of the metals for the advancement of technology and subsequent betterment of human living be reduced, steps can be advanced to enact policies of pollution prevention and abatement to alleviate the adverse effects of the metals on human health. The heavy metals and REEs are our useful resources as well as potential pollutants. We must use them wisely and dispose of them properly after their applications. For readers' convenience, Table 1.19 is included to show REE applications, and potential supply issues for clean-energy technologies.

REFERENCES

1. The facts on copper, Dartmouth College, Hanover, NH, http://www.dartmouth.edu/, 2015.
2. Canadian soil quality guidelines for copper environmental and human health, Canadian Council of Environmental Ministers, Winnipeg, Manitoba, 1997.
3. US EPA. Toxics Release Inventory. U.S. Environmental Protection Agency, Washington, DC. www.epa. gov, 1999.

4. Copper fact sheet. Bureau of Reclamation, Technical Service center, Water Treatment Engineering and Research Group, D-8230, Denver, CO, 2001.

5. US EPA. *Environmental Technology Verification Report Environmental Bio-Detection Products Inc. Toxi-chromotest.* Washington, DC: US Environmental Protection Agency. June 2006. www.epa.gov/etv/pubs/600etv06055.pdf

6. Audi, G. Nubase evaluation of nuclear and decay properties. *Nucl. Phys. A* 729, 3–128, 2003. Atomic mass data center. Doi: 10.1016/j:nuclphysa.200311.001.

7. Supply and Services Canada, Health and Welfare Canada. *Nutrition Recommendations—The Report of the Scientific Review Committee.* Ottawa: Supply and Services Canada, Health and Welfare Canada, 1990.

8. Held, K.D. Role of Fenton chemistry in thiol-induced toxicity and apoptosis. *Radiat. Res.* 145(5), 542–53, 1996.

9. CDC (Centers for Disease Control and Prevention). Surveillance for waterborne-disease outbreaks—United States, 1993–1994. *MMWR* 45(SS-1), 12–13, 1996.

10. Knobeloch, L., Ziarnik, M., Howard, J., Theis, B., Farmer, D., Anderson, H., and Proctor, M. Gastrointestinal upsets associated with ingestion of copper-contaminated water. *Environ. Health Perspect.* 102(11), 958–961, 1994.

11. Pizarro, F., Olivares, M., Uauy, R., Contreras, P., Rebelo, A., and V. Gidi. Acute gastrointestinal effects of graded levels of copper in drinking water. *Environ. Health Perspect.* 107(2), 117–121, 1999.

12. Wapnir, R.A. Copper absorption and bioavailability. *Am. J. Clin. Nutr.* 67(5 Suppl.), 1054S–1060S, 1998.

13. O'Donohue, J., Reid, M.A., Varghese, A., Portmann, B., and Williams, R. Micronodular cirrhosis and acute liver failure due to chronic copper self-intoxication. *Eur. J. Gastroenterol. Hepatol.* 5, 561–562, 1993.

14. Brewer, G.J. and Yuzbasiyan-Gurkan, V. Wilson disease. *Medicine* 71(3), 139–164, 1992.

15. Mori, M., Hattori, A., Sawaki, M., Tsuzuki, N., Sawada, N., Oyamada, M., Sugawara, N., and Enomoto, K. The LEC rat: A model for human hepatitis, liver cancer, and much more. *Am. J. Pathol.* 144(1), 200–204, 1994.

16. Kitaura, K., Chone, Y., Satake, N., Akagi, A., Ohnishi, T., Suzuki, Y., and Izumi, K. Role of copper accumulation in spontaneous renal carcinogenesis in Long-Evans Cinnamon rats. *Jpn. J. Cancer Res.* 90(4), 385–392, 1999.

17. Keen, C.L. Teratogenic effects of essential trace metals: Deficiencies and excesses. In: *Toxicology of Metals.* Chang, L.W., Magos, L., and Suzuki, T., eds. New York: CRC Press, pp. 977–1001, 1996.

18. Hanna, L.A., Peters, J.M., Wiley, L.M., Clegg, M.S., and Keen, C.L. Comparative effects of essential and nonessential metals on preimplantation mouse embryo development in vitro. *Toxicology* 116, 123–131, 1997.

19. Becker, T.W., Krieger, G., and Witte, I. DNA single and double strand breaks induced by aliphatic and aromatic aldehydes in combination with copper (II). *Free Radic. Res.* 24(5), 325–332, 1996.

20. Glass, G.A. and Stark, A.A. Promotion of glutathione-gamma-glutamyl transpeptidase-dependent lipid peroxidation by copper and ceruloplasmin: The requirement for iron and the effects of antioxidants and antioxidant enzymes. *Environ. Mol. Mutagen.* 29(1), 73–80, 1997.

21. US EPA. Monitoring requirements for lead and copper in tap water. *Fed. Regist.* 56(110), 26555–26557, 1991. US Environmental Protection Agency, Washington, DC.

22. US EPA. Drinking water maximum contaminant level goals and national primary drinking water regulations for lead and copper. *Fed. Regist.* 59125, 33860–33864, 1994. US Environmental Protection Agency, Washington, DC.

23. IPCS (International Programme on Chemical Safety). *Copper. Environmental Health Criteria 200.* Geneva, Switzerland: WHO, 1998.

24. Galal-Gorchev, H. and Herrman, J.L. Letter to A.C. Kolbye, Jr., editor of *Regulatory and Pharmacology*, on the evaluation of copper by the Joint FAO/WHO Expert Committee on Food Additives from WHO. Sept. 12, 1996.

25. Stilman, M.L. Metallothioneins. *Coor. Chem. Rev.* 144, 461–511, 1995.

26. Walsh, C.T., Sandstead, H., and Prasad, A. Zinc: Health effects and research priorities for the 1990s. *Environ. Health Perspect.* 102(Suppl 2), 5–46, 1994.

27. Yadrick, M.K., Kenney, M.A., and Winterfeldt, E.A. Iron, copper, and zinc status: Response to supplementation with zinc or zinc and iron in adult females. *Am. J. Clin. Nutr.* 49, 145–150, 1989.

28. Ogra, Y. and Suzuki, K.T. Targeting of tetrathiomolybdate on the copper accumulating in the liver of LEC rats. *J. Inorg. Biochem.* 70(1), 49–55, 1998.

29. Yu, S., Wests, C.E., and Beynen, A.C. Increasing intakes of iron reduces status, absorption and biliary excretion of copper in rats. *Br. J. Nutr.* 71, 887–895, 1994.
30. Pekelharing, H.L.M., Lemmens, A.G., and Beynen, A.C. Iron, copper and zinc status in rats fed on diets containing various concentrations of tin. *Br. J. Nutr.* 71, 103–109, 1994.
31. Aburto, E.M., Cribb, A.E., and Fuentealba, I.C. Effect of chronic exposure to excess dietary copper and dietary selenium supplementation on liver specimens from rats. *Am. J. Vet. Res.* 62(9), 1423–1427, 2001.
32. Aburto, E.M., Cribb, A.E., and Fuentealba, I.C. Morphological and biochemical assessment of the liver response to excess dietary copper in Fischer 344 rats. *Can. J. Vet. Res.* 65(2), 97–103, 2001.
33. ATSDR Toxicological profile for Zinc. U.S. Department of health and human services, Public Health Service Agency for Toxic Substances and Disease Registry, 2005.
34. Anderson, M.B., Lepak, K., and Farinas, V. Protective action of zinc against cobalt-induced testicular damage in the mouse. *Reprod. Toxicol.* 7, 49–54, 1993.
35. Goodwin, F. Zinc and zinc alloys. In: *Kirk-Othmer's Encyclopedia of Chemical Technology.* Kroschwitz, J., ed. New York: John Wiley & Sons, pp. 789–839, 1998.
36. Ohnesorge, F.K. and Wilhelm, M. Zinc. In: *Metals and Their Compounds in the Environment. Occurrence, Analysis, and Biological Relevance.* Merian, E., ed. Weinheim, VCH, pp. 1309–1342, 1991.
37. WHO. *Environmental Health Criteria; 221, Zinc.* Geneva: World Health Organization, 2001.
38. Lewis, R.J., Sr., ed. *Hawley's Condensed Chemical Dictionary.* 12th ed. New York: Van Nostrand Reinhold Co., p. 1242, 1993.
39. USGS. Zinc. U.S. Geological Survey, Mineral Commodity Summaries, 2001. http://minerals.usgs.gov/minerals/pubs/commodity/zinc
40. USGS. *Trace Elements and Organic Compounds in Streambed Sediment and Fish Tissue of Coastal New England Streams, 1998–99.* Denver, CO: U.S. Geological Survey. Water-Resources Investigations Report 02-4179, 2002.
41. Sweet, C.W., Vermette, S.J., and Landsberger, S. Sources of toxic trace elements in urban air in Illinois. *Environ. Sci. Technol.* 27, 2502–2510, 1993.
42. Gundersen, P. and Steinnes, E. Influence of pH and TOC concentration on Cu, Zn, Cd, and Al speciation in rivers. *Water Res.* 37, 307–318, 2003.
43. Dudka, S. and Chlopecka, A. Effect of solid-phase speciation on metal mobility and phytoavailability in sludge-amended soil. *Water Air Soil Pollut.* 51, 153–160, 1990.
44. King, L.M., Banks, W.A., and George, W.J. Differential zinc transport into testis and brain of cadmium-sensitive and -resistant murine strains. *J. Androl.* 21, 656–663, 2000.
45. Knudsen, E., Jensen, M., Solgaard, P. et al. Zinc absorption estimated by fecal monitoring of zinc stable isotopes validated by comparison with whole-body retention of zinc radioisotopes in humans. *J. Nutr.* 125, 1274–1282, 1995.
46. Hunt, J.R., Matthys, L.A., and Johnson, L.K. Zinc absorption, mineral balance, and blood lipids in women consuming controlled lactoovovegetarian and omnivorous diets for 8 wk. *Am. J. Clin. Nutr.* 67, 421–430, 1998.
47. Samman, S. and Roberts, D. C. The effect of zinc supplements on plasma zinc and copper levels and the reported symptoms in healthy volunteers. *Med. J. Aust.* 146, 246–249, 1987.
48. Samman, S. and Roberts, D.C. The effect of zinc supplements on lipoproteins and copper status. *Atherosclerosis* 70, 247–252, 1988.
49. Hale, W.E., May, F.E., and Thomas, R.G. Effect of zinc supplementation on the development of cardio-vascular disease in the elderly. *J. Nutr. Elder.* 8, 49–57, 1988.
50. Prasad, A., Bao, B., and Beck, F.W. Antioxidant effect of zinc in humans. *Free Radic. Biol. Med.* 37, 1182–1190, 2004.
51. Davis, C.D., Milne, D.B., and Nielsen, F.H. Changes in dietary zinc and copper affect zinc-status indicators of postmenopausal women, notably, extracellular superoxide dismutase and amyloid precursor proteins. *Am. J. Clin. Nutr.* 71, 781–788, 2000.
52. Milne, D.B., Davis, C.D., and Nielsen, F.H. Low dietary zinc alters indices of copper function and status in postmenopausal women. *Nutrition* 17, 701–708, 2001.
53. Chandra, R.K. Excessive intake of zinc impairs immune responses. *JAMA* 252, 1443–1446, 1984.
54. Brown, J.J. Zinc fume fever. *Br. J. Radiol.* 61, 327–329, 1988.
55. Malo, J.L., Malo, J., and Cartier, A. Acute lung reaction due to zinc inhalation. *Eur. Respir. J.* 3, 111–114, 1990.
56. Malo, J.L., Cartier, A., and Dolovich, J. Occupational asthma due to zinc. *Eur. Respir. J.* 6, 447–450, 1993.

57. Gordon, T., Chen, L.C., and Fine, J.M. Pulmonary effects of inhaled zinc oxide in human subjects, Guinea pigs, rats, and rabbits. *Am. Ind. Hyg. Assoc. J.* 53, 503–509, 1992.

58. Fine, J.M., Gordon, T., and Chen, L.C. Metal fume fever: Characterization of clinical and plasma IL-6 responses in controlled human exposures to zinc oxide fume at and below the threshold limit value. *J. Occup. Environ. Med.* 39, 722–726, 1997.

59. Martin, C.J., Le, X.C., and Guidotti, T.L. Zinc exposure in Chinese foundry workers. *Am. J. Ind. Med.* 35, 574–580, 1999.

60. Zerahn, B., Kofoed-Enevoldsen, A., and Jensen, B.V. Pulmonary damage after modest exposure to zinc chloride smoke. *Respir. Med.* 93, 885–890, 1999.

61. Pettilä, V., Takkunen, O., and Tukiainen, P. Zinc chloride smoke inhalation: A rare cause of severe acute respiratory distress syndrome. *Intensive Care Med.* 26, 215–217, 2000.

62. Zaporowska, H. and Wasilewski, W. Combined effect of vanadium and zinc on certain selected haematological indices in rats. *Comp. Biochem. Physiol. C* 103, 143–147, 1992.

63. Bentley, P.J. and Grubb, B.R. Effects of a zinc-deficient diet on tissue zinc concentrations in rabbits. *J. Anim. Sci.* 69, 4876–4882, 1991.

64. de Oliveira, F.S. Viana, M.R., and Antoniolli, A.R. Differential effects of lead and zinc on inhibitory avoidance learning in mice. *Braz. J. Med. Biol. Res.* 34, 117–120, 2001.

65. Evenson, D.P., Emerick, R.J., and Jost, L.K. Zinc-silicon interactions influencing sperm chromatin integrity and testicular cell development in the rat as measured by flow cytometry. *J. Anim. Sci.* 71, 955–962, 1993.

66. US EPA. *Toxicological Review of Zinc and Compounds (CAS No. 7440-66-6) in Support of Summary Information on the Integrated Risk Information System (iris).* Washington, DC: US Environmental Protection Agency, 2005.

67. US EPA. *Toxicological Review of Zinc and Compound.* Washington, DC: US Environmental Protection Agency, 2005. EPA/635/R-05/002.

68. Lönnerdal, B. Dietary factors influencing zinc absorption. *J. Nutr.* 130, 1378S–1383S, 2000.

69. Hwang, S.J., Chang, J.M., and Lee, S.C. Short- and long-term uses of calcium acetate do not change hair and serum zinc concentrations in hemodialysis patients. *Scand. J. Clin. Lab. Invest.* 59, 83–87, 1999.

70. O'Brien, K.O., Zavaleta, N., and Caulfield, L.E. Prenatal iron supplements impair zinc absorption in pregnant Peruvian women. *J. Nutr.* 130, 2251–2255, 2000.

71. Bougle, D., Isfaoun, A., and Bureau, F. Long-term effects of iron: Zinc interactions on growth in rats. *Biol. Trace Elem. Res.* 67, 37–48, 1999.

72. Coogan, T.P., Bare, R.M., and Waalkes, M.P. Cadmium-induced DNA strand damage in cultured liver cells: Reduction in cadmium genotoxicity following zinc pretreatment. *Toxicol. Appl. Pharmacol.* 113, 227–233, 1992.

73. Brzoska, M.M., Moniuszko-Jakoniuk, J., and Jurczuk, M. The effect of zinc supply on cadmium-induced changes in the tibia of rats. *Food Chem. Toxicol.* 39, 729–737, 2001.

74. Lasley, S.M. and Gilbert, M.E. Lead inhibits the rat N-methyl-D-aspartate receptor channel by binding to a site distinct from the zinc allosteric site. *Toxicol. Appl. Pharmacol.* 159, 224–233, 1999.

75. Bebe, F. and Panemangalore, M. Modulation of tissue trace metal concentrations in weaning rats fed different levels of zinc and exposed to oral lead and cadmium. *Nutr. Res.* 16, 1369–1380, 1996.

76. Anderson, M.B., Lepak, K., and Farinas, V. Protective action of zinc against cobalt-induced testicular damage in the mouse. *Reprod. Toxicol.* 7, 49–54, 1993.

77. ATSDR. *Toxicological Profile for Silver.* Atlanta, GA: US Department of Health and Human Services, Public Health Service, Agency for Toxic Substances and Disease Registry (TP-90-24), 1990.

78. Eisler R. *Silver Hazards to Fish, Wildlife and Invertebrates: A Synoptic Review.* Washington, DC: US Department of the Interior, National Biological Service, 44 pp. (Biological Report 32 and Contaminant Hazard Reviews Report 32), 1997.

79. Morse J., Presley, B., Taylor, R., Benoit, G., and Santschi, P. Trace metal chemistry of Galveston Bay: Water, sediment, and biota. *Mar. Environ. Res.* 36, 1–37, 1993.

80. Szefer, P., Szefer, K., Pempkowiak, J., Skwarzec, B., Bojanowski, R., and Holm, E. Distribution and coassociations of selected metals in seals of the Antarctic. *Environ. Pollut.* 83, 341–349, 1994.

81. Falandysz, J. and Danisiewicz, D. Bioconcentration factors (BCF) of silver in wild *Agaricus campestris*. *Bull. Environ. Contam. Toxicol.* 55, 122–129, 1995.

82. Bell, R. and Kramer, J. Structural chemistry and geochemistry of silver-sulfur compounds: Critical review. *Environ. Toxicol. Chem.* 18(1), 9–22, 1999.

83. Silver Institute. *World Silver Survey 2000.* Washington, DC: The Silver Institute, 2000.

84. Sanders, J., Riedel, G., and Abbe, G. Factors controlling the spatial and temporal variability of trace metal concentrations in *Crassostrea virginica* (Gmelin). In: *Estuaries and Coasts: Spatial and Temporal Intercomparisons*. Elliot, M. and Ducrotoy, J., eds. *Proceedings of the Estuarine and Coastal Sciences Association Symposium*, 4–8 September 1989, University of Caen, France. Fredensborg, Olsen & Olsen, pp. 335–339 (ECSA Symposium 19), 1991.

85. Webb, N. and Wood, C. Bioaccumulation and distribution of silver in four marine teleosts and two marine elasmobranchs: Influence of exposure duration, concentration, and salinity. *Aquat. Toxicol.*, 49(1–2), 111–129, 2000.

86. Buhl, K. and Hamilton, S. Relative sensitivity of early life stages of Arctic grayling, coho salmon, and rainbow trout to nine inorganics. *Ecotoxicol. Environ. Saf.*, 22, 184–197, 1991.

87. Bryan, G. and Langston, W. Bioavailability, accumulation and effects of heavy metals in sediments with special reference to United Kingdom estuaries: A review. *Environ. Pollut.*, 76, 89–131, 1992.

88. Ratte, H. Bioaccumulation and toxicity of silver compounds: A review. *Environ. Toxicol. Chem.*, 18(1), 89–108, 1999.

89. Hirsch, M. Availability of sludge-borne silver to agricultural crops. *Environ. Toxicol. Chem.*, 17(4), 610–616, 1998.

90. Hirsch, M., Ritter, M., Roser, K., Garrisi, P., and Forsythe, S. The effect of silver on plants grown in sludge-amended soils. In: *Transport, Fate, and Effects of Silver in the Environment*. Andren, A., Bober, T., Crecelius, E., Kramer, J., Luoma, S., Rodgers, J., and Sodergren, A., eds. *Proceedings of the 1st International Conference*. 8–10 August 1993. Madison, WI: University of Wisconsin Sea Grant Institute, pp. 69–73, 1993.

91. Nalecz-Jawecki, G., Demkowicz-Dobrzanski, K., and Sawicki, J. Protozoan *Spirostomum ambiguum* as a highly sensitive bioindicator for rapid and easy determination of water quality. *Sci. Total Environ. Suppl.* 2, 1227–1234, 1993.

92. Berthet, B., Amiard, J., Amiard-Triquet, C., Martoja, M., and Jeantet, A. Bioaccumulation, toxicity and physico-chemical speciation of silver in bivalve molluscs: Ecotoxicological and health consequences. *Sci. Total Environ.*, 125, 97–122, 1992.

93. Diamond, J., Mackler, D., Collins, M., and Gruber, D. Derivation of freshwater silver criteria for the New River, Virginia, using representative species. *Environ. Toxicol. Chem.*, 9, 1425–1434, 1990.

94. Hook, S. and Fisher, N. Sublethal effects of silver in zooplankton: Importance of exposure pathways and implications for toxicity testing. *Environ. Toxicol. Chem.*, 20(3), 568–574, 2001.

95. Berry, W., Cantwell, M., Edwards, P., Serbst, J., and Hansen, D. Predicting toxicity of sediments spiked with silver. *Environ. Toxicol. Chem.*, 18(1), 40–48, 1999.

96. Ferguson, E. and Hogstrand, C. Acute silver toxicity to seawater-acclimated rainbow trout: Influence of salinity on toxicity and silver speciation. *Environ. Toxicol. Chem.*, 17(4), 589–593, 1998.

97. Shaw, J., Wood, C., Birge, W., and Hogstrand, C. Toxicity of silver to the marine teleost (*Oligocottus maculosus*): Effects of salinity and ammonia. *Environ. Toxicol. Chem.*, 17(4), 594–600, 1998.

98. Hogstrand, C., Galvez, F., and Wood, C. Toxicity, silver accumulation and metallothionein induction in freshwater rainbow trout during exposure to different silver salts. *Environ. Toxicol. Chem.*, 15, 1102–1108, 1996.

99. Grosell, M., Hogstrand, C., Wood, C., and Hansen, H. A nose-to-nose comparison of the physiological effects of exposure to ionic silver versus silver chloride in the European eel (*Anguilla anguilla*) and the rainbow trout (*Oncorhynchus mykiss*). *Aquat. Toxicol.*, 48(2–3), 327–342, 2000.

100. Birge, W. and Zuiderveen, J. The comparative toxicity of silver to aquatic biota. In: Transport, fate and effects of silver in the environment. Andren, W. and Bober, T. eds. *Abstracts of the 3rd International Conference*. 6–9 August 1995, Washington, DC. Madison, WI: University of Wisconsin Sea Grant Institute, pp. 79–85, 1996.

101. Lide, D. R., ed. *"Molybdenum", CRC Handbook of Chemistry and Physics*. Boca Raton, FL: CRC Press, vol. 4, pp. 18, 1994. ISBN 0-8493-0474-1.

102. Considine, G.D., ed. "Molybdenum", Van Nostrand's Encyclopedia of Chemistry. New York: Wiley-Interscience, pp. 1038–1040, 2005. 0-471-61525-0.

103. Emsley, J. Nature's Building Blocks. Oxford: Oxford University Press, pp. 262–266, 2001. 0-19-850341-5.

104. Lide, D. R., ed. *CRC Handbook of Chemistry and Physics*. Boca Raton, FL: CRC Press, vol. 11, pp. 87–88, 2006. ISBN 0-8493-0487-3.

105. Momčilovic, B. Acute human molybdenum toxicity. *Arh. Hig. Radr. Toksikol.* 50(3), 289–297, 1999.

106. Sullivan, J.B. and Krieger, G.R. *Hazardous Materials Toxicology*. Baltimore, MD: Williams and Wilkins, pp. 905–907, 1992.

107. WHO. *Guidelines for Drinking Water Quality.* 2nd edition. Geneva: World Health Organization, 1993.
108. Lesser, S.H. and Weiss, S.J. *Art Hazards.* New York: Saunders Co., pp. 451–458, 1995.
109. Castorph, H.R. and Walker, M. *Toxic Metal Syndrome.* Garden City Park, NY: Avery Publication Group, 1995.
110. Cantone, M.C., De Bartolo, D., and Giussani, A. A methodology for biokinetic studies using stable isotopes: Results of reported molybdenum investigations on a healthy volunteer. *Appl. Radiat. Isot.* 48, 333–338, 1997.
111. DeStasio, G., Perfetti, P., and Oddo, N. Metal uptake in neurone: A systemic study. *Neuroreport* 3, 965–968, 1992.
112. Ding, S., Liang, T., Zhang, C., Yan, J., and Zhang, Z. Accumulation and fractionation of rare earth elements (REEs) in wheat: Controlled by phosphate precipitation, cell wall absorption and solution complexation. *J. Exp. Bot.* 56(420), 2765–2775, 2005.
113. Zhang, H., Zhu, W.F., and Feng, J. Subchronic toxicity of rare earth elements and estimated daily intake allowance. *Ninth Annual V. M. Goldschmidt Conference*, August 22–27, Cambridge, Massachusetts, Vol. 1, p. 7025, 1999.
114. Fu, F.F. and Tasuku, A. Distribution of rare earth elements in seaweed: Implication of two different sources of rare earth elements and silicon in seaweed. *J. Phycol.* 36, 62–70, 2000.
115. Kataoka, T., Strkalenburg, A., Nakanishi, T.M., Delhaize, E., and Ryan, P.R. Several lanthanides activate malate efflux from roots of aluminium-tolerant wheat. *Plant Cell Environ.* 25, 453–460, 2002.
116. Wang, Z.J., Liu, D.F., Lu, P., and Wang, C.X. Accumulation of rare earth elements in corn after agricultural application. *J. Environ. Qual.* 30, 37–45, 2001.
117. Fu, F.F., Akage, T., and Shinotsuka, K. Distribution patterns of rare earth elements in fern: Implication for intake of fresh silicate particles by plants. *Biol. Trace Elem. Res.* 64, 13–26, 1998.
118. Miekeley, N., Casartelli, E.A., and Dotto, R.M. Concentration levels of rare-earth elements and thorium in plants from the Morro Do Ferro environment as an indicator for the biological availability of transuranium elements. *J. Radioanal. Nucl. Chem.* 182, 75–89, 1994.
119. Wyttenbach, A., Schleppi, P., Bucher, J., Furrer, V., and Tobler, L. The accumulation of the rare earth elements and of scandium in successive needle age classes of Norway spruce. *Biol. Trace Elem. Res.* 41, 13–29, 1994.
120. Wyttenbach, A., Furrer, V., Schleppi, P., and Tobler, L. Rare earth elements in soil and in soil-grown plants. *Plant Soil* 199, 267–273, 1998.
121. UNEP. *Recycling of Copper, Lead and Zinc Bearing Wastes*, Environment Monographs No. 109. Paris, France: United Nations Environment Programme, 1995.
122. Kundig, K.J.A. *Copper's Role in the Safe Disposal of Radioactive Wastes: Copper's Relevant Properties.* Copper Development Association, New York, 1998. www.copper.org.
123. US EPA. *Copper Mining and Production Wastes.* Washington, DC: US Environmental Protection Agency, 2015.
124. NREPC. *How to Manage Silver-Bearing Hazardous Wastes.* Frankfort, KY: Natural Resources and Environmental Protection Cabinet, March 1997.
125. Kodak. *Waste Management.* Rochester, NY: Kodak Environmental Services, 2015.
126. Molybdenum.com. *Molybdenum MSDS.* Willowbrook, IL: Molybdenum.com, p. 60527, 2015.
127. Wang, L.K., Hung, Y.T., Lo, H.H., and Yapijakis, C. *Handbook of Industrial and Hazardous Wastes Treatment.* New York: Marcel Dekker, Inc., 1345p, 2004.
128. Wang, L.K., Hung, Y.T., and Shammas, N.K. *Handbook of Advanced Industrial and Hazardous Wastes Treatment.* Boca Raton, FL: CRC Press, 1378p, 2010.
129. Feng, J., Chuanhua, L., Jinhui, P., Libo, Z., and Shaohua, J. Solvent extraction of copper with laminar flow of microreactor from leachant containing copper and iron. In: *Rare Metal Technology 2015.* Neelameggham, N.R., Alam, S., Oosterhof, I., Jha, A.A., and Wang, S. eds. Hoboken, NJ: John Wiley and Sons, Inc., pp. 45–54, 2015.
130. Yang, Y., Yin, W., Jiang, T., Xu, B., and Li, Q. Research on process of hydrometallurgical extracting Au, Ag, and Pd from decopperized anode slime. In: *Rare Metal Technology 2015.* Neelameggham, N.R., Alam, S., Oosterhof, I., Jha, A.A., and Wang, S., eds. Hoboken, NJ: John Wiley and Sons, Inc., pp. 107–116, 2015.
131. Kim, J., Kim, H., Kim, M., Lee, J., and Kumar, J. Status of separation and purification of rare earth elements from Korean ore. In: *Rare Metal Technology 2015.* Neelameggham, N.R., Alam, S., Oosterhof, I., Jha, A.A., and Wang, S., eds. Hoboken, NJ: John Wiley and Sons, Inc., pp. 117–126, 2015.

132. Nakanishi, B., Lambotte, G., and Allanore, A. Ultra high temperature rare earth metal extraction by electrolysis. In: *Rare Metal Technology 2015*. Neelameggham, N.R., Alam, S., Oosterhof, I., Jha, A.A., and Wang, S., eds. Hoboken, NJ: John Wiley and Sons, Inc., pp. 117–189, 2015.

133. Thriveni, T., Kumar, J., Ramakrishna, C., Jegal, Y., and Ahn, J. Rare earth elements gallium and yttrium recovery from (KC) Korean red mud samples by solvent extraction and heavy metals removal/ stabilization by carbonation. In: *Rare Metal Technology 2015*. Neelameggham, N.R., Alam, S., Oosterhof, I., Jha, A.A., and Wang, S., eds. Hoboken, NJ: John Wiley and Sons, Inc., pp. 157–168, 2015.

134. US EPA. *Rare Earth Elements: A Review of Production, Processing, Recycling, and Associated Environmental Issues*. Cincinnati, OH: US Environmental Protection Agency. EPA/600/R-12/572, Dec. 2012.

135. Chen, J.P., Yang, L., Ng, W.J., Wang, L.K., and Thong, S.L. Ion exchange. In: *Advanced Physicochemical Treatment Processes*. Wang, L.K., Hung, Y.T., and Shammas, N.K., eds. NJ: Humana Press, pp. 261–292, 2006.

136. Shammas, N.K. and Wang, L.K. Alternative and membrane filtration technologies. In: *Water Engineering: Hydraulics, Distribution and Treatment*. Hoboken, NJ: John Wiley and Sons, Inc., pp. 513–544, 2016.

137. Shammas, N.K. and Wang, L.K. Chemical precipitation and water softening. In: *Water Engineering: Hydraulics, Distribution and Treatment*. Hoboken, NJ: John Wiley and Sons, Inc., pp. 593–604, 2016.

138. Wang, L.K., Wu, J.S., Shammas, N.K., and Vaccari, D.A. Recarbonation and softening. In: *Physicochemical Treatment Processes*. Wang, L.K., Hung, Y.T., and Shammas, N.K., eds. NJ: Humana Press, pp. 199–228, 2005.

139. Wang, L.K., Chen, J.P., Hung, Y.T., and Shammas, N.K. *Membrane and Desalination Technologies*. Totowa, NJ: Humana Press, 716p, 2011.

140. Krofta, M. and Wang, L.K. *Flotation Engineering*. Lenox, MA: Lenox Institute of Water Technology, 2000. Technical Manual No. Lenox/1-06-2000/368.

141. Wang, L.K., Shammas, N.K., Selke, W.A., and Aulenbach, D.B. *Flotation Technology*. Totowa, NJ: Humana Press, 680p, 2010.

2 Nano-Bioremediation
Applications of Nanotechnology for Bioremediation

S. Raj Kumar and P. Gopinath

CONTENTS

ABSTRACT

Owing to the rapid growth of industrialization, urbanization, and modern agricultural practices, pollution of stream water or groundwater and soil is on the rise. The biggest challenge to researchers is the removal of contaminants. To fulfill the human desire for energy generation and other needs, natural resources have been exploited resulting in the degradation of water quality and environmental pollution leading to ecological imbalance. Pollution is defined as the presence of pollutants in the environment that causes instability, disorder, harm, and discomfort to the ecosystem, that is, physical systems or living organisms. Present treatment technologies, though efficient, cause several problems which make remediation processes complex. Among these technologies, bioremediation has been prominently practiced as an efficient cost-effective technology for controlling hazardous pollutants like heavy metals in soil and water. This chapter reviews the treatment technologies currently available for removing heavy metals and some of the nanotechnology applications in water treatment. A novel method of nano-bioremediation is effective and more significant for heavy metal removal in all aspects in which the drawbacks of bioremediation can be possibly avoided by the application of nanotechnology.

2.1 INTRODUCTION

Currently, pollution due to heavy metals in wastewater, groundwater, lakes, and streams has caused serious long-term health impacts in human beings. Industrialization makes the conventional methods unsuitable for the treatment due to their nonspecificity, decrease in efficiency, and cost expensive. To overcome these difficulties, few studies have been reported by combining the biological methods with other remediation techniques such as biophysical methods, biochemical methods, physiochemical methods, and nano-based physiochemical methods. This chapter reviews the prevailing treatment methods such as physical, chemical, physiochemical, and biological methods, and mainly focuses on bioremediation, its mechanism, and the applications of nanotechnology in bioremediation. Recent studies had reported that the metal oxide-based nanoparticles can be used as an effective nano-adsorbents for the removal of heavy metals and organic pollutants from water. However, the applications of polymer templated nanoparticles and functionalized nanoparticles for the removal of heavy metals are gaining much attention owing to their promising superiority over the other methods of water treatment. Moreover, in this chapter, several nano-based materials have been discussed in detail which will give an overall knowledge regarding the nano-bioremediation for removing heavy metal contaminants from ground water and industrial effluents. Henceforth, the proposed nanobio approach will be offering the most promising and reliable treatment technology in terms of efficiency and cost in accordance with the socio-economic conditions of the developing nations.

2.2 CURRENT TREATMENT TECHNOLOGIES FOR THE REMOVAL OF HEAVY METALS AND OTHER CONTAMINANTS

2.2.1 PHYSICAL TREATMENT METHODS

1. Precipitation: It is the process in which a suitable anion is added to precipitate the metal salts. The chemicals generally used in this technique are manganese sulfate, copper sulfate, ammonium sulfate, alum, ferric salts, etc. The efficiency is affected by low pH and the presence of salts (ions) and sludge disposal makes the process even more expensive (1). Precipitation using bisulfide, lime, or ion exchange lacks the specificity and is not effective in removing metal ions at lower concentration.

2. Ion exchange: It is the method used in industries for the recovery of heavy metals from effluents. This process consists of a solid-phase ion-exchange material that has the ability to exchange cations or anions. The commonly used ion-exchange matrix material is the synthetic ion-exchange resin. It is relatively expensive and has the ability to achieve parts per billion (ppb) levels even when a relatively large volume is treated. The difficulty in this method is that it cannot handle high metal concentration because of foul formation in the matrix due to the presence of solids and organic materials in the wastewater. In addition, it is not selective in nature and is also highly sensitive to pH variations of the solution (2).

3. Electrowinning: It is a method prominently used in industrial metallurgical and mining processes, such as metal transformation, acid mine drainage, electronics, and electrical industries, and heap leaching for the recovery and removal of heavy metals (2).

4. Electrocoagulation: It is a similar electrochemical-based approach where electric current is used to remove metals from solution. In other words, in this method the contaminants in wastewater are retained in the solution by application of external voltage. When the ions and charged particles are neutralized with ions of opposite electrical charges by the electrocoagulation system, they become destabilized and are precipitated (2).

5. Cementation: It is one of the types of precipitation methods involving an electrochemical approach in which metals with greater oxidation capability flows into the solution. Copper is most frequently separated by cementation and also some of the metals, such as gallium (Ga), lead (Pb), gold (Au), silver (Ag), antimony (Sb), cadmium (Cd), tin (Sn), and arsenic

(As) can be recovered by this method. In adsorption, the adsorbent surface is attached with the metal species by the physisorption and chemisorption process. The metal ion removal efficiency is influenced by various factors, such as pH, surface area of the adsorbent and its surface energy, etc. Commonly used adsorbents are activated alumina, activated carbon, potassium permanganate ($KMnO_4$)-coated glauconite, granular ferric hydroxide, iron-oxide-coated sand, copper–zinc granules, etc. (2).

6. Membrane filtration: It involves the separation of metals from water by passing the metal through a semipermeable membrane with pressure gradient which acts as the driving force. The main drawback of this approach is fouling which occurs due to the coprecipitation of Fe^{2+} and Mn^{2+} ions present in the water. Apart from these, factors such as pressure difference monitoring and water pretreatment make the process more expensive (2).

7. Electrodialysis: It is analogous to the reverse osmosis (RO) process except for the driving force, where an electric field is applied across a semipermeable membrane for separation of charged metal ions from contaminated water. It demonstrates greater efficiency in removing heavy metals from groundwater and is based on parameters such as porosity, pH, groundwater flow rate, texture, ionic conductivity, and water content. This treatment method can be allied with other processes, such as treatment for reactive zones, membrane filtration, flushing of the surfactant, reactive zonal treatment, permeable reactive barriers (PRBs), and bioaugmentation to achieve prosperous remediation goals (2,3).

2.2.2 Chemical Treatment Methods

Large volumes of heavy metal contaminants are dispersed in the groundwater which is very difficult for conventional treatment technologies to handle. Some of the available chemical treatment methods and their drawbacks are discussed below.

1. Reduction: Reductants like gaseous hydrogen sulfide and dithionites are injected deeply into the polluted regions having high permeability with alkaline pH in the soil. Degradation or immobilization of the contaminants takes place in these polluted regions. Formation of toxic intermediates is the drawback associated with the reduction process (4–6). One such example is the colloidal zerovalent ion (ZVI) that can be deeply inserted into the aquifer where it undergoes quick deterioration and toxic substances are produced as byproducts (7,8).

2. Chemical washing: It is the direct method for heavy metal removal by means of strong extractants like acids. This method deteriorates the soil quality which is hazardous to the surroundings. *Ex situ* treatment of the polluted soil is risky, and management problems and hazardous waste treatment is very complex.

3. Chelate flushing: It is the method for extracting a huge quantity of heavy metals, as the active agents being used in the process can be rejuvenated and recycled. Solvent loaded resins are very effective which is 100% regenerative and are used in PRBs. A drawback with this method is that chelates such as ethylenediaminetetraacetic acid (EDTA) and diethylenetriaminepentaacetic acid (DTPA) are expensive and carcinogenic in nature (9–11).

2.2.3 Biological Treatment Methods

The field of biotechnology extends its vast application into the environmental field for water treatment, and is termed bioremediation. In this section, biotechnology-based water treatment technologies so far available and extensively followed are discussed. Bioremediation is now considered as the eco-friendly and cost-effective treatment technology for the elimination of metal pollutants mainly in water and soil. Although bioremediation is preferred, sometimes the contaminants themselves become toxic to the microorganisms involved in the process. These problems led researchers to find an alternative solution by extending bioremediation techniques to obtain high resistance under extreme conditions and stable remediation properties to maintain a high rate of bioremediation.

1. Bioremediation: Bioremediation is the productive utilization of living systems, such as bacteria, fungi, algae, and some plants to degrade, detoxify, transform, immobilize, or stabilize toxic environmental contaminants into an innocuous state or to levels below the concentration limits acceptable by the regulatory authorities. *Escherichia, Citrobacteria, Klebsiella, Rhodococcus, Staphylococcus, Alcaligenes, Bacillus,* and *Pseudomonas* are the organisms that are commonly used in bioremediation. Bioremediation consists of various remediation strategies, such as bioaugmentation, that is, natural attenuation process by using indigenous microorganisms, stimulated process by adding nutrients (biostimulation), use of genetically modified organisms, phytoremediation involving the use of some plants, and biomineralization involving the thorough biodegradation of organic substances into inorganic components (12). Different bioremediation approaches are schematically shown in Figure 2.1.

2. Biofiltration: The biofilter consists of a porous medium in which the surface is covered with water and microorganisms. It is based on the mechanism of complex formation between the contaminants and organic substances in water. The porous medium used in this process adsorbs and is finally biotransformed into metabolic byproducts, biomass, carbon dioxide, and water. The three important activities that take place in a biofilter are the attachment, growth, and degradation and detachment of microorganisms (13,14).

3. Biosorption: It is widely followed biological method based on the materials of biological origin such as biomass obtained from dead or inactive bacteria. It is a passive process where no external energy is required and has many advantages such as highly efficient regeneration of biosorbents, metal recovery, minimal sludge formation and cost effective (15). The biomass acting as a ion exchange matrix binds and exhibits its intrinsic properties in order to remove heavy metals from very dilute aqueous solutions. Biosorbent materials like the cell-wall structure of certain algae, fungi, and bacteria are responsible for this phenomenon (16). On account of all these advantages biosorption is considered to be a better solution for the remediation of metal-containing effluents. Most sorption-based remediation is carried out in single metal ion and only limited studies have been carried out in mixed metal

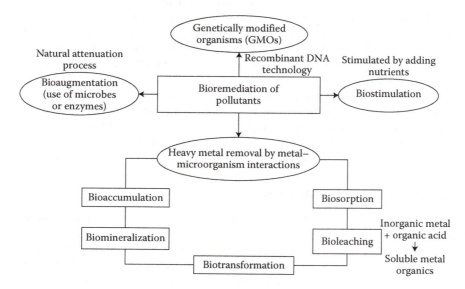

FIGURE 2.1 Overview of bioremediation approaches. (From Joutey, N.T, Bahafid, W., Sayel, H., Ghachtouli, N.E., *Agricultural and Biological Sciences*, 2013. ISBN 978-953-51-1154-2, doi: 10.5772/56194.)

solutions. Several biomass types of bacteria, algae, and yeasts having biosorptive potential have been identified. These low cost biosorbents make the process highly economical and competitive for environmental applications (17). This process consists of several features including the selective removal of metals over a broad range of pH and temperature, rapid kinetics of adsorption and desorption, and low capital and operation cost. The development of such biosorbents is found to have high metal affinity toward the different types of hazardous heavy metals released from the industrial wastes. Subsequently, biosorbents made by the combination of the nonliving biomass comprising distinct types of microorganisms is also used. Apart from these, biosorbents made by the combination of nonliving biomass comprising distinct types of microorganisms is used. The immobilized biomass is recommended for large-scale application compared to native biomass but complete investigation on several immobilization techniques is needed to analyze its efficacy, ease of use, and cost effectiveness. Several methodologies have been carried out to remove the heavy methods such as formation of complexes, using chelating agents, electrostatic interactions, and ion exchange using materials derived from agricultural activities and other natural sources. Also preliminary treatment with chemical agents is prerequisite in order to increase the sorption efficiency and stability. The biosorption approach is lucrative due to the greater adsorption rate, easy availability of biosorbents, low cost and free from toxicity.

4. Biophysiochemical method: It is the method in which the biological process is coupled with an adsorption or coagulation technique. It is considered to be a good alternative remediation technique because of its several advantages over other conventional physiochemical treatment methods. Further, biological processes hold much promise in sludge disposal protocols and also act as an integral component of any arsenic treatment technology (18). Acidithiobacillus ferrooxidans BY-3, a chemolithotrophic bacterium, acts as a natural biosorbent which is isolated from the mines, has been extensively used for the removal of organic and inorganic arsenic compounds from the aqueous solutions (19). Srivastava et al. (20) isolated five different fungal strains and used them for the removal of arsenic from contaminated sites. Another approach is the biotic oxidation of iron using microorganisms *Gallionella ferruginea* and *Leptothrix ochracea*. It is based on the mechanism that makes for the deposition of iron oxides in the filter medium along with the microorganism that provides the supportive environment for metal adsorption and removal from the aqueous solution. It avoids the use of chemical reagents for the oxidation of trivalent arsenic, does not require monitoring of a breakthrough point, as in the sorption processes, because the iron oxides are continuously produced *in situ*. Another important advantage is that it is a combination of biological oxidation–filtration–sorption processes that can be applied for the simultaneous removal of several inorganic contaminants, such as iron, manganese, and arsenic from groundwater (21).

5. Novel biosorbents: Novel biosorbents have been developed to enhance the selectivity and accumulating properties of the microorganisms used for the bioremediation process. A genetic engineering technique-based biosorption process has the ability to increase the remedial activity of the microorganisms. Future study involves the development of engineered microorganisms having higher adsorption capacity and specificity for the toxic metal ions. Few investigations have been carried out for the determination of compatibility of these biosorbents for treating industrial effluents. Overall, it is not a small feat to replace well-established conventional techniques. Although it is costly, the technique has huge potential as many biosorbents are known to perform well. Therefore, extensive research on pilot and full-scale biosorption process is necessary (22).

6. Bioaugmentation: It is an *in situ* bioremediation technique in which genetic engineering has been used to increase the metabolizing ability of microorganisms. Kostal et al. (23) worked on the genetic manipulation of *E. coli* strain and over expressed ArsR genes which results in the accumulation of As. It is considered to be an effective method to increase the accumulation and binding of arsenic in selective ligands and for its removal (23). Chauhan

et al. (24) developed a collection of metagenomes from the sludge of industrial effluent treatment plants, and discovered a new As(V) resistance gene (arsN) which encodes a protein similar to acetyl transferase. Over expression of this protein leads to higher arsenic resistance in *E. coli*. Similarly new innovative approaches in biology, such as metagenomic studies, directed evolution, and genome shuffling can be used for developing new arsenic-resistant pathways which are suitable for arsenic remediation (25). This was well explained by the modification of an arsenic-resistance operon by DNA shuffling (26).

7. Bacterial sulfate reduction (BSR): Jong and Parry (27) treated arsenic and other acidic metals, such as magnesium (Mg), aluminum (Al), copper (Cu), iron (Fe), zinc (Zn), nickel (Ni), and sulfate contaminants in an upflow anaerobic packed-bed reactor using sulfate-reducing bacteria. More than 77.5% of the initial concentrations of arsenic metal were removed and it was well supported by the work carried out by Simonton et al. (28) who demonstrated the effective removal for chromium (Cr) and arsenic (>60%–80%) metals from solution using sulfate-reducing bacteria *Desulfovibrio desulfuricans*. In a similar work, Steed et al. (29) also developed a sulfate-reducing biological process with the aim of removing heavy metals from acid mine drainage. Apart from these, Fukushi et al. (30) used another indirect method to remove arsenic by sequestering the metal into insoluble sulfides by the metabolic action of the sulfate-reducing bacteria, using a wide range of organic substrates with SO_4^{2-} as the terminal electron acceptor under anaerobic conditions. Microbial mediated conversion between As(III) and As(V) increases the arsenic mobility because As(III) is more mobile and toxic compared to As(V).

8. Phytoremediation: There are several plant species available for soil remediation which can uptake contaminants from soil, surface water, groundwater, and sediments. Phytofiltration employs the ability of plants in accumulating and showing effective tolerance to heavy metals like arsenic. Usually hydrilla plants are used in phytofiltration and an improvement in this process can be sought by growing these plants in actual field conditions in the presence of contaminated water. Phytoextraction is the uptake of contaminants by plant roots and translocation within the plants. This method is applied to metal-contaminated soil but is often associated with several disadvantages, such as removal of plant biomass, metal reclamation, disposal of biomass, and phototoxic effect of the metals. Phytodegradation, also known as phytotransformation, stands for the breakdown of contaminants by plants through metabolic processes within the plant, or *ex situ* breakdown of contaminants through the effect of enzymes produced by the plants. The degradation of the products and formation of toxic intermediates are the main problem associated with this process. Another plant-mediated heavy metal sequestration process is phytovolatilization where plants take up the contaminants which are then transpired with release of the contaminant or modified forms of the contaminants from the plants to the atmosphere (31–34).

Various conventional water treatment methods presently followed are summarized in Table 2.1.

2.3 NANOTECHNOLOGY

Nanomaterials (NMs) are defined as the materials having size ranging from 1 through 100 nm with a minimum of one dimension. In this small scale, NMs possess unique properties compared to the other materials. Many of these materials have been explored for application in wastewater treatment. They utilize the size-dependent properties of NMs, such as high surface area, high reactivity, strong sorption, and faster dissolution. Although many nano-based technologies are successful on the laboratory scale, only few technologies have been used for small-scale testing or commercialization. Such nano-based technologies include nanotechnology-associated membranes, nano-adsorbents, and nano-photocatalysts. Although products based on these three approaches have been commercialized, they are not successful in large-scale wastewater treatment.

TABLE 2.1

Classification of Conventional Water Treatment Methods for Heavy Metal Remediation

Methods	Mechanism of Action	Significance	Drawbacks	Targets	References
Precipitation	Generation of iron and other metal precipitates with reduction and precipitation of other metals	Perform similar to natural process	Clogging and corrosion of ZVIs	Zn, Cu, Pb, Se, Ca, Mn, Ni, Cd, Al, Mg, Cr, As, Sr, Co	(35–40)
Denitrification and BSR	Formation of sulfides from divalent metals and hydroxides from trivalent metals	95% metal removal in PRBs	Continuous supply of nutrients is required. Steady supply of nutrients should be provided	Fe, Ni, Zn, Al, Mn, Cu, U, Se, As, V, Cr	(41–44)
Absorption:					
1. Inorganic surfactants	Sorption of metal is dependent on the charge carried by the surfactant	Complex formation with surfactants	Aquifer with maximum permeability is required	Cd, Pb, Zn, As, Cu, Ni	(9,11,45)
2. Industrial byproducts	Surface site adsorption	Raw materials from industries	Field application is necessary	As, Cd, Pb	(46–49)
3. Ferrous materials	Metal sorption of iron oxide and its derivatives	As(V) with iron form inner sphere complexes	Oxidation is difficult and materials to be replaced frequently	As(V), Cr, Hg, Cu, Cd, Pb	(50–54)
Membrane and filtration technology	Slow electric charges, complexation, dialysis, arresting of micelles in three-dimensional structure	very high removing efficiency	Blockage of filter, rejuvenate or restoration of filter materials	Tc, Hg, Cu, Pb, Cr, Zn, As, U, Cd	(10,55,56)
Reduction:					
1. Using dithionites	Precipitation at alkaline pH	Active over larger area	Formation of toxic gas intermediate and handling is difficult	Cr, U, Th	(4–6)
2. Using ZVI and colloidal Fe	Precipitation and sorption on ZVI	No toxic exposure in deep aquifers		CrO_4^{2-}, TcO_4^{-}, UO_2^{2+}, As	(7,8,57,58)
Chelate flushing	Formation of stable complexes	Action of ligands at very low dose and regeneration	Persistent, toxic, and expensive	Fe, Cu, Cr, As, Hg, Pb, Zn, Cd	(59–62)

(Continued)

TABLE 2.1 (*Continued*)

Classification of Conventional Water Treatment Methods for Heavy Metal Remediation

Methods	Mechanism of Action	Significance	Drawbacks	Targets	References
Ion exchange	Liquid–liquid extraction and separation of solid phase	Selectivity in removing low level of metal ions	High cost and contaminant specific	Heavy metals and transition metals	(62)
Biological methods: Activity in the subsurface	Oxidation, precipitation, and bioaccumulation	Very low cost and effective over long time	Not applicable for aquifers, slow method, modeling is impossible	Cu, Ni, Cr, Cd, Zn, Co, Pb	(63–66)
BSR	Reduced to precipitates, catalyzed by the SRB	Onsite treatment, offsite application in bioreactors and also applied in PRBs	Limited rate of the reaction and residence time is required	Divalent metal cations	(67–71)
Biosorption	Plants, fungus, bacteria, and DNA aptamers recover metals from the cytoplasm	Absorption of metals, such as Fe, Zn, Ni, and As. Anionic interference is absent	Due to the high acidic conditions desorption of heavy metals occurs	Fe, Zn, Ni, As, Cr, Cd	(34,72–74)
Cellulosic materials and agricultural wastes	Heavy metals are adsorbed at pH 4–6 in the cellulose material	High concentration of metals are treated using cheap cellulose materials	Field study has not been done	Pb, Ni, Cu, Cd, Zn	(75–80)

2.3.1 Applications of Nanotechnology for Heavy Metal Remediation

In the recent past, applications of nanotechnology for feasible solutions in wastewater treatment have been adopted. The different nanoparticles with desirable properties and their application in wastewater treatment for the removal of heavy metals and other contaminants are discussed (81). NMs having unique properties combined with conventional treatment techniques provide wide opportunities to make dramatic changes in wastewater treatment methodologies. The combination of both nanoscience and engineering offers better opportunities in the restoration of heavy metal-contaminated groundwater.

2.3.2 Need for New Technology

Amid all the water contaminants, heavy metals play a major role in causing severe health-associated complications in human beings and animals. This is due to their nonbiodegradable characteristics and extreme toxicity. Itai-Itai is a well-known Japanese disease caused due to the prolonged exposure of the heavy metal chromium. The severe and prolonged disorders caused by chromium are hypertension, skeletal malformation in the fetus, testicular atrophy, renal damage, and emphysema. Therefore, it is essential that new innovative approaches are required in order to eradicate these hazardous metals from water.

Different varieties of novel materials, such as graphene derivatives (82), carbon based sorbents (83), chelates (84), activated carbons (85), chitosan/natural zeolites (86), and clay minerals (87) are being investigated with the aim of adsorbing heavy metal ions from aqueous solutions. Due to lower adsorption capabilities and efficiencies, the conventional adsorbents commonly used have been limited in heavy metal remediation (88). The applications of various NMs for the removal of heavy metals are briefly explained below.

2.3.3 APPLICATIONS OF VARIOUS NANOMATERIALS FOR THE REMOVAL OF HEAVY METALS FROM WASTEWATER

NMs have wide applications in various fields. Recently they are being applied in the area of water purification in order to reduce the bulk concentrations of toxic substances, such as radionuclides, metal ions, and organic and inorganic compounds to the ppb levels (89). Magnetite nanoparticles (Fe_3O_4) impregnated with a silica compound is used for the removal of a large amount of toxic substances usually existing in the environment, also for the biological separation of cells and remediation purposes (90). Apart from this, nanostructured silica alone can also be applied in wastewater treatment in order to eliminate heavy metal ions (91).

2.3.4 NANOPARTICLES AS ADSORBENTS

NMs exhibit good adsorbent properties due to their larger surface area and high affinity toward the target compound/compounds when functionalized with various chemical groups. For the removal of heavy metals, the adsorption process is considered to be the better remediation option compared to conventional treatment technologies. The adsorption process has several advantages, such as greater efficiency, operational simplicity, and cost effectiveness (92). Several efforts have been made to apply these unique materials for developing selective sorbents with high capacity for removing heavy toxic metal ions from groundwater. Some of the effective adsorbents are discussed below.

1. Polymers as nano-adsorbents: Polymers such as dendrimers are good adsorbents capable of removing both heavy metals and organic compounds. The sorption mechanism of polymers includes electrostatic interactions, hydrophobic effects, bond formation between hydrogen atoms, and complexation (93). The interior portion of dendrimers is hydrophobic in nature for adsorption of organic compounds and the exterior portion is linked either with a hydroxyl or amine group for heavy metal adsorption. The dendrimer is associated with an ultrafiltration system in order to recover Cu^{2+} ions from aqueous solutions and remove the metal ions at an initial concentration of 10 ppm (parts per million) (94). After the adsorption process, the dendrimers along with heavy metal ions were recovered and regenerated by a filtration system with a pH as low as 4.

   The case study on the adsorption mechanism of chitosan biopolymer for arsenic is discussed below. When the chitosan-grafted biopolymer adsorbs metal ions, it undergoes any one of the following mechanisms, such as chelation of metals, formation of ion pairs, and electrostatic interactions (95). In certain cases, it also undergoes complexation and diffusion mechanisms (96) as a result of van der Waals and hydrophobic interactions, hydrogen bond formation, and the ion-exchange process (97). The promising adsorbents, such as iron-oxide-coated sand (98) and magnetic nanoparticles (99) encapsulated with polymer exhibits very high potential for heavy metal remediation.

2. Other nanosorbents: The adsorption of Pb(II), Cd(II), and Ni(II) ions from aqueous solution using chitosan methacrylic acid (MAA) nanoparticles was studied (100). Two types of akaganeite materials were prepared and tested for the sorption of heavy metals such as antimony (Sb) and arsenic derivatives (101). It has also been reported that the nanocrystalline akaganeite (–FeOOH)-coated quartz sand (CACQS) can be used for bromate removal from aqueous solutions (101). Some of the nano-adsorbents used for the deduction of arsenic are already in commercial use. Their costs and performance have been compared with former adsorbents in preliminary studies. One such adsorbent that has been commercialized includes ArsenXnp which is a medium based on hybrid ion-exchange material consisting of polymers and oxidized nanoparticles. Similarly, ADSORBSIA is a medium made up of nanocrystalline titanium dioxide beads where the size ranges from 0.25 to 1.2 mm in diameter. These two adsorbents are highly efficient in removing heavy metals like arsenic whereas slight backwash is required for the ArsenXnp (103). Both ArsenXnp and ADSORBSIA are proved as being cost effective and have been put to use for small scale to medium scale drinking water treatment systems.

   Researchers are continually giving their efforts in developing new adsorbents by scientifically exploring the design of nanoparticles and their adsorption capacity in order to remove hazardous metals present in drinking water (104). The combined effect of two compounds namely iron metals and carbonaceous substances show enhanced removal of pollutants as compared to the individual compounds. In a similar way, by means of chemical reactions a magnetite/graphene oxide nano composite has been synthesized with a particle size of about 10–15 nm to remove cobalt ions in the aqueous solution of about 22.70 mg/g at the temperature of 343 K (105).

3. Metal-based nano-adsorbents: Metallic nanoparticles are being investigated with the aim of removing heavy metals, such as mercury, nickel, copper, arsenic, cadmium, chromium, lead, etc. Calcium-doped zinc oxide nanoparticles as a selective adsorbent for the extraction of lead ion was explored (106). Many metal oxidized NMs along with nanosized magnetite and titanium dioxide outcompetes the adsorption capacity of activated carbon (107). In another study, metal hydroxide nanoparticles have been loaded on the activated carbon in order to remove arsenic and other organic pollutants accordingly (108,109).

   Nanoparticles produced from metal or metal oxides that are extensively used in wastewater treatment for eradicating the heavy metals are manganese oxides (110), copper oxides (111), cerium oxide (112), magnesium oxides (113), titanium oxides (114), silver nanoparticles (115), and ferric oxides (116).

   Huang et al. (117) synthesized the titanate nanoflowers with a large surface area and demonstrated the heavy metal removal capability of these titanate nanoflowers. Relative experiments show that the synthesized nanoflowers exhibit greater adsorption capacity as compared to titanate nanotubes and nanowires. Also the titanate nanoflowers exhibit high selectivity for toxic metal ions. The results were found to conform to the standard adsorption Langmuir model and pseudo-second-order kinetics (117). Further, Zhang et al. prepared arrays of magnesium hydroxide nanotubes to form $Mg(OH)_2/Al_2O_3$ composites in order to remove nickel ions from contaminated water (118).

4. Iron-based NMs: The selection of a suitable process for wastewater treatment is an intricate assignment, because the selection is based on several factors, such as standard quality, cost, and efficiency. Iron nanoparticles and polymer coated nanoparticles plays a significant role in the removal of heavy metals such as Cr(VI) and As(III) (119,120). The common iron-based NMs used for remediation are nanosized zero-valent ion (NZVI), iron sulfide nanoparticles, bimetallic Fe nanoparticles, and nanosized FeO (121). Studies carried out by Kanel et al. (122) suggested nano ZVI to be a better alternative for arsenic remediation. The removal of Cr(VI) and Pb(II) from aqueous solutions using supported, nanoscale ZVI was carried out (123). Zhong et al. (124) synthesized iron-oxide nanoparticles by using the

evolution process in ethylene glycol medium and demonstrated the outstanding ability to remove heavy metals and other contaminants in wastewater. A detailed comparative study of field applications of nano-, micro-, and millimetric ZVI for the remediation of contaminated aquifers was carried out to determine the efficiency of size-based ZVI NMs (125).

Applications of iron-oxide NMs in water treatment are divided into two groups, that is, nano-adsorbent or immobilized carriers to increase removal efficiency and as photo catalysts to break down the hazardous toxic contaminants into a less toxic material. A review on the applications of iron-oxide NMs is schematically shown in Figure 2.2. A few iron oxide constructed methodologies had been suggested and implemented in wastewater treatment plants; however, several methods are still at the experimental stage. The iron-oxide NMs are considered to be the effective methodology for the absorption of heavy metals and organic pollutants. Iron oxide NMs face some impending problems when applied to *in vitro* and *in vivo* studies. The impact of these NMs is becoming critical due to the discharges occurring in the environment. The increase in these toxic discharges along with industrial growth will encourage researchers to assess future risks. In addition to all these, a few other controversial issues over the human health and environmental outcomes of these unique materials need to be addressed.

5. Photocatalytic NMs: Photocatalysis is an advanced oxidation process (AOP) which is used in the deterioration of organic contaminants in a modest and effective way. Oxidation process through photocatalysis is an innovative method to remove trace amounts of pathogens and pollutants. It is considered to be the significant pretreatment method for the eradication of nonbiodegradable and toxic pollutants and thereby increases their decontamination activity. Modification of the nanoparticles through catalyzation and by other means had been produced to increase the remediation speed and efficiency (127). The chief obstacles for their broad application in decontamination process are diminished photocatalytic activity and slower kinetic reaction (81). In these photocatalytic materials, nanosized semiconductor materials, such as zinc oxide (ZnO), titanium dioxide (TiO$_2$), tungsten oxide (WO$_3$), and cadmium sulfide (CdS) are categorized under various processes, such as conjugated adsorption along the electrical double layers, high-adsorption surface area, and photochemical activity. Also these materials are immediately available, low cost, and have low toxicity (128–130).

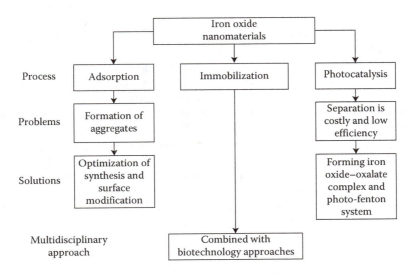

FIGURE 2.2 Applications of iron-oxide NMs for removal of heavy metals. (From Xu, P. et al. *Science of the Total Environment.* 424, 1–10, 2012.)

Electrochemical processes and technologies in wastewater treatment processes and titanium-based photocatalytic interactions have been investigated for application on a large scale to mitigate the problems faced in wastewater treatment (131–133). Many researchers worked on the interaction of TiO_2-based nanoparticles in the form of filters, membranes, or in colloidal form with biomolecules, to understand their remediation strategy for various applications (132,134–141).

6. Nanobiomaterials for heavy metal remediation: Studies were conducted on some bacteria to produce an iron sulfide compound which acts as an adsorbent for several toxic metal ions (142). Apart from bacteria, *Noaea mucronata* is a plant species used for the accumulation of heavy metals, such as lead, copper, cadmium, zinc, iron, and nickel. The nanoparticles obtained from this plant are used for the bioremediation of heavy metal contaminants from groundwater, streams, and rivers. The study results envisage that the initial concentrations of the above mentioned heavy metals decreased relatively after 3 days of remediation (143). The study conducted on plant species such as *Centaurea virgata*, *Scariola orientalis*, *Noaea mucronata*, *Chenopodium album*, *Cydonia oblonga*, *Reseda lutea*, and *Salix excelsa* revealed that these plants are very good heavy metal accumulators. Specifically *Noaea mucronata* is a suitable accumulator for lead to a level more than 1000 ppm (144).

7. Carbon-based nanoparticles: Carbon-based NMs are extensively used for the removal of heavy metals because of its nontoxicity and greater adsorption capacity (145). The first used adsorbent commonly used for metal ion removal is activated carbon, but it is difficult for activated carbon to reduce up to ppb levels. After advances in the emerging field of nanotechnology several innovative unique materials, such as fullerenes, graphenes, and carbon nanotubes (CNTs) (145) are used as adsorbents.

Though activated carbon is a better adsorbent for organic and inorganic pollutants, it has some limitations in the applicability to most heavy metals, specifically to arsenic As(V) (146). The two main features necessary for an effective adsorbent are large surface area and the presence of functional groups. Inorganic adsorbents generally do not possess both properties simultaneously. CNT sheets have been used as an adsorbent for divalent heavy metals such as Cu^{2+}, Zn^{2+}, Pb^{2+}, Cd^{2+}, and Co^{2+} (147). Carbon based polymeric NMs such as polystyrene and acrylics are having high surface area as well as intrinsic functional groups requisite for the adsorption of inorganic contaminants (148,149). CNTs exhibit greater efficiency in adsorbing a wide variety of organic compounds compared to activated carbon (150). This property is mainly due to the interactions between the pollutant and the CNTs due to their large surface area.

CNTs exhibit rapid kinetic potential due to their higher adsorption capacity in the case of metal ions. CNTs having surface functional groups, such as carboxyl, hydroxyl, and phenol are the most important adsorption sites for metal ions. This adsorption process generally occurs through chemical bond formation and electrostatic attraction. Hence the surface oxidation could extensively increase the adsorption ability of the CNTs. Several works have demonstrated that CNTs are good adsorbents in comparison to activated carbon for the removal of heavy metals, such as copper (Cu^{2+}), cobalt (Co^{2+}), cadmium (Cd^{2+}), and (Zn^{2+}) (151,152). The adsorption potential of CNTs is faster because of their large accessible surface area and mesoporous structure. In general, however, adsorbents like CNTs cannot be completely used as a good substitute in the place of activated carbon. The surface chemistry of these CNTs can also be modified according to the target specific pollutants; so that it can be used for the removals of intractable compounds or trace organic contaminants.

There are two types of CNTs, that is, single-walled carbon nanotubes (SWCNTs) and multiwalled carbon nanotubes (MWCNTs). SWCNTs possess high antimicrobial activity whereas MWCNTs have both antimicrobial properties and the adsorption ability for the removal of heavy metals. CNTs are extensively used in wastewater treatment for heavy metal removal. Firstly, CNTs are used independently as adsorbents for divalent metal ions. Pyrzyñska and Bystrzejewski (153) conducted adsorption studies on metals, such as cobalt and copper, and found that NMs like CNTs and carbon-encapsulated magnetic nanoparticles show greater adsorption ability compared to activated carbon and also suggested the restrictions and benefits of heavy metal adsorption of these materials. Also, the study conducted by Stafiej and Pyrzynska (154) suggested that the adsorption capacity is affected by factors like metal ion concentrations and pH. This result has been cross checked with the Freundlich adsorption model and positive results were obtained. Functionalization of CNTs for adsorption is discussed below.

CNTs oxidized with acids consists of several functional groups, such as carboxyl (—COOH), carbonyl (—C=O), and hydroxyl (—OH) groups on the CNTs' surfaces (155–157). The noncovalent functional methods are usually practiced in the CNTs. These functionalized CNTs are developed with the aim of increasing water solubility which can be utilized for several field applications (158,159). The adsorption capacity of MWCNTs is enhanced by the process of oxidation and functionalization of amino group that results in the adsorption of cadmium removal. This action gives the opportunity that the amino-functionalized multiwall carbon nanotubes (MWCNTs) can be used for the development of filtration membranes required for the removal of heavy metal ions from industrial wastewater even at a very high temperature (160). In order to verify the heavy metal ion removal efficiency of MWCNTs, an extensive study has been conducted by comparing the heavy metal removing potential of MWCNTs in two samples, one from Red Sea water (RSW) and another from the King Abdulaziz University wastewater (KAUWW) treatment plant. The results showed that heavy metal removal efficiency of MWCNTs is higher in the samples of KAUWW compared to the RSW samples. This is due to the surplus concentrations of metals, such as magnesium, sodium, calcium, and potassium. These metal ions exhibit a screening property that shrinks the adsorption capacity of MWCNTs (161,162).

Another carbon-based materials used as an adsorbent is graphene. It is made up of single or multiple layered atomic graphites, consisting of a two-dimensional structure which possesses significant thermal and mechanical properties. Graphene oxide (GO) based layered nanosheets was developed by Zhao et al. (2011) (82) using Hummer's method for the removal of heavy metals such as cobalt and cadmium where the sorption activity depend upon various parameters such as pH, ionic strength and the functional group etc. Chandra et al. (2010) showed that magnetite–graphene nanoparticles with size of 10 nm exhibit greater binding capability for arsenic owing to the enhanced sites of adsorption in the graphene composite materials (163).

8. Nanofibers: Electrospinning is a plain, proficient, and cheaper method for the production of ultra-fine nanofibers by means of resources such as metals, polymers, or ceramics (164,165). These nanofibers form mats with complex pore structures due to their porosity and higher surface area. The physical properties such as morphology, composition, diameter, spatial arrangement, and secondary derivatives of these electrospun nanofibers are fabricated depending upon particular fields of application (165). Nanofiber membranes have been commercially used as air filters, but their application in treating wastewater has not yet been evaluated. The membranous nanofibers removes microparticles at a higher elimination rate but there is no foul formation from the aqueous solutions. Hence, it has been deployed as the preliminary treatment preceding the ultrafiltration or RO processes (166). NMs with certain functional groups are mixed with the spinning solutions in order to produce nanofibers in situ or nanoparticles associated with nanofibers (165). These electrospun nanofibers are used for the building up of membranous filters for multipurpose use either by means of materials like titanium dioxide (TiO_2) or by using the functionalized

NMs. Electrospun polyacrylonitrile nanofiber mats are used for heavy metal ion removal because of their tremendous potential as a heterogeneous adsorbent for metal ions (167). Carbon nanofibers grown on iron (Fe) are being used for the removal of arsenic (V) in wastewater (168). The fabrication and characterization of poly(ethylene oxide) templated nickel oxide nanofibers for dye degradation was studied (169).

2.4 NANO-BIOREMEDIATION OF HEAVY METALS IN WATER

There is urgency for the development of innovative treatment technologies because new water quality standards have been promulgated for water treatment. Remarkably, nanotechnology is the science and art of manipulating matter at the atomic and molecular level, which has the potential to enhance environmental water quality and sustainability through various routes, such as water treatment, prevention of pollution, and remediation processes. It is being evolved as a green technology that can enhance the environmental performance and economic development of industries, and reduce resource consumption and energy requirements. Hence, much attention has been focused on the development and potential benefits of NMs in water treatment processes. However, concerns regarding their potential effects on human and environmental toxicity have been raised. If these gaps are assessed carefully, these NMs will play a cardinal role in ensuring good quality water and soil to meet the ever-increasing demand for potable water and safe soil for agricultural activities (170,171). The United States Environmental Protection Agency (USEPA) supports research on novel remediation approaches for the removal of heavy metals and contaminants which are efficient and cost effective compared to conventional water treatment techniques. This chapter therefore, emphasized on the novel technique of "nano-bioremediation" that has the potential not only to reduce the overall costs of cleaning up large-scale contaminated sites, but can also reduce processing time.

2.5 CONCLUSION

Heavy metals that are predominantly considered as toxic materials that need immediate remediation are lead, mercury, copper, arsenic, cadmium, chromium, etc. Emerging applications of NMs will endeavor to find an effective remediation solution for removing these heavy metals. Furthermore, specific control and design of NMs at the molecular level will give increased affinity, capacity, and selectivity of pollutants which lead to the reduced releases of hazardous substances into the air and water, providing safe drinking water. This chapter finally concludes that the proposed nano-bioremediation can be termed nano-renovogen which is a future nanotechnology-based bioremediation process for contaminant removal.

ACKNOWLEDGMENTS

We give sincere thanks to the Department of Science and Technology (Water Technology Initiative—Project No. DST/TM/WTI/2K13/94 (G)), Government of India, for financial support. SR is thankful to the Ministry of Human Resource Development, Government of India, for his fellowship.

REFERENCES

1. Mondal, P., Majumdar, C.B., Mohanty, B. Laboratory based approaches for arsenic remediation from contaminated water: Recent developments. *Journal of Hazardous Materials*. B137, 464–479, 2006.
2. Ahluwalia, S.S., Goyal, D. Microbial and plant derived biomass for removal of heavy metals from wastewater. *Bioresource Technology*. 98, 2243–2257, 2007.
3. Gray, N.F. *Water Technology*. John Wiley & Sons, New York, 473–474, 1999.

4. Amonette, J.E., Szecsody, J.E., Schaef, H.T., Templeton, J.C., Gorby, Y.A., Fruchter, J.S. Abiotic reduction of aquifer materials by dithionite: A promising in situ remediation technology. *Proceedings of the 33rd Hanford Symposium on Health and the Environment*. Battelle Press, Columbus, OH, Pasco, Washington, vol. 2, 851–881, 1994.

5. Fruchter, J.S., Cole, C.R., Williams, M.D., Vermeul, V.R., Teel, S.S., Amonette, J.E., Szecsody, J.E., Yabusaki, S.B. *Creation of a Subsurface Permeable Treatment Barrier Using In Situ Redox Manipulation*. Pacific Northwest National Laboratory, Richland, WA, 1997.

6. Sevougian, S.D., Steefel, C.I., Yabusaki, S.B. Enhancing the design of in situ chemical barriers with multicomponent reactive transport modeling, in situ remediation: Scientific basis for current and future technologies. *Proceedings of the 33rd Hanford Symposium on Health and the Environment*. Battelle Press, Columbus, OH, Pasco, Washington, 1–28, 1994.

7. Cantrell, K.J., Kaplan, D.I., Wietsma, T.W. Zero-valent iron for the in situ remediation of selected metals in groundwater. *Journal of Hazardous Materials*. 42, 201–212, 1995.

8. Manning, B.A., Hunt, M.L., Amrhein, C., Yarmoff, J.A. Arsenic (III) and Arsenic (V) reactions with zerovalent iron corrosion products. *Environmental Science and Technology*. 36, 5455–5461, 2002.

9. Mulligan, C.N., Yong, R.N., Gibbs, B.F. Remediation technologies for metal contaminated soils and groundwater: An evaluation. *Engineering Geology*. 60, 193–207, 2001.

10. Sikdar, S.K., Grosse, D., Rogut, I. Membrane technologies for remediating contaminated soils: A critical review. *Journal of Membrane Science*. 151, 75–85, 1998.

11. Torres, L.G., Lopez, R.B., Beltran, M. Removal of As, Cd, Cu, Ni, Pb, and Zn from a highly contaminated industrial soil using surfactant enhanced soil washing. *Physics and Chemistry of the Earth, Parts A/B/C*. 37–39, 30–36, 2012.

12. Joutey, N.T., Bahafid, W., Sayel, H., Ghachtouli, N.E. Biodegradation: Involved microorganisms and genetically engineered microorganisms. *Biodegradation-Life of Sciences*, eds., R. Chamy and F. Rosenkranz, 290–320, 2013. ISBN 978-953-51-1154-2, doi: 10.5772/56194.

13. Devinny, J.S., Deshusses, M.A., Webster, T.S. *Biofiltration for Air Pollution Control*. CRC press, Boca Raton, FL, 1–320, 1999.

14. Srivastava, N.K., Majumdar, C.B. Novel biofiltration methods for the treatment of heavy metals from industrial wastewater. *Journal of Hazardous Materials*. 151, 1–8, 2008.

15. Hashim, M.A., Mukhopadhyay, S., Sahu, J.N., Sengupta, B. Remediation technologies for heavy metal contaminated groundwater. *Journal of Environmental Management*. 92, 2355–2388, 2011.

16. Volesky, B., Holan, Z.R. Biosorption of heavy metals. *Biotechnology Progress*. 11, 235–250, 1995.

17. Wang, J., Chen, C. Biosorbents for heavy metals removal and their future. *Biotechnology Advances*. 27, 195–226, 2009.

18. Jain, C.K. and Singh, R.D. Technological options for the removal of arsenic with special reference to South East Asia. *Journal of Environmental Management*. 107, 1–18, 2012.

19. Yana, L., Yina, H., Zhang, S., Lenga, F., Nana, W., Li, H. Biosorption of inorganic and organic arsenic from aqueous solution by *Acidithiobacillus ferrooxidans* BY-3. *Journal of Hazardous Materials*. 178, 209–217, 2010.

20. Srivastava, P.K., Vaish, A., Dwivedi, S., Chakrabarty, D., Singh, N., Tripathi, R.D. Biological removal of arsenic pollution by soil fungi. *Science of the Total Environment*. 409, 2430–2442, 2011.

21. Pokhrel, D., Viraraghavan, T. Biological filtration for removal of arsenic from drinking water. *Journal of Environmental Management*. 90, 1956–1961, 2009.

22. Vijayaraghavan, K., Yun, Y.S. Bacterial biosorbents and biosorption. *Biotechnology Advances*. 26, 266–291, 2008.

23. Kostal, J.R.Y., Wu, C.H., Mulchandani, A., Chen, W. Enhanced arsenic accumulation in engineered bacterial cells expressing ArsR. *Applied and Environmental Microbiology*. 70, 4582–4587, 2004.

24. Chauhan, N.S., Ranjan, R., Purohit, H.J., Kalia, V.C., Sharma, R. Identification of genes conferring arsenic resistance to *Escherichia coli* from an effluent treatment plant sludge metagenomic library. *FEMS Microbiology Ecology*. 67, 130–139, 2009.

25. Dai, M.H., Copley, S.D. Genome shuffling improves degradation of the anthropogenic pesticide pentachlorophenol by *Sphingobium chlorophenolicum* ATCC 39723. *Applied and Environmental Microbiology*. 70, 2391–2397, 2004.

26. Crameri, A., Dawes, G., Rodriguez, E., Silver, S., Stemmer, W.P.C. Molecular evolution of an arsenate detoxification pathway DNA shuffling. *Nature Biotechnology*. 15, 436–438, 1997.

27. Jong, T., Parry, D.L. Removal of sulphate and heavy metals by sulphate reducing bacteria in short-term bench scale up flow anaerobic packed bed reactor runs. *Water Research*. 37, 3379–3389, 2003.

28. Simonton, S., Dimsha, M., Thomson, B., Barton, L.L., Cathey, G. Long-term stability of metals immobilized by microbial reduction. *Proceedings of the 2000 Conference on Hazardous Waste Research: Environmental Challenges and Solutions to Resource Development, Production and Use*, Southeast Denver, CO, 394–403, 2000.

29. Steed, V.S., Suidan, M.T., Gupta, M., Miyahara, T., Acheson, C.M., Sayles, G.D. Development of a sulfate-reducing biological process to remove heavy metals from acid mine drainage. *Water Environment Research*. 72, 530–535, 2000.

30. Fukushi, K., Sasaki, M., Sato, T., Yanase, N., Amano, H., Ikeda, H. A natural attenuation of arsenic in drainage from an abandoned arsenic mine dump. *Applied Geochemistry*. 18, 1267–1278, 2003.

31. Jadia, C.D., Fulekar, M.H. Phytoremediation of heavy metals: Recent techniques. *African Journal of Biotechnology*. 8(6), 921–928, 2009.

32. Pivetz, B.E. *Phytoremediation of Contaminated Soil and Ground Water at Hazardous Waste Sites*. Environmental Research Services Corporation. EPA/540/S 01/500, 2001.

33. Ali, H., Khan, E., Sajad, M.A. Phytoremediation of heavy metals—Concepts and applications. *Chemosphere*. 91, 869–888, 2013.

34. Srivastava, S., Shrivastava, M., Suprasanna, P., D'Souza, S.F. Phytofiltration of arsenic from simulated contaminated water using *Hydrilla verticillata* in field conditions. *Ecological Engineering*. 37, 1937–194, 2011.

35. Faulkner, D.W.S., Hopkinson, L., Cundy, A.B. Electro kinetic generation of reactive iron-rich barriers in wet sediments: Implications for contaminated land management. *Mineralogical Magazine*. 69, 749–757, 2005.

36. Hopkinson, L., Cundy, A.B. FIRS (ferric iron remediation and stabilization): A novel electrokinetic technique for soil remediation and engineering. *CL: AIRE Research Bulletin*. 2003. Available from: http://www.claire.co.uk/.

37. Jeen, S.W., Gillham, R.W., Przepiora, A. Predictions of long-term performance of granular iron permeable reactive barriers: Field-scale evaluation. *Journal of Contaminant Hydrology*. 123, 50–64, 2011.

38. Jun, D., Yongsheng, Z., Weihong, Z., Mei, H. Laboratory study on sequenced permeable reactive barrier remediation for landfill leachate-contaminated groundwater. *Journal of Hazardous Materials*. 161, 224–230, 2009.

39. Puls, R.W., Paul, C.J., Powell, R.M. The application of *in situ* permeable reactive (zero-valent iron) barrier technology for the remediation of chromate contaminated groundwater: A field test. *Applied Geochemistry*. 14, 989–1000, 1999.

40. USEPA, *Permeable Reactive Barrier Technologies for Contaminant Remediation*, EPA 600/R-98/125, Washington, DC 20460, 94, 1998.

41. Benner, S.G., Blowes, D.W., Gould, W.D., Herbert, R.B., Ptacek, C.J. Geochemistry of a permeable reactive barrier for metals and acid mine drainage. *Environmental Science and Technology*. 33, 2793–2799, 1999.

42. Jarvis, A.P., Moustafa, M., Orme, P.H.A., Younger, P.L. Effective remediation of grossly polluted acidic, and metal-rich, spoil heap drainage using a novel, low cost, permeable reactive barrier in Northumberland, UK. *Environmental Pollution*. 143, 261–268, 2006.

43. Jeyasingh, J., Somasundaram, V., Philip, L., Bhallamudi, S.M. Pilot scale studies on the remediation of chromium contaminated aquifer using biobarrier and reactive zone technologies. *Chemical Engineering Journal*. 167, 206–214, 2011.

44. Thiruvenkatachari, R., Vigneswaran, S., Naidu, R. Permeable reactive barrier for groundwater remediation. *Journal of Industrial and Engineering Chemistry*. 14, 145–156, 2008.

45. Scherer, M.M., Richter, S., Valentine, R.L., Alvarez, P.J.J. Chemistry and microbiology of permeable reactive barriers for *in situ* groundwater cleanup. *Critical Reviews in Environmental Science and Technology*. 30, 363–411, 2000.

46. Amin, M.N., Kaneco, S., Kitagawa, T., Begum, A., Katsumata, H., Suzuki, T., Ohta, K. Removal of arsenic in aqueous solutions by adsorption onto waste rice husk. *Industrial and Engineering Chemistry Research*. 45, 8105–8110, 2006.

47. Mohan, D., Chander, S. Removal and recovery of metal ions from acid mine drainage using lignite a low cost sorbent. *Journal of Hazardous Materials*. 137, 1545–1553, 2006.

48. Mohan, D., Pittman, J.C.U., Bricka, M., Smith, F., Yancey, B., Mohammad, J., Steele, P.H., Alexandre-Franco, M.F., Serrano, V.G., Gong, H. Sorption of arsenic, cadmium, and lead by chars produced from fast pyrolysis of wood and bark during bio-oil production. *Journal of Colloid and Interface Science*. 310(1), 57–73, 2007.

49. Sneddon, I.R., Garelick, H., Valsami-Jones, E. An investigation into arsenic(V) removal from aqueous solutions by hydroxyl apatite and bone-char. *Mineralogical Magazine.* 69, 769–780, 2005.
50. Chowdhury, S.R., Yanful, E.K. Arsenic and chromium removal by mixed Magnetite-maghemite nanoparticles and the effect of phosphate on removal. *Journal of Environmental Management.* 91, 2238–2247, 2010.
51. Rao, T.S., Karthikeyan, J. Removal of As(V) from water by adsorption on to low-cost and waste materials. *Progress in Environmental Science and Technology.* Science Press, Beijing, China, 684–691, 2007.
52. Ruiping, L., Lihua, S., Jiuhui, Q., Guibai, L. Arsenic removal through adsorption, sand filtration and ultrafiltration: *In situ* precipitated ferric and manganese binary oxides as adsorbents. *Desalination.* 249, 1233–1237, 2009.
53. Smedley, P.L., Kinniburgh, D.G. A review of the source, behaviour and distribution of arsenic in natural waters. *Applied Geochemistry.* 17, 517–568, 2002.
54. Sylvester, P., Westerhoff, P., Möller, T., Badruzzaman, M., Boyd, O. A hybrid sorbent utilizing nanoparticles of hydrous iron oxide for arsenic removal from drinking water. *Environmental Engineering Science.* 24, 104–112, 2007.
55. Hsieh, L.H.C., Weng, Y.-H., Huang, C.-P., Li, K.-C. Removal of arsenic from groundwater by electro-ultrafiltration. *Desalination.* 234, 402–408, 2008.
56. Sang, Y., Li, F., Gu, Q., Liang, C., Chen, J. Heavy metal-contaminated groundwater treatment by a novel nanofiber membrane. *Desalination.* 223, 349–360, 2008.
57. Gillham, R.W., OíHannesin, S.F., Orth, W.S. Metal enhanced abiotic degradation of halogenated aliphatics: Laboratory tests and field trials. *Proceedings of the 6th Annual Environmental Management and Technical Conference/Haz. Mat. Central Conference.* Advanstar Exposition, Glen Ellyn, IL, Rosemont, Illinois, 440–461, 1993.
58. Su, C., Puls, R.W. Arsenate and arsenite removal by zerovalent iron: Kinetics, redox transformation, and implications for *in situ* groundwater remediation. *Environmental Science and Technology.* 35, 1487–1492, 2001.
59. Blue, L.Y., Van Aelstyn, M.A., Matlock, M., Atwood, D.A. Low-level mercury removal from groundwater using a synthetic chelating ligand. *Water Research.* 42, 2025–2028, 2008.
60. Hong, P.K.A., Cai, X., Cha, Z. Pressure-assisted chelation extraction of lead from contaminated soil. *Environmental Pollution.* 153, 14–21, 2008.
61. Lim, T.T., Tay, J.-H., Wang, J.-Y. Chelating-agent-enhanced heavy metal extraction from a contaminated acidic soil. *Journal of Environmental Engineering.* 130, 59–66, 2004.
62. Warshawsky, A., Strikovsky, A.G., Vilensky, M.Y., Jerabek, K. Interphase mobility and migration of hydrophobic organic metal extractant molecules in solvent-impregnated resins. *Separation Science and Technology.* 37, 2607–2622, 2002.
63. Baker, A.J.M. Metal hyper accumulation by plants: Our present knowledge of eco physiological phenomenon. In: Abstract Book: *Current Topics in Plant Biochemistry, Physiology and Molecular Biology: Annual Symposium.* 14, 55, 1995.
64. Salati, S., Quadri, G., Tambone, F., Adani, F. Fresh organic matter of municipal solid waste enhances phytoextraction of heavy metals from contaminated soil. *Environmental Pollution.* 158, 1899–1906, 2010.
65. Wilson, B.H., Smith, G.B., Rees, J.F. Biotransformation of selected alkylbenzenes and halogenated aliphatic hydrocarbons in methanogenic aquifer material: A microcosm study. *Environmental Science and Technology.* 20, 997–1002, 1986.
66. Yong, R.N., Mulligan, C.N. *Natural Attenuation of Contaminants in Soils.* CRC Press, Boca Raton (FL), 2004.
67. Dvorak, D.H., Hedin, R.S. Treatment of metal contaminated water using bacterial sulfate reduction: Results from pilot-scale reactors. *Biotechnology and Bioprocess Engineering.* 40, 609–616, 1992.
68. Gibert, O., de Pablo, J., Cortina, J.L., Ayora, C. Treatment of acid mine drainage by sulphate-reducing bacteria using permeable reactive barriers: A review from laboratory to full-scale experiments. *Reviews in Environmental Science and Biotechnology.* 1, 327–333, 2002.
69. Hammack, R.W., Edenborn, H.M. The removal of nickel from mine waters using bacterial sulfate reduction. *Applied Microbiology and Biotechnology.* 37, 674–678, 1992.
70. Hammack, R.W., Edenborn, H.M., Dvorak, D.H. Treatment of water from an open-pit copper mine using biogenic sulfide and limestone: A feasibility study. *Water Research.* 28, 2321–2329, 1994.
71. Waybrant, K.R., Blowes, D.W., Ptacek, C.J. Selection of reactive mixtures for use in permeable reactive walls for treatment of mine drainage. *Environmental Science and Technology.* 32, 1972–1979, 1998.

72. Kim, M., Um, H.-J., Bang, S., Lee, S.-H., Oh, S.-J., Han, J.-H., Kim, K.-W., Min, J., Kim, Y.-H. Arsenic removal from Vietnamese groundwater using the arsenic binding DNA aptamer. *Environmental Science and Technology.* 43, 9335–9340, 2009.

73. Pandey, P.K., Verma, Y., Choubey, S., Pandey, M., Chandrasekhar, K. Biosorptive removal of cadmium from contaminated groundwater and industrial effluents. *Bioresource Technology.* 99, 4420–4427, 2008.

74. Prakasham, R.S., Merrie, J.S., Sheela, R., Saswathi, N., Ramakrishna, S.V. Biosorption of chromium VI by free and immobilized *Rhizopus arrhizus*. *Environmental Pollution.* 104, 421–427, 1999.

75. Han, D., Gary, H.P., Spalding, B., Scott, B.C. Electrospun and oxidized cellulosic materials for environmental remediation of heavy metals in groundwater, model cellulosic surfaces. *ACS Symposium Series.* 1019, 243–257, 2009.

76. Hasan, S., Hashim, M.A., Gupta, B.S. Adsorption of Ni(SO$_4$) on Malaysian rubber-wood ash. *Bioresource Technology.* 72, 153–158, 2000.

77. Kamel, S., Hassan, E.M., El-Sakhawy, M. Preparation and application of acrylonitrile-grafted cyanoethyl cellulose for the removal of copper (II) ions. *Journal of Applied Polymer Science.* 100, 329–334, 2006.

78. Sahu, J.N., Acharya, J., Meikap, B.C. Response surface modeling and optimization of chromium(VI) removal from aqueous solution using tamarind wood activated carbon in batch process. *Journal of Hazardous Materials.* 172, 818–825, 2009.

79. Sud, D., Mahajan, G., Kaur, M.P. Agricultural waste material as potential adsorbent for sequestering heavy metal ions from aqueous solutions—A review. *Bioresource Technology.* 99, 6017–6027, 2008.

80. Tabakci, M., Erdemir, S., Yilmaz, M. Preparation, characterization of cellulose grafted with calix[4] arene polymers for the adsorption of heavy metals and dichromate anions. *Journal of Hazardous Materials.* 148, 428–435, 2007.

81. Qu, X., Alvarez, P.J.J., Li, Q. Applications of nanotechnology in water and wastewater treatment. *Water Research.* 47, 3931–3946, 2013.

82. Zhao, G., Li, J., Ren, X., Chen, C., Wang, X. Few-layered graphene oxide nanosheets as superior sorbents for heavy metal ion pollution management. *Environmental Science and Technology.* 45, 10454–10462, 2011.

83. Moreno-Castilla, C., Álvarez-Merino, M.A., López-Ramón, M.V., Rivera-Utrilla, J. Cadmium ion adsorption on different carbon adsorbents from aqueous solutions. Effect of surface chemistry, pore texture, ionic strength, and dissolved natural organic matter. *Langmuir.* 20, 8142–8148, 2004.

84. Sun, S., Wang, L., Wang, A. Adsorption properties of crosslinked carboxymethyl-chitosan resin with Pb (II) as template ions. *Journal of Hazardous Materials.* 136, 930–937, 2006.

85. Kobya, M., Demirbas, E., Senturk, E., Ince, M. Adsorption of heavy metal ions from aqueous solutions by activated carbon prepared from apricot stone. *Bioresource Technology.* 96, 1518–1521, 2005.

86. Wang, X., Zheng, Y., Wang, A. Fast removal of copper ions from aqueous solution by chitosan-g-poly(acrylic acid)/attapulgite composites. *Journal of Hazardous Materials.* 168, 970–977, 2009.

87. Oubagaranadin, J.U.K., Murthy, Z.V.P. Adsorption of divalent lead on a montmorillonite – Illite type of clay. *Industrial and Engineering Chemistry Research.* 48, 10627–10636, 2009.

88. Wang, X., Guo, Y., Yang, L., Han, M., Zhao, J., Cheng, X. Nanomaterials as sorbents to remove heavy metal ions in wastewater treatment. *Journal of Environmental and Analytical Toxicology.* 2(7), 1–7, 2012. doi: 10.4172/2161-0525.1000154.

89. Savage, N., Diallo, M.S. Nanomaterials and water purification: Opportunities and challenges. *Journal of Nanoparticle Research.* 7, 331–342, 2005.

90. Wu, P.G., Zhu, J.H., Xu, Z.H. Template-assisted synthesis of mesoporous magnetic nano composite particles. *Advanced Functional Materials.* 14, 345–351, 2004.

91. Xin Rong, Z., Ping, Y., and Meng Yeu, Z. Research on photocatalytic degradation of organophosphorous pesticides using TiO$_2$.SiO$_2$/beads. *Industrial Water Treatment.* 21(3), 13–39, 2001.

92. Zamboulis, D., Peleka, E.N., Lazaridis, N.K., Matis, K.A. Metal ion separation and recovery from environmental sources using various flotation and sorption techniques. *Journal of Chemical Technology and Biotechnology.* 86, 335–344, 2011.

93. Wang, D., Deraedt, C., Ruiz, J., Astruc, D. Magnetic and dendritic catalysts. *Accounts of Chemical Research.* 48, 1871–1880, 2015.

94. Rao Kotte, M., Kuvarega, A.T., Cho, M., Mamba, B.B., Diallo, M.S. Mixed matrix PVDF membranes with in situ synthesized PAMAM dendrimer-like particles: A new class of sorbents for Cu(II) recovery from aqueous solutions by ultrafiltration. *Environmental Science and Technology.* 49, 9431–9442, 2015.

95. Guibal, E. Interactions of metal ions with chitosan-based sorbents. *Separation and Purification Technology.* 38, 43–74, 2004.

96. Varma, A.J., Deshpande, S.V., Kennedy, J.F. Metal complexation by chitosan and its derivatives: A review. *Carbohydrate Polymers.* 55, 77–93, 2004.

97. Saha, S., Sarkar, P. Arsenic remediation from drinking water by synthesized nano-alumina dispersed in chitosan-grafted polyacrylamide. *Journal of Hazardous Materials.* 227–228, 68–78, 2012.

98. Devi, R.R., Umlong, I.M., Das, B., Borah, K., Thakur, A.J., Raul, P.K., Banerjee, S., Singh, L. Removal of iron and arsenic (III) from drinking water using iron oxide-coated sand and limestone. *Applied Water Science.* 4, 175–182, 2014.

99. Ekramul Mahmud, H.N.M., Obidul Huq, A.K., Yahya, R.B. The removal of heavy metal ions from wastewater/aqueous solution using polypyrrole-based adsorbents: A review. *RSC Advances.* 6, 14778–14791, 2016.

100. Heidari, A., Younesi, H., Mehrabanb, Z., Heikkinen, H. Selective adsorption of Pb(II), Cd(II), and Ni(II) ions from aqueous solution using chitosan–MAA nanoparticles. *International Journal of Biological Macromolecules.* 61, 251–263, 2013.

101. Kolbe, F., Weiss, H., Morgenstern, P., Wennrich, R., Lorenz, W., Schurk, K., Stanjek, H., Daus, B. Sorption of aqueous antimony and arsenic species onto akaganeite. *Journal of Colloid and Interface Science.* 357, 460–465, 2011.

102. Xu, C., Shi, J., Zhou, W., Gao, B., Yue, Q., Wang, X. Bromate removal from aqueous solutions by nano crystalline akaganeite ($-FeOOH$)-coated quartz sand (CACQS). *Chemical Engineering Journal.* 187, 63–68, 2012.

103. Aragon, M., Kottenstette, R., Dwyer, B., Aragon, A., Everett, R., Holub, W., Siegel, M., Wright, J. *Arsenic Pilot Plant Operation and Results.* Sandia National Laboratories, Anthony, New Mexico, 2007.

104. Zhu, J., Wei, S., Chen, M., Gu, H., Rapole, S.B., Pallavkar, S., Ho, T.C., Hopper, J., Guo, Z. Magnetic nanocomposites for environmental remediation. *Advanced Powder Technology.* 24, 459–467, 2013.

105. Liu, M., Chen, C., Hu, J., Wu, X., Wang, X. Synthesis of magnetite/graphene oxide composite and application for cobalt(II) removal. *The Journal of Physical Chemistry C.* 115, 25234–25240, 2011.

106. Khan, S.B., Marwani, H.M., Asiri, A.M., Bakhsh, E.M. Exploration of calcium doped zinc oxide nanoparticles as selective adsorbent for extraction of lead ion. *Desalination and Water Treatment*, 1–10, 2015. doi: 10.1080/19443994.2015.1109560.

107. Mayo, J.T., Yavuz, C., Yean, S., Cong, L., Shipley, H., Yu, W., Falkner, J., Kan, A., Tomson, M., Colvin, V.L. The effect of nanocrystalline magnetite size on arsenic removal. *Science and Technology of Advanced Materials.* 8(1–2), 71–75, 2007.

108. Hristovski, K.D., Nguyen, H., Westerhoff, P.K. Removal of arsenate and 17-ethinyl estradiol (EE2) by iron (hydr) oxide modified activated carbon fibers. *Journal of Environmental Science and Health Part A—Toxic/Hazardous Substances and Environmental Engineering.* 44(4), 354–361, 2009.

109. Hristovski, K.D., Westerhoff, P.K., Moller, T., Sylvester, P. Effect of synthesis conditions on nano-iron (hydr) oxide impregnated granulated activated carbon. *Chemical Engineering Journal.* 146(2), 237–243, 2009.

110. Gupta, K., Bhattacharya, S., Chattopadhyay, D., Mukhopadhyay, A., Biswas, H, Dutta, J., Ray, N.R., Ghosh, U.C. Ceria associated manganese oxide nanoparticles: Synthesis, characterization and arsenic (V) sorption behavior. *Chemical Engineering Journal.* 172, 219–229, 2011.

111. Goswami, A., Raul, P.K., Purkait, M.K. Arsenic adsorption using copper (II) oxide nanoparticles. *Chemical Engineering Research and Design.* 90(9), 1387–1396, 2012.

112. Cao, C.Y., Cui, Z.M., Chen, C.Q., Song, W.G., Cai, W. Ceria hollow nanospheres produced by a template-free microwave-assisted hydrothermal method for heavy metal ion removal and catalysis. *Journal of Physical Chemistry.* 114, 9865–9870, 2010.

113. Gao, C., Zhang, W., Li, H., Lang, L., Xu, Z. Controllable fabrication of mesoporous MgO with various morphologies and their absorption performance for toxic pollutants in water. *Crystal Growth and Design.* 8, 3785–3790, 2008.

114. Luo, T., Cui, J., Hu, S., Huang, Y., Jing, C. Arsenic removal and recovery from copper smelting wastewater using TiO_2. *Environmental Science and Technology.* 44, 9094–9098, 2010.

115. Fabrega, J., Luoma, S.N., Tyler, C.R., Galloway, T.S., Lead, J.R. Silver nanoparticles: Behaviour and effects in the aquatic environment. *Environmental International.* 37, 517–531, 2011.

116. Feng, L., Cao, M., Ma, X., Zhu, Y., Hu, C. Super paramagnetic high-surface area Fe_3O_4 nanoparticles as adsorbents for arsenic removal. *Journal of Hazardous Materials.* 217–218, 439–446, 2012.

117. Huang, J., Cao, Y., Liu, Z., Deng, Z., Tang, F., Wang, W. Efficient removal of heavy metal ions from water system by titanate nanoflowers. *Chemical Engineering Journal.* 180, 75–80, 2012.

118. Zhang, S., Cheng, F., Tao, Z., Gao, F., Chen J. Removal of nickel ions from wastewater by $Mg(OH)_2$/MgO nanostructures embedded in Al_2O_3 membranes. *Journal of Alloys and Compounds.* 426, 281–285, 2006.

119. Abdollahi, M., Zeinali, S., Nasirimoghaddam, S., Sabbaghi, S. Effective removal of As(III) from drinking water samples by chitosan-coated magnetic nanoparticles. *Desalination and Water Treatment.* 56, 2092–2104, 2015.

120. Mu, Y., Ai, Z., Zhang, L., Song, F. Insight into core–shell dependent anoxic Cr(VI) removal with Fe @ Fe_2O_3 nanowires: Indispensable role of surface bound Fe(II). *ACS Applied Materials and Interfaces,* 7, 1997–2005, 2015.

121. Ludwig, R.D., Su, C., Lee, T.R., Wilkin, R.T., Acree, S.D. Ross, R.R., Keeley, A. In situ chemical reduction of Cr(VI) in groundwater using a combination of ferrous sulfate and sodium dithionite: A field investigation. *Environmental Science and Technology.* 41, 5299–5305, 2007.

122. Kanel, S.R., Greneche, J.M., Choi, H. Arsenic (V) removal from groundwater using nano scale zerovalent iron as a colloidal reactive barrier material. *Environmental Science and Technology.* 40(6), 2045–2050, 2006.

123. Ponder, S.M., Darab, J.G., Mallouk, T.E. Remediation of Cr(VI) and Pb(II) aqueous solutions using supported, nanoscale zerovalent iron. *Environmental Science and Technology.* 34(12), 2564–2569, 2000.

124. Zhong, L.S., Hu, J.S., Liang, H.P., Cao, A.M., Song, W.G., Wan, L.J. Self-assembled 3D flowerlike iron oxide nanostructures and their application in water treatment. *Journal of Advanced Materials.* 18(18), 2426–2431, 2006.

125. Comba, S., Di Molfetta, A., Sethi, R. A comparison between field applications of nano-, micro-, and millimetric zero-valent iron for the remediation of contaminated aquifers. *Water Air and Soil Pollution.* 215, 595–607, 2011.

126. Xu, P., Zeng, G.M., Huang, D.L., Hu, S., Zhao, M.H., Lai, C., Wei, Z., Huang, C., Xie, G.X., Liu, Z.F. Use of iron oxide nanomaterials in wastewater treatment: A review. *Science of the Total Environment.* 424, 1–10, 2012.

127. Shan, G., Yan, S., Tyagi, R.D., Surampalli, R.Y., Zhang, T.C. Applications of nanomaterials in environmental science and engineering: Review. The *Practice Periodical of Hazardous Toxic and Radioactive Waste Management.* 13, 110–119, 2009.

128. Lin, H.F., Liao, S.C., Hung, S.W. The dc thermal plasma synthesis of ZnO nanoparticles for visible-light photocatalyst. *Journal of Photochemistry and Photobiology A.* 174(1), 82–87, 2005.

129. Jing, L., Yichun, Q., Baiqi, W., Shudan, L., Baojiang, J., Libin, Y., Wei, F., Honggang, F., Jiazhong, S. Review of photoluminescence performance of nano-sized semiconductor materials and its relationships with photocatalytic activity. *Solar Energy Materials and Solar Cells.* 90(12), 1773–1787, 2006.

130. Nakataa, K., Fujishima, A. TiO_2 photocatalysis: Design and applications. *Journal of Photochemistry and Photobiology C: Photochemistry Reviews.* 13, 169–189, 2012.

131. Zhang, C., Jiang, Y., Li, Y., Hu, Z., Zhou, L. Zhou, M. Three-dimensional electrochemical process for wastewater treatment: A general review. *Chemical Engineering Journal.* 228, 455–467, 2013.

132. Goel, M., Chovelon, J.M., Ferronato, C., Bayard, R., Sreekrishnan, T.R. The remediation of wastewater-containing 4-chlorophenol-using integrated photocatalytic and biological treatment. *Journal of Photochemistry and Photobiology B: Biology,* 98(1), 1–6, 2010.

133. Augugliano, V., Litter, M., Palmisano, L., Soria, J. The combination of heterogeneous photocatalysis with chemical and physical operations: A tool for improving the photo process performance. *Journal of Photochemistry and Photobiology C: Photochemical Reviews.* 7(4), 127–144, 2006.

134. Zhang, X.J., Ma, T.Y., Yuan, Z.Y. Titania–phosphonate hybrid porous materials: Preparation, photocatalytic activity and heavy metal ion adsorption. *Journal of Materials Chemistry.* 18, 2003–2010, 2008.

135. Li, D., Zhua, Q., Hana, C., Yanga, Y., Jiang, W., Zhang, Z. Photocatalytic degradation of recalcitrant organic pollutants in water using a novel cylindrical multi-column photoreactor packed with TiO_2-coated silica gel beads. *Journal of Hazardous Materials.* 285, 398–408, 2015.

136. Murgoloa, S., Petronella, F., Ciannarellaa, R., Comparelli, R., Agostiano, A., Curri, M.L., Mascolo, M. UV and solar-based photocatalytic degradation of organic pollutants by nano-sized TiO_2 grown on carbon nanotubes. *Catalysis Today.* 240, 114–124, 2015.

137. Kenanakis, G., Vernardoua, D., Dalamagkas, A., Katsarakis, N. Photocatalytic and electrooxidation properties of TiO_2 thin films deposited by sol–gel. *Catalysis Today.* 240, 146–152, 2015.

138. López-Munoz, M.J., Revilla, A., Alcalde, G. Brookite TiO_2-based materials: Synthesis and photocatalytic performance in oxidation of methyl orange and As(III) in aqueous suspensions. *Catalysis Today.* 240, 138–145, 2015.

139. Han, F., Kambala, V.S.R., Srinivasan, M., Rajarathnam, D., Naidu, R. Tailored titanium dioxide photocatalysts for the degradation of organic dyes in wastewater treatment: A review. *Applied Catalysis A: General.* 359(1–2), 25–40, 2009.

140. Nawrocki, J., Kasprzyk-Hordernb, B. The efficiency and mechanisms of catalytic ozonation. *Applied Catalysis B: Environmental.* 99, 27–42, 2010.

141. Kusic, H., Koprivanac, N., Srsan, L. Azo dye degradation using Fenton type processes assisted by UV irradiation: A kinetic study. *Journal of Photochemistry and Photobiology A: Chemistry.* 181(2–3), 195–202, 2006.

142. Mahmoud, M.E., Yakut, A.A., Abdel-Aal, H., Osman, M.M. Enhanced biosorptive removal of cadmium from aqueous solutions by silicon dioxide nano-powder, heat inactivated and immobilized *Aspergillus ustus. Desalination.* 279(1–3), 291–297, 2011.

143. Mohsenzadeh, F., Rad, A.C. Bioremediation of heavy metal pollution by nano-particles of *Noaea Mucronata. International Journal of Bioscience, Biochemistry and Bioinformatics.* 2, 2, 2012.

144. Rizwan, M., Singh, M., Mitra, C.K., Morve, R.K. Ecofriendly application of nanomaterials: Nanobioremediation. *Journal of Nanoparticles.* 2014, 1–7, 2014.

145. Mauter, M.S., Elimelech, M. Environmental applications of carbon-based nanomaterials. *Environmental Science and Technology.* 42, 5843–5859, 2008.

146. Daus, B., Wennrich, R., Weiss, H. Sorption materials for arsenic removal from water: A comparative study. *Water Research.* 38(12), 2948–2954, 2004.

147. Tofighy, M.A., Mohammadi, T. Adsorption of divalent heavy metal ions from water using carbon nanotube sheets. *Journal of Hazardous Materials.* 185, 140–147, 2011.

148. Pan, B., Pan B, Zhang, W., Lv, L., Zhang, Q., Zheng, S. Development of polymeric and polymer-based hybrid adsorbents for pollutants removal from waters. *Chemical Engineering Journal.* 151(1–3), 19–29, 2009.

149. Zhao, X., Lv, L., Pan, B., Zhang, W., Zhang, S., Zhang, Q. Polymer-supported nanocomposites for environmental application: A review. *Chemical Engineering Journal.* 170(2–3), 381–394, 2011.

150. Pan, B., Xing, B.S. Adsorption mechanisms of organic chemicals on carbon nanotubes. *Environmental Science and Technology.* 42(24), 9005–9013, 2008.

151. Bystrzejewski, M., Pyrzyn´ ska, K., Huczko, A., Lange, H. Carbon-encapsulated magnetic nanoparticles as separable and mobile sorbents of heavy metal ions from aqueous solutions. *Carbon.* 47, 1189–1206, 2009.

152. Lu, C., Chiu, H., Bai, H. Comparisons of adsorbent cost for the removal of zinc(II) from aqueous solution by carbon nanotubes and activated carbon. *Journal of Nanoscience and Nanotechnology,* 7(4–5), 1647–1652, 2007.

153. Pyrzyñska, K., Bystrzejewski, M. Comparative study of heavy metal ions sorption onto activated carbon, carbon nanotubes, and carbon-encapsulated magnetic nanoparticles. *Colloids and Surfaces A: Physicochemical and Engineering Aspects,* 362, 102–109, 2010.

154. Stafiej, A., Pyrzynska, K. Adsorption of heavy metal ions with carbon nanotubes. *Separation and Purification Technology.* 58, 49–52, 2007.

155. Ma, P.C., Siddiqui, N.A., Marom, G., Kim, J.K. A dispersion and functionalization of carbon nanotubes for polymer-based nanocomposites: A review. *Composites: Part A.* 41, 1345–1367, 2010.

156. Liua, S., Shena, Q., Caoa, Y., Gana, L., Wanga, Z., Steigerwaldb, M.L., Guo, X. Chemical functionalization of single-walled carbon nanotube field-effect transistors as switches and sensors. *Coordination Chemistry Reviews.* 254, 1101–1116, 2010.

157. Veličković, Z., Vuković, G.D., Marinković, A.D., Moldovand, M.S., Peric-Gruji, A.A., Petar S. Uskoković, P.S., Ristić, M.D. Adsorption of arsenate on iron(III) oxide coated ethylenediamine functionalized multiwall carbon nanotubes. *Chemical Engineering Journal.* 181–182, 174–181, 2012

158. Chen, C.Y., Zepp, R.G. Probing photosensitization by functionalized carbon nanotubes. *Environmental Science and Technology.* 49, 13835–13843, 2015.

159. Upadhyayula, V.K.K., Deng, S., Mitchell, M.C., Smith, G.B. Application of carbon nanotube technology for removal of contaminants in drinking water: A review. *Science of the Total Environment.* 408(1), 1–13, 2009.

160. Hoek, E.M.V., Ghosh, A.K. Nanotechnology-based membranes for water purification. In: *Nanotechnology Applications for Clean Water.* Street, A., Sustich, R., Duncan, J., and Savage, N. (Eds.), William Andrew, Inc., Norwich, NY, 47 – 58, 2009.

161. Tripathy, S., Kanungo, S. Adsorption of Co^{2+}, Ni^{2+}, Cu^{2+} and Zn^{2+} from 0.5 M NaCl and major ion sea water on a mixture of δ-MnO_2 and amorphous FeOH, *Journal of Colloid Interface Science* 284, 30–38, 2005.

162. Swedlund, P.J., Webster, J.G., Miskelly, G.M. Goethite adsorption of Cu(II), Pb(II), Cd(II), and Zn(II) in the presence of sulfate: Properties of the ternary complex. *Geochimica et Cosmochimica Acta.* 73, 1548–1562, 2009.

163. Chandra, V., Park, J., Chun, Y., Lee, J.W., Hwang, I.C., Kim, K.S. Water-dispersible magnetite-reduced graphene oxide composites for arsenic removal. *ACS Nano.* 4, 3979–3986, 2010.
164. Cloete, T.E., Kwaadsteniet, M.D., Botes, M., Lopez-Romero, J.M. *Nanotechnology in Water Treatment Applications.* Caister Academic Press, UK, 2010.
165. Malwal, D., Gopinath, P. Fabrication and applications of ceramic nanofibers in water remediation: A review. *Critical Reviews in Environmental Science and Technology.* 46(5), 500–534, 2016.
166. Peng, C., Zhang, J., Xiong, Z., Zhao, B., Liu, P. Fabrication of porous hollow g-Al_2O_3 nanofibers by facile electrospinning and its application for water remediation. *Microporous and Mesoporous Materials.* 215, 133–142, 2015.
167. Kampalanonwat, P., Supaphol, P. Preparation and adsorption behavior of aminated electro spun polyacrylonitrile nanofibre mats for heavy metal ion removal. *Applied Materials and Interfaces.* 2(2), 3619–3627, 2010.
168. Gupta, A.K., Deva, D., Sharma, A., Verma, N. Fe-grown carbon nanofibers for removal of arsenic (V) in wastewater. *Industrial and Engineering Chemistry Research.* 49(15), 7074–7084, 2010.
169. Malwal, D., Gopinath, P. Fabrication and characterization of poly (ethylene oxide) templated nickel oxide nanofibers for dye degradation. *Environmental Science: Nano.* 2, 78–85, 2015.
170. Theron, J., Walker, J.A., Cloete, T.E. Nanotechnology and water treatment: Applications and emerging opportunities. *Critical Reviews in Microbiology.* 34, 43–69, 2008.
171. Bhatt, I., Tripathi, B.N. Interaction of engineered nanoparticles with various components of the environment and possible strategies for their risk assessment. *Chemosphere.* 82, 308–317, 2011.

3 Recycling of Filter Backwash Water and Alum Sludge from Water Utility for Reuse

Mu-Hao Sung Wang, Lawrence K. Wang,
Nazih K. Shammas, and Milos Krofta

CONTENTS

ABSTRACT

The feasibility of recycling filter backwash water and alum sludge generated from water purification plants has been investigated. Actual wastewater and alum sludge used in this study were collected from a water plant employing water treatment processes including chemical addition, mixing, flocculation, clarification, filtration, and chlorination. Wastewater and sludge are generated mainly from the clarifier and the filter backwash. The waste recycle system presented here consists of (a) recycling the filter backwash water to the intake system for the reproduction of potable water, (b) dividing the combined sludge into two fractions for alum solubilization, separately, in an acid reactor and an alkaline reactor, (c) removing the inert silts from alum solutions by two separate water–solids separators for ultimate disposal, and (d) returning the solubilized alums from the two separate water–solids separators in proper proportions for reuse as flocculants.

The proposed recycle process was designed to provide a cost-effective system for achieving "zero" wastewater discharge and alum recovery from a water purification plant. Recommended process design parameters necessary to achieve the above stated goals have been established. Experimental results tend to suggest that practical designs based on the proposed water recycle and sludge thickening and alum recovery (STAR) system are technically feasible. Additional conclusions drawn from this research are (a) discharging raw alum sludge from a water treatment plant to a nearby wastewater treatment plant through a sewer system is a viable means of sludge disposal for the water utility; (b) the thickened raw alum sludge can be disposed on land as a soil amendment without adverse effect on soil if the pH of the disposed alum sludge is near neutral; (c) recycling the recovered alum for water purification within the water treatment plant is technically feasible the problem of impurity concentration (heavy metals and soluble organics) can be met by a scheduled recycling application or an automatic blowdown; (d) the U.S. Federal and the Commonwealth of Massachusetts prohibit recycling of the recovered alum for water purification within the water treatment plant because its long-term health effect is unknown; (e) employing recovered alum from a water treatment sludge as a precipitant for phosphate removal in a wastewater treatment plant is technically feasible although its economic feasibility needs to be studied; and (f) direct recycle of filter backwash water from a dissolved air flotation–filtration (DAFF) water treatment plant to the plant's intake unit is both technically and economically feasible. The recovered alum (either aluminum sulfate or sodium aluminate) can be effectively used for removal of heavy metals and phosphorus from wastewater.

3.1 INTRODUCTION

3.1.1 Typical Physicochemical Treatment Plant

The flow diagram of a typical physicochemical water treatment plant with direct recycling of filter backwash water is shown in Figure 3.1 (1–3). The physicochemical treatment system is mainly used for water purification. Raw water is treated by flash mixing, flocculation, clarification, filtration, and disinfection. Clarification can be accomplished by either conventional sedimentation (1,2), or innovative dissolved air flotation (DAF). The unit operation of filtration can be sand filtration and/or granular-activated carbon (GAC) filtration. The filter effluent is disinfected and stored in a clear well where the water is ready to be pumped to the water distribution system for domestic and industrial consumption. The most common coagulants used in water purification are alum, sodium aluminate, and ferric

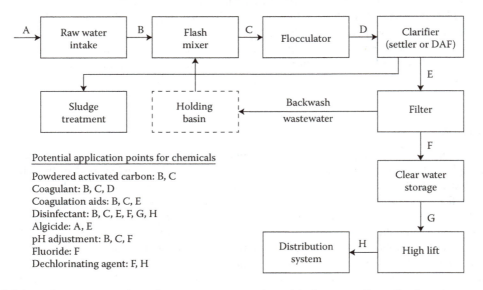

FIGURE 3.1   Flow diagram of a typical water treatment plant with direct recycling of backwash water.

sulfate. As shown in Figure 3.1, the most common water treatment plant produces mainly two waste streams: (a) backwash wastewater from filters and (b) waste sludge from clarifiers. Disposal of waste sludge and wastewater becomes an important concern to environmental engineers and government officials (1–51). It is important to note that the same flow diagram shown in Figure 3.1 can also be used for wastewater treatment. The Niagara Falls Wastewater Treatment Plant, Niagara Falls, New York, consisting of chemical mixing, flocculation, clarification (sedimentation), GAC filtration, and disinfection, is a complete physical–chemical plant for treating combined industrial and municipal wastewater.

3.1.2 FILTER BACKWASH WATER RECYCLE AND SLUDGE THICKENING

In a water treatment plant, a holding tank shown in Figure 3.1 is required for the recycling of backwash wastewater generated from a conventional water treatment plant in which sedimentation basins are used as clarifiers (1,2). If the filter backwash wastewater from a conventional plant is intended to be recycled for reuse, a huge holding tank is required to equalize the wastewater flow and to settle and separate the sludge there. Accordingly total recycle and reuse of filter backwash water in conventional plants is technically feasible but economically unattractive. Besides, the consistency of settled waste sludge from sedimentation clarifiers is in the range of 0.2%–0.5%. A separate sludge thickener is generally required if alum sludge recovery is intended.

An innovative water treatment plant using DAF cells for clarification can directly recycle its filter backwash wastewater to a mixer–flocculator for reprocessing, thus eliminating the need for a huge holding tank (4,5,8,9). Besides, the consistency of the DAF floated sludge can be as high as 2.6% if desired thus, eliminating the requirement for a separate sludge thickener if alum recovery is intended (34–37).

3.2 SLUDGE THICKENING AND ALUM RECOVERY SYSTEM

A STAR system stands for the sludge thickening and alum recovery system (6). Alum sludge from either conventional or innovative water treatment plants contains mainly aluminum hydroxide, which is an amphoteric species capable of reacting with both acid and alkaline reagents, as indicated in the following two reactions:

$$2Al(OH)_3 + 3H_2SO_4 \rightarrow Al_2(SO_4)_3 + 6H_2O$$
$$Al(OH)_3 + NaOH \rightarrow NaAlO_2 + 2H_2O$$

Sulfuric acid (H_2SO_4) and sodium hydroxide (NaOH) are the most common acid and base, respectively, used in the STAR system.

Krofta and Wang (6) successfully used nitric acid (HNO_3) and hydrochloric acid (HCl) for recovery of alum sludge as aluminum nitrate and aluminum chloride, respectively.

Potassium hydroxide (KOH) is also an effective alkaline chemical for alum recovery in the form of potassium aluminate (6).

A demonstrated alum sludge recovery scheme is presented as shown in Figure 3.2. The major source of alum sludge comes from a DAF clarifier. A small portion of alum sludge could be contributed by backwashing the filters. Route A in Figure 3.2 shows that alum can be recovered as aluminum sulfate (i.e., filter alum) by adding sulfuric acid. Route B shows that alum can be recovered as sodium aluminate (i.e., soda alum). Routes A and B have been demonstrated to be feasible, but a pH adjustment procedure is generally needed when either recovered alum is being recycled for reuse. This is due to the fact that the pH of the acid reactor effluent is extremely low, and the pH of the alkaline reactor effluent is extremely high. The optimum pH for alum coagulation, however, is about 6.3 (3).

An effective alum recycle alternative is that part of the alum sludge can be regenerated by adding a strong acid (Route A in Figure 3.2) and the remaining portion of the alum sludge can be

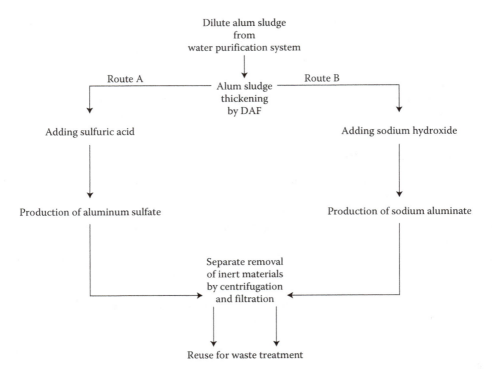

FIGURE 3.2 Recovery and reuse of alum sludge. Alum recycling system can be route A, route B, or the combination of routes A and B.

regenerated by adding a strong base (Route B in Figure 3.2). Recycling both aluminum sulfate and sodium aluminate (or the like), at appropriate ratios, to the intake system for reuse would eliminate the additional pH adjustment requirement (3).

Figure 3.3 shows a proposed STAR operation. The alum sludge from a DAF-filtration (DAFF) clarifier is already thickened by DAF. Any commercial DAF thickener will be equally feasible for sludge thickening. Part of the thickened raw alum sludge can be converted to aluminum sulfate by the addition of sulfuric acid in an acid mixing reactor, and the remaining part of the sludge can be converted to aluminate by adding caustic soda in a base mixing reactor. After acid and alkaline treatments are over, the residual solid sludge is composed of mainly inert materials which can be separated by a separation unit, such as centrifugation or filtration. An effective water–solids separator manufactured by Krofta Engineering Corporation and the Lenox Institute of Water Technology is shown in Figure 3.4. The two liquid streams containing high concentrations of recovered alums can then be withdrawn for reuse. The parts of the Krofta water–solids separator shown in Figure 3.4 are noted below:

1. Centrifuge tank
2. Window frame
3. Inspection window
4. Pressurized air inlet
5. Tank breather
6. Sludge inlet
7. Sludge outlet
8. Gasket
9. Gasket

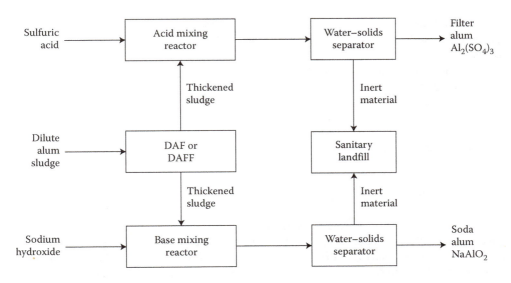

FIGURE 3.3 STAR system.

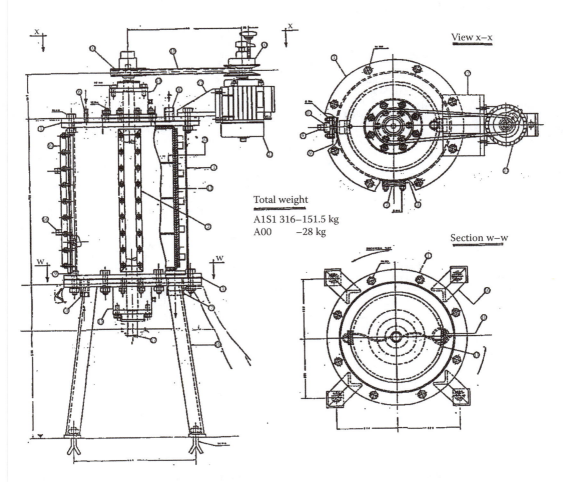

FIGURE 3.4 A Krofta water–solids separator.

10. Rotaring basket
11. Filtering cloth
12. Water outlet
13. Header spray-washers
14. Pressurized water inlet
15. Mechanical group
16. Tank support
17. Motor support
18. Variable shave
19. Driving V belt
20. Variable sheave
21. Electrical motor

Figure 3.5 shows the total waste recycle system of an improved water purification plant using alum as primary coagulant and using a DAFF clarifier for water treatment. Hundred percent of filter backwash water is recycled to DAFF clarifier's mixing and flocculation chamber for reproduction of potable water. The alum sludge thickened by DAFF clarifier's flotation goes to the remaining STAR units (acid mixing reactor, base mixing reactor, and water–solids separators shown in Figure 3.4) for alum recovery. The only waste produced from this plant is a small amount of inert material suitable for sanitary landfill.

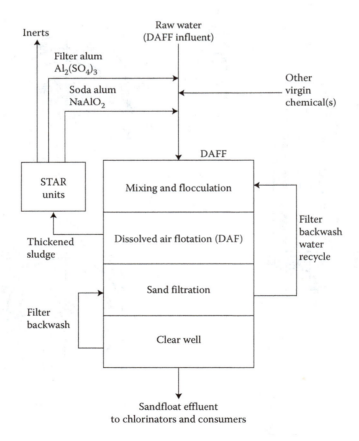

FIGURE 3.5 Innovative potable water treatment plant (DAFF plant) with filter backwash water recycle, STAR system.

3.3 DEMONSTRATION PROGRAM

An extensive research program was conducted in 1984–1988 and 1998–1999 at the Lenox Water Treatment Plant (LWTP) (5,6) and the Lenox Institute of Water Technology for optimization and demonstration of total filter backwash water recycle and the STAR system. The following were the specific objectives of the investigation:

1. To demonstrate the feasibility of the total filter backwash water recycle
2. To study the chemical reactions and the reaction temperature of the STAR system considering cold weather conditions
3. To study the reaction time of the STAR system
4. To optimize the sulfuric acid dosage for alum recovery, in turn to determine the sulfuric acid cost for this STAR application
5. To determine the metals and organic contents of recovered aluminum sulfate solution
6. To compare the recovered alum with commercial liquid alum

New testing data for alum recovery are presented in Tables 3.1 through 3.24. A measure of 36 N concentrated sulfuric acid was used throughout this entire study. The recovery of alum sludge in the form of sodium aluminate is reported elsewhere in 1988 (31).

3.4 FILTER BACKWASH WATER RECYCLE AND SLUDGE THICKENING

The total recycle and reuse of filter backwash wastewater has been successfully practiced at the Lenox Water Treatment Plant, Lenox, Massachusetts, from July 1982 to 1989. The heart of the LWTP at the time of this study was a 22-ft diameter DAFF package plant with a design capacity of 1.2 MGD (million gallons per day), or 4.542 MLD (million liters per day). The DAFF water treatment plant was mainly composed of mixing–flocculation, DAF, sand filtration, clear well, and chlorination, of which DAF was responsible for water clarification, filter backwash water treatment, and sludge thickening. The alum sludge was concentrated by DAF to about 2.6%.

TABLE 3.1

Recovery of Aluminum from Lenox Alum Sludge (Test No. 1; TSS = 15,923 mg/L) Using Concentrated Sulfuric Acid

Reaction Temperature (°C)	Reaction Time (h)	mL Sludge Plus mL Conc. H_2SO_4	Recovered Soluble Aluminum (mg/L)	Aluminum Recovery (%)
18	0	10,000 + 0	0.05	0
23	0.25	10,000 + 100	3,240	72
22	0.50	10,000 + 100	4,180	92.8
21	1	10,000 + 100	4,060	90.2
19	2	10,000 + 100	4,060	90.2
19	4	10,000 + 100	4,060	90.2
16	20	10,000 + 100	3,940	87.6
17	23	10,000 + 100	3,600	80.0
17	26	10,000 + 100	3,660	81.3
17	92	10,000 + 100	3,890	86.4

Note: TSS of thickened alum sludge = 15,923 mg/L; initial A1 of thickened sludge = 4,500 mg/L; influent flow rate at LWTP = 560 gpm; sludge flow rate at LWTP = 3 gpm; alum dosage at LWTP = 3.9 mg/L as Al_2O_3; mixing rate in the period 0–26 h = 80 rpm; mixing rate in the period 26–92 h = 0 rpm; and settled sludge volume after 92 h = 150 mL sludge per 650 mL of total volume.

TABLE 3.2

Recovery of Aluminum from Lenox Alum Sludge (Test No. 2; TSS = 17,283 mg/L) Using Concentrated Sulfuric Acid

Reaction Temperature (°C)	Reaction Time (h)	mL Sludge Plus mL Conc. H₂SO₄	Recovered Soluble Aluminum (mg/L)	Aluminum Recovery (%)
18	0	20,000 + 0	0.05	0
27	0.25	20,000 + 500	3,600	75
27	0.50	20,000 + 500	4,020	83.8
25	1	20,000 + 500	4,340	90.4
23	2	20,000 + 500	4,400	91.7
22	5	20,000 + 500	3,660	76.3
17	21	20,000 + 500	3,940	82.1
17	24	20,000 + 500	3,940	82.1
17	93	20,000 + 500	3,950	82.3
18	105	20,000 + 500	3,760	78.3

Note: TSS of thickened alum sludge = 17,283 mg/L; initial Al of thickened sludge = 4,800 mg/L; influent flow rate at LWTP = 560 gpm; sludge flow rate at LWTP = 3 gpm; alum dosage at LWTP = 3.9 mg/L as Al₂O₃; mixing rate in the period 0–24 h = 80 rpm; mixing rate in the period 34–105 h = 0 rpm; and settled sludge volume after 105 h = 150 mL sludge per 830 mL of total volume.

TABLE 3.3

Recovery of Aluminum from Lenox Alum Sludge (Test No. 3; TSS = 20,508 mg/L) Using Concentrated Sulfuric Acid

Reaction Temperature (°C)	Reaction Time (h)	mL Sludge Plus mL Conc. H₂SO₄	Recovered Soluble Aluminum (mg/L)	Aluminum Recovery %
Test No. 3A				
19	0	2,000 + 0	0.05	0
NA	0.25	2,000 + 50	NA	NA
27	0.50	2,000 + 50	4,840	86.4
24.5	1	2,000 + 50	4,730	84.5
22	2	2,000 + 50	4,620	82.5
17	69	2,000 + 50	5,090	90.8
Test No. 3B				
19	0	2,000 + 0	0.05	0
32	0.25	2,000 + 70	4,770	85.2
29	0.50	2,000 + 70	4,830	86.3
24.5	1	2,000 + 70	4,700	83.9
21.5	2	2,000 + 70	4,720	84.3
17	69	2,000 + 70	4,900	87.5

Note: TSS of thickened alum sludge = 20,508 mg/L; initial Al of thickened sludge = 5,600 mg/L; influent flow rate at LWTP = 560 gpm; sludge flow rate at LWTP = 3 gpm; and alum dosage at LWTP = 3.9 mg/L as Al₂O₃.

TABLE 3.4

Recovery of Aluminum from Lenox Alum Sludge (Test No. 4; TSS = 23,527 mg/L) Using Concentrated Sulfuric Acid

Reaction Temperature (°C)	Reaction Time (h)	mL Sludge Plus mL Conc. H_2SO_4	Recovered Soluble Aluminum (mg/L)	Aluminum Recovery (%)
Test No. 4A				
19	0	2,000 + 0	0.05	0
MA	0.25	2,000 + 20	NA	NA
24.5	0.50	2,000 + 20	3,730	62.9
24	1	2,000 + 20	3,980	67.1
22	2	2,000 + 20	3,200	53.96
17	69	2,000 + 20	4,430	74.7
Test No. 4B				
19	0	2,000 + 0	0.05	0
28	0.25	2,000 + 70	4,030	68.0
27	0.50	2,000 + 70	4,130	69.6
23.5	1	2,000 + 70	4,730	79.8
23	2	2,000 + 70	4,640	78.2
17	69	2,000 + 70	4,680	78.9

Note: TSS of thickened alum sludge = 23,527 mg/L; initial Al of thickened sludge = 5,930 mg/L; influent flow rate at LWTP = 560 gpm; sludge flow rate at LWTP = 3 gpm; and alum dosage at LWTP = 3.9 mg/L as Al_2O_3.

TABLE 3.5

Recovery of Aluminum from Lenox Alum Sludge (Test No. 5; TSS = 31,276 mg/L) Using Concentrated Sulfuric Acid

Reaction Temperature (°C)	Reaction Time (h)	mL Sludge Plus mL Conc. H_2SO_4	Recovered Soluble Aluminum (mg/L)	Aluminum Recovery (%)
Test No. 5A				
19	0	2,000 + 0	0.05	0
NA	0.25	2,000 + 20	NA	NA
25	0.50	2,000 + 20	3,460	64.0
24	1	2,000 + 20	4,230	78.33
22.5	2	2,000 + 20	4,570	84.6
17	69	2,000 + 20	NA	NA
Test No. 5B				
19	0	2,000 + 0	0.05	0
31	0.25	2,000 + 70	4,340	80.4
28	0.50	2,000 + 70	4,320	80.0
23.5	1	2,000 + 70	NA	NA
22	1.5	2,000 + 70	4,030	74.6
22	2	2,000 + 70	4,360	80.7
17	69	2,000 + 70	NA	NA

Note: TSS of thickened alum sludge = 31,276 mg/L; initial Al of thickened sludge = 5,400 mg/L; influent flow rate at LWTP = 560 gpm; sludge flow rate at LWTP = 3 gpm; and alum dosage at LWTP = 3.9 mg/L as Al_2O_3.

TABLE 3.6
Effect of Initial Raw Alum Sludge TSS Concentration and Acid Dosage on Aluminum Recovery

Test No.	Initial TSS of Raw Sludge (mg/L)	Conc. H_2SO_4 Dosage mL/L Sludge	Aluminum Recovery (%)	Recovered Soluble Al (mg/L)
1	15,923	10	92.8–90.2	4,180–4,060
2	17,283	25	83.8–90.4	4,020–4,340
3A	20,508	25	86.4–84.5	4,840–4,730
3B	20,508	35	86.3–83.9	4,830–4,700
4A	23,527	10	62.9–67.1	3,730–3,980
4B	23,527	35	69.6–79.8	4,130–4,730
5A	31,276	10	64.0–78.33	3,460–4,230
5B	31,276	35	80.0–	4,320–

TABLE 3.7
Centrifugation of Alum Sludge Pretreated by Concentrated Sulfuric Acid

Centrifugation Time (min)	Sludge Volume (mL)	Sludge TSS (mg/L)	Centrifugate	
			Volume (mL)	A1 (mg/L)
0	7,200	11,606	0	NA
5	1,000	71,312	6,200	3,750
10	900	72,779	6,300	3,750
15	800	87,518	6,400	3,760

Note: Initial raw alum sludge TSS = 17,283 mg/L before acid treatment; acid treatment = 250-mL concentrated sulfuric acid per 10 L of raw alum sludge at room temperature 17–27°C; TSS after acid treatment before centrifugation = 11,606 mg/L; and centrifuge operation = 1,725 rpm.

TABLE 3.8
Monitoring of TOC of Recovered Alum Solutions

Test No.	Test Conditions	Raw Water TOC (mg/L)	Effluent TOC (mg/L)	Solution TOC (mg/L)
1	TSS = 15,923 mg/L 10 mL/L acid	5.8	0.78	1,080
2	TSS = 17,283 mg/L 25 mL/L acid	5.8	0.78	1,351
3A	TSS = 20,508 mg/L 25 mL/L acid	6.0	0.78	1,528
4B	TSS = 23,528 mg/L 35 mL/L acid	6.0	0.78	1,545
5B	TSS = 31,276 mg/L 35 mL/L acid	6.0	0.78	1,630

Note: Acid = 36 N concentrated sulfuric acid.

TABLE 3.9

Recovery of Aluminum from Lenox Alum Sludge (Test No. 9; TSS = 24,077 mg/L) Using Concentrated Sulfuric Acid

Reaction Temperature (°C)	Reaction Time (h)	mL Sludge Plus mL Conc. H_2SO_4	Recovered Soluble Aluminum (mg/L)	Aluminum Recovery (%)
3	0	10,000 + 0	0.05	0
4	0.25	10,000 + 250	3,680	64.7
5	0.50	10,000 + 250	3,590	63.1
3	1	10,000 + 250	3,780	66.4
2	2	10,000 + 250	4,060	71.4
1	4	10,000 + 250	4,080	71.7
1	7	10,000 + 250	4,020	70.6
5	143	10,000 + 250	3,900	68.5

Note: TSS of thickened alum, sludge = 24,077 mg/L; initial A1 of thickened sludge = 5,690 mg/L; influent flow rate at LWTP = 560 gpm; sludge flow rate at LWTP = 3 gpm; alum dosage at LWTP = 3.9 mg/L as Al_2O_3; mixing rate in the period 0–7 h = 100 rpm; mixing rate in the period 7–143 h = 0 rpm; and settled sludge volume after 143 h = no noticeable sludge per 500 mL of total volume.

TABLE 3.10

Recovery of Aluminum from Lenox Alum Sludge (Test No. 10; TSS = 25,753 mg/L) Using Concentrated Sulfuric Acid

Reaction Temperature (°C)	Reaction Time (h)	mL Sludge Plus mL Conc. H_2SO_4	Recovered Soluble Aluminum (mg/L)	Aluminum Recovery (%)
18	0	10,000 + 0	0.05	0
20	0.75	10,000 + 170	1,387	24.86
20	1.0	10,000 + 170	1,718	30.79
19	1.5	10,000 + 170	1,945	34.86
19	2	10,000 + 170	1,647	29.52
19	126	10,000 + 170	3,180	56.98
18	150	10,000 + 170	3,814	68.4

Note: TSS of thickened alum sludge = 25,753 mg/L; initial Al of thickened sludge = 5,580 mg/L; influent flow rate at LWTP = 560 gpm; sludge flow rate at LWTP = 3 gpm; alum dosage at LWTP = 3.9 mg/L as Al_2O_3; mixing rate in the period 0–150 h = 0 rpm; and settled sludge volume after 150 h = approximately 25 mL sludge per 190 mL of total volume. Floated sludge volume after 150 h = approximately 30 mL sludge per 190 mL of total volume.

The DAFF plant used mainly poly aluminum chloride (66 mg/L average) in the winter and filter alum (73.6 mg/L) in the other three seasons. Sodium aluminate was only used occasionally for pH control. All three chemicals produced aluminum hydroxide sludge as end products.

While the entire full scale operation was successful, a complete 12-month operational data (July 1, 1986–June 30, 1987) was presented below to indicate the fact that the total recycle of filter backwash water for reuse would not adversely affect the plant effluent's water quality:

LWTP (DAFF) influent

Flow = 148–760 gpm (gallon per minute) = 560–2877 Lpm (liter per minute) (average 377 gpm = 1427 Lpm)

TABLE 3.11

Recovery of Aluminum from Lenox Alum Sludge (Test No. 11; TSS = 25,830 mg/L) Using Concentrated Sulfuric Acid

Reaction Temp. (°C)	Reaction Time (h)	mL Sludge Plus mL Conc. H$_2$SO$_4$	Recovered Soluble Al (mg/L)	Aluminum Recovery (%)	Color Unit
14	0	10,000 + 0	0.05	0	0
14	0.25	10,000 + 10	18.5	0.35	5
14	0.5	10,000 + 10	47.4	0.90	8
15	1.0	10,000 + 10	56.1	1.10	10
18	24	10,000 + 10	84	1.6	15
17	45	10,000 + 10	101	1.9	18

Note: TSS of thickened alum sludge = 25,830 mg/L; initial Al of thickened sludge = 5,279 mg/L; influent flow rate at LWTP = 560 gpm; sludge flow rate at LWTP = 3 gpm; alum dosage at LWTP = 3.9 mg/L as Al$_2$O$_3$; mixing rate in the period 0–24 h = 30 rpm; mixing rate in the period 24–45 h = 0 rpm; settled sludge volume after 45 h = 972 mL sludge per 1000 mL of total volume.

TABLE 3.12

Recovery of Aluminum from Lenox Alum Sludge (Test No. 12; TSS = 25,830 mg/L) Using Concentrated Sulfuric Acid

Reaction Temp. (°C)	Reaction Time (h)	mL Sludge Plus mL Conc. H$_2$SO$_4$	Recovered Soluble Al (mg/L)	Aluminum Recovery (%)	Color Unit
14	0	10,000 + 0	0.05	0	0
14	0.25	10,000 + 20	NA	NA	20
14	0.5	10,000 + 20	250	4.7	20
15	1.0	10,000 + 20	270	5.1	20
18	24	10,000 + 20	270	5.1	20
17	45	10,000 + 20	311	5.9	23

Note: TSS of thickened alum sludge = 25,830 mg/L; initial Al of thickened sludge = 5,279 mg/L; influent flow rate at LWTP = 560 gpm; sludge flow rate at LWTP = 3 gpm; alum dosage at LWTP = 3.9 mg/L as Al$_2$O$_3$; mixing rate in the period 0–24 h = 30 rpm; mixing rate in the period 24–45 h = 0 rpm; and settled sludge volume after 45 h = 943 mL sludge per 1000 mL of total volume.

Temperature = 37–75°F = 2.8–24°C (average 51.8°F = 11°C)
Turbidity = 0.65–7.35 NTU (nephelometric turbidity unit) (average 1.6 NTU)
pH = 6.7–8.6 unit (average 7.6 unit)
Alkalinity = 60–92 mg/L CaCO$_3$ (average 73.5 mg/L CaCO$_3$)
Color = 0–15 unit (average 6 unit)
Aluminum = 0.01–0.08 mg/L Al (average 0.06 mg/L Al)

LWTP (DAFF) effluent

Turbidity = 0.02–0.53 NTU (average 0.08 NTU)
pH = 6.6–8.0 unit (average 7.1 unit)

TABLE 3.13

Recovery of Aluminum from Lenox Alum Sludge (Test No. 13; TSS = 25,830 mg/L) Using Concentrated Sulfuric Acid

Reaction Temp. (°C)	Reaction Time (h)	mL Sludge Plus mL Conc. H_2SO_4	Recovered Soluble Al (mg/L)	Aluminum Recovery (%)	Color Unit
14	0	10,000 + 0	0.05	0	0
14	0.25	10,000 + 40	470	8.9	15
14	0.5	10,000 + 40	580	11	25
15	1.0	10,000 + 40	400	7.6	20
18	24	10,000 + 40	400	7.6	20
17	45	10,000 + 40	460	8.7	23

Note: TSS of thickened alum sludge = 25,830 mg/L; initial Al of thickened sludge = 5,279 mg/L; influent flow rate at LWTP = 560 gpm; sludge flow rate at LWTP = 3 gpm; alum dosage at LWTP = 3.9 mg/L as Al_2O_3; mixing rate in the period 0–24 h = 30 rpm; mixing rate in the period 24–45 h = 0 rpm; and settled sludge volume after 45 h = 932 mL sludge per 1000 mL of total volume.

TABLE 3.14

Recovery of Aluminum from Lenox Alum Sludge (Test No. 14; TSS = 25,830 mg/L) Using Concentrated Sulfuric Acid

Reaction Temp. (°C)	Reaction Time (h)	mL Sludge Plus mL Conc. H_2SO_4	Recovered Soluble Al (mg/L)	Aluminum Recovery (%)	Color Unit
14	0	10,000 + 0	0.05	0	0
14.5	0.25	10,000 + 60	840	15.9	30
14.5	0.5	10,000 + 60	910	17.3	35
16	1.0	10,000 + 60	790	15.0	35
18	24	10,000 + 60	790	15.0	50
17	45	10,000 + 60	1,128	21.4	74
17	45	10,000 + 60	1,670	31.6	85

Note: TSS of thickened alum sludge = 25,830 mg/L; initial Al of thickened sludge = 5,279 mg/L; influent flow rate at LWTP = 560 gpm; sludge flow rate at LWTP = 3 gpm; alum dosage at LWTP = 3.9 mg/L as Al_2O_3; mixing rate in the period 0–24 h = 30 rpm; mixing rate in the period 24–45 h = 0 rpm; and settled sludge volume after 45 h = 932 mL sludge per 1,000 mL of total volume.

Alkalinity = 48–86 mg/L $CaCO_3$ (average 66 mg/L $CaCO_3$)
Color = 0 unit (average 0 unit)
Aluminum = 0.01–0.10 mg/L Al (average 0.05 mg/L Al)

It can be seen that accomplishment of water purification, filter backwash recycle, and sludge thickening by the DAFF clarifier is technically feasible. The DAFF effluent quality was excellent (average effluent turbidity = 0.08 NTU; average effluent color = 0 unit). There was no accumulation of aluminum residual (average effluent Al = 0.05 mg/L) in the effluent even though the filter backwash wastewater was 100% recycled for 7 years.

TABLE 3.15

Recovery of Aluminum from Lenox Alum Sludge (Test No. 15; TSS = 25,830 mg/L) Using Concentrated Sulfuric Acid

Reaction Temp. (°C)	Reaction Time (h)	mL Sludge Plus mL Conc. H₂SO₄	Recovered Soluble Al (mg/L)	Aluminum Recovery (%)	Color Unit
14	0	10,000 + 0	0.05	0	0
15	0.25	10,000 + 80	1,150	21.8	40
15	0.5	10,000 + 80	1,290	24.4	60
16	1.0	10,000 + 80	1,980	37.5	90
18	24	10,000 + 80	2,750	52.1	125
17	45	10,000 + 80	3,300	62.5	150

Note: TSS of thickened alum sludge = 25,830 mg/L; initial Al of thickened sludge = 5,279 mg/L; influent flow rate at LWTP = 560 gpm; sludge flow rate at LWTP = 3 gpm; alum dosage at LWTP = 3.9 mg/L as Al₂O₃; mixing rate in the period 0–24 h = 30 rpm; mixing rate in the period 24–45 h = 0 rpm; and settled sludge volume after 45 h = 950 mL sludge per 1,000 mL of total volume.

TABLE 3.16

Recovery of Aluminum from Lenox Alum Sludge (Test No. 16; TSS = 25,830 mg/L) Using Concentrated Sulfuric Acid

Reaction Temp. (°C)	Reaction Time (h)	mL Sludge Plus mL Conc. H₂SO₄	Recovered Soluble Al (mg/L)	Aluminum Recovery (%)	Color Unit
14	0	10,000 + 0	0.05	0	0
16	0.25	10,000 + 100	1,250	23.7	50
16	0.5	10,000 + 100	1,400	26.5	60
17	1.0	10,000 + 100	1,900	36.0	100
18	24	10,000 + 100	2,280	43.2	120
17	45	10,000 + 100	3,040	57.6	160

Note: TSS of thickened alum sludge = 25,830 mg/L; initial Al of thickened sludge = 5,279 mg/L; influent flow rate at LWTP = 560 gpm; sludge flow rate at LWTP = 3 gpm; alum dosage at LWTP = 3.9 mg/L as Al₂O₃; mixing rate in the period 0–24 h = 30 rpm; mixing rate in the period 24–45 h = 0 rpm; and settled sludge volume after 45 h = 949 mL sludge per 1,000 mL of total volume.

3.5   REACTION TEMPERATURE AND REACTION TIME OF ACID REACTOR

Tests nos. 1–5 were conducted in the acid reactor under room temperatures using 36 N concentrated sulfuric acid for alum recovery. The test results are documented in Tables 3.1 through 3.5.

From the data in Tables 3.1 through 3.5, one can conclude that with adequate mixing (at 80 rpm) the reaction time of 30–60 min would be sufficient for alum recovery at room temperature. The percent aluminum recovery actually reduced with further increase of reaction time when raw alum sludge concentration was below 20,000 mg/L (see Tables 3.1 and 3.2).

Table 3.6 indicates that at the reaction time of 30–60 min, the higher the initial TSS (total suspended solids) of raw alum sludge (up to 20,508 mg/L), the higher the soluble aluminum concentration in the recovered solution. A further increase in raw alum sludge concentration (23,527–31,276 mg/L) did not increase the recovered soluble aluminum concentration.

The data in Table 3.7 clearly show that centrifugation is an efficient unit operation for the separation of residual inert sludge from the recovered aluminum solution. Further investigations were

TABLE 3.17

Recovery of Aluminum from Lenox Alum Sludge (Test No. 17; TSS = 25,830 mg/L) Using Concentrated Sulfuric Acid

Reaction Temp. (°C)	Reaction Time (h)	mL Sludge Plus mL Conc. H_2SO_4	Recovered Soluble AL (mg/L)	Aluminum Recovery (%)	Color Unit
15	0	10,000 + 0	0.05	0	0
20.5	0.25	10,000 + 150	2,240	42.4	230
20.5	0.5	10,000 + 150	2,530	47.9	250
20	1.0	10,000 + 150	3,350	63.5	500
18	24	10,000 + 150	4,070	76.2	600
17	45	10,000 + 150	4,288	81.2	640

Note: TSS of thickened alum sludge = 25,830 mg/L; initial Al of thickened sludge = 5,279 mg/L; influent flow rate at LWTP = 560 gpm; sludge flow rate at LWTP = 3 gpm; alum dosage at LWTP = 3.9 mg/L as Al_2O_3; mixing rate in the period 0–24 h = 30 rpm; mixing rate in the period 24–45 h = 0 rpm; and settled sludge volume after 45 h = 462 mL sludge per 1,000 mL of total volume. Floated sludge volume after 45 h = 31 mL sludge per 1,000 mL of total volume.

TABLE 3.18

Recovery of Aluminum from Lenox Alum Sludge (Test No. 18; TSS = 25,830 mg/L) Using Concentrated Sulfuric Acid

Reaction Temp. (°C)	Reaction Time (h)	mL Sludge Plus mL Conc. H_2SO_4	Recovered Soluble Al (mg/L)	Aluminum Recovery (%)	Color Unit
15	0	10,000 + 0	0.05	0	0
22	0.25	10,000 + 200	2,930	55.5	1,000
21	0.5	10,000 + 200	3,960	75.0	1,500
21	1.0	10,000 + 200	4,070	77.1	2,500
18	24	10,000 + 200	4,151	78.6	2,550
17	45	10,000 + 200	4,477	84.8	2,750

Note: TSS of thickened alum sludge = 25,830 mg/L; initial Al of thickened sludge = 5,279 mg/L; influent flow rate at LWTP = 560 gpm; sludge flow rate at LWTP = 3 gpm; alum dosage at LWTP = 3.9 mg/L as Al_2O_3; mixing rate in the period 0–24 h = 30 rpm; mixing rate in the period 24–45 h = 0 rpm; and settled sludge volume after 45 h = 200 mL sludge per 1,000 mL of total volume. Floated sludge volume after 45 h = 29 mL sludge per 1,000 mL of total volume.

being conducted using vacuum filtration, pressure filtration, sedimentation, slow filtration, absorption, etc. for separation of residual inert solids.

The total organic carbon (TOC) concentration in the recovered alum solutions were monitored and reported in Table 3.8. At the raw sludge concentration range of 15,923–31,276 mg/L, the TOC of recovered alum solution was in the range of 1080–1630 mg/L, which was considered to be OK. After recycle and reuse of the alum, the TOC of treated water would only be slightly increased. Although TOC accumulation in the effluent was expected, it might be prevented by adequate operational procedures. For instance, after sludge recycle for a determined long period of time, DAFF plant should be fed with virgin alum and the STAR system should be fed with all fresh raw alum sludge again.

Various tests were also conducted under refrigerator-controlled temperatures to simulate operational conditions in winter.

TABLE 3.19

Recovery of Aluminum from Lenox Alum Sludge (Test No. 19; TSS = 25,830 mg/L) Using Concentrated Sulfuric Acid

Reaction Temp. (°C)	Reaction Time (h)	mL Sludge Plus mL Conc. H₂SO₄	Recovered Soluble Al (mg/L)	Aluminum Recovery (%)	Color Unit
15	0	10,000 + 0	0.05	0	0
23	0.25	10,000 + 250	4,380	82.97	2,500
22	0.5	10,000 + 250	4,460	84.49	2,650
22	1.0	10,000 + 250	4,660	88.3	2,600
18	24	10,000 + 250	4,640	87.9	2,600
17	45	10,000 + 250	4,907	92.9	2,750

Note: TSS of thickened alum sludge = 25,830 mg/L; initial Al of thickened sludge = 5,279 mg/L; influent flow rate at LWTP = 560 gpm; sludge flow rate at LWTP = 3 gpm; alum dosage at LWTP = 3.9 mg/L as Al₂O₃; mixing rate in the period 0–24 h = 30 rpm; mixing rate in the period 24–45 h = 0 rpm; and settled sludge volume after 45 h = 161 mL sludge per 1,000 mL of total volume. Floated sludge volume after 45 h = 32 mL sludge per 1,000 mL of total volume.

TABLE 3.20

Recovery of Aluminum from Lenox Alum Sludge (Test No. 20; TSS = 25,830 mg/L) Using Concentrated Sulfuric Acid

Reaction Temp. (°C)	Reaction Time (h)	mL Sludge Plus mL Conc. H₂SO₄	Recovered Soluble Al (mg/L)	Aluminum Recovery (%)	Color Unit
15	0	10,000 + 0	0.05	0	0
24	0.25	10,000 + 300	4,130	78.2	2,250
23	0.5	10,000 + 300	4,120	78.1	2,250
23	1.0	10,000 + 300	4,577	86.7	2,500
18	24	10,000 + 300	4,761	90.2	2,600
17	45	10,000 + 300	5,035	95.4	2,750

Note: TSS of thickened alum sludge = 25,830 mg/L; initial Al of thickened sludge = 5,279 mg/L; influent flow rate at LWTP = 560 gpm; sludge flow rate at LWTP = 3 gpm; alum dosage at LWTP = 3.9 mg/L as Al₂O₃; mixing rate in the period 0–24 h = 30 rpm; mixing rate in the period 24–45 h = 0 rpm; and settled sludge volume after 45 h = 229 mL sludge per 1,000 mL of total volume. Floated sludge volume after 45 h = 29 mL sludge per 1,000 mL of total volume.

Tests no. 9 (Table 3.9) and no. 19 (Table 3.19) are the simulations of winter operation and warm weather operation, respectively. In both tests, the LWTP's sludge was thickened by DAF to 24,077–25,830 mg/L of TSS before acid treatment using 36 N concentrated sulfuric acid. In both cases, 250 mL of sulfuric acid was dosed to every 10,000 mL of thickened sludge (TSS = 24,077–25,830 mg/L). It can be seen from Tables 3.9 and 3.19 that warm-temperature operation required shorter reaction time (0.5–1.0 h), produced more recovered soluble aluminum (4460–4660 mg/L Al), and had a higher percentage of aluminum recovery (84.49%–88.3%) in comparison with the cold weather operation. The following is a brief summary:

Cold temperature at 1–5°C

Reaction time = 2–4 h
Soluble Al = 4060–4080 mg/L
Aluminum recovery = 71.4%–71.7%

TABLE 3.21
Recovery of Aluminum from Lenox Alum Sludge (Test No. 21; TSS = 34,690 mg/L) Using Concentrated Sulfuric Acid

Reaction Temperature (°C)	Reaction Time (h)	mL Sludge Plus mL Conc. H_2SO_4	Recovered Soluble Aluminum (mg/L)	Aluminum Recovery (%)
Test 21A				
17	0	10,000 + 0	0.05	0
18	0.25	10,000 + 10	56	0.86
17.5	0.50	10,000 + 10	84	1.3
17.5	1	10,000 + 10	105	1.6
Test 21b				
17	0	10,000 + 0	0.05	0
18	0.25	10,000 + 20	104	1.6
18	0.50	10,000 + 20	237	3.7
18	1	10,000 + 20	216	3.3
Test 21C				
17	0	10,000 + 0	0.05	0
18.5	0.25	10,000 + 40	466	7.2
18.5	0. 50	10,000 + 40	467	7.2
18.5	1	10,000 + 40	483	7.48
Test 21D				
17	0	10,000 + 0	0.05	0
20	0.25	10,000 + 60	750	11.6
19	0.50	10,000 + 60	798	12.4
19	1	10,000 + 60	1,130	17.0
Test 21E				
17	0	10,000 + 0	0.05	0
20	0.25	10,000 + 80	1,163	18.0
19	0.50	10,000 + 80	1,132	17.5
18.5	1	10,000 + 80	1,220	18.9
Test 21F				
17	0	10,000 + 0	0.05	0
21	0.25	10,000 + 100	1,577	24.4
21	0.50	10,000 + 100	1,625	25.2
20	1	10,000 + 100	1,936	30.0

Note: TSS of thickened alum sludge = 34,690 mg/L; initial Al of thickened sludge = 6454 mg/L; influent flow rate at LWTP = 560 gpm; sludge flow rate at LWTP = 3 gpm; alum dosage at LWTP = 3.9 mg/L as Al_2O_3; and mixing rate in the period 0–1 h = 80 rpm.

Warm temperature at 15–23°C

Reaction time = 0.5–1.0 h
Soluble Al = 4460–4660 mg/L
Aluminum recovery = 84.5

3.6   SULFURIC ACID REQUIREMENT IN ACID REACTOR

A comparison between Table 3.9 (Test No. 9) and Table 3.10 (Test no. 10) clearly indicates that 170 mL of 36 N sulfuric acid was insufficient for treatment of 10,000 mL of thickened Lenox alum sludge (TSS = 25,753 mg/L) even at a warm temperature (18–20°C).

TABLE 3.22

Settling Velocity of Alum Sludge Which was Pretreated by Concentrated Sulfuric Acid

Settling Time (h)	Sludge Volume (mL)
0	400
0.25	400
0.5	400
1	395
2	395
3	395
4	390
20	390
144	215

Note: 500 mL graduated cylinder was used for this test; initial raw alum sludge TSS = 25,830 mg/L; acid treatment = 250 mL concentrated sulfuric acid per 10 L of raw alum sludge; temperature = 17°C; and reaction time = 45 h (Test no. 19).

TABLE 3.23
TOC of Recovered Alum Solutions

Test No.	Test Conditions	Raw Water TOC (mg/L)	Effluent TOC (mg/L)	Soluble TOC (mg/L)
9	TSS = 24,077 mg/L			
	25 mL/L acid	6.0	0.78	1,508
10	TSS = 25,753 mg/L			
	17 mL/L acid	6.5	2.3	1,238
	TSS = 25,830 mg/L			
11	1 mL/L acid	6.7	2.4	48.4
12	2 mL/L acid	6.7	2.4	395.3
13	4 mL/L acid	6.7	2.4	474.3
14	6 mL/L acid	6.7	2.4	652.2
15	8 mL/L acid	6.7	2.4	731.2
16	10 mL/L acid	6.7	2.4	853.4
17	15 mL/L acid	6.7	2.4	1185.8
18	20 mL/L acid	6.7	2.4	1230.0
19	25 mL/L acid	6.7	2.4	1422.9
20	30 mL/L acid	6.7	2.4	1462.5
	TSS = 34,690 mg/L			
21A	1 mL/L acid	6.5	2.3	46.1
2 IB	2 mL/L acid	6.5	2.3	500.0
21C	4 mL/L acid	6.5	2.3	653.8
2 ID	6 mL/L acid	6.5	2.3	730.5
21E	8 mL/L acid	6.5	2.3	854.6
21F	10 mL/L acid	6.5	2.3	883.8

TABLE 3.24

Acid Treatment of Lenox Sludge for Alum Recovery

Parameters	Quality of Recovered Alum
pH (unit)	<2
Total suspended solid (mg/L)	2,038
Volatile suspended solid (mg/L)	1,078
Fixed suspended solid (mg/L)	960
Aluminum (mg/L)	4,660
Arsenic (mg/L)	0
Barium (mg/L)	NA
Cadmium (mg/L)	0.01
Chromium (mg/L)	0
Copper (mg/L)	3.24
Iron (mg/L)	136.5
Lead (mg/L)	1.3
Manganese (mg/L)	8.86
Mercury (mg/L)	0
Nickel (mg/L)	1.5
Platinum (mg/L)	0
Potassium (mg/L)	18
Selenium (mg/L)	0
Sodium (mg/L)	43
Titanium (mg/L)	0
Zinc (mg/L)	0.35
Total coliform, #/100 mL	0
THMFP (mg/L)	6100.6
THM (mg/L)	0
TOC (mg/L)	1,528
COD (mg/L)	1,400
Color (unit)	2,600

Note: 10,000 mL of Lenox sludge (initial TSS = 25,830 mg/L) was treated with 250 mL of concentrated sulfuric acid at room temperature (22°C) and 30 rpm of mixing for 1 h. See Test no. 19.

Accordingly an extensive study (Test nos. 11–20) was conducted to determine the optimum sulfuric acid dosage for alum recovery. Results are presented in Tables 3.11 through 3.20. For every 10,000 mL of thickened alum sludge (TSS = 25,830 mg/L), the volume of 36 N sulfuric acid dosages were dosed with an increasing trend from 10 to 20, 40, 60, 80, 100, 150, 200, 250, and 300 mL. Results are presented in Tables 3.11 through 3.20, respectively. Apparently, the optimum dosage was 150–300 mL of concentrated sulfuric acid per 10,000 mL of thickened alum sludge (TSS = 25,830 mg/L). Figure 3.6 illustrates the effect of sulfuric acid dosage on alum recovery.

It is important to note that the aluminum recovery efficiency of each acid treatment can be visually observed in accordance with the color of the recovered alum solution. The higher the color, the higher the concentration of recovered soluble aluminum (see Tables 3.11 through 3.20).

3.7 EFFECT OF THICKENED SLUDGE CONCENTRATION ON ALUM RECOVERY

In Test nos. 21A–21F (see Table 3.21), 10–100 mL of 36 N sulfuric acid was dosed to 10,000 mL of 3.469% thickened sludge in the acid reactor. On the other hand, in Test nos. 11–16 (Tables 3.11

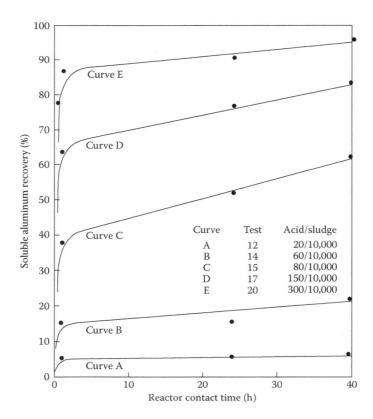

Curve	Test	Acid/sludge
A	12	20/10,000
B	14	60/10,000
C	15	80/10,000
D	17	150/10,000
E	20	300/10,000

FIGURE 3.6   Effect of reactor contact time and acid-to-sludge volumetric ratio.

through 3.16), 10–100 mL of 36 N sulfuric acid was dosed to 10,000 mL of 2.583% thickened sludge. The results indicate that when sulfuric acid was underdosed, an increase in the concentration of thickened sludge (from 2.583% to 3.469%) did not increase the alum recovery efficiency. DAF alone thickened the alum sludge to 2.583%. An evaporator was earlier used for further sludge thickening to 3.469%.

It is obvious if sulfuric acid is overdosed, an increase in concentration of thickened sludge will increase linearly until the optimum ratio of sulfuric acid to thickened sludge is reached.

3.8   SEPARATION OF INERT SILTS FROM RECOVERED LIQUID ALUM

It was demonstrated previously that centrifugation was an efficient unit operation for separation of inert sludge from the acid/base-treated liquid solutions from two reactors (Figure 3.3). Results were reported in Table 3.7.

At a centrifugation detention time of 15 min at 172.5 rpm, the TSS of inert sludge was concentrated to 8.75% which was good. For a cost-effective operation, the centrifugation time could be set at 5 min. Figure 3.4 shows that the process unit was feasible for water–solids separation.

In a supplemental study, the use of gravity sedimentation for the same purpose was attempted, and the sedimentation results are presented in Table 3.22 for Test no. 19.

Based on the results in Table 3.22, it is concluded that plain gravity sedimentation is not feasible for the alum sludge's solid–liquid separation. After settling of 144 h, the sludge volume was only reduced from 400 to 215 mL, which was not acceptable.

3.9   ANALYSIS OF RECOVERED LIQUID ALUM

The TOC of recovered raw liquid alum (after centrifugation but before filtration) was extensively analyzed and reported in Table 3.23. The raw water TOC and plant effluent TOC of the LWTP are also listed in the same table for the purpose of comparison. The raw water TOC and effluent TOC were 6.0–6.7 and 0.78–2.4 mg/L, respectively.

The TOC of recovered liquid alum was indeed high, and increased with increasing acid dosage. It is also expected that the higher the recovered soluble aluminum concentration, the higher the released soluble TOC (see Tables 3.21 and 3.23).

Much more detailed chemical and microbiological examinations were performed on typical recovered liquid alum. Table 3.24 indicates the results. The recovered liquid alum had extremely how pH (<2), high soluble aluminum (4660 mg/L), no coliform bacteria, no THM (trihalomethane), no arsenic, high color (2600 units), high COD (chemical oxygen demand) (1400 mg/L), high TOC (1528 mg/L), high THMFP (trihalomethane formation potential) (6100 mg/L), and very low heavy metals, such as zinc, titanium, selenium, platinum, nickel, mercury, iron, manganese, lead, copper, chromium, cadmium, and barium.

It is encouraging to note that the volatile suspended solid (VSS) was high (1078 mg/L) which means that not all organics are solubilized by acid treatment. Such volatile suspended solids (VSSs) and fixed suspended solids (FSS) can be further reduced by a physical operation, such as the built-in filtration mechanism of the tested Krofta water–solids separator (Figure 3.4).

3.10   COMPARISON OF RECOVERED LIQUID ALUM AND COMMERCIAL ALUM

Commercial alum supplied by Holland Co., Adams, Massachusetts, was also analyzed for TOC. A comparison of the commercial alum and the recovered raw liquid alum is given below:

Recovered liquid alum

COD = 1400 mg/L
Soluble aluminum = 4660 mg/L Al
Al_2O_3 = 8.820 mg/L

Commercial liquid alum

COD = 420 mg/L
Soluble aluminum = 43,941 mg/L Al
Al_2O_3 = 83,000 mg/L

3.11   SUMMARY OF DEMONSTRATION PROJECT

3.11.1   Filter Backwash Water Recycle

The 1.2-MGD (4.542 MLD) LWTP had one 22-ft diameter DAFF unit consisting of mixing-flocculation, DAF, sand filtration, clear well, and postchlorination. The DAFF unit has faithfully served the town of Lenox's 6500 residents and 3500 tourists since July 1982. With permission from the Commonwealth of Massachusetts and the U.S. Environmental Protection Agency (USEPA), in the demonstration period 100% of filter backwash wastewater was recycled to the DAFF plant's flocculation chamber for reproduction of potable water. Therefore, the total recycle of filter backwash water in a potable flotation plant is definitely feasible and cost effective, and has been practiced at the LWTP since 1982, and at the Pittsfield Water Treatment Plant (PWTP) since 1986 (34–37).

3.11.2 STAR System

For liquid alum recovery in Lenox, Massachusetts, the Lenox alum sludge was thickened by a DAF clarifier (DAFF or DAF, or equivalent) from approximately 2500 to 25,830 mg/L or higher. The thickened sludge, if gently mixed (at 30 rpm) with 36 N sulfuric acid at a ratio of 250 mL acid to 10,000 mL thickened sludge at room temperature for 1 h contact time, produced a recovered liquid alum with 4660 mg/L of soluble aluminum or 8802 mg/L in terms of Al_2O_3.

Although the COD of the recovered liquid alum was about 1400 mg/L, and the other organic parameters are high (THMFP = 6100 ppb; TOC = 1528 mg/L), they were all diluted because only a very small amount of recovered liquid alum was needed to treat raw water for water purification. Considering an average alum dosage of 2.5 mg/L as Al_2O_3, the mixture of raw water and the recovered liquid alum would have the following characteristics:

COD = 6.34 mg/L
TOC = 6.92 mg/L
THMFP = 27.65 ppb

which are all very reasonable. The Lenox raw water TOC was measured to be 6.0–6.7 mg/L. After alum sludge was recycled for reuse, the TOC of Lenox raw water containing chemicals only increased by 6.5%, which was negligible.

Winter operation of an alum sludge recycle system is technically feasible although the higher the temperature, the better alum recovery efficiency. For winter operation, a reactor reaction time of 2–4 h was required during the demonstration experiments. For warm weather operation, a short reaction time of 0.5–1.0 h was sufficient.

After the chemical reaction was over, the inert substances were separated from the reactor effluent for ultimate disposal by sanitary landfill. It has been demonstrated that centrifugation is a better unit operation than sedimentation for removal of inert substances. A Krofta water–solids separator which incorporates both centrifugation and filtration was ideal for this operation. Other commercial water–solids separator including centrifugation and filtration will also be acceptable.

Although the USEPA has approved all materials and chemicals used in the STAR system for the demonstration project, the STAR system has not been approved for routine long-term water treatment. The major problem associated with the recycling of alum sludge within a water utility is that organic impurities and heavy metals may be recycled as well. The impurities include inert soil materials, organic substances, and convertible mineral matters. The inert soil materials become the bulk of the sludge remaining after acidification. Organic substances especially the color-causing substances may be resolubilized by acidification, thus requiring actions for their removal. Similarly convertible mineral matter, particularly iron, manganese, and other heavy metals, are subject to dissolution. When the impurities are recycled together with the recycled alum, they may increase in concentration until the acidified supernatant from alum sludge becomes too rich in impurities to perform satisfactorily. The problem of impurity concentration can be technically met by the automatic blowdown due to only 45%–55% recovery of alum coagulant. Nevertheless both the federal and state governments must approve the recovered alum for water purification within a water plant. The recovered alum, however, can be easily approved for wastewater treatment by the governments, especially for the removal of phosphorus and heavy metals (38–43).

3.12 DISCUSSIONS AND CONCLUSIONS OF THIS RESEARCH

3.12.1 Economical Analysis of Filter Backwash Water Recycle and Sludge Thickening

The water loss of a majority of conventional water treatment plants (including mainly mixing-flocculation, sedimentation, filtration, clear well, and disinfection) is about 9% of total raw water pumpage due to the fact that the filter backwash wastewater is totally wasted (i.e., without recycle) and the settled sludge is bulky and dilute in TSS. A comparable innovative Lenox DAFF water treatment plant (including mainly mixing-flocculation, DAF, filtration, clear well, and disinfection) recycles its filter backwash wastewater and chemical flocs for reproduction of drinking water, thus its water loss is only about 0.5%, contributed by floated sludge. (*Note*: If STAR system is used for alum recovery in a DAFF plant, even the 0.5% water can be saved.) The rates of water treatment by the two types of plants can be estimated as follows:

Conventional plant

$$\text{Water production (effluent flow)} = \text{Plant influent flow} \times 0.91$$

$$\text{Plant influent flow} = 1.0989 \text{ water production (effluent flow)}$$

Innovative DAFF Plant (DAFF Plant)

$$\text{Water production (effluent flow)} = \text{Plant influent flow} \times 0.995$$

$$\text{Plant influent flow} = 1.005 \text{ water production (effluent flow)}$$

Assuming the coagulant dosages (mg coagulant per liter of influent water) for both conventional and innovative plants are identical, the conventional plant requires much more coagulants by weight (ton/day) because the conventional plant must treat about 9% more water (i.e., factor 1.0989 vs. factor 1.005) in order to supply the same volume of effluent for community consumption. The added advantage of innovative DF-filtration plant is that it conserves about 9% of water which can be very precious in drought areas.

If a conventional plant does recycle and reuse its filter backwash water, it needs a huge holding tank (see Figure 3.1) for backwash water equalization and sludge separation. The capital cost for the holding tank is high.

The settled sludge from a conventional plant's sedimentation clarifiers is low in concentration and requires a separate sludge thickener if alum recovery or other sludge treatment is intended. The requirement of a separate sludge thickener signifies another added capital cost for a conventional water treatment plant.

3.12.2 Economical Analysis of STAR System

The daily chemical treatment costs can be significantly reduced if the newly developed STAR system can be adopted.

The purpose of sludge recovery is to solve a sludge problem. Coagulant recovery offers added economic benefits.

These benefits include less coagulation chemical cost, and smaller amounts of inert solid carried to disposal by a sanitary landfill. Most of the chemical cost saving involves the acid and/or alkaline treatment. The design engineer can be assured that there will always be a cost difference between sulfuric acid and alum because the acid is required to manufacture the alum. There will be a big cost difference between sodium hydroxide and sodium aluminate, because the former is the raw chemical and the latter is the product.

White and White (7) presented an abstract of annual operating costs from their investigations. Raw water no. 3 was considered by White and White (7) to be the typical raw water source with no unusual problems, so the economics were typical of what was to be expected. Annual costs included: solubilization of alum sludge in acid reactor, dewatering costs on a stationary horizontal vacuum filtration bed for water–solids separation, and hauling and disposal of the residue. These annual costs showed a saving in favor of alum recovery of some 20% more than the cost of commercial alum itself if recovery was not practiced.

In another study in Germany (32), a computer program was developed which compared the capital and operational costs for water treatment plants with and without aluminum recovery from precipitation sludge. Annual costs of chemical consumption in a water treatment plant with aluminum recovery is at least 25% lower than in those with no coagulant recovery.

The STAR system is economically worthy of the design engineer's consideration. Such a system can be properly designed and safely operated. With the extreme variability from one raw water or wastewater to another, it is highly recommended that pilot testing be undertaken before such a design is attempted.

3.12.3   REUSE OF RAW ALUM SLUDGE

Pittsfield Water Treatment Plant (PWTP), Pittsfield, Massachusetts, is a DAFF water purification plant. Its DAF thickened raw alum sludge is discharged to the Pittsfield Wastewater Treatment Plant (PWWTP) for phosphorus removal (8). It is then concluded that discharge of raw alum sludge to wastewater treatment plants is a viable means of sludge disposal for the water utility.

The Lenox Water Treatment Plant (LWTP), Lenox, Massachusetts, is also a DAFF water purification plant. Its DAF thickened raw alum sludge was disposed on land for many years as a soil amendment (5,9,37,44). No adverse effect on soil was discovered when the pH of the disposed alum sludge was near neutral.

Elangovan and Subramanian (45) have concluded that raw alum sludge can be reused in clay brick manufacturing.

Recent studies (46–47) show that an alum sludge-based constructed wetland system can significantly remove organic matter and nutrients from the high-strength wastewater.

3.12.4   REUSE OF RECOVERED ALUM

The recovered alum from the Lenox Water Treatment Plant (LWTP; DAFF plant) in this research was directly applied to the LWTP for water purification for a very short period of time to demonstrate its technical feasibility under the condition that at least 50% virgin alum had to be used during alum recycle operation, and the recovered alum could not be continuously used for treating water over 3 days. After using 100% virgin alum for three consecutive days, then the combination of 50% recovered alum and 50% virgin alum could be applied together again for 3 days. In the entire research period, no adverse effect on water quality in terms of TOC, THM, THMFP, residual aluminum, turbidity, color, heavy metals, etc. was discovered. Nevertheless, continuous recycling of the recovered alum for water purification is not allowed by the federal and local governments because its long term health effects are unknown.

The LWTP's recovered alum was shipped to the nearby Lenox Wastewater Treatment Plant (LWWTP), Lenox, Massachusetts, for wastewater treatment as a part of this demonstration project. It was discovered that over 85% phosphorus removal could be achieved consistently in the entire research period. Under these situations, buildup caused by recycling of impurities could not occur, while advantage would accrue to both water utility and wastewater utility. It is concluded that employing recovered alum from water treatment sludge as a precipitant for phosphate removal in a wastewater treatment plant is technically feasible. The problem is economics. For instance, the LWWTP cannot possibly consume all acidified sludge supernatant (containing recovered alum) from the LWTP within its own town of Lenox boundary. The economic feasibility of this option needs to be further studied.

The AquaCritox process is a supercritical water oxidation process in which alum sludge is heated to between 374°C and 500°C at 221 bar pressure in the presence of oxygen (43). All of the organic matter is completely oxidized in an exothermic reaction producing carbon dioxide, water, aluminum hydroxide, and iron hydroxide as a water insoluble precipitate mixture. The pure precipitated coagulant hydroxide mixture is readily reacted with sulfuric acid to form fresh aluminum or iron sulfate that is capable of meeting the USEPA specifications for coagulants. This is another option for disposal and reuse of alum generated from the water utility. Additional research data and current practices of recycling of backwash wastewater and alum sludge can be found from the literature (44–49).

REFERENCES

1. Pallo, P. E., Schwartz, V. J., and Wang, L. K. Recycling and reuse of filter backwash water containing alum sludge, *Water and Sewage Works*, 115 (5), 123, 1972.
2. Wang, L. K., Pallo, P. E., Schwartz, V. J., and Kown, B. T. Continuous pilot plant study of recycling of filter backwash water, *Journal American Water Works Association*, 65 (5), 355–358, 1973.
3. Wang, L. K. and Yang, J. Y. Total waste recycle system for water purification plant using alum as primary coagulant, *Resource Recovery and Conservation*, 1, 67–84, 1975.
4. Krofta, M. and Wang, L. K. Application of dissolved air flotation to the Lenox Massachusetts water supply: Water purification by flotation, *Journal of New England Water Works Association*, 99, 249–264, 1985.
5. Krofta, M. and Wang, L. K. Application of dissolved air flotation to the Lenox Massachusetts water supply: Sludge thickening by flotation or lagoon, *Journal of New England Water Works Association*, 99, 265–284, 1985.
6. Krofta, M. and Wang, L. K. Municipal waste treatment by supracell flotation, chemical oxidation and star system, Technical Report No. LIR/l0–86/214, Lenox Institute of Water Technology, 22pp., October 1986. NTIS-PB88 2005481/AS
7. White, A. R. and White, P. M. N. Alum recovery, an aid to the disposal of water plant solids, *Technical Paper Presented at the ASCE 1984 Spring Convention*, Atlanta, GA, May 1984.
8. Krofta, M. and Wang, L. K. Winter operation of nation's largest potable flotation plants, *Technical Papers Presented at the 1987 Joint Conference of American Water Works Association and Water Pollution Control Federation*, Cheyenne, WY, September 20–23, 1987. NTIS-PB8820056 3/AS.
9. Krofta, M. and Wang, L. K. Winter operation of the nation's first potable flotation plant, *Technical Papers Presented at the 1987 Joint Conference of American Water Works Association and Water Pollution Control Federation*, Cheyenne, WY, September 20–23, 1987. NTIS-P88 200555/AS.
10. Roberts, J. M. and Roddy, C. P. Recovery and reuse of alum sludge at Tampa, *Journal American Water Works Association*, 52 (7), 91, 1960.
11. Fulton, C. R. Alum recovery for filter plant waste treatment, *Water and Waste Engineering*, 7, 78, 1974.
12. Grurmiiinger, R. M. Disposal of waste alum sludge from water treatment plants, *Journal of Water Pollution Control Federation*, 47, 543, 1975.
13. Fulton, C. P. Recover alum to reduce waste disposal costs, *Journal American Water Works Association*, 66 (5), 312, 1974.
14. Chen, B. H. H. Alum recovery from representative water treatment plant sludge, *Journal American Water Works Association*, 68 (4), 204, 1976.
15. Cornwell, D. A. and Zoltek Jr., J. Recycling of alum used for phosphorous removal in domestic wastewater treatment, *Journal of Water Pollution Control Federation*, 49 (4), 600, 1977.
16. Westerhoff, C. P. Water treatment plant sludges, Part I, *Journal American Water Works Association*, 70 (9), 498, 1978.
17. Westerhoff, C. P. Water treatment plant sludges, Part II, *Journal American Water Works Association*, 70 (10), 548, 1978.
18. Webster, 3. A. Operation and experiences at Doer water treatment works, with special reference to use of activated silica and the recovery of alum from sludge, *Journal Institute Water Engineering*, 20 (2), 167, 1966.
19. AWWA Editor. Disposal of wastes from water treatment plants, Part I, *Journal American Water Works Association*, 61 (10), 541, 1969.
20. AWWA Editor. Disposal of wastes from water treatment plants, Part II, *Journal American Water Works Association*, 61 (11), 619, 1969.
21. AWWA Editor. Disposal of wastes from water treatment plants, Part III, *Journal American Water Works Association*, 61 (12), 681, 1969.
22. AWWA Editor. Disposal of water treatment plant wastes, *Journal American Water Works Association*, 64 (12), 814, 1972.
23. O'Connor, J. L. *Management of Water Treatment Residues*, McGraw-Hill, Water Quality and Treatment, American Water Works Association, New York, 1971.

24. Cornwell, D. A. and Susan, J. A. Characteristics of acid-treated sludges, *Journal American Water Works Association*, 71 (10), 604, 1979.
25. Palm, A. T. The treatment and disposal of alum sludge, *Proceedings of Society of Water Treatment Examination*, 3, 131, 1954.
26. Issac, P. E. and Vahidi, I. The recovery of alum sludge, *Proceedings of Society of Water Treatment Examination*, 10, 19, 1969.
27. Westerhoff, C. P. and Daly, M. P. Water treatment wastes and disposal, Part I, *Journal American Water Works Association*, 66 (5), 319, 1974.
28. Westerhoff, C. P. and Daly, M. P. Water treatment wastes and disposal, Part II, *Journal American Water Works Association*, 66 (6), 378, 1974.
29. Westerhoff, C. P. and Daly, M. P. Water treatment wastes and disposal, Part III, *Journal American Water Works Association*, 66 (7), 441, 1974.
30. Lindsey, E. E. and Longsame, O. Recovery and reuse of alum from water filtration plant sludge by ultra-filtration, *American Institute of Chemical Engineers Symposium Series of Water*, 71, 185, 1975.
31. Wang, L. K. and Wu, B. C. Recovery and reuse of filter alum and soda alum in a water treatment plant, Technical Report No. LIR/06-88/309, Lenox Institute of Water Technology, Lenox, MA, June 8, 1988.
32. Kempa, J. and Juslyna, J. Methods and feasibilities of aluminum recovery from precipitation sludge, *Recovery Energy Matter, Residue Waste, Recycling Institute*, 792–799, 1982.
33. Goldman, M. L. and Watwon, F. Feasibility of alum sludge reclamation, Water Resources Research Center, Washington Technical Institute, Washington, DC, July 1975.
34. Wang, L. K., Hung, Y. T., and Shammas, N. K. (eds.). *Physicochemical Treatment Processes*, Humana Press, Totowa, NJ, 723pp., 2005.
35. Wang, L. K., Hung, Y. T., and Shammas, N. K. (eds.). *Advanced Physicochemical Treatment Processes*, Humana Press, Totowa, NJ, 690pp., 2006.
36. Wang, L. K., Hung, Y. T., and Shammas, N. K. (eds.), *Advanced Physicochemical Treatment Technologies*, Humana Press, Totowa, NJ, 710pp., 2007.
37. Wang, L. K., Shammas, N. K., Selke, W. A., and Aulenbach, D. B. (eds.). *Flotation Technology*, Humana Press, Totowa, NJ, 680pp., 2010.
38. Bugbee, G. J. and Frink, C. R., *Alum Sludge as a Soil Amendment: Effects on Soil Properties and Plant Growth*, Connecticut Agricultural Experiment Station, New Haven, CT, Bulletin 827, November 1985.
39. Faust, S. D. and Aly, O. M. *Chemistry of Water Treatment*, CRC Press, New York, 256pp, 1998.
40. Chu, W. Lead metal removal by recycled alum sludge, *Water Research*, 33 (13), 3019–3025, 1999.
41. Georgantas, D. A., Matsis, V. M., and Grigoropoulou, H. P. Soluble phosphorus removal through adsorption on spent alum sludge, *Environmental Technology*, 27 (10), 1093–1095, 2006.
42. Sanin, F. D., Clarkson, W. W., and Veslind, P. A. *Sludge Engineering: The Treatment and Disposal of Wastewater Sludges*, Technology and Engineering, New York, 393pp., 2010.
43. SCFI Group, Ltd. Alum Sludge: From Waste to Valuable Product with AquaCritox Recovery Process, SCFI Group, Ltd., Rubicon, CIT Campus, Cork, Ireland. April 2011.
44. State of Kentucky. Recycling of Water Treatment Waste Alum Sludge, Frankfort, KY, 2011.
45. Elangovan, C. and Subramanian, K. Reuse of alum sludge in clay brick manufacturing, *Water Science and Technology*, 11 (3), 333–341, 2011.
46. Hu, Y. Y., Zhao, Y., Zhao, X., and Kumar, J. L. High rate nitrogen removal in an alum sludge based inter-mittent aeration constructed wetland, *Environmental Science and Technology*, 46 (8), 4583–4590, 2012.
47. Ireland Environmental Protection Agency (IEPA). Development of an Alum Sludge-Based Constructed Wetland System for Improving Organic Matter and Nutrients Removal in High-Strength Wastewater, Ireland Environmental Protection Agency. Dublin, Ireland. 2012.
48. Wang, L. K. and Shammas, N. K. Waste sludge management in water utilities, in: Hung, Y. T., Wang, L. K., and Shammas, N. K. (eds.), *Handbook of Environment and Waste Management: Land and Groundwater Pollution Control*, Vol. 2, World Scientific, Singapore, pp. 1061–1091, 2014.
49. Shammas, N. K. and Wang, L. K. *Water Engineering: Hydraulic, Distribution and Treatment*, John Wiley & Sons, Inc., Hoboken, NJ, pp. 661–682, 2016.
50. Shammes, N.K. and Wang, L.K. *Water and Wastewater Engineering: Water Supply and Wastewater Removal*. Third edition. John Wiley & Sons, Inc. Hoboken, NJ, 824 pp, 2011.
51. Shammas, N.K., Wang, L.K., Queiroz Faria, L.C. and Chaves Ferro, M.A. Abastecimento de Água e Remoc ã de Resĺduos. Gropo Editorial Nacional LTC, Rio de Janeiro. 750 pp, 2013.

4 Selection of Remedial Alternatives for Soil Contaminated with Heavy Metals

Nazih K. Shammas

CONTENTS

ABSTRACT

Metals account for much of the contamination found at hazardous waste sites. They are present in the soil and ground water at approximately 65% of the Superfund sites. The metals most frequently identified are lead, arsenic, chromium, cadmium, nickel, and zinc. Other metals often identified as contaminants include copper and mercury.

This chapter provides remedial project managers, engineers, on-scene coordinators, contractors, and other state or private remediation managers and their technical support personnel with information to facilitate the selection of appropriate remedial alternatives for soil contaminated with arsenic (As), cadmium (Cd), chromium (Cr), mercury (Hg), and lead (Pb).

Common compounds, transport, and fate are discussed for each of the five elements. A general description of metal-contaminated Superfund soils is provided. The technologies covered are containment (immobilization), solidification/stabilization (S/S), vitrification, soil washing, soil flushing, pyrometallurgy, electrokinetics, and phytoremediation. Use of treatment trains and remediation costs are also addressed.

4.1 INTRODUCTION

Metals account for much of the contamination found at hazardous waste sites. They are present in the soil and ground water at approximately 65% of the Superfund or CERCLA (Comprehensive Environmental Response, Compensation, and Liability Act) (1) sites for which the U.S. Environmental Protection Agency (USEPA) has signed records of decisions (RODs) (2). The metals most frequently identified are lead, arsenic, chromium, cadmium, nickel, and zinc. Other metals often identified as contaminants include copper and mercury. In addition to the Superfund program, metals make up a significant portion of the contamination requiring remediation under the Resource Conservation and Recovery Act (RCRA) (3) and contamination present at federal facilities, notably those that are the responsibility of the Department of Defense (DoD) and the Department of Energy (DOE).

This chapter provides remedial project managers, engineers, on-scene coordinators, contractors, and other state or private remediation managers and their technical support personnel with information to facilitate the selection of appropriate remedial alternatives for soil contaminated with arsenic (As), cadmium (Cd), chromium (Cr), mercury (Hg), and lead (Pb) (4–6).

Common compounds, transport, and fate are discussed for each of the five elements. A general description of metal-contaminated Superfund soils is provided. The technologies covered are containment (immobilization), solidification/stabilization (S/S), vitrification, soil washing, soil flushing, pyrometallurgy, electrokinetics, and phytoremediation. Use of treatment trains and remediation costs are also addressed.

It is assumed that users of this chapter will, as necessary, familiarize themselves with (1) the applicable or relevant and appropriate regulations pertinent to the site of interest; (2) applicable health and safety regulations and practices relevant to the metals and compounds discussed; and (3) relevant sampling, analysis, and data interpretation methods. Information on Pb battery (Pb and As), wood preserving (As and Cr), pesticide (Pb, As, and Hg), and mining sites have been addressed in USEPA Superfund documents (7–12). The greatest emphasis is on remediation of inorganic forms of the metals of interest. Organometallic compounds, organic–metal mixtures, and multimetal mixtures are briefly addressed.

4.2 OVERVIEW OF METAS AND THEIR COMPOUNDS

This section provides a brief, qualitative overview of the physical characteristics and mineral origins of the five metals, and factors affecting their mobility. More comprehensive and quantitative reviews of the behavior of these five metals in soil can be found in readily available USEPA Superfund documents (4,13,14).

4.2.1 OVERVIEW OF PHYSICAL CHARACTERISTICS AND MINERAL ORIGINS

Arsenic is a semimetallic element or metalloid that has several allotropic forms. The most stable allotrope is a silver-gray, brittle, crystalline solid that tarnishes in air. As compounds, mainly As_2O_3, can be recovered as a byproduct of processing complex ores mined mainly for copper, lead, zinc, gold, and silver. As occurs in a wide variety of mineral forms, including arsenopyrite, $FeAsS_4$, which is the main commercial ore of As worldwide.

Cadmium is a bluish-white, soft, ductile metal. Pure Cd compounds rarely are found in nature, although occurrences of greenockite (CdS) and otavite ($CdCO_3$) are known. The main sources of Cd are sulfide ores of lead, zinc, and copper. Cd is recovered as a byproduct when these ores are processed.

Chromium is a lustrous, silver-gray metal. It is one of the less common elements in the Earth's crust, and occurs only in compounds. The chief commercial source of Cr is the mineral chromite, $FeCr_2O_4$. Cr is mined as a primary product and is not recovered as a byproduct of any other mining operation. There are no chromite ore reserves, nor is there primary production of chromite in the United States.

Mercury is a silvery, liquid metal. The primary source of Hg is cinnabar (HgS), a sulfide ore. In a few cases, Hg occurs as the principal ore product; it is more commonly obtained as the byproduct of processing complex ores that contain mixed sulfides, oxides, and chloride minerals (these are usually associated with base and precious metals, particularly gold). Native or metallic Hg is found in very small quantities in some ore sites. The current demand for Hg is met by secondary production (i.e., recycling and recovery).

Lead is a bluish-white, silvery, or gray metal that is highly lustrous when freshly cut but tarnishes when exposed to air. It is very soft and malleable, has a high density (11.35 g/cm^3) and low-melting point (327.4°C), and can be cast, rolled, and extruded. The most important Pb ore is galena, PbS. Recovery of Pb from the ore typically involves grinding, flotation, roasting, and smelting. Less common forms of the mineral are cerussite, $PbCO_3$, anglesite, $PbSO_4$, and crocoite, $PbCrO_4$.

4.2.2 OVERVIEW OF BEHAVIOR OF AS, CD, CR, PB, AND HG

Since metals cannot be destroyed, remediation of metal-contaminated soil consists primarily of manipulating (i.e., exploiting, increasing, decreasing, or maintaining) the mobility of metal contaminant(s) to produce a treated soil that has an acceptable total or leachable metal content. Metal mobility depends upon numerous factors. Metal mobility in soil–waste systems is determined by (13)

1. Type and quantity of soil surfaces present
2. Concentration of metal of interest
3. Concentration and type of competing ions and complexing ligands, both organic and inorganic
4. pH
5. Redox status

"Generalization can only serve as rough guides of the expected behavior of metals in such systems. Use of literature or laboratory data that do not mimic the specific site soil and waste system will not be adequate to describe or predict the behavior of the metal. Data must be site-specific. Long term effects must also be considered. As organic constituents of the waste matrix degrade, or as pH or redox conditions change, either through natural processes of weathering or human manipulation, the potential mobility of the metal will change as soil conditions change" (13).

Cd, Cr(III), and Pb are present in cationic forms under natural environmental conditions (13). These cationic metals generally are not mobile in the environment and tend to remain relatively close to the point of initial deposition. The capacity of soil to adsorb cationic metals increases with increasing pH, cation exchange capacity, and organic carbon content. Under the neutral to basic conditions typical of most soils, cationic metals are strongly adsorbed on the clay fraction of soils and can be adsorbed by hydrous oxides of iron, aluminum, or manganese present in soil minerals. Cationic metals will precipitate as hydroxides, carbonates, or phosphates. In acidic, sandy soils, the cationic metals are more mobile. Under conditions that are atypical of natural soils (e.g., pH <5 or >9; elevated concentrations of oxidizers or reducers; high concentrations of soluble organic or inorganic complexing or colloidal substances), but that may be encountered as a result of waste disposal or remedial processes, the mobility of these metals may be substantially increased. Also, competitive adsorption between various metals has been observed in experiments involving various solids with oxide surfaces (γ-FeOOH, α-SiO$_2$, and γ-Al$_2$O$_3$). In several experiments, Cd adsorption was decreased by the addition of Pb or Cu for all three of these solids. The addition of zinc resulted in the greatest decrease of Cd adsorption. Competition for surface sites occurred when only a few percent of all surface sites were occupied (15).

As, Cr(VI), and Hg behaviors differ considerably from Cd, Cr(III), and Pb. As and Cr(VI) typically exist in anionic forms under environmental conditions. Hg, although it is a cationic metal, has unusual properties (e.g., liquid at room temperature, easily transforms among several possible valence states).

In most As-contaminated sites, As appears as As$_2$O$_3$ or as anionic As species leached from As$_2$O$_3$, oxidized to As(V), and then sorbed onto iron-bearing minerals in the soil. As may be present also in organometallic forms, such as methylarsenic acid, H$_2$AsO$_3$CH$_3$, and dimethylarsenic acid, (CH$_3$)$_2$AsO$_2$H, which are active ingredients in many pesticides, as well as the volatile compounds arsine (AsH$_3$) and its methyl derivatives [i.e., dimethylarsine HAs(CH$_3$)$_2$ and trimethylarsine, As(CH$_3$)$_3$]. These As forms illustrate the various oxidation states that As commonly exhibits (–III, 0, III, and V) and the resulting complexity of its chemistry in the environment.

As(V) is less mobile and less toxic than As(III). As(V) exhibits anionic behavior in the presence of water, and hence its aqueous solubility increases with increasing pH, and it does not complex or precipitate with other anions. As(V) can form low-solubility metal arsenates. Calcium arsenate, Ca$_3$(AsO$_4$)$_2$, is the most stable metal arsenate in well-oxidized and alkaline environments, but it is unstable in acidic environments. Even under initially oxidizing and alkaline conditions, absorption of CO$_2$ from the air will result in the formation of CaCO$_3$ and release of arsenate. In sodic soils, sufficient sodium is available, such that the mobile compound Na$_3$AsO$_4$ can form. The slightly less stable manganese arsenate, Mn$_2$(AsO$_4$)$_2$, forms in both acidic and alkaline environments, while iron arsenate is stable under acidic soil conditions. In aerobic environments, HAsO$_4$ predominates at pH <2 and is replaced by H$_2$AsO$_4^-$, HAsO$_4^{2-}$, and AsO$_4^{3-}$ as pH increases to about 2, 7, and 11.5, respectively. Under mildly reducing conditions, H$_3$AsO$_3$ is a predominant species at low pH, but is replaced by H$_2$AsO$_3^-$, HAsO$_3^{2-}$, and AsO$_3^{3-}$ as pH increases. Under still more reducing conditions and in the presence of sulfide, As$_2$S$_3$ can form. As$_2$S$_3$ is a low-solubility, stable solid. AsS$_2$ and AsS$_2^-$ are thermodynamically unstable with respect to As$_2$S$_3$ (16). Under extreme reducing conditions, elemental As and volatile arsine, AsH$_3$, can occur. Just as competition between cationic metals affects mobility in soil, competition between anionic species (chromate, arsenate, phosphate, sulfate, etc.) affects anionic fixation processes and may increase mobility.

The most common valence states of Cr in the Earth's surface and near-surface environment are +3 (trivalent or Cr(III)) and +6 (hexavalent or Cr(VI)). The trivalent Cr (discussed above) is the most thermodynamically stable form under common environmental conditions. Except in leather tanning, industrial applications of Cr generally use the Cr(VI) form. Due to kinetic limitations, Cr(VI) does not always readily reduce to Cr(III) and can remain present over an extended period of time.

Cr(VI) is present as the chromate, CrO_4^{2-}, or dichromate, $Cr_2O_7^{2-}$, anion, depending on pH and concentration. Cr(VI) anions are less likely to be adsorbed to solid surfaces than Cr(III). Most solids in soils carry negative charges that inhibit Cr(VI) adsorption. Although clays have high capacity to adsorb cationic metals, they interact little with Cr(VI) because of the similar charges carried by the anion and clay in the common pH range of soil and groundwater. The only common soil solid that adsorbs Cr(VI) is iron oxyhydroxide. Generally, a major portion of Cr(VI) and other anions adsorbed in soils can be attributed to the presence of iron oxyhydroxide. The quantity of Cr(VI) adsorbed onto the iron solids increases with decreasing pH.

At metal-contaminated sites, Hg can be present in mercuric form (Hg^{2+}) mercurous form (Hg_2^{2+}), elemental form (Hg), or alkylated form (e.g., methyl and ethyl Hg). Hg_2^{2+} and Hg^{2+} are more stable under oxidizing conditions. Under mildly reducing conditions, both organically bound Hg and inorganic Hg compounds can convert to elemental Hg, which then can be readily converted to methyl or ethyl Hg by biotic and abiotic processes. Methyl and ethyl Hg are mobile and toxic forms.

Hg is moderately mobile, regardless of the soil. Both the mercurous and mercuric cations are adsorbed by clay minerals, oxides, and organic matter. Adsorption of cationic forms of Hg increases with increasing pH. Mercurous and mercuric Hg are also immobilized by forming various precipitates. Mercurous Hg precipitates with chloride, phosphate, carbonate, and hydroxide. At concentrations of Hg commonly found in soil, only the phosphate precipitate is stable. In alkaline soils, mercuric Hg precipitates with carbonate and hydroxide to form a stable (but not exceptionally insoluble) solid phase. At lower pH and high chloride concentration, soluble $HgCl_2$ is formed. Mercuric Hg also forms complexes with soluble organic matter, chlorides, and hydroxides that may contribute to its mobility (13). In strong reducing conditions, HgS, a very low-solubility compound is formed.

4.3   DESCRIPTION OF SUPERFUND SOILS CONTAMINATED WITH METALS

Soils can become contaminated with metals from direct contact with industrial plant waste discharges; fugitive emissions; or leachate from waste piles, landfills, or sludge deposits. The specific type of metal contaminant expected at a particular Superfund site would obviously be directly related to the type of operation that had occurred there. Table 4.1 lists the types of operations that are directly associated with each of the five metal contaminants (5).

Wastes at CERCLA sites are frequently heterogeneous on a macro and micro scale. The contaminant concentration and the physical and chemical forms of the contaminant and matrix usually are complex and variable. Of these, waste disposal sites collect the widest variety of waste types; therefore concentration profiles vary by orders of magnitude through a pit or pile. Limited volumes of high-concentration "hot spots" may develop due to variations in the historical waste disposal patterns or local transport mechanisms. Similar radical variations frequently occur on the particle-size scale as well. The waste often consists of a physical mixture of very different solids, for example, paint chips in spent abrasive.

Industrial processes may result in a variety of solid metal-bearing waste materials, including slags, fumes, mold sand, fly ash, abrasive wastes, spent catalysts, spent-activated carbon, and refractory bricks (17). These process solids may be found above ground as waste piles or below ground in landfills. Solid-phase wastes can be dispersed by well-intended but poorly controlled reuse projects. Waste piles can be exposed to natural disasters or accidents causing further dispersion.

4.4   SOIL CLEANUP GOALS AND TECHNOLOGIES FOR REMEDIATION

Table 4.2 provides an overview of cleanup goals (actual and potential) for both total and leachable metals. Based on inspection of the total metals cleanup goals, one can see that they vary considerably both within the same metal and between metals.

TABLE 4.1

Principle Sources of As, Cd, Cr, Hg, and Pb-Contaminated Soils

Contaminant	Principle Sources
As	Wood preserving
	As-waste disposal
	Pesticide production and application
	Mining
Cd	Plating
	Ni–Cd battery manufacturing
	Cd-waste disposal
Cr	Plating
	Textile manufacturing
	Leather tanning
	Pigment manufacturing
	Wood preserving
	Cr-waste disposal
Hg	Chloralkali manufacturing
	Weapons production
	Copper and zinc smelting
	Gas line manometer spills
	Paint application
	Hg-waste disposal
Pb	Ferrous/nonferrous smelting
	Pb-acid battery breaking
	Ammunition production
	Leaded paint waste
	Pb-waste disposal
	Secondary metals production
	Waste oil recycling
	Firing ranges
	Ink manufacturing
	Mining
	Pb-acid battery manufacturing
	Leaded glass production
	Tetraethyl Pb production
	Chemical manufacturing

Source: USEPA. *Technology Alternatives for the Remediation of Soils Contaminated with AS, Cd, Cr, Hg, and Pb.* EPA/540/S-97/500, U.S. Environmental Protection Agency, Cincinnati, OH, August 1997.

Similar variation is observed in the actual or potential leachate goals. The observed variation in cleanup goals has at least two implications with regard to technology alternative evaluation and selection. First, the importance of identifying the target metal(s), contaminant state (leachable vs. total metal), the specific type of test and conditions, and the numerical cleanup goals early in the remedy evaluation process is made apparent. Depending on which cleanup goal is selected, the required removal or leachate reduction efficiency of the overall remediation can vary by several orders of magnitude (5,18). Second, the degree of variation in goals both within and between the metals, plus the many factors that affect mobility of the metals, suggest that generalizations about effectiveness of a technology for meeting total or leachable treatment goals should be viewed with some caution.

TABLE 4.2

Cleanup Goals (Actual and Potential) for Total and Leachable Metals

Description	As	Cd	Cr (Total)	Hg	Pb
Total Metals Goals (mg/kg)					
Background (mean)	5	0.06	100	0.03	10
Background (range)	1–50	0.01–0.70	1–1,000	0.01–0.30	2–200
Superfund site goals from TRD	5–65	3–20	6.7–375	1–21	200–500
Theoretical minimum total metals to ensure TCLP Leachate < threshold (i.e., TCLP × 20)	100	20	100	4	100
California total threshold limit concentration	500	100	500	20	1,000
Leachable Metals (µg/L)					
TCLP threshold for RCRA waste	5,000	1,000	5,000	200	5,000
Extraction procedure toxicity test	5,000	1,000	5,000	200	5,000
Synthetic precipitate leachate	—	—	—	—	—
Multiple extraction procedure	—	—	—	—	—
California soluble threshold leachate concentration	5,000	1,000	5,000	200	5,000
Maximum contaminant level[a]	50	5	100	2	15
Superfund site goals from TRD	50	—	50	0.05–2	50

Source: USEPA. *Technology Alternatives for the Remediation of Soils Contaminated with AS, Cd, Cr, Hg, and Pb.* EPA/540/S-97/500, U.S. Environmental Protection Agency, Cincinnati, OH, August 1997.

Note: —, No specified level and no example cases identified.

[a] Maximum contaminant level = the maximum permissible level of contaminant in water delivered to any user of a public system.

Technologies potentially applicable to the remediation of soils contaminated with the five metals or their inorganic compounds are listed below (2,5):

Technology Class	Specific Technology
Containment	Caps
	Vertical barriers
	Horizontal barriers
Solidification/stabilization	Cement based
	Polymer microencapsulation
	Vitrification
Separation/concentration	Soil washing
	Soil flushing
	Pyrometallurgy
	Electrokinetics
	Phytoremediation

The best demonstrated available technology (BDAT) status refers to the determination under the RCRA of the BDAT for various industry-generated hazardous wastes that contain the metals of interest. Whether the characteristics of a Superfund metal-contaminated soil (or fractions derived from it) are similar enough to the RCRA waste to justify serious evaluation of the BDAT for a specific Superfund soil must be made on a site-specific basis. Other limitations relevant to BDATs include (a) the regulatory basis for BDAT standards focus BDATs on proven, commercially available technologies at the time of the BDAT determination, (b) a BDAT may be identified, but that does not necessarily preclude the use of other technologies, and (c) a technology identified as BDAT may not necessarily be the current technology of choice in the RCRA hazardous waste treatment industry.

The USEPA's Superfund Innovative Technology Evaluation (SITE) program evaluates many emerging and demonstrated technologies in order to promote the development and use of innovative technologies to cleanup Superfund sites across the country. The major focus of SITE is the Demonstration Program, which is designed to provide engineering and cost data for selected technologies.

Cost is not discussed in each technology narrative; however, a summary table is provided at the end of the technology discussion section that illustrates technology cost ranges and treatment train options.

4.5 CONTAINMENT

Containment technologies for application at Superfund sites include landfill covers (caps), vertical barriers, and horizontal barriers (4). For metal remediation, containment is considered an established technology except for in situ installation of horizontal barriers.

4.5.1 PROCESS DESCRIPTION

Containment ranges from a surface cap that limits infiltration of uncontaminated surface water to subsurface vertical or horizontal barriers that restrict lateral or vertical migration of contaminated groundwater. The material provided here is primarily from USEPA (5,9).

4.5.1.1 Caps

Capping systems reduce surface water infiltration; control gas and odor emissions; improve esthetics; and provide a stable surface over the waste. Caps can range from a simple native soil cover to a full RCRA Subtitle C composite cover.

Cap construction costs depend on the number of components in the final cap system (i.e., costs increase with the addition of barrier and drainage components). Additionally, cost escalates as a function of topographic relief. Side slopes steeper than 3 horizontal to 1 vertical can cause stability and equipment problems that dramatically increase the unit cost (4,19).

4.5.1.2 Vertical Barriers

Vertical barriers minimize the movement of contaminated groundwater off-site or limit the flow of uncontaminated groundwater onsite. Common vertical barriers include slurry walls in excavated trenches; grout curtains formed by injecting grout into soil borings; vertically injected, cement–bentonite grout-filled borings or holes formed by withdrawing beams driven into the ground; and sheet-pile walls formed of driven steel.

Certain compounds can affect cement–bentonite barriers. The impermeability of bentonite may significantly decrease when it is exposed to high concentrations of creosote, water-soluble salts (copper, Cr, As), or fire retardant salts (borates, phosphates, and ammonia). Specific gravity of salt solutions must be >1.2 to impact bentonite (20,21). In general, soil–bentonite blends resist chemical attack best if they contain only 1% bentonite and 30%–40% natural soil fines. Treatability tests should evaluate the chemical stability of the barrier if adverse conditions are suspected.

Carbon steel used in pile walls quickly corrodes in dilute acids, slowly corrodes in brines or salt water, and remains mostly unaffected by organic chemicals or water. Salts and fire retardants can reduce the service life of a steel sheet pile; corrosion-resistant coatings can extend their anticipated life. Major steel suppliers will provide site-specific recommendations for cathodic protection of piling.

Construction costs for vertical barriers are influenced by the soil profile of the barrier material used and by the method of placing it. The most economical shallow vertical barriers are soil–bentonite trenches excavated with conventional backhoes; the most economical deep vertical barriers consist of a cement–bentonite wall placed by a vibrating beam.

4.5.1.3 Horizontal Barriers

In situ horizontal barriers can underlie a sector of contaminated materials onsite without removing the hazardous waste or soil. Established technologies use grouting techniques to reduce the permeability of underlying soil layers. Studies performed by the U.S. Army Corps of Engineers (22) indicate that conventional grout technology cannot produce an impermeable horizontal barrier because it cannot ensure uniform lateral growth of the grout. These same studies found greater success with jet grouting techniques in soils that contain fines sufficient to prevent collapse of the wash hole and that present no large stones or boulders that could deflect the cutting jet.

Since few in situ horizontal barriers have been constructed, accurate costs have not been established. Work performed by Corps of Engineers for USEPA has shown that it is very difficult to form effective horizontal barriers. The most efficient barrier installation used a jet wash to create a cavity in sandy soils into which cement–bentonite grouting was injected. The costs relate to the number of borings required. Each boring takes at least one day to drill.

4.5.2 Site Requirements

In general, the site must be suitable for a variety of heavy construction equipment including bulldozers, graders, backhoes, multishaft drill rigs, various rollers, vibratory compactors, forklifts, and seaming devices (23,24). When capping systems are being utilized, onsite storage areas are necessary for the materials to be used in the cover. If site soils are adequate for use in the cover, a borrow area needs to be identified and the soil tested and characterized. If site soils are not suitable, it may be necessary to truck in other low-permeability soils (23). In addition, an adequate supply of water may also be needed in order to achieve the optimum soil density.

The construction of vertical containment barriers, such as slurry walls, requires knowledge of the site, the local soil and hydrogeologic conditions, and the presence of underground utilities (25). Preparation of the slurry requires batch mixers, hydration ponds, pumps, hoses, and an adequate supply of water. Therefore, onsite water storage tanks and electricity are necessary. In addition, areas adjacent to the trench need to be available for the storage of trench spoils (which could potentially be contaminated) and the mixing of backfill. If excavated soils are not acceptable for use as backfill, suitable backfill must be trucked to the site (25).

4.5.3 Applicability

Containment is most likely to be applicable to (5)

1. Wastes that are low-hazard (e.g., low toxicity or low concentration) or immobile
2. Wastes that have been treated to produce low-hazard or low-mobility wastes for onsite disposal
3. Wastes whose mobility must be reduced as a temporary measure to mitigate risk until a permanent remedy can be tested and implemented

Situations where containment would not be applicable include

1. Wastes for which there is a more permanent and protective remedy that is cost-effective
2. Where effective placement of horizontal barriers below existing contamination is difficult
3. Where drinking water sources will be adversely affected if containment fails, and if there is inadequate confidence in the ability to predict, detect, or control harmful releases due to containment failure

Important advantages of containment are (5)

1. Surface caps and vertical barriers are relatively simple and rapid to implement at low cost and can be more economical than excavation and removal of waste
2. Caps and vertical barriers can be applied to large areas or volumes of waste
3. Engineering control (containment) is achieved, and may be a final action if metals are well immobilized and potential receptors are distant
4. A variety of barrier materials are available commercially
5. In some cases it may be possible to create a land surface that can support vegetation and/ or be applicable for other purposes

Disadvantages of containment include (5)

1. Design life is uncertain
2. Contamination remains onsite, available to migrate should containment fail
3. Long-term inspection, maintenance, and monitoring is required
4. Site must be amenable to effective monitoring
5. Placement of horizontal barriers below existing waste is difficult to implement successfully

4.5.4 PERFORMANCE AND BDAT STATUS

Containment is widely accepted as a means of controlling the spread of contamination and preventing the future migration of waste constituents. Table 4.3 shows a list of selected sites where containment has been selected for remediating metal-contaminated solids.

The performance of capping systems, once installed, may be difficult to evaluate (23). Monitoring well systems or infiltration monitoring systems can provide some information, but it is often not possible to determine whether the water or leachate originated as surface water or groundwater.

With regard to slurry walls and other vertical containment barriers, performance may be affected by a number of variables including geographic region, topography, and material availability. A thorough characterization of the site and a compatibility study are highly recommended (25).

Containment technologies are not considered "treatment technologies" and hence no BDATs involving containment have been established.

4.5.5 SITE PROGRAM DEMONSTRATION PROJECTS

Ongoing SITE demonstrations applicable to soils contaminated with the metals of interest include

- Morrison Knudsen Corporation (high clay grouting technology)
- RKK, Ltd. (frozen soil barriers)

TABLE 4.3
Containment Applications at Selected Superfund Sites with Metal Contamination

Site Name/State	Specific Technology	Key Metal Contaminants	Associated Technology
Ninth Avenue Dump, IN	Containment-slurry wall	Pb	Slurry wall/capping
Industrial Waste Control, AK	Containment-slurry wall	As, Cd, Cr, Pb	Capping/French drain
E.H. Shilling Landfill, OH	Containment-slurry wall	As	Capping/clay berm
Chemtronic, NC	Capping	Cr, Pb	Capping
Ordnance Works Disposal, WV	Capping	As, Pb	Capping
Industriplex, MA	Capping	As, Pb, Cr	Capping

Source: USEPA. *Technology Alternatives for the Remediation of Soils Contaminated with AS, Cd, Cr, Hg, and Pb.* EPA/540/S-97/500, U.S. Environmental Protection Agency, Cincinnati, OH, August 1997.

4.6   SOLIDIFICATION/STABILIZATION TECHNOLOGIES

The term "solidification/stabilization" refers to a general category of processes that are used to treat a wide variety of wastes, including solids and liquids. Solidification and stabilization are each distinct technologies, as described below (26):

Solidification—refers to processes that encapsulate a waste to form a solid material and to restrict contaminant migration by decreasing the surface area exposed to leaching and/or by coating the waste with low-permeability materials. Solidification can be accomplished by a chemical reaction between a waste and binding (solidifying) reagents or by mechanical processes. Solidification of fine waste particles is referred to as microencapsulation, while solidification of a large block or container of waste is referred to as macroencapsulation.

Stabilization—refers to processes that involve chemical reactions that reduce the leachability of a waste. Stabilization chemically immobilizes hazardous materials (such as heavy metals) or reduces their solubility through a chemical reaction. The physical nature of the waste may or may not be changed by this process.

S/S aims to accomplish one or more of the following objectives (4):

1. Improve the physical characteristics of the waste by producing a solid from liquid or semi-liquid wastes
2. Reduce the contaminant solubility by formation of sorbed species or insoluble precipitates (e.g., hydroxides, carbonates, silicates, phosphates, sulfates, or sulfides)
3. Decrease the exposed surface area across which mass transfer loss of contaminants may occur by formation of a crystalline, glassy, or polymeric framework which surrounds the waste particles
4. Limit the contact between transport fluids and contaminants by reducing the material's permeability

S/S technology usually is applied by mixing contaminated soils or treatment residuals with a physical binding agent to form a crystalline, glassy, or polymeric framework surrounding the waste particles. In addition to the microencapsulation, some chemical fixation mechanisms may improve the waste's leach resistance. Other forms of S/S treatment rely on macroencapsulation, where the waste is unaltered but macroscopic particles are encased in a relatively impermeable coating (27), or on specific chemical fixation, where the contaminant is converted to a solid compound resistant to leaching. S/S treatment can be accomplished primarily through the use of either inorganic binders (e.g., cement, fly ash, and/or blast furnace slag) or by organic binders such as bitumen (4). Additives may be used, for example, to convert the metal to a less mobile form or to counteract adverse effects of the contaminated soil on the S/S mixture (e.g., accelerated or retarded setting times, and low physical strength). The form of the final product from S/S treatment can range from a crumbly, soil-like mixture to a monolithic block. S/S is more commonly done as an ex situ process, but the in situ option is available. The full range of inorganic binders, organic binders, and additives is too broad; the emphasis in this chapter is on ex situ, cement-based S/S, which is widely used; in situ, cement-based S/S, which has been applied to metals at full-scale; and polymer microencapsulation, which appears applicable to certain wastes that are difficult to treat via cement-based S/S.

Additional information and references on S/S of metals can be found in USEPA (4,28–30). Innovative S/S technologies (e.g., sorption and surfactant processes, bituminization, emulsified asphalt, modified sulfur cement, polyethylene extrusion, soluble silicate, slag, lime, and soluble phosphates) are addressed in USEPA reports (26,31–35).

4.6.1   Process Description

4.6.1.1   Ex Situ, Cement-Based S/S

Ex situ, cement-based S/S is performed on contaminated soil that has been excavated and classified to reject oversize. Cement-based S/S involves mixing contaminated materials with an appropriate ratio of cement or similar binder/stabilizer, and possibly water and other additives. A system is also necessary for delivering the treated wastes to molds, surface trenches, or subsurface injection. Off-gas treatment (if volatiles or dust are present) may be necessary. The fundamental materials used to perform this technology are Portland-type cements and pozzolanic materials. Portland cements are typically composed of calcium silicates, aluminates, aluminoferrites, and sulfates. Pozzolans are very small spheroidal particles that are formed in combustion of coal (fly ash) and in lime and cement kilns, for example. Pozzolans of high silica content are found to have cement-like properties when mixed with water. Cement-based S/S treatment may involve using only Portland cement, only pozzolanic materials, or blends of both. The composition of the cement and pozzolan, together with the amount of water, aggregate, and other additives, determines the set time, cure time, pour characteristics, and material properties (e.g., pore size and compressive strength) of the resulting treated waste. The composition of cements and pozzolans, including those commonly used in S/S applications, are classified according to American Society for Testing and Materials (ASTM) standards. S/S treatment usually results in an increase (>50% in some cases) in the treated waste volume. Ex situ treatment provides high throughput (100–200 m^3/day mixer^{-1}).

Cement-based S/S reduces the mobility of inorganic compounds by formation of insoluble hydroxides, carbonates, or silicates; substitution of the metal into a mineral structure; sorption; physical encapsulation; and perhaps other mechanisms. Cement-based S/S involves a complex series of reactions, and there are many potential interferences (e.g., coating of particles by organics, excessive acceleration or retardation of set times by various soluble metal and inorganic compounds; excessive heat of hydration; pH conditions that solubilize anionic species of metal compounds, etc.) that can prevent attainment of S/S treatment objectives for physical strength and leachability. While there are many potential interferences, Portland cement is widely used and studied, and a knowledgeable vendor may be able to identify, and confirm via treatability studies, approaches to counteract adverse effects by the use of appropriate additives or other changes in formulation.

4.6.1.2   In Situ, Cement-Based S/S

In situ, cement-based S/S has only two steps: (1) mixing and (2) off-gas treatment. The processing rate for in situ S/S is typically considerably lower than for ex situ processing. In situ S/S is demonstrated to depths of 10 m and may be able to extend to 50 m. The most significant challenge in applying S/S in situ for contaminated soils is achieving complete and uniform mixing of the binder with the contaminated matrix (36). Three basic approaches are used for in situ mixing of the binder with the matrix (5):

1. Vertical auger mixing.
2. In-place mixing of binder reagents with waste by conventional earthmoving equipment, such as draglines, backhoes, or clamshell buckets.
3. Injection grouting, which involves forcing a binder containing dissolved or suspended treatment agents into the subsurface, allowing it to permeate the soil. Grout injection can be applied to contaminated formations lying well below the ground surface. The injected grout cures in place to produce an in situ treated mass.

4.6.1.3   Polymer Microencapsulation S/S

Polymer microencapsulation S/S can include application of thermoplastic or thermosetting resins. Thermoplastic materials are the most commonly used organic-based S/S treatment materials.

Potential candidate resins for thermoplastic encapsulation include bitumen, polyethylene, and other polyolefins, paraffins, waxes, and sulfur cement. Of these candidate thermoplastic resins, bitumen (asphalt) is the least expensive and by far the most commonly used (37). The process of thermoplastic encapsulation involves heating and mixing the waste material and the resin at elevated temperature, typically 130–230°C in an extrusion machine. Any water or volatile organics in the waste boil off during extrusion and are collected for treatment or disposal. Because the final product is a stiff, yet plastic resin, the treated material typically is discharged from the extruder into a drum or other container.

S/S process quality control requires information on the range of contaminant concentrations; potential interferences in waste batches awaiting treatment; and treated product properties such as compressive strength, permeability, leachability, and in some instances, toxicity (28).

4.6.2 Site Requirements

The site must be prepared for the construction, operation, maintenance, decontamination, and decommissioning of the equipment. The size of the area required for the process equipment depends on several factors, including the type of S/S process involved, the required treatment capacity of the system, and site characteristics, especially soil topography and load-bearing capacity. A small mobile ex situ unit occupies space for two, standard flatbed trailers. An in situ system requires a larger area to accommodate a drilling rig as well as a larger area for auger decontamination.

4.6.3 Applicability

This section addresses expected applicability based on the chemistry of the metal and the S/S binders. The soil-contaminant-binder equilibrium and kinetics are complicated, and many factors influence metal mobility, so there may be exceptions to the generalizations presented below.

4.6.3.1 Cement-Based S/S

For cement-based S/S, if a single metal is the predominant contaminant in soil, then Cd and Pb are the most amenable to cement-based S/S. The predominant mechanism for immobilization of metals in Portland and similar cements is precipitation of hydroxides, carbonates, and silicates. Both Pb and Cd tend to form insoluble precipitates in the pH ranges found in cured cement. They may resolubilize, however, if the pH is not carefully controlled. For example, Pb in aqueous solutions tends to resolubilize as $Pb(OH)^{3-}$ around pH 10 and above. Hg, while it is a cationic metal like Pb and cadmium, does not form low-solubility precipitates in cement, so it is difficult to stabilize reliably by cement-based processes, and this difficulty would be expected to be greater with increasing Hg concentration and with organomercury compounds. As, due to its formation of anionic species, also does not form insoluble precipitates in the high pH cement environment, and cement-based solidification is generally not expected to be successful. Cr(VI) is difficult to stabilize in cement due to the formation of anions that are soluble at high pH. However, Cr(VI) can be reduced to Cr(III), which does form insoluble hydroxides. Although Hg and As(III and V) are particularly difficult candidates for cement-based S/S, this should not necessarily eliminate S/S (even cement-based) from consideration since (a) as with Cr(VI) it may be possible to devise a multistep process that will produce an acceptable product for cement-based S/S; (b) a noncement-based S/S process (e.g., lime and sulfide for Hg; oxidation to As(V) and coprecipitation with iron) may be applicable; or (c) the leachable concentration of the contaminant may be sufficiently low that a highly efficient S/S process may not be required to meet treatment goals.

The discussion of applicability above also applies to in situ, cement-based S/S. If in situ treatment introduces chemical agents into the ground, this chemical addition may cause a pollution problem in itself, and may be subject to additional requirements under the Land Disposal Restrictions.

4.6.3.2 Polymer Microencapsulation

Polymer microencapsulation has been mainly used to treat low-level radioactive wastes. However, organic binders have been tested or applied to wastes containing chemical contaminants such as As, metals, inorganic salts, polychlorinated biphenyls (PCBs), and dioxins (37). Polymer microencapsulation is particularly well suited to treating water-soluble salts such as chlorides or sulfates that are generally difficult to immobilize in a cement-based system (38). Characteristics of the organic binder and extrusion system impose compatibility requirements on the waste material. The elevated operating temperatures place a limit on the quantity of water and volatile organic chemicals (VOCs) in the waste feed. Low volatility organics will be retained in the bitumen but may act as solvents causing the treated product to be too fluid. The bitumen is a potential fuel source so the waste should not contain oxidizers such as nitrates, chlorates, or perchlorates. Oxidants present the potential for rapid oxidation, causing immediate safety concerns, as well as slow oxidation, which results in waste form degradation.

Cement-based S/S of multiple metal wastes is particularly difficult if a set of treatment and disposal conditions cannot be found that simultaneously produces low mobility species for all the metals of concern. For example, the relatively high pH conditions that favor Pb immobilization would tend to increase the mobility of As. On the other hand, the various metal species in a multiple metal waste may interact (e.g., formation of low-solubility compounds by combination of Pb and arsenate) to produce a low mobility compound.

Organic contaminants are often present with inorganic contaminants at metal-contaminated sites. S/S treatment of organic-contaminated waste with cement-based binders is more complex than treatment of inorganics alone. This is particularly true with VOCs where the mixing process and heat generated by cement hydration reactions can increase vapor losses (39–42). However, S/S can be applied to wastes that contain lower levels of organics, particularly when inorganics are present and/or the organics are semivolatile or nonvolatile. Also, recent studies indicate the addition of silicates or modified clays to the binder system may improve S/S performance with organics (27).

4.6.4 PERFORMANCE AND BDAT STATUS

Year 2000 information about the use of S/S at Superfund remedial sites indicates that S/S has been used at 167 sites since FY 1982 (26). Figure 4.1 shows the number of projects by status for the following stages: predesign/design, design completed/being installed, operational, and completed. Data are shown for in situ and ex situ S/S projects. In addition, information about all source control technologies is provided. With respect to S/S projects, the majority of both in situ and ex situ projects (62%) are completed, followed by projects in the predesign/design stage (21%). Overall, completed S/S projects represent 30% of all completed Superfund projects in which treatment technologies have been used for source control.

Figure 4.2 shows the types of binder materials used for S/S projects at Superfund remedial sites, including inorganic binders, organic binders, and combination organic and inorganic binders. Many of the binders used include one or more proprietary additives. Examples of inorganic binders include cement, fly ash, lime, soluble silicates, and sulfur-based binders, while organic binders include asphalt, epoxide, polyesters, and polyethylene. More than 90% of the S/S projects used inorganic binders. In general, inorganic binders are less expensive and easier to use than organic binders. Organic binders are generally used to solidify radioactive wastes or specific hazardous organic compounds.

Figure 4.3 shows the types of contaminant groups and combination of contaminant groups treated by S/S at Superfund remedial sites. S/S was used to treat metals only in 56% of the projects, and used to treat metals alone or in combination with organics or radioactive metals at approximately 90% of the sites. S/S was used to treat organics only at 6% of the sites (26). Figure 4.4 provides a further breakdown of the metals treated by S/S at Superfund remedial sites. The top five metals treated by S/S are Pb, Cr, As, Cd, and Cu.

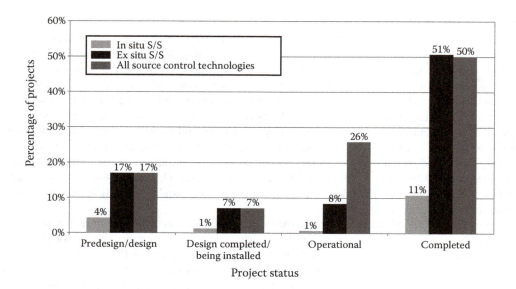

FIGURE 4.1 Percentage of Superfund remedial projects by status. Number of projects: source control = 682, ex situ S/S = 139, in situ S/S = 28. (From USEPA. *Solidification/Stabilization Use at Superfund Sites. EPA-542-R-00-010,* U.S. Environmental Protection Agency, Washington, DC, September 2000.)

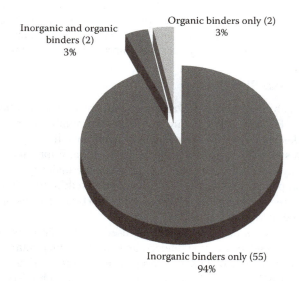

FIGURE 4.2 Binder materials used for S/S projects. Total number of projects = 59. (From USEPA. *Solidification/Stabilization Use at Superfund Sites. EPA-542-R-00-010,* U.S. Environmental Protection Agency, Washington, DC, September 2000.)

S/S with cement-based and pozzolan binders is a commercially available, established technology (5). Table 4.4 shows a selected list of sites where S/S has been selected for remediating metal-contaminated solids. Note that S/S has been used to treat all five metals (Cr, Pb, As, Hg, and Cd). Although it would not generally be expected that cement-based S/S would be applied to As- and Hg-contaminated soils, it was beyond the scope of the project to examine in detail

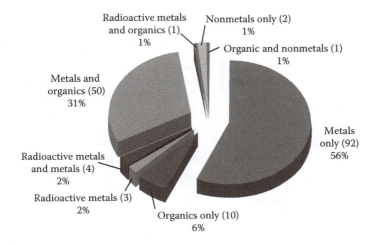

FIGURE 4.3 Contaminant types treated by S/S. Total number of projects = 163. (From USEPA. *Solidification/Stabilization Use at Superfund Sites. EPA-542-R-00-010,* U.S. Environmental Protection Agency, Washington, DC, September 2000.)

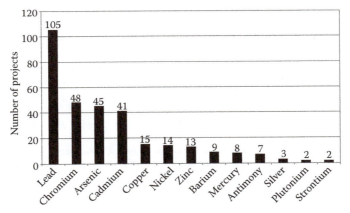

FIGURE 4.4 Number of S/S projects treating specific metals. (From USEPA. *Solidification/Stabilization Use at Superfund Sites. EPA-542-R-00-010,* U.S. Environmental Protection Agency, Washington, DC, September 2000.)

the characterization data, S/S formulations, and performance data upon which the selections were based, so the selection/implementation data are presented without further comment.

Applications of polymer microencapsulation have been limited to special cases where the specific performance features are required for the waste matrix, and contaminants allow reuse of the treated waste as a construction material (43).

S/S is a BDAT for the following waste types (5):

- Cd nonwastewater other than Cd-containing batteries
- Cr nonwastewater following reduction to Cr(III)
- Pb nonwastewater
- Wastes containing low concentrations (<260 mg/kg) of elemental Hg-sulfide precipitation
- Plating wastes and steelmaking wastes

TABLE 4.4

Solidification/Stabilization Applications at Superfund Sites with Metal Contamination

Site Name/State	Specific Technology	Key Metal Contaminants	Associated Technology
DeRewal Chemical, NJ	Solidification	Cr, Cd, Pb	GW pump and treatment
Marathon Battery Co., NY	Chemical fixation	Cd, Ni	Dredging, off-site disposal
Nascolite, Millville, NJ	Stabilization of wetland soils	Pb	On-site disposal of stabilized soils; excavation and off-site disposal of wetland soils
Roebling Steel, NJ	Solidification/ stabilization	As, Cr, Pb	Capping
Waldick Aerospace, NJ	S/S	Cd, Cr	Off-site disposal
Aladdin Plating, PA	Stabilization	Cr	Off-site disposal
Palmerton Zinc, PA	Stabilization, fly ash, lime, potash	Cd, Pb	—
Tonolli Corp., PA	S/S	As, Pb	In situ chemical limestone barrier
Whitmoyer Laboratories, PA	Oxidation/fixation	As	GW pump and treatment, capping, grading, and revegetation
Bypass 601, NC	S/S	Cr, Pb	Capping, regrading, revegetation, GW pump and treatment
Flowood, MS	S/S	Pb	Capping
Independent Nail, SC	S/S	Cd, Cr	Capping
Pepper's Steel and Alloys, FL	S/S	As, Pb	On-site disposal
Gurley Pit, AR	In situ S/S	Pb	
Pesses Chemical, TX	Stabilization	Cd	Concrete capping
E.I. Dupont de Nemours, IA	S/S	Cd, Cr, Pb	Capping, regrading, and revegetation
Shaw Avenue Dump, IA	S/S	As, Cd	Capping, groundwater monitoring
Frontier Hard Chrome, WA	Stabilization	Cr	
Gould Site, OR	S/S	Pb	Capping, regrading, and revegetation

Source: USEPA. *Technology Alternatives for the Remediation of Soils Contaminated with AS, Cd, Cr, Hg, and Pb.* EPA/540/S-97/500, U.S. Environmental Protection Agency, Cincinnati, OH, August 1997.

Although vitrification, not S/S, was selected as BDAT for RCRA As-containing nonwastewater, USEPA does not preclude the use of S/S for treatment of As (particularly inorganic As) wastes but recommends that its use be determined on a case-by-case basis. A variety of stabilization techniques including cement, silicate, pozzolan, and ferric coprecipitation were evaluated as candidate BDATs for As. Due to concerns about long-term stability and the waste volume increase, particularly with ferric coprecipitation, stabilization was not accepted as BDAT.

4.6.5 SITE PROGRAM DEMONSTRATION PROJECTS

Completed SITE demonstrations applicable to soils contaminated with the metals of interest include (5)

- Advanced Remediation Mixing, Inc. (ex situ S/S)
- Funderburk and Associates (ex situ S/S)
- Geo-Con, Inc. (in situ S/S)
- Soliditech, Inc. (ex situ S/S)
- STC Omega, Inc. (ex situ S/S)

- WASTECH Inc. (ex situ S/S)
- Separation and Recovery Systems, Inc. (ex situ S/S)
- Wheelabrator Technologies Inc. (ex situ S/S)

4.6.6   Cost of S/S

Information about the cost of using S/S to treat wastes at Superfund remedial sites was reported by USEPA for 29 completed projects in 2000 (26). Total costs in terms of 2014 USD (44) for S/S projects ranged from USD 100,000 to USD 21,000,000 including the cost of excavation, treatment, and disposal (if ex situ). The cost ranged from USD 14/m^3 to approximately USD 2,100/m^3. The average cost for these projects was USD 396/m^3, including two projects with relatively high costs (approximately USD 2,100/m^3). Excluding those two projects, the average cost per cubic meter was USD 338 (26).

4.7   VITRIFICATION

Vitrification applies high-temperature treatment aimed primarily at reducing the mobility of metals by incorporation into a chemically durable, leach resistant, vitreous mass. Vitrification can be carried out on excavated soils as well as in situ.

4.7.1   Process Description

During the vitrification process, organic wastes are pyrolyzed (in situ) or oxidized (ex situ) by the melt front, whereas inorganics, including metals, are incorporated into the vitreous mass. Off-gases released during the melting process, containing volatile components and products of combustion and pyrolysis, must be collected and treated (4,45,46). Vitrification converts contaminated soils to a stable glass and crystalline monolith (46). With the addition of low-cost materials such as sand, clay, and/or native soil, the process can be adjusted to produce products with specific characteristics, such as chemical durability. Waste vitrification may be able to transform the waste into useful, recyclable products such as clean fill, aggregate, or higher valued materials such as erosion-control blocks, paving blocks, and road dividers.

4.7.1.1   Ex Situ Vitrification

Ex situ vitrification (ESV) technologies apply heat to a melter through a variety of sources such as combustion of fossil fuels (coal, natural gas, and oil) or input of electric energy by direct joule heat, arcs, plasma torches, and microwaves. Combustion or oxidation of the organic portion of the waste can contribute significant energy to the melting process, thus reducing energy costs. The particle size of the waste may need to be controlled for some of the melting technologies. For wastes containing refractory compounds that melt above the unit's nominal processing temperature, such as quartz or alumina, size reduction may be required to achieve acceptable throughputs and a homogeneous melt. For high-temperature processes using arcing or plasma technologies, size reduction is not a major factor. For the intense melters using concurrent gas-phase melting or mechanical agitation, size reduction is needed for feeding the system and for achieving a homogeneous melt.

4.7.1.2   In Situ Vitrification

In situ vitrification (ISV) technology is based on electric melter technology, and the principle of operation is joule heating, which occurs when an electrical current is passed through a region that behaves as a resistive heating element. Electrical current is passed through the soil by means of an array of electrodes inserted vertically into the surface of the contaminated soil zone. Because dry soil is not conductive, a starter path of flaked graphite and glass frit is placed in a small trench between the electrodes to act as the initial flow path for electricity. Resistance heating in the starter path transfers heat to the soil that then begins to melt. Once molten, the soil becomes conductive. The melt grows outward and downward as power is gradually increased to the full constant operating power level. A single melt can treat a region of up to 1,000 T. The maximum treatment depth has been demonstrated to be about 6 m. Large contaminated areas are treated in multiple settings that fuse the blocks together to form one large monolith (4). Further information on ISV can be found in References 47–50.

4.7.2   Site Requirements

The site must be prepared for the mobilization, operation, maintenance, and demobilization of the equipment. Site activities such as clearing vegetation, removing overburden, and acquiring backfill material are often necessary for ESV as well as ISV. Ex situ processes will require areas for storage of excavated, treated, and possibly pretreated materials. The components of one ISV system are contained in three transportable trailers: an off-gas and process control trailer, a support trailer, and an electrical trailer. The trailers are mounted on wheels sufficient for transportation to and over a compacted ground surface (51).

The field-scale ISV system evaluated in the SITE program required three-phase electrical power at either 12,500 or 13,800 V, which is usually taken from a utility distribution system (52). Alternatively, the power may be generated onsite by means of a diesel generator. Typical applications require 800–1,000 kW h/T (47).

4.7.3   Applicability

Setting cost and implementability aside, vitrification should be most applicable where nonvolatile metal contaminants have glass solubilities exceeding the level of contamination in the soil. Cr-contaminated soil should pose the least difficulties for vitrification, since it has low volatility, and glass solubility between 1% and 3%. Vitrification may or may not be applicable for Pb, As, and Cd, depending on the level of difficulty encountered in retaining the metals in the melt, and controlling and treating any volatile emissions that may occur. Hg clearly poses problems for vitrification due to high volatility and low glass solubility (<0.1%) but may be allowable at very low concentrations.

Chlorides present in the waste in excess of about 0.5% by weight (wt) typically will not be incorporated into and discharged with the glass but will fume off and enter the off-gas treatment system. If chlorides are excessively concentrated, salts of alkali, alkaline earths, and heavy metals will accumulate in solid residues collected by off-gas treatment. Separation of the chloride salts from the other residuals may be required before or during return of residuals to the melter. When excess chlorides are present, there is also a possibility that dioxins and furans may form and enter the off-gas treatment system.

Waste matrix composition affects the durability of the treated waste. Sufficient glass-forming materials, SiO_2 (>30% wt.%), and combined alkali, Na + K (>1.4% wt), are required for vitrification of wastes. If these conditions are not met, frit and/or flux additives typically are needed. Vitrification is also potentially applicable to soils contaminated with mixed metals and metal–organic wastes.

Specific situations where ESV would not be applicable or would face additional implementation problems include (5)

1. Wastes containing >25% moisture content cause excessive fuel consumption
2. Wastes where size reduction and classification are difficult or expensive
3. Volatile metals, particularly Cd and Hg, will vaporize and must be captured and treated separately
4. Arsenic-containing wastes may require pretreatment to produce less volatile forms

5. Metal concentrations in soil that exceed their solubility in glass
6. Sites where commercial capacity is not adequate or transportation cost to a fixed facility is unacceptable

Specific situations, in addition to those cited above, where ISV would not be applicable or would face additional implementation problems include (5)

1. Metal-contaminated soil where a less costly and adequately protective remedy exists
2. Projects that cannot be undertaken because of limited commercial availability
3. Contaminated soil <2 m and >6 m below the ground surface
4. Presence of an aquifer with high hydraulic conductivity (e.g., soil permeability >1 × 10^{-5} cm/s) limits economic feasibility due to excessive energy required
5. Contaminated soil mixed with buried metal that can result in a conductive path causing short circuiting of electrodes
6. Contaminated soil mixed with loosely packed rubbish or buried coal can start underground fires and overwhelm off-gas collection and treatment systems
7. Volatile heavy metals near the surface can be entrained in combustion product gases and not retained in the melt
8. Sites where surface slope >5% may cause the melt to flow
9. In situ voids >150 m^3 interrupt conduction and heat transfer
10. Underground structures and utilities <6 m from the melt zone must be protected from heat or avoided

Where it can be successfully applied, advantages of vitrification include (5)

1. Vitrified product is an inert, impermeable solid that should reduce leaching for long periods of time
2. Volume of vitrified product will typically be smaller than initial waste volume
3. Vitrified product may be usable
4. A wide range of inorganic and organic wastes can be treated
5. There is both an ex situ and an in situ option available

A particular advantage of ex situ treatment is better control of processing parameters. Also, fuel costs may be reduced for ESV by the use of combustible waste materials. This fuel cost-saving option is not directly applicable for ISV, since combustibles would increase the design and operating requirements for gas capture and treatment.

4.7.4 Performance and BDAT Status

ISV has been implemented at metal-contaminated Superfund sites and was evaluated under the SITE program (53). Some improvements are needed with melt containment and air emission control systems. ISV has been operated at a large scale 10 times, including two demonstrations on radioactively contaminated sites at the DOE's Hanford Nuclear Reservation (45,54). Pilot-scale tests have been conducted at Oak Ridge National Laboratory, Idaho National Engineering Laboratory, and Arnold Engineering Development Center. More than 150 tests and demonstrations at various scales have been performed on a broad range of waste types in soils and sludges. The technology has been selected as a preferred remedy at 10 private, Superfund, and DOD sites (55). Table 4.5 provides a summary of ISV technology selection/application at metal-contaminated Superfund sites. A number of ESV systems are under development. The technical resource document identified one full-scale ex situ melter that was reported to be operating on RCRA organics and inorganics.

Vitrification is a BDAT for the As-containing wastes.

TABLE 4.5

In Situ Vitrification Applications at Superfund Sites with Metal Contamination

Site Name/State	Key Metal Contaminants
Parsons Chemical, MI	Hg (low)
Rocky Mountain Arsenal, CO	As, Hg

Source: USEPA. *Technology Alternatives for the Remediation of Soils Contaminated with AS, Cd, Cr, Hg, and Pb.* EPA/540/S-97/500, U.S. Environmental Protection Agency, Cincinnati, OH, August 1997.

4.7.5 SITE PROGRAM DEMONSTRATION PROJECTS

Completed SITE demonstrations applicable to soils contaminated with the metals of interest include (5)

- Babcock & Wilcox Co. (cyclone furnace—ESV)
- Retech, Inc. (plasma arc—ESV)
- Geosafe Corporation (ISV)
- Vortec Corporation (ex situ oxidation and vitrification process)

4.8 SOIL WASHING

Soil washing is an ex situ remediation technology that uses a combination of physical separation and aqueous-based separation unit operations to reduce contaminant concentrations to site-specific remedial goals (56). Although soil washing is sometimes used as a stand-alone treatment technology, it is more often combined with other technologies to complete site remediation. Soil washing technologies have successfully remediated sites contaminated with organic, inorganic, and radioactive contaminants (56). The technology does not detoxify or significantly alter the contaminant but transfers the contaminant from the soil into the washing fluid or mechanically concentrates the contaminants into a much smaller soil mass (57) for subsequent treatment (see Figure 4.5).

Further information on soil washing can be found in USEPA innovative technology reports and programs (58,59).

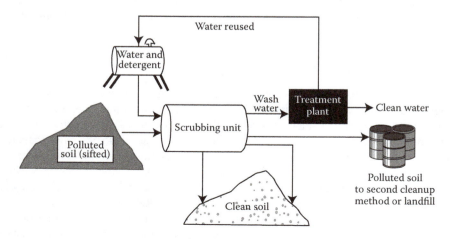

FIGURE 4.5 Soil washing operation. (From USEPA. *A Citizen's Guide to Soil Washing. EPA 542-F-01-008*, U.S. Environmental Protection Agency, Washington, DC, May 2001.)

4.8.1 PROCESS DESCRIPTION

Soil washing systems are quite flexible in terms of the number, type, and order of processes involved. Soil washing is performed on excavated soil and may involve some or all of the following, depending on the contaminant-soil matrix characteristics, cleanup goals, and specific process employed (5,57):

1. Mechanical screening to remove various oversize materials
2. Crushing to reduce applicable oversize to suitable dimensions for treatment
3. Physical processes (e.g., soaking, spraying, tumbling, and attrition scrubbing) to liberate weakly bound agglomerates (e.g., silts and clays bound to sand and gravel) followed by size classification to generate coarse-grained and fine-grained soil fraction(s) for further treatment
4. Treatment of the coarse-grained soil fraction(s)
5. Treatment of the fine-grained fraction(s)
6. Management of the generated residuals

Treatment of the coarse-grained soil fraction typically involves additional application of physical separation techniques and possibly aqueous-based leaching techniques. Physical separation techniques (e.g., sorting, screening, elutriation, hydrocyclones, spiral concentrators, and flotation) exploit physical differences (e.g., size, density, shape, color, and wetability) between contaminated particles and soil particles in order to produce a clean (or nearly clean) coarse fraction and one or more metal-concentrated streams. Many of the physical separation processes listed above involve the use of water as a transport medium, and if the metal contaminant has significant water solubility, then some of the coarse-grained soil cleaning will occur as a result of transfer to the aqueous phase. If the combination of physical separation and unaided transfer to the aqueous phase cannot produce the desired reduction in the soil's metal content, which is frequently the case for metal contaminants, then solubility enhancement is an option for meeting cleanup goals for the coarse fraction. Solubility enhancement can be accomplished in several ways (5,60,61):

1. Converting the contaminant into a more soluble form (e.g., oxidation–reduction and conversion to soluble metal salts)
2. Using an aqueous-based leaching solution (e.g., acidic, alkaline, oxidizing, and reducing) in which the contaminant has enhanced solubility
3. Incorporating a specific leaching process into the system to promote increased solubilization via increased mixing, elevated temperatures, higher solution/soil ratios, efficient solution/soil separation, multiple stage treatment, etc.
4. A combination of the above

After the leaching process is completed on the coarse-grained fraction, it will be necessary to separate the leaching solution and the coarse-grained fraction by settling. A soil rinsing step may be necessary to reduce the residual leachate in the soil to an acceptable level. It may also be necessary to readjust soil parameters such as pH or redox potential before replacement of the soil on the site. The metal-bearing leaching agent must also be treated further to remove the metal contaminant and permit reuse in the process or discharge, and this topic is discussed below under management of residuals.

Treatment of fine-grained soils is similar in concept to the treatment of coarse-grained soils, but the production rate would be expected to be lower and hence more costly than for the coarse-grained soil fraction. The reduced production rate arises from factors including (a) the tendency of clays to agglomerate, thus requiring time, energy, and high water/clay ratios to produce leachable slurry and (b) slow settling velocities that require additional time and/or capital equipment to produce acceptable soil–water separation for multibatch or countercurrent treatment, or at the end of treatment. A site-specific determination needs to be made whether the fines should be treated to produce clean fines or whether they should be handled as a residual waste stream.

Management of generated residuals is an important aspect of soil washing. The effectiveness, implementability, and cost of treating each residual stream are important to the overall success of soil washing for the site. Perhaps the most important of the residual streams is the metal-loaded leachant that is generated, particularly if the leaching process recycles the leaching solution. Furthermore, it is often critical to the economic feasibility of the project that the leaching solution be recycled. For these closed or semi-closed-loop leaching processes, successful treatment of the metal-loaded leachant is imperative to the successful cleaning of the soil. The leachant must (a) have adequate solubility for the metal so that the metal reduction goals can be met without using excessive volumes of leaching solution and (b) be readily, economically, and repeatedly adjustable (e.g., pH adjustment) to a form in which the metal contaminant has very low solubility so that the recycled aqueous phase retains a favorable concentration gradient compared to the contaminated soil. Also, efficient soil–water separation is important prior to recovering metal from the metal-loaded leachant in order to minimize contamination of the metal concentrate. Recycling the leachant reduces logistical requirements and costs associated with makeup water, storage, permitting, compliance analyses, and leaching agents. It also reduces external coordination requirements and eliminates the dependence of the remediation on the ability to meet publicly owned treatment works (POTW) discharge requirements.

Other residual streams that may be generated and require proper handling include (5)

1. Untreatable, uncrushable oversize
2. Recyclable metal-bearing particulates, concentrates, or sludges from physical separation or leachate treatment
3. Nonrecyclable metal-bearing particulates, concentrates, soils, sludges, or organic debris that fail toxicity characteristic leaching procedure (TCLP) thresholds for RCRA hazardous waste
4. Soils or sludges that are not RCRA hazardous wastes but are also not sufficiently clean to permit return to the site
5. Metal-loaded leachant from systems where leachant is not recycled
6. Rinsate from treated soil

4.8.2 SITE REQUIREMENTS

The area required for a unit at a site will depend on the vendor system selected, the amount of soil storage space, and/or the number of tanks or ponds needed for washwater preparation and wastewater storage and treatment. Typical utilities required are water, electricity, steam, and compressed air; the quantity of each is vendor- and site-specific. It may be desirable to control the moisture content of the contaminated soil for consistent handling and treatment by covering the excavation, storage, and treatment areas. Climatic conditions such as annual or seasonal precipitation cause surface runoff and water infiltration; therefore, runoff control measures may be required. Since soil washing is an aqueous-based process, cold weather impacts include freezing as well as potential effects on leaching rates.

4.8.3 APPLICABILITY

Soil washing is potentially applicable to soils contaminated with all five metals of interest. Conditions that particularly favor soil washing include (5)

1. A single principal contaminant metal that occurs in dense, insoluble particles that report to a specific, small mass fraction(s) of the soil
2. A single contaminant metal and species that is very water or aqueous leachant soluble and has a low soil/water partition coefficient
3. Soil containing a high proportion (e.g., >80%) of soil particles >2 mm are desirable for efficient contaminant-soil and soil–water separation.

Conditions that clearly do not favor soil washing include (5)

1. Soils with a high (i.e., >40%) silt and clay fraction
2. Soils that vary widely and frequently in significant characteristics such as soil type, contaminant type and concentration, and where blending for homogeneity is not feasible
3. Complex mixtures (e.g., multicomponent, solid mixtures where access of leaching solutions to contaminant is restricted; mixed anionic and cationic metals where pH of solubility maximums are not close)
4. High clay content, cation exchange capacity, or humic acid content, which would tend to interfere with contaminant desorption
5. Presence of substances that interfere with the leaching solution (e.g., carbonaceous soils would neutralize extracting acids; similarly, high humic acid content will interfere with an alkaline extraction)
6. Metal contaminants in a very low-solubility, stable form (e.g., PbS) may require long contact times and excessive amounts of reagent to solubilize

4.8.4 PERFORMANCE AND BDAT STATUS

Soil washing has been used at waste sites in Europe, especially in Germany, the Netherlands, and Belgium (62). Table 4.6 lists selected Superfund sites where soil washing has been selected and/or implemented.

Acid leaching, which is a form of soil washing, is the BDAT for Hg.

TABLE 4.6
Soil Washing Applications at Selected Superfund Sites with Metal Contamination

Site Name/State	Specific Technology	Key Metal Contaminants	Associated Technology
Ewan Property, NJ	Water washing	As, Cr, Cu, Pb	Pretreatment by solvent extraction to remove organics
GE Wiring Devices, PR	Water with KI solution additive	Hg	Treated residues disposed onsite and covered with clean soil
King of Prussia, NJ	Water with washing agent additives	Ag, Cr, Cu	Sludges to be land disposed
Zanesville Well Field, OH	Soil washing	Hg, Pb	SVE to remove organics
Twin Cities Army Ammunition Plant, MN	Soil washing	Cd, Cr, Cu, Hg, Pb	Soil leaching
Sacramento Army Depot Sacramento, CA	Soil washing	Cr, Pb	Off-site disposal of wash liquid

Source: USEPA. *Technology Alternatives for the Remediation of Soils Contaminated with AS, Cd, Cr, Hg, and Pb*. EPA/540/S-97/500, U.S. Environmental Protection Agency, Cincinnati, OH, August 1997.

4.8.5 SITE DEMONSTRATIONS AND EMERGING TECHNOLOGIES PROGRAM PROJECTS

SITE demonstrations applicable to soils contaminated with the metals of interest include (5)

- Bergmann USA (physical separation/leaching) BioGenesis[SM] (physical separation/leaching)
- Biotrol, Inc. (physical separation)
- Brice Environmental Services Corp. (physical separation)
- COGNIS, Inc. (leaching)
- Toronto Harbor Commission (physical separation/leaching)

Four SITE Emerging Technologies Program projects have been completed that are applicable to soils contaminated with the metals of interest.

4.9 SOIL FLUSHING

Soil flushing is the in situ extraction of contaminants from the soil via an appropriate washing solution. Water or an aqueous solution is injected into or sprayed onto the area of contamination, and the contaminated elutriate is collected and pumped to the surface for removal, recirculation, or onsite treatment and reinjection. The technology is applicable to both organic and inorganic contaminants, and metals in particular (4). For the purpose of metals remediation, soil flushing has been operated at full scale, but for a small number of sites.

4.9.1 PROCESS DESCRIPTION

Soil flushing uses water, a solution of chemicals in water, or an organic extractant to recover contaminants from the in situ material. The contaminants are mobilized by solubilization, formation of emulsions, or a chemical reaction with the flushing solutions. After passing through the contamination zone, the contaminant-bearing fluid is collected by strategically placed wells or trenches and brought to the surface for disposal, recirculation, or onsite treatment and reinjection. During elutriation, the flushing solution mobilizes the sorbed contaminants by dissolution or emulsification.

One key to efficient operation of a soil-flushing system is the ability to reuse the flushing solution, which is recovered along with groundwater. Various water treatment techniques can be applied to remove the recovered metals and render the extraction fluid suitable for reuse. Recovered flushing fluids may need treatment to meet appropriate discharge standards prior to release to a POTW or receiving waters. The separation of surfactants from recovered flushing fluid, for reuse in the process, is a major factor in the cost of soil flushing. Treatment of the flushing fluid results in process sludges and residual solids, such as spent carbon and spent ion exchange resin, which must be appropriately treated before disposal. Air emissions of volatile contaminants from recovered flushing fluids should be collected and treated, as appropriate, to meet applicable regulatory standards. Residual flushing additives in the soil may be a concern and should be evaluated on a site-specific basis (63). Subsurface containment barriers can be used in conjunction with soil-flushing technology to help control the flow of flushing fluids.

Further information on soil flushing can be found in References 58,63–65.

4.9.2 SITE REQUIREMENTS

Stationary or mobile soil-flushing systems are located onsite. The exact area required will depend on the vendor system selected and the number of tanks or ponds needed for washwater preparation and wastewater treatment. Certain permits may be required for operation, depending on the system being utilized. Slurry walls or other containment structures may be needed along with hydraulic controls to ensure capture of contaminants and flushing additives. Impermeable membranes may be necessary to limit infiltration of precipitation, which could cause dilution of the flushing solution and loss of hydraulic control. Cold weather freezing must also be considered for shallow infiltration galleries and aboveground sprayers (66).

4.9.3   APPLICABILITY

Soil flushing may be easy or difficult to apply, depending on the ability to wet the soil with the flushing solution and to install collection wells or subsurface drains to recover all the applied liquids. The achievable level of treatment varies and depends on the contact of the flushing solution with the contaminants and the appropriateness of the solution for contaminants, and the hydraulic conductivity of the soil. Soil flushing is most applicable to contaminants that are relatively soluble in the extracting fluid, and that will not tend to sorb onto soil as the metal-laden flushing fluid proceeds through the soil to the extraction point. Based on the earlier discussion of metal behavior, some potentially promising scenarios for soil flushing would include Cr(VI), As(III or V) in permeable soil with low iron oxide, low clay, and high pH; Cd in permeable soil with low clay, low cation exchange capacity, and moderately acidic pH; and, Pb in acid sands. A single target metal would be preferable to multiple metals, due to the added complexity of selecting a flushing fluid that would be reasonably efficient for all contaminants. Also, the flushing fluid must be compatible with not only the contaminant, but also the soil. Soils that counteract the acidity or alkalinity of the flushing solution will decrease its effectiveness. If precipitants occur due to interaction between the soil and the flushing fluid, then this could obstruct the soil pore structure and inhibit flow to and through sectors of the contaminated soil. It may take long periods of time for soil flushing to achieve cleanup standards.

A key advantage of soil flushing is that the contaminant is removed from the soil. Recovery and reuse of the metal from the extraction fluid may be possible in some cases, although the value of the recovered metal would not be expected to fully off-set the costs of recovery. The equipment used for the technology is relatively easy to construct and operate. It does not involve excavation, treatment, and disposal of the soil, which avoids the expense and hazards associated with these activities.

4.9.4   PERFORMANCE AND **BDAT** STATUS

Table 4.7 lists the Superfund sites where soil flushing has been selected and/or implemented. Soil flushing has a more established history for removal of organics but has been used for Cr removal (e.g., United Chrome Products Superfund Site, near Corvallis, Oregon). In situ technologies, such as soil flushing, are not considered RCRA BDAT for any of the five metals (5).

Soil flushing techniques for mobilizing contaminants can be classified as conventional and unconventional. Conventional applications employ water only as the flushing solution. Unconventional

---

TABLE 4.7
Soil-Flushing Applications at Selected Superfund Sites with Metal Contamination

Site Name/State	Specific Technology	Key Metal Contaminants	Associated Technology
Lipari Landfill, NJ	Soil flushing of soil and wastes contained by slurry wall	Cr, Hg, Pb	Slurry wall and cap
United Chrome Products, OR	Cap; excavation from impacted wetlands	Cr	Electrokinetic pilot test, Considering in situ reduction

Source: USEPA. *Technology Alternatives for the Remediation of Soils Contaminated with AS, Cd, Cr, Hg, and Pb.* EPA/540/S-97/500, U.S. Environmental Protection Agency, Cincinnati, OH, August 1997.

applications that are currently being researched include the enhancement of the flushing water with additives, such as acids, bases, and chelating agents to aid in the desorption/dissolution of the target contaminants from the soil matrix to which they are bound.

Researchers are also investigating the effects of numerous soil factors on heavy metal sorption and migration in the subsurface. Such factors include pH, soil type, soil horizon, particle size, permeability, specific metal type and concentration, and type and concentrations of organic and inorganic compounds in solutions. Generally, as the soil pH decreases, cationic metal solubility and mobility increase. In most cases, metal mobility and sorption are likely to be controlled by the organic fraction in topsoils, and clay content in the subsoils.

4.9.5   SITE DEMONSTRATION AND EMERGING TECHNOLOGIES PROGRAM PROJECTS

There are no in situ soil flushing projects reported to be completed either as SITE demonstration or Emerging Technologies Program Projects (66).

4.10   PYROMETALLURGY

Pyrometallurgy is used here as a broad term encompassing elevated temperature techniques for extraction and processing of metals for use or disposal. High-temperature processing increases the rate of reaction and often makes the reaction equilibrium more favorable, lowering the required reactor volume per unit output (4). Some processes that clearly involve both metal extraction and recovery include roasting, retorting, or smelting. While these processes typically produce a metal-bearing waste slag, metal is also recovered for reuse. A second class of pyrometallurgical technologies included here is a combination of high-temperature extraction and immobilization. These processes use thermal means to cause volatile metals to separate from the soil and report to the fly ash, but the metal in the fly ash is then immobilized, instead of recovered, and there is no metal recovered for reuse. A third class of technologies are those that are primarily incinerators for mixed organic–inorganic wastes, but which have the capability of processing wastes containing the metals of interest by either capturing volatile metals in the exhaust gases or immobilizing the nonvolatile metals in the bottom ash or slag. Since some of these systems may have applicability to some cases where metals contamination is the primary concern, a few technologies of this type are noted that are in the SITE program. Vitrification is addressed in Section 4.7. It is not considered pyrometallurgical treatment since there is typically neither a metal extraction nor a metal recovery component in the process.

4.10.1   PROCESS DESCRIPTION

Pyrometallurgical processing usually is preceded by physical treatment (5) to produce a uniform feed material and upgrade the metal content.

Solids treatment in a high-temperature furnace requires efficient heat transfer between the gas and solid phases while minimizing particulate in the off-gas. The particle-size range that meets these objectives is limited and is specific to the design of the process. The presence of large clumps or debris slows heat transfer, so pretreatment to either remove or pulverize oversize material is normally required. Fine particles also are undesirable because they become entrained in the gas flow, increasing the volume of dust to be removed from the flue gas. The feed material is sometimes pelletized to give a uniform size. In many cases a reducing agent and flux may be mixed in prior to pelletization to ensure good contact between the treatment agents and the contaminated material and to improve gas flow in the reactor (4).

Due to its relatively low boiling point (357°C) and ready conversion at elevated temperature to its metallic form, Hg is commonly recovered through roasting and retorting at much lower temperatures than the other metals. Pyrometallurgical processing to convert compounds of the other four metals to elemental metal requires a reducing agent, fluxing agents to facilitate melting and to slag off impurities, and a heat source. The fluid mass often is called a melt, but the operating temperature, although quite high, is often still below the melting points of the refractory compounds being processed. The fluid forms as a lower-melting-point material due to the presence of a fluxing agent such as calcium. Depending on processing temperatures, volatile metals such as Cd and Pb may fume off and be recovered from the off-gas as oxides. Nonvolatile metals, such as Cr or nickel, are tapped from the furnace as molten metal. Impurities are scavenged by formation of slag (4). The effluents and solid products generated by pyrometallurgical technologies typically include solid, liquid, and gaseous residuals. Solid products include debris, oversized rejects, dust, ash, and the treated medium. Dust collected from particulate control devices may be combined with the treated medium or, depending on analyses for carryover contamination, recycled through the treatment unit.

4.10.2   SITE REQUIREMENTS

Few pyrometallurgical systems are available in mobile or transportable configurations. Since this is typically an off-site technology, the distance of the site from the processing facility has an important influence on transportation costs. Off-site treatment must comply with USEPA's off-site treatment policies and procedures. The off-site facility's environmental compliance status must be acceptable, and the waste must be of a type allowable under their operating permits. In order for pyrometallurgical processing to be technically feasible, it must be possible to generate a concentrate from the contaminated soil that will be acceptable to the processor. The processing rate of the off-site facility must be adequate to treat the contaminated material in a reasonable amount of time. Storage requirements and responsibilities must be determined. The need for air discharge and other permits must be determined on a site-specific basis.

4.10.3   APPLICABILITY

With the possible exception of Hg, or a highly contaminated soil, pyrometallurgical processing where metal recovery is the goal would not be applied directly to the contaminated soil, but rather to a concentrate generated via soil washing. Pyrometallurgical processing in conventional rotary kilns, rotary furnaces, or arc furnaces is most likely to be applicable to large volumes of material containing metal concentrations (particularly, Pb, Cd, or Cr) higher than 5%–20%. Unless a very concentrated feed stream can be generated (e.g., approximately 60% for Pb), there will be a charge, in addition to transportation, for processing the concentrate. Lower metal concentrations can be acceptable if the metal is particularly easy to reduce and vaporize (e.g., Hg) or is particularly valuable (e.g., gold or platinum). Arsenic is the weakest candidate for pyrometallurgical recovery, since there is almost no recycling of arsenic in the U.S. Arsenic is also the least valuable of the metals. The price ranges for the five metals (4) are reported here in terms of 2014 USD (44):

	2014 USD/T
As	350–700 (as As trioxide)
Cd	8,500
Cr	11,200
Pb	1,000–1,100
Hg	7,600–13,000

4.10.4   PERFORMANCE AND **BDAT** STATUS

The USEPA technical document (4) contains a list of approximately 35 facilities/addresses/contacts that may accept concentrates of the five metals of interest for pyrometallurgical processing. Sixteen of the 35 facilities are Pb recycling operations, 7 facilities recover Hg, and the remainders address a range of RCRA wastes that contain the metals of interest. Due to the large volume of electric arc furnace emission control waste, extensive processing capability has been developed to recover Cd, Pb, and Zn from solid waste matrices. The available process technologies include (5)

- Waelz kiln process (Horsehead Resource Development Company, Inc.)
- Waelz kiln and calcination process (Horsehead Resource Development Company, Inc.)
- Flame reactor process (Horsehead Resource Development Company, Inc.)
- Inclined rotary kiln (Zia Technology)

Plasma arc furnaces are successfully treating waste at two steel plants. These are site-dedicated units that do not accept outside material for processing.

Pyrometallurgical recovery is a BDAT for the following waste types (5)

- Cd-containing batteries
- Pb nonwastewater in the noncalcium sulfate subcategory
- Hg wastes prior to retorting
- Pb acid batteries
- Zinc nonwastewater
- Hg from wastewater treatment sludge

4.10.5   SITE DEMONSTRATION AND EMERGING TECHNOLOGIES PROGRAM PROJECTS

SITE demonstrations applicable to soils contaminated with the metals of interest include (5)

- RUST Remedial Services, Inc. (X-Trax Thermal Desorption)
- Horsehead Resource Development Company, Inc. (flame reactor)

4.11   ELECTROKINETICS

Electrokinetic remediation relies on the application of low-intensity direct current between electrodes placed in the soil. Contaminants are mobilized in the form of charged species, particles, or ions (2). Attempts to leach metals from soils by electro-osmosis date back to the 1930s. In the past, research focused on removing unwanted salts from agricultural soils. Electrokinetics has been used for dewatering of soils and sludges since the first recorded use in the field in 1939 (67). Electrokinetic extraction has been used in the former Soviet Union since the early 1970s to concentrate metals and to explore for minerals in deep soils. By 1979, research had shown that the content of soluble ions increased substantially in electro-osmotic consolidation of polluted dredgings, while metals were not found in the effluent (68). By the mid-1980s, numerous researchers had realized independently that electrokinetic separation of metals from soils was a potential solution to contamination (69).

Several organizations are developing technologies for the enhanced removal of metals by transporting contaminants to the electrodes where they are removed and subsequently treated above ground. A variation of the technique involves treatment without removal by transporting contaminants through specially designed treatment zones that are created between electrodes. Electrokinetics also can be used to slow or prevent migration of contaminants by configuring cathodes and anodes in a manner that causes contaminants to flow toward the center of a contaminated area of soil. Performance data illustrate the potential for achieving removals >90% for some metals (2).

The range of potential metals is broad. The commercial applications in Europe treated copper, lead, zinc, arsenic, cadmium, chromium, and nickel. There is also potential applicability for radionuclides and some types of organic compounds. The electrode spacing and duration of remediation is site-specific. The process requires adequate soil moisture in the vadose zone, so the addition of a conducting pore fluid may be required (particularly due to a tendency for soil drying near the

TABLE 4.8

Overview of Electrokinetic Remediation Technology

General Characteristics

- Depth of soil that is amenable to treatment depends on electrode placement
- Best used in homogeneous soils with high moisture content and high permeability

Approach #1 Enhanced Removal	Approach #2 Treatment without Removal
Description	*Description*
Electrokinetic transport of contaminants toward the polarized electrodes to concentrate the contaminants for subsequent removal and ex situ treatment	Electro-osmotic transport of contaminants through treatment zones placed between the electrodes. The polarity of the electrodes is reversed periodically, which reverses the direction of the contaminants back and forth through treatment zones. The frequency with which electrode polarity is reversed is determined by the rate of transport of contaminants through the soil
Status	*Status*
Demonstration projects using full-scale equipment are reported in Europe. Bench- and pilot-scale laboratory studies are reported in the United States and at least two full-scale field studies are ongoing in the United States	Demonstrations are ongoing
Applicability	*Applicability*
Pilot scale: lead, arsenic, nickel, mercury, copper, zinc *Lab scale:* lead, cadmium, chromium, mercury, zinc, iron, magnesium, uranium, thorium, radium	Technology developed for organic species and metals
Comments	*Comments*
Field studies are under evaluation by USEPA, DOE, DoD, and Electric Power Research Institute (EPRI) The technique primarily would require addition of water to maintain the electric current and facilitate migration; however, there is ongoing work in application of the technology in partially saturated soils	This technology is being developed for deep clay formations

Source: USEPA. *Recent Developments for In Situ Treatment of Metal Contaminated Soils.* Contract # 68-W5-0055 U.S. Environmental Protection Agency, Washington, DC, March 1997.

anode). Specially designed pore fluids also are added to enhance the migration of target contaminants. The pore fluids are added at either the anode or cathode, depending on the desired effects.

Table 4.8 presents an overview of two variations of electrokinetic remediation technology. Geokinetics International, Inc.; Battelle Memorial Institute; Electrokinetics, Inc.; and Isotron Corporation all are developing variations of technologies categorized under Approach #1, Enhanced Removal. The consortium of Monsanto, E.I. du Pont de Nemours and Company, General Electric, DOE, and the USEPA Office of Research and Development is developing the Lasagna Process, which is categorized under Approach #2, Treatment without Removal (2).

4.11.1 Process Description

Electrokinetic remediation, also referred to as electrokinetic soil processing, electromigration, electrochemical decontamination, or electroreclamation, can be used to extract radionuclides, metals, and some types of organic wastes from saturated or unsaturated soils, slurries, and sediments (70). This in situ soil processing technology is primarily a separation and removal technique for extracting contaminants from soils.

The principle of electrokinetic remediation relies upon application of a low-intensity direct current through the soil between two or more electrodes. Most soils contain water in the pores between the soil particles and have an inherent electrical conductivity that results from salts present in the soil (71). The current mobilizes charged species, particles, and ions in the soil by the following processes (72):

1. Electromigration (transport of charged chemical species under an electric gradient)
2. Electro-osmosis (transport of pore fluid under an electric gradient)
3. Electrophoresis (movement of charged particles under an electric gradient)
4. Electrolysis (chemical reactions associated with the electric field)

Figure 4.6 presents a schematic diagram of a typical conceptual electrokinetic remediation application.

Electrokinetics can be efficient in extracting contaminants from fine-grained, high-permeability soils. A number of factors determine the direction and extent of the migration of the contaminant. Such factors include the type and concentration of the contaminant, the type and structure of the soil, and the interfacial chemistry of the system (73). Water or some other suitable salt solution may be added to the system to enhance the mobility of the contaminant and increase the effectiveness of the technology (e.g., buffer solutions may change or stabilize pore fluid pH). Contaminants arriving at the electrodes may be removed by any of several methods, including electroplating at the electrode, precipitation or coprecipitation at the electrode, pumping of water near the electrode, or complexing with ion exchange resins (73).

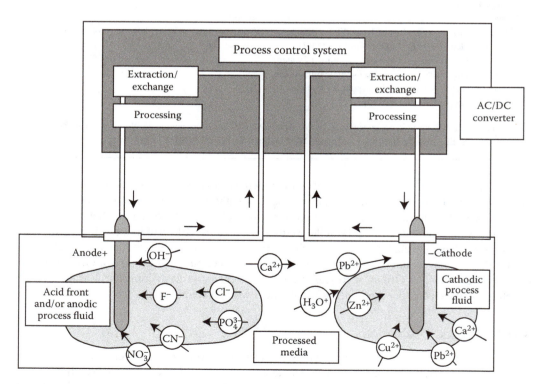

FIGURE 4.6   Diagram of one electrode configuration used in field implementation of electrokinetics. (From USEPA. *Recent Developments for In Situ Treatment of Metal Contaminated Soils.* Contract # 68-W5-0055 U.S. Environmental Protection Agency, Washington, DC, March 1997.)

Electrochemistry associated with this process involves an acid front that is generated at the anode if water is the primary pore fluid present. The variation of pH at the electrodes results from the electrolysis of the water. The solution becomes acidic at the anode because hydrogen ions are produced and oxygen gas is released, and the solution becomes basic at the cathode, where hydroxyl ions are generated and hydrogen gas is released (74). At the anode, the pH could drop to below 2, and it could increase at the cathode to above 12, depending on the total current applied. The acid front eventually migrates from the anode to the cathode. Movement of the acid front by migration and advection results in the desorption of contaminants from the soil (70). The process leads to temporary acidification of the treated soil, and there are no established procedures for determining the length of time needed to reestablish equilibrium. Studies have indicated that metallic electrodes may dissolve as a result of electrolysis and introduce corrosion products into the soil mass. However, if inert electrodes, such as carbon, graphite, or platinum, are used, no residue will be introduced in the treated soil mass as a result of the process (2).

4.11.2 SITE REQUIREMENTS

Before electrokinetic remediation is undertaken at a site, a number of different field and laboratory screening tests must be conducted to determine whether the particular site is amenable to the treatment technique.

1. *Field conductivity surveys:* The natural geologic spatial variability should be delineated because buried metallic or insulating material can induce variability in the electrical conductivity of the soil and, therefore, the voltage gradient. In addition, it is important to assess whether there are deposits that exhibit very high electrical conductivity, in which case the technique may be inefficient.
2. *Chemical analysis of water:* The pore water should be analyzed for dissolved major anions and cations, as well as for the predicted concentration of the contaminant(s). In addition, electrical conductivity and pH of the pore water should be measured.
3. *Chemical analysis of soil:* The buffering capacity and geochemistry of the soil should be determined at each site.
4. *pH effects:* The pH values of the pore water and the soil should be determined because they have a great effect on the valence, solubility, and sorption of contaminant ions.
5. *Bench-scale test:* The dominant mechanism of transport, removal rates, and amounts of contamination left behind can be examined for different removal scenarios by conducting bench-scale tests. Because many of these physical and chemical reactions are interrelated, it may be necessary to conduct bench-scale tests to predict the performance of electrokinetics remediation at the field scale (69,70).

4.11.3 APPLICABILITY AND DEMONSTRATION PROJECTS

Various methods, developed by combining electrokinetics with other techniques, are being applied for remediation. This section describes different types of electrokinetic remediation methods for use at contaminated sites. The methods discussed were developed by Electrokinetics, Inc.; Geokinetics International, Inc.; Isotron Corporation; Battelle Memorial Institute; a consortium effort; and P&P Geotechnik GmbH (2).

4.11.3.1 Electrokinetics, Inc.

Electrokinetics, Inc. operates under a licensing agreement with Louisiana State University. The technology is patented by and assigned to Louisiana State University (75) and a complementing process patent is assigned to Electrokinetics, Inc. (76). As depicted in Figure 4.5, groundwater and/or a processing fluid (supplied externally through the boreholes that contain the electrodes) serves as the conductive medium. The additives in the processing fluid, the products of electrolysis reactions at the electrodes, and the dissolved chemical entities in the contaminated soil are transported across the contaminated soil by conduction under electric fields. This transport, when coupled with sorption, precipitation/dissolution, and volatilization/complexation, provides the fundamental mechanism that can affect the electrokinetic remediation process. Electrokinetics, Inc. accomplishes extraction and removal by electrodeposition, evaporation/condensation, precipitation, or ion exchange, either at the electrodes or in a treatment unit that is built into the system that pumps the processing fluid to and from the contaminated soil. Pilot-scale testing was carried out with support from the USEPA that also developed a design and analysis package for the process (77).

4.11.3.2   Geokinetics International, Inc.

Geokinetics International, Inc. (GII) obtained a patent for an electroreclamation process. The key claims in the patent are the use of electrode wells for both anodes and cathodes and the management of the pH and electrolyte levels in the electrolyte streams of the anode and the cathode. The patent also includes claims for the use of additives to dissolve different types of contaminants (78). Fluor Daniel is licensed to operate GII's metal removal process in the United States.

GII has developed and patented electrically conductive ceramic material (EBONEX®) that has an extremely high resistance to corrosion. It has a lifetime in soil of at least 45 years and is self-cleaning. GII also has developed a batch electrokinetic remediation (BEK®) process. The process which incorporates electrokinetic technology normally requires 24–48 h for complete remediation of the substrate. BEK is a mobile unit that remediates ex situ soils on site. GII also has developed a solution treatment technology (EIX®) that allows removal of contamination from the anode and the cathode solutions up to a thousand times faster than can be achieved through conventional means (2).

4.11.3.3   Isotron Corporation

Isotron Corporation participated in a pilot-scale demonstration of electrokinetic extraction supported by DOE's Office of Technology Development. The demonstration took place at the Oak Ridge K-25 facility in Tennessee. Completed laboratory tests showed that the Isotron process could affect the movement and capture of uranium present in soil from the Oak Ridge site (79).

Isotron Corporation also was involved with Westinghouse Savannah River Company in a demonstration of electrokinetic remediation. The demonstration, supported by DOE's Office of Technology Development, took place at the old TNX basin at the Savannah River site in South Carolina. Isotron used the Electrosorb® process with a patented cylinder to control buffering conditions in situ. An ion exchange polymer matrix called Isolock® was used to trap metal ions. The process was tested for the removal of lead and chromium (79).

4.11.3.4   Battelle Memorial Institute

Another method that uses electrokinetic technology is electroacoustical soil decontamination. This technology combines electrokinetics with sonic vibration. Through application of mechanical vibratory energy in the form of sonic or ultrasonic energy, the properties of a liquid contaminant in soil can be altered in a way that increases the level of removal of the contaminant. Battelle Memorial Institute of Columbus, OH developed the in situ treatment process that uses both electrical and acoustical forces to remove floating contaminants, and possibly metals, from subsurface zones of contamination. The process was selected for USEPA's SITE program (80).

4.11.3.5   Consortium Process

Monsanto Company has coined the name Lasagna to identify its products and services that are based on the integrated in situ remediation process developed by a consortium. The proposed technology combines electro-osmosis with treatment zones that are installed directly in the contaminated soils to form an integrated in situ remedial process, as shown in Figure 4.7. The consortium consists of Monsanto, E.I. du Pont de Nemours and Company (DuPont), and General Electric (GE), with participation by the USEPA Office of Research and Development and DOE.

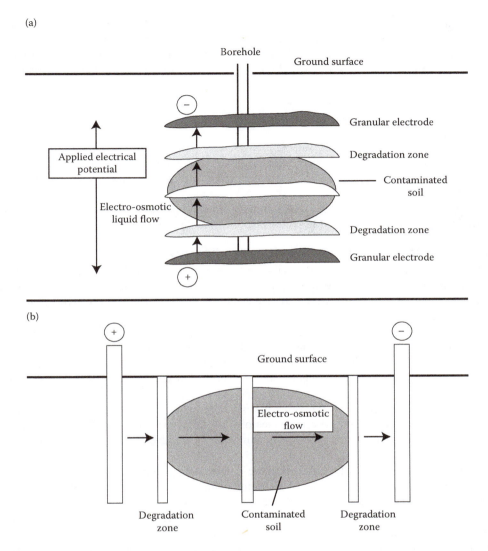

FIGURE 4.7 Schematic diagram of the Lasagna™ process: (a) horizontal configuration and (b) vertical configuration. Electro-osmotic flow is reversed upon switching electrical polarity. (From USEPA. *Recent Developments for In Situ Treatment of Metal Contaminated Soils.* Contract # 68-W5-0055 U.S. Environmental Protection Agency, Washington, DC, March 1997.)

The in situ decontamination process occurs as follows (2):

1. Creates highly permeable zones in close proximity sectioned through the contaminated soil region and turns them into sorption-degradation zones by introducing appropriate materials (sorbents, catalytic agents, microbes, oxidants, buffers, and others).
2. Uses electro-osmosis as a liquid pump to flush contaminants from the soil into the treatment zones of degradation.
3. Reverses liquid flow, if desired, by switching the electrical polarity, a mode that increases the efficiency with which contaminants are removed from the soil; allows repeated passes through the treatment zones for complete sorption.

Initial field tests of the consortium process were conducted at DOE's gaseous diffusion plant in Paducah, Kentucky. The experiment tested the combination of electro-osmosis and in situ sorption in treatment zones. Technology development for the degradation processes and their integration into the overall treatment scheme were carried out at bench and pilot scales, followed by field experiments of the full-scale process (81).

4.11.4 Performance and Cost

Work sponsored by USEPA, DOE, the National Science Foundation, and private industry, when coupled with the efforts of researchers from academic and public institutions have demonstrated the feasibility of moving electrokinetics remediation to pilot-scale testing and demonstration stages (70).

This section describes testing and cost summary results reported by Louisiana State University, Electrokinetics, Inc., GII, Battelle Memorial Institute, and the consortium (2).

4.11.4.1 Louisiana State University—Electrokinetics, Inc.

The Louisiana State University (LSU)—Electrokinetics, Inc. Group has conducted bench-scale testing on radionuclides and on organic compounds. Test results have been reported for lead, cadmium, chromium, mercury, zinc, iron, and magnesium. Radionuclides tested include uranium, thorium, and radium.

In collaboration with USEPA, the LSU—Electrokinetics, Inc. Group has completed pilot-scale studies of electrokinetic soil processing in the laboratory. Electrokinetics, Inc. carried out a site-specific pilot-scale study of the Electro-Klean™ electrical separation process. Pilot field studies also have been reported in the Netherlands on soils contaminated with lead, arsenic, nickel, mercury, copper, and zinc.

A pilot-scale laboratory study investigating the removal of 2,000 mg/kg of lead loaded onto kaolinite was completed. Removal efficiencies of 90%–95% were obtained. The electrodes were placed one inch apart in a 2-ton kaolinite specimen for 4 months, at a total energy cost of about 2014 USD 26/T (80).

With the support of DOD, Electrokinetics, Inc. carried out a comprehensive demonstration study of lead extraction from a creek bed at a U.S. Army firing range in Louisiana. USEPA took part in independent assessments of the results of that demonstration study under the SITE program. The soils are contaminated with levels as high as 4500 mg/kg of lead; pilot-scale studies have demonstrated that concentrations of lead decreased to <300 mg/kg in 30 weeks of processing. The TCLP values dropped from >300 mg/L to <40 mg/L within the same period. At the site of the demonstration study, Electrokinetics, Inc. used the CADEX™ electrode system that promotes transport of species into the cathode compartment, where they are precipitated and/or electrodeposited directly. Electrokinetics, Inc. used a special electrode material that is cost-effective and does not corrode. Under the supervision and support of the Electric Power Research Institute and power companies in the southern United States, a treatability and a pilot-scale field testing study of soils in sites contaminated with arsenic has been performed, in a collaborative effort between Southern Company Services Engineers and Electrokinetics, Inc. (2).

The processing cost of a system designed and installed by Electrokinetics, Inc. consists of energy cost, conditioning cost, and fixed costs associated with installation of the system. Power consumption is related directly to the conductivity of the soil across the electrodes. Electrical conductivity of soils can span orders of magnitude, from 30 mhos/cm to more than 3,000 µmhos/cm, with higher values being in saturated, high-plasticity clays. A mean conductivity value is 500 µmhos/cm. The voltage gradient is held to approximately 1 V/cm in an attempt to prevent adverse effects of temperature increases and for other practical reasons (70). It may be cost-prohibitive to attempt to remediate high-plasticity soils that have high electrical conductivities. However, for most deposits having conductivities of 500 µmhos/cm, the daily energy consumption will be approximately 12 kW h/m^3 day^{-1} or about USD 1.20/m^3 day^{-1}, (USD 0.10/kWh) and USD 36/m^3 month^{-1}. The processing time will depend upon several factors, including the spacing of the electrodes, and the type of conditioning scheme that will be used. If an electrode spacing of 4 m is selected, it may be necessary to process the site over several months.

Pilot-scale studies using "real-world" soils indicate that the energy expenditures in extraction of metals from soils may be 500 kW h/m^3 or more at electrode spacing of 1.0–1.5 m (77). The vendor estimates that the direct cost of about USD 50/m^3 (USD 0.10/kW h) suggested for this energy expenditure, together with the cost of enhancement, could result in direct costs of USD 100/m^3. If no other efficient in situ technology is available to remediate fine-grained and heterogeneous subsurface deposits contaminated with metals, this technique would remain potentially competitive.

4.11.4.2 Geokinetics International, Inc.

GII has successfully demonstrated in situ electrochemical remediation of metal-contaminated soils at several sites in Europe. Geokinetics, an associate company of GII, also has been involved in the electrokinetics arena in Europe. Table 4.9 summarizes the physical characteristics of five of the sites, including the size, the contaminant(s) present, and the overall performance of the technology at each site. GII estimates its typical costs for "turn key" remediation projects are in the range of 2007 USD 160–260/m^3 (2).

4.11.4.3 Battelle Memorial Institute

The technology demonstration through the SITE program was completed (80). The results indicate that the electroacoustical technology is technically feasible for the removal of inorganic species from clay soils (66).

4.11.4.4 Consortium Process

The Phase I field test of the Lasagna™ process has been completed. Scale-up from laboratory units was successfully achieved with respect to electrical parameters and electro-osmotic flow. Soil samples taken throughout the test site before and after the test indicate a 98% removal of trichloroethylene (TCE) from a tight clay soil (i.e., hydraulic conductivity less than 1×10^{-7} cm/s). TCE soil levels

---

TABLE 4.9
Performance of Electrochemical Soil Remediation Applied at Five Field Sites in Europe

Site Description	Soil Volume (m^3)	Soil Type	Contaminant	Initial Concentration (mg/kg)	Final Concentration (mg/kg)
Former paint factory	230	Peat/clay soil	Cu	1,220	<200
			Pb	>3,780	<280
Operational galvanizing plant	40	Clay soil	Zn	>1,400	600
Former timber plant	190	Heavy clay soil	As	>250	<30
Temporary landfill	5,440	Argillaceous sand	Cd	>180	<40
Military air base	1,900	Clay	Cd	660	47
			Cr	7,300	755
			Cu	770	98
			Ni	860	80
			Pb	730	108
			Zn	2,600	289

Source: USEPA. *Recent Developments for In Situ Treatment of Metal Contaminated Soils.* Contract # 68-W5-0055 U.S. Environmental Protection Agency, Washington, DC, March 1997.

were reduced from the 100–500 mg/kg range to an average concentration of 1 mg/kg (82). Various treatment processes are being investigated in the laboratory to address other types of contaminants, including heavy metals (82).

4.11.5 SUMMARY OF ELECTROKINETIC REMEDIATION

Electrokinetic remediation may be applied to both saturated and partially saturated soils. One problem to overcome when applying electrokinetic remediation to the vadose zone is the drying of soil near the anode. When an electric current is applied to soil, water will flow by electro-osmosis in the soil pores, usually toward the cathode. The movement of the water will deplete soil moisture adjacent to the anode, and moisture will collect near the cathode. However, processing fluids may be circulated at the electrodes. The fluids can serve both as a conducting medium and as a means to extract or exchange the species and introduce other species. Another use of processing fluids is to control, depolarize, or modify either or both electrode reactions. The advance of the process fluid (acid or the conditioning fluid) across the electrodes assists in desorption of species and dissolution of carbonates and hydroxides. Electro-osmotic advection and ionic migration lead to the transport and subsequent removal of the contaminants. The contaminated fluid is then recovered at the cathode.

Spacing of the electrode will depend upon the type and level of contamination and the selected current voltage regime. When higher voltage gradients are generated, the efficiency of the process might decrease because of increases in temperature. A spacing that will generate a potential gradient in the order of 1 V/cm is preferred. The spacing of electrodes generally will be as much as 3 m. The duration of the remediation will be site-specific. The remediation process should be continued until the desired removal is achieved. However, it should be recognized that, in cases in which the duration of treatment is reduced by increasing the electrical potential gradient, the efficiency of the process will decrease (83,84).

The advantage of the technology is its potential for cost-effective use for both in situ and ex situ applications. The fact that the technique requires the presence of a conducting pore fluid in a soil mass may have site-specific implications. Also, heterogeneities or anomalies found at sites, such as submerged foundations, rubble, large quantities of iron or iron oxides, large rocks, or gravel; or submerged cover material, such as seashells, are expected to reduce removal efficiencies (70).

4.12 PHYTOREMEDIATION

This technology is in the stage of commercialization for treatment of soils contaminated with metals, and in the future may provide a low-cost option under specific circumstances. At the current stage of development, this process is best suited for sites with widely dispersed contamination at low concentrations where only treatment of soils at the surface (in other words, within depth of the root zone) is required (2).

Phytoremediation is the use of plants to remove, contain, or render harmless environmental contaminants. This definition applies to all biological, chemical, and physical processes that are influenced by plants and that aid in the cleanup of contaminated substances (85). Plants can be used in site remediation, both to mineralize and immobilize toxic organic compounds at the root zone and to accumulate and concentrate metals and other inorganic compounds from soil into aboveground shoots (86). Although phytoremediation is a relatively new concept in the waste management community, techniques, skills, and theories developed through the application of well-established agroeconomic technologies are easily transferable. The development of plants for restoring sites contaminated with metals will require the multidisciplinary research efforts of agronomists, toxicologists, biochemists, microbiologists, pest management specialists, engineers, and other specialists (85,86). Table 4.10 presents an overview of phytoremediation technology.

TABLE 4.10

Overview of Phytoremediation Technology

General Characteristics

Best used at sites with low to moderate disperse metals content and with soil media that will support plant growth

Applications limited to depth of the root zone

Longer times required for remediation compared with other technologies

Different species have been identified to treat different metals

Approach #1—Phytoextraction (Harvest)	**Approach #2—Phytostabilization (Root Fixing)**
Description	*Description*
Uptake of contaminants from soil into aboveground plant tissue, which is periodically harvested and treated	Production of chemical compounds by the plant to immobilize contaminants at the interface of roots and soil. Additional stabilization can occur by raising the pH level in the soil
Status	*Status*
Field testing for effectiveness on radioactive metals is ongoing in the vicinity of the damaged nuclear reactor in Chernobyl, Ukraine	Research is ongoing
Field testing also is being conducted in Trenton, NJ and Butte, MT and by the Idaho National Engineering Laboratory (INEL) in Fernald, OH	
Applicability	*Applicability*
Potentially applicable for many metals. Nickel and zinc appear to be most easily absorbed. Preliminary results for absorption of copper and cadmium are encouraging	Potentially applicable for many metals, especially lead, chromium, and mercury
Comments	*Comments*
Cost affected by volume of biomass produced that may require treatment before disposal. Cost affected by concentration and depth of contamination and number of harvests required	Long-term maintenance is required

Source: USEPA. *Recent Developments for In Situ Treatment of Metal Contaminated Soils.* Contract # 68-W5-0055 U.S. Environmental Protection Agency, Washington, DC, March 1997.

Two basic approaches for metals remediation include phytoextraction and phytostabilization. Phytoextraction relies on the uptake of contaminants from the soil and their translocation into aboveground plant tissue, which is harvested and treated. Although hyperaccumulating trees, shrubs, herbs, grasses, and crops have potential, crops seem to be most promising because of their greater biomassproduction. Nickel and zinc appear to be the most easily absorbed, although tests with copper and cadmium are encouraging (2). Significant uptake of lead, a commonly occurring contaminant, has not been demonstrated on a large scale. However, some researchers are experimenting with soil amendments that would facilitate uptake of lead by plants.

4.12.1 Process Description

Metals considered essential for at least some forms of life include vanadium (V), chromium (Cr), manganese (Mn), iron (Fe), cobalt (Co), nickel (Ni), copper (Cu), zinc (Zn), and molybdenum (Mo) (86). Because many metals are toxic in concentrations above minute levels, an organism must regulate the cellular concentrations of such metals. Consequently, organisms have evolved transport systems to regulate the uptake and distribution of metals. Plants have remarkable metabolic and absorption capabilities, as well as transport systems that can take up ions selectively from the soil.

Plants have evolved a great diversity of genetic adaptations to handle potentially toxic levels of metals and other pollutants that occur in the environment. In plants, the uptake of metals occurs primarily through the root system, in which the majority of mechanisms to prevent metal toxicity are found (87). The root system provides an enormous surface area that absorbs and accumulates the water and nutrients essential for growth. In many ways, living plants can be compared to solar-powered pumps that can extract and concentrate certain elements from the environment (88).

Plant roots cause changes at the soil-root interface as they release inorganic and organic compounds (root exudates) in the area of the soil immediately surrounding the roots (the rhizosphere) (89). Root exudates affect the number and activity of microorganisms, the aggregation and stability of soil particles around the root, and the availability of elements. Root exudates can increase (mobilize) or decrease (immobilize) directly or indirectly the availability of elements in the rhizosphere. Mobilization and immobilization of elements in the rhizosphere can be caused by (90,91)

1. Changes in soil pH
2. Release of complexing substances, such as metal-chelating molecules
3. Changes in oxidation–reduction potential
4. Increase in microbial activity

Phytoremediation technologies can be developed for different applications in environmental cleanup and are classified into three types:

1. Phytoextraction
2. Phytostabilization
3. Rhizofiltration

4.12.1.1 Phytoextraction

Phytoextraction technologies use hyperaccumulating plants to transport metals from the soil and concentrate them into the roots and aboveground shoots that can be harvested (85,86,89). A plant containing more than 0.1% of Ni, Co, Cu, Cr, or 1% Zn and Mn in its leaves on a dry weight basis is called a hyperaccumulator, regardless of the concentration of metals in the soil (86,92,93).

Almost all metal-hyperaccumulating species known today were discovered on metal-rich soils, either natural or artificial, often growing in communities with metal excluders (86,94). Actually, almost all metal-hyperaccumulating plants are endemic to such soils, suggesting that hyperaccumulation is an important ecophysiological adaptation to metal stress and one of the manifestations of resistance to metals. The majority of hyperaccumulating species discovered so far are restricted to a few specific geographical locations (86,92). For example, Ni hyperaccumulators are found in New Caledonia, the Philippines, Brazil, and Cuba. Ni and Zn hyperaccumulators are found in southern and central Europe and Asia Minor.

Dried or composted plant residues or plant ashes that are highly enriched with metals can be isolated as hazardous waste or recycled as metal ore (95). The goal of phytoextraction is to recycle as "bio-ores" metals reclaimed from plant ash in the feed stream of smelting processes. Even if the plant ashes do not have enough concentration of metal to be useful in smelting processes, phytoextraction remains beneficial because it reduces by as much as 95% the amount of hazardous waste to be landfilled (2). Several research efforts in the use of trees, grasses, and crop plants are being pursued to develop phytoremediation as a cleanup technology. The following paragraphs briefly discuss these three phytoextraction techniques.

The use of trees can result in the extraction of significant amounts of metal because of their high biomass production. However, the use of trees in phytoremediation requires long-term treatment and may create additional environmental concerns about falling leaves. When leaves containing metals fall or blow away, recirculation of metals to the contaminated site and migration off-site by wind transport or through leaching can occur (2).

Some grasses accumulate surprisingly high levels of metals in their shoots without exhibiting toxic effects. However, their low biomass production results in relatively low yield of metals. Genetic breeding of hyperaccumulating plants that produce relatively large amounts of biomass could make the extraction process highly effective (96).

It is known that many crop plants can accumulate metals in their roots and aboveground shoots, potentially threatening the food chain. For example, in May 1980 regulations proposed under RCRA for hazardous waste include limits on the amounts of cadmium and other metals that can be applied to crops. Recently, however, the potential use of crop plants for environmental remediation has been under investigation. Using crop plants to extract metals from the soil seems practical because of their high biomass production and relatively fast rate of growth. Other benefits of using crop plants are that they are easy to cultivate and exhibit genetic stability (97).

4.12.1.2   Phytostabilization

Phytostabilization uses plants to limit the mobility and bioavailability of metals in soils. Ideally, phytostabilizing plants should be able to tolerate high levels of metals and to immobilize them in the soil by sorption, precipitation, complexation, or the reduction of metal valences. Phytostabilizing plants also should exhibit low levels of accumulation of metals in shoots to eliminate the possibility that residues in harvested shoots might become hazardous wastes (88). In addition to stabilizing the metals present in the soil, phytostabilizing plants also can stabilize the soil matrix to minimize erosion and migration of sediment. Dr. Gary Pierzynski of Kansas State University is studying phytostabilization in poplar trees, which were selected for the study because they can be deep-planted and may be able to form roots below the zone of maximum contamination (2).

Since most sites contaminated with metals lack established vegetation, metal-tolerant plants are used to revegetate such sites to prevent erosion and leaching (98). However, that approach is a containment rather than a remediation technology. Some researchers consider phytostabilization an interim measure to be applied until phytoextraction becomes fully developed. However, other researchers are developing phytostabilization as a standard protocol of metal remediation technology, especially for sites at which the removal of metals does not seem to be economically feasible. After field applications conducted by a group in Liverpool, England, varieties of three grasses were made commercially available for phytostabilization (88):

- *Agrostis tenuis,* cv *Parys* for copper wastes
- *Agrosas tenuis,* cv *Coginan* for acid lead and zinc wastes
- *Festuca rubra,* cv *Merlin* for calcareous lead and zinc wastes

4.12.1.3   Rhizofiltration

One type of rhizofiltration uses plant roots to absorb, concentrate, and precipitate metals from wastewater (88), which may include leachate from soil. Rhizofiltration uses terrestrial plants instead of aquatic plants because the terrestrial plants develop much longer, fibrous root systems covered with root hairs that have extremely large surface areas. This variation of phytoremediation uses plants that remove metals by sorption, which does not involve biological processes. Use of plants to translocate metals to the shoots is a slower process than phytoextraction (98).

Another type of rhizofiltration, which is more fully developed, involves construction of wetlands or reed beds for the treatment of contaminated wastewater or leachate. The technology is cost-effective for the treatment of large volumes of wastewater that have low concentrations of metals (98). Since rhizofiltration focuses on the treatment of contaminated water, it is not discussed further in this chapter.

Table 4.11 presents the advantages and disadvantages of each of the types of phytoremediation currently being researched that are categorized as either phytoextraction on phytostabilization (88).

TABLE 4.11

Types of Phytoremediation Technology: Advantages and Disadvantages

Type of Phytoremediation	Advantages	Disadvantages
Phytoextraction by trees	High biomass production	Potential for off-site migration and leaf transportation of metals to surface Metals are concentrated in plant biomass and must be disposed of eventually
Phytoextraction by grasses	High accumulation	Low biomass production and slow growth rate Metals are concentrated in plant biomass and must be disposed of eventually
Phytoextraction by crops	High biomass and increased growth rate	Potential threat to the food chain through ingestion by herbivores Metals are concentrated in plant biomass and must be disposed of eventually
Phytostabilization	No disposal of contaminated biomass required	Remaining liability issues, including maintenance for indefinite period of time (containment rather than removal)
Rhizofiltration	Readily absorbs metals	Applicable for treatment of water only Metals are concentrated in plant biomass and must be disposed of eventually

Source: USEPA. *Recent Developments for In Situ Treatment of Metal Contaminated Soils.* Contract # 68-W5-0055 U.S. Environmental Protection Agency, Washington, DC, March 1997.

4.12.1.4   Future Development

Faster uptake of metals and higher yields of metals in harvested plants may become possible through the application of genetic engineering and/or selective breeding techniques. Recent laboratory-scale testing has revealed that a genetically altered species of mustard weed can uptake mercuric ions from the soil and convert them to metallic mercury, which is transpired through the leaves (2). Improvements in phytoremediation may be attained through research and a better understanding of the principles governing the processes by which plants affect the geochemistry of their soils. In addition, future testing of plants and microflora may lead to the identification of plants that have metal accumulation qualities that are far superior to those currently known.

4.12.2   Applicability

Plants have been used to treat wastewater for more than 300 years, and plant-based remediation methods for slurries of dredged material and soils contaminated with metals have been proposed since the mid-1970s (85,99). Reports of successful remediation of soils contaminated with metals are rare, but the suggestion of such application is more than two decades old, and progress is being made at a number of pilot test sites (94). Successful phytoremediation must meet cleanup standards in order to be approved by regulatory agencies.

No full-scale applications of phytoremediation have been reported. One vendor, Phytotech, Inc., is developing phytostabilization for soil remediation applications. Phytotech also has patented strategies for phytoextraction and is conducting several field tests in Trenton, New Jersey and in Chernobyl, Ukraine (97). Also, as was previously mentioned, a group in Liverpool, England has made three grasses commercially available for the stabilization of lead, copper, and zinc wastes (88).

4.12.3   Performance and Cost

A variety of new research approaches and tools are expanding understanding of the molecular and cellular processes that can be employed through phytoremediation (100).

4.12.3.1   Performance

Potential for phytoremediation (phytoextraction) can be assessed by comparing the concentration of contaminants and volume of soil to be treated with the particular plant's seasonal productivity of biomass and ability to accumulate contaminants. Table 4.6 lists selected examples of plants identified as metal hyperaccumulators and their native countries (92,101). If plants are to be effective remediation systems, 1 ton of plant biomass, costing from several hundred to a few thousand dollars to produce, must be able to treat large volumes of contaminated soil. For metals that are removed from the soil and accumulated in aboveground biomass, the total amount of biomass per hectare required for soil cleanup is determined by dividing the total weight of metal per hectare to be remediated by the accumulation factor, which is the ratio of the accumulated weight of the metal to the weight of the biomass containing the metal. The total biomass per hectare (T/ha) then can be divided by the productivity of the plant (T/ha year^{-1}) to determine the number of years (year) required to achieve cleanup standards—a major determinant of the overall cost and feasibility of phytoremediation (100).

As discussed earlier, the amount of biomass is one of the factors that determine the practicality of phytoremediation. Under the best climatic conditions, with irrigation, fertilization, and other factors, total biomass productivity can approach 100 T/ha/year. One unresolved issue is the tradeoff between accumulation of toxic elements and productivity (102). In practice, a maximum harvest biomass yield of 10–20 T/ha/year is likely, particularly for plants that accumulate metals.

These values for productivity of biomass and the metal content of the soil would limit annual capacity for removal of metals to approximately 10–400 kg/ha/year, depending on the pollutant, species of plant, climate, and other factors. For a target soil depth of 30 cm (4000 T/ha), this capacity amounts to an annual reduction of 2.5–100 mg/kg of soil contaminants. This rate of removal of contamination often is acceptable, allowing total remediation of a site over a period of a few years to several decades (100).

4.12.3.2   Cost

The practical objective of phytoremediation is to achieve major reductions in the cost of cleanup of hazardous sites. Salt and others (88) note the cost-effectiveness of phytoremediation with an example: Using phytoremediation to cleanup 1 acre of sandy loam soil to a depth of 50 cm typically will cost 2014 USD 70,000 to 120,000, compared with a cost of at least USD 470,000 for excavation and disposal storage without treatment (88). One objective of field tests is to use commercially available agricultural equipment and supplies for phytoremediation to reduce costs. Therefore, in addition to their remediation qualities, the agronomic characteristics of the plants must be evaluated.

The processing and ultimate disposal of the biomass generated is likely to be a major percentage of overall costs, particularly when highly toxic metals and radionuclides are present at a site. Analysis of the costs of phytoremediation must include the entire cycle of the process, from the growing and harvesting of the plants to the final processing and disposal of the biomass. It is difficult to predict costs of phytoremediation, compared with overall cleanup costs at a site. Phytoremediation also may be used as a follow-up technique after areas having high concentrations of pollutants have been mitigated or in conjunction with other remediation technologies, making cost analysis more difficult.

4.12.3.3   Future Directions

Because metal hyperaccumulators generally produce small quantities of biomass, they are unsuited agronomically for phytoremediation. Nevertheless, such plants are a valuable store of genetic and physiologic material and data (85). To provide effective cleanup of contaminated soils, it is essential to find, breed, or engineer plants that absorb, translocate, and tolerate levels of metals in the range of 0.1%–1.0%. It also is necessary to develop a methodology for selecting plants that are native to the area.

Three grasses are commercially available for the stabilization of lead, copper, and zinc wastes (88). An integrated approach that involves basic and applied research, along with consideration of safety, legal, and policy issues, will be necessary to establish phytoremediation as a practicable cleanup technology (85).

According to a DOE report three broad areas of research and development can be identified for the in situ treatment of soil contaminated with metals (100):

1. *Mechanisms of uptake, transport, and accumulation:* Research is needed to develop better understanding of the use of physiological, biochemical, and genetic processes in plants. Research on the uptake and transport mechanisms is providing improved knowledge about the adaptability of those systems and how they might be used in phytoremediation.
2. *Genetic evaluation of hyperaccumulators:* Research is being conducted to collect plants growing in soils that contain high levels of metals and screen them for specific traits useful in phytoremediation. Plants that tolerate and colonize environments polluted with metals are a valuable resource, both as candidates for use in phytoremediation and as sources of genes for classical plant breeding and molecular genetic engineering.
3. *Field evaluation and validation:* Research is being conducted to employ early and frequent field testing to accelerate implementation of phytoremediation technologies and to provide data to research programs. Standardization of field-test protocols and subsequent application of test results to real problems are also needed.

Research in these areas is expected to grow as many of the current engineering technologies for cleaning surface soil of metals are costly and physically disruptive. Phytoremediation, when fully developed, could result in significant cost savings and in the restoration of numerous sites by a relatively noninvasive, solar-driven, in situ method that, in some forms, can be esthetically pleasing (85).

4.12.4 SUMMARY OF PHYTOREMEDIATION TECHNOLOGY

Phytoremediation is in the early stage of development and is being field tested at various sites in the United States and overseas for its effectiveness in capturing or stabilizing metals, including radioactive wastes. Limited cost and performance data are currently available. Phytoremediation has the potential to develop into a practicable remediation option at sites at which contaminants are near the surface, are relatively nonleachable, and pose little imminent threat to human health or the environment (85). The efficiency of phytoremediation depends on the characteristics of the soil and the contaminants; these factors are summarized in the sections that follow.

4.12.4.1 Site Conditions

The effectiveness of phytoremediation generally is restricted to surface soils within the rooting zone. The most important limitation to phytoremediation is rooting depth, which can be 20, 50, or even 100 cm, depending on the plant and soil type. Therefore, one of the favorable site conditions for phytoremediation is contamination with metals that is located at the surface (100).

The type of soil, as well as the rooting structure of the plant relative to the location of contaminants can have a strong influence on the uptake of any metal substance by the plant. Amendment of soils to change soil pH, nutrient compositions, or microbial activities must be selected in treatability studies to govern the efficiency of phytoremediation. Certain generalizations can be made about such cases; however, much work is needed in this area (85). Since the amount of biomass that can be produced is one of the limiting factors affecting phytoremediation, optimal climatic conditions, with irrigation and fertilization of the site, should be considered for increased productivity of the best plants for the site (100).

TABLE 4.12
Typical Treatment Trains

	Containment	S/S	Vitrification	Soil Washing	Pyrometallurgical	Soil Flushing
Pretreatment						
Excavation	•	E,P	I,E	•	•	
Debris removal		E,P	E	•	•	
Oversize reduction		E,P	E	•	•	
Adjust pH	•	I,E,P				
Reduction [e.g., Cr(VI) to Cr(III)]	•	I,E				
Oxidation [e.g., As(III) to As(V)]	•	I,E				
Treatment to remove or destroy organics		I,E				
Physical separation of rich and lean fractions		I,E,P	E	•	•	
Dewatering and drying for wet sludge	•	P	E		•	
Conversion of metals to less volatile forms [e.g., As_2O_3 to $Ca_3(AsO_4)_2$]			E			
Addition of high-temperature reductants					•	
Pelletizing					•	
Flushing fluid delivery and extraction system						•
Containment barriers	•	I,E,P	I	•		•
Posttreatment/Residuals Management						
Disposal of treated solid residuals (preferably below the frost line and above the water table)		I,E,P	E		•	
Containment barriers		I,E,P	I,E			•
Off-gas treatment		I,E,P	I,E		•	
Reuse for onsite paving		P				
Metal recovery from extraction fluid by aqueous processing (ion exchange, electrowinning, etc.)						
Pyrometallurgical recovery of metal from sludge				•		
Processing and reuse of leaching solution				•	•	
S/S treatment of leached residual				•		
Disposal of solid process residuals (preferably below the frost line and above the water table)				•		
Disposal of liquid process residuals				•		•
S/S treatment of slag or fly ash					•	
Reuse of slag/vitreous product as construction material			E	•		
Reuse of metal or metal compound					•	
Further processing of metal or metal compound					•	
Flushing liquid/groundwater treatment/ disposal						•

Source: USEPA. *Technology Alternatives for the Remediation of Soils Contaminated with AS, Cd, Cr, Hg, and Pb.* EPA/540/S-97/500, U.S. Environmental Protection Agency, Cincinnati, OH, August 1997.

Note: Technology has been divided into the following categories: I = in situ process; E = ex situ process; P = polymer (microencapsulation ex situ).

4.12.4.2 Waste Characteristics

Sites that have low to moderate contamination with metals might be suitable for growing hyperaccumulating plants, although the most heavily contaminated soils do not allow plant growth without the addition of soil amendments. Unfortunately, one of the most difficult metal cations for plants to translocate is lead, which is present at numerous sites in need of remediation. Although significant uptake of lead has not yet been demonstrated, one researcher is experimenting with soil amendments that make lead more available for uptake (88).

Capabilities to accumulate lead and other metals are dependent on the chemistry of the soil in which the plants are growing. Most metals, and lead in particular, occur in numerous forms in the soil, not all of which are equally available for uptake by plants (85,103). The maximum removal of lead requires a balance between the nutritional requirements of plants for biomass production and the bioavailability of lead for uptake by plants. Maximizing availability of lead requires low pH and low levels of available phosphate and sulfate. However, limiting the fertility of the soil in such a manner directly affects the health and vigor of plants (85).

4.13 USE OF TREATMENT TRAINS

Several of the metal remediation technologies discussed are often enhanced through the use of treatment trains. Treatment trains use two or more remedial options applied sequentially to the contaminated soil and often increase the effectiveness while decreasing the cost of remediation. Processes involved in treatment trains include soil pretreatment, physical separation designed to decrease the amount of soil requiring treatment, additional treatment of process residuals or off-gases, and a variety of other physical and chemical techniques, which can greatly improve the performance of the remediation technology (104–114). Table 4.12 provides examples of treatment trains used to enhance each of the proved and commercialized metal remediation technology (5).

TABLE 4.13
Estimated Cost Ranges of Metals Remediation Technologies

Type of Remediation	Cost Range 2014 USD/T
Containment[a]	15–140
Solidification/stabilization	93–440
Vitrification	600–1,330
Soil washing	93–370
Soil flushing[b]	93–300
Pyrometallurgical	383–850
Electrokinetics[b]	70–190
Phytoremediation[c]	35–60

Source: USEPA. *Recent Developments for In Situ Treatment of Metal Contaminated Soils.* Contract # 68-W5-0055 U.S. Environmental Protection Agency, Washington, DC, March 1997; USEPA. *Technology Alternatives for the Remediation of Soils Contaminated with AS, Cd, Cr, Hg, and Pb.* EPA/540/S-97/500, U.S. Environmental Protection Agency, Cincinnati, OH, August 1997.

[a] Includes landfill caps and slurry walls. A slurry wall depth of 6 m is assumed.

[b] Costs reported in USD/m^3, assumed soil specific gravity of 1.6.

[c] Costs reported per acre for a soil depth of 0.50 m.

4.14 COST RANGES OF REMEDIAL TECHNOLOGIES

Estimated cost ranges for the basic operation of the technology are presented in Table 4.13. The reader is cautioned that the cost estimates generally do not include pretreatment, site preparation, regulatory compliance costs, costs for additional treatment of process residuals (e.g., stabilization of incinerator ash or disposal of metals concentrated by solvent extraction), or profit (5,115). Since the actual cost of employing a remedial technology at a specific site may be significantly different than these estimates, data are best used for order-of-magnitude cost evaluations.

REFERENCES

1. Federal Register. *Comprehensive Environmental Response, Compensation, and Liability Act (CERCLA or Superfund)*, 42 U.S.C. s/s 9601 et seq. (1980), United States Government, Public Laws. Available at: www.access.gpo.gov/uscode/title42/chapter103_.html, 2016.
2. USEPA. *Recent Developments for In Situ Treatment of Metal Contaminated Soils*. Contract # 68-W5-0055 U.S. Environmental Protection Agency, Washington, DC, March 1997.
3. Federal Register. *Resource Conservation and Recovery Act (RCRA)*, 42 US Code s/s 6901 et seq. (1976), U.S. Government, Public Laws. www.federalregister.gov/resource-conservation-and-recovery-act-rcra, 2016.
4. USEPA. *Contaminants and Remedial Options at Selected Metal-Contaminated Sites*. EPA/540/R-95/512, U.S. Environmental Protection Agency, Washington, DC, July 1995.
5. USEPA. *Technology Alternatives for the Remediation of Soils Contaminated with AS, Cd, Cr, Hg, and Pb*. EPA/540/S-97/500, U.S. Environmental Protection Agency, Cincinnati, OH, August 1997.
6. USEPA. *In Situ Technologies for the Remediation of Soils Contaminated with Metals—Status Report*. U.S. Environmental Protection Agency, Cincinnati, OH, July 1996.
7. USEPA. *Selection of Control Technologies for Remediation of Lead Battery Recycling Sites*. EPA/540/2-91/014, U.S. Environmental Protection Agency, Cincinnati, OH, 1991.
8. USEPA. *Engineering Bulletin: Selection of Control Technologies for Remediation of Lead Battery Recycling Site*. EPA/540/S-92/011, U.S. Environmental Protection Agency, Cincinnati, OH, 1992.
9. USEPA. *Contaminants and Remedial Options at Wood Preserving Sites*. EPA 600/R-92/182, U.S. Environmental Protection Agency, Washington, DC, 1992.
10. USEPA. *Presumptive Remedies for Soils, Sediments, and Sludges at Wood Treater Sites*. EPA/540/R-95/128, U.S. Environmental Protection Agency, Washington, DC, 1995.
11. USEPA. *Contaminants and Remedial Options at Pesticide Sites*. EPA/600/R-94/202, U.S. Environmental Protection Agency, Washington, DC, 1994.
12. USEPA. *Separation/Concentration Technology Alternatives for the Remediation of Pesticide-Contaminated Soil*. EPA/540/S-97/503, U.S. Environmental Protection Agency, Washington, DC, 1997.
13. Deka, J. and Sarma, H. P. Heavy metal contamination in soil in an industrial zone and its relation with some soil properties. *Archives of Applied Science Research*, 4 (2), 831–836, 2015.
14. Palmer, C. D. and Puls, R. W. *Natural Attenuation of Hexavalent Chromium in Ground Water and Soils*. EPA/540/S-94/505, U.S. Environmental Protection Agency, Washington, DC, 1994.
15. Benjamin, M. M. and Leckie, J. D. Adsorption of metals at oxide interfaces: Effects of the concentrations of adsorbate and competing metals. In: *Contaminant sand Sediments, Vol. 2: Analysis, Chemistry, Biology*, Baker R. A. (Ed.), Chapter 16. Ann Arbor Science Publishers, Inc., Ann Arbor, MI, 1980.
16. Wagemann, R. Some theoretical aspects of stability and solubility of inorganic As in the freshwater environment. *Water Research*, 12, 139–145, 1978.
17. Zimmerman, L. and Coles, C. Cement industry solutions to waste management—The utilization of processed waste by-products for cement manufacturing. In: *Proceedings of the 1st International Conference for Cement Industry Solutions to Waste Management*, Calgary, Alberta, Canada, pp. 533–545, 1992.
18. Earth Platform. *Contaminated Soil Remediation*. http://www.earthplatform.com/contaminated/soil/remediation, 2015.
19. Sharma, H. D. and Reddy, K. R. *Geoenvironmental Engineering: Site Remediation, Waste Containment, and Emerging Waste Management Technologies*. John Wiley & Sons, Hoboken, NJ, 2004.
20. Weston R. F. *Installation Restoration General Environmental Technology Development Guidelines for In-Place Closure of Dry Lagoons*. U.S. Army Toxic and Hazardous Materials, Washington, DC, May 1985.

21. USEPA. *Slurry Trench Construction for Pollution Migration Control.* EPA/540/2-84/001, U.S. Environmental Protection Agency, Washington, DC, February 1984.
22. USEPA. *Grouting Techniques in Bottom Sealing of Hazardous Waste Sites.* EPA/600/2-86/020, U.S. Environmental Protection Agency, Washington, DC, 1986.
23. USEPA. *Engineering Bulletin: Landfill Covers.* EPA/540/S-93/500, U.S. Environmental Protection Agency, Cincinnati, OH, February 1993.
24. FRTR. Physical barriers. In: *Remediation Technologies Screening Matrix and Reference Guide.* http://www.frtr.gov/matrix2/section4/4-53.html, 2015.
25. USEPA. *Engineering Bulletin: Slurry Walls.* EPA/540/S–92/008, U.S. Environmental Protection Agency, Cincinnati, OH, October 1992.
26. USEPA. *Solidification/Stabilization Use at Superfund Sites.* EPA-542-R-00-010, U.S. Environmental Protection Agency, Washington, DC, September 2000.
27. USEPA. *Technical Resource Document: Solidification/Stabilization and Its Application to Waste Materials.* EPA/530/R-93/012, U.S. Environmental Protection Agency, Cincinnati, OH, June 1993.
28. USEPA. *Engineering Bulletin: Solidification/Stabilization of Organics and Inorganics.* EPA/540/S-92/015, U.S. Environmental Protection Agency, Cincinnati, OH, 1992.
29. Conner, J. R. *Chemical Fixation and Solidification of Hazardous Wastes.* VanNostrand Reinhold, New York, 1990.
30. USEPA. *Solidification/Stabilization and Its Application to Waste Materials.* EPA/530/R-93/012, U.S. Environmental Protection Agency, Washington, DC, June 1993.
31. Anderson, W. C. (Ed.). *Innovative Site Remediation Technology: Solidification/Stabilization,* Vol. 4. American Academy of Environmental Engineers and Scientists (AAEES), Annapolis, MD.
32. WASTECH. American Academy of Environmental Engineers (EPA printed under license No. EPA/542-B-94-001), June 1994.
33. USEPA. *A Citizen's Guide to Solidification/Stabilization.* EPA 542-F-01-024, U.S. Environmental Protection Agency, Washington, DC, December 2001.
34. USACE. *Solidification/Stabilization of Contaminated Material, Unified Facility Guide Specification.* UFGS-02160a, U.S. Army Corps of Engineers, Washington, DC, October 2000.
35. ANL. *Fact Sheet—Solidification/Stabilization.* Drilling Waste Management Information System, Argonne National Laboratory. http://web.ead.anl.gov/dwm/techdesc/solid/index.cfm, 2015.
36. USEPA. *Handbook on In Situ Treatment of Hazardous Waste-Contaminated Soils.* EPA/540/2-90/002, U.S. Environmental Protection Agency, Cincinnati, OH, 1990.
37. Arniella, E. F. and Blythe L. J. Solidifying traps hazardous waste. *Chemical Engineering,* 97 (2), 92–102, 1990.
38. Kalb, P. D., Burns, H. H., and Meyer, M. Thermo-plastic encapsulation treatability study for a mixed waste incinerator off-gas scrubbing solution. In: *Third International Symposium on Stabilization/Solidification of Hazardous, Radioactive, and Mixed Wastes,* Gilliam, T. M. (Ed.). ASTM STP 1240, American Society for Testing and Materials, Philadelphia, PA, 1993.
39. Ponder, T. G. and Schmitt, D. Field assessment of air emission from hazardous waste stabilization operation. In: *Proceedings of the 17th Annual Hazardous Waste Research Symposium.* EPA/600/9-91/002, Cincinnati, OH, 1991.
40. Shukla, S. S., Shukla, A. S., and Lee, K. C. Solidification/stabilization study for the disposal of pentachlorophenol. *Journal of Hazardous Materials,* 30, 317–331, 1992.
41. USEPA. *Evaluation of Solidification/Stabilization as a Best Demonstrated Available Technology for Contaminated Soils.* EPA/600/2-89/013, U.S. Environmental Protection Agency, Cincinnati, OH, 1989.
42. Weitzman, L. and Hamel L. E. Volatile emissions from stabilized waste. In: *Proceedings of the 15th Annual Research Symposium.* EPA/600/9-90/006, U.S. Environmental Protection Agency, Cincinnati, OH, 1990.
43. Means, J. L., Nehring, K. W., and Heath, J. C. Abrasive blast material utilization in asphalt roadbed material. In: *Third International Symposium on Stabilization/Solidification of Hazardous, Radioactive, and Mixed Wastes.* ASTM STP 1240, American Society for Testing and Materials, Philadelphia, PA, 1993.
44. US ACE. Yearly average cost index for utilities. In: *Civil Works Construction Cost Index System Manual.* 110-2-1304, U.S. Army Corps of Engineers, Washington, DC, 44pp. PDF file is available at www.publications.usace.army.mil/USAC-publications/Engineer-Manuals/udt_43544_param_page/5/, 2016.
45. Buelt, J. L., Timmerman, C.L., Oma, K. H., FitzPatrick, V. F., and Carter J. G. *In Situ Vitrification of Transuranic Waste: An Updated Systems Evaluation and Applications Assessment.* PNL-4800, Pacific Northwest Laboratory, Richland, WA, 1987.

46. USEPA. *Vitrification Technologies for Treatment of Hazardous and Radioactive Waste.* EPA/625/R-92/002, U.S. Environmental Protection Agency, Cincinnati, OH, May 1992.

47. USEPA. *Engineering Bulletin-In Situ Vitrification Treatment.* EPA/540/S-94/504, U.S. Environmental Protection Agency, Cincinnati, OH, October 1994.

48. USEPA. *Engineering Bulletin: In Situ Vitrification Treatment.* EPA/540/S-94/504, U.S. Environmental Protection Agency, Washington, DC, Revised May 2002.

49. FRTR. Solidification/stabilization—In situ soil remediation technology. In: *Remediation Technologies Screening Matrix and Reference Guide.* http://www.frtr.gov/matrix2/section4/4-8.html, 2015.

50. USEPA. *Geosafe Corporation In Situ Vitrification Innovative Technology Evaluation Report.* EPA/540/R-94/520, U.S. Environmental Protection Agency, Washington, DC, March 1995.

51. Cocârţă, D. M., Dinu, R. N., Dumitrescu, C., Reşetar-Deac, A.M., and Tanasiev, V. Risk-based approach for thermal treatment of soils contaminated with heavy metals. In: *Proceedings of the 16th International Conference on Heavy Metals in the Environment*, Article # 01005, E3S Web of Conferences, Rome, Italy, Vol. 1, April 2013.

52. Timmerman, C. L. *In Situ Vitrification of PCB Contaminated Soils.* EPRI CS-4839, Electric Power Research Institute, Palo Alto, CA, 1986.

53. USEPA. *The Superfund Innovative Technology Evaluation Program: Technology Profiles*, 4th Edition. EPA/540/5-91/008, U.S. Environmental Protection Agency, Washington, DC, 1991.

54. Luey, J., Koegler, S. S., Kuhn, W. L., Lowery, P. S., and Winkelman, R. G. *In Situ Vitrification of a Mixed-Waste Contaminated Soil Site.* The 116-B-6A Crib at Hanford, PNL-8281, Pacific Northwest Laboratory, Richland, WA, 1992.

55. Hansen, J. E. and FitzPatrick. V. F. *In Situ Vitrification Applications.* Geosafe Corporation, Richland, WA, 1991.

56. USEPA. *Engineering Bulletin: Soil Washing Treatment.* EPA/540/2-90/017, U.S. Environmental Protection Agency, Cincinnati, OH, 1996.

57. USEPA. *A Citizen's Guide to Soil Washing.* EPA 542-F-01-008, U.S. Environmental Protection Agency, Washington, DC, May 2001.

58. William C. A. (Ed.). *Innovative Site Remediation Technology: Soil Washing/Flushing*, Vol. 3. American Academy of Environmental Engineers (Published by EPA under EPA 542-B-93-012), November 1993.

59. USEPA. *Technology Focus—Soil Washing.* Technology Innovation Program, U.S. Environmental Protection Agency, Washington, DC. http://clu-in.org/techfocus/default.focus/sec/Soil_Washing/cat/Overview, 2015.

60. Ehsan, S., Prasher, S. O., and Marshall, W. D. A washing procedure to mobilize mixed contaminants from soil. II. Heavy metals. *Journal of Environmental Quality*, 35, 2084–2091, 2006.

61. Fischer, K. and Bipp, H. P. Removal of heavy metals from soil components and soils by natural chelating Agents. Part II. Soil Extraction by Sugar Acids. *Water, Air, and Soil Pollution*, 38 (1–4), 271–288, 2002.

62. USEPA. *Citizens Guide to Soil Washing.* EPA/542/F-92/003, U.S. Environmental Protection Agency, Washington, DC. March 1992.

63. USEPA. *Engineering Bulletin: In Situ Soil Flushing.* EPA/540/2-91/021, U.S. Environmental Protection Agency, Cincinnati, OH, October 1991.

64. FRTR. Soil flushing—In situ soil remediation technology. *Remediation Technologies Screening Matrix and Reference Guide.* http://www.frtr.gov/matrix2/section4/4-6.html, 2015.

65. CPEO. *Soil Flushing.* Center for Public Environmental Oversight (CPEO), San Francisco, CA http://www.cpeo.org/techtree/ttdescript/soilflus.htm, 2015.

66. USEPA. *Superfund Innovative Technology Evaluation Program: Technology Profiles*, 7th Edition. EPA/540/R-94/526, U.S. Environmental Protection Agency, Washington, DC, November 1994.

67. Pamukcu, S. and Wittle, J. K. Electrokinetic removal of selected metals from soil. *Environment Progress* II, 3, 241–250, 1992.

68. Acar, Y. B. Electrokinetic cleanups. *Civil Engineering*, October, 58–60, 1992.

69. Mattson, E. D. and Lindgren, E. R. Electrokinetics: An innovative technology for in situ remediation of metals. In: *Proceedings, National Groundwater Association, Outdoor Acnon Conference*, Minneapolis, MN, May 1994.

70. Acar, Y. B. and Gale R. J. Electrokinetic remediation: Basics and technology status. *Journal of Hazardous Materials*, 40, 117–137, 1995.

71. Will, F. Removing toxic substances from the soil using electrochemistry. *Chemistry and Industry,* May 15, 376–379, 1995.

72. Rodsand, T. and Acar, Y. B. Electrokinetic extraction of lead from spiked Norwegian marine clay. *Geoenvironment 2000*, 2, 1518–1534, 1995.

73. Lindgren, E. R., Kozak, M. W., and Mattson E. D. Electrokinetic remediation of contaminated soils: An update. *Waste Management'92*. Tucson, Arizona, p.1309, 1992.
74. Jacobs, R. A. and Sengun, M. Z. Model of experiences on soil remediation by electric fields. *Journal of Environmental Science and Health*, 29A, 9, 1994.
75. Acar, Y. B. and Gale, R. J. *Electrochemical Decontamination of Soils and Slurries*. U.S. Patent No.: 5,137, 608, Commissioner of Patents and Trademarks, Washington, DC, August 15, 1992.
76. Marks, R., Acar, Y. B., and Gale, R. J. *In situ Bioelectrokinetic Remediation of Contaminated Soils Containing Hazardous Mixed Wastes*. U.S. Patent No. 5,458,747, Commissioner of Patents and Trademarks, Washington, DC, October 17, 1995.
77. Acar, Y. B. and Alshawabkeh, A. N. Electrokinetic remediation: I. Pilot-scale tests with lead spiked kaolinite, II. Theoretical model. *Journal of Geotechnical Engineering*, 122 (3), 173–196, 1996.
78. W. Pool. *Process for the Electroreclamation of Soil Material*. Patent #5,433,829, U.S. Patent Office, July 18, 1995.
79. USEPA. *In Situ Remediation Technology Status Report: Electrokinetics*. EPA 542-K-94-007, U.S. Environmental Protection Agency, Washington, DC, 1995.
80. Editor. Innovative in situ cleanup processes. *The Hazardous Waste Consultant*, September/October 1992.
81. DOE. *Development of an Integrated In-Situ Remediation Technology*. Technology Development Data Sheet, DE-AR21-94MC31185, U.S. Department of Energy, 1995.
82. USEPA. *Lasagna™ Public-Private Partnership*. EPA 542-F-96-010A, U.S. Environmental Protection Agency, Washington, DC, 1996.
83. Szpyrkowicz, L., Radaelli, M., Bertini, S., Daniele, S., and Casarin, F. Simultaneous removal of metals and organic compounds from a heavily polluted soil. *Electrochimica Acta*, 52 (10), 3386–3392, 2007.
84. CPEO. *Electrokinetics*. Center for Public Environmental Oversight (CPEO), San Francisco, CA http://www.cpeo.org/techtree/ttdescript/elctro.htm, 2015.
85. Cunningham, S. D. and Berti, W. R. Remediation of contaminated soils with green plants: An overview. *In Vitro Cellular and Developmental Biology* (Tissue Culture Association), 29, 207–212, 1993.
86. Raskin, I. Bioconcentration of metals by plants. *Environmental Biotechnology*, 5, 285–290, 1994.
87. Goldsbrough, P. Phytochelatins and metallothioneins: Complementary mechanisms for metal tolerance. In: *Fourteenth Annual Symposium 1995 in Current Topics in Plant Biochemistry, Physiology and Molecular Biology*, Columbia, MO, 1995.
88. Salt, D. E. Phytoremediation: A novel strategy for the removal of toxic metals from the environment using plants. *Biotechnology*, 13, 468–474, 1995.
89. Kumar, P. B. A. Phytoextraction: The use of plants to remove metals from soils. *Environmental Science and Technology*, 29, 1232–1238, 1995.
90. Durham, S. *Using Plants to Clean Up Soil*. U.S. Department of Agriculture (USDA). http://www.ars.usda.gov/is/pr/2007/070123.htm, 2015.
91. Morel, I. L. Root exudates and metal mobilization. In: *Fourteenth Annual Symposium 1995 in Current Topics in Plant Biochemistry, Physiology and Molecular Biology*, Columbia, MO, 1995.
92. Baker, A. J. M. and Brooks, R. R. Terrestrial higher plants which hyperaccumulate metallic elements—A review of their distribution, ecology, and phytochemistry. *Biorecovery*, 1, 81–126, 1989.
93. Hyperaccumulation in the genus *Alyssum*. In: *Fourteenth Annual Symposium 1995 in Current Topics in Plant Biochemistry, Physiology and Molecular Biology*, Columbia, MO, 1995.
94. Baker, A. J. M. Metal hyperaccumulation by plants: Our present knowledge of ecophysiological phenomenon. In: *Fourteenth Annual Symposium 1995 in Current Topics in Plant Biochemistry, Physiology and Molecular Biology*, Columbia, MO, 1995.
95. Greger, M. and Landberg, M. T. Improving removal of metals from soil by *Salix*. In: *Proceedings of the 7th International Conference on the Biogeochemistry of Trace Elements*, Uppsala, Sweden, June 15–19, 2003.
96. Chaney, R.L., M. Malik, Y.M. Li, S.L. Brown, J.S. Angle, and A.J.M. Baker. Phytoremediation of soil metals. *Current Opinions in Biotechnology*, 8, 279–284, 1997.
97. King Communications Group, Inc. Promise of heavy metal harvest lures venture funds. *The Bioremediation Report*, Vol. 4, p. 1, Washington, DC, January 1995.
98. Ensley, B. D. Will plants have a role in bioremediation? In: *Fourteenth Annual Symposium 1995 in Current Topics in Plant Biochemistry, Physiology and Molecular Biology*, Columbia, MO, 1995.
99. Cunningham, S. D. and Lee, C. R. Phytoremediation: Plant-based remediation of contaminated soil and sediments. In: *Proceedings of a Symposium of the Soil Science Society of America*, Chicago, IL, November, 1994.

100. DOE. *Summary Report of a Workshop on Phytoremediation Research Needs*. U.S. Department of Energy, Santa Rosa, CA. July 24–26, 1994.
101. Baker, A. J. M., Brooks, R. R., and Reeves, R. D. Growing for gold … and copper … and zinc. *New Scientist*, 1603, 44–48, 1989.
102. Dávila, O.G., Gómez-Bernal, J.M., and Ruíz-Huerta, E.A., *Plants and Soil Contamination with Heavy Metals in Agricultural Areas of Guadalupe, Zacatecas, Mexico*. Academia. Edu http://www.academia.edu/1439828/Plants_and_Soil_Contamination_with_Heavy_Metals_in_Agricultural_Areas_of_Guadalupe_Zacatecas_Mexico, 2015.
103. USDA. Acidifying soil helps plant remove cadmium, zinc metals. Agricultural research service. *Science Daily*. http://www.sciencedaily.com/releases/2005/06/050619192657.htm, 2015.
104. de Souza, R.B., Maziviero, T.G., Christofoletti, C.A., Pinheiro, T.G., and Fontanetti, C.S., Soil contamination with heavy metals and petroleum derivates: Impact on edaphic fauna and remediation strategies. In: *Soil Processes and Current Trends in Quality Assessment*, Soriano, M. C. H. (Ed.). Inteck Publications, Rijeka, Croatia, February, 2013.
105. Wang, L. K., Hung, Y. T., and Shammas, N. K. (Eds.). *Handbook of Advanced Industrial and Hazardous Wastes Treatment*. CRC Press, Taylor & Francis, Boca Raton, VA, 2010.
106. Basta, N. T. and Gradwohl, R. *Remediation of Heavy Metal-Contaminated Soil Using Rock Phosphate*. http://www.ipni.net/ppiweb/bcrops.nsf/$webindex/87396A7CCC0E3C4D852568F00066A560/$file/98-4p29.pdf, 2015.
107. Wang, L. K., Shammas, N. K., Selke, W. A., and Aulenbach, D. B. (Eds.). *Flotation Technology*. Humana Press, Totowa, NJ, 2010.
108. Lee, K.-Y. and Kim, K.-W., Heavy metal removal from shooting range soil by hybrid electrokinetics with bacteria and enhancing agents. *Environmental Science Technology*, 44 (24), 9482–9487, 2010.
109. Mohanty, B. and Mahindrakar, A. B. Removal of heavy metal by screening followed by soil washing from contaminated soil. *International Journal of Technology and Engineering System*, 2 (3), 290–293, 2011.
110. Carbtrol[R]. *Heavy Metal Removal System*. http://www.carbtrol.com/heavy_metal.html, 2015.
111. Wastech Control and Engineering, Heavy Metal Contamination Removal. http://www.wastechengineering.com/heavy-metal-removal-systems.html, 2015.
112. Christian, D., Wong, E., Crawford, R. L., Cheng, I. F., and Hess. T. F. Heavy metals removal from mine runoff using compost bioreactors. *Environmental Technology*, 31 (14), 1533–1546, 2010.
113. Wuana, R. A., Okieimen, F. E., and Imborvungu, J. A. Removal of heavy metals from a contaminated soil using organic chelating acids, *International Journal of Environmental Science and Technology*, 7 (3), 485–496, 2010.
114. Veolia. *Heavy Metal Removal, Ceramic Membranes*. Available at: www.wateronline.com/doc/heavy-metals-with-ceramem-ceramic-0001, 2016.
115. Hyman, M. and Dupont, R. R. *Groundwater and Soil Remediation: Process Design and Cost Estimating of Proven Technologies*. ASCE Publications, Reston, VA, 534, 2001.

5 Removal of Heavy Metals by Low-Cost Adsorption Materials

Siew-Teng Ong, Sie-Tiong Ha, Pei-Sin Keng,
Siew-Ling Lee, and Yung-Tse Hung

CONTENTS

ABSTRACT

Owing to the environmental impact as well as the growing awareness among the public, it is imperative to remove or reduce the concentration of heavy metals to environmentally acceptable levels
before being discharged to open stream. The conventional methods for removing heavy metals
suffer from many drawbacks such as high cost, sludge disposal problem, complex technology, and
limited applicability. Therefore, intensive research has been carried out using low-cost materials to
remove these heavy metals at an affordable cost. This chapter examines (i) some commonly found
heavy metals in wastewater, (ii) main treatment technologies and their limitations, (iii) various
studies using waste materials from agriculture and industry or naturally occurring biosorbents, (iv)
chemical properties and characterization studies on the low-cost adsorbents, (v) influential parameters in affecting the removal efficiency, and (vi) equilibrium, kinetic models, and process design
used in the adsorption process.

5.1   HEAVY METALS

Wastewater may be defined as a combination of liquid and water-transported wastes from homes, commercial buildings, industrial facilities, and institutions along with any groundwater infiltration, surface
water, and stormwater inflow that may enter the sewer system. The rapid growth of human population
and industrialization in the world has resulted in increased wastewater generation. This kind of waste
may contain various pollutants such as heavy metals, toxic organic compounds, phosphorus, detergents,
biodegradable organics, nutrients, dissolved inorganic solids, and refractory organics.

Amongst all, heavy metals pose one of the most serious environmental problems and one of the most
difficult to solve. The term "heavy metals" is misleading because they are not all "heavy" in terms of
atomic weight, density, or atomic number. Besides, they are not even entirely metallic in character, for
example, arsenic. As a rough generalization, the heavy metals include all the metals in the periodic table
except those in Groups I and II (1). Heavy metals such as mercury, lead, arsenic, chromium, copper, cadmium, and nickel are widely used in industry, particularly in metal finishing or metal-plating industries
and in products such as batteries and electronic devices. Nevertheless, the technologically important
heavy metals also cause increasing environmental hazards. Table 5.1 shows the concentrations of leachate contaminants found in the petroleum, calcium fluoride, and metal finishing industrial sludges.

Wastewater containing heavy metals has been of great concern due to their toxicity and carcinogenic effect. Even very small amounts can cause severe physiological or neurological damage. Thus,
numerous ways have been attempted to prevent or minimize this kind of potential health hazard.
This includes government regulations, research to develop methods for waste treatment by scientists, and revision of the technologies used in industries to produce degradable wastes or disposal of
wastes in ways less damaging to the environment and human beings.

5.1.1   CHROMIUM

Chromium (Cr) was discovered in 1979 by the French chemist Louis N. Vauquelin in the rare mineral crocoite ($PbCrO_4$). It was named for the varied colors of its compounds (chroma = color) (3).

TABLE 5.1

Concentrations of Specific Cations, Anions, and Organics in the Three Industrial Sludge Leachates (m/L)

Measured[a] Pollutant	Acidic Petroleum Sludge Leachate	Neutral Calcium Fluoride Sludge Leachate	Basic Metal Finishing Sludge Leachate
Ca	34–50	180–318	31–38
Cu	0.09–0.17	0.10–0.16	0.45–0.53
Mg	27–50	4.8–21	24–26
Ni	—[b]	—	—
Zn	0.13–0.17	—	—
F	0.95–1.2	6.7–11.6	1.2–1.5
Total CN	0.20–1.2	—	—
COD	251–340	44–49	45–50

Source: US Environmental Protection Agency. 1980. *Evaluation of Sorbents for Industrial Sludge Leachate Treatment*, EPA-600/2-80-052. US EPA, Cincinnati, OH.

[a] Fe, Cd, Cr, and Pb contents were analyzed, but found to be below measurable levels.

[b] Dashed line indicated amounts below measurable levels.

It is a naturally occurring element which is commonly found in rocks, minerals, and sources of geologic emissions such as volcanic dusts and gases. Chromium has atomic number 24. There are 13 known isotopes of chromium (mass number 45–47) in which four are stable, giving chromium the relative atomic mass 51.9961. Although chromium can exist in several chemical forms displaying oxidation numbers from 0 to VI, only two of them: trivalent chromium, Cr(III) and hexavalent chromium Cr(VI), are stable enough to occur in the environment (4).

Cr(VI), a Lewis base, is water soluble and always exists in solution as a component of a complex anion. Basically, the speciation of Cr(VI) is concentration and pH dependent. At pH <1, the dominant species is chromic acid (H_2CrO_4); while the equilibrium between monohydrogen chromate ion ($HCrO_4^-$) and dichromate ion ($Cr_2O_7^{2-}$) occurs at pH 2–6. Meanwhile, chromate ion (CrO_4^{2-}) presents as the major component with a pH above 6.

Chromium is usually found in industrial effluents because of their widespread usage in a variety of commercial processes. Chromium and its compounds are used in metal alloys such as stainless steel; protective coatings on metal; magnetic tapes; and pigments for paints, cement, paper, rubber, composite floor covering and other materials. Other uses include chemical intermediate for wood preservatives, organic chemical synthesis, photochemical processing, and industrial water treatment. In medicine, chromium compounds are used in astringents and antiseptics whereas they serve as catalysts and fungicides in the leather tanning industry. Chromium is also found application in brewery processing and brewery warmer water where it acts as an algaecide against slime forming bacteria and yeasts (5).

Since Cr(VI) is able to penetrate through cell membranes efficiently and undergoes strong oxidization, making it a serious environmental pollutant which may represent a considerable health risk (4). Acute high exposure levels cause skin ulceration, perforation of the nasal septum, gastrointestinal irritation, kidney and liver damage as well as internal hemorrhage (5,6). Cr(VI) compounds are also found to produce a variety of genotoxic effects, including DNA damage, mutations, and chromosomal aberrations, in both *in vitro* and *in vivo* test systems (7). The United State Public Health Service has estimated the upper limit from lifetime exposure to 1 mg/L Cr(VI) to result in 120 additional cases of cancer in a population of 10,000 (6).

5.1.2 COPPER

Copper (Cu) is a crystalline reddish metal, with an atomic number of 29 and atomic weight of 63.55. It exists mainly in four valence states, that is, Cu(0), Cu(I), Cu(II), and Cu(III) of which Cu(II) is the most common and stable ion (8). It is easily complexed and is involved in many metabolic processes in living organisms. Copper is among the 25 most abundant elements in the Earth's crust, occurring at about 50–100 g/ton, and has played an important role in human technological, industrial, and cultural development since primitive times. Copper is also distinguished by several properties which contribute to its extensive use: (i) a combination of mechanical workability with corrosion resistance to many substances, (ii) excellent electrical conductivity, (iii) superior thermal conductivity, (iv) efficient as an ingredient of alloys to improve their physical and chemical properties, (v) capable as catalysts for several kinds of chemical reaction, (vi) nonmagnetic characteristics, advantageous in electrical and magnetic apparatus, and (vii) nonsparking characteristics, mandatory for tools for use in explosive atmosphere (9).

Copper is one of the few common metals that find greater commercial applications as pure metal rather than in alloys. The major uses of copper are building construction (roofing parts and gutters) and plumbing installation (valves and pipe fittings), electrical and electronics products (wire, motors, generators, and cable), and household appliances (radios and televisions sets). Apart from these, it is also used in the production of alloys with zinc, nickel, and tin, as catalysts and in the electrochemical industry. Copper salts are useful as pigments, fungicides, and biocides as well as in various pharmaceutical uses. For instance, copper chromate is used as pigments, catalysts for liquid-phase hydrogenation, and as potato fungicides (10).

Copper is also an essential element nutritionally, being among the most abundant metallic elements in the human body, which is needed in many protein and enzymes (i.e., ferroxidases, cytochrome oxidase, superoxide dismutase, and amine oxidases). However, like all heavy metals, intake of excessively large doses of copper by humans will cause severe health disorders such as liver and renal damage, gastrointestinal irritation, anemia, and central nervous system irradiation. Long-term exposure can lead to copper poisoning, especially in people whose bodies have trouble regulating copper because of certain genetic disorders or illness, such as Wilson's disease (11).

5.1.3 CADMIUM

Cadmium (Cd) is a soft, bluish-white metal with an atomic number of 48. It is similar in many respects to zinc (prefers the oxidation state of +2) and mercury (shows low-melting point compared with other transition metals). Cadmium is a metal widely used in industries such as cadmium plating, alkaline batteries, copper alloys, paints, and plastics. Its high resistance to corrosion makes it applicable as a protective layer when it is deposited on other metals.

Most of the Cd compounds released to the environment are contained in solid wastes form (e.g., coal ash, sewage sludge, flue dust, and fertilizers). Cd has been well recognized for its negative effect on the environment where it accumulates throughout the food chain, posing a serious threat to human health. The extremely long biological half-life of Cd also causes a major concern.

Toxic effects of cadmium on humans include both chronic and acute disorders such as testicular atrophy, hypertension, damage to kidneys and bones, anemia, itai-itai, and so on. It has been recorded that the intake of Cd-contaminated rice led to itai-itai disease and renal abnormalities, including proteinuria and glucosuria.

Cd is also found in cigarette smoke and long-term inhalation of CdO dust could cause a syndrome characterized by damage to the pulmonary and renal systems. Acute Cd poisoning may lead to lung edema, in some cases with lethal outcome.

5.2 TREATMENT OF HEAVY METALS

5.2.1 CHEMICAL PRECIPITATION

Chemical precipitation is perhaps the oldest and the most widely used method for the removal of heavy metals from wastewater. This method can be considered as a low-cost and effective process for the removal of large quantities of metal ion. Precipitation involves the formation of an insoluble compound from a solution upon addition of a properly selected reagent. The most commonly used chemicals are lime or caustic for hydroxide precipitation, sodium sulfide or sodium hydrosulfide for sulfide precipitation and sodium bicarbonate for carbonate precipitation. Figure 5.1 illustrates the different designs of hydroxide precipitation, soluble sulfide precipitation (SSP), and insoluble sulfide precipitation (ISP) processes in the wastewater treatment systems. The precipitate can then be separated from the wastewater using some physical separation process, such as sedimentation, coagulation, and filtration. Table 5.2 presents the comparison of metal per liter of raw feed before treatment and wastewater after treatment using five variations of chemical precipitation techniques.

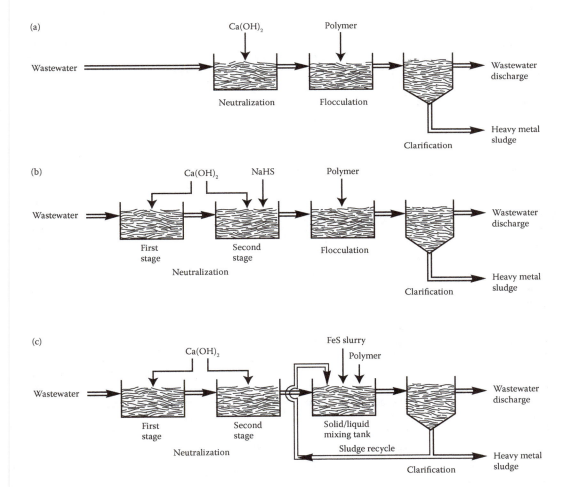

FIGURE 5.1 Wastewater treatment processes for removing heavy metals in the electroplating industry: (a) hydroxide precipitation, (b) SSP, and (c) ISP. (From US Environmental Protection Agency. 1980. *Control and Treatment Technology for the Metal Finishing Industry Sulfide Precipitation*, EPA-625/8–80–003. US EPA, Cincinnati, OH.)

TABLE 5.2

Chemical Analysis of Raw and Treated Wastewater Used in Pilot Tests

Contaminant (µg/L)	Raw Feed Before Treatment	Wastewater After Treatment[a]				
		LO-C	LO-CF	LWS-C	LWS-CF	LSPF
Pilot Test 1						
Cadmium	45	15	8	11	7	20
Total chromium	163,000	3,660	250	1,660	68	159
Copper	4,700	135	33	82	18	3
Nickel	185	30	38	33	31	18
Zinc	2,800	44	10	26	2	11
Lead	119	119	88	104	59	120
Pilot Test 2						
Cadmium	58	7	12	<5	<5	<5
Total chromium	6,300	4	2	5	7	3
Hexavalent chromium	<5	<1	<1	<1	<1	<1
Copper	1,100	860	848	13	13	132
Nickel	160	30	34	33	23	34
Zinc	650,000	2,800	2,300	104	19	242
Mercury	<1	NA	NA	NA	NA	NA
Silver	16	NA	NA	NA	NA	NA
Pilot Test 3						
Cadmium	34	21	21	1	1	1
Total chromium	3	NA	NA	NA	NA	NA
Copper	20	7	8	2	1	4
Nickel	64	29	29	72	34	31
Zinc	440,000	37,000	29,000	730	600	2,000
Mercury	<10	NA	NA	NA	NA	NA
Lead	45	13	14	9	11	13
Silver	61	4	4	1	3	4
Tin	200	<10	<10	<10	<10	<10
Ammonium	([b])	NA	NA	NA	NA	NA
Pilot Test 4						
Cadmium	58,000	1,130	923	26	<10	<10
Total chromium	5,000	138	103	49	50	37
Copper	2,000	909	943	60	160	929
Nickel	3,000	2,200	2,300	1,800	1,900	2,600
Zinc	290,000	1,200	510	216	38	12
Iron	740,000	2,000	334	563	229	305
Mercury	<0.3	<0.3	<0.3	<0.3	<0.3	<0.3
Silver	14	14	10	7	7	8
Tin	5,000	129	81	71	71	71
Pilot Test 5						
Cadmium	<40	<1	<1	<1	<1	<1
Total chromium	1,700	109	39	187	17	20
Copper	21,000	1,300	367	2,250	169	11
Nickel	119,000	12,000	9,400	11,000	3,500	5,300

(Continued)

TABLE 5.2 (*Continued*)
Chemical Analysis of Raw and Treated Wastewater Used in Pilot Tests

Contaminant (µg/L)	Raw Feed Before Treatment	Wastewater After Treatment[a]				
		LO-C	LO-CF	LWS-C	LWS-CF	LSPF
Zinc	13,000	625	10	192	8	5
Iron	NA	2	<2	5	<2	<2
Lead	13	7	5	4	3	3
Silver	6	NA	NA	NA	NA	NA

Source: US Environmental Protection Agency. 1980. *Control and Treatment Technology for the Metal Finishing Industry Sulfide Precipitation*, EPA-625/8–80–003. US EPA, Cincinnati, OH.

Note: Wastewater by pilot test: 1—high chromium rinse from aluminum cleaning, anodizing, and electroplating; 2—chromium, copper, and zinc rinse from electroplating; 3—high zinc rinse from electroplating; 4 and 5—mixed heavy metal rinse from electroplating.

[a] LO-C = lime only, clarified; LO-CF = lime only, clarified, filter; LWS-C = lime with sulfide, clarified; LWS-CF = lime with sulfide, clarified, filtered; LSPF = lime, sulfide polished, filtered; and NA = not applicable.

[b] Qualitative tests indicated the presence of significant amounts of ammonium.

Although this process has wide applicability in the removal of toxic metals from aqueous waste, still, there are limitations need to be addressed. For instance, chemical precipitation is not applicable when the metal of interest is highly soluble and does not precipitate out of solution at any pH such as Cr(VI). Consequently, treatment of Cr(VI) usually consists of a two-stage process: the reduction of Cr(VI) to Cr(III) using sulfur dioxide gas from sodium bisulfate solution, followed by the precipitation of Cr(III) (13). This method is not favorable since it does not allow complete recovery of chromium in the desired hexavalent oxidation state.

For hydroxide precipitation, it requires certain pH in order to reduce metal concentration to that below the level required by standards. This is very difficult to achieve if the solution contains multiple metals as the pH of minimum solubility varies from metal to metal. Figure 5.2 shows the theoretical minimum solubilities for different metals occur at different pH values. For sulfide precipitation, the limitations are the evolution of sulfide gas and discharge of excess soluble sulfide. Nevertheless, sulfide precipitation still appears to be a better alternative compared with hydroxide for removing heavy metals from wastewater. This is mainly attributed to the attractive features of sulfide such as high reactivity (reaction between S^{2-}/HS^{-} with heavy metal ions) and insolubility of metal sulfides over a broad pH range (12). Other limitation of chemical precipitation is the need to use excess amounts of chemical for precipitation to avoid resolubilization of any precipitated compound after filtration thereby implying it is costly. Besides, the disposal of sludge produced during chemical precipitation has created another environmental problem. The generated sludges are hazardous and require a special storage facility and specific treatment before disposal. Table 5.3 lists the characteristics of the wastewater before treatment (hydroxide precipitation, SSP), the volume of sludge generated and the amount of chemical reagents consumed in the treatment. The ultimate disposal of these significant quantities of sludges and large amounts of reagents consumed may be very expensive and indirectly increase the cost of treatment.

5.2.2 Ion Exchange

Ion exchange is a chemical treatment process used to remove the dissolved ionic species from contaminated aqueous streams. It involves the reversible exchange of ions in solution with the ions held

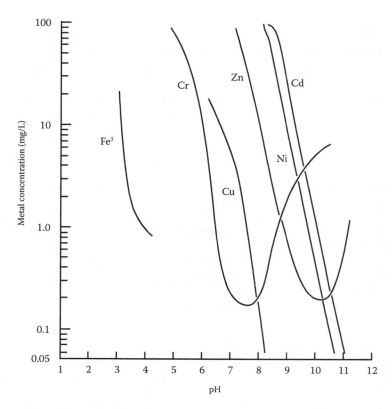

FIGURE 5.2 Metal solubility as a function of pH. (From US Environmental Protection Agency. 1973. *Waste Treatment: Upgrading Metal-Finishing Facilities to Reduce Pollution*, EPA-625/3–73-002. US EPA, Cincinnati, OH.)

by a solid ion-exchanging material, in which there is no directly perceptible permanent change in the structure of the solid. Ion exchangers are generally utilized in column reactors so that a high degree of exchanger utilization is achieved. They can be characterized by a number of physical properties including particle size, density, degree of cross-linking, resistance to oxidation, and thermal stability.

Ion exchange resin can be broadly classified as strong or weak cation exchangers and strong or weak anion exchangers. Table 5.4 shows the capacity of ion exchangers and cost of ion-exchange operation for metal recovery. The classification of the resins is based on the active ion-exchange sites of the resin, for example, strong acid cation exchange resin possesses sulfonic groups; weak acid cation exchange resin generally contains carboxylic acid groups; strong base anion exchange posseses quaternary ammonium groups while weak base anion exchange resin contains functional groups that are derived from weak base amines, such as tertiary ($-NR_2$), secondary ($-NHR$), or primary ($-NH_2$) amino groups. Chelating resins behave similar to weak acid cation resins but exhibit a high degree of selectivity for metal cations over sodium, calcium, or magnesium.

Soluble heavy metals, which are amenable to treatment by ion exchange include arsenic, barium, cadmium, chromium, cyanide, mercury, selenium, and silver. The advantage of ion-exchange technique is put to use in the treatment of wastewater without generating sludge. Besides, it permits the reuse of rinse water in a close cycle and recovery of metal in the wastewater. However, regardless of the efficiency of ion-exchange resins for heavy metal removal, the cost incurred (Table 5.4) prohibits the treatment of highly concentrated wastewater; it is thereby typically used as a polishing step after precipitation.

TABLE 5.3
Wastewater Treatment Process Details of Pilot Tests

	Pilot Test[a]				
Characteristic	1	2[b]	3	4	5
Raw feed before treatment					
pH	1.7	1.2	6.4	2.4	7.1
Conductivity (μmho/cm)	10,600 at 70°F	149,000 at 68°F	12,100 at 77°F	5,600 at 66°F	1,500 at 70°F
Color	Yellow	Colorless	Colorless	Colorless	Pale green
Precipitation pH for LO and LWS processes	8.5	6.2/9.0	9.0	10.0	8.5
Sludge volume (%)[c]					
LO process	18	78/23	[d]	43	5
LWS process	16	78/13	[d]	37	6
Process consumables (mg/L)					
Sulfuric acid for Cr^{6+} reduction	0	0	0	0	339
Sodium sulfite for Cr^{6+} reduction	226	31	0	41	25
Calcium oxide for neutralization	1,530	14,380	911	2,680	145
Sulfide for LWS process	8	381		400	91
Sulfide for LPSF process	1	5		141	67

Source: US Environmental Protection Agency. 1980. *Control and Treatment Technology for the Metal Finishing Industry Sulfide Precipitation*, EPA-625/8–80–003. US EPA, Cincinnati, OH.

Note: LO = lime only; LWS = lime with sulfide; and LSPF = lime, sulfide polished, filtered.

[a] Wastewater by pilot test: 1—high chromium rinse from aluminum cleaning, anodizing, and electroplating; 2—chromium, copper, and zinc rinse from electroplating; 3—high zinc rinse from electroplating; 4 and 5—mixed heavy metal rinse from electroplating.

[b] Because of the exceptionally large volume of sludge generated by this wastewater, precipitation was accomplished in two stages. First- and second-stage values are separated by a diagonal line; single values apply to the total process.

[c] Sludge volume per solution volume, percent after 1 h settling.

[d] Data not available.

5.2.3 MEMBRANE SYSTEM

One of the growing interests in the reduction and/or recycling of hazardous waste involves the use of membrane separation processes. These processes include reverse osmosis, electrodialysis, hyperfiltration, and ultrafiltration. Reverse osmosis is a pressure-driven membrane process in which a feed stream containing inorganic ions under the pressure is separated into a purified permeated stream and a concentrate stream. The pure water is forced through a semipermeable membrane into the less concentrated solution and the flow stops when equal concentrations are attained on both sides of the membrane, at which point the solvent molecules pass through the membrane in both directions at equal rates. The most commonly used membrane materials are cellulose acetate, aromatic polyamides, and thin film composites. One of the major applications of reverse osmosis has been in the recovery of metals from the effluents generated by the electroplating plants, which have been engaged in electroplating nickel, copper, brass, and cadmium.

Ultrafiltration and hyperfiltration utilize pressure and a semipermeable membrane to separate nonionic materials from the solvent. These membrane separation techniques are particularly effective for the removal of suspended solid, oil, and grease, large organic molecules, and heavy metal complexes from the wastewater stream.

Electrodialysis is used for the separation, removal, or concentration of ionized species in aqueous solutions by the selective transport of ions through ion-exchange membranes under the influence

TABLE 5.4

Ion Exchange Capacity and Cost of Ion Exchange Operation for Metal Recovery

	Cation Exchange			Anion Exchange	
Metal Form	Capacity (lb/ft³)	Cost (cents/lb)	Metal Form	Capacity (lb/ft³)	Cost (cents/lb)
Al_2O_3	1.1	14	Sb	4.5	6.7
BeO	0.5	30	Bi	3.1	9.7
Cd	6.7	2.3	Cr_2O_3	1.9	16
Ce_2O_3	5.6	2.7	Ga	5.2	5.8
CsCl	16.0	9.4	Ge	5.4	5.6
CoO	3.6	4.2	Au	7.3	4.1
Cu	3.8	3.9	Ha	6.6	4.9
Pb	12.4	1.2	Ir	7.1	4.2
LiO	0.8	18	Mo	3.6	8.4
Mg	1.5	10	Nb	3.4	8.8
MgO	1.5	10	Pd	3.9	7.8
Mn	3.3	4.6	Pt	7.2	4.2
Hg	12	13	Re	13.8	2.2
Ni	3.5	4.3	Rh	2.9	10
Ra	13.6	11	Ta	6.7	4.5
Rare earths	6.3	2.4	ThO_2	8.6	3.5
Ag	13	1.2	W_2O3	6.8	4.4
Sn	7.1	2.1	V_2O_5	3.8	7.9
Zn	3.9	38	UO_2	8.8	3.4
			Zr	3.4	8.8

Source: US Environmental Protection Agency. 1973. *Traces of Heavy Metals in Water Removal Processes and Monitoring*, EPA-902/9-74-001. US EPA, Cincinnati, OH.

of an electrical potential across the membrane. Depending on the ion-exchange material, the membranes are permeable to either anions or cations, but not both. These membranes allow the ions to transfer through them from a less concentrated to a more concentrated solution.

A membrane system can be used for the removal of heavy metal ions but the concentration of metal ions in the feed stream has to be reasonably low for a successful operation of a membrane process. With the increasing concentration of metals in the feed streams, the rejection of the membrane is lowered and a membrane scaling is often noted. This shows an increase in the process cost but a decline in process efficiency. In addition, membranes used in the process are considerably expensive materials, a fact that is aggravated by their relative short operation life. Membranes are subjected to deterioration in the presence of microorganism, compaction, scaling, and loss of productivity with time. As such, this system remains as an expensive treatment option and requires a high level of technical expertise to operate.

5.2.4 ADSORPTION

Adsorption is an attachment of the molecules of a gas or a liquid to the surface of another substance (usually solid); these molecules form a closely adherent film or layer held in place by different attractive forces. The three defined forces are physical, chemical, and electrostatic interactions. Physical adsorption results from the action of van der Waals forces; chemical adsorption involves electronic interactions between specific surface sites and solute molecules; an electrostatic interaction is generally reserved for Coulombic attractive forces between ions and charged functional groups.

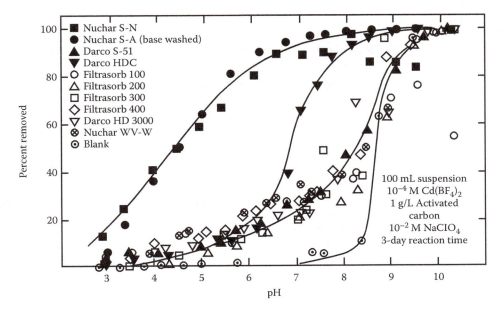

FIGURE 5.3 Typical cadmium(II) removal of different types of activated carbons as affected by pH. (From US Environmental Protection Agency. 1983. *Activated Carbon Process for the Treatment of Cadmium(II)-Containing Wastewaters*, EPA/600/S2-83-061. US EPA, Cincinnati, OH.)

Activated carbon, both in granular and powder form, is recognized as one of the most well-known adsorbents. Granular-activated carbons are widely used in flow through column reactors for carbon adsorption systems. Figure 5.3 exemplifies the effect of 17 different types of commercial activated carbons on Cd(II) removal. The adsorption properties of activated carbon are primarily a result of its highly porous structure, or equivalently the high specific surface area of the finished product. This kind of adsorption process is reversible and it is usually used in removing the adsorbed contaminants after the adsorption capacity of the carbon has been exhausted.

The applications of activated carbon adsorption for heavy metals have also been well documented. However, it is ineffective for very low concentrations. Another drawback of activated carbon adsorption in heavy metal removal is its high affinity toward organic molecules. Thus, in the presence of any high molecular-weight compounds, the internal pores in the deep regions of the bed are blocked and unavailable to adsorb contaminants. Besides, the activation process and regeneration of activated carbon require high capital investment. The heat treatment and activation process must be repeated after every regeneration process following the elution of saturated carbon. Apart from that, the carbon suffers from weight loss and reduction in adsorption capacity by approximately 10%–15% after each regeneration process. Another problem associated with the carbon adsorbent is the development of excessive head loss as a result of suspended solid accumulation, biological growth in the bed, or fouling of the influent screen.

5.2.5 Biosorption

Generally, all biological materials have certain biosorptive ability. In this case, biosorption can be considered as a new sorption process developed for the removal of toxic metal ions from wastewater. This kind of sorption process involves the removal of metal or metalloids species, compounds, and particulates by biological materials through passive sorption. Tables 5.5 through 5.7 present the removal capacities of biosorbents, nonbiosorbents, and activated carbons for treating sludge leachates from petroleum, calcium fluoride, and metal finishing industries, respectively.

TABLE 5.5

Net Sorbent Removal Capacities for Treating Acidic Petroleum Sludge Leachate (μg/g)[a]

Pollutant	Acidic Fly Ash	Basic Fly Ash	Zeolite	Vermiculite	Illite	Kaolinite	Activated Alumina	Activated Carbon
Ca	0	0	1,390	686	721	10.5	200	128
Cu	2.4	1.9	5.2	1.1	0	0	0.35	0
Mg	0	102	746	67	110	595	107	8.6
Zn	1.6	1.7	10.8	4.5	0	0	0.40	1.1
F-	8.7	6.2	4.1	0	9.3	3.5	3.4	1.2
CN-	2.7	2.5	4.7	7.6	12.1	3.1	0	2.4
COD	3,818	3,998	468	6,654	4,807	541	411	3,000
TOC	1,468	737	170	2,545	2,175	191	176	1,270

Source: US Environmental Protection Agency. 1980. *Evaluation of Sorbents for Industrial Sludge Leachate Treatment*, EPA-600/2-80-052. US EPA, Cincinnati, OH.

Note: +Cl-, Cd, Cr, Fe, Ni, and Pb were measured and found in low concentrations.

[a] μg of contaminant removed/g of sorbent used.

TABLE 5.6

Net Sorbent Removal Capacities for Treating Neutral Calcium Fluoride Sludge Leachate (μg/g)[a]

Pollutant	Acidic Fly Ash	Basic Fly Ash	Zeolite	Vermiculite	Illite	Kaolinite	Activated Alumina	Activated Carbon
Ca	261	0	5054	0	0	857	6140	357
Cu	2.1	0.36	8.2	0	0	6.7	2.9	2.0
Mg	230	155	0	0	0	0	214	3.0
F-	102	51.8	27.7	0	175	132	348	0
COD	690	203	171	0	108	185	0	956
TOC	153	44.7	93	0	26.1	71	0	325

Source: US Environmental Protection Agency. 1980. *Evaluation of Sorbents for Industrial Sludge Leachate Treatment*, EPA-600/2-80-052. US EPA, Cincinnati, OH.

Note: +Cl-, CN-, Cd, Cr, Cu, Fe, Ni, Pb, and Zn were measured and found in low concentrations.

[a] μg of contaminant removed/g of sorbent used.

Biomass raw materials (e.g., seaweed and algae) or wastes from other industrial operations (e.g., fungi from fermentation process) serve as attractive sources of biosorbents. Table 5.8 illustrates the Freundlich constants for sorption of four heavy metals, which adequately described the removal efficiency of the filamentous fungi. The cell wall of the biosorbents which consists of mainly polysaccharides, proteins, and lipids is capable of concentrating heavy metal ions, known as bioaccumulation. Furthermore, the presence of many functional groups such as carboxylate, hydroxyl, sulfate, phosphate, and amino groups which can bind metal ions is also considered as an added feature for this kind of biosorbent.

The interaction between biosorbents and metal can occur via complexation, coordination, chelation, ion exchange, adsorption, and inorganic microprecipitation. Any one or combination of the mentioned metal binding mechanisms may be subjected to various degrees in immobilizing one or more metallic species on the biosorbent.

Some of the advantages of biosorption worth mentioning include its (i) ability to bind heavy metal ions in the presence of commonly encountered ions such as calcium, magnesium, sodium, chloride, sulfate, and potassium without interference; (ii) efficiency in metal removal and is often

TABLE 5.7
Net Sorbent Removal Capacities for Treating Basic Metal Finishing Sludge Leachate (μg/g)[a]

Pollutant	Acidic Fly Ash	Basic Fly Ash	Zeolite	Vermiculite	Illite	Kaolinite	Activated Alumina	Activated Carbon
Ca	87.3	97.8	1,240	819	1280	735	737	212
Cu	13.0	6.1	85.4	15.2	43.1	23.7	6.2	16.8
Mg	296	176	1,328	344	1,122	494	495	188
Ni	3.8	1.7	13.5	2.3	5.1	4.6	2.3	4.7
F-	0	0	2.1	0	2.2	2.6	11.4	0
COD	1,080	259	0	618	1,744	0	0	1,476
TOC	430	115	0	244	729	0	0	589

Source: US Environmental Protection Agency. 1980. *Evaluation of Sorbents for Industrial Sludge Leachate Treatment,* EPA-600/2-80-052. US EPA, Cincinnati, OH.

Note: +CI, CN, Cd, Cr, Fe, Pb, and Zn were measured and found in low concentrations.

[a] μg of contaminant removed/g of sorbent used.

comparable with commercial ion exchangers; and (iii) role in improving a zero-waste economic policy especially in the case of reuse of agricultural and industrial byproducts.

However, the capability of biosorbents in metal removal is greatly affected by several factors. These include the specific surface properties of the biosorbents and physicochemical parameters of the solution, for instances, temperature, pH, initial metal ion concentration, and biomass concentration. If there is more than one metal to be bound simultaneously, the combined effects would depend on metal ion combination, levels of metal concentration, and on the order of metal addition (18,19).

Some of the advantages and disadvantages of the main conventional removal of heavy metals technologies discussed above are summarized in Table 5.9. Cost of treatment is always the main consideration in choosing a suitable type of heavy metal treatment. Figure 5.4 shows the estimated total annual cost for various treatment processes in Cd(II) wastewater.

TABLE 5.8
Freundlich Constants for Heavy Metal Sorption by Filamentous Fungi

Metal	Fungus	K	N	r²
Ag	*A. niger*	1.096	0.892	0.953
	M. rouxii	3.373	0.641	0.806
Cd	*A. niger*	0.156	0.679	0.861
	M. rouxii	0.039	0.875	0.994
Cu	*A. niger*	0.889	0.495	0.921
	M. rouxii	0.746	0.551	0.963
La	*A. niger*	2.877	0.426	0.971
	M. rouxii	5.702	0.314	0.968

Source: US Environmental Protection Agency. 1990. *Sorption of Heavy Metals by Intact Microorganisms, Cell Walls, and Clay-Wall Composites,* EPA/600/M-90/004. US EPA, Cincinnati, OH.

Note: The constant K represents the amount of metal sorbed in μmol/g at an equilibrium concentration of 1 μM and n is the slope of the log transformed isotherm.

TABLE 5.9

Comparison of Main Convectional Heavy Metals Removal Technologies

Technology	Advantages	Disadvantages
Chemical precipitation	• Simple and inexpensive treatment • Chemicals used are easily available	• High sludge production • Disposal problem • pH sensitive • Moderate metal selectivity (sulfide) • Nonmetal selectivity (hydroxide)
Ion exchange	• High regeneration of materials • Effective pure effluent • Metal selectivity • Metal recovery	• High cost • Sensitive to suspended solids
Membrane systems	• Less solid waste production • Pure effluent • Metal recovery • Minimal chemical consumption	• High initial and running cost • Membrane scaling • Sensitive to suspended solids • High pressures • Efficiency decreases with the presence of other metals
Adsorption by activated carbon	• Removes most of the heavy metals • High efficiency	• Pores blockage in the presence of high molecular-weight compounds • High regeneration cost • Weight loss and reduction in adsorption capacity
Biosorption	• Economically attractive, utilization of nature resources, regeneration is avoidable	• Temperature, pH, initial metal ion concentration and biomass concentration dependent

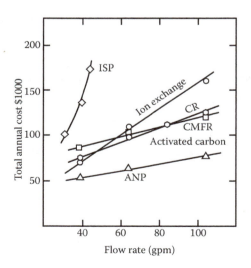

FIGURE 5.4　Estimated total cost of various cadmium(II) treatment processes. (From US Environmental Protection Agency. 1983. *Activated Carbon Process for the Treatment of Cadmium(II)-Containing Wastewaters*, EPA/600/S2-83-061. US EPA, Cincinnati, OH.) *Note*: CMFR = completely mixed flow reactor-activated carbon; CR = column reactor-activated carbon; ANP = alkaline neutralization precipitation; and ISP = insoluble sulfide precipitation.

5.3 APPLICATION OF LOW-COST ADSORBENTS FOR HEAVY METALS REMOVAL

The need of industries to lessen the pollution loads before the discharge enters the surface water and the limitations of existing conventional methods for heavy metals removal have led to the search in developing low-cost treatment methods. Numerous techniques have been attempted; but still, adsorption is constantly viewed as a highly effective method for this purpose. The application of low-cost adsorbent materials makes this approach even more attractive and feasible. In this context, low-cost adsorbent materials can be defined as those that are generally available at free cost and are abundant in nature. Utilization of naturally occurring material or locally available agricultural waste materials, or industrial byproducts as the adsorbents for the removal of heavy metals from wastewaters offers not only an economical approach for heavy metal removal, but also other advantages such as the possibility of attaining a zero-waste situation in the environment.

A lot of investigations have been reported on using these low-cost adsorbent materials for the adsorption of individual or multiple heavy metals in an aqueous solution. Some of these adsorbents have shown excellent performance in the removal of heavy metals from industrial wastewater. In this section, some of the selected materials from industrial byproducts, agriculture waste, and biosorbents were discussed in terms of their efficiency for heavy metals removal. Recent reported adsorption capacities of the selected adsorbents are presented in Tables 5.10 through 5.15 to provide some idea of adsorbent effectiveness. However, the reported adsorption capacities must be taken as values that can be attained only under specific conditions since the adsorption capacities of the adsorbents would be varied, depending on the characteristics of the adsorbent, the experimental conditions as well as the extent of chemical modifications. Thus, the reader is encouraged to refer to the original articles for a detailed information on the experimental conditions.

5.3.1 FLY ASH

Fly ash, a waste from the perspective of power generation is generally gray in color, abrasive, mostly alkaline, and refractory in nature. The primary components of fly ash have been identified as alumina (Al_2O_3), silica (SiO_2), calcium oxide (CaO), and iron oxide (Fe_2O_3), with varying amounts of carbon, calcium, magnesium, and sulfur. The chemical composition and physical properties of fly ash may vary due to the variations in coals from different sources as well as differences in the design of coal-fired boilers. However, an empirical formula for fly ash based on the dominance of certain key elements has been proposed as (20)

$$Si_{1.0}Al_{0.45}Ca_{0.51}Na_{0.047}Fe_{0.039}Mg_{0.020}K_{0.013}Ti_{0.011}$$

Generally, fly ash can be classified into two types: (i) type C, which is normally produced from the burning of low-rank coals (lignites or subbituminous coals) and has cementitious properties (self-hardening upon reaction with H_2O) and (ii) type F which is commonly produced from the burning of higher-rank coals (bituminous coals or anthracites) that is pozzolanic in nature (hardening when reacted with $Ca(OH)_2$ and H_2O). The main difference between these two types lies on the sum of SiO_2, Al_2O_3, and Fe_2O_3.

Most of the fly ash generated is disposed of as landfill, a practice which is under examination for environmental concerns. Therefore, continuing research efforts have been made to utilize this waste material into new products rather than land disposal to lessen the environmental burden. The potential applications of fly ash include as raw material in cement and brick production and as filler in road works. The conversion of fly ash into zeolite has gained considerable interest as well. Another attractive possibility might be to make it into a low-cost adsorbent for gas and water treatment provided production could match industrial needs. A lot of investigations have been reported

TABLE 5.10

Adsorption Capacities of Metals by Fly Ash[a]

Metals	Adsorbent	Adsorption Capacity[b]	Temperature (°C)	References
As(III)	Fly ash coal-char	3.7–89.2	25	21
As(V)	Fly ash	7.7–27.8	20	22
	Fly ash coal-char	0.02–34.5	25	21
Cd(II)	Fly ash	1.6–8.0	–	23
	Fly ash zeolite	95.6	20	23
	Fly ash	0.67–0.83	20	24
	Afsin-Elbistan fly ash	0.08–0.29	20	25
	Seyitomer fly ash	0.0077–0.22	20	25
	Fly ash	198.2	25	26
	Fly ash-washed	195.2	25	26
	Fly ash-acid	180.4	25	26
	Bagasse fly ash	1.24–2.0	30–50	27
	Fly ash	0.05	25	28
	Coal fly ash	18.98	25	29
	Coal fly ash pellets	18.92	–	29
	Bagasse fly ash	6.19	–	30
	Fly ash zeolite X	97.78	–	31
Co(II)	Fly ash zeolite 4A	13.72	–	32
Cr(III)	Fly ash	52.6–106.4	20–40	33
	Bagasse fly ash	4.35	–	34
	Fly ash zeolite 4A	41.61	–	32
Cr(VI)	Fly ash + wollastonite	2.92	–	35
	Fly ash + China clay	0.31	–	35
	Fly ash	1.38	30–60	36
	Fe impregnated fly ash	1.82	30–60	36
	Al impregnated fly ash	1.67	30–60	36
	Afsin-Elbistan fly ash	0.55	20	25
	Seyitomer fly ash	0.82	20	25
	Bagasse fly ash	4.25–4.35	30–50	34
	Fly ash	23.86	–	37
Cs(I)	Fly ash zeolite	443.9	25	38
Cu(II)	Fly ash	1.39	30	39
	Fly ash +wollastonite	1.18	30	39
	Fly ash	1.7–8.1	–	23
	Afsin-Elbistan fly ash	0.34–1.35	20	40
	Seyitomer fly ash	0.09–1.25	20	40
	Fly ash	207.3	25	41
	Fly ash-washed	205.8	25	41
	Fly ash-acid	198.5	25	41
	Fly ash	0.63–0.81	25	42
	Bagasse fly ash	2.26–2.36	30–50	43
	Fly ash	0.76	32	44
	Fly ash	7.5	–	41
	Coal fly ash pellets	20.92	25	29
	Fly ash zeolite 4A	50.45	–	32
	Fly ash	7.0	–	45
	Coal fly ash (CFA)	178.5–249.1	30–60	46
	CFA-600	126.4–214.1	30–60	46
	CFA–NAOH	76.7–137.1	30–60	46

(Continued)

TABLE 5.10 (*Continued*)
Adsorption Capacities of Metals by Fly Ash[a]

Metals	Adsorbent	Adsorption Capacity[b]	Temperature (°C)	References
	Fly ash zeolite X	90.86	–	31
	Fly ash	7.0	–	47
Hg(II)	Fly ash	2.82	30	48
	Fly ash	11.0	30–60	36
	Fe impregnated fly ash	12.5	30–60	36
	Al impregnated fly ash	13.4	30–60	36
	Sulfo-calcic fly ash	5.0	30	41
	Silico-aluminous ashes	3.2	30	41
	Fly ash-C	0.63–0.73	5–21	49
Ni(II)	Fly ash	9.0–14.0	30–60	50
	Fe impregnated fly ash	9.8–14.93	30–60	50
	Al impregnated fly ash	10–15.75	30–60	50
	Afsin-Elbistan fly ash	0.40–0.98	20	40
	Seyitomer fly ash	0.06–1.16	20	40
	Bagasse fly ash	1.12–1.70	30–50	27
	Fly ash	3.9	–	41
	Fly ash zeolite 4A	8.96	–	32
	Afsin-Elbistan fly ash	0.98	–	25
	Seyitomer fly ash	1.16	–	25
	Bagasse fly ash	6.48	–	30
	Fly ash	0.03	–	51
Pb(II)	Fly ash zeolite	70.6	20	52
	Fly ash	444.7	25	53
	Fly ash-washed	483.4	25	53
	Fly ash-acid	437.0	25	53
	Fly ash	753	32	53
	Bagasse fly ash	285–566	30–50	54
	Fly ash	18.8	–	22
	Fly ash zeolite X	420.61	–	31
Zn(II)	Fly ash	6.5–13.3	30–60	50
	Fe impregnated fly ash	7.5–15.5	30–60	50
	Al impregnated fly ash	7.0–15.4	30–60	50
	Fly ash	0.25–2.8	20	24
	Afsin-Elbistan fly ash	0.25–1.19	20	40
	Seyitomer fly ash	0.07–1.30	20	40
	Bagasse fly ash	2.34–2.54	30–50	43
	Bagasse fly ash	13.21	30	55
	Fly ash	4.64	23	56
	Fly ash	0.27	25	28
	Fly ash	0.068–0.75	0–55	57
	Fly ash	3.4	–	41
	Fly ash zeolite 4A	30.80	–	32
	Bagasse fly ash	7.03	–	30
	Fly ash	11.11	–	47
	Rice husk ash	14.30	–	58
	Fly ash	7.84	–	47

[a] These reported adsorption capacities are values obtained under specific conditions. Readers are encouraged to refer to the original articles for information on experimental conditions.

[b] In mg/g.

TABLE 5.11
Adsorption Capacities of Metals by Rice Husk[a]

Metals	Adsorbent	Adsorption Capacity[b]	Temperature (°C)	References
As(III)	Copolymer of iron and aluminum impregnated with active silica derived from rice husk ash	146	–	61
As(V)	Rice husk	615.11	–	62
	Quaternized rice husk	18.98	–	63
Au(I)	Rice husk	64.10	40	64
	Rice husk	50.50	30	64
	Rice husk	39.84	20	64
	Rice husk ash	21.2	–	65
Cd(II)	Partial alkali digested and autoclaved rice husk	16.7	–	66
Cd(II)	Phosphate-treated rice husk	103.09	20	67
	Rice husk	73.96	–	68
	Rice husk	21.36	–	62
	Rice husk	4	–	69
	Rice husk	8.58 ± 0.19	–	70
	Rice husk	0.16	–	71
	Rice husk	0.32	–	72
	NaOH treated rice husk	125.94	–	68
	NaOH treated rice husk	7	–	69
	NaOH treated rice husk	20.24 ± 0.44	–	70
	NaHCO$_3$ treated rice husk	16.18 ± 0.35	–	70
	Epichlorohydrin treated rice husk	11.12 ± 0.24	–	70
	Rice husk ash	3.04	–	73
	Polyacrylamide grafted rice husk	0.889	–	74
	HNO$_3$, K$_2$CO$_3$ treated rice husk	0.044 ± 0.1[c]	30	75
	Partial alkali digested and autoclaved rice husk	9.57	–	66
Cr(III)	Rice husk	1.90	–	72
	Rice husk ash	240.22	–	76
Cr(VI)	Rice husk	164.31	–	62
	Rice husk	4.02	–	71
	Rice husk ash	26.31	–	37
	Rice husk-based activated carbon	14.2–31.5	–	77
	Formaldehyde treated rice husk	10.4	–	78
	Preboiled rice husk	8.5	–	78
Cu(II)	Tartaric acid modified rice husk	29	27	79
	Tartaric acid modified rice husk	22	50	79
	Tartaric acid modified rice husk	18	70	79
	Tartaric acid modified rice husk	31.85	–	80
	Rice husk heated to 500°C (RHA500)	16.1	–	81
	Rice husk	1.21	–	72
	Rice husk	0.2	–	73
	Rice husk	7.1	–	81
	Rice husk ash	11.5191	–	82
	RH-cellulose	7.7	–	81
	Rice husk heated to 300°C (RHA300)	6.5	–	81
	Microwave incinerated rice husk ash (800°C)	3.497	–	83
	Microwave incinerated rice husk ash (500°C)	3.279	–	83
	HNO$_3$, K$_2$CO$_3$ treated rice husk	0.036 ± 0.2[c]	30	75

(Continued)

TABLE 5.11 (*Continued*)

Adsorption Capacities of Metals by Rice Husk [a]

Metals	Adsorbent	Adsorption Capacity[b]	Temperature (°C)	References
	Partial alkali digested and autoclaved rice husk	10.9	–	66
Fe(II)	Copolymer of iron and aluminum impregnated with active silica derived from rice husk ash	222	–	61
Hg(II)	Rice husk ash	6.72	30	84
	Rice husk ash	9.32	15	84
	Rice husk ash	40.0–66.7	–	85
	Polyaniline/rice husk ash nanocomposite	Not determined	–	86
	Partial alkali digested and autoclaved rice husk	36.1	–	66
Mn	Copolymer of iron and aluminum impregnated with active silica derived from rice husk ash	158	–	61
	Partial alkali digested and autoclaved rice husk	8.30	–	66
Ni(II)	Rice husk	0.23	–	72
	Rice husk ash	4.71	–	87
	Microwave-irradiated rice husk (MIRH)	1.17	30	88
	Partial alkali digested and autoclaved rice husk	5.52	–	66
Pb(II)	Rice husk ash	12.61	30	84
	Rice husk ash	12.35	15	84
	HNO_3, K_2CO_3 treated rice husk	0.058 ± 0.1[c]	30	75
	Rice husk ash	207.50	–	89
	Rice husk ash	91.74	–	90
	Copolymer of iron and aluminum impregnated with active silica derived from rice husk ash	416	–	61
	Tartaric acid modified rice husk	120.48	–	79
	Tartaric acid modified rice husk	108	27	79
	Tartaric acid modified rice husk	105	50	79
	Tartaric acid modified rice husk	96	70	79
	Partial alkali digested and autoclaved rice husk	58.1	–	66
	Tartaric acid modified rice husk	21.55	–	69
	Rice husk	6.385	25	91
	Rice husk	5.69	30	92
	Rice husk	45	–	69
	Rice husk	11.40	–	62
Zn(II)	HNO_3, K_2CO_3 treated rice husk	0.037 ± 0.2[c]	30	75
	Rice husk	30.80	50	93
	Rice husk	29.69	40	93
	Rice husk	28.25	30	93
	Rice husk	26.94	20	93
	Rice husk ash	14.30	–	58
	Rice husk ash	7.7221	–	82
	Rice husk ash	5.88	–	73
	Partial alkali digested and autoclaved rice husk	8.14	–	66
	Rice husk	0.75	–	72
	Rice husk	0.173	–	71

[a] These reported adsorption capacities are values obtained under specific conditions. Readers are encouraged to refer to the original articles for information on experimental conditions.

[b] In mg/g except in footnote c.

[c] In mmol/g.

TABLE 5.12

Adsorption Capacities of Metals by Wheat-Based Materials[a]

Metals	Adsorbent	Adsorption Capacity[b]	References
Cd(II)	Wheat straw	14.56	94
	Wheat straw	11.60	95
	Wheat straw	40.48	96
	Wheat bran	51.58	97
	Wheat bran	15.71	98
	Wheat bran	21.0	99
	Wheat bran	101	100
Cr(III)	Wheat straw	21.0	101
	Wheat bran	93.0	99
Cr(VI)	Wheat straw	47.16	96
	Wheat bran	35	102
	Wheat bran	40.8	103
	Wheat bran	310.58	104
	Wheat bran	0.942	105
Cu(II)	Wheat straw	11.43	94
	Wheat straw-citric acid treated	78.13	106
	Wheat bran	12.7	107
	Wheat bran	17.42	108
	Wheat bran	8.34	109
	Wheat bran	6.85	110
	Wheat bran	51.5	111
	Wheat bran	15.0	99
Hg(II)	Wheat bran	70.0	99
Ni(II)	Wheat straw	41.84	96
	Wheat bran	12.0	99
Pb(II)	Wheat bran	87.0	112
	Wheat bran	62.0	99
	Wheat bran	79.4	100
Zn(II)	Wheat bran	16.4	107
U(VI)	Wheat straw	19.2–34.6	113

[a] These reported adsorption capacities are values obtained under specific conditions. Readers are encouraged to refer to the original articles for information on experimental conditions.

[b] In mg/g.

on the use of fly ash in the adsorption of individual pollutants in an aqueous solution or flue gas. The results obtained when using these particular fly ashes are encouraging for the removal of heavy metals and organics from industrial wastewater. Adsorption capacities of fly ash for the removal of metals are provided in Table 5.10.

5.3.2 RICE HUSK

Rice is grown on every continent except Antarctica and ranks second only to wheat in terms of worldwide area and production. When rough rice or paddy rice is husked, rice husk is generated as a waste and generally, every 100 kg of paddy rice produces 20 kg of husk. Of course, the rice husk

TABLE 5.13

Adsorption Capacities of Metals by Chitosan and Chitosan Composites[a]

Adsorbent	Metal	Adsorption Capacity[b]	Temperature (°C)	References
Chitosan/cotton fibers (via Schiff base bond)	Hg(II)	104.31	35	114
Chitosan/cotton fibers (via C–N single bond)	Hg(II)	96.28	25	114
Chitosan/cotton fibers (via Schiff base bond)	Cu(II)	24.78	25	115
Chitosan/cotton fibers (via Schiff base bond)	Ni(II)	7.63	25	115
Chitosan/cotton fibers (via Schiff base bond)	Pb(II)	101.53	25	115
Chitosan/cotton fibers (via Schiff base bond)	Cd(II)	15.74	25	115
Chitosan/cotton fibers (via Schiff base bond)	Au(III)	76.82	25	116
Chitosan/cotton fibers (via C–N single bond)	Au(III)	88.64	25	116
Magnetic chitosan	Cr(VI)	69.40	–	117
Chitosan/magnetite	Pb(II)	63.33	–	118
Chitosan/magnetite	Ni(II)	52.55	–	118
Chitosan/cellulose	Cu(II)	26.50	25	119
Chitosan/cellulose	Zn(II)	19.81	25	119
Chitosan/cellulose	Cr(VI)	13.05	25	119
Chitosan/cellulose	Ni(II)	13.21	25	119
Chitosan/cellulose	Pb(II)	26.31	25	119
Chitosan/perlite	Cu(II)	196.07	–	120
Chitosan/perlite	Ni(II)	114.94	–	120
Chitosan/perlite	Cd(II)	178.6	25	121
Chitosan/perlite	Cr(VI)	153.8	25	122
Chitosan/perlite	Cu(II)	104.0	25	123
Chitosan/ceramic alumina	As(III)	56.50	25	124
Chitosan/ceramic alumina	As(V)	96.46	25	124
Chitosan/ceramic alumina	Cu(II)	86.20	25	125
Chitosan/ceramic alumina	Ni(II)	78.10	25	125
Chitosan/ceramic alumina	Cr(VI)	153.8	25	126
Chitosan/montmorillonite	Cr(VI)	41.67	25	127
Chitosan/alginate	Cu(II)	67.66	–	128
Chitosan/calcium alginate	Ni(II)	222.2	–	129
Chitosan/silica	Ni(II)	254.3	–	129
Chitosan/PVC	Cu(II)	87.9	–	130
Chitosan/PVC	Ni(II)	120.5	–	130
Chitosan/PVA	Cd(II)	142.9	50	131
Chitosan/PVA	Cu(II)	47.85	–	132
Chitosan/sand	Cu(II)	10.87	–	133
Chitosan/sand	Cu(II)	8.18	–	134
Chitosan/sand	Pb(II)	12.32	–	134
Chitosan/clinoptilolite	Cu(II)	574.49	–	135
Chitosan/clinoptilolite	Cu(II)	719.39	25	136
Chitosan/clinoptilolite	Co(II)	467.90	25	136
Chitosan/clinoptilolite	Ni(II)	247.03	25	136
Chitosan/nano-hydroxyapatite	Fe(III)	6.75	–	137
Poly(methacrylic acid) grafted-chitosan/bentonite	Th(IV)	110.5	30	138
Chitosan-coated acid-treated oil palm shell charcoal (CCAB)	Cr(VI)	60.25	–	139
Chitosan-coated oil palm shell charcoal (CCB)	Cr(VI)	52.68	–	139
Acid-treated oil palm shell charcoal (AOPSC)	Cr(VI)	44.68	–	139

[a] These reported adsorption capacities are values obtained under specific conditions. Readers are encouraged to refer to the original articles for information on experimental conditions.

[b] In mg/g.

TABLE 5.14
Adsorption Capacities of Metals by Untreated and Pretreated Algae-Based Materials[a]

Algae	Metals	Adsorption Capacity[b]	References
Ascophyllum nodosum (B)	Cd(II)	0.338–1.913	143
Ascophyllum nodosum	Ni(II)	1.346–2.316	144
Ascophyllum nodosum	Pb(II)	1.313–2.307	144
Ascophyllum nodosum-CaCl₂ treated	Cd(II)	0.930	145
Ascophyllum nodosum-CaCl₂ treated	Cu(II)	1.090	145
Ascophyllum nodosum-CaCl₂ treated	Pb(II)	1.150	145
Ascophyllum nodosum-Bis(ethenil)sulfone treated	Pb(II)	1.733	144
Ascophyllum nodosum-divinil sulfone treated	Cd(II)	1.139	143
Ascophyllum nodosum-formaldehyde treated	Cd(II)	0.750	146
Ascophyllum nodosum-formaldehyde treated	Cd(II)	0.750	147
Ascophyllum nodosum-formaldehyde treated	Cd(II)	0.854	147
Ascophyllum nodosum-formaldehyde treated	Cu(II)	0.990	146
Ascophyllum nodosum-formaldehyde treated	Cu(II)	1.306	147
Ascophyllum nodosum-formaldehyde treated	Cu(II)	1.432	147
Ascophyllum nodosum-formaldehyde treated	Pb(II)	1.3755	147
Ascophyllum nodosum-formaldehyde treated	Ni(II)	1.618	147
Ascophyllum nodosum-formaldehyde treated	Ni(II)	1.431	147
Ascophyllum nodosum-formaldehyde treated	Zn(II)	0.680	146
Ascophyllum nodosum-formaldehyde treated	Zn(II)	0.719	147
Ascophyllum nodosum-formaldehyde treated	Zn(II)	0.8718	147
Ascophyllum nodosum-formaldehyde (3CdSO₄, H₂O) treated	Cd(II)	1.121	143
Ascophyllum nodosum-formaldehyde + CH₃COOH treated	Ni(II)	0.409	144
Ascophyllum nodosum-formaldehyde + CH₃COOH treated	Pb(II)	1.308	144
Ascophyllum nodosum-formaldehyde + urea treated	Cd(II)	1.041	143
Ascophyllum nodosum-formaldehyde + urea treated	Ni(II)	0.511	144
Ascophyllum nodosum-formaldehyde + urea treated	Pb(II)	0.854	144
Ascophyllum nodosum-formaldehyde Cd(CH₃COO)₂ treated	Cd(II)	1.326	143
Ascophyllum nodosum-glutaraldehyde treated	Cd(II)	1.259	143
Ascophyllum nodosum-glutaraldehyde treated	Cd(II)	0.480	147
Ascophyllum nodosum-glutaraldehyde treated	Cd(II)	0.4626	147
Ascophyllum nodosum-glutaraldehyde treated	Cu(II)	0.8497	147
Ascophyllum nodosum-glutaraldehyde treated	Cu(II)	0.803	147
Ascophyllum nodosum-glutaraldehyde treated	Ni(II)	0.9199	147
Ascophyllum nodosum-glutaraldehyde treated	Ni(II)	1.959	147
Ascophyllum nodosum-glutaraldehyde treated	Pb(II)	1.318	144
Ascophyllum nodosum-glutaraldehyde treated	Pb(II)	0.898	147
Ascophyllum nodosum-glutaraldehyde treated	Pb(II)	0.8157	147
Ascophyllum nodosum-glutaraldehyde treated	Zn(II)	0.3671	147
Ascophyllum nodosum-glutaraldehyde treated	Zn(II)	0.138	147
Caulerpa lentillifera (G)-dried macroalgae	Cu(II)	0.042–0.088	148
Caulerpa lentillifera (G)-dried macroalgae	Cd(II)	0.026–0.042	148
Caulerpa lentillifera (G)-dried macroalgae	Pb(II)	0.076–0.139	148
Caulerpa lentillifera (G)-dried macroalgae	Zn(II)	0.021–0.141	148
Caulerpa lentillifera (G)-dried macroalgae	Cu(II)	0.112	149
Caulerpa lentillifera (G)-dried macroalgae	Cd(II)	0.0381	149
Caulerpa lentillifera (G)-dried macroalgae	Pb(II)	0.142	149
Chaetomorpha linum (G)	Cd(II)	0.48	150
Chlorella miniata (G)	Cu(II)	0.366	151

(Continued)

TABLE 5.14 (*Continued*)

Adsorption Capacities of Metals by Untreated and Pretreated Algae-Based Materials[a]

Algae	Metals	Adsorption Capacity[b]	References
Chlorella miniata	Ni(II)	0.237	151
Chlorella vulgaris (G)	Cd(II)	0.30	152
Chlorella vulgaris	Ni(II)	0.205–1.017	152
Chlorella vulgaris	Pb(II)	0.47	152
Chlorella vulgaris	Zn(II)	0.37	152
Chlorella vulgaris	Cr(VI)	0.534	153
Chlorella vulgaris	Cr(VI)	1.525	154
Chlorella vulgaris	Cu(II)	0.295	151
Chlorella vulgaris	Cu(II)	0.254–0.549	153
Chlorella vulgaris	Cu(II)	0.758	154
Chlorella vulgaris	Fe(III)	0.439	153
Chlorella vulgaris	Ni(II)	1.017	154
Chlorella vulgaris	Ni(II)	0.205	151
Chlorella vulgaris-artificial cultivation	Cr(IV)	1.525	154
Chlorella vulgaris-artificial cultivation	Cu(II)	0.759	154
Chlorella vulgaris-artificial cultivation	Ni(II)	1.017	154
Cladophora glomerata (G)	Pb(II)	0.355	155
Chondrus crispus (R)	Ni(II)	0.443	144
Chondrus crispus treated with 1-chloro-2,3-epoxipropane	Pb(II)	1.009	144
Chondrus crispus	Pb(II)	0.941	144
Codium fragile (G)	Cd(II)	0.0827	156
Codium taylori (G)	Ni(II)	0.099	144
Codium taylori	Pb(II)	1.815	144
Corallina officinalis (R)	Cd(II)	0.2642	156
Durvillaea potatorum (B)-CaCl$_2$ treated	Cd(II)	0.260	157
Durvillaea potatorum-CaCl$_2$ treated	Cd(II)	1.130	157
Durvillaea potatorum-CaCl$_2$ treated	Cd(II)	1.100	157
Durvillaea potatorum-CaCl$_2$ treated	Cd(II)	1.100	157
Durvillaea potatorum-CaCl$_2$ treated	Cd(II)	1.120	157
Durvillaea potatorum-CaCl$_2$ treated	Cu(II)	0.040	158
Durvillaea potatorum-CaCl$_2$ treated	Cu(II)	0.180	158
Durvillaea potatorum-CaCl$_2$ treated	Cu(II)	0.990	158
Durvillaea potatorum-CaCl$_2$ treated	Cu(II)	1.210	158
Durvillaea potatorum-CaCl$_2$ treated	Cu(II)	1.310	158
Durvillaea potatorum-CaCl$_2$ treated	Ni(II)	0.17	159
Durvillaea potatorum-CaCl$_2$ treated	Ni(II)	0.68	159
Durvillaea potatorum-CaCl$_2$ treated	Ni(II)	1.13	159
Durvillaea potatorum-CaCl$_2$ treated	Pb(II)	0.020	158
Durvillaea potatorum-CaCl$_2$ treated	Pb(II)	0.760	158
Durvillaea potatorum-CaCl$_2$ treated	Pb(II)	1.290	158
Durvillaea potatorum-CaCl$_2$ treated	Pb(II)	1.470	158
Durvillaea potatorum-CaCl$_2$ treated	Pb(II)	1.550	158
Ecklonia maxima (B)-CaCl$_2$ treated	Cd(II)	1.150	145
Ecklonia maxima-CaCl$_2$ treated	Cu(II)	1.220	145
Ecklonia maxima-CaCl$_2$ treated	Pb(II)	1.400	145
Ecklonia radiata (B)-CaCl$_2$ treated	Cd(II)	1.040	145
Ecklonia radiata-CaCl$_2$ treated	Cu(II)	0.070	158
Ecklonia radiata-CaCl$_2$ treated	Cu(II)	0.450	158

(Continued)

TABLE 5.14 (*Continued*)

Adsorption Capacities of Metals by Untreated and Pretreated Algae-Based Materials[a]

Algae	Metals	Adsorption Capacity[b]	References
Ecklonia radiata-CaCl₂ treated	Cu(II)	0.950	158
Ecklonia radiata-CaCl₂ treated	Cu(II)	1.060	158
Ecklonia radiata-CaCl₂ treated	Cu(II)	1.110	158
Ecklonia radiata-CaCl₂ treated	Pb(II)	0.050	158
Ecklonia radiata-CaCl₂ treated	Pb(II)	0.420	158
Ecklonia radiata-CaCl₂ treated	Pb(II)	0.990	158
Ecklonia radiata-CaCl₂ treated	Pb(II)	1.170	158
Ecklonia radiata-CaCl₂ treated	Pb(II)	1.260	158
Fucus vesiculosus (B)	Cd(II)	0.649	143
Fucus vesiculosus	Ni(II)	0.392	144
Fucus vesiculosus	Pb(II)	1.105–2.896	144
Fucus vesiculosus-formaldehyde treated	Ni(II)	0.559	144
Fucus vesiculosus-formaldehyde treated	Pb(II)	1.752	144
Fucus vesiculosus-formaldehyde + HCl treated	Pb(II)	1.453	144
Galaxaura marginata (R)	Ni(II)	0.187	144
Galaxaura marginata	Pb(II)	0.121	144
Galaxaura marginata-CaCO₃ treated	Ni(II)	0.187	144
Galaxaura marginata-CaCO₃ treated	Pb(II)	1.530	144
Gracilaria corticata (R)	Pb(II)	0.2017–0.2606	155
Gracilaria edulis (R)	Cd(II)	0.24	150
Gracilaria salicornia (R)	Cd(II)	0.16	150
Laminaria hyperbola (B)-treated CaCl₂	Cd(II)	0.820	145
Laminaria hyperbola-treated CaCl₂	Cu(II)	1.220	145
Laminaria hyperbola-treated CaCl₂	Pb(II)	1.350	145
Laminaria japonica (B)-treated CaCl₂	Cd(II)	1.110	145
Laminaria japonica-treated CaCl₂	Cu(II)	1.200	145
Laminaria japonica-treated CaCl₂	Pb(II)	1.330	145
Lessonia flavicans (B)-treated CaCl₂	Cd(II)	1.160	145
Lessonia flavicans-treated CaCl₂	Cu(II)	1.250	145
Lessonia flavicans-treated CaCl₂	Pb(II)	1.450	145
Lessonia nigrescens (B)-treated CaCl₂	Cd(II)	1.110	145
Lessonia nigrescens-treated CaCl₂	Cu(II)	1.260	145
Lessonia nigrescens-treated CaCl₂	Pb(II)	1.460	145
Padina sp. (B)	Cd(II)	0.53	160
Padina sp.-CaCl₂ treated	Cd(II)	0.52	160
Padina sp.-CaCl₂ treated	Cu(II)	0.8	161
Padina gymnospora (B)	Ni(II)	0.170	144
Padina gymnospora	Pb(II)	0.314	144
Padina gymnospora-CaCO₃ treated	Ni(II)	0.238	144
Padina gymnospora-CaCO₃ treated	Pb(II)	0.150	144
Padina tetrastromatica (B)	Pb(II)	1.049	155
Padina tetrastromatica	Cd(II)	0.53	150
Polysiphonia violacea (R)	Pb(II)	0.4923	155
Porphyra columbina (R)	Cd(II)	0.4048	156
Sargassum sp. (B)	Cd(II)	1.40	162
Sargassum sp.	Cr(VI)	1.3257	163
Sargassum sp.	Cr(VI)	1.30	164
Sargassum sp.	Cu(II)	1.08	164

(Continued)

TABLE 5.14 (*Continued*)
Adsorption Capacities of Metals by Untreated and Pretreated Algae-Based Materials[a]

Algae	Metals	Adsorption Capacity[b]	References
Sargassum baccularia (B)	Cd(II)	0.74	150
Sargassum fluitans (B)	Ni(II)	0.409	144
Sargassum fluitans	Pb(II)	1.594	144
Sargassum fluitans-epichlorohyridin treated	Pb(II)	0.975	144
Sargassum fluitans-epichlorohyridin treated	Ni(II)	0.337	144
Sargassum fluitans-formaldehyde treated	Cd(II)	0.9519	147
Sargassum fluitans-formaldehyde treated	Cu(II)	1.7938	147
Sargassum fluitans-formaldehyde treated	Ni(II)	1.9932	147
Sargassum fluitans-formaldehyde treated	Pb(II)	1.8244	147
Sargassum fluitans-formaldehyde treated	Zn(II)	0.9635	147
Sargassum fluitans-formaldehyde + HCl treated	Ni(II)	0.749	144
Sargassum fluitans-glutaraldehyde treated	Cd(II)	1.0676	147
Sargassum fluitans-glutaraldehyde treated	Cu(II)	1.574	147
Sargassum fluitans-glutaraldehyde treated	Ni(II)	0.7337	147
Sargassum fluitans-glutaraldehyde treated	Pb(II)	1.6603	147
Sargassum fluitans-glutaraldehyde treated	Zn(II)	0.9942	147
Sargassum fluitans-NaOH treated	Al(III)	0.950	165
Sargassum fluitans-NaOH treated	Al(III)	1.580	165
Sargassum fluitans-NaOH treated	Al(III)	3.740	165
Sargassum fluitans-NaOH treated	Cu(II)	0.650	165
Sargassum fluitans-NaOH treated	Cu(II)	1.350	165
Sargassum fluitans-NaOH treated	Cu(II)	1.540	165
Sargassum fluitans-protonated biomass	Cd(II)	0.710	166
Sargassum fluitans-protonated biomass	Cu(II)	0.800	166
Sargassum hystrix (B)	Pb(II)	1.3755	155
Sargassum natans (B)	Cd(II)	1.174	143
Sargassum natans	Ni(II)	0.409	144
Sargassum natans	Pb(II)	1.221	144
Sargassum natans	Pb(II)	1.1487	155
Sargassum siliquosum (M)	Cd(II)	0.73	150
Sargassum vulgare (M)	Ni(II)	0.085	144
Sargassum vulgare	Pb(II)	1.100	144
Sargassum vulgare-protonated biomass	Cd(II)	0.790	166
Sargassum vulgare-protonated biomass	Cu(II)	0.930	166
Scenedesmus obliquus (G)	Cu(II)	0.524	154
Scenedesmus obliquus	Ni(II)	0.5145	154
Scenedesmus obliquus	Cr(VI)	1.131	154
Scenedesmus obliquus-artificial cultivation	Cr(VI)	1.131	154
Scenedesmus obliquus-artificial cultivation	Cu(II)	0.524	154
Scenedesmus obliquus-artificial cultivation	Ni(II)	0.514	154
Ulva lactuca (G)	Pb(II)	0.61	155
Undaria pinnatifida (B)	Pb(II)	1.945	167

Note: (B): brown alga; (G): green alga; and (R): red alga.

[a] These reported adsorption capacities are values obtained under specific conditions. Readers are encouraged to refer to the original articles for information on experimental conditions.

[b] In mmol/g.

TABLE 5.15

Adsorption Capacities of Metals by Bacterials[a]

Metals	Adsorbent	Adsorption Capacity[b]	References
Cd(II)	*Aeromonas caviae*	155.3	169
	Enterobacter sp.	46.2	170
	Ochrobactrum anthropi	–	171
	Pseudomonas aeruginosa	42.4	172
	Pseudomonas putida	8.0	173
	Pseudomonas putida	500.00	174
	Pseudomonas sp.	278.0	175
	Sphingomonas paucimobilis	–	176
	Staphylococcus xylosus	250.0	175
	Streptomyces pimprina	30.4	177
	Streptomyces rimosus	64.9	178
Cr(VI)	*Aeromonas caviae*	284.4	169
	Bacillus coagulans	39.9	179
	Bacillus megaterium	30.7	179
	Bacillus coagulans	39.9	179
	Bacillus licheniformis	69.4	180
	Bacillus megaterium	30.7	179
	Bacillus thuringiensis	83.3	181
	Pseudomonas sp.	95.0	175
	Pseudomonas fluorescens	111.11	174
	Staphylococcus xylosus	143.0	175
	Zoogloea ramigera	2	182
Cu(II)	*Bacillus firmus*	381	183
	Bacillus sp.	16.3	184
	Bacillus subtilis	20.8	185
	Enterobacter sp.	32.5	170
	Micrococcus luteus	33.5	185
	Pseudomonas aeruginosa	23.1	172
	Pseudomonas cepacia	65.3	186
	Pseudomonas putida	6.6	173
	Pseudomonas putida	96.9	187
	Pseudomonas putida	15.8	188
	Pseudomonas putida	163.93	174
	Pseudomonas stutzeri	22.9	185
	Sphaerotilus natans	60	189
	Sphaerotilus natans	5.4	189
	Streptomyces coelicolor	66.7	190
	Thiobacillus ferrooxidans	39.8	191
Fe(III)	*Streptomyces rimosus*	122.0	192
Ni(II)	*Bacillus thuringiensis*	45.9	193
	Pseudomonas putida	556	174
	Streptomyces rimosus	32.6	194
Pb(II)	*Bacillus* sp.	92.3	184
	Bacillus firmus	467	183
	Corynebacterium glutamicum	567.7	195
	Enterobacter sp.	50.9	170

(Continued)

TABLE 5.15 (*Continued*)
Adsorption Capacities of Metals by Bacterials[a]

Metals	Adsorbent	Adsorption Capacity[b]	References
	Pseudomonas aeruginosa	79.5	172
	Pseudomonas aeruginosa	0.7	196
	Pseudomonas putida	270.4	187
	Pseudomonas putida	56.2	173
	Streptomyces rimosus	135.0	197
Pd(II)	*Desulfovibrio desulfuricans*	128.2	198
	Desulfovibrio fructosivorans	119.8	198
	Desulfovibrio vulgaris	106.3	198
Pt(IV)	*Desulfovibrio desulfuricans*	62.5	198
	Desulfovibrio fructosivorans	32.3	198
	Desulfovibrio vulgaris	40.1	198
Th(IV)	*Arthrobacter nicotianae*	75.9	199
	Bacillus licheniformis	66.1	199
	Bacillus megaterium	74.0	199
	Bacillus subtilis	71.9	199
	Corynebacterium equi	46.9	199
	Corynebacterium glutamicum	36.2	199
	Micrococcus luteus	77.0	199
	Zoogloea ramigera	67.8	199
U(VI)	*Arthrobacter nicotianae*	68.8	199
	Bacillus licheniformis	45.9	199
	Bacillus megaterium	37.8	199
	Bacillus subtilis	52.4	199
	Corynebacterium equi	21.4	199
	Corynebacterium glutamicum	5.9	199
	Micrococcus luteus	38.8	199
	Nocardia erythropolis	51.2	199
	Zoogloea ramigera	49.7	199
Zn(II)	*Streptomyces rimosus*	30	200
	Bacillus firmus	418	183
	Aphanothece halophytica	133.0	201
	Pseudomonas putida	6.9	173
	Pseudomonas putida	17.7	188
	Streptomyces rimosus	30.0	200
	Streptomyces rimosus	80.0	200
	Streptoverticillium cinnamomeum	21.3	202
	Thiobacillus ferrooxidans	82.6	206
	Thiobacillus ferrooxidans	172.4	191

[a] These reported adsorption capacities are values obtained under specific conditions. Readers are encouraged to refer to the original articles for information on experimental conditions.

[b] In mg/g.

production may vary with different rice species. Therefore, in many rice producing countries, the utilization of this abundant scaly residue is of great significance.

Rice husk is considered as a lignocellulosic agricultural byproduct that contains approximately 32.24% cellulose, 21.34% hemicelluloses, 21.44% lignin, and 15.05% mineral ash (59). The percentage of silica in its mineral ash is about 96.34% (60). Such a high percentage of silica coupled with a large amount of lignin, a structural polymer, is very unusual in nature. It has made rice husk not only resistant to water penetration and fungal decomposition, but also resistant to the best efforts of man to dispose it since the rice husk does not biodegrade easily.

Of all cereal byproducts, rice husk has the lowest percentage of total digestible nutrients (<10%). It also contains very low protein and available carbohydrates, and yet, at the same time, high in crude fiber and crude ash. Owing to its abrasive character, poor nutritive value, low bulk density, and high ash content which would sometimes cause harmful effects, the husk is not widely used as animal feed.

Rice husk is a waste from a rice cultivation perspective. From an agricultural byproducts utilization perspective, however, rice husk is a resource yet to be fully utilized and exploited. The researchers are thus looking for ways to valorize rice husk. Efforts have been made to utilize rice husk as a building material. In this regard, rice husk is used to insulate walls, floors, and roof cavities because of its excellent properties, such as good heat insulation, does not emit smell or gases, and it is not corrosive. Unfortunately, the cost of building materials manufactured using rice husk as the aggregate is not competitive with that using other aggregates.

Thus, another interesting possibility for utilizing this cheap and readily available resource might be as a low-cost adsorbent in the removal of heavy metals from aqueous environment. The excellent characteristics of rice husk such as its insolubility in water, good chemical stability, high mechanical strength, and its granular structure, make this likelihood to be higher. Considerable researches have been attempted on the use of rice husk, either untreated or modified, to remove heavy metals using different methods. Adsorption capacities of metals by untreated and treated rice husk are presented in Table 5.11.

5.3.3 WHEAT STRAW AND WHEAT BRAN

Every year, large amounts of straw and bran from *Triticum aestivum* (wheat), a major food crop of the world, are produced as byproducts/waste materials. Wheat straw has been used as fodder and in paper industry to produce low-quality boards or packing materials. The stems are burnt directly in some parts of the world for energy purposes, adding seriously to atmospheric pollution and wastage of resources.

The main components found in wheat straw are cellulose (37%–39%), hemicellulose (30%–35%), lignin (~14%), and sugars. Considering its chemical properties, wheat straw normally consists of different functional groups such as carboxyl, hydroxyl, sulfhydryl, amide, amine, and so on. The percentage composition of different substances varies in different parts of the world, although the substances are almost similar.

Both wheat straw and wheat bran have been investigated for their adsorption behavior toward metal ions (Table 5.12). The reported variations in metal capacities of wheat-based materials correspond to the variation in the structure of wheat bran used in different studies, along with other parameters. Apart from this, the discrepancies in the origin, area, soil, and kind of wheat from where wheat-based material is obtained may explain such a variation in results.

5.3.4 CHITIN, CHITOSAN, AND CHITOSAN COMPOSITES

The utilization of byproducts, chitin, generated from crustacean processing could be helpful in addressing the environmental problem as the biodegradation of this waste is very slow in nature. As a matter of fact, the application of biopolymers such as chitin and chitosan can be seen as one of the emerging techniques for the removal of certain hazardous pollutants from the environment.

Chitin is the second most abundant polymer in nature after cellulose. It is a kind of natural biopolymer which has a chemical structure similar to cellulose and is generally found in a wide range of natural sources such as in the exoskeletons of crustaceans, cell wall of fungi, insects, annelids, and molluscs. It contains 2-acetamido-2-deoxy-β-D-glucose through a β (1 $\rightarrow$ 4) linkage. Chitosan is a type of natural poly(aminosaccharide) consisting mainly of a poly($1 \rightarrow 4$)-2 amino-2-deoxy-D-glucose unit, synthesized from the deacetylation of chitin. Chitosan is known as an excellent biomaterial because of its special characteristics, for instance, hydrophilicity, biocompatibility, biodegradability, nontoxicity, and good adsorption properties.

Apart from the mentioned physicochemical characteristics, the possibility of using chitin in a variety of forms, from flake types to gels, beads, and fibers is also the contributing factor as to why this waste material has drawn particular attention. It has been demonstrated that chitin can provide readily available binding sites for a wide range of molecules due to its high contents of amino and hydroxyl functional groups. Nevertheless, the adsorption properties would still depend strongly on the sources of chitin, the degree of N-acetylation, and on variations in crystallinity and amino content.

Chitosan is very sensitive to pH as it can either form gel or dissolve depending on the pH values. Apparently, this characteristic has limited chitosan's performance as a biosorbent in wastewater treatment. To overcome this problem, cross-linking reagents such as glyoxal, formaldehyde, glutaraldehyde, epichlorohydrin, ethylene glycon diglycidyl ether, and isocyanates have been used to stabilize chitosan in acidic media. Cross-linking agents do not only prevent chitosan from becoming soluble under these conditions but also enhance its mechanical properties. As a result, cross-linked chitosan not only has stronger mechanical properties compared with its parent biopolymer, but might also has higher affinity for the targeted pollutants.

Biosorption using chitosan-based materials, such as chitosan derivatives and chitosan composites have been extensively investigated for the removal of heavy metals. Among them are chitosan derivatives containing nitrogen, phosphorus, and sulfur as heteroatoms, and other derivatives such as chitosan-crown ethers and chitosan ethylenediaminetetraacetic acid (EDTA)/diethylenetriaminepentaacetic acid (DTPA) complexes. As for chitosan composites, various kinds of substances have been used to form composites with chitosan, which include montmorillonite, polyurethane, activated clay, bentonite, polyvinyl alcohol, polyvinyl chloride, kaolinite, oil palm ash, and perlite. Table 5.13 presents the heavy metal removal capacities through adsorption process by chitosan and chitosan composites.

5.3.5 ALGAE

Algae are a large and diverse group of simple plant-like organisms, ranging from unicellular to multicellular forms, which can be seen in aquatic habitats, freshwater, marine, and moist soil. Algae contain chlorophyll and carry out oxygenic photosynthesis. This biosorbent has been extensively studied due to its ubiquitous occurrence in nature. Algae have found applications as fertilizer, energy sources, pollution control, stabilizing substances, in nutrition, etc. Figure 5.5 presents efficiency of heavy metals uptake by various algae.

Several characteristics are used to classify algae, including the nature of the chlorophyll(s) present, the carbon reserve polymers produced, the cell-wall structure, and the type of motility. Although all algae contain chlorophyll a, there are some, which contain other chlorophylls that differ in minor ways from chlorophylls a. The presence of these additional chlorophylls is characteristic of particular algal groups. The major groups of algae include Chrysophyta (golden-brown algae, diatoms), Euglenophyta (euglenoids is also considered as protozoa), Pyrrophyta (dino-flagellates), Chlorophyta (green algae), Phaeophyta (brown algae), and Rhodophyta (red algae). Adsorption capacities of metals by untreated and treated algae are provided in Table 5.14. From the published literatures, brown algae are the most widely studied among the three groups of algae (red, green, and brown algae). This could be related to sorption capability of the algae, whereby brown algae emerges to offer better sorption than red or green algae (141,142). Researchers have used mainly brown algae treated in different ways to improve their sorption capacity (141).

The algal cell is surrounded by a thin, rigid cell wall that contains pores of about 3–5 nm wide to allow low molecular-weight constituents such as water, ions, gases, and other nutrients to pass through freely for metabolism and growth. However, the cell walls are essentially impermeable to

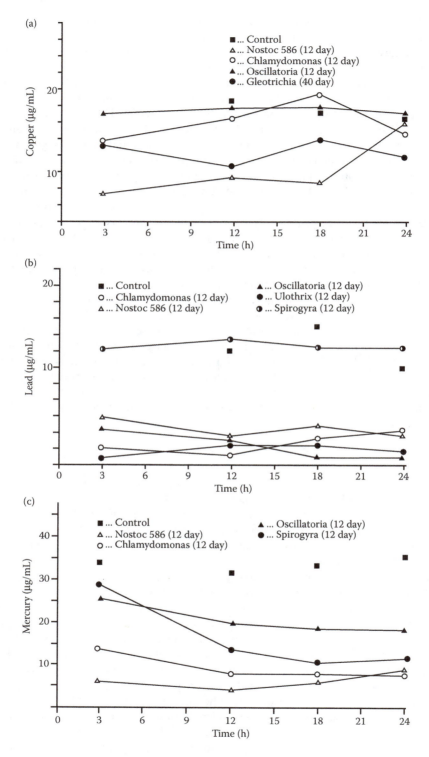

FIGURE 5.5 Uptake of heavy metals (a) Cu, (b) Pb, and (c) Hg by various algae. (From US Environmental Protection Agency. 1983. *Factors Influencing Metal Accumulation by Algae*, EPA-600/S2-82-100. US EPA, Cincinnati, OH.)

larger molecules or to macromolecules. It is usually made of a multilayered microfibrillar framework generally consisting of cellulose and interspersed with amorphous material (168).

In biosorption, various algae have been used and investigated for heavy metal removal in aqueous solutions by a number of researchers. The metal biosorption by algae mainly depend on the components on the cell, especially through cell surface and the spatial structure of the cell wall. Various functional groups, such as carboxyl, hydroxyl, sulfate, and amino groups in algal cell-wall polysaccharides have been proven to play a very important role in metal binding. The biomass characteristics, physicochemical properties of the targeted metals, and solution pH also have a significant impact on the biosorption performance.

5.3.6 BACTERIA

Bacteria are microscale organisms whose single cells have neither a membrane-bound nucleus nor other membrane-bound organelles such as the mitochondria and chloroplasts. They have simple morphology and commonly present in three basic shapes: spherical or ovoid (coccus), rod (bacillus, with a cylindrical shape), and spiral (spirillum). Bacteria vary in size as much as in shape due to differences in genetics and ecology. The smallest bacteria are about 0.3 μm, and a few bacteria become fairly large, for example, some spirochetes occasionally reach 500 μm in length, and cyanobacterium *Oscillatoria* is about 7 μm in diameter.

A "typical" bacterial cell (e.g., *Escherichia coli*) contains cell wall, cell membrane, and cytoplasmic matrix consisting of several constituents, which are not membrane-enclosed: inclusion bodies, ribosomes, and the nucleoid with its genetic material. Some bacteria have special structure, such as flagella and S-layer. The major function of the cell wall is to (i) provide the cell shape and protect it from osmotic lysis, (ii) protect cell from toxic substances, and (iii) to offer the site of action for several antibiotics. Moreover, it is a necessary component for normal cell division. Cellular wall shape and strength are primarily due to peptidoglycan. The amount and exact composition of peptidoglycan are only found in cell walls and vary among the major bacterial groups.

Bacteria are of special interest in search for and the development of new biosorbent materials due to their availability, small size, ubiquity, ability to grow under controlled conditions, and resiliency to a wide range of environmental situations. Adsorption capacities of metals by bacterial surfaces are given in Table 5.15.

5.4 CHEMICAL PROPERTIES AND CHARACTERIZATION STUDIES

Fourier-transform infrared spectroscopy (FTIR) spectrum analysis is usually used to study the functional groups on the adsorbents. UV–Vis spectroscopy is used to investigate whether the removal of Cr(VI) involves the reduction of Cr(VI) to Cr(III) by measuring the absorbance of the purple–violet complex of Cr(VI) with 1,5-diphenylcarbazide acidic solution at 540 nm. The difference between the total and Cr(VI) concentrations was taken to represent the Cr(III) concentration.

To elucidate the surface morphology of the adsorbents before and after sorption, several techniques can be used which include scanning electron microscopy (SEM), transmission electron microscopy (TEM), and atomic force microscopy (AFM). Both SEM and TEM involved the use of focused beam of electrons instead of light to "image" the materials of interest and gain information as to its structure and composition. Whereas for AFM, it is a stylus-type instrument, in which a sharp probe, scanned raster-fashion across the sample, is used to detect changes in the surface structure on the atomic scale. As the interaction force between the cantilever tip and surface varies, deflections are produced in the cantilever. These deflections are measured, and used to compile a topographic image of the surface. Color mapping is the usual method used for displaying the data where light color indicates high features or high topography and lower topography is shown by darker color. And often, if the adsorbents were subjected to chemical modifications, the resulting materials become more intense and display a higher topography.

5.5   INFLUENCE OF OPERATIONAL PARAMETERS

5.5.1   Effect of pH

Since the efficiency of the adsorption process is strongly dependent on pH, in most of the adsorption process of heavy metals by various low-cost adsorbents, pH is one of the commonly examined parameters. Generally, the prominent effect of this parameter is because the solution pH influences the metal chemistry as well as the surface binding sites of the biosorbents. From the literature, it is evident that at certain pH, the metal ions could be precipitated out as hydroxides. Therefore, in most of the studies, the solution pH at which precipitation occurred will not be investigated since the dominant removal process was due to precipitation and not of experimental interest. In most of the lignocellulosic adsorbents, the presence of carboxyl functional groups has been well documented. It is suggested that at low pH (<2.0), the carboxyl groups on the surface of the adsorbents were predominantly protonated (–COOH), and hence incapable of binding the cationic species. With increasing pH, adsorption became favorable as the adsorption sites were made available for binding positively charged metal ions.

In the adsorption of Cr(VI) using natural rice hull (NRH) and ethylenediamine-modified rice hull (enRH), the modified adsorbent exhibited greater uptake capability for Cr(VI) and the adsorption decreased with increasing pH (203). This is due to the distribution of Cr(VI) species which is controlled by the ion equilibria and the total Cr(VI) concentration used. Under the experimental condition, it is postulated that $HCrO_4^-$ was the major species and played an important role in association with the adsorbents. At low pH, the amine groups on the surface of enRH was protonated by H^+, rendering it favorable for electrostatic attraction between $HCrO_4^-$ and positively charged binding sites. The lower uptake at pH 1 is closely related to the reduction of Cr(VI) to Cr(III). It has been well documented that under acidic conditions, Cr(VI) demonstrates a very high positive redox potential which denotes it is strongly oxidizing and unstable in the presence of electron donors (204). The absence of lone pair in NRH as compared with those present in enRH explained the low reduction capability of NRH, and thereby, adsorption decreased with increasing pH.

Generally, an adsorption process is accompanied by a decrease in pH due to the release of H^+. However, exception cases were observed in the adsorption involving Cr(VI) and As(V). The increase in pH implies the release of OH^- ions into the solution upon protonation of the adsorbents.

5.5.2   Effect of Initial Concentration of Heavy Metals and Contact Time

The nature of the adsorbent and its available binding sites played a crucial role in determining the time needed for the attainment of equilibrium. Nevertheless, the typical adsorption pattern exhibited by various adsorbents in adsorbing heavy metals is a rapid ion-exchange process followed by chemisorption. The fast initial metal uptake is attributed to the rapid attachment of heavy metals onto the surface of the adsorbents, whereas the following slower adsorption is related to the interior penetration (intraparticle diffusion). In terms of initial heavy metals concentrations, the trend of uptake usually followed the normal course of adsorption process; the least concentrated showing the highest percentage uptake while the amount of heavy metals adsorbed decreased. Adsorption process involving a mixture of heavy metals sometimes reached equilibrium faster than those metals that present singly. The faster adsorption rate in this kind of systems could be due to the higher total metal ion concentration in the system which in turn gives rise to a greater driving force and collision probability between metal ions and the adsorbent. By comparing the uptake of heavy metal ions that are present in a mixture or single metal ion solution showed that the effect could be synergistic or antagonistic. Different explanations have been given regarding the sorption affinity of the adsorbents and these include competitive effect, ionic size, stability of the bond between the metal ions and the adsorbents, nature of metal-ion sorbents, interaction, and the distribution of the reaction group on the adsorbents (205).

5.5.3 Effect of the Chelator

One of the common problems associated with heavy metals removal in the conventional treatment method is the presence of a chelator. The chelators could mask the presence of metal ions, rendering their removal from the solution difficult or impossible. Owing to this, the effect of chelators that are commonly found in the environment, such as ethylenediamine tetraacetic acid (EDTA), nitrilotriacetic acid (NTA), and salicylic acid (SA) were often tested for their influence on the adsorption of heavy metals. NTA is chosen because it is a substitute used for polyphosphate in the detergent whereas SA is representing humic acid which is reported to be present in natural wastes. For the adsorption of Cu(II) and Pb(II), the results have shown that both NTA and EDTA inhibit the metals uptake by the modified adsorbent (79). This is because NTA and EDTA formed stable complexes with Cu(II) and Pb(II) and they compete more effectively with the binding sites of both metal ions. The effectiveness of a chelator is expressed in chelator stability constants, $\log K_1$ where the larger $\log K_1$ value will give higher efficiency of the chelating effect. The results obtained were in accordance with the $\log K_1$ values of 5.55, 9.80, and 16.28, respectively. Therefore, it is of utmost important to assess critically and differently the adsorption of heavy metals by various adsorbents if chelators are known to be present in the same system because it could be a significant suppressing effect.

5.6 EXPERIMENTAL METHODS AND MODELING OF HEAVY METALS ADSORPTION

5.6.1 Batch Adsorption Experiments

In a batch adsorption experiment, the adsorbent must be in contact with the adsorbate for a period of time to ensure that the concentration of the adsorbate in solution is in equilibrium with the adsorbate on the surface. Usually, the time required for the attainment of equilibrium is pH, concentration, agitation and is particle size dependent. For the batch equilibrium operations, a porous adsorbent with a smaller particle size is generally favored for its higher surface area, resulting in a more effective adsorbent–adsorbate contact and in a reduction of diffusional resistance inside the pores. After the adsorption process, the solid (adsorbent and adsorbate absorbed) and liquid phases (adsorbate residue in solution) are separated via several methods, for example, settling, filtration, or centrifugation. Owing to the cost involved, the used adsorbent is either discarded or regenerated. The most common applicability of batch adsorption studies will be in adsorption isotherm and kinetics modeling.

5.6.2 Equilibrium Modeling of Biosorption in a Batch System

The adsorption properties and equilibrium data are usually known as adsorption isotherms. They are considered as the basic, yet the key requirements in adsorption system design. The good enough description of the adsorbate–adsorbent interaction provided by these data can optimize the application of the adsorbents. Apart from establishing an appropriate and correct correlation for the equilibrium data, the compliance of the data to a suitable mathematical model is also equally important. An accurate mathematical description is crucial for a reliable prediction on the adsorption parameters. It is also essential to allow a quantitative comparison on the adsorption behavior of different adsorption systems under a variety operating conditions.

Adsorption equilibrium is achieved when the amount of adsorbate being adsorbed onto the adsorbent is equal to the amount being desorbed. The equilibrium condition can be represented by plotting the adsorbate concentration in solid phase versus that in liquid phase. The position of equilibrium in the adsorption process is measured from the distribution of adsorbate molecules between the adsorbent and the liquid phase, which can generally be expressed by one or more of a series of isotherm models. The shape generated from an isotherm is usually used to predict the "favorable" behavior of an adsorption system. Besides, the isotherm shape provides qualitative information on the nature of the solute–surface interaction. The adsorption isotherms are also applied extensively in the determination of the maximum adsorption capacity of adsorbents for a particular adsorbate. This information is important as a fundamental and convenient tool to evaluate the performance of different adsorbents and select the most appropriate one for a particular adsorption application under certain conditions.

On the other hand, two- and three-parameter models, originally used for gas-phase adsorption, are available and readily adopted to correlate adsorption equilibria in liquid-phase adsorption. The experimental adsorption data are well described by the equilibrium isotherm equations generated from each model. The different equation parameters and the underlying thermodynamic assumptions of these models often provide insight into the adsorption mechanism, surface properties, and affinity of the adsorbent. Apparently, establishing the most appropriate correlation of equilibrium curves is crucial in optimizing the adsorption condition, subsequently contributing to an improvement of the adsorption system.

5.6.2.1 Two-Parameter Isotherms

Langmuir, Freundlich, and Brunauer–Emmet–Teller (BET) models are some of the widely met isotherms. Meanwhile, Dubinin–Radushkevich (D–R) and Temkin isotherms appear to be gaining less popularity among the two-parameter models. Other seldom used two-parameter models such as Halsey and Hurkins–Jura (H–J) are also discussed briefly. The application of each model for an adsorption system is often limited by assumptions made within the model.

5.6.2.1.1 Langmuir Isotherm

The Langmuir model is one of most popular isotherm models used to quantifying the amount of the adsorbed adsorbate on an adsorbent as a function of concentration at a particular temperature (207). Inherent within this model, some assumptions are valid for a biosorption process, including monolayer coverage of the adsorbate over a homogeneous adsorbent surface. All the sites on the adsorbent are equivalent and once an adsorbate molecule occupies a site, no further adsorption can take place at that site. Therefore, this model assumes occurrence of adsorption takes place at specific homogeneous sites on the surface of the adsorbent. Graphically, a plateau in the plot of q_e versus C_e characterizes the Langmuir isotherm. This explains why no further adsorption is allowed at equilibrium where a saturation point is reached. In addition, the Langmuir equation is applicable to homogeneous adsorption where the adsorption of each molecule has equal adsorption activation energy. Thus, this isotherm model is always utilized to describe adsorption of an adsorbate molecule from a liquid solution as

$$q_e = \frac{q_{max}K_LC_e}{1+K_LC_e} \tag{5.1}$$

where
q_{max} = mass of the adsorbate adsorbed/mass of adsorbent for a complete monolayer
K_L = Langmuir constant related to the enthalpy of adsorption

The Langmuir equation can be written in different linear forms as

$$\frac{C_e}{q_e} = \frac{1}{q_{max}}C_e + \frac{1}{K_Lq_{max}} \tag{5.2}$$

$$\frac{1}{q_e} = \left(\frac{1}{K_Lq_{max}}\right)\frac{1}{C_e} + \frac{1}{q_{max}} \tag{5.3}$$

$$q_e = q_{max} - \left(\frac{1}{K_L}\right)\frac{q_e}{C_e} \qquad (5.4)$$

$$\frac{q_e}{C_e} = K_L q_{max} - K_L q_e \qquad (5.5)$$

In some cases, different isotherm parameters are obtained using the four Langmuir linear equations (Equations 5.2 through 5.5), but they are identical when the nonlinear method is applied. Hence, the nonlinear method exists as a better approach to obtain the isotherm parameters (208). Despite better result provided by the nonlinear method, the linear least-square method is still more favorable among the researchers due to its simplicity and convenience.

The Langmuir isotherm is considered as the conventional method used in quantifying the maximum uptake and estimating the adsorption capacity q_{max} of different adsorbents. The obtained q_{max} should logically be temperature independent as it is supposed to coincide with saturation of a fixed number of identical surface sites that possess equal affinity for the adsorbate. However, small to modest changes in adsorption capacity with temperature is usually detected in real experimental conditions. The divergence from its formulation strongly indicates the presence of the surface functional groups on the adsorbent rather than a set of identical surface sites that are related to the saturation limit. Practically, the adsorption capacity is always influenced by the number of active sites on the adsorbent, the chemical state of the sites, the affinity between the sites (i.e., binding strength), and by the sites accessible to the adsorbate.

The Langmuir adsorption model suffers from the disadvantage of failure to account for the surface roughness of the adsorbate. Availability of multiple site-type that has arisen from rough inhomogeneous surfaces and changing of some parameters from site to site, such as the heat of adsorption has made this model to deviate drastically in many cases. Other than that, adsorbate–adsorbent interactions are ignored in this model. It has been proven experimentally that the existence of adsorbate–adsorbent interactions in heat of adsorption data, namely direct interaction and indirect interaction must be taken into consideration. In direct interactions, the adjacent adsorbed molecules can make adsorbing near another adsorbate molecule more or less favorable. Meanwhile, indirect interaction is referred to as the tendency of the adsorbate to change the surface around the adsorbed site, subsequently affecting the adsorption behavior of the nearby sites.

The decrease of K_L value with elevating temperature is an indicator for the exothermal nature of the adsorption process (209–212). In a physical adsorption, the bonding between adsorbates and the surface was primarily by physical forces, which become weaken at higher temperatures. Meanwhile, the endothermic process of the binding of adsorbates to active sites needs thermal energy; thus the elevation in temperature was more favorable for chemisorption (endothermic) (213). Alternatively, the exothermal or endothermal nature of the adsorption process can be further confirmed using the van't Hoff plots. An integrated van't Hoff equation provides the thermodynamic property and it relates the Langmuir constant, K_L to the temperature as

$$K_L = K_0 \exp\left(-\frac{\Delta H}{RT}\right) \qquad (5.6)$$

where
 K_0 = parameter of the van't Hoff equation
 ΔH = enthalpy of adsorption

5.6.2.1.2   Freundlich Isotherm

Freundlich isotherm (214) is another most frequently used isotherm for description of heterogeneous systems. In fact, this isotherm model is the oldest of the nonlinear isotherms. It assumes neither homogenous site energies nor limited levels of adsorption. Therefore, concentration of adsorbate on the adsorbent surface increases with increasing adsorbate concentration in the system. The exponential equation is expressed in following form:

$$q_e = K_F C_e^{1/n} \tag{5.7}$$

where

    q_e = mass of the adsorbate adsorbed/mass adsorbent
    C_e = adsorbate concentration in solution, mass/volume
    K_F = Freundlich constant related to adsorption capacity at a particular temperature
    n = Freundlich constant related to adsorption intensity at a particular temperature ($n > 1$)

Equation 5.1 can also be written in a linearized logarithmic form

$$\log q_e = \log K_F + \frac{1}{n} \log C_e \tag{5.8}$$

By plotting $\log q_e$ versus $\log C_e$, values of $1/n$ and $\log K_F$ can be obtained from the graph slope and intercept, respectively. Log K_F is equivalent to $\log q_e$ when C_e equals unity. The K_F value depends on the units upon which q_e and C_e are expressed if $1/n \neq 1$. Usually, Freundlich constant n ranges from 1 to 10 for a favorable adsorption. Larger value of n may indicate a stronger interaction between the adsorbent and the adsorbate. On the contrary, linear adsorption leading to identical adsorption energies for all sites is observed when $1/n$ equals 1 (215). Obviously, Freundlich isotherm is widely used in the study of due to its ability to fit nearly all experimental adsorption–desorption data. In particular, this isotherm provides excellently fitting data of highly heterogeneous adsorbent systems. The limitation of Freundlich isotherm of being inappropriate over a wide concentration range is always ignored by researchers since a moderate concentration range is normally used in most biosorption studies.

Adsorption capacity is the most significant property of an adsorbent. It is defined as the value of amount of a specific adsorbate taken up by an adsorbent per unit mass of the adsorbent. This variable is governed by the nature of the adsorbent, such as pore and particle size distribution, specific surface area, cation exchange capacity, and surface functional groups. Besides, pH and temperature of the system may also affect the adsorption capacity of an adsorbent. In general, the adsorption capacities of most of the biosorbents (obtained from K_F) are considerably low as compared with the commercially available activated carbons. Nevertheless, different types of biosorbents are still receiving intensive attraction from the researchers in view of their biosorption advantages and cost-effectiveness.

5.6.2.1.3   Temkin Isotherm

The Temkin model (216) takes into accounts of indirect interactions between the adsorbate molecules on adsorption isotherms. The derivation of Temkin isotherm assumes that as the surface of the adsorbent is occupied by the adsorbate, the heat of adsorption of all molecules in the layer would decrease linearly with coverage due to the indirect interactions. It makes the Temkin model differ from Freundlich model which implies a logarithmical decrease in the heat of adsorption. The Temkin equation proposes a linear decrease of adsorption energy as an increase in the degree of completion of the adsorption centers on an adsorbent. The equation is expressed as

$$q_e = \frac{RT}{b} \ln a C_e \qquad (5.9)$$

where
  a = the Temkin isotherm constant
  b = the Temkim constant related to the heat of adsorption

The linear form of the Temkin equation (Equation 5.10) is applicable to analyze the adsorption data at moderate concentrations. Both constants a and b can be determined from a plot of q_e versus $\ln C_e$:

$$q_e = \frac{RT}{b} \ln a + \frac{RT}{b} \ln C_e \qquad (5.10)$$

The simple assumptions made within the Temkin equation cause the derivation for this equation not well suited for a complex phenomenon involved in liquid-phase adsorption. Unlike gas-phase adsorption, the adsorbed molecules are not necessarily organized in a tightly packed structure with identical orientation in liquid-phase adsorption. In addition, the formation of micelles from the adsorbed molecules and the presence of solvent molecules add to the complexity of adsorption in liquid phase. In fact, liquid-phase adsorption is also greatly impacted by other factors such as pH, solubility of the adsorbate in the solvent, and temperature and surface chemistry of the adsorbent. For this reason, this equation is rarely used for the representation of experimental data of complex systems.

5.6.2.1.4 BET Model

The first isotherm for multimolecular layer adsorption was derived by Brunauer, Emmer, and Teller (217). This major advance in adsorption theory, the so-called BET theory, has solved the constraint found in Langmuir isotherm. Assuming the adsorbent surface is composed of fixed individual sites and molecules can be adsorbed more than one layer thick on the surface of the adsorbent, this model suggests a random distribution of sites covered by one, two, three, or more adsorbate molecules. Besides, the model is made based on the assumptions that there is no interaction between each adsorption layer, and the Langmuir theory can be applied to each layer. In other words, the same kinetics concept proposed by Langmuir is applied to this multiple layering process, that is, the rate of adsorption on any layer is equal to the rate of desorption from that layer. The simplified form of the BET equation is written as

$$q_e = q_{max} \frac{K_B C_e}{(C_e - C_s)[1 + (K_B - 1)(C_e / C_s)]} \qquad (5.11)$$

where
  q_{max} = mass of the adsorbate adsorbed/mass of the adsorbent for a complete monolayer
  C_s = concentration of the adsorbate at saturation of all layers
  K_B = constant related to energy of adsorption

Equation 5.11 can be converted into a linear form:

$$\frac{C_e}{(C_s - C_e)q} = \frac{1}{K_B q_{max}} + \left(\frac{K_B - 1}{K_B q_{max}} \right) \left(\frac{C_e}{C_s} \right) \qquad (5.12)$$

The BET model is based on an ideal assumption that all sites are energetically identical along with no horizontal interaction between the adsorbed molecules. As a result, it may be applicable for systems involving heterogeneous materials and simple nonpolar gases, but it is not valid for complex systems dealing with heterogeneous adsorbent such as biosorbents and adsorbates. Consequently, it has lost its popularity in the interpretation of liquid-phase adsorption data for complex solids.

5.6.2.1.5 D–R Isotherm

By not assuming a homogeneous surface or constant adsorption potential, Dubinin and Radushkevich (218) have proposed another equation used in the analysis of isotherms. This model suggests the close relationship between characteristic adsorption curve and porous structure of the biosorbent. Apart from estimating the porosity and the characteristics of adsorption, this model can also be used to determine the apparent free energy of the adsorption process. The D–R isotherm is expressed as

$$q_e = Q_m \exp(-K\varepsilon^2)$$
(5.13)

where
K = the constant related to the adsorption energy
Q_m = the adsorption capacity of the adsorbent per unit mass
ε = Polanyi potential which is correlated to temperature

The D–R equation can be rearranged into a linear form:

$$\ln q_e = \ln Q_m - K\varepsilon^2$$
(5.14)

The slope of the plot $\ln q_e$ versus ε^2 gives K and the intercept yields the adsorption capacity, Q_m. The constant K is related to the mean free energy of adsorption (E) per mole of the adsorbate during the transportation process from infinite distance in solution to the surface of the solid. Thus, E can be calculated from the K value using the relation

$$E = \frac{1}{\sqrt{2K}}$$
(5.15)

In fact, this energy E can be computed using the following relationship (219):

$$\varepsilon = RT \ln\left(1 + \frac{1}{C_e}\right)$$
(5.16)

Since the D–R isotherm is temperature dependent, a characteristic curve with all the suitable data lying on the same curve can be obtained by plotting the adsorption data at different temperatures ($\ln q_e$ versus ε^2). In other words, the applicability of the D–R equation in expressing the adsorption equilibrium data is confirmed if the identity curve is obtained. Apparently, the validity of the ascertained parameters would be questionable when the fitting procedure gives high correction values, but the characteristic curve generated from the analyzed data shows deviation. Nevertheless, the characteristic curve of biosorption systems is rarely examined as the experiments were usually conducted at one temperature. The disadvantage of the D–R isotherm is its suitability for only an intermediate range of adsorbate concentrations as it may exhibit unrealistic asymptotic behavior.

5.6.2.1.6 Hasley Isotherm

Like the Freundlich isotherm, the Hasley model (220) is suitable for multilayer adsorption. The advantage of this isotherm is its usage to confirm the heteroporous nature of the adsorbent by excellent fitting of the experimental data to this model. The Hasley equation is expressed as

$$q_e = \mathrm{Exp}\left(\frac{\ln k_H - \ln C_e}{n} \right) \tag{5.17}$$

where
  k_H = the Hasley isotherm constant
  n = the Hasley isotherm exponent

5.6.2.1.7 H–J Isotherm

The H– adsorption isotherm (221) is suitable for multilayer adsorption. This model suggests the existence of a heterogeneous pore distribution in the adsorbent. The H–J isotherm is given as follows:

$$q_e = \sqrt{\frac{A_H}{B_2 + \log C_e}} \tag{5.18}$$

where
  A_H = isotherm parameter
  B_2 = isotherm constant

5.6.2.2 Three-Parameter Isotherms

There are cases when the two-parameter models are not competent enough to correlate and describe the equilibrium data. For this reason, models involving more than two parameters are needed to interpret the data. A particular model might be inapplicable in a certain situation, while in some cases more than one model can explain the biosorption mechanism. Some available three-parameter isotherms for the prediction of biosorption experimental data are presented.

5.6.2.2.1 Redlich–Peterson Isotherm

By combining elements from both the Langmuir and Freundlich equations, the Redlich–Peterson (R–P) isotherm model (222) suggests that the adsorption mechanism is a hybrid of the two and does not follow ideal monolayer adsorption. The isotherm model is capable to characterize adsorption equilibrium over a wide concentration range:

$$q_e = \frac{K_{RP}C_e}{1 + a_{RP}C_e^{\beta}} \tag{5.19}$$

where K_{RP}, a_{RP}, and β are the R– parameters. The exponent β lies between 0 and 1.

Its limiting behavior is summarized here: when $\beta = 1$, the R–P equation resembles the Langmuir equation:

$$q_e = \frac{K_{RP}C_e}{1 + a_{RP}C_e} \tag{5.20}$$

If $\beta = 0$, the equation represents Henry's law:

$$q_e = \frac{K_{RP}C_e}{1 + a_{RP}} \tag{5.21}$$

Since the β values are close to unity in most biosorption cases, the adsorption data are rather be fitted with the Langmuir model.

The linearized form of Equation 5.19 is written as

$$\ln\left(K_R \frac{C_e}{q_e} - 1\right) = \ln a_{RP} + \beta \ln C_e \qquad (5.22)$$

The linear forms of the equations allow determination of the parameters of the Langmuir and Freundlich models. However, it is not possible to obtain the parameters of the R–P isotherms from the linear equation because R–P isotherm incorporates three parameters. To solve this problem, a minimization procedure has to be adopted to verify the parameters of Equation 5.22 by maximizing the correlation coefficients between the experimental data points and those from theoretical model predictions with the solver add-in function for Microsoft Excel.

5.6.2.2.2 Sips Isotherm

To avoid the problem of continuing increase in the adsorbed amount with rising concentration as observed in the Freundlich model, Sips isotherm was proposed (223). In fact, the Sips expression (Equation 5.21) is similar to the Freundlich isotherm, and differs only on the finite limit of the adsorbed amount at sufficiently high concentration:

$$q_e = q_{max} \frac{(K_S C_e)^\gamma}{1 + (K_S C_e)^\gamma} \qquad (5.23)$$

where K_S = Sips isotherm constant

Besides, Equation 5.23 is akin to the Langmuir equation, Equation 5.1. The distinctive feature in Equation 5.23 is the presence of an additional parameter, γ. The parameter γ characterizes heterogeneity of the system, which could stem from the biosorbent or the adsorbate, or a combination of both. In the case γ is unity, Equation 5.23 is equivalent to Equation 5.3.

5.6.2.2.3 Toth Equation

Both Freundlich and Sips equations have their limitations in describing an adsorption data. As discussed previously, Freundlich equation is not able to predict adsorption equilibria data at intense concentration, while Sips equation is invalid at the low concentration end. Obviously, both mentioned equations are not reduced to the correct Henry law type at the low concentration limit. To overcome this, Toth isotherm (224) which obeys Henry's law at low concentration and reaches an adsorption maximum at high concentration is proposed. The Toth isotherm is derived from the potential theory and it is capable to describe adsorption for heterogeneous systems. It assumes an asymmetrical quasi-Gaussian energy distribution with its left-hand side widened, that is, most sites have adsorption energy less than the mean value:

$$q_e = q_{max} \frac{C_e}{\left[a_t + C_e^t\right]^{1/t}} \qquad (5.24)$$

where
a_T = adsorptive potential constant
t = heterogeneity coefficient of the adsorbent ($0 < t \leq 1$)

Toth equation possesses a parameter to characterize the heterogeneity of the system. The Toth equation reduces to the Langmuir equation when a surface is homogeneous, $t = 1$.

5.6.3 Kinetic Modeling of Biosorption in a Batch System

High adsorption capacity and fast adsorption rate are two important criteria for an ideal adsorbent. As the efficiency of the adsorption process is strongly dependent on the rate of the adsorbate to attach onto the surface of the adsorbent, kinetic studies appear as an important step in the selection of a suitable adsorbent. Apart from reflecting the factors affecting the adsorption process, results from kinetic studies also provide prediction on the adsorption rate. In adsorption processes, the three commonly used kinetic models are the intraparticle diffusion model, pseudo-first-order kinetic model, and pseudo-second-order kinetic model. These kinetic models are applicable to examine the rate determining mechanism of the adsorption process as well as the role of the adsorption surface, the chemical reaction involved, and/or diffusion mechanisms. In practice, kinetic studies were carried out in batch reactions using various adsorbent doses and particle sizes, initial adsorbate concentrations, agitation speeds, pH values, and temperatures along with different adsorbent and adsorbate types. Subsequently, the best-fitting kinetic rate equation is determined using linear regression. To confirm that the experimental data is in good agreement with the kinetic rate equations using the coefficients of determination, the linear least-square method is always applied to the linearly transformed kinetic rate equations.

Generally, the mechanism of adsorbate removal by adsorption is postulated as in the following steps:

1. Bulk diffusion: transport of adsorbate from the bulk solution to the surface of the adsorbent
2. Film diffusion: diffusion of adsorbate through the boundary layer to the surface of the adsorbent
3. Pore diffusion or intraparticle diffusion: migration of adsorbate from the surface to within the particle's pores
4. Adsorption: adsorption of adsorbate on the active sites that are available on the internal surface of the pores

It has been demonstrated in many studies that the bulk diffusion can be ignored providing sufficient stirring to avoid particle and solute gradients in the batch system. Therefore, the adsorption dynamics can be approximated by three consecutive steps 2 through 4 only. A rapid uptake which is immeasurably fast occurs in the adsorption process, in the last step of the mechanism. It is suggested that this step contributes no resistance and it can be considered as an instantaneous process especially in the case of physical adsorption. As a result, the overall rate of the adsorption process is controlled by either film or intraparticle diffusion, or by a combination of both.

In the case of chemical reactions, the adsorption rate may be controlled by its own kinetic rates. Not only the diffusion equations but also the boundary conditions and the adsorption isotherm equation for a complete modeling of kinetics should be taken into account since the adsorption kinetics provide valuable insights into the practical application of the process design and operation control. It has hence led to a complicated system of equations. However, the system is often possible to be simplified by separating the diffusion steps. Based on the assumptions that the initial adsorption rate was characterized by external diffusion and was controlled by intraparticle diffusion, the diffusion mechanisms were considered independently.

The film diffusion is an important rate-controlling step in the first step of adsorption. The change in adsorbate concentration with respect to time is presented as follows:

$$\frac{dC}{dt} = -k_L A(C - C_s) \tag{5.25}$$

where
C = bulk liquid phase concentration of the adsorbate at any time t
C_s = surface concentration of the adsorbate
k_L = external mass transfer coefficient
A = specific surface area for mass transfer

It is assumed that during the initial stage of adsorption, the intraparticle resistance is negligible and the transport is mainly due to film diffusion mechanism. The surface concentration of the adsorbate, C_s can be ignored and $C = C_0$ at $t = 0$. With these assumptions Equation 5.25 can be written in a simplified form:

$$\left[\frac{d(C/C_0)}{dt}\right] = -k_L A \tag{5.26}$$

5.6.3.1 Intraparticle Diffussion Model

Weber and Morris (225) developed the intraparticle diffussion model to describe the intraparticle diffusion by correlating adsorption capacity to effective diffusivity of the adsorbate within the particle. The model is expressed as

$$q_t = f\left(\frac{Dt}{r_p^2}\right)^{1/2} = K_{WM}t^{1/2} \tag{5.27}$$

where
r_p = particle radius
D = effective diffusivity of solutes within the particle
q_t = adsorption capacity at time t
K_{WM} = intraparticle diffusion rate constant

Intraparticle diffusion is the only rate determining step, the plot of q versus $t^{1/2}$ should give a straight-line passing through the origin. The intraparticle diffusion rate constant K can be obtained from the slope of the straight-line. However, the adsorption process may involve some other mechanisms if the adsorption data exhibit multilinear plots. The first shaper portion is a good evidence of a significant external resistance to mass transfer surrounding the particles in the early stage of adsorption. The intraparticle diffusion dominates in the second linear portion, which is a gradual adsorption stage. Eventually, the intraparticle diffusion starts to slow down due to the extremely low solute concentration in solution in the third portion. The third portion is also recognized as the final equilibrium stage. Apparently, the adsorption mechanism can be rationalized by a good correlation of rate data in this model and K values can be determined by linearization of the curve $q = f(t^{0.5})$.

Owing to reasons such as (i) the greater mechanical obstruction to movement presented by the surface molecules or surface layers and (ii) the restraining chemical attractions between the adsorbate and the adsorbent, diffusion within the particle is a much slower process compared with the movement of the adsorbate from the solution to the external solid surface. During adsorption of the adsorbate in a batch system, adsorbate molecules reach at the adsorbent surface more quickly than they can diffuse into the solid. Accumulation of the adsorbate at the surface tends to establish a (pseudo)-equilibrium. Since the surface concentration is depleted by inward adsorption, further adsorption of the adsorbate can take place only at the same rate.

Pseudo-first- and pseudo-second orders are two simplified kinetic models which have been applied to test the adsorption kinetics of adsorbents. Basically, these two models take account of all the steps of adsorption including external film diffusion, intraparticle diffusion, and adsorption.

5.6.3.2 Pseudo-First-Order Kinetic Model

Pseudo-first-order kinetic model is also known as Lagergren model (226). In this model, adsorption is considered to be first order in adsorption capacity and chemisorption is the rate-limiting step, and hence it only predicts the behavior over the "whole" range of studies supporting the validity. In spite of its limitation, this model has been widely used to characterize the adsorption behavior of an adsorbate. The Lagergren first-order rate expression based on solid capacity is generally written as

$$\frac{dq}{dt} = k_1(q_e - q_t) \tag{5.28}$$

where

q_e = adsorption capacity at equilibrium state
q_t = adsorption capacity at time t
k_1 = rate constant of pseudo-first-order adsorption

Integration of Equation 5.28 with the boundary conditions at $t = 0$, $q_t = 0$, and at $t = t$, $q_t = q_t$ results in

$$\ln(q_e - q_t) = \ln q_e - k_1 t \tag{5.29}$$

The nonlinear form of Equation 5.27 is given as

$$q_t = q_e(1 - \exp(-k_1 t)) \tag{5.30}$$

Hypothetically, the straight-line plots of $\ln(q_e - q_t)$ against t of Equation 5.29 should be made at different initial adsorbate concentrations to verify the rate constant and equilibrium adsorbate uptake. A straight-line of $\ln(q_e - q_t)$ versus t confirms the applicability of this kinetic model. The q_e value obtained by this method is always compared with the experimental value. Even though the least-square fitting process yields a high correlation coefficient, a reaction cannot be classified as first order if a large discrepancy in the q_e values is observed. A time lag resulted from external mass transfer or boundary layer diffusion at the beginning of the adsorption process could be the reason for the difference in q_e values. In this case, nonlinear procedure fitting of Equation 5.30 appears as an alternative way to predict q_e and k_1, although this is not a common exercise.

5.6.3.3 Pseudo-Second-Order Kinetic Model

Since the system's kinetics determines adsorbate residence time and the reactor dimensions, predicting the rate of adsorption for a given system is among the most important factors in adsorption system design. Although the adsorption capacity is strongly dependent on various factors such as the nature of the adsorbate, initial adsorbate concentration, temperature, pH of solution and adsorbent particle size, a kinetic model is only concerned with the effect of observable parameters on the overall rate.

Ho and McKay's pseudo-second-order model (227) is derived on the basis of the adsorption capacity of the solid phase. This model can be expressed as

$$\frac{dq}{dt} = k_2(q_e - q_t)^2 \tag{5.31}$$

where k_2 = rate constant of pseudo-second-order adsorption

Integration of Equation 5.31 with the boundary conditions at $t = 0$, $q = 0$, and at $t = t$, $q_t = q_t$, yields

$$\frac{1}{q_e - q} = \frac{1}{q_e} + k_2 t \tag{5.32}$$

Equation 4.32 can be converted into linear form as

$$\frac{t}{q} = \frac{t}{q_e} + \frac{1}{k_2 q_e^2}$$

(5.33)

Ho and McKay equation is applicable to most adsorption systems for the entire experimental duration of adsorption using different adsorbate concentrations and adsorbent dosages. Most importantly, it allows determination of adsorption capacity, pseudo-second-order rate constant, and initial adsorption rate without prior knowledge of experimental parameters.

5.6.4 Continuous Packed-Bed System in the Biosorption of Heavy Metals

The batch adsorption method is feasible to adopt for an adsorption system involving small volumes of adsorbate. However, for large-scale application of biosorption process, continuous flow treatments would be the better choice. In this method, adsorbates in solution are fed continuously to either the top or the bottom of a stationary bed of solid adsorbent. The amount of the adsorbate being adsorbed increases as a function of time and an unsteady-state condition prevails. In the adsorption process under continuous flow conditions, the equilibrium between adsorption and desorption is rarely achieved. The adsorbent is usually regenerated for reuse when the adsorptive capacity of the adsorbent is approached. Since this type of test conditions provides a closer simulation of commercial systems, it is commonly applied in the assessment of the suitability of an adsorbent for a particular adsorbate. Among all the different experimental setups, the packed-bed column is perhaps the most effective device for continuous operations.

In a downflow packed-bed column, initially, when the feed adsorbate solution moves through the column, it is in contact with the fresh adsorbent at the top of the column. As the solution flows down the column, most of the adsorbate is adsorbed progressively from the liquid onto the adsorbent. The concentration of the adsorbate in the effluent remains either very low or even untraceable or as the adsorbate solution passes through the adsorption zone, the adsorbate is either being removed partially or completely. The length of the adsorption zone is somewhat arbitrary as it is dependent of the value of the adsorbate concentration selected for its lower boundary. Adsorbate concentration in the effluent rises slowly if more adsorbate solution enters the column due to equilibrium and kinetic factors. When the upper portion of packing adsorbent is saturated with the adsorbate, the adsorption zone will move down the column like a slowly moving wave. Finally, the lower edge of the adsorption zone arrives at the bottom of the column, leading to a remarkable increase in adsorbate concentration in the effluent. With this rapid rise, the flow is stopped as little additional adsorption takes place with the entire bed approaching an equilibrium state with the feed. This point is referred to as the breakthrough point. The plot of adsorbate effluent concentration versus time is known as the breakthrough curve and it can be used to describe the performance of a continuous packed bed.

There are several factors that affect the breakthrough point and the breakthrough curve, such as the nature of the adsorbate and the adsorbent, geometry of the column, and the operating conditions. The breakthrough point usually increases with increasing bed height, reducing adsorbent's particle size, and with decreasing flow rate. The general position of the breakthrough curve along the time or volume axis may indicate the loading behavior of the adsorbate to be removed from a solution in a fixed bed. It is often expressed in terms of normalized concentration defined as the ratio of effluent adsorbate concentration to inlet adsorbate concentration (C/C_0) as a function of time or volume of the effluent (V_{eff}) for a given bed height. The breakthrough curve would approach a straight vertical line if the adsorption isotherm were favorable and if the adsorption rate were infinite. As the mass transfer rate decreases, the breakthrough curve becomes less sharp. It is noteworthy that the breakthrough curves are diffuse and exhibit an S-shape since the mass transfer is always finite.

A number of simple mathematical models have been developed to predict the dynamic behavior of the column. Various models that are used to characterize the fixed-bed performance for the biosorption process are presented here.

5.6.4.1 Adams–Bohart Model

The Adams–Bohart model (228) is originally developed for gas adsorption. The adsorption is an equation used to characterize the relationship between C/C_0 and t for the adsorption of chlorine on charcoal in a fixed-bed column. It assumes that the adsorption rate is proportional to both the residual capacity of the adsorbent and the concentration of the adsorbing species. Regardless of the phase of the adsorbate, its overall approach can be applied to quantitative description of other systems. The solution of the differential equations for mass transfer rate in solid and liquid phases makes the Adams–Bohart model applicable to fixed-bed column of different biosorption applications. The linear form of the model is shown in Equation 5.32:

$$\ln \frac{C}{C_0} = k_{AB} C_0 t - k_{AB} N \frac{Z}{U_0} \tag{5.34}$$

where
  C = adsorbate concentration remaining at each contact time
  C_0 = initial adsorbate concentration
  k_{AB} = Adams–Bohart kinetic constant
  N = metal concentration in the bulk liquid
  Z = bed depth of column
  U_0 = linear velocity calculated by dividing the flow rate by the column's sectional area

It is noteworthy that when $t \to \infty$, $N \to N_0$, where N_0 is the saturation concentration. Equation 5.32 is derived based on the assumption of low concentration field where $C < 0.15 C_0$ and it is generally valid in the initial part of the breakthrough. Therefore, this model is often utilized in describing the initial part of the breakthrough curve only. Values describing the characteristic operational parameters of the column can be determined from a plot of $\ln C/C_0$ against t at a given bed height and flow rate.

5.6.4.2 Bed Depth–Service Time Model

Starting from the Adams and Bohard model, the bed depth–service time (BDST) model (228) correlates the service time (t) with the process variables by ignoring intraparticle mass resistance and external film resistance. This model is commonly used for determining the capacity of fixed bed at different breakthrough values. By assuming that the adsorbate is adsorbed onto the adsorbent surface directly, this model states that the service time for a column is given by

$$t = \frac{N_0}{C_0 U_0} Z - \frac{1}{K_a C_0} \ln \left(\frac{C_0}{C} - 1 \right) \tag{5.35}$$

where
  K_a = rate constant in BDST
  N_0 = adsorption capacity

The equation is reduced to Equation 5.35 at 50% breakthrough (C_o/C) = 2 and $t = t_{0.5}$

$$t_{0.5} = \left(\frac{N_o}{C_o U_o} \right) Z \tag{5.36}$$

or

$$t_{0.5} = \text{constant} \times Z \tag{5.37}$$

If the adsorption data fits the model, a straight-line passing through the origin should be obtained in a plot of BDST at 50% breakthrough against bed depth using Equation 5.36.

5.6.4.3 Yoon–Nelson Model

Yoon–Nelson model (229) is a relatively simple theoretical model as it does not require detailed information on the adsorbent and solute characteristics, adsorbent type, and on the physical properties of adsorption bed adsorbent. It assumes the rate of decrease in the probability of adsorption for each adsorbate molecule is proportional to the probability of adsorption of the adsorbate and the probability of adsorbate breakthrough on the adsorbent. The Yoon and Nelson equation regarding a single-component system is given by

$$\ln \frac{C}{C_0 - C} = k_{YN} t - \tau k_{YN} \tag{5.38}$$

where

k_{YN} = Yoon and Nelson rate constant
t = time required for 50% adsorbate breakthrough
τ = breakthrough (sampling) time

Calculation of theoretical breakthrough curves for a single-component system requires the determination of the parameters k_{YN} and τ for the adsorbate of interest. These values may be determined from the available experimental data. If the model adequately describes the experimental data, a straight-line should be obtained by a plot of $\ln C/(C_0 - C)$ versus sampling time (t), the slope and intercept of which are k_{YN} and τk_{YN}, respectively.

5.6.4.4 Thomas Model

The Thomas model (230) appears as one of the most commonly used approximate models based on the assumption of Langmuir kinetics of adsorption–desorption and no axial dispersion. This model is usually used to obtain information on the maximum adsorption capacity of an adsorbate in column design. By considering the rate driving force obeys second-order reversible reaction kinetics, the expression of Thomas model for an adsorption column is given as follows:

$$\frac{C}{C_0} = \frac{1}{1 + \exp(k_{Th}/Q(q_0 X - C_0 V_{eff}))} \tag{5.39}$$

where

k_{Th} = Thomas rate constant
Q = Flow rate
q_0 = Maximum solid-phase concentration of the solute
X = Amount of adsorbent in the column
V_{eff} = Effluent volume

The Thomas model can be converted into linear form as follows:

$$\ln\left(\frac{C_0}{C} - 1\right) = \frac{k_{Th} q_0 X}{Q} - \frac{k_{Th} C_0}{Q} V_{eff} \tag{5.40}$$

A plot of $\ln[(C_0/C) - 1]$ against t at a given flow rate allows determination of the kinetic coefficient k_{Th} and the adsorption capacity of the bed q_0.

5.6.4.5 Clark Model

Clark (231) defined a new simulation of breakthrough curves which combined the Freundlich equation and the mass transfer concept. The equation generated based on this model has the following form:

$$\frac{C}{C_0} = \left(\frac{1}{1 + Ae^{-rt}}\right)^{1/n-1} \tag{5.41}$$

with

$$A = \left(\frac{C_0^{n-1}}{C_{break}^{n-1}} - 1\right)e^{rt_{break}} \tag{5.42}$$

and

$$R(n-1) = r \quad \text{and} \quad R = \frac{k_{Cl}}{U_0}v \tag{5.43}$$

where

C_{break} = Outlet concentration at breakthrough (or limit effluent concentration)
t_{break} = Time at breakthrough
k_{Cl} = Clark rate constant
v = Migration rate

For a particular adsorption process on a fixed bed and a chosen treatment objective, values of A and r can be determined using Equation 5.43 by nonlinear regression analysis, enabling the prediction of the breakthrough curve according to the relationship between C/C_0 and t in Equation 5.43.

5.6.5 Response Surface Methodology

Response surface methodology (RSM) is a collection of mathematical and statistical techniques for designing experiments, building models, evaluating the effects of variables, and searching optimum conditions of variables to predict targeted responses. It can be considered as an important branch of experimental design and a critical technology particularly in developing new processes, optimizing their performance, and improving design and formulation of new products. Its great applications would be in situations that involve a large number of variables influencing the performance measure or quality characteristic of the product or process. This kind of performance measure or quality characteristic is termed as the response. Most real-world applications for RSM will involve more than one response.

As such, identifying and fitting an appropriate response surface model in heavy metal treatment process can be seen as an attractive approach to improve the removal rate, reduced process variability, time, and overall costs. Moreover, the factors that influence the experiments are identified, optimized, and possible synergic or antagonistic interactions that may exist between factors can be evaluated. There are three main steps involved in the development and optimization process: (i) experimental design, (ii) modeling, and (iii) optimization.

Optimization of a process could be performed either by empirical or statistical methods. However, the empirical method is time consuming and does not necessarily enable an effective optimization.

This could be solved through the statistics-based procedure, RSM. The optimization process by RSM involves three major steps:

1. Performing statistically designed experiments
2. Estimating the coefficients in a mathematical model
3. Predicting the response and checking the adequacy of the model

RSM represents the independent process variables in this quantitative form (232):

$$Y = f(A_1, A_2, A_3, \ldots, A_n) \tag{5.44}$$

where
   Y = the amount of metal adsorbed (mg/L)
   f = response function
   $A_1, A_2, A_3, \ldots, A_n$ = the independent variables

Response surface is obtained by plotting the expected response but the value of f is unknown and can be very complicated. So RSM approximates its value by a suitable lower-order polynomial. If response varies in a linear manner, the response can be represented by this inear function equation as

$$Y = b_o + b_1 A_1 + b_2 A_2 + \cdots + b_n A_n \tag{5.45}$$

But if curvature is there in the system, a higher-order polynomial sush as the quadratic model is used which can be stated in the form of the following equation:

$$Y = b_o + \Sigma b_i A_i + \Sigma b_{ii} A^2_i + \Sigma b_{ij} A_i A_j \tag{5.46}$$

where
   b_o = offset term
   A_i = first-order main effect
   A_{ii} = second-order main effect
   A_{ij} = interaction effect

The application of RSM in the adsorption studies for heavy metals removal can minimize the number of experiments involved and optimize the effective parameters collectively (233–235).

5.7 CONCLUSIONS

The application of low-cost adsorbents in heavy metals removal will make the process highly economical and competitive particularly for environmental applications in detoxifying effluents from metal-plating and metal-finishing operations, mining and ore processing operations, battery and accumulator manufacturing operations, thermal power generation (coal-fired plants in particular), nuclear power generation, and so on. A number of investigations have demonstrated that biosorption is a useful alternative to the conventional systems for the removal of heavy metals from aqueous solution. This technology need not necessarily replace the conventional treatment routes but may complement them.

The adsorption capacity of low-cost materials normally can be improved by pretreatment or modification using physical or chemical methods. Chemical modification in general improved the adsorption capacity of adsorbents probably due to the higher number of active binding sites after modification, better ion-exchange properties, and due to the formation of new functional groups that favors metal uptake. Although chemically modified low-cost adsorbents can enhance its adsorptivity toward heavy metals, the cost of chemicals used and methods of modification also have to be taken into consideration in order to produce "low-cost" adsorbents.

Although excellent removal capabilities were apparent for several low-cost adsorbents, the utilization of these materials in industrial-scale applications is still far from reality. All these arguments converge into one conclusion: more effort is required to implement low-cost materials as adsorbents for removal of heavy metals. The researchers from various scientific backgrounds, from engineering to biochemistry, working together, will make a significant contribution to elucidating the biosorption mechanisms. Further testing in real wastewater should be conducted, and at the same time, appropriate mathematical models need to be developed. It is desirable to have a low-cost adsorbent with a wide range of metal affinities as this will be particularly useful for industrial effluents that carry more than one type of metals.

REFERENCES

1. Volesky, B. 1990. Biosorption and biosorbents. In: Volesky, B. (ed.) *Biosorption of Heavy Metals*, 3–44. Florida: CRC Press, Inc.
2. US Environmental Protection Agency, 1980. *Evaluation of Sorbents for Industrial Sludge Leachate Treatment*, EPA-600/2-80-052. US EPA, Cincinnati, OH.
3. Theopold, K.H. 1994. Chromium: Inorganic and coordination chemistry. In: King, R.B. (ed.) *Encyclopedia of Inorganic Chemistry*, 666–677. New York: John Wiley and Sons.
4. Stasicka, Z. and Kotaś, J. 2000. Chromium occurrence in the environment and methods of its speciation. *Environ Pollut*, 107:263–283.
5. US Environmental Protection Agency, *National Primary Drinking Water Regulations: Technical Fact Sheet on Chromium*. http://www.epa.gov/ogwdw/pdfs/factsheets/ioc/tech/chromium.pdf
6. Salem, H. 1989. The chromium paradox in modern life: Introductory address to the symposium. *Sci Total Environ*, 86:1–3.
7. Wetterhahn, K.E. and Hamilton, J.W. 1989. Molecular basis of hexavalent chromium carcinogenicity: Effect on gene expression. *Sci Total Environ*, 86:113–129.
8. Cotton, F.A., Wilkinson, G., Murillo, C.A., and Bochmanm, M. 1999. *Advanced Inorganic Chemistry*. Sixth edition, New York: John Wiley and Sons, Inc.
9. Considine, D.M. and Considine, G.D. 1984. *Encyclopedia of Chemistry*. Fourth Edition, 287–293. New York: Van Nostrand Reinhold Company.
10. Scheinberg, H. 1991. Copper. In: Merian E. (ed.) *Metals and Their Compounds in the Environment*, 893–908. Weinheim: VCH Publishers.
11. US Environmental Protection Agency. *National Primary Drinking Water Regulations: Consumer Fact Sheet on Copper*. http://www.epa.gov/ogwdw/pdfs/factsheets/ioc/copper.pdf
12. US Environmental Protection Agency, 1980. *Control and Treatment Technology for the Metal Finishing Industry Sulfide Precipitation*, EPA-625/8–80–003. US EPA, Cincinnati, OH.
13. Beszedits, S. 1988. Chromium removal from industrial wastewaters. *Adv Env Sci Technol*, 20:231–261.
14. US Environmental Protection Agency. 1973. *Waste Treatment: Upgrading Metal-Finishing Facilities to Reduce Pollution*, EPA-625/3–73-002. US EPA, Cincinnati, OH.
15. US Environmental Protection Agency. 1973. *Traces of Heavy Metals in Water Removal Processes and Monitoring*, EPA-902/9-74-001. US EPA, Cincinnati, OH.
16. US Environmental Protection Agency. 1983. *Activated Carbon Process for the Treatment of Cadmium(II)-Containing Wastewaters*, EPA/600/S2-83-061. US EPA, Cincinnati, OH.
17. US Environmental Protection Agency. 1990. *Sorption of Heavy Metals by Intact Microorganisms, Cell Walls, and Clay-Wall Composites*, EPA/600/M-90/004. US EPA, Cincinnati, OH.
18. Ting, Y.P., Lawson, F., and Prince, I.G. 1991. Uptake of cadmium and zinc by the alga *Chlorella vulgaris*: II. Multi ion situation. *Biotechnol Bioeng*, 37:445–455.
19. Pascucci, P.R. 1993. Simultaneous multielement study on the binding of metals in solution by algal biomass *Chlorella vulgaris*. *Anal Lett*, 26:1483–1493.
20. Iyer, R.S. and Scott, J.A. 2001. Power station fly ash – a review of value-added utilization outside of the construction industry. *Resour Conserv Recy*, 31:217–228.

21. Pattanayak, J., Mondal, K., Mathew, S., and Lalvani, S.B. 2000. A parametric evaluation of the removal of As(V) and As(III) by carbon-based adsorbents. *Carbon*, 38:589–596.
22. Diamadopoulos, E., Loannidis, S., and Sakellaropoulos, G.P. 1993. As(V) removal from aqueous solutions by fly ash. *Water Res*, 27:1773–1777.
23. Ayala, J., Blanco, F., Garcia, P., Rodriguez, P., and Sancho, J. 1998. Asturian fly ash as a heavy metals removal material. *Fuel*, 77:1147–1154.
24. Bayat, B. 2002. Combined removal of zinc (II) and cadmium (II) from aqueous solutions by adsorption onto high calcium Turkish Fly Ash. *Water Air Soil Pollut*, 136:6992.
25. Bayat, B. 2002. Comparative study of adsorption properties of Turkish fly ashes: II. The case of chromium (VI) and cadmium (II). *J Hazard Mater*, 95:275–290.
26. Apak, R., Tutem, E., Hugul, M., and Hizal, J. 1998 Heavy metal cation retention by unconventional sorbents (red muds and fly ashes). *Water Res*, 32:430–440.
27. Gupta, V.K., Ali, I., Jain, C.K., Sharma, M., and Saini, V.K. 2003. Removal of cadmium and nickel from wastewater using bagasse fly ash—A sugar industry waste. *Water Res*, 37:4038.
28. Weng, C.H. and Huang, C.P. 1994. Treatment of metal industrial wastewater by fly ash and cement fixation. *J Environ Eng*, 120:1470–1487.
29. Papandreou, A., Stournaras, C.J., and Panias, D. 2007. Copper and cadmium adsorption on pellets made from fired coal fly ash. *J Hazard Mater*, 148:538–547.
30. Ho, G.E., Mathew, K., and Newman, P.W.G. 1989. Leachate quality from gypsum neutralized red mud applied to sandy soils. *Water Air Soil Pollut*, 47:1–18.
31. Apiratikul, R. and Pavasant, P. 2008. Sorption of Cu^{2+}, Cd^{2+}, and Pb^{2+} using modified zeolite from coal fly ash. *Chem Eng J*, 144:245–258.
32. Hui, K.K., Chao, C.Y., and Kot, S.C. 2005. Removal of mixed heavy metal ions in wastewater by zeolite 4A and residual products from recycled coal fly ash. *J Hazard Mater*, 127:89–101.
33. Cetin, C. and Pehlivan, E. 2007. The use of fly ash as a low cost, environmentally friendly alternative to activated carbon for the removal of heavy metals from aqueous solutions. *Colloids Surf A: Physicochem Eng Asp*, 298:83–87.
34. Gupta, V.K. and Ali, I. 2004. Removal of lead and chromium from wastewater using bagasse fly ash—A sugar industry waste. *J Colloid Interface Sci*, 271:321–328.
35. Panday, K.K., Prasad, G., and Singh, V.N. 1984. Removal of Cr(VI) from aqueous solutions by adsorption on fly ash-wollastonite. *J Chem Technol Biotechnol*, 34A: 367–374.
36. Banerjee, S.S., Joshi, M.V., and Jayaram, R.V. 2004. Removal of Cr(VI) and Hg(II) from aqueous solution using fly ash and impregnated fly ash. *Sep Sci Technol*, 39:1611–1629.
37. Bhattacharya, A.K., Naiya, T.K., Mandal, S.N., and Das, S.K. 2008. Adsorption, kinetics and equilibrium studies on removal of Cr(VI) from aqueous solutions using different low-cost adsorbents. *Chem Eng J*, 137: 529–541.
38. Mimura, H., Yokota, K., Akiba, K., and Onodera, Y. 2001. Alkali hydrothermal synthesis of zeolites from coal fly ash and their uptake properties of cesium ion. *J Nucl Sci Technol*, 38:766–772.
39. Panday, K.K., Prasad, G., and Singh, V.N. 1985. Copper(II) removal from aqueous solutions by fly ash. *Water Res*, 19:869–873.
40. Bayat, B. 2002. Comparative study of adsorption properties of Turkish fly ashes: I. The case of nickel(II), copper(II) and zinc(II). *J Hazard Mater*, 95:251–273.
41. Ricou, P., Lecuyer, I., and Cloirec, P.L. 1999. Removal of Cu^{2+}, Zn^{2+} and Pb^{2+} by adsorption onto fly ash and fly ash/lime mixing. *Water Sci Technol*, 39:239–247.
42. Lin, C.J. and Chang, J.E. 2001. Effect of fly ash characteristics on the removal of Cu(II) from aqueous solution. *Chemosphere*, 44:1185–1192.
43. Gupta, V.K. and Ali, I. 2000. Utilisation of bagasse fly ash (a sugar industry waste) for the removal of copper and zinc from wastewater. *Sep Purif Technol*, 18:131–140.
44. Rao, M., Parwate, A.V., Bhole, A.G., and Kadu, P.A. 2003. Performance of low-cost adsorbents for the removal of copper and lead. *J Water Supply Res Technol*, 52:49–58.
45. Hossain, M.A., Kumita, M., Michigami, Y., and More, S. 2005. Optimization of parameters for Cr(VI) adsorption on used black tea leaves. *Adsorption*, 11:561–568.
46. Hsu, T.C., Yu, C.C., and Yeh, C.M. 2008. Adsorption of Cu^{2+} from water using raw and modified coal fly ashes. *Fuel*, 87:1355–1359.
47. Gupta, V.K. 1998. Equilibrium uptake, sorption dynamics, process development, and column operations for the removal of copper and nickel from aqueous solution and wastewater using activated slag, a low-cost adsorbent. *Ind Eng Chem Res*, 37:192–202.

48. Sen, A.K. and De, A.K. 1987. Adsorption of mercury(II) by coal fly ash. *Water Res*, 21:885–888.

49. Kapoor, A. and Viraraghavan, T. 1992. Adsorption of mercury from wastewater by fly ash. *Adsorpt Sci Technol*, 9:130–147.

50. Banerjee, S.S., Jayaram, R.V., and Joshi, M.V. 2003. Removal of nickel and zinc(II) from wastewater using fly ash and impregnated fly ash. *Sep Sci Technol*, 38:1015–1032.

51. Rao, M., Parwate, A.V., and Bhole, A.G. 2002. Removal of Cr^{6+} and Ni^{2+} from aqueous solution using bagasse and fly ash. *Waste Manage*, 22:821–830.

52. Gan, Q. 2000. A case study of microwave processing of metal hydroxide sediment sludge from printed circuit board manufacturing wash water. *Waste Manage*, 20:695–701.

53. Yadava, K.P., Tyagi, B.S., Panday, K.K., and Singh, V.N. 1987. Fly ash for the treatment of Cd(II) rich effluents. *Environ Technol Lett*, 8:225–234.

54. Goswami, D. and Das, A.K. 2000. Removal of arsenic from drinking water using modified fly-ash bed. *Inter J Water*, 1:61–70.

55. Gupta, V.K. and Sharma, S. 2003. Removal of zinc from aqueous solutions using bagasse fly ash—A low cost adsorbent. *Ind Eng Chem Res*, 42:6619–6624.

56. Weng, C.H. and C.P. Huang, 1990. Removal of Trace Heavy Metals by Adsorption onto Fly Ash. *Proceedings of the 1990 ASCE Environmental Engineering Specialty Conference.* 923–924, Arlington, VA.

57. Weng, C.H. and Huang, C.P. 2004. Adsorption characteristics of Zn(II) from dilute aqueous solution by fly ash. *Colloids Surf A: Physicochem Eng Asp*, 247:137–143.

58. Bhattacharya, A.K., Mandal, S.N., and Das, S.K. 2006. Adsorption of Zn(II) from aqueous solution by using different adsorbents. *Chem Eng J*, 123:43–51.

59. Rahman, I.A., Ismail, J., and Osman, H. 1997. Effect of nitric acid digestion on organic materials and silica in rice husk. *J Mater Chem*, 7:1505–1509.

60. Rahman, I.A. and Ismail, J., 1993. Preparation and characterization of a spherical gel from a low-cost material. *J Mater Chem*, 3:931–934.

61. Abo-El-Enein, S.A., Eissa, M.A., Diafullah, A.A., Rizk, M.A., and Mohamed, F.M. 2009. Removal of some heavy metals ions from wastewater by copolymer of iron and aluminum impregnated with active silica derived from rice husk ash. *J Hazard Mater*, 172:574–579.

62. Roy, D., Greenlaw, P.N., and Shane, B.S. 1993. Adsorption of heavy metals by green algae and ground rice hulls. *J Environ Sci Health*, A28:37–50.

63. Lee, C.K., Low, K.S., Liew, S.C., and Choo, C.S. 1999. Removal of arsenic(V) from aqueous solution by quaternized rice husk. *Environ Technol*, 20:971–978.

64. Nakbanpote, W., Thiravavetyan, P., and Kalambaheti, C. 2002. Comparison of gold adsorption by *Chlorella vulgaris*, rice husk and activated carbon. *Miner Eng*, 15:549–552.

65. Nakbanpote, W., Thiravavetyan, P., and Kalambaheti, C. 2000. Preconcentration of gold by rice husk ash. *Miner Eng*, 13:391–400.

66. Krishnani, K.K., Meng, X., Christodoulatos C., and Boddu, V.M. 2008. Biosorption mechanism of nine different heavy metals onto biomatrix from rice husk. *J Hazard Mater*, 153:1222–1234.

67. Ajmal, M., Rao, R.A.K., Anwar, S., Ahmad, J., and Ahmad, R. 2003. Adsorption studies on rice husk: Removal and recovery of Cd(II) from wastewater. *Biores Technol*, 86:147–149.

68. Ye, H., Zhu, Q., and Du, D. 2010. Adsorptive removal of Cd(II) from aqueous solution using natural and modified rice husk. *Biores Technol*, 101:5175–5179.

69. Tarley, C.R.T. Ferreira, S.L.C., and Arruda, M.A.Z. 2004. Use of modified rice husks as a natural solid adsorbent of trace metals: Characterisation and development of an on-line preconcentration system for cadmium and lead determination by FAAS. *Microchem J*, 77:163–175.

70. Kumar, U. and Bandyopadhyay, M. 2006. Sorption of cadmium from aqueous solution using pretreated rice husk. *Biores Technol*, 97:104–109.

71. Munaf, E. and Zein, R. 1997. The use of rice husk for removal of toxic metals from waste water. *Environ Technol*, 18:359–362.

72. Marshall, W.E., Champagne, E.T., and Evans, W.J. 1993. Use of rice milling byproducts (hulls and bran) to remove metal ions from aqueous solution. *J Environ Sci Health*, A28:1977–1992.

73. Srivastava, V.C., Mall, I.D., and Mishra, I.M. 2008. Removal of cadmium(II) and zinc(II) metal ions from binary aqueous solution by rice husk ash. *Colloids Surf A: Physicochem Eng Asp*, 312:172–184.

74. Sharma, N., Kaur, K., and Kaur, S. 2009. Kinetic and equilibrium studies on the removal of Cd^{2+} ions from water using polyacrylamide grafted rice (*Oryza sativa*) husk and (*Tectona grandis*) saw dust. *J Hazard Mater*, 163:1338–1344.

75. Akhtar, M., Iqbal, S., Kausar, A., Bhanger, M.I., and Shaheen, M.A. 2010. An economically viable method for the removal of selected divalent metal ions from aqueous solutions using activated rice husk. *Colloids Surf B: Biointerfaces*, 75:149–155.

76. Wang, L.H. and Lin, C.I. 2008. Adsorption of chromium (III) ion from aqueous solution using rice hull ash. *J Chin Inst Chem Eng*, 39:367–373.

77. Guo, Y., Qi, J., Yang, S., Yu, K., Wang, Z., and Xu, H. 2002. Adsorption of Cr(VI) on micro- and meso-porous rice husk-based active carbon. *Mater Chem Phys*, 78:132–137.

78. Bansal, M., Garg, U., Singh, D., and Garg, V.K. 2009. Removal of Cr(VI) from aqueous solutions using pre-consumer processing agricultural waste: A case study of rice husk. *J Hazard Mater*, 162:312–320.

79. Wong, K.K., Lee, C.K., Low, K.S., and Haron, M.J. 2003. Removal of Cu and Pb by tartaric acid modified rice husk from aqueous solutions. *Chemosphere*, 50:23–28.

80. Wong, K.K., Lee, C.K., Low, K.S., and Haron, M.J. 2003. Removal of Cu and Pb from electroplating wastewater using tartaric acid modified rice husk. *Process Biochem*, 39:437–445.

81. Nakbanpote, W., Goodman, B.A., and Thiravetyan, P. 2007. Copper adsorption on rice husk derived materials studied by EPR and FTIR. *Colloids Surf A: Physicochem Eng Asp*, 304:7–13.

82. Feroze, N., Ramzan, N., Khan, A., and Cheema, I.I. 2011. Kinetic and equilibrium studies for Zn (II) and Cu (II) metal ions removal using biomass (Rice Husk) Ash. *J Chem Soc Pak*, 33:139–146.

83. Johan, N.A., Kutty, S.R.M., Isa, M.H., Muhamad, N.S., and Hashim, H. 2011. Adsorption of copper by using microwave incinerated rice Husk Ash (MIRHA). *Int J Civil Environ Eng*, 3:211–215.

84. Feng, Q., Lin, Q., Gong, F., Sugita, S., and Shoya, M. 2004. Adsorption of lead and mercury by rice husk ash. *J Colloid Interface Sci*, 278:1–8.

85. Tiwari, D.P., Singh, D.K., and Saksena, D.N. 1995. Hg(II) adsorption from solutions using rice-husk ash. *J Environ Eng*, 121:479–481.

86. Ghorbani, M., Lashkenari, M.S., and Eisazadeh, H. 2011. Application of polyaniline nanocomposite coated on rice husk ash for removal of Hg(II) from aqueous media. *Synthetic Met*, 161:1430–1433.

87. Srivastava, V.C., Mall, I.D., and Mishra, I.M. 2009. Competitive adsorption of cadmium(II) and nickel(II) metal ions from aqueous solution onto rice husk ash. *Chem Eng Processing: Process Intens*, 48:370–379.

88. Pillai, M.G., Regupathi, I., Kalavathy, M.H., Murugesan, T., and RoseMiranda, L. 2009. Optimization and analysis of nickel adsorption on microwave irradiated rice husk using response surface methodology (RSM). *J Chem Technol Biotechnol*, 84:291–301.

89. Wang, L.H. and Lin, C.I. 2008. Adsorption of lead(II) ion from aqueous solution using rice hull ash. *Ind Eng Chem Res*, 47:4891–4897.

90. Naiya, T.K., Bhattacharya, A.K., Mandal, S., and Das, S.K. 2009. The sorption of lead(II) ions on rice husk ash. *J Hazard Mater*, 163:1254–1264.

91. Surchi, K.M.M. 2011. Agricultural wastes as low cost adsorbents for Pb removal: Kinetics, equilibrium and thermodynamics. *Int J Chem*, 3:103–112.

92. Zulkali, M.M.D., Ahmad, A.L., Norulakmal, N.H., and Sharifah, N.S. 2006. Comparative studies of *Oryza sativa* L. husk and chitosan as lead adsorbent. *J Chem Technol Biotechnol*, 81:1324–1327.

93. Mishra, S.P., Tiwari, D., and Dubey, R.S. 1997. The uptake behaviour of rice (Jaya) husk in the removal of Zn(II) ions—A radiotracer study. *Appl Radiat Isotopes*, 48:877–882.

94. Dang, V.B.H., Doan, H.D., Dang-Vu, T., and Lohi, A., 2009. Equilibrium and kinetics of biosorption of cadmium (II) and copper (II) ions by wheat straw. *Biores Technol*, 100:211–219.

95. Tan, G. and Xiao, D. 2009. Adsorption of cadmium ion from aqueous solution by ground wheat stems. *J Hazard Mater*, 164:1359–1363.

96. Dhir, B. and Kumar, R. 2010. Adsorption of heavy metals by *Salvinia* biomass and algricultural residues. *Int J Environ Res*, 4:427–432.

97. Nouri, L. and Hamdaoui, O. 2007. Ultrasonication-assisted sorption of cadmium from aqueous phase by wheat bran. *J Phys Chem A*, 111:8456–8463.

98. Nouri, L., Ghodbane, I., Hamdaoui, O., and Chiha, M. 2007. Batch sorption dynamics and equilibrium for the removal of cadmium ions from aqueous phase using wheat bran. *J Hazard Mater*, 149:115–125.

99. Farajzadeh, M.A. and Monji, A.B. 2004. Adsorption characteristics of wheat bran towards heavy metal cations. *Sep Sci Technol*, 38:197–207.

100. Özer, A. and Pirincci, H.B. 2006. The adsorption of Cd(II) ions on sulfuric acid-treated wheat bran. *J Hazard Mater* B, 137:849–855.

101. Chojnacka, K. 2006. Biosorption of Cr(III) ions by wheat straw and grass: a systematic characterization of new biosorbents. *Polish J Environ Studies*, 15:845–852.

102. Dupont, L. and Guillon, E. 2003. Removal of hexavalent chromium with a lignocellulosic substrate extracted from wheat bran. *Environ Sci Technol*, 37:4235–4241.

103. Wang, X.S., Li, Z.Z., and Sun, C. 2008. Removal of Cr(VI) from aqueous solutions by lowcost biosorbents: Marine macroalgae and agricultural by-products. *J Hazard Mater*, 153:1176–1184.

104. Singh, K.K., Hasan, H.S., Talat, M., Singh, V.K., and Gangwar, S.K. 2009. Removal of Cr(VI) from aqueous solutions using wheat bran. *Chem Eng J*, 151:113–121.

105. Nameni, M., Moghadam, M.R.A., and Aram, M. 2008. Adsorption of hexavalent chromium from aqueous solutions by wheat bran. *Intern J Environ Sci Technol*, 5:161–168.

106. Gong, R., Guan, R., Zhao, J., Liu, X., and Ni, S. 2008. Citric acid functionalizing wheat straw as sorbent for copper removal from aqueous solution. *J Health Sci*, 54:174–178.

107. Dupont, L., Bouanda, J., Dumonceau, J., and Aplincourt, M. 2005. Biosorption of Cu(II) and Zn(II) onto a lignocellulosic substrate extracted from wheat bran. *Environ Chem Lett*, 2:165–168.

108. Aydın, H., Bulut, Y., and Yerlikaya, C. 2008. Removal of copper (II) from aqueous solution by adsorption onto low-cost adsorbents. *J Environ Manage*, 87:37–45.

109. Basci, N., Kocadagistan, E., and Kocadagistan, B. 2004. Biosorption of copper (II) from aqueous solutions by wheat shell. *Desalination*, 164:135–140.

110. Wang, X.S., Li, Z.Z., and Sun, C. 2009. A comparative study of removal of Cu(II) from aqueous solutions by locally low-cost materials: Marine macroalgae and agricultural by-products. *Desalination*, 235:146–159.

111. Özer, A., Özer, D., and Özer, A. 2004. The adsorption of copper (II) ions on to dehydrate wheat bran (DWB): Determination of the equilibrium and thermodynamic parameters. *Process Biochem*, 39:2183–2191.

112. Bulut, Y. and Baysal, Z. 2006. Removal of Pb(II) from wastewater using wheat bran. *J Environ Manage*, 78:107–113.

113. Wang, X., Xia, L., Tan, K., and Zheng, W. 2011. Studies on adsorption of uranium(VI) from aqueous solution by wheat straw. *Environ Prog Sustain Energy*. doi: 10.1002/ep.10582.

114. Qu, R.J., Sun, C.M., Fang, M., Zhang, Y., Ji, C.N. Xu, Q. et al. 2009. Removal of recovery of Hg(II) from aqueous solution using chitosan-coated cotton fibers. *J Hazard Mater*, 167:717–727.

115. Zhang, G.Y., Qu, R.J., Sun, C.M., Ji, C.N., Chen, H. Wang, C.H. et al. 2008. Adsorption for metal ions of chitosan coated cotton fiber. *J Appl Polym Sci*, 110:2321–2327.

116. Qu, R.J., Sun, C.M., Wang, M.H., Ji, C.N., Xu, Q. Zhang, Y. et al. 2009. Adsorption of Au(III) from aqueous solution using cotton fiber/chitosan composite adsorbents. *Hydrometallurgy*, 100:65–71.

117. Huang, G.L., Zhang, H.Y., Jeffrey, X.S., and Tim, A.G.L. 2009. Adsorption of chromium(VI) from aqueous solutions using cross-linked magnetic chitosan beads. *Ind Eng Chem Res*, 48:2646–2651.

118. Tran, H.V., Tran, L.D., and Nguyen, T.N. 2010. Preparation of chitosan/magnetite composite beads and their application for removal of Pb(II) and Ni(II) from aqueous solution. *Mater Sci Eng C*, 30:304–310.

119. Sun, X.Q., Peng, B., Jing, Y., Chen, J., and Li, D.Q. 2009. Chitosan(chitin)/cellulose composite biosorbents prepared using ionic liquid for heavy metal ions adsorption. *Separations*, 55:2062–2069.

120. Kalyani, S., Ajitha, P.J., Srinivasa, R.P., and Krishnaiah, A. 2005. Removal of copper and nickel from aqueous solutions using chitosan coated on perlite as biosorbent. *Sep Sci Technol*, 40:1483–1495.

121. Shameem, H., Abburi, K., Tushar, K.G., Dabir, S.V., Veera, M.B., and Edgar, D.S. 2006. Adsorption of divalent cadmium (Cd(II)) from aqueous solutions onto chitosan-coated perlite beads. *Ind Eng Chem Res*, 45:5066–5077.

122. Shameem, H., Abburi, K., Tushar, K.G., Dabir, S.V., Veera, M.B., and Edgar, D.S. 2003. Adsorption of chromium(VI) on chitosan-coated perlite. *Sep Sci Technol*, 38:3775–3793.

123. Shameem, H., Tushar, K.G., Dabir, S.V., and Veera, M.B. 2008. Dispersion of chitosan on perlite for enhancement of copper(II) adsorption capacity. *J Hazard Mater*, 152:826–837.

124. Veera, M.B., Krishnaiah, A., Jonathan, L.T., Edgar, D.S., and Richard, H. 2008. Removal of arsenic (III) and arsenic (V) from aqueous medium using chitosan-coated biosorbent. *Water Res*, 42:633–642.

125. Veera, M.B., Krishnaiah, A., Ann, J.R., and Edgar, D.S. 2008. Removal of copper(II) and nickel(II) ions from aqueous solutions by a composite chitosan biosorbent. *Sep Sci Technol*, 43:1365–1381.

126. Veera, M.B., Krishnaiah, A., Jonathan, L.T., and Edgar, D.S. 2003. Removal of hexavalent chromium from wastewater using a new composite chitosan biosorbent. *Environ Sci Technol*, 37:4449–4456.

127. Fan, D.H., Zhu, X.M., Xu, M.R., and Yan, J.L. 2006. Adsorption properties of chromium (VI) by chitosan coated montmorillonite. *J Biol Sci*, 6:941–945.

128. Wan Ngah, W.S. and Fatinathan, S. 2008. Adsorption of Cu(II) ions in aqueous solution using chitosan beads, chitosan–GLA beads and chitosan–alginate beads. *Chem Eng J*, 143:62–72.

129. Vijaya, Y., Srinivasa, R.P., Veera, M.B., and Krishnaiah, A. 2008. Modified chitosan and calcium alginate biopolymer sorbents for removal of nickel (II) through adsorption. *Carbohydr Polym*, 72:261–271.

130. Srinivasa, R.P., Vijaya, Y., Veera, M.B., and Krishnaiah, A. 2009. Adsorptive removal of copper and nickel ions from water using chitosan coated PVC beads. *Bioresour Technol*, 100:194–199.
131. Kumar, M., Bijay, P.T., and Vinod, K.S. 2009. Crosslinked chitosan/polyvinyl alcohol blend beads for removal and recovery of Cd(II) from wastewater. *J Hazard Mater*, 172:1041–1048.
132. Wan Ngah, W.S., Kamari, A., and Koay, Y.J. 2004. Equilibrium kinetics studies of adsorption of copper (II) on chitosan and chitosan/PVA beads. *Int J Biol Macromol*, 34:155–161.
133. Wan, M.W., Kan, C.C., Lin, C.H., Buenda, D.R., and Wu, C.H. 2007. Adsorption of copper (II) by chitosan immobilized on sand. *Chia-Nan Annual Bulletin*, 33:96–106.
134. Wan, M.W., Kan, C.C., Buenda, D.R., and Maria, L.P.D. 2010. Adsorption of copper(II) and lead(II) ions from aqueous solution on chitosan-coated sand. *Carbohydr Polym*, 80:891–899.
135. Dragan, E.S., Dinu, M.V., and Timpu, D. 2010. Preparation and characterization of novel composites based on chitosan and clinoptilolite with enhanced adsorption properties for Cu^{2+}. *Bioresour Technol*, 101:812–817.
136. Dinu, M.V. and Dragan, E.S. 2010. Evaluation of Cu^{2+}, Co^{2+}, and Ni^{2+} ions removal from aqueous solution using a novel chitosan/clinoptilolite composites: Kinetics and isotherms. *Chem Eng J*, 160:157–163.
137. Kousalya, G.N., Muniyappan, R.G., Sairam, S.C., and Meenakshi, S. 2010. Synthesis of nano-hydroxy-apatite chitin/chitosan hybrid biocomposites for the removal of Fe(III). *Carbohydr Polym*, 82:549–599.
138. Thayyath, S.A., Sreenivasan, R., and Abdul Rauf, T. 2010. Adsorptive removal of thorium(IV) from aqueous solutions using poly(methacrylic acid)-grafted chitosan/bentonite composite matrix: Process design and equilibrium studies. *Colloids Surf A: Physicochem Eng Asp*, 368:13–22.
139. Nomanbhay, S.M. and Palanisamy, K. 2005. Removal of heavy metal from industrial wastewater using chitosan coated oil palm shell charcoal. *Electron J Biotechnol*, 8:43–53.
140. US Environmental Protection Agency. 1983. *Factors Influencing Metal Accumulation by Algae*, EPA-600/S2-82-100. US EPA, Cincinnati, OH.
141. Romera, E., Gonzalez, F., Ballester, A., Blazquez, M.L., and Munoz, J.A. 2006. Biosorption with algae: A statistical review. *Crit Rev Biotechnol*, 26:223–35.
142. Brinza, L., Dring, M.J., and Gavrilescu, M. 2007. Marine micro- and macro-algal species as biosorbents for heavy metals. *Environ Eng Manage J*, 6:237–251.
143. Holan, Z.R., Volesky, B., and Prasetyo, I. 1993. Biosorption of cadmium by biomass of marine algae. *Biotechnol Bioeng*, 41:819–825.
144. Holan, Z.R. and Volesky, B. 1994. Biosorption of lead and nickel by biomass of marine algae. *Biotechnol Bioeng*, 43:1001–1009.
145. Yu, Q., Matheickal, J.T., Yin, P., and Kaewsarn, P. 1999. Heavy metal uptake capacities of common marine macro algal biomass. *Water Res*, 33:1534–1537.
146. Chong, K.H. and Volesky, B. 1995. Description of two-metal biosorption equilibria by Langmuir-type models. *Biotechnol Bioeng*, 47:451–460.
147. Leusch, A., Holan, Z., and Volesky, B. 1995. Biosorption of heavy metals (Cd, Cu, Ni, Pb, Zn) by chemically-reinforced biomass of marine algae. *J Chem Technol Biotechnol*, 62:279–288.
148. Pavasant, P., Apiratikul, R., Sungkhum, V., Suthiparinyanont, P., Wattanachira, S., and Marhaba, T.F. 2006. Biosorption of Cu2+, Cd2+, Pb2+, and Zn2+ using dried marine green macroalga *Caulerpa lentillifera*. *Bioresour Technol*, 97:2321–2329.
149. Apiratikul, R. and Pavasant, P. 2008. Batch and column studies of biosorption of heavy metals by *Caulerpa lentillifera*. *Bioresour Technol*, 99:2766–2777.
150. Hashim, M.A. and Chu, K.H. 2004. Biosorption of cadmium by brown, green, and red seaweeds. *Chem Eng J*, 97:249–255.
151. Lau, P.S., Lee, H.Y., Tsang, C.C.K., Tam, N.F.Y., and Wong, Y.S. 1999. Effect of metal interference, pH and temperature on Cu and Ni biosorption by *Chlorella vulgaris* and *Chlorella miniata*. *Environ Technol*, 20:953–961.
152. Klimmek, S., Stan, H.-J., Wilke, A., Bunke, G., and Buchholz, R. 2001. Comparative analysis of the biosorption of cadmium, lead, nickel, and zinc by algae. *Environ Sci Technol*, 35:4283–4288.
153. Aksu, Z., Açikel, Ü., and Kutsal, T. 1997. Application of multicomponent adsorption isotherms to simultaneous biosorption of iron(III) and chromium(VI) on *C. vulgaris*. *J Chem Technol Biotechnol*, 70:368–378.
154. Dönmez, G.C., Aksu, Z., Öztürk, A., and Kutsal, T. 1999. A comparative study on heavy metal biosorption characteristics of some algae. *Process Biochem*, 34:885–892.
155. Jalali, R., Ghafourian, H., Asef, Y., Davarpanah, S.J., and Sepehr, S. 2002. Removal and recovery of lead using nonliving biomass of marine algae. *J Hazard Mater*, 92:253–262.

156. Basso, M.C., Cerrella, E.G., and Cukierman, A.I. 2002. Empleo de algas marinas para la biosorción de metales pesados de aguas contaminadas. *Avances en Energías Renovables y Medio Ambiente*, 6:69–74.

157. Matheickal, J.T., Yu, Q., and Woodburn, G.M. 1999. Biosorption of cadmium(II) from aqueous solutions by pre-treated biomass of marine alga *Durvillaea potatorum*. *Water Res*, 33:335–342.

158. Matheickal, J.T. and Yu, Q. 1999. Biosorption of lead(II) and copper(II) from aqueous solutions by pre-treated biomass of Australian marine algae. *Bioresour Technol*, 69:223–229.

159. Yu, Q. and Kaewsarn, P. 2000. Adsorption of Ni²⁺ from aqueous solutions by pretreated biomass of marine macroalga *Durvillaea potatorum*. *Separ Sci Technol*, 35:689–701.

160. Kaewsarn, P. and Yu, Q. 2001. Cadmium(II) removal from aqueous solutions by pre-treated biomass of marine alga *Padina* sp. *Environ Pollut*, 112:209–213.

161. Kaewsarn, P. 2002. Biosorption of copper(II) from aqueous solutions by pre-treated biomass of marine algae *Padina* sp. *Chemosphere*, 47:1081–1085.

162. Tobin, J.M., Cooper, D.G., and Neufeld, R.J. 1984. Uptake of metal ions by *Rhizopus arrhizus* biomass. *Appl Environ Microbiol*, 47:821–824.

163. Cossich, E.S., Tavares, C.R.G., and Ravagnani, T.M.K. 2002. Biosorption of chromium(III) by Sargassum sp. Biomass. *Electron J Biotechnol*, 5:44–52.

164. Silva, E.A., Cossich, E.S., Tavares, C.G., Cardozo Filho, L., and Guirardello, R. 2003. Biosorption of binary mixtures of Cr(III) and Cu(II) ions by *Sargassum* sp. *Braz J Chem Eng*, 20:213–227.

165. Lee, H.S., and Volesky, B. 1999. Interference of aluminum in copper biosorption by an algal biosorbent. *Water Qual Res J Can*, 34:519–533.

166. Davis, T.A., Volesky, B., and Vieira, R.H.S.F. 2000. Sargassum seaweed as biosorbent for heavy metals. *Water Res*, 34:4270–7278.

167. Kim, Y.H., Yeon Park, J., Yoo, Y.J., and Kwak, J.W. 1999. Removal of lead using xanthated marine brown alga, *Undaria pinnatifida*. *Process Biochem*, 34:647–652.

168. Madigan, M.T., Martinko, J.M., and Parker, J. 2000. *Brock Biology of Microorganisms*. Nineth edition, Upper Saddle River, NJ: Pearson Prentice Hall.

169. Loukidou, M.X., Karapantsios, T.D., Zouboulis, A.I., and Matis, K.A. 2004. Diffusion kinetic study of cadmium(II) biosorption by *Aeromonas caviae*. *J Chem Technol Biotechnol*, 79:711–719.

170. Lu, W.-B., Shi, J.-J., Wang, C.-H., and Chang, J.-S. 2006. Biosorption of lead, copper and cadmium by an indigenous isolate Enterobacter sp. J1 possessing high heavy-metal resistance. *J Hazard Mater*, 134:80–86.

171. Ozdemir, G., Ozturk, T., Ceyhan, N., Isler, R., and Cosar, T. 2003. Heavy metal biosorption by biomass of *Ochrobactrum anthropi* producing exopolysaccharide in activated sludge. *Bioresour Technol*, 90:71–74.

172. Chang, J.-S., Law, R., and Chang, C.-C. 1997. Biosorption of lead, copper and cadmium by biomass of Pseudomonas aeruginosa PU21. *Water Res*, 31:1651–1658.

173. Pardo, R., Herguedas, M., Barrado, E., and Vega, M. 2003. Biosorption of cadmium, copper, lead and zinc by inactive biomass of *Pseudomonas putida*. *Anal Bioanal Chem*, 376:26–32.

174. Hussein, H., Ibrahim, S.F., Kandeel, K., and Moawad, H. 2004. Biosorption of heavy metals from waste water using *Pseudomonas* sp. *Electron J Biotechnol*, 17:38–46.

175. Ziagova, M., Dimitriadis, G., Aslanidou, D., Papaioannou, X., Litopoulou Tzannetaki, E., and Liakopoulou-Kyriakides, M. 2007. Comparative study of Cd(II) and Cr(VI) biosorption on *Staphylococcus xylosus* and *Pseudomonas* sp. in single and binary mixtures. *Bioresour Technol*, 98:2859–2865.

176. Tangaromsuk, J., Pokethitiyook, P., Kruatrachue, M., and Upatham, E.S. 2002. Cadmium biosorption by *Sphingomonas paucimobilis* biomass. *Bioresour Technol*, 85:103–105.

177. Puranik, P.R., Chabukswar, N.S., and Paknikar, K.M. 1995. Cadmium biosorption by *Streptomyces pimprina* waste biomass. *Appl Microbiol Biotechnol*, 43:1118–1121.

178. Selatnia, A., Bakhti, M.Z., Madani, A., Kertous, L., and Mansouri, Y. 2004. Biosorption of Cd²⁺ from aqueous solution by a NaOH-treated bacterial dead *Streptomyces rimosus* biomass. *Hydrometallurgy*, 75:11–24.

179. Srinath, T., Verma, T., Ramteke, P.W., and Garg, S.K. 2002. Chromium (VI) biosorption and bioaccumulation by chromate resistant bacteria. *Chemosphere*, 48:427–435.

180. Zhou, M., Liu, Y., Zeng, G., Li, X., Xu, W., and Fan, T. 2007. Kinetic and equilibrium studies of Cr(VI) biosorption by dead *Bacillus licheniformis* biomass. *World J Microbiol Biotechnol*, 23:43–48.

181. Şahin, Y. and Öztürk, A. 2005. Biosorption of chromium(VI) ions from aqueous solution by the bacterium *Bacillus thuringiensis*. *Process Biochem*, 40:1895–1901.

182. Nourbakhsh, M., Sağ, Y., Özer, D., Aksu, Z., Kutsal, T., and Çağlar, A. 1994. A comparative study of various biosorbents for removal of chromium(VI) ions from industrial waste waters. *Process Biochem*, 29:1–5.

183. Salehizadeh, H. and Shojaosadati, S.A. 2003. Removal of metal ions from aqueous solution by polysaccharide produced from *Bacillus firmus*. *Water Res*, 37:4231–4235.

184. Tunali, S., Çabuk, A., and Akar, T. 2006. Removal of lead and copper ions from aqueous solutions by bacterial strain isolated from soil. *Chem Eng J*, 115:203–211.

185. Nakajima, A., Yasuda, M., Yokoyama, H., Ohya-Nishiguchi, H., and Kamada, H. 2001. Copper biosorption by chemically treated *Micrococcus luteus* cells. *World J Microbiol Biotechnol*, 17:343–347.

186. Savvaidis, I., Hughes, M.N., and Poole, R.K. 2003. Copper biosorption by *Pseudomonas cepacia* and other strains. *World J Microbiol Biotechnol*, 19:117–121.

187. Uslu, G. and Tanyol, M. 2006. Equilibrium and thermodynamic parameters of single and binary mixture biosorption of lead (II) and copper (II) ions onto *Pseudomonas putida*: Effect of temperature. *J Hazard Mater*, 135:87–93.

188. Chen, X.C., Wang, Y.P., Lin, Q., Shi, J.Y., Wu, W.X., and Chen, Y.X. 2005. Biosorption of copper(II) and zinc(II) from aqueous solution by *Pseudomonas putida* CZ1. *Colloids Surf B: Biointerfaces*, 46:101–107.

189. Beolchini, F., Pagnanelli, F., Toro, L., and Vegliò, F. 2006. Ionic strength effect on copper biosorption by *Sphaerotilus natans*: Equilibrium study and dynamic modelling in membrane reactor. *Water Res*, 40:144–152.

190. Öztürk, A., Artan, T., and Ayar, A. 2004. Biosorption of nickel(II) and copper(II) ions from aqueous solution by *Streptomyces coelicolor* A3(2). *Colloids Surf B: Biointerfaces*, 34:105–111.

191. Liu, H.-L., Chen, B.-Y., Lan, Y.-W., and Cheng, Y.-C. 2004. Biosorption of Zn(II) and Cu(II) by the indigenous *Thiobacillus thiooxidans*. *Chem Eng J*, 97:195–201.

192. Selatnia, A., oukazoula, A., Kechid, N., Bakhti, M.Z., and Chergui, A. 2004b. Biosorption of Fe^{3+} from aqueous solution by a bacterial dead *Streptomyces rimosus* biomass. *Process Biochem*, 39:1643–1651.

193. Öztürk, A. 2007. Removal of nickel from aqueous solution by the bacterium *Bacillus thuringiensis*. *J Hazard Mater*, 147:518–523.

194. Selatnia, A., Madani, A., Bakhti, M.Z., Kertous, L., Mansouri, Y., and Yous, R. 2004d. Biosorption of Ni^{2+} from aqueous solution by a NaOH-treated bacterial dead *Streptomyces rimosus* biomass. *Miner Eng*, 17:903–911.

195. Choi, S.B. and Yun, Y.-S. 2004. Lead biosorption by waste biomass of *Corynebacterium glutamicum* generated from lysine fermentation process. *Biotechnol Lett*, 26:331–336.

196. Lin, C.-C. and Lai, Y.-T. 2006. Adsorption and recovery of lead(II) from aqueous solutions by immobilized *Pseudomonas Aeruginosa* PU21 beads. *J Hazard Mater*, 137:99–105.

197. Selatnia, A., Boukazoula, A., Kechid, N., Bakhti, M.Z., Chergui, A., and Kerchich, Y. 2004c. Biosorption of lead (II) from aqueous solution by a bacterial dead *Streptomyces rimosus* biomass. *Biochem Eng J*, 19:127–135.

198. De Vargas, I., Macaskie, L.E., and Guibal, E. 2004. Biosorption of palladium and platinum by sulfate-reducing bacteria. *J Chem Technol Biotechnol*, 79:49–56.

199. Nakajima, A. and Tsuruta, T. 2004. Competitive biosorption of thorium and uranium by *Micrococcus luteus*. *J Radioanaly Nucl Chem*, 260:13–18.

200. Mameri, N., Boudries, N., Addour, L., Belhocine, D., Lounici, H., Grib, H., and Pauss, A. 1999. Batch zinc biosorption by a bacterial nonliving *Streptomyces rimosus* biomass. *Water Res*, 33:1347–1354.

201. Incharoensakdi, A. and Kitjaharn, P. 2002. Zinc biosorption from aqueous solution by a halotolerant cyanobacterium *Aphanothece halophytica*. *Curr Microbiol*, 45:261–264.

202. Puranik, P.R. and Paknikar, K.M. 1997. Biosorption of lead and zinc from solutions using *Streptoverticillium cinnamoneum* waste biomass. *J Biotechnol*, 55:113–124.

203. Tang, P. L., Lee, C.K., Low, K.S., and Zainal, Z. 2003. Sorption of Cr(VI) and Cu(II) in aqueous solution by ethylenediamine modified rice hull. *Environ Technol*, 24:1243–1251.

204. Kotas, J. and Stasicka, Z. 2000. Chromium occurrence in the environment and methods of its speciation. *Environ Pollut*, 107:263–283.

205. Low, K.S., Lee, C.K. and Leo, A.C. 1995. Removal of metals from electroplating wastes using banan pith. *Bioresour Technol*, 51:227–231.

206. Celaya, R.J., Noriega, J.A., Yeomans, J.H., Ortega, L.J. and Ruiz-Manríquez, A. 2000. Biosorption of Zn(II) by *Thiobacillus ferrooxidans*. *Bioprocess Eng*, 22:539–542.

207. Langmuir, I. 1918. The adsorption of gases on plane surfaces of glass, mica and platinum. *J Am Chem Soc*, 40:1361–1368.

208. Ho, Y. 2006. Isotherms for the sorption of lead onto peat: Comparison of linear and non-linear methods. *Polish J Environ Stud*, 15:81–86.
209. Ho, Y. and Ofomaja, A.E. 2006. Biosorption thermodynamics of cadmium on coconut copra meal as biosorbent. *Biochem Eng J*, 30:117–123.
210. Padmavathy, V. 2008. Biosorption of nickel(II) ions by baker's yeast: Kinetic, thermodynamic and desorption studies. *Biores Technol*, 99:3100–3109.
211. Djeribi, R. and Hamdaoui, O. 2008. Sorption of copper(II) from aqueous solutions by cedar sawdust and crushed brick. *Desalination*, 225:95–112.
212. Shaker, M.A. 2007. Thermodynamic profile of some heavy metal ions adsorption onto biomaterial surfaces, *Am J Appl Sci*, 4:605–612.
213. Febrianto, J., Kosasih, A.N., Sunarso, J., Ju, Y.S., Indraswati, N., and Ismadi, S. 2009. Equilibrium and kinetic studies in adsorption of heavy metals using biosorbent: A summary of recent studies. *J Hazard Mater*, 162:616–645.
214. Freundlich, H. 1906. Adsorption in solution. *Phys Chem Soc*, 40:1361–1368.
215. Site, A.D. 2001. Factors affecting sorption of organic compounds in natural sorbent/water systems and sorption coefficients for selected pollutants: A review. *J Phys Chem Ref Data*, 30:187–439.
216. Temkin, M.I. 1941. Adsorption equilibrium and the kinetics of processes on nonhomogeneous surfaces and in the interaction between adsorbed molecules. *Zh Fiz Chim*, 15:296–332.
217. Brunauer, S., Emmett, P.H., and Teller, E. 1938. Adsorption of gases in multimolecular layers. *J Am Chem Soc*, 60:309–319.
218. Dubinin, M.M. and Radushkevich, L.V. 1947. Equation of the characteristic curve of activated charcoal. *Proc Acad Sci Phys Chem Sec*, USSR, 55:331–333.
219. Hasany, S.M. and Chaudhary, M.H. 1996. Sorption potential of hare river sand for the removal of antimony from acidic aqueous solution. *Appl Rad Isot*, 47:467–471.
220. Halsey, G. 1948. Physical adsorption on non-uniform surfaces. *J Chem Phys*, 16:931–937.
221. Harkins, W.D. and Jura, E.J. 1944. The decrease of free surface energy as a basis for the development of equations for adsorption isotherms; and the existence of two condensed phases in films on solids. *J Chem Phys*, 12:112–113.
222. Redlich, O.J. and Peterson, D.L. 1959. A useful adsorption isotherm. *J Phys Chem*, 63:1024.
223. Sips, R.J. 1948. On the structure of a catalyst surface. *Chem Phys*, 16:490–495.
224. Toth, J. 1971. State equations of the solid gas interface layer. *Acta Chem Acad Hung*, 69:311–317.
225. Weber, W.J. and Morris, J.C. 1963. Kinetic of adsorption on carbon from solution. *J Sanit Eng Div ASCE*, 89SA2:31–59.
226. Lagergren, S. 1898. Zur theorie der sogenannten adsorption gelöster stoffe, Kungliga Svenska Vetenskapsakademiens. *Handlingar*, 24:1–39.
227. Ho, Y.S. and McKay, G. 1999. Pseudo-second order model for sorption processes. *Process Biochem*, 34:451–465.
228. Bohart, G. and Adams, E.Q. 1920. Some aspects of the behaviour of charcoal with respect to chlorine. *J Am Chem Soc*, 42:523–544.
229. Yoon, Y.H. and Nelson, J.H. 1984. Application of gas adsorption kinetics. I. A. theoretical model for respirator cartridge service time. *Am Ind Hyg Assoc J*, 45:509–516.
230. Thomas, H.C. 1944. Heterogeneous ion exchange in a flowing system. *J Am Chem Soc*, 66:1664–1666.
231. Clark, R.M. 1987. Evaluating the cost and performance of field-scale granular activated carbon systems. *Environ Sci Technol*, 21:573–580.
232. Kiran, B. and Kaushik, A. 2008. Chromium binding capacity of *Lyngbya putealis* exopolysaccharides. *Biochem Eng J*, 38:47–54.
233. Rahimi, S., Moattari, R.M., Rajabi, L., and Derakhshan, A.A. 2015. Optimization of lead removal from aqueous solution using goethite/chitosan nanocomposite by response surface methodology. *Colloids and Surf A: Physiochem Eng Asp*, 484:216–225.
234. Srivastava, V., Sharma, Y.C., and Sillanpää, M. 2015. Application of response surface methodology for optimization of Co(II) removal from synthetic wastewater by adsorption on NiO nanoparticles. *J Mol Liq*, 211:613–620.
235. Davarnejad, R. and Panahi, P. 2016. Cu(II) and Ni(II) removal from aqueous solutions by adsorption on Henna and optimization of effective parameters by using the response surface methodology. *J Indus Eng Chem*, 33:270–275.

6 Sulfide Precipitation for Treatment of Metal Wastes

Nazih K. Shammas and Lawrence K. Wang

CONTENTS

ABSTRACT

Electroplating and other metal finishing operations discharge their spent process water to either waterways or publicly owned treatment works (POTWs) and they comprise more individual wastewater discharges than any other industrial category. The pollutants contained in these discharges are potentially toxic; therefore, to comply with the Clean Water Act, the water must be treated

185

before being discharged to a waterway or a POTW. The regulations require oxidation of cyanide, reduction of hexavalent chromium, removal of heavy metals, and control of pH.

Sulfide precipitation is one among many methods available for removing metals from metal finishing process wastewater. This chapter presents information on various technologies that have been demonstrated. By providing process descriptions, advantages and disadvantages, and economic characteristics of each system, this chapter can facilitate the evaluation of effective means of pollution control by those involved in metal finishing wastewater pollution control.

6.1 INTRODUCTION

The pollutants contained in the electroplating and other metal finishing operations discharges are potentially toxic; hence, their process water is one of the many industrial wastes subject to regulation under the Resource Conservation and Recovery Act (RCRA) (1,2) and the Hazardous and Solid Waste Amendments (HSWA) (3). The metal finishing industry has also been subject to extensive regulation under the Clean Water Act (CWA) (4). Therefore, to comply with these federal regulations, the metal finishing process water must be treated before being discharged to a waterway or a POTW (publicly owned treatment works). The regulations require oxidation of cyanide, reduction of hexavalent chromium, removal of heavy metals, and control of pH.

Metals are usually removed by adding an alkali; such as hydrated lime [$Ca(OH)_2$] or caustic soda (NaOH) to adjust the pH of the wastewater to the point where the metals exhibit minimum solubilities (5–11). The metals precipitate as metal hydroxides (12) and can be removed from the wastewater by flocculation and clarification (13–16); In many cases, the addition of a postfiltration (17) step can further reduce the total metal concentration in the effluent by removing any metal hydroxide carryover. Some common limitations of the hydroxide process are as follows (18):

1. The theoretical minimum solubilities for different metals occur at different pH values (Figure 6.1). For mixtures of metal ions, it must be determined whether a single pH can produce sufficiently low, though not minimum, solubilities for the metal ions present in the wastewater (19).
2. Because hydroxide precipitates tend to resolubilize if the solution pH is increased or decreased from their minimum solubility points, maximum removal efficiency will not be achieved unless the pH is controlled within a narrow range.
3. The presence of complexing ions—such as phosphates, tartrates, EDTA, and ammonia—that are commonly found in cleaner and plating formulations may have an adverse effect on metal removal efficiencies when hydroxide precipitation is used. Figure 6.2 shows the solubility of nickel ions as a function of pH when precipitated with other metal ions in the presence of certain complexing ions used in a proprietary electroless nickel plating bath.

Despite these limitations, hydroxide precipitation (particularly when followed by flocculation and filtration) produces a high-quality effluent when applied to many waste streams. Often coprecipitation of a mixture of metal ions will result in residual metal solubilities lower than those that could be achieved by precipitating each metal at its optimum pH. In other cases, modification of the hydroxide process has improved its performance in treating waste streams containing complexed heavy metals. This improved performance is usually realized by dissolving another positively charged ion such as Fe^{2+} or Ca^{2+} into the wastewater and then precipitating the metals. High-pH lime treatment and ferrous sulfate ($FeSO_4$) precipitation techniques use this principle.

Sulfide precipitation has been demonstrated to be an effective alternative to hydroxide precipitation for removing various heavy metals from industrial wastewater (20–27). The high reactivity of

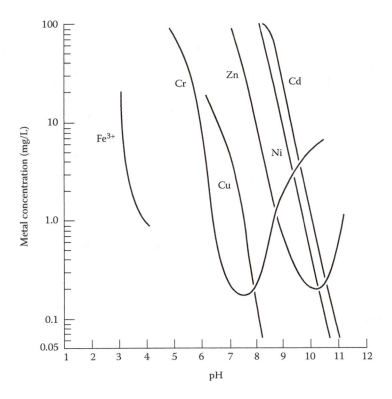

FIGURE 6.1 Metal solubility as a function of pH. (From USEPA, *Waste Treatment: Upgrading Metal-Finishing Facilities to Reduce Pollution*, EPA 625/3-73-002. U.S. Environmental Protection Agency, Washington, DC, July 1973.)

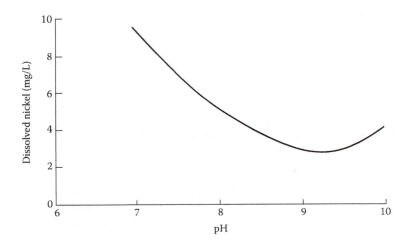

FIGURE 6.2 Solubility of complexed nickel when precipitated with caustic soda. (From USEPA, *Control and Treatment Technology for the Metal Finishing Industry: Sulfide Precipitation, Summary Report,* EPA 625/8-80-003, Environmental Protection Agency, The Industrial Environmental Research Laboratory, Cincinnati, OH, April 1980.)

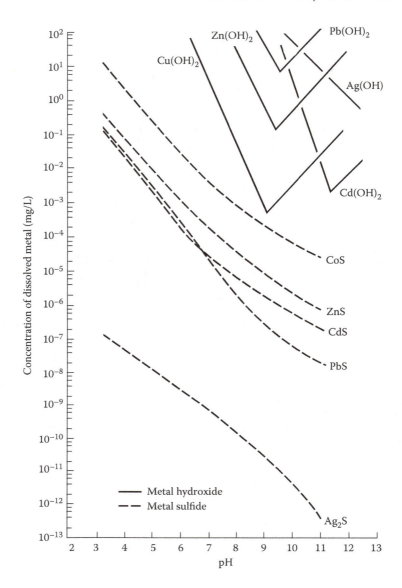

FIGURE 6.3 Solubilities of metal hydroxides and sulfides as a function of pH. Plotted data for metal sulfides based on experimental data listed in Seidell's solubilities. (From USEPA, *Control and Treatment Technology for the Metal Finishing Industry: Sulfide Precipitation, Summary Report,* EPA 625/8-80-003, Environmental Protection Agency, The Industrial Environmental Research Laboratory, Cincinnati, OH, April 1980.)

sulfides (S^{2-}, HS^-) with heavy metal ions and the insolubility of heavy metal sulfides over a broad pH range are attractive features compared with the hydroxide precipitation process (Figure 6.3). Sulfide precipitation can also achieve low metal solubilities in the presence of certain complexing and chelating agents.

The main difference between the two processes that use sulfide precipitation is the means of introducing the sulfide ion into the wastewater. In the soluble sulfide precipitation (SSP) process, the sulfide is added in the form of a water-soluble sulfide reagent such as sodium sulfide (Na_2S) or sodium hydrosulfide (NaHS). A more recently developed process adds a slightly soluble ferrous sulfide (FeS) slurry to the wastewater to supply the sulfide ions needed to precipitate the heavy metals.

In the past, operational difficulties prevented more than minimal application of the SSP process. Later investigations, however, have eliminated or reduced these problems. Technological advances in the area of selective-ion electrodes have provided a probe that has proven successful in pilot-scale evaluations for controlling the addition of soluble sulfide reagent to match reagent demand. Eliminating sulfide reagent overdose can prevent the odor problem commonly associated with these systems. In soluble sulfide systems that do not automatically adjust reagent dosage to match demand, the process tanks must be enclosed and vacuum evacuated to minimize sulfide odor problems in the work area. The formulation of polyelectrolyte conditioners that effectively flocculate the fine metal sulfide particles has eliminated the difficulty in separating the precipitants from the discharge and has resulted in sludges that are easily dewatered (18).

A patented sulfide precipitation process called Sulfex™ has proven effective in separating heavy metals from plating waste streams. The process uses a freshly prepared ferrous sulfide slurry (prepared by reacting $FeSO_4$ and NaHS) as the source of the sulfide ions needed to precipitate the metals from the wastewater. The process operates on the principle that FeS will dissociate into ferrous ions and sulfide ions to the degree predicted by its solubility product. As sulfide ions are consumed, additional FeS will dissociate to maintain the equilibrium concentration of sulfide ions. In alkaline solutions, the ferrous ions will precipitate as ferrous hydroxides. Because most heavy metals have sulfides less soluble than ferrous sulfide, they will precipitate as metal sulfides.

An advantage of the insoluble sulfide precipitation (ISP) process is the absence of any detectable hydrogen sulfide (H_2S) odor—a problem historically associated with SSP treatment systems. Another advantage is that the ISP process will reduce hexavalent chromium to the trivalent state under the same process conditions required for metal precipitation, thus eliminating the need to segregate and pretreat chromium waste streams. Disadvantages of the ISP process include considerably higher than stoichiometric reagent consumption and significantly higher sludge generation factors than either the hydroxide or soluble sulfide treatment processes.

Figure 6.4 compares typical process flow diagrams of a hydroxide treatment system and both types of sulfide systems. Most of the elements of the sulfide systems are common to the hydroxide precipitation treatment sequence. The sulfide treatment processes also can be used as a polishing system after a conventional hydroxide precipitation/clarification process to significantly reduce the consumption of sulfide reagent.

The final selection of a hydroxide or sulfide process should also consider any different constraints for disposal of the resulting sludge. Preliminary studies have indicated that metal ion leachability is lower for metal sulfide sludges than for hydroxide sludges. However, the long-term impacts of weathering and of bacterial and air oxidation of sulfide sludges have not been evaluated.

The importance of design safeguards to avoid the potential hazards associated with sulfide precipitation processes cannot be overemphasized. For example, a sulfide reagent coming into contact with an acidic waste stream can result in the evolution of toxic H_2S fumes in the work area. The potential danger can be minimized by fairly conventional design safeguards, but the safeguards must be well maintained to be effective. Another potential problem for plants discharging to enclosed sewers is the danger associated with residual levels of sulfide in the wastewater. This problem occurs primarily with the SSP processes because the low solubility of FeS in the ISP process controls the residual sulfide concentration at a very low level. Elimination of the H_2S hazard to sewer workers would require either oxidation of the wastewater before discharge or process controls to ensure a low sulfide residual in the discharge.

This chapter is intended to promote an understanding of the use of sulfide precipitation for the removal of heavy metals from industrial waste streams. The chapter includes a general discussion of the sulfide precipitation process theory and an evaluation of both soluble and insoluble sulfide treatment systems in terms of performance, cost, and operating reliability (28–32).

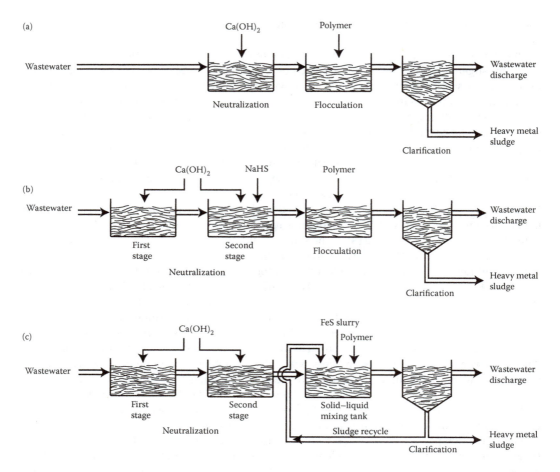

FIGURE 6.4 Wastewater treatment processes for removing heavy metals: (a) hydroxide precipitation, (b) SSP, and (c) ISP. (From USEPA, *Control and Treatment Technology for the Metal Finishing Industry: Sulfide Precipitation, Summary Report,* EPA 625/8-80-003, Environmental Protection Agency, The Industrial Environmental Research Laboratory, Cincinnati, OH, April 1980.)

6.2 PROCESS THEORY

The precipitation of a dissolved metal ion as a metal sulfide (MS) occurs when the metal ion (M^{2+}) contacts a sulfide ion (S^{2-}):

$$M^{2+} + S^{2-} \rightarrow MS \tag{6.1}$$

Most heavy metals encountered in electroplating wastewater will form stable metal sulfides; common exceptions include the trivalent chromic and ferric ions.

The two processes employed to precipitate metals as sulfides differ mainly in the method used to introduce the sulfide ions into the wastewater. The SSP process uses a water-soluble sulfide compound; consequently, the concentration of dissolved sulfide depends on the quantity of reagent added. The ISP process mixes the wastewater with a slurry of slightly soluble FeS, which will dissociate to satisfy its solubility product, yielding a dissolved sulfide concentration of approximately 0.02 µg/L in the wastewater. Use of FeS as the source of sulfide ions controls the level of dissolved sulfide at a concentration low enough to eliminate any detectable emission of H_2S but still provide an inventory of undissolved sulfide that automatically replaces the sulfide consumed in precipitation reactions.

TABLE 6.1

Solubilities of Sulfides That Automatically Replaces the Sulfide Consumed in Precipitation Reactions

Metal Sulfide	K_{sp} (64–77°F)[a]	Sulfide Concentration (mol/L)
Manganous sulfide	1.4×10^{-15}	3.7×10^{-8}
Ferrous sulfide	3.7×10^{-19}	6.1×10^{-10}
Zinc sulfide	1.2×10^{-23}	3.5×10^{-12}
Nickel sulfide	1.4×10^{-24}	1.2×10^{-12}
Stannous sulfide	1.0×10^{-25}	3.2×10^{-13}
Cobalt sulfide	3.0×10^{-26}	1.7×10^{-13}
Lead sulfide	3.4×10^{-28}	1.8×10^{-14}
Cadmium sulfide	3.6×10^{-29}	6.0×10^{-15}
Silver sulfide	1.6×10^{-49}	3.4×10^{-17}
Bismuth sulfide	1.0×10^{-97}	4.8×10^{-20}
Copper sulfide	8.5×10^{-45}	9.2×10^{-23}
Mercuric sulfide	2.0×10^{-49}	4.5×10^{-25}

Source: USEPA, *Control and Treatment Technology for the Metal Finishing Industry: Sulfide Precipitation, Summary Report,* EPA 625/8-80-003, Environmental Protection Agency, The Industrial Environmental Research Laboratory, Cincinnati, OH, April 1980.

[a] Solubility product of a metal sulfide, K_{sp}, equals the product of the molar concentrations of the metal and sulfide.

In the ISP process, the dissolved sulfide ions will precipitate as a metal sulfide any metal with a sulfide solubility less than that of FeS. As shown in Table 6.1, the only heavy metal with a sulfide more soluble than FeS is manganese. In an alkaline solution, the ferrous ions generated in the dissociation of the FeS will precipitate as hydroxides. Maintaining low levels of ferrous ions in the effluent requires that the pH be controlled between 8.5 and 9.5.

One advantage of the ISP process is the ability of the sulfide and ferrous ions to reduce hexavalent chromium to its trivalent state, which eliminates the need to segregate and treat chromium wastes separately. Under alkaline conditions, the chromium will then precipitate as chromium hydroxide $[Cr(OH)_3]$. The overall reduction reaction is

$$H_2CrO_4 + FeS + 4H_2O \rightarrow Cr(OH)_3 + Fe(OH)_3 + S + 2H_2O \qquad (6.2)$$

In the SSP process, the sulfide ion is capable of reducing hexavalent chromium as follows:

$$2H_2CrO_4 + 3NaHS + 8H_2O \rightarrow 2Cr(OH)_3 + 3S + 7H_2O + 3NaOH \qquad (6.3)$$

The question of whether a soluble sulfide reagent can reduce and precipitate hexavalent chromium in one step was addressed in a study conducted for the U.S. Navy (33). The study concluded that the reduction could be accomplished in the presence of ferrous ions (or conceivably some other suitable secondary metal). The ferrous ion acts principally as a catalyst for chromium reduction. Less than stoichiometric dosages of iron are required to effect reduction of most of the chromium. Nearly stoichiometric dosages, however, are required to achieve levels typical of other reduction processes.

6.2.1 SSP PROCESS CHEMISTRY

The addition of a sulfide reagent that has a high solubility in wastewater will yield a relatively high concentration of dissolved sulfide, compared with the ISP process. This high concentration of dissolved sulfide causes a rapid precipitation of the metals dissolved in the water as metal sulfides, which often results in the generation of small particle fines and hydrated colloidal particles. The rapid precipitation reaction tends more toward discrete particle precipitation than toward nucleation precipitation (the precipitation of a particle from solution onto an already existing particle). The resulting poor-settling or -filtering floc is difficult to separate from the wastewater discharges. This problem has been solved by the effective use, separately or combined, of coagulants and flocculants to aid in the formation of large, fast-settling particle flocs (13).

Another disadvantage of an SSP system is the H_2S odor often associated with it (21). The odor detection level of hydrogen sulfide—0.1–1.0 ppmv—is very low compared with the workplace H_2S concentration limit of 10 ppmv specified by the Occupational Safety and Health Administration (OSHA) for worker safety.

The rate of H_2S formation in a water solution is a function of pH (concentration of hydrogen ions) and sulfide ion concentration. The formation of H_2S from dissolved sulfide ions proceeds as follows:

$$S^{2+} + H^+ \rightarrow HS^- \tag{6.4}$$

$$HS^- + H^+ \rightarrow H_2S \tag{6.5}$$

Actually, the strong base S^{2-} is not present in any significant amount except at high pH. For example, at a pH of 11, less than 0.05% of the dissolved sulfide is in the S^{2-} form; the remainder is in either the HS^- or H_2S form. Figure 6.5 is a graph for determining the percentage of the dissolved sulfide in the form of H_2S as a function of the pH of the solution. The relationship shows that at a pH of 9, H_2S accounts for only 1% of the free sulfide in solution. The rate of evolution of H_2S from a sulfide solution per unit of water–air interface will depend on the temperature of the solution (which determines the H_2S solubility), the dissolved sulfide concentration, and the pH. In practice, considering typical response lags of instruments and incremental reagent addition, control of the level of dissolved sulfide and pH would require fine tuning and rigorous maintenance to prevent an H_2S odor problem in the work area. In operating treatment systems, the H_2S odor problem is eliminated by enclosing and vacuum evacuating the process vessels.

Adding a sulfide reagent to wastewater containing precipitated metal hydroxides will result in the resolubilization of the metal hydroxides. The dissolving of the metal hydroxides occurs because the dissolved metal ion concentration is now lower than the equilibrium level predicted by the hydroxide solubility. These newly liberated metal ions will be precipitated by any excess sulfide present. The following reactions occur:

$$M^{2+} + S^{2-} \rightarrow MS \tag{6.6}$$

$$M(OH)_2 \rightarrow M^{2+} + 2(OH)^- \tag{6.7}$$

$$M^{2+} + S^- \rightarrow MS \tag{6.8}$$

Normally, the precipitated solids are in contact with the wastewater long enough to result in an almost complete conversion of metal hydroxides to metal sulfides. Therefore, the sulfide reagent demand depends on the total metal concentration contained in the wastewater. Consequently, a significant reduction in sulfide reagent consumption could be achieved by separating the precipitated metal hydroxides from the wastewater before adding the sulfide reagent.

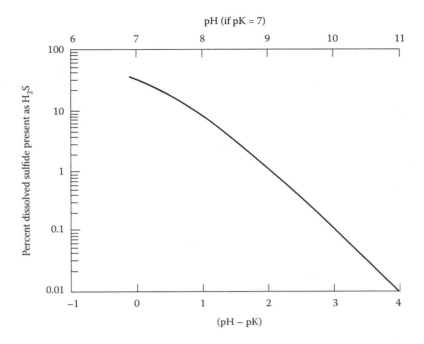

Specific electrical conductance	Value of pK		
of solution at 77°F ($\mu\Omega$/cm)	50°F	68°F	104°F
0[a]	7.24	7.10	6.82
100	7.22	7.08	6.80
1,000	7.18	7.04	6.76
50,000[b]	7.09	6.95	6.67

[a] Distilled H_2O.
[b] Seawater.

FIGURE 6.5 Percent of dissolved sulfide in the H_2S form. pK (logarithmic practical ionization constant) is used to measure the degree of dissociation of weak acids, in this case H_2S. (From USEPA, *Control and Treatment Technology for the Metal Finishing Industry: Sulfide Precipitation, Summary Report,* EPA 625/8-80-003, Environmental Protection Agency, The Industrial Environmental Research Laboratory, Cincinnati, OH, April 1980.)

6.2.2 ISP Process Chemistry

The Sulfex process precipitates dissolved metals as sulfides by mixing the wastewater with an FeS slurry in a solid–liquid contact chamber. The FeS dissolves to maintain the sulfide ion concentration at a level of 0.02 μg/L.

The following reactions occur when FeS is introduced into a solution containing dissolved metals and metal hydroxide (18):

$$FeS \rightarrow Fe^{2+} + S^{2-} \tag{6.9}$$

$$M^{2+} + S^{2-} \rightarrow MS \tag{6.10}$$

$$M(OH)_2 \rightarrow M^{2+} + 2(OH)^- \tag{6.11}$$

$$Fe^{2+} + 2(OH)^- \rightarrow Fe(OH)_2 \qquad\qquad (6.12)$$

The addition of ferrous ions to the wastewater and their precipitation as ferrous hydroxide [Fe(OH)$_2$] results in a considerably larger quantity of solid waste from this process than from a conventional hydroxide precipitation process.

As with SSP, the ISP process achieves an almost complete conversion of previously precipitated metal hydroxides to metal sulfides. The reaction goes to completion because of the long residence time of the solids in the treatment system before discharge.

Figure 6.6 shows the three different factors that affect the ability of FeS to precipitate copper from a solution containing metal complexing compounds. Hence, the design criteria that must be addressed are based on these three factors (18):

1. A dense sludge blanket must be maintained in the solid–liquid contact zone.
2. Adequate mixing time is required for the precipitation reaction to reach equilibrium.
3. From 2 to 4 times the required quantity is needed to realize the low levels of dissolved copper achievable by sulfide precipitation.

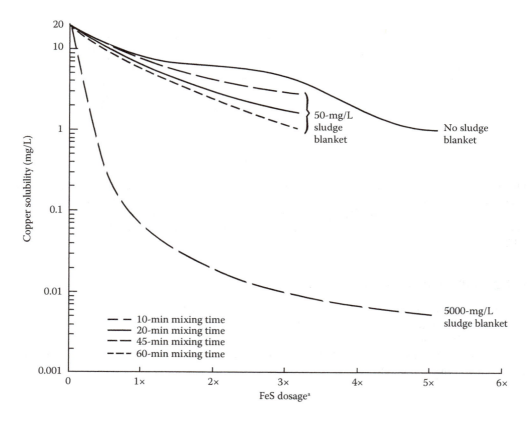

FIGURE 6.6 Influence of FeS dosage, sludge blanket concentration, and mixing time on copper solubility. [a]X = stoichiometic equivalent concentration of FeS required to precipitate 20 mg Cu^{+2}/L = 27.7 mg FeS/L. Results of jar tests with complexed Cu influent; pH values maintained between 7 and 8 during tests. (From USEPA, *Waste Treatment: Upgrading Metal-Finishing Facilities to reduce Pollution*, EPA 625/3-73-002. U.S. Environmental Protection Agency, Washington, DC, July 1973.)

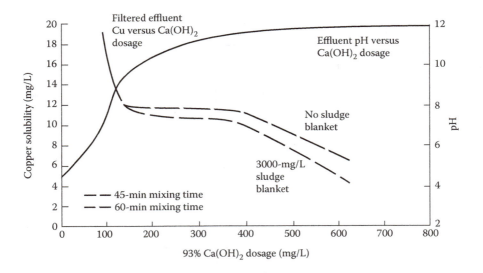

FIGURE 6.7 Influence of Ca(OH)$_2$ dosage, sludge blanket concentration, and pH on copper solubility. Results of hydroxide process jar tests with completed Cu influent. (From USEPA, *Waste Treatment: Upgrading Metal-Finishing Facilities to reduce Pollution*, EPA 625/3-73-002. U.S. Environmental Protection Agency, Washington, DC, July 1973.)

To illustrate the relative effectiveness of sulfide precipitation, Figure 6.7 represents the solubility of copper in the same complexing compound solution as a function of pH. Even at a pH of 12, the level of dissolved copper cannot be reduced below 2 mg/L.

6.3 SOLUBLE SULFIDE PRECIPITATION

Use of a water-soluble sulfide compound to reduce the solubility of heavy metals in a wastewater discharge is an effective method of improving the performance of a hydroxide precipitation treatment system. This section describes the results of an investigation of the use of SSP and presents information on systems using the technology (18).

6.3.1 PILOT PLANT EVALUATION

6.3.1.1 Test Description

The U.S. Environmental Protection Agency's (USEPA's) Industrial Environmental Research Laboratory funded a pilot study to compare and evaluate five treatment systems using variations of SSP and hydroxide precipitation processes to treat metal finishing wastewater. The pilot tests were designed to simulate the three basic process systems (shown in Figure 6.8) in order to provide a source of the data needed by firms interested in using the SSP treatment process. The five process variations tested were (18)

1. Lime only, clarified (LO-C)—the conventional process using lime as a neutralizing agent to precipitate the dissolved metals and clarification (9,10) to separate the suspended solids from the discharge (System A)
2. Lime only, clarified, filtered (LO-CF)—the LO-C process with a filtration step (9,10) downstream of clarification to improve the suspended solids removal (System A)
3. Lime with sulfide, clarified (LWS-C)—the LO-C process with controlled addition of a soluble sulfide reagent in the neutralizing chamber (System B)

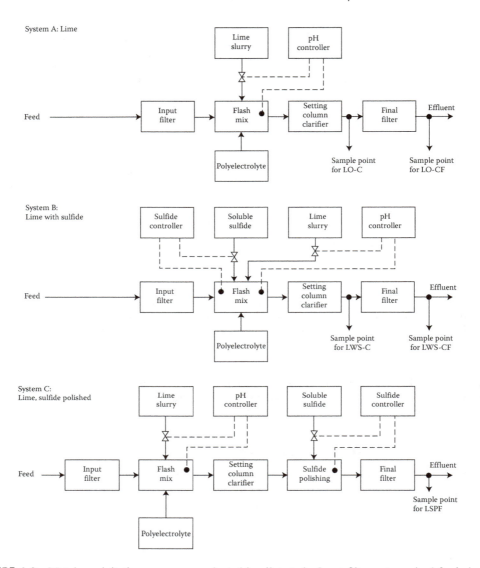

FIGURE 6.8 Metal precipitation processes evaluated in pilot study. Input filter not required for industrial applications. Abbreviations: LOC, lime only, clarified; LOCF, lime only, clarified, filtered; LWSC, lime with sulfide, clarified; LWSCF, lime with sulfide, clarified, filtered; LSPF, lime, sulfide polished, filtered. (From USEPA, *Control and Treatment Technology for the Metal Finishing Industry: Sulfide Precipitation, Summary Report,* EPA 625/8-80-003, Environmental Protection Agency, The Industrial Environmental Research Laboratory, Cincinnati, OH, April 1980.)

4. Lime with sulfide, clarified, filtered (LWS-CF)—the LWS-C process with a filtration step downstream of clarification to improve the suspended solids removal (System B)
5. Lime, sulfide polished, filtered (LSPF)—a polishing sulfide precipitation process featuring lime neutralization and clarification to remove the metal hydroxides followed by addition of a soluble sulfide reagent to reduce the metal solubility and a filtration step to remove the precipitated solids (System C)

These process variations were evaluated with 14 actual raw wastewater feed samples obtained from various industrial firms engaged in electroplating and metal finishing. The pilot plant could operate in any of the five modes and could process 0.034 gal/min (130 mL/min) of wastewater in a continuous treatment sequence. Samples were pretreated as required for chromium reduction and cyanide oxidation. Attempts were not made to reduce hexavalent chromium with sulfide reagent.

In the sulfide process variations, the soluble sulfide reagent addition was controlled automatically by a specific-ion sulfide reference electrode pair to maintain a preselected potential of −550 mV with respect to the reference electrode. The value of −550 mV corresponds to about 0.5 mg/L of free sulfide, which was selected as the control point because at that concentration: (a) the curve of electrical potential versus sulfide concentration has its maximum gradient and (b) the wastewater solution has no detectable sulfide odor.

The study reported that the dependability of the sulfide specific-ion electrode was excellent during the 6-month test period.

6.3.1.2 Test Results

Results of five of the pilot tests are presented in Tables 6.2 and 6.3. Table 6.2 lists the characteristics of the wastewater before treatment, the volume of sludge generated, and the amount of reagents consumed in the treatment. Table 6.3 compares the amount of metal per liter of raw feed before treatment and wastewater after treatment using the five process variations.

TABLE 6.2
Wastewater Treatment Process Details of Pilot Tests

| | Pilot Test[a] | | | | |
Characteristic	1	2[b]	3	4	5
Raw feed before treatment:					
pH	1.7	1.2	6.4	2.4	7.1
Conductivity (μmho/cm)	10,600 at 72°F	149,000 at 68°F	12,100 at 77°F	5,600 at 66°F	1,500 at 70°F
Color	Yellow	Colorless	Colorless	Colorless	Pale green
Precipitation pH for LO and LWS processes	8.5	6.2/9.0	9.0	10.0	8.5
Sludge volume (%)[c]:					
LO process	18	78/23	([d])	43	5
LWS process	16	78/13	([d])	37	6
Process consumables (mg/ L):					
Sulfuric acid for Cr^{6+} reduction	0	0	0	0	339
Sodium sulfite for Cr^{6+} reduction	226	31	0	41	25
Calcium oxide for neutralization	1,530	14,380	911	2.680	145
Sulfide for LWS process	8	381	([d])	400	91
Sulfide for LSPF process	1	5	([d])	141	67

Source: USEPA, *Control and Treatment Technology for the Metal Finishing Industry: Sulfide Precipitation, Summary Report,* EPA 625/8-80-003, Environmental Protection Agency, The Industrial Environmental Research Laboratory, Cincinnati, OH, April 1980.

Note: LO = lime only; LWS = lime with sulfide; and LSPF = lime, sulfide polished, filtered.

[a] Wastewater by pilot test: 1—high-chromium rinse from aluminum cleaning, anodizing, and electroplating; 2—chromium, copper, and zinc rinse from electroplating; 3—high-zinc rinse from electroplating; 4 and 5—mined heavy metal rinse from electroplating.

[b] Because of the exceptionally large volume of sludge generated by this wastewater, precipitation was accomplished in two stages. First- and second-stage values are separated by a diagonal line; single values apply to the total process.

[c] Sludge volume per solution volume, percent after 1 h settling.

[d] Data not available.

TABLE 6.3
Chemical Analysis of Raw and Treated Wastewater Used in Pilot Tests

Contaminant (µg/L)	Row Feed Before Treatment	Wastewater After Treatment[a]				
		LO-C	LO-CF	LWS-C	LWS-CF	LSPF
Pilot Test 1						
Cadmium	45	15	8	11	7	20
Total chromium	163,000	3,660	250	1,660	68	159
Copper	4,700	135	33	82	18	3
Nickel	185	30	38	33	31	18
Zinc	2,800	44	10	26	2	11
Load	119	119	88	104	59	120
Pilot Test 2						
Cadmium	58	7	12	<5	<5	<5
Total chromium	6,300	4	2	5	7	3
Hexavalent chromium	<5	<1	<1	<1	<1	<1
Copper	1,100	860	848	13	13	132
Nickel	160	30	34	33	23	34
Zinc	650,000	2,800	2,300	104	19	242
Mercury	<1	NA	NA	NA	NA	NA
Silver	16	NA	NA	NA	NA	NA
Pilot Test 3						
Cadmium	34	21	21	1	1	1
Total chromium	3	NA	NA	NA	NA	NA
Copper	20	7	8	2	1	4
Nickel	64	29	29	72	34	31
Zinc	440,000	37,000	29,000	730	600	2,000
Mercury	<10	NA	NA	NA	NA	NA
Load	45	13	14	9	11	13
Silver	61	4	4	1	3	4
Tin	200	<10	<10	<10	<10	<10
Ammonium	([b])	NA	NA	NA	NA	NA
Pilot Test 4						
Cadmium	58,000	1,130	923	26	<10	<10
Total chromium	5,000	138	103	49	50	37
Copper	2,000	909	943	60	160	929
Nickel	3,000	2,200	2,300	1,800	1,900	2.600
Zinc	290,000	1,200	510	216	38	12
Iron	740,000	2,000	334	563	229	305
Mercury	<0.3	<0.3	<0.3	<0.3	<0.3	<0.3
Silver	14	14	10	7	7	8
Tin	5,000	129	81	71	71	71
Pilot Test 5						
Cadmium	<40	<1	<1	<1	<1	<1
Total chromium	1,700	109	39	187	17	20
Copper	21,000	1,300	367	2,250	169	11
Nickel	119,000	12,000	9,400	11,000	3,500	5,300

(Continued)

TABLE 6.3 (*Continued*)
Chemical Analysis of Raw and Treated Wastewater Used in Pilot Tests

Contaminant (µg/L)	Row Feed Before Treatment	Wastewater After Treatment[a]				
		LO-C	LO-CF	LWS-C	LWS-CF	LSPF
Zinc	13,000	625	10	192	8	5
Iron	NA	2	<2	5	<2	<2
Load	13	7	5	4	3	3
Silver	6	NA	NA	NA	NA	NA

Source: USEPA, *Control and Treatment Technology for the Metal Finishing Industry: Sulfide Precipitation, Summary Report,* EPA 625/8-80-003, Environmental Protection Agency, The Industrial Environmental Research Laboratory, Cincinnati, OH, April 1980.

Note: 2—chromium, copper, and zinc rinse from electroplating; 3—high-zinc rinse from electroplating; 4 and 5—mixed heavy metal rinse from electroplating.

[a] LO-C = time only, clarified; LO-CF = lime only, clarified, filtered; LWS-C = limo with sulfide, clarified; LWS-CF = lime with sulfide, clarified, filtered; LSPF = lime, sulfide polished, filtered; NA = not applicable.

[b] Qualitative tests indicated the presence of significant amounts of ammonium.

Pilot Test 1 simulated treatment of wastewater containing a high concentration of chromium and moderate levels of copper and zinc. As can be seen from the effluent quality of the LO-CF process, the hydroxide solubilities of the metals in this wastewater were quite low and use of a sulfide reagent to achieve lower metal solubilities was not required. The significant reduction in the chromium concentration across the filter can be seen by comparing the effluent quality of the LO-C and LO-CF processes. This situation points out how poor solids removal can have significant adverse effects on an otherwise effective metal precipitation treatment system.

Pilot Tests 2 and 3 were performed with wastewater that was not effectively treated by hydroxide precipitation. In these tests, significantly improved effluent quality was achieved by the sulfide precipitation treatment. In *Pilot Test 2*, the effluent produced by the LO-CF process contained relatively high levels of zinc and copper, 2.3 and 0.8 mg/L, respectively. Treatment with a soluble sulfide compound considerably reduced the effluent concentration of these metals. In *Pilot Test 3*, soluble sulfide treatment of wastewater with a high zinc concentration was significantly more effective than hydroxide precipitation.

Tests also were conducted on wastewater containing an assortment of heavy metals at relatively high concentrations. The results of *Pilot Tests 4 and 5* (shown in Table 6.3) indicate that low levels of all metal pollutants could not be achieved by treatment of these particular wastewater with either hydroxide or sulfide precipitation. In *Pilot Test 4*, sulfide precipitation removed the cadmium, copper, and zinc to considerably lower levels than the hydroxide precipitation process, but both processes had a high residual nickel concentration in the effluent. A similar situation occurred with nickel in *Pilot Test 5*.

The data on effluent quality from this study suggest the following general conclusions about the treatment of wastewater with either hydroxide or sulfide precipitation for removal of heavy metals:

1. In most cases, metal removal can be improved by precipitating metals as sulfides rather than as hydroxides.
2. Some wastewater can be effectively treated to low residual concentrations of all metals present by either hydroxide or sulfide precipitation processes; some wastewater cannot be effectively treated by either hydroxide or sulfide precipitation.

3. Consistent removal of metals to effluent concentrations of less than 1 mg/L requires filtration to remove residual suspended solids. Because fine particles (which include precipitated metals) are only minimally different in density from water, they cannot be effectively separated by clarification and therefore contribute to the effluent metal concentration.

Another significant finding of the study is the quantity of sulfide reagent consumed in precipitating the metals as sulfides. In the LWS processes, the bulk of the test runs consumed between 1.0 and 2.5 times the stoichiometric sulfide reagent demand based on the total mass of metals that form sulfides in the wastewater. This reagent demand factor supports the belief that all metals are precipitated as sulfides and that any metals initially precipitated as hydroxides are converted to metal sulfides.

In the LSPF process, the metals precipitated as hydroxides are separated by clarification before addition of the sulfide reagent. The sulfide reagent demand for most of the LSPF process tests ranged from 2 to 6 times the stoichiometric sulfide reagent demand. The stoichiometric demand in this case can be calculated from the concentration of metals in the LO-C effluent. The study contained no conclusions as to the cause of the significantly higher sulfide reagent demand relative to the stoichiometric requirements.

6.3.2 SSP System Description and Performance

Treatment of wastewater by the SSP process has proved effective for precipitation of many of the metals typically encountered in electroplating wastewater. The primary application of SSP has been for waste streams containing low concentrations of metals and complexing agents, which interfere with effective metal removal by hydroxide precipitation.

Figure 6.9a is a schematic of a continuous SSP system used to treat a heavy metal waste stream discharged from a large mechanical equipment manufacturer. Part of the wastewater results from electroplating land surface finishing operations. The wastewater pH is adjusted to 7.5 in the first-stage neutralizer and is maintained at approximately 8.5 in the second-stage neutralizer. If the pH falls below 7 in the first stage, a low-pH alarm sounds and the pump feeding the second-stage neutralizer is shut off. Consequently, a surge volume is required in the system to store the wastewater until the pH returns to the control set-point. Sodium hydrosulfide is added in the second-stage neutralizer at a rate set to maintain a dosage of 5–10 mg of free sulfide/L of wastewater. Automatic controls are not used to adjust sulfide reagent feed rate to account for changes in demand. The required sulfide reagent addition rate is determined by periodic testing.

The system shown in Figure 6.9a uses a separate hexavalent chromium reduction system, although the free sulfide can potentially accomplish the reduction. This approach was not evaluated because performing chromium reduction in the second-stage neutralizer would increase sulfide reagent demand to approximately 35–50 mg/L of feed (based on consumption equal to twice the stoichiometric reagent demand) and would make sulfide reagent demand considerably more variable. Without an automatic sulfide reagent addition system to match supply with demand, the increased variability in reagent demand would reduce the reliability of the treatment system. The existing chromium reduction unit, which uses sodium bisulfite ($NaHSO_3$) as the reducing agent, reduced the hexavalent chromium to the required level. Therefore, sulfide precipitation was used only to achieve the superior metal removal required by the discharge permit.

The reduction in the metal solubility achieved by adding NaHS to this plant's wastewater is shown in Table 6.4. The data indicate that the metal solubility decreases as the sulfide reagent dosage increases.

Table 6.4 also shows the solubilities of the metal hydroxides after pH adjustment to 8.5. Effective metal removal is achieved by this treatment system with a sulfide reagent in the sulfide dosage range of 5–10 mg/L.

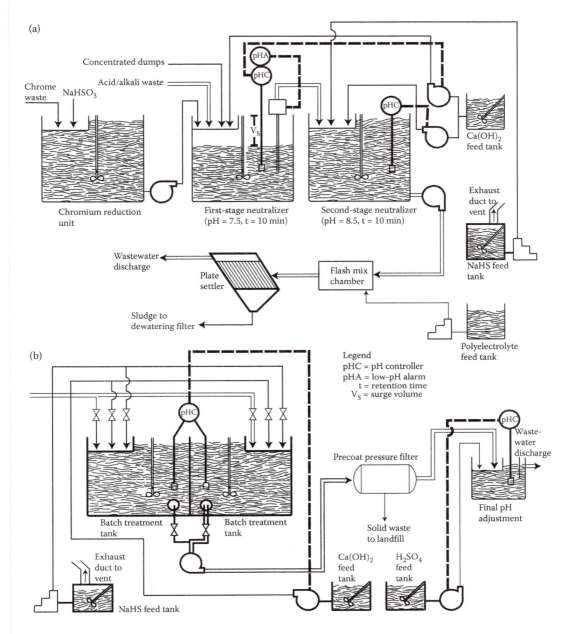

FIGURE 6.9 SSP treatment systems: (a) continuous and (b) batch. (From USEPA, *Control and Treatment Technology for the Metal Finishing Industry: Sulfide Precipitation, Summary Report,* EPA 625/8-80-003, Environmental Protection Agency, The Industrial Environmental Research Laboratory, Cincinnati, OH, April 1980.)

Figure 6.9b shows a commercially operated batch wastewater treatment system using a soluble sulfide reagent. The system includes two batch treatment tanks, each sized to hold 1 day's wastewater flow. The sequence of treatment follows:

1. The pH of the full, off-stream tank is raised automatically to a value of 11 by the addition of hydrated lime.
2. Depending on the volume of wastewater in the tank, a quantity of NaHS is metered into the tank.

TABLE 6.4

Sulfide Precipitation of Cadmium, Zinc, and Mercury

Metal (mg/L)	Raw Waste	Hydroxide Solubility at pH of 8.5	Supernatant[a] Sulfide Addition (mg/L)		
			1	5	10
Cadmium	2.1	2.0	1.6	0.39	0.06
Zinc	3.0	2.25	1.8	1.5	1.1
Mercury	0.006	0.0027	0.0013	0.001	0.0008

Source: USEPA, *Control and Treatment Technology for the Metal Finishing Industry: Sulfide Precipitation, Summary Report,* EPA 625/8-80-003, Environmental Protection Agency, The Industrial Environmental Research Laboratory, Cincinnati, OH, April 1980.

Note: Stoichiometric sulfide requirement to precipitate mixture given is 2.1 mg/L of sulfide based on raw waste composition.

[a] Polyelectrolyte dose = 1 mg/L; settling time of 2 h.

3. The tank is agitated for approximately 30 min and a sample is taken, filtered, and analyzed for the metal that is characteristically most difficult to remove.
4. If the metal concentration is low enough, the contents of the tank are pumped through a diatomaceous earth precoat pressure filter (34) and, after final pH polishing , are discharged. If the reference metal level is not low enough, additional NaHS is added and steps 3 and 4 are repeated.

The performance of the batch system in reducing the level of total metals in the wastewater discharge is presented in Table 6.5. As shown, the pH of the wastewater is raised to 11 before the NaHS is added. Experimentally, it was found that the sulfide addition would reduce the dissolved

TABLE 6.5

Removal of Complexed Copper and Other Metals from Electroplating Wastewater

Metal (mg/L)	Untreated Wastewater	Filtrate
Copper	17	0.4
Nickel	0.3	<0.2
Lead	1.85	<0.2
Zinc	0.86	0.4
Tin	4.29	<1.0

Source: USEPA, *Control and Treatment Technology for the Metal Finishing Industry: Sulfide Precipitation, Summary Report,* EPA 625/8-80-003, Environmental Protection Agency, The Industrial Environmental Research Laboratory, Cincinnati, OH, April 1980.

Note: Batch treatment sequence: lime added to raise pH to 11; NaHS added to equivalent sulfide ion concentration of 20 mg/L (stoichiometric requirement = 10 mg/L; filtered through diatomaceous earth filter; final pH adjustment to 8 before discharge).

metal concentration to equally low levels at a pH of 8.5. Removal of fluorides present in the plant's wastewater, however, required elevating the pH to 11.

The continuous and batch SSP systems described in this section are located in segregated waste treatment areas. Despite careful control of the wastewater pH and sulfide addition rate, the H_2S odor in the area was a nuisance. To reduce the ambient level of H_2S, the open-top treatment tanks where the sulfide reagent is added to the wastewater were modified into closed-top, vacuum-evacuated tanks. In the batch system shown in Figure 6.9b, the final pH adjustment tank contributed to the odor problem and was modified similarly. The exhaust from these tanks, which contains a low level of H_2S, is vented outdoors. These changes, plus rigid control of pH and sulfide dosage levels, have resulted in an almost undetectable H_2S odor in the waste treatment area.

6.3.3 SSP POLISHING TREATMENT SYSTEM

Sulfide reagent demand for the SSP treatment system shown in Figure 6.9a is a function of the total metal concentration of the raw wastewater. Sufficient reagent must be supplied to convert all entering metals to metal sulfides. In treating wastewater containing high metal loadings, significant sulfide reagent cost savings can be realized by using SSP to polish the effluent after a conventional pH adjustment/clarification treatment sequence (Figure 6.10). The LSPF process evaluated in the pilot studies discussed earlier simulated the use of SSP as a polishing system.

In addition to reducing sulfide reagent consumption, using sulfide precipitation as a polishing system will reduce the variability of reagent demand. The reagent demand for the polishing system will be a function of wastewater flow and the concentration of metals in the overflow from the first-stage clarifier. The metal concentration in the wastewater at this point should not be subject to the wide variability that often characterizes the raw wastewater feed metal concentration. Without an automatic reagent addition control loop, dosing the wastewater with a predetermined amount of sulfide reagent would be considerably more reliable in a polishing treatment application.

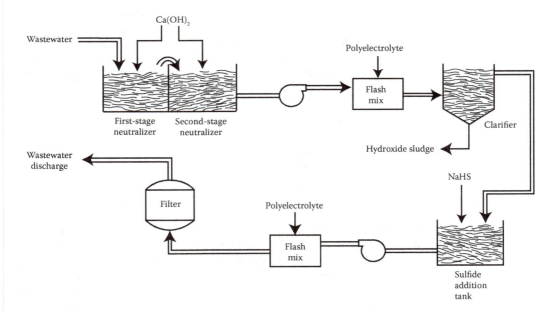

FIGURE 6.10 SSP polishing system. (From USEPA, *Control and Treatment Technology for the Metal Finishing Industry: Sulfide Precipitation, Summary Report,* EPA 625/8-80-003, Environmental Protection Agency, The Industrial Environmental Research Laboratory, Cincinnati, OH, April 1980.)

The plant operating the treatment system shown in Figure 6.9a evaluated the use of SSP as a polishing treatment to reduce the variability of sulfide reagent demand. It was found that clarifying the wastewater before adding the sulfide reagent resulted in the formation of poor-settling particles that were difficult to remove from the wastewater. The current treatment sequence, in which the sulfide reagent is added in the second-stage neutralizer, removes precipitated metal more effectively. It was concluded that the presence of the precipitated metal hydroxides and lime solids in the wastewater entering the second-stage neutralizer provided nucleation sites, which promoted the coagulation (13) of the precipitated metal sulfides.

An SSP pilot study reports success in forming metal sulfide particles that were easily removed from the wastewater despite precipitation in a solution lacking nucleation sites. The researchers found that conditioning the colloidal metal sulfide precipitants with a cationic coagulant to increase the particle size and then adding an anionic flocculent to link the particles produced large, fast-settling particles when flocculated. In the pilot study discussed previously, the sulfide polishing process precipitated metals as sulfides after the wastewater had been clarified to remove suspended solids. The study indicated that the metal sulfide solids were removed effectively by filtration.

The additional equipment requirements of a polishing treatment system include a second mixing tank to add the sulfide reagent and a second solids separation unit (using either a clarifier or a filter) installed downstream of the metal hydroxide clarification step. A second polyelectrolyte addition system also may be required to enhance the efficiency of the metal sulfide solids separation step.

6.3.4 Hydroxide System Modifications for SSP

Augmenting a hydroxide precipitation wastewater treatment system with SSP to achieve a lower level of metals in the effluent can be a cost-effective means of achieving compliance. The cost of using soluble sulfide treatment will be significantly affected by the reliability and dependability of using the specific-ion sulfide reference electrode to control the sulfide reagent addition. If the residual sulfide concentration can be maintained consistently at a level of 0.3–0.5 mg/L in the wastewater, it should not be necessary to modify existing treatment tanks to eliminate sulfide odor in the work area. Because the reliability of the control system has not been established, two alternative approaches emerge for converting a hydroxide system to use SSP.

With no automatic control of the level of residual sulfide in the wastewater, converting the conventional hydroxide precipitation system (Figure 6.11a) to an SSP system (Figure 6.11b) requires several process modifications. The modifications, which are discussed in the following paragraphs, include (18)

1. NaHS reagent feed tank and feed pump
2. Second-stage neutralizer/soluble sulfide treatment tank
3. Clarifier enclosure and vacuum evacuation
4. Control system
5. Sand filter or other polishing filtration unit
6. Aeration system

The NaHS feed tank should have a closed top with a vent connecting to an exhaust system. In installations where venting any odor is considered a public nuisance, the vent can be connected to a scrubber system. Using a scrubber eliminates the discharge of any odor, whereas simply venting outdoors eliminates any hazard to the worker during reagent preparation. The feed pump should be a positive displacement pump with a variable stroke to facilitate the metering of reagent into the system.

The second-stage neutralizer/soluble sulfide treatment tank is used for adding the sulfide reagent to the wastewater. The tank also provides improved pH control to ensure that the sulfide reagent does not come into contact with acidic wastewater. The tank contents should be agitated. The tank should be sized to provide a minimum retention time of 20 min, and it should be equipped with a

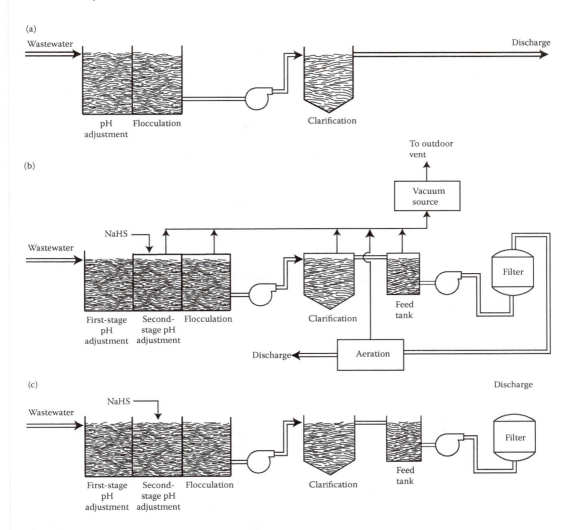

FIGURE 6.11 Conversion of hydroxide treatment system to use S;SP: (a) hydroxide precipitation system, (b) SSP system, and (c) SSP system with automatic control of sulfide residual. (From USEPA, *Control and Treatment Technology for the Metal Finishing Industry: Sulfide Precipitation, Summary Report,* EPA 625/8-80-003, Environmental Protection Agency, The Industrial Environmental Research Laboratory, Cincinnati, OH, April 1980.)

pH control loop and alkali neutralizing reagent feed system. To minimize any H_2S odor associated with the treatment, the tank should be totally enclosed and vacuum evacuated.

To convert the conventional hydroxide precipitation system to an SSP system, it is also necessary to totally enclose and vacuum evacuate the clarifier.

A control system is needed to avoid mixing of the sulfide reagent with low-pH wastewater. An instrumentation loop that interrupts the wastewater feed to the sulfide treatment tank if the pH of this stream falls below set-point is one way of minimizing the potential hazard. Low-pH conditions also should sound an alarm and interrupt the sulfide feed to the system. This type of control will result in the need for surge volume upstream of the sulfide treatment tank to store the volume buildup until the pH is brought back above the set-point.

A sand filter or other polishing filtration unit (17) that removes suspended solids in the clarifier overflow to very low levels is recommended for any treatment system that must achieve very low levels of metals in the effluent. The significance of reducing the solubility of a metal pollutant by means of sulfide precipitation will be lost unless the level of suspended solids, which include insoluble metals, is also controlled at a low level.

An aeration system may be needed to oxidize residual sulfide before wastewater discharge. If wastewater is discharged into a sewer system, precautions must be taken to ensure that the discharge does not contain high levels of sulfide. Discharge of wastewater containing significant quantities of sulfide could be hazardous to individuals working in a poorly vented sewer system. No specific limit exists for direct discharge of sulfide, but its presence contributes to the biochemical oxygen demand (BOD) of the wastewater. The easily oxidized sulfide compounds can be treated in an air sparged tank with a retention time of approximately 30 min. If indoors, this tank also should be totally enclosed and vacuum evacuated.

For a process using automatic control of the sulfide reagent addition (Figure 6.11c), the required modifications to convert the hydroxide system to an SSP system would include the following (18):

1. NaHS reagent feed tank and feed pump—identical to the tank and pump required for the previous case, except the feed pump is actuated by a signal from the sulfide reagent control system to maintain a constant residual sulfide concentration in the wastewater
2. Second-stage neutralizer/soluble sulfide treatment tank—for addition of the sulfide reagent to the wastewater, but in this case the residual free sulfide ion concentration is maintained at a level below 0.5 mg/L by means of a sulfide ion control loop
3. Control system to avoid mixing of the sulfide reagent with low-pH wastewater
4. Sand filter

The second-stage neutralizer/sulfide treatment tank and the downstream process tanks will not need to be enclosed and vacuum evacuated if careful control of pH (between 8 and 9.5) and sulfide ion concentration is maintained. Control of sulfide ion concentration also should eliminate the need to aerate the wastewater before discharge. The other elements of the sulfide system shown in Figure 6.11—first-stage pH adjustment, polyelectrolyte conditioning, and clarification—are common to hydroxide precipitation systems.

For batch treatment SSP systems, a two-tank system for alternately collecting and treating the wastewater would be required. The treatment sequence for a batch system was presented earlier. If the residual level of sulfide cannot be controlled, aeration of the wastewater after chemical treatment may be required in addition to enclosing and vacuum evacuating the tanks during treatment. The wastewater could be aerated in the treatment tank before flocculation (if required) and solid–liquid separation.

Retrofitting a hydroxide system to use soluble sulfide polishing would require a mixing tank to add the sulfide reagent to the wastewater downstream of the existing clarifier and a second solids separation unit. Because the solids generation rate in the soluble sulfide polishing step should be low, a sand or mixed-media filter should be suitable for removing the suspended solids from the wastewater before discharge.

Polyelectrolyte conditioning and flocculation may be required between the sulfide reagent addition tank and the solids removal filter. Without instrumentation for reliable control of the residual sulfide concentration, the sulfide reagent mixing tank and downstream equipment would need to be enclosed and ventilated, and aeration of the effluent might be required.

6.3.5 SSP Cost Estimating

Improving the performance of a hydroxide precipitation system through the use of SSP will require investment capital to modify the treatment system and will increase the cost to operate the system.

There is some uncertainty in predicting the extent of the modifications needed to convert a hydroxide system to use SSP. Demonstration of the reliability of automatic control of the sulfide reagent feed

TABLE 6.6
Equipment Cost Factors for SSP Treatment Systems

Equipment Component	Installed Cost ($1,000),[a] by Wastewater Flow Rate (gal/min)		
	30	60	90
Sodium hydrosulfide feed tank and metering pump	8	8	8
Automatic sulfide reagent addition control	8.5	8.5	8.5
Low-pH prevention control loop	3.7–4.9	3.7–4.9	3.7–4.9
Second-stage pH adjustment and sulfide reagent mixing tank:			
Open top	44	54	59
Totally enclosed and vented	56	68	73
Suspended solids polishing filter	59	80	100
Aerator	10	17	22

Source: USEPA, *Control and Treatment Technology for the Metal Finishing Industry: Sulfide Precipitation, Summary Report*, EPA 625/8-80-003, Environmental Protection Agency, The Industrial Environmental Research Laboratory, Cincinnati, OH, April 1980; USEPA, *Environmental Pollution Control Alternatives: Economics of Wastewater Treatment Alternatives for the Electroplating Industry*, EPA 625/5-79-01 6 U.S. Environmental Protection Agency, June 1979.

Note: Costs escalated to 2012 USD. (From US ACE, Yearly average cost index for utilities. In: *Civil Works Construction Cost Index System Manual*, 110-2-1304, U.S. Army Corps of Engineers, Washington, DC, 44pp. PDF file is available at http://www.nww.usace.army.mil/cost, 2015.)

[a] Installed costs of different components are presented. Engineering and design costs, site preparation, and equipment freight charges are not included.

is needed to eliminate this uncertainty. Table 6.6 presents the costs (including hardware and installation) of the different equipment components that may be required (18,35). All costs have been escalated to 2012 USD using U.S. Army Corps of Engineers Yearly Average Cost Index for Utilities (36):

1. NaHS feed tank and metering pump
2. Automatic control of sulfide reagent addition
3. Low-pH prevention control loop
4. Mixing tank
5. Suspended solids polishing filter
6. Aerator

The cost for a sodium hydrosulfide feed tank is based on a 400-gal (1514-L), closed-top, carbon-steel tank that has a removable lid, exhaust vent, and appropriate nozzles. The diaphragm metering pump is rated to deliver 0–20 gal/h (0–76 L/h).

A specific-ion sulfide reference electrode pair automatically controls the sulfide reagent feed pump. A control loop prevents low-pH conditions in the sulfide treatment tank by automatically shutting down the wastewater feed pump and sulfide reagent feed pump if the wastewater pH falls below the control set-point. The cost presented assumes the prior existence of a pH probe and a surge volume to hold the diverted flow.

Second-stage pH adjustment and sulfide reagent addition occur in an agitated tank sized for 20-min retention of wastewater. Costs are given for both an open-top and a totally enclosed and vented tank.

The suspended solids polishing filter costs presented are for dual mixed-media filters (17), skid mounted and sized so that one filter can process the maximum flow during backwash. The unit is equipped with a blower for low-pressure air scouring, a backwash storage tank, and a pump to bleed the wash back into the treatment system.

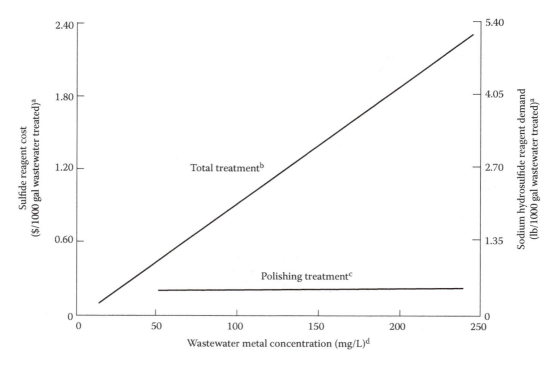

FIGURE 6.12   Soluble sulfide reagent cost. [a]Based on NaHS (72% flake) at $900/ton. [b]Total treatment at 2 times the stoichiometric reagent demand. [c]Polishing treatment at 4 times the stoichiometric reagent demand and a total metal hydroxide solubility of 10 mg/L. [d]Includes all metals that form sulfides, based on metal with molecular weight of 62.6 (average of Ni, Cu, and Zn). Cost 2012 USD. (From USEPA, *Control and Treatment Technology for the Metal Finishing Industry: Sulfide Precipitation, Summary Report,* EPA 625/8-80-003, Environmental Protection Agency, The Industrial Environmental Research Laboratory, Cincinnati, OH, April 1980; USEPA, *Environmental Pollution Control Alternatives: Economics of Wastewater Treatment Alternatives for the Electroplating Industry,* EPA 625/5-79-01 6 U.S. Environmental Protection Agency, June 1979.)

The aerator cost is based on an enclosed, vacuum-evacuated tank sized for 30-min retention of wastewater and equipped with an air sparger.

Higher operating costs—for operating labor and treatment reagents—will result from incorporating SSP into an existing treatment system. Additional operating labor will be required to prepare the sulfide reagent and to maintain and operate the additional equipment components. Additional expense will result from the consumption of the sulfide reagent. The consumption rate will depend on the volume of wastewater treated and the required dosage. The dosage per volume of wastewater treated will be a function of the wastewater metal concentration. Figure 6.12 presents the sulfide reagent cost per 1,000 gal (3800 L) of wastewater treated as a function of metal concentration for an SSP system used to treat the total metal load as well as for polishing treatment.

Sludge generation rates will increase with the use of SSP compared with a conventional hydroxide treatment system because of improved metal removal, but the increase should be insignificant. For example, precipitating an additional 5 mg/L of dissolved metals from a waste stream will increase the clarifier underflow rate by less than 1 gal of sludge per 1,000 gal of wastewater treated, based on an underflow concentration of 1% solids by weight. Also, the dewatering properties of sulfide sludges are believed to be superior to those of hydroxide sludges.

If the pH of the neutralized wastewater is increased to minimize odor, more alkali will be consumed, causing an increase in cost. The increased cost of alkali should not be significant except for high-volume treatment systems. Use of a pH above 10 would necessitate a final adjustment to lower the pH to the acceptable discharge range.

6.4 INSOLUBLE SULFIDE PRECIPITATION

A commercially available ISP wastewater treatment system was developed to provide a treatment process that offers the superior metal removal of sulfide precipitation systems without the unpleasant H_2S odor often associated with soluble sulfide systems. Since the first commercial demonstration of the process in 1978, many additional installations have become operational. The process is patented, and its use requires payment of a licensing fee to the patent holder. This section describes the process, presents performance data on three currently operating systems, and evaluates use of the process for treatment of electroplating wastewater.

6.4.1 PROCESS DESCRIPTION

6.4.1.1 Process Equipment Components

A hydroxide neutralization/ISP treatment system for control of pH and precipitation of heavy metals is depicted in Figure 6.13. In this system, the hexavalent chromium is reduced to its trivalent state by the sulfide and ferrous ions present in the mixer/clarifier, thus eliminating the need for a separate chromium reduction unit. With the exception of chromium and iron, all other heavy metals in the wastewater precipitate as sulfides. The key elements of the system are (18)

1. pH control
2. Mixer/clarifier
3. Reagent addition to mixer/clarifier
4. FeS feed rate control
5. Sand filter

Effective metal removal by sulfide or hydroxide precipitation requires that the pH of the wastewater be controlled within the neutral to slightly alkaline range. Although the dependence of metal solubility on pH is not critical for sulfide precipitation systems, it still affects metal removal (see Figure 6.3). It is more important to eliminate the danger of the FeS slurry coming into contact with acidic wastewater; FeS is soluble in acidic solutions, and mixing it with low-pH wastewater would result in the emission of toxic H_2S fumes in the work area. The risk is minimized by installing a recycle control on the feed to the mixer/clarifier. If the pH of the feed stream drops below 7, valves automatically reroute the feedback to the second-stage neutralizer. For this reason, a surge volume, shown as V_s in Figure 6.13, is required to store the accumulated wastewater until the control setpoint is reestablished.

The mixer/clarifier shown in Figure 6.13 serves two purposes. First, it provides the solid–liquid contact volume between the wastewater and the FeS slurry necessary to maintain the wastewater sulfide ion concentration at its saturation point. As illustrated in Figure 6.6, both mixing time and sludge blanket density in the solid–liquid contact zone affect metal removal. Second, it clarifies the effluent of suspended solids.

To achieve low concentrations of dissolved metals, which are characteristic of metal sulfides, the liquid residence time in the solid–liquid contact zone of the mixer/clarifier must be sufficient for the metal precipitation reaction to reach completion. Proper agitation in the contact zone will enhance the degree of reaction completion achieved as well as promote particle growth of the precipitated metal sulfides. The formation of large, rapid-settling particles facilitates the removal of the solids by clarification.

Reagent addition to the mixer/clarifier is controlled by a flow-measuring device that monitors the feed to the mixer/clarifier and sends a signal to a counter, which computes the cumulative flow. The additions of fresh FeS and polymer are controlled to provide a set quantity of each when the counter records a set volumetric throughput. The dosage rate is determined for both reagents by performing a series of jar tests. A sample is taken from the second-stage neutralizer and tested to determine the required addition of FeS.

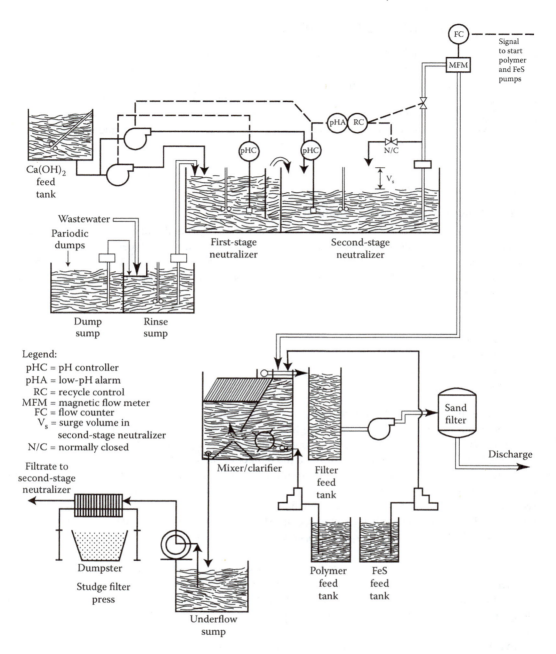

FIGURE 6.13 Sulfex ISP treatment system. (From USEPA, *Control and Treatment Technology for the Metal Finishing Industry: Sulfide Precipitation, Summary Report,* EPA 625/8-80-003, Environmental Protection Agency, The Industrial Environmental Research Laboratory, Cincinnati, OH, April 1980.)

Jar tests are conducted on approximately four samples to determine the lowest FeS dosage that provides optimum metal removal. Because polyelectrolyte demand should be proportional to the demand for FeS, it is fed at a constant ratio of the demand for FeS. Jar tests are normally conducted once or twice per shift to determine the required addition rate.

The FeS feed rate control loop automatically adds a preset amount of reagent each time an increment of wastewater enters the mixer/clarifier. The amount of reagent added is set manually based on the results of the jar tests. The inability to adjust the FeS reagent dosage automatically in response to changes in the reagent demand complicates the operation of ISP treatment systems. To compensate for the lack of automatic control, two features must be considered in design of the system:

1. FeS reagent demand averaging
2. Maintaining an inventory of unreacted FeS in the mixer/clarifier

Reagent demand averaging requires the elimination of sharp deviations in wastewater flow rate and pollutant concentration entering the treatment system. Flow variability normally is eliminated by providing a surge volume upstream of the treatment process and treating the wastewater at a constant average rate. The variability of pollutant concentration can be reduced by use of an averaging tank—an agitated tank that stores and blends the treatment system feed before processing. The impact of averaging tank volume and retention time on reagent demand variability is presented graphically in Figure 6.14. As shown, with 1 h of retention time in upstream process tanks, the variability of the mixer/clarifier (blended feed) reagent demand is equal to 54% of the plant feed reagent demand variability; with 4-h retention time in upstream process tanks, the mixer/clarifier reagent demand variability is reduced to 15% of the plant feed variability. The graph presents an idealized situation of reagent demand fluctuating around a constant average demand. In actual practice, however, the deviations may be long term and may not average out to a constant demand rate. The relationship between retention time in upstream blending tanks and demand fluctuations is a key to operating any treatment process that does not automatically adjust reagent supply to changes in demand.

Maintaining an inventory of unreacted FeS in the mixer/clarifier is needed to provide the sulfide reagent when reagent demand exceeds supply. Because demand fluctuations are inevitable, an

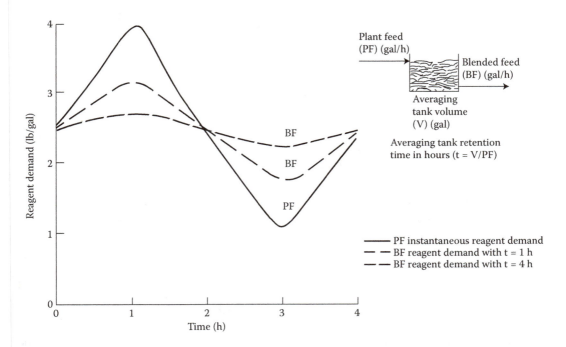

FIGURE 6.14 Impact of averaging tank volume on reagent demand variability. (From USEPA, *Control and Treatment Technology for the Metal Finishing Industry: Sulfide Precipitation, Summary Report*, EPA 625/8-80-003, Environmental Protection Agency, The Industrial Environmental Research Laboratory, Cincinnati, OH, April 1980.)

inventory of reagent is essential to consistently achieve maximum removal of metals. The quantity of FeS stored in the mixer/clarifier is proportional to the quantity of solids maintained in the unit and to the concentration of FeS in those solids.

A sand filter is included in the system to ensure that the wastewater discharge contains a minimum concentration of suspended solids. To meet strict metal discharge requirements, the level of dissolved and insoluble metals in the effluent discharge must be reduced to a minimum. For both sulfide and hydroxide precipitation systems, a sand filter ensures that upsets in the treatment system causing turbidity in the clarifier overflow will not jeopardize effluent quality.

6.4.1.2 FeS Reagent Consumption

As shown in Figure 6.6, precipitation of dissolved metals to the low-solubility level characteristic of metal sulfides normally requires 2–4 times the stoichiometric amount of FeS. The ratio of the amount of reagent added to the stoichiometric demand establishes the equilibrium concentration of FeS in the sludge blanket solids. The FeS added in excess of the stoichiometric demand provides the inventory of unreacted reagent that is consumed when reagent demand exceeds supply.

The concentration of FeS in the sludge blanket as a function of the ratio of reagent addition to stoichiometric reagent demand is shown in Figure 6.15. The quantity of reagent consumed as a function of this ratio also is shown. Because the underflow rate is set to balance the solids loading rate, the concentration of FeS in the sludge blanket also determines the amount lost in the sludge underflow.

By defining the volume of the solid–liquid contact zone and the density of the sludge blanket in this zone, the amount of FeS stored can be approximated. The larger the quantity of unreacted

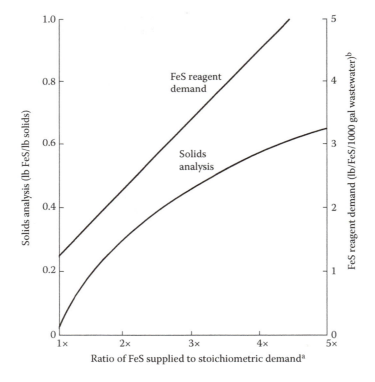

FIGURE 6.15 Sludge blanket FeS concentration and associated reagent demand. [a]x = the stoichiometric FeS reagent requirement. [b]Based on treatment of wastewater containing 100 mg/L Cu^{+2}. (From USEPA, *Control and Treatment Technology for the Metal Finishing Industry: Sulfide Precipitation, Summary Report,* EPA 625/8-80-003, Environmental Protection Agency, The Industrial Environmental Research Laboratory, Cincinnati, OH, April 1980.)

FeS maintained in the blanket, the greater the ability of the system to compensate automatically for increases in reagent demand. The FeS supply can be increased by

1. Increasing the FeS reagent feed rate
2. Designing larger solid–liquid contact volume into the system
3. Maintaining the maximum sludge blanket solids concentration in the solid–liquid contact volume that is compatible with good clarification in the settling zone of the mixer/clarifier

The first two methods of increasing the FeS inventory have economic penalties: the reagent cost and sludge volume rise as dosage is increased, and the initial cost and space requirements increase as larger mixing volume is designed into the system. Therefore, maintaining a dense sludge blanket in the mixing zone is the most efficient way to achieve good reagent use and to provide the inventory of FeS needed for reagent demand increases. In practice, this requires monitoring the blanket level and adjusting the sludge draw off rate to match the solids accumulation rate in the system.

6.4.1.3 Operating Procedure

The ISP system shown in Figure 6.13 required a full-time operator during one shift and approximately 2–4 h of operator attention during other shifts. Operator duties are as follows (18):

1. Once each shift, a sample of mixer/clarifier feed is removed from the second-stage neutralizer for jar testing to determine the required FeS addition rate.
2. Based on the jar test results, the FeS and polyelectrolyte addition control system is set to feed the needed quantity of reagents each time a set feed increment has entered the mixer/clarifier.
3. The timer that controls the sludge blowdown is adjusted to reflect any change in the solids loading rate. (This relates to the jar test performed in the first step.)
4. The level of solids in the mixer/clarifier is monitored periodically (normally every 1 or 2 h) by performing a settling test on samples removed from the mixing zone of the mixer/clarifier. The sludge blowdown rate is adjusted to maintain the maximum solids concentration in the mixing zone that is compatible with low levels of turbidity in the clarified effluent.

Other operator duties generally required for operation of this system and most treatment systems include

1. Preparation of treatment reagents—in this case, reagents include lime slurry, Sulfex reagent (Figure 6.16), and polyelectrolyte
2. Operation of sludge dewatering filter
3. Periodic back-flush cleaning of the sand filter
4. Periodic calibration of pH probes
5. Collection of samples required for discharge permit
6. Regularly scheduled lubrication of system elements

6.4.2 ISP Polishing Treatment System

The FeS reagent demand for the system shown in Figure 6.13 is a function of the total metal load entering the mixer/clarifier. Sufficient FeS must be added not only to precipitate the dissolved metals but also to convert the precipitated metal hydroxides to metal sulfides. For treatment systems with a high mass flow of metals, FeS consumption will be high and considerable waste solids (a combination of metal sulfides, metal hydroxides, and unreacted FeS) will be generated. For these applications, the reduction in reagent consumption and solid waste disposal charges may justify

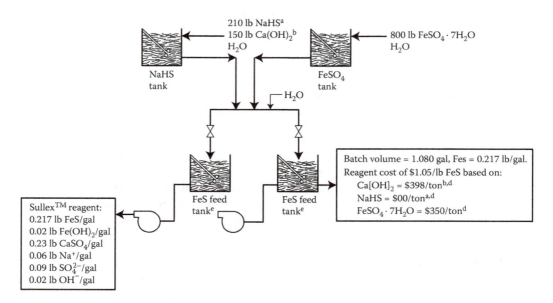

FIGURE 6.16 FeS feed system. [a]70% to 72% flake. [b]93% pure. [c]Includes shipping and palletizing. [d]Includes shipping. [e]One in use, one for batch preparation. (From USEPA, *Control and Treatment Technology for the Metal Finishing Industry: Sulfide Precipitation, Summary Report,* EPA 625/8-80-003, Environmental Protection Agency, The Industrial Environmental Research Laboratory, Cincinnati, OH, April 1980; USEPA, *Environmental Pollution Control Alternatives: Economics of Wastewater Treatment Alternatives for the Electroplating Industry,* EPA 625/5-79-01 6 U.S. Environmental Protection Agency, June 1979.)

using ISP to polish the clarified overflow after a conventional hydroxide precipitation/clarification treatment sequence (Figure 6.17).

In this polishing system, the FeS demand is based on the metals contained in first clarifier overflow. If hexavalent chromium is present in the wastewater, it will be reduced in the second-stage mixer/clarifier and precipitated along with the dissolved metals. Two advantages of this approach, compared with the system shown in Figure 6.13, are reduced FeS reagent demand and reduced sludge generation, which is a function of metal loading and reagent consumption. Another advantage is that the

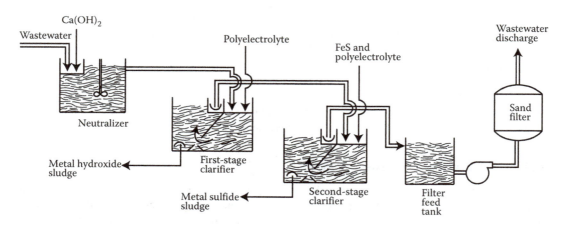

FIGURE 6.17 ISP polishing system. (From USEPA, *Control and Treatment Technology for the Metal Finishing Industry: Sulfide Precipitation, Summary Report,* EPA 625/8-80-003, Environmental Protection Agency, The Industrial Environmental Research Laboratory, Cincinnati, OH, April 1980.)

concentration of metals in the first-stage clarifier overflow will not be subject to the wide variation that often characterizes the wastewater feed metal concentration. The metal hydroxide equilibrium solubility will determine the concentration of dissolved metals in the overflow; this concentration will establish reagent demand. Again, because reagent supply is not adjusted automatically for changes in demand, this feature increases reliability. The concentration of hexavalent chromium, which is unaffected by the hydroxide treatment, will still be subject to variation, but the variability should be reduced because of the larger volume of upstream process tanks in a polishing treatment system.

Identification of the optimum system—polishing sulfide precipitation or treatment of the total metal load—requires determining whether the operating cost savings of the polishing system offset the additional cost of a second mixer/clarifier and polyelectrolyte feeder.

6.4.3 ISP System Performance

Three plants that use the Sulfex system to remove heavy metals from wastewater discharge were placed in plating shops where no wastewater treatment systems existed. Two of the plants (plants A and B) treat the total metal load with FeS, whereas the third (plant C) employs ISP as a polishing step after hydroxide precipitation/clarification.

Plant A performs copper, nickel, and chromium plating (both electroplating and electroless plating) of plastic components. The heavy metals in the wastewater are complexed with a variety of chelating agents. During the pilot evaluation, it was apparent that hydroxide precipitation would not remove the metals to the levels required in the discharge permit (Table 6.7). After a pilot evaluation showed that ISP could achieve the required discharge limitations, the firm hired a vendor to design a treatment system using this technology. The vendor guaranteed that the system would meet all discharge regulations.

The system was designed to treat 40 gal/min (151 L/min) of wastewater and is essentially identical to the system shown in Figure 6.13. The performance of the system in removing copper, nickel, total chromium, and hexavalent chromium (Cr^{6+}) during a 60-h test period is shown in Figures 6.18 and 6.19. Figure 6.20 shows the corresponding sample point locations.

The performance in chromium removal shows a deviation from normal removal efficiency between hours 16 and 28 that corresponds to an increase in the level of hexavalent chromium in the mixer/clarifier feed during hours 8 through 28. By comparing the stoichiometric FeS demand

TABLE 6.7
Plant A Discharge Permit Requirements

| | Discharge Limits[a] | | | |
| | Mass (lb/d) | | Concentration (mg/L) | |
Item	Average[b]	Maximum[c]	Average[b]	Maximum[c]
Suspended solids	35.3	53.0	NA[d]	NA[d]
Total copper	0.89	1.77	1.0	1.5
Total nickel	0.89	1.77	1.0	1.5
Total chromium	0.89	1.77	1.0	1.5
Hexavalent chromium	0.089	0.177	0.05	0.10

Source: USEPA, *Control and Treatment Technology for the Metal Finishing Industry: Sulfide Precipitation, Summary Report,* EPA 625/8-80-003, Environmental Protection Agency, The Industrial Environmental Research Laboratory, Cincinnati, OH, April 1980.

[a] Required pH level is between 6.0 and 9.5.
[b] Monthly average of daily 24-h composite samples.
[c] Highest daily 24-h composite in the month.
[d] Not applicable.

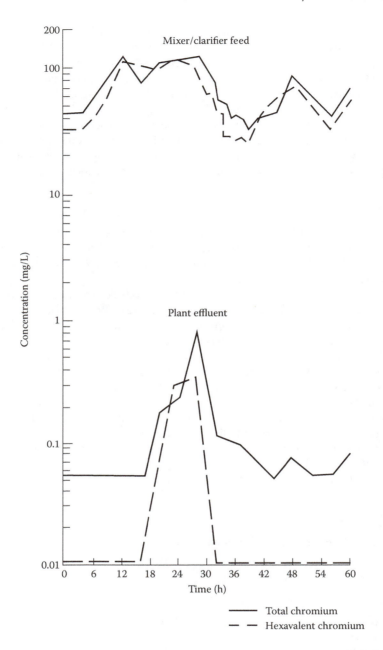

FIGURE 6.18 Plant A's performance in removing chromium. (From USEPA, *Control and Treatment Technology for the Metal Finishing Industry: Sulfide Precipitation, Summary Report,* EPA 625/8-80-003, Environmental Protection Agency, The Industrial Environmental Research Laboratory, Cincinnati, OH, April 1980.)

with the quantity supplied and the associated mixer/clarifier removal efficiency (Figure 6.21), it is obvious that the FeS feed was not increased sufficiently to compensate for the increased demand. Consequently, the level of unreacted FeS in the sludge blanket was gradually depleted, and at hour 16 insufficient FeS was present in the blanket to achieve the normal high level of removal. This condition persisted until hour 28. The FeS stored in the sludge blanket maintained the high removal efficiency between hours 8 and 16, despite a low FeS reagent supply/demand ratio.

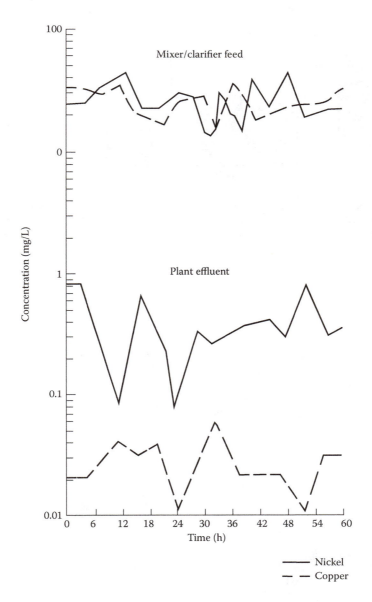

FIGURE 6.19 Plant A's performance in removing nickel and copper. (From USEPA, *Control and Treatment Technology for the Metal Finishing Industry: Sulfide Precipitation, Summary Report,* EPA 625/8-80-003, Environmental Protection Agency, The Industrial Environmental Research Laboratory, Cincinnati, OH, April 1980.)

Figure 6.21 shows that optimum removal efficiency for the chromium is achieved with an FeS dosage of approximately 3 times the stoichiometric demand. The stoichiometric demand was determined by laboratory analysis of mixer/clarifier feed samples. The removal efficiencies for nickel and copper were relatively constant and showed no discernible trends over the dosage ratios encountered during the test period.

Based on an FeS dosage rate of 3 times the stoichiometric demand and the observed consumption of other treatment reagents, the cost of treatment chemicals and sludge generation factors for the ISP system at this facility are shown in Table 6.8.

Plant B manufactures parts for the automotive industry. Wastewater from the metal finishing portion of the process contains varying quantities of chromium (hexavalent and trivalent), zinc,

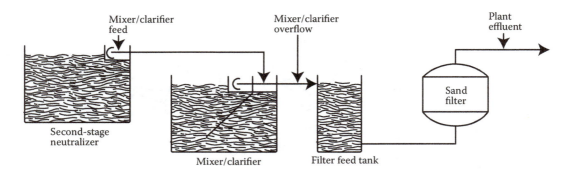

FIGURE 6.20 Sample points anionic and cationic polymer feed systems. (From USEPA, *Control and Treatment Technology for the Metal Finishing Industry: Sulfide Precipitation, Summary Report,* EPA 625/8-80-003, Environmental Protection Agency, The Industrial Environmental Research Laboratory, Cincinnati, OH, April 1980.)

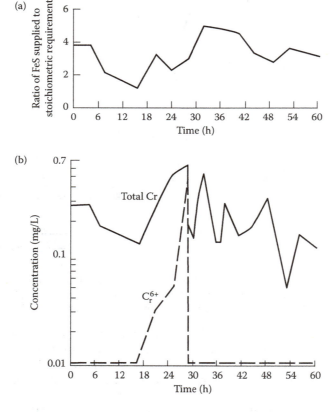

FIGURE 6.21 Impact of FeS supply/demand ratio on reduction of hexavalent chromium at plant A: (a) FeS supply versus stoichiometric requirement and (b) mixer/clarifier overflow chromium concentration. (From USEPA, *Control and Treatment Technology for the Metal Finishing Industry: Sulfide Precipitation, Summary Report,* EPA 625/8-80-003, Environmental Protection Agency, The Industrial Environmental Research Laboratory, Cincinnati, OH, April 1980.)

TABLE 6.8
Wastewater Treatment Process Characteristics for Plants A, B, and C[a]

Characteristic	Value		
	Plant A	Plant B	Plant C
Wastewater			
Average flow rate (gal/min)	39	21	16
pH:			
Feed	2.0–4.0	4.5–6.0	2.5–3.0
Effluent	9.0–10.0	8.5–9.5	7.5–8.5
Average feed concentration (mg/L)			
Nickel	31	NA	NA
Copper	28	NA	NA
Hexavalent chromium	76	27	0.07
Total chromium	88	39	8
Zinc	NA	48	24
Iron	NA	1.4	127
Phosphorus	NA	NA	289
Treatment chemicals			
Lime:[b]			
lb/h	8.8	2.0	8.1
Calcium chloride (for phosphate removal):[b]			
lb/h	NA	NA	17.0
Canonic polymor:[b]			
lb/h	0.1	0.17	0.02
Anionic polymer:[b]			
lb/h	NA	NA	0.01
Ferrous sulfide:			
lb/h	12.5[c]	4.5[d]	0.30[b]
Total chemicals ($/h)	5.78	2.23	2.48
Chemical cost ($/1,000 gal)	6.03	4.32	6.30[e]
Sludge generation factors			
Dry solids generation:			
lb/h	23.7	7.2	16.4
First stage	NA	NA	16
Second stage	NA	NA	0.4
lb/1,000 gal wastewater	10.1	5.7	17[e]
Underflow volume (gal/h at 0.75% solids)	380	114	262
Filter cake volume (gal/h at 30% solids)	7.9	2.4	5.3

Source: USEPA, *Control and Treatment Technology for the Metal Finishing Industry: Sulfide Precipitation, Summary Report,* EPA 625/8-80-003, Environmental Protection Agency, The Industrial Environmental Research Laboratory, Cincinnati, OH, April 1980; USEPA, *Environmental Pollution Control Alternatives: Economics of Wastewater Treatment Alternatives for the Electroplating Industry,* EPA 625/5-79-01 6 U.S. Environmental Protection Agency, June 1979.

Note: NA = not applicable. Costs escalated to 2012 USD. (From US ACE, Yearly Average Cost Index for Utilities. In: *Civil Works Construction Cost Index System Manual,* 110-2-1304, U.S. Army Corps of Engineers, Washington, DC, 44pp. PDF file is available at http://www.nww.usace.army.mil/cost, 2015.)

[a] All three plants use an ISP process to remove metals from wastewater, but plant C uses ISP as a polishing system.

[b] Observed rates.

[c] Based on 3 times the stoichiometric requirement.

[d] Based on 4 times the stoichiometric requirement.

[e] Without the presence of phosphates, treatment cost equals $2.0/1,000 gal, solids generation equals 6.4 lb/1,000 gal.

and iron in solution with phosphates, organic chelating agents, and assorted chemicals used in the process baths. The wastewater is treated in a neutralization/ISP/clarification treatment sequence similar to that shown in Figure 6.13. Then it is mixed with the remainder of the wastewater from the plant and is discharged to the city wastewater treatment system.

The wastewater flow rate to the system averaged 20 gal/min (76 L/min). The performance of the system during a 2-day test in removing chromium (total and hexavalent), zinc, and iron from the wastewater is shown in Figures 6.22 and 6.23. The same sample location designation used in

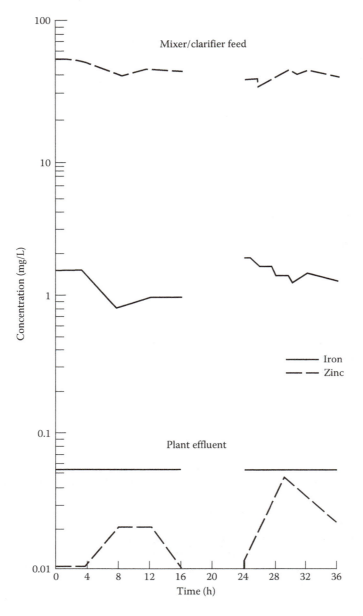

FIGURE 6.22 Plant B's performance in removing iron and zinc. The plant operates two shifts per day; there was no waste water flow between hours 16 and 24. (From USEPA, *Control and Treatment Technology for the Metal Finishing Industry: Sulfide Precipitation, Summary Report,* EPA 625/8-80-003, Environmental Protection Agency, The Industrial Environmental Research Laboratory, Cincinnati, OH, April 1980.)

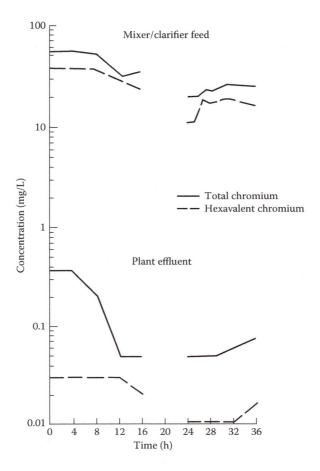

FIGURE 6.23 Plant B's performance in removing chromium. There was no wastewater flow to the system between hours 16 and 24. (From USEPA, *Control and Treatment Technology for the Metal Finishing Industry: Sulfide Precipitation, Summary Report,* EPA 625/8-80-003, Environmental Protection Agency, The Industrial Environmental Research Laboratory, Cincinnati, OH, April 1980.)

Figure 6.20 applies. Figure 6.24 defines the ratio of FeS supply to stoichiometric demand for the same test period. The ratio varied from 3 to 5 times the stoichiometric demand during the test period. The quality of the effluent, which contained lower pollutant levels than those specified in both local and state guidelines, showed no discernible trends within this range of reagent supply/demand ratios.

The cost of treatment chemicals and the sludge generation factors for the ISP system at this facility are shown in Table 6.8. Chemical costs were approximately USD 4.32/1,000 gal of wastewater treated.

Plant C uses the ISP process to polish the clarified overflow from a conventional hydroxide precipitation/clarification treatment sequence. The system treats approximately 15–18 gal/min (57–68 L/min) of wastewater from a programmed, barrel-dip, zinc-phosphatizing plating line. The system is similar to the one shown in Figure 6.17; it has a second mixer/clarifier and polymer feed system, installed after the second-stage neutralizer, to remove the precipitated metal hydroxides and phosphates. Dual polyelectrolyte feed systems are needed because an anionic polymer is used in the hydroxide removal clarifier and a cationic polymer is used to enhance the settling of the precipitated metal sulfides. The sludge production and FeS consumption are reduced considerably compared with a system treating the total metal load with sulfide precipitation. Less than 5% of the waste solids removed from the system are attributed to the sulfide precipitation step.

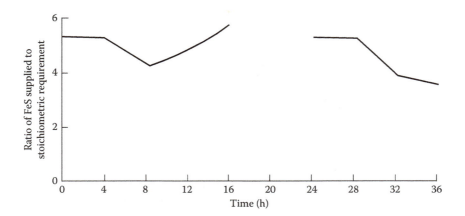

FIGURE 6.24 FeS supplied versus stoichiometric requirement at plant B. There was no wastewater flow to the system between hours 16 and 24. (From USEPA, *Control and Treatment Technology for the Metal Finishing Industry: Sulfide Precipitation, Summary Report,* EPA 625/8-80-003, Environmental Protection Agency, The Industrial Environmental Research Laboratory, Cincinnati, OH, April 1980.)

TABLE 6.9
Influent and Effluent Wastewater Characteristics for ISP Polishing System

	Wastewater Analysis		
Item	Influent	Effluent	Permit Requirements[a]
pH	2.9	8.5	6.6–9.5
Phosphorus (mg/L)	289	0.3	<1.2
Total suspended solids (mg/L)	320	6	<23
Total chromium (mg/L)	8	<0.10	<0.6
Hexavalent chromium (mg/L)	0.07	<0.02	<0.06
Nickel (mg/L)	0.77	<0.1	<0.6
Zinc (mg/L)	24	0.12	<0.6
Iron (mg/L)	127	0.60	<1.2

Source: USEPA, *Control and Treatment Technology for the Metal Finishing Industry: Sulfide Precipitation, Summary Report,* EPA 625/8-80-003, Environmental Protection Agency, The Industrial Environmental Research Laboratory, Cincinnati, OH, April 1980.

[a] Monthly average of daily composite samples.

Table 6.8 presents the chemical consumption and sludge generation rates for plant C. Treatment of the phosphates in the wastewater accounts for a large percentage of the treatment cost, and the phosphate solids constitute the bulk of the sludge generated. The chemical cost associated with removal of the heavy metals contained in the wastewater was estimated at USD 1.98/1,000 gal. Without the presence of phosphates, the solids generation rate would equal 6.4 lb/1,000 gal (0.76 kg/m^3) of wastewater.

Table 6.9 presents the pollutant concentrations in plant C's raw waste and effluent discharge and shows the effluent quality required by the discharge permit.

In this polishing application, FeS is fed into the second-stage mixer/clarifier to yield a concentration of approximately 40 mg/L in the wastewater. The dosage rates for the insoluble solids systems treating the total metal load for plants A and B are approximately 640 and 430 mg/L, respectively.

6.4.4 HYDROXIDE SYSTEM MODIFICATIONS FOR ISP

The metal removal efficiency of a hydroxide precipitation system can be improved by incorporating ISP into the system. Sulfide precipitation can be used either to convert the metals to metal sulfides before the clarifier or as a polishing system to precipitate dissolved metals from wastewater after the insoluble metal hydroxides have been removed by clarification.

6.4.4.1 Equipment Requirements

The key component of an ISP system is the solid–liquid contact chamber where the wastewater is mixed thoroughly with the insoluble sulfide contained in the sludge blanket. Three design criteria must be addressed in specifying this piece of equipment:

1. Liquid residence time in the mixing zone
2. Sludge blanket volume and density
3. Mixing efficiency

Figure 6.25 is a schematic of the mixer/clarifier designed specifically for this application. In the systems currently using ISP, the unit is sized to provide approximately 1 h of liquid residence time in the mixing zone. Because the mixing zone volume is equal to the solids retention volume, a large inventory of unreacted FeS can be maintained in the unit. The agitator is designed to maintain a dense fluidized sludge in the mixing zone. Sample ports are located in the different zones of the unit to check the sludge density. The unit also has a timed sludge drawoff valve that can be set to balance the blowdown to the solids accumulation rate automatically.

Other elements needed to augment a treatment system with ISP include

1. FeS reagent preparation tanks, reagent storage, and feed pumps
2. A reagent feed control system that matches reagent dosage to wastewater flow rate
3. A control loop to interrupt the wastewater feed during low-pH conditions

In converting a hydroxide system to use sulfide precipitation, the addition of a polishing filtration system to remove residual suspended solids from the clarifier overflow could significantly reduce effluent metal concentrations. Meeting strict effluent metal discharge limits will require an effluent with low levels of both suspended and dissolved metals.

6.4.4.2 Treatment System Evaluation

The cost advantages of using ISP as a polishing system must be weighed against the higher equipment costs and space requirements of a second clarifier. It might be more cost effective for plants with small metal loadings to incorporate ISP upstream of the existing clarifier and thus avoid the expense of a second clarifier.

Retrofitting a hydroxide treatment system that already has a flocculation zone to enhance the settling properties of the precipitated metals before clarification can be accomplished simply and with minimum investment. Many existing systems include a flocculation chamber (13) either in a separate vessel or as part of the clarifier itself. As shown in Figure 6.26, sulfide precipitation can be incorporated into this type of treatment system by installing:

1. An FeS reagent addition system and feed control system to feed FeS into the flocculation chamber in proportion to the volume of wastewater processed

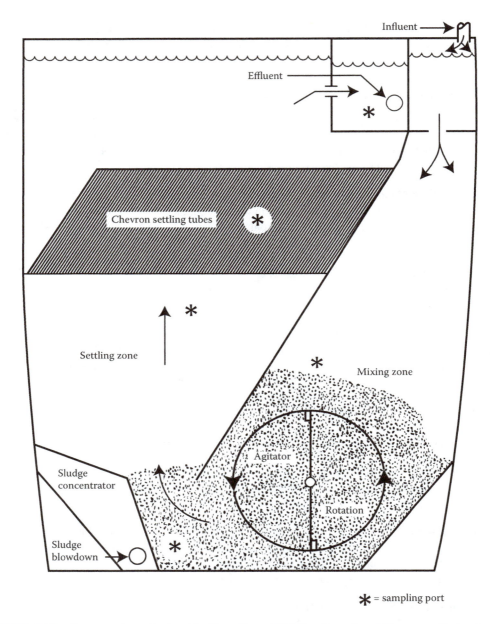

FIGURE 6.25 Cross section of mixer/clarifier. (From USEPA, *Control and Treatment Technology for the Metal Finishing Industry: Sulfide Precipitation, Summary Report,* EPA 625/8-80-003, Environmental Protection Agency, The Industrial Environmental Research Laboratory, Cincinnati, OH, April 1980.)

2. A sludge recirculation loop (if not already existing) to recycle solids from the clarifier underflow back to the flocculator
3. A low-pH feed interrupt control loop to stop the feed to the flocculator if the pH of this stream falls below the set-point

Pilot tests must be performed to determine if the residence time, agitation, and blanket density in the flocculation chamber are conducive to effective metal removal. Figure 6.6 defined the different

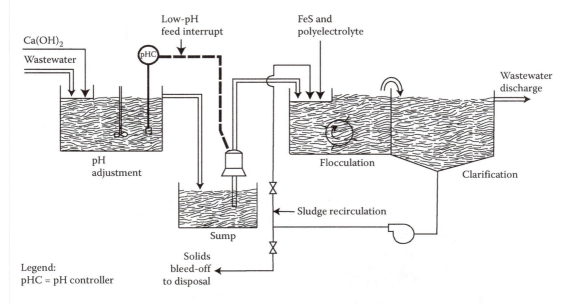

FIGURE 6.26 Retrofit of a hydroxide system with insoluble sulfide treatment. (From USEPA, *Control and Treatment Technology for the Metal Finishing Industry: Sulfide Precipitation, Summary Report,* EPA 625/8-80-003, Environmental Protection Agency, The Industrial Environmental Research Laboratory, Cincinnati, OH, April 1980.)

variables for evaluation by pilot testing or jar testing. Deficiencies in the flocculator residence time, mixing efficiency, and the like can be tolerated, although they generally result in increased reagent consumption.

An approach for treatment systems that do not have flocculation zones is either to add a flocculator or to replace the existing clarifier with the mixer/clarifier designed for this application (Figure 6.25). The most reliable approach to using ISP as a polishing system would be to install a mixer/clarifier downstream of the existing clarifier.

6.4.4.3 ISP Batch Treatment Systems

As with continuous treatment systems, batch treatment using ISP would require contact between the wastewater and a dense sludge blanket to achieve maximum metal removal. Consequently, a large volume of solids would be needed for each batch, necessitating storage of the settled sludge after batch treatment. Figure 6.27 shows a configuration of an ISP batch treatment system and the associated treatment sequence. The major process components of the system are

1. Two tanks equipped with mechanical agitation
2. A precipitation tank
3. Reagent storage and feed systems to add the lime (or caustic soda), FeS, and polymer

The two agitated tanks alternate as the wastewater collection tank and pretreatment tank. Pretreatment is required to neutralize the acidic wastewater before mixing it with the metal sulfide sludge. A precipitation tank is needed to bring the wastewater into contact with the FeS slurry and to provide storage volume for maintaining an inventory of sludge solids in the system. Gentle agitation is required to suspend the sludge solids during mixing and to promote particle growth of the precipitated solids.

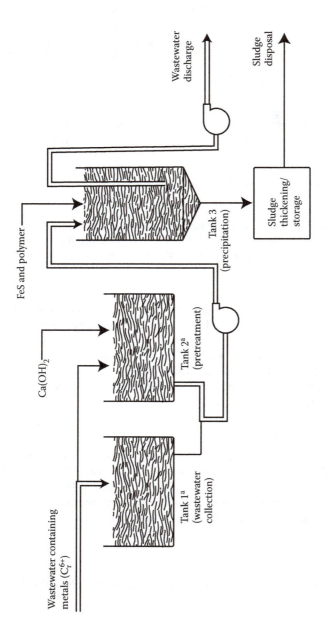

FIGURE 6.27 Batch wastewater treatment using ISP. [a]Tanks 1 and 2 alternate in process function. Treatment sequence: When Tank 2 is filled to capacity, incoming wastewater is diverted to Tank 1. The pH of the wastewater in Tank 2 is adjusted to 8.5. A sample is removed from Tank 2 and analyzed by jar test procedure to determine required FeS dosage. The wastewater in Tank 2, along with the required amount of FeS and polymer, is charged into Tank 3. The wastewater/sludge mixture in Tank 3 is agitated for 1 hour. Agitation is stopped and the solids are allowed to settle. A sample of the clarified wastewater is analyzed to check water quality. The wastewater in Tank 3 is decanted and discharged. A portion of the settled sludge is discharged to sludge disposal to maintain a constant sludge inventory. (From USEPA, *Control and Treatment Technology for the Metal Finishing Industry: Sulfide Precipitation, Summary Report*, EPA 625/8-80-003, Environmental Protection Agency, The Industrial Environmental Research Laboratory, Cincinnati, OH, April 1980.)

6.4.5 ISP Treatment Costs

6.4.5.1 Operating Costs

The following costs associated with using ISP are in addition to the operating costs of a conventional hydroxide precipitation system (18):

1. Reagent costs for FeS and polyelectrolyte
2. Labor cost of additional operational duties described earlier
3. Disposal cost of any additional solid waste generated
4. Licensing fee charged by the patent holder to use the process

Reagent costs for FeS depend on the quantity of metals to be precipitated (or, in the case of hexavalent chromium, the quantity to be reduced chemically) and the ratio of reagent needed for effective removal to the stoichiometric reagent requirement. Figure 6.28a shows the FeS consumption rates and reagent cost for various metal concentrations in the wastewater and typical ratios

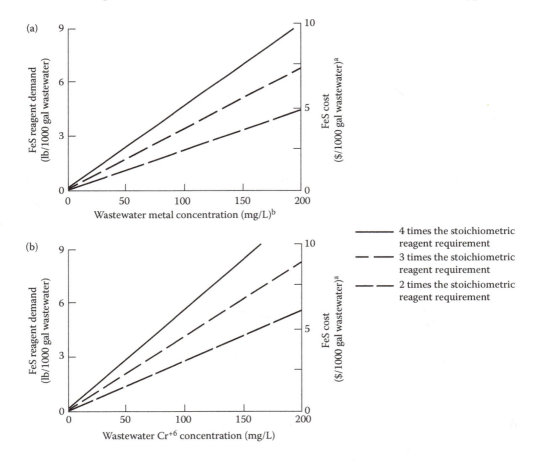

FIGURE 6.28 FeS consumption and cost factors for: (a) precipitation of metals and (b) hexavalent chromium reduction. [a]Based on FeS at \$1.05/lb. [b]Only includes those metals, other than iron, that form sulfides; based on metal with molecular weight of 62.5 (average of Ni, Cu, and Zn). Cost in 2012 USD. (From USEPA, *Control and Treatment Technology for the Metal Finishing Industry: Sulfide Precipitation, Summary Report,* EPA 625/8-80-003, Environmental Protection Agency, The Industrial Environmental Research Laboratory, Cincinnati, OH, April 1980; USEPA, *Environmental Pollution Control Alternatives: Economics of Wastewater Treatment Alternatives for the Electroplating Industry,* EPA 625/5-79-01 6 U.S. Environmental Protection Agency, June 1979.)

of reagent demand to stoichiometric requirement. The wastewater metal concentration is defined as the metals other than iron that will form sulfides. To compute reagent consumption rates, it was assumed that the metals have a "plus 2" valence and a molecular weight equal to the average molecular weight of copper, nickel, and zinc. Although determination of the optimum dosage ratio requires testing, wastewater with no heavy metal complexing agents generally requires 1.5–2 times the stoichiometric reagent requirements, whereas wastewater containing complexed heavy metals will require 3–4 times the stoichiometric reagent dosage. Figure 6.28b presents the FeS reagent demand and cost for wastewater treatment over a range of hexavalent chromium concentrations.

At three operating plants, labor requirements for the ISP systems varied only slightly. Each plant employed a full-time operator for one shift and required 2–6 h of operator attention on other shifts.

ISP systems generate considerably more sludge in treating a volume of wastewater than the conventional hydroxide precipitation scheme. The additional sludge results from precipitation as hydroxides of the ferrous and ferric ions liberated as the sulfide reagent is consumed and from the

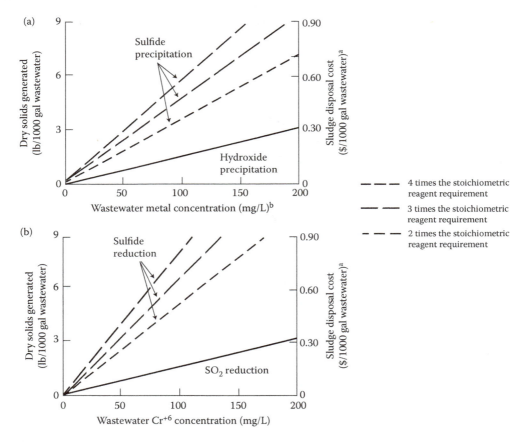

FIGURE 6.29 Sludge generation factors for: (a) precipitation of metals and (b) hexavalent chromium reduction. [a]Only includes metal hydroxide and metal sulfide solids. [b]Only includes metals, other than iron, that form metal sulfides; based on a metal with a molecular weight of 62.5 (average of Ni, Cu, and Zn); ferrous ions in wastewater will generate 1.34 lb solids/1000 gal at a concentration of 100 mg/L ferric ions will generate 1.6 lb solids/1000 gal at a concentration of 100 mg/L. [c]Based on disposal at 25% solids by weight and $0.25/gal sludge. Cost in 2012 USD. (From USEPA, *Control and Treatment Technology for the Metal Finishing Industry: Sulfide Precipitation, Summary Report,* EPA 625/8-80-003, Environmental Protection Agency, The Industrial Environmental Research Laboratory, Cincinnati, OH, April 1980; USEPA, *Environmental Pollution Control Alternatives: Economics of Wastewater Treatment Alternatives for the Electroplating Industry,* EPA 625/5-79-01 6 U.S. Environmental Protection Agency, June 1979.)

excess FeS that is used in treatment. Figure 6.29 compares the solids generation rates for ISP systems with those for treatment systems using hydroxide precipitation for metal removal and sulfur dioxide (SO_2) for chromium reduction. The graph also shows solid waste disposal charges, assuming the sludge is disposed of at 25% solids by weight and at a cost of USD 0.25/gal. For plants with different sludge disposal cost formulas, the disposal cost can be calculated by multiplying the cost indicated in Figure 6.29 by the ratio of the actual disposal cost to the assumed rate of USD 0.25/gal.

Owing to the high cost of sludge disposal—normally from USD 0.12/gal to USD 0.50/gal—it is cost effective to invest in mechanical dewatering equipment to reduce the sludge volume. At the three operating plants, recessed plate filter presses were installed to dewater the sludge before transport to the disposal site. The presses dewatered the underflow from less than 1% solids by weight to 25%–30% solids by weight.

Total sludge generation for both hydroxide and sulfide systems will be somewhat higher than the rates shown in Figure 6.29. The additional solids are caused by the presence of lime solids, suspended solids in the wastewater feed, and insoluble byproducts resulting from neutralization. For treating waste streams to remove heavy metals, the additional solids should be approximately the same for insoluble sulfide and hydroxide systems. For chromium reduction, SO_2 reduction systems often require the wastewater to be acidified, and the quantity of alkali for subsequent neutralization is larger than that required with sulfide reduction. Consequently, the additional lime required for neutralization with SO_2 reduction will result in more lime solids in the sludge.

Licensing fees for the use of ISP to treat wastewater are charged annually and are determined by the flow rate of wastewater treated. This fee is small, however, compared with other costs typically associated with wastewater treatment.

6.4.5.2 Equipment Costs

The actual total installation costs for the three ISP treatment systems described earlier are presented in Table 6.10. All three systems were installed in plants that had no existing treatment systems. The systems in plants A and B are similar to the one illustrated in Figure 6.13. The costs presented also include duplexing of many of the pumps and reagent storage tanks, a control panel, and additional instrumentation not shown on the flow diagram. Plant C is a sulfide polishing system similar to the one shown in Figure 6.17. The installed cost of this system includes the additional equipment required by a polishing system—a second clarifier (to separate the insoluble compounds resulting from hydroxide neutralization) and a second polyelectrolyte feed system.

Much of the equipment in an ISP system is common to hydroxide systems. Cost data on wastewater treatment equipment for the metal finishing industry are presented in the USEPA report, *Economics of Wastewater Treatment Alternatives for the Electroplating Industry* (35). Converting a hydroxide system to use ISP in many cases will require only the installation of a mixer/clarifier downstream of the existing clarifier and a feed system to meter the FeS and polyelectrolyte into the wastewater.

Table 6.11 presents the cost (including installation and hardware) of installing the following ISP process equipment components in an existing treatment system:

1. Mixer/clarifier
2. FeS reagent preparation and feed system
3. Polymer feed system
4. Control loops
5. Suspended solids polishing filters

The installed costs presented for a mixer/clarifier are for a preassembled, skid-mounted component requiring only piping and electrical connections for installation. The FeS reagent preparation and feed system includes two FeS feed tanks with low-level alarms, two reagent pumps, a mixing tank, and a transfer pump; the costs are for skid-mounted, preassembled units, constructed of carbon steel (see Figure 6.16).

TABLE 6.10

Installation Costs for Three Sulfex ISP Treatment Systems

Cost Component	ISP System Cost ($1,000)		
	Plant A	Plant B	Plant C
Installation costs			
Process equipment	492	258	NA
Underground tanks	101	135	NA
Shipping end installation	81	62	NA
Additional building space	56	NA	NA
Startup expenses	8	NA	NA
Engineering	NA	48	NA
Other	NA	3	NA
Total installation costs	738[a]	506[b]	412[c]

Source: USEPA, *Control and Treatment Technology for the Metal Finishing Industry: Sulfide Precipitation, Summary Report,* EPA 625/8-80-003, Environmental Protection Agency, The Industrial Environmental Research Laboratory, Cincinnati, OH, April 1980; USEPA, *Environmental Pollution Control Alternatives: Economics of Wastewater Treatment Alternatives for the Electroplating Industry,* EPA 625/5-79-01 6 U.S. Environmental Protection Agency, June 1979.

Note: NA = not available. Costs escalated to 2012 USD. (From US ACE. Yearly Average Cost Index for Utilities. In: *Civil Works Construction Cost Index System Manual,* 110-2-1304, U.S. Army Corps of Engineers, Washington, DC, 44pp, 2015 (36,37).

[a] ISP system design flow = 40 gal/min
[b] ISP polishing system design flow = 35 gal/min
[c] ISP polishing system design flow = 15 gal/min

The costs presented for the polymer feed system are based on a system with two plastic polymer feed tanks and two positive displacement pumps with adjustable stroke. The skid-mounted, preassembled components are equipped with a low-level alarm and dilution water-mixing apparatus. Costs are given for two control loops: a reagent addition control system with a magnetic flow meter and flow counter (to match the addition of FeS and polymer with wastewater volumetric throughput) and a low-pH feed interruption control. The costs for suspended solids polishing filters are for dual mixed-media filters, skid mounted and sized so that one filter can process the maximum flow during backwash. The filters are equipped with a blower for low-pressure air scouring, a backwash storage tank, and a pump to bleed the wash back into the system.

6.4.5.3 Cost Comparison of Conventional Chemical Reduction and ISP Chromium Reduction

Replacing a conventional chromium reduction system with reduction by FeS can be advantageous. In some cases an operating cost benefit will result. Another advantage of reducing chromium with FeS is that the hexavalent chromium wastewater does not need to be segregated for individual treatment; it can be treated in the common neutralization/precipitation treatment sequence. Figure 6.30 defines typical treatment sequences for reduction of chromium by chemical means and using FeS. The FeS treatment process eliminates the need to lower and raise the pH of the wastewater and results in a significant saving in acid and alkali reagent. Table 6.12 presents treatment and sludge disposal costs for the two chromium reduction systems shown in Figure 6.30. The chemical

TABLE 6.11

Equipment Cost Factors for ISP Treatment System Components

Equipment Component	Installed Cost ($1,000)
Mixer/clarifier	
30-gal/min wastewater flow rate	44
60-gal/min wastewater flow rate	54
90-gal/min wastewater flow rate	59
Ferrous sulfide reagent preparation and feed system	
5-lb/h FeS feed rate[a]	39
10-lb/h FeS feed rate	49
15-lb/h FeS feed rate	59
Polymer feed system	12
Control loops	
Reagent addition system	11
Low-pH feed interruption control	5
Suspended solids polishing filters	
30-gal/min wastewater flow rate	59
60-gal/min wastewater flow rate	80
90-gal/min wastewater flow rate	100

Source: USEPA, *Control and Treatment Technology for the Metal Finishing Industry: Sulfide Precipitation, Summary Report,* EPA 625/8-80-003, Environmental Protection Agency, The Industrial Environmental Research Laboratory, Cincinnati, OH, April 1980; USEPA, *Environmental Pollution Control Alternatives: Economics of Wastewater Treatment Alternatives for the Electroplating Industry*, EPA 625/5-79-01 6 U.S. Environmental Protection Agency, June 1979.

Note: Costs are basic installed costs of different components. Engineering and design costs, site preparation, and equipment freight charges are not included. Costs escalated to 2012 USD. (US ACE. Yearly Average Cost Index for Utilities. In: *Civil Works Construction Cost Index System Manual*, 110-2-1304, U.S. Army Corps of Engineers, Washington, DC, 44pp. PDF file is available at http://www.nww.usace.army.mil/cost, 2015.)[a] For lower feed rates, less automated systems are available for approximately $12,000.

consumption factors assume that the lime consumption is twice the stoichiometric amount required to neutralize the wastewater and precipitate the dissolved metals. The excess lime is needed to overcome buffering normally encountered when neutralizing waste streams. It is further assumed that lime solids equal to 50% of the mass of lime used in neutralization are present in the sludge. These lime solids result from precipitation of insoluble byproducts in the neutralization reaction as well as from the tendency for some portion of the lime used not to dissolve and add to the sludge volume. Consequently, the lime required in the chemical reduction treatment sequence to raise the pH from 2 to 8 results in considerable sludge generation.

Figure 6.31 compares the cost of treatment chemicals and sludge disposal for the two chromium reduction systems shown in Figure 6.30 over a range of hexavalent chromium concentrations in the wastewater. A cost saving can be realized for FeS reduction compared with conventional chemical reduction. For wastewater requiring twice the stoichiometric FeS dosage, a treatment cost advantage exists over treatment of wastewater containing less than 50 mg/L Cr^{6+} by SO_2 reduction and that containing less than 100 mg/L Cr^{6+} by $NaHSO_3$ reduction. For FeS reduction systems requiring twice the stoichiometric dosage rate, a savings in solid waste disposal costs also would be realized for treatment of wastewater containing less than 150 mg/L Cr^{6+}. At higher FeS dosage requirements,

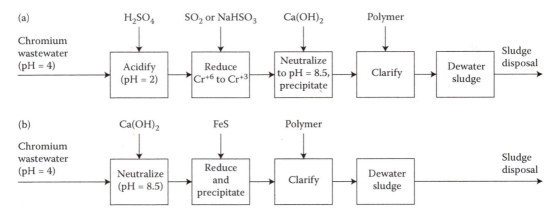

FIGURE 6.30 Comparison of chromium reduction treatment sequences: (a) chemical and (b) insoluble sulfide. Table 6.12 presents cost basis for comparison of chemical and insoluble sulfide chromium reduction systems. (From USEPA, *Control and Treatment Technology for the Metal Finishing Industry: Sulfide Precipitation, Summary Report,* EPA 625/8-80-003, Environmental Protection Agency, The Industrial Environmental Research Laboratory, Cincinnati, OH, April 1980.)

TABLE 6.12

Cost Basis for Comparison of Chemical and Insoluble Sulfide Chromium Reduction Treatment Systems Shown in Figure 6.30

| | Cost in 2012 USD[a] | | | |
| | Treatment[b] | | Sludge Disposal[c] | |
Parameter	$/lb Cr⁶⁺	$/1,000 gal Wastewater	S/lb Cr⁶⁺	$/1,000 gal Wastewater
Chemical reduction				
Sulfur dioxide	1.05	1.39	0 39	0.29
Sodium bisulfite	2.00	1.66	0.39	0.29
Insoluble sulfide reduction				
Ferrous sulfide at dosage equal to 2 times stoichiometric requirement	3.86	0.07	0.51	0.02
Ferrous sulfide at dosage equal to 4 times stoichiometric requirement	7.61	0.07	0.81	0.02

Source: USEPA, *Control and Treatment Technology for the Metal Finishing Industry: Sulfide Precipitation, Summary Report,* EPA 625/8-80-003, Environmental Protection Agency, The Industrial Environmental Research Laboratory, Cincinnati, OH, April 1980; USEPA, *Environmental Pollution Control Alternatives: Economics of Wastewater Treatment Alternatives for the Electroplating Industry,* EPA 625/5-79-01 6 U.S. Environmental Protection Agency, June 1979.

Note: 2012 cost basis. Sulfur dioxide and sodium bisulfite consumption is equal to 2 times the stoichiometric requirement at a hexavalent chromium (Cr^{6+}) concentration of 50 mg/L. Lime consumption is equal to 2 times the stoichiometric requirement for unbuffered waste streams. Lime solids are 50% of lime dosage and contribute to sludge volume. Costs escalated to 2012 USD. (From US ACE. Yearly Average Cost Index for Utilities. In: *Civil Works Construction Cost Index System Manual,* 110-2-1304, U.S. Army Corps of Engineers, Washington, DC, 44pp, 2015. See references 36 and 37.)

[a] Total treatment cost is based on both mass of chromium reduced and volume of wastewater treated.

[b] Based on lime at $0.035/lb, sulfur dioxide at $0.15/lb, sodium bisulfite at $0.20/lb, sulfuric acid at $0.05/lb, and ferrous sulfide at $0.43/lb.

[c] Based on disposal at 25% solids by weight at a cost of $0.10/gal sludge.

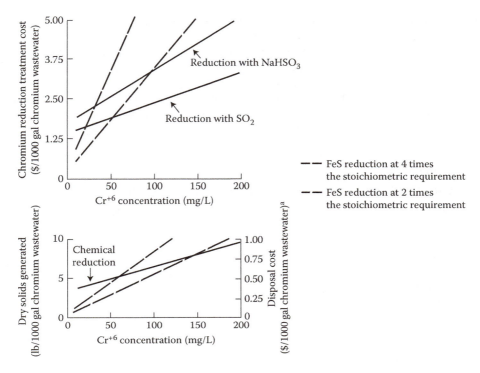

FIGURE 6.31 Costs of treatment chemicals and sludge disposal for chemical and insoluble sulfide chromium reduction. [a]Based on disposal at 25% solids by weight at $0.25/gal sludge. Based on treatment parameters defined in Table 6.12. Cost in 2012 USD. (From USEPA, *Control and Treatment Technology for the Metal Finishing Industry: Sulfide Precipitation, Summary Report,* EPA 625/8-80-003, U.S. Environmental Protection Agency, The Industrial Environmental Research Laboratory, Cincinnati, OH, April 1980; USEPA, *Environmental Pollution Control Alternatives: Economics of Wastewater Treatment Alternatives for the Electroplating Industry,* EPA 625/5-79-01 6 U.S. Environmental Protection Agency, June 1979.)

such as 4 times the stoichiometric demand, chromium reduction using FeS is more economical for the treatment of dilute chromium waste streams.

It is important to point out that the preceding comparisons are based on typical operating conditions and reagent costs; a comparative analysis for a specific plant should use actual operating data (e.g., reagent consumption and sludge generation).

6.4.5.4 Cost Comparison of ISP Polishing and Total Metal Treatment

Converting all metals in a waste stream to metal sulfides via sulfide precipitation uses considerable FeS and results in a large volume of waste solids. Separation of the precipitated metal hydroxides from the wastewater before polishing with sulfide precipitation can reduce both reagent consumption and solid waste generation. In a polishing application, the FeS reagent demand is a function of the dissolved metal concentration in the wastewater after hydroxide precipitation/clarification. Conversion of a sulfide precipitation system to a polishing system requires installation of a second clarifier and polyelectrolyte feed system to separate the precipitated metal hydroxides from the neutralized wastewater before adding the sulfide reagent.

The reagent consumption and solid waste generation factors associated with treatment of the total metal load were presented in Figure 6.28. To estimate reagent requirements for a sulfide polishing system, it is necessary to determine the concentration of metals in the wastewater after hydroxide neutralization/precipitation/clarification. Reagent consumption ranges between 1.5 and 4 times stoichiometric demand for polishing systems. Compared with the reagent consumption factors presented in

Figure 6.28, the sulfide precipitation polishing system at plant C required an FeS dosage rate of 40 mg/L in the wastewater. Note, however, that this system did not have a significant level hexavalent chromium in the wastewater; hexavalent chromium is not removed by hydroxide precipitation, and reagent demand for chromium reduction will be the same for sulfide polishing or sulfide precipitation systems.

Plant A uses ISP for total treatment of the metals in the wastewater. Table 6.13 presents the costs of wastewater treatment using ISP as a polishing step compared with its use to precipitate the total metal load at plant A. The major cost saving results from reduced FeS consumption; the required FeS dosage is reduced by separation of precipitated metal hydroxides before the addition of the sulfide reagent.

TABLE 6.13
Potential Benefits for Use of ISP Polishing System at Plant A

Item	Value	
Wastewater characteristics		
Average flow rate (gal/min)	39	
pH		
Feed	2–4	
Effluent	9–10	
Average feed concentration (mg/L)		
Nickel	31	
Copper	28	
Hexavalent chromium	76	
Total chromium	88	
	Current system	**Polishing system**
Treatment chemical costs ($/h)		
Lime[a]	0.68	0.68
Polyelectrolyte[b]	1.02	0.85
Ferrous sulfide[c]	13 13	8.71
Total	14.83	10.24
Cost saving	NA	4.59
Sludge generation factors		
Dry solids generation (lb/h):		
First stage	NA	6.2
Second stage	NA	13.1
Total	23.6	19.3
Sludge cake volume (gal/h at 30% solids)	7.9	6.4
Disposal cost at $0.46/gal sludge ($/h)	3.63	2.94
Disposal cost saving ($/h)	NA	0.69
Net savings: treatment chemical cost savings plus disposal cost savings ($/h)	NA	5.28
Annual saving based on 6,000 h/year operation ($/year)	NA	31,700

Source: USEPA, *Control and Treatment Technology for the Metal Finishing Industry: Sulfide Precipitation, Summary Report,* EPA 625/8-80-003, Environmental Protection Agency, The Industrial Environmental Research Laboratory, Cincinnati, OH, April 1980; USEPA, *Environmental Pollution Control Alternatives: Economics of Wastewater Treatment Alternatives for the Electroplating Industry,* EPA 625/5-79-01 6 U.S. Environmental Protection Agency, June 1979.

Note: 2012 cost basis. NA = not applicable.

[a] Observed rates.
[b] Design rate.
[c] Based on 3 times the stoichiometric requirement.

Based on the savings indicated in Table 6.13, a profitability analysis of the investment required to convert to a polishing system is presented in Table 6.14. The USD 63,000 investment required for the conversion would have an average after-tax return on investment of 13%.

The costs of FeS reagent and solid waste disposal for ISP systems and sulfide polishing systems are compared further in Figure 6.32 for each 1,000 gal (3785 L) of wastewater treated at various metal concentrations. The solid waste disposal cost estimate assumed disposal of the sludge at 25% solids by weight at a cost of UDS 0.25/gal of waste and that the sludge from both systems would

TABLE 6.14
Economics of Converting Plant A ISP Treatment System to ISP Polishing System Operating 6,000 h/year

Item	Value
Installation costs ($)	
Equipment:	
40 gal/min mixer/clarifier	44,000
Polyelectrolyte feeder	12,000
Total equipment installation	56,000
Additional installation: estimated freight, site preparation, and miscellaneous	7,000
Total installation costs:	63,000
Additional annual operating costs ($/year)	
Labor (100 h/year at $20/h)	2,000
Supervision	0
Maintenance (6% of investment)	3,800
General plant overhead	2,000
Utilities	
Electricity	500
Water (polymer feeder)	500
Total operating costs	8,800
Annual fixed costs ($/year)	
Depreciation (10% of investment)	6,300
Taxes and insurance (1% of investment)	630
Total fixed costs	6,930
Total operating and fixed costs ($/year)	15,730
Annual savings ($/year)	
Chemicals	27,550
Sludge disposal	4,150
Total annual savings	31,700
Not savings: annual savings minus operating and fixed costs ($/year)	15,970
Not savings after taxes, 48% tax rate ($/year)	8,300
After-tax average return on investment (%)	13.0
Cash flow from investment: net savings after taxes plus depreciation ($/year)	14,600
Payback period: total investment/cash flow (year)	4.3

Source: USEPA, *Control and Treatment Technology for the Metal Finishing Industry: Sulfide Precipitation, Summary Report,* EPA 625/8-80-003, Environmental Protection Agency, The Industrial Environmental Research Laboratory, Cincinnati, OH, April 1980; USEPA, *Environmental Pollution Control Alternatives: Economics of Wastewater Treatment Alternatives for the Electroplating Industry,* EPA 625/5-79-01 6 U.S. Environmental Protection Agency, June 1979.

Note: 2012 cost basis.

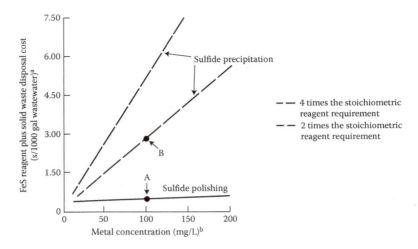

FIGURE 6.32 Treatment cost of ISP versus insoluble sulfide polishing. [a]Solid waste disposal at 25% solids by weight and $0.25/gal. [b]Based on total metal concentration in wastewater; includes only metals, other than iron, that form sulfides; based on a metal with a molecular weight of 62.5 (average of Ni, Cu, and Zn). Cost in 2012 USD. (From USEPA, *Control and Treatment Technology for the Metal Finishing Industry: Sulfide Precipitation, Summary Report,* EPA 625/8-80-003, U.S. Environmental Protection Agency, The Industrial Environmental Research Laboratory, Cincinnati, OH, April 1980; USEPA, *Environmental Pollution Control Alternatives: Economics of Wastewater Treatment Alternatives for the Electroplating Industry*, EPA 625/5-79-01 6 U.S. Environmental Protection Agency, June 1979.)

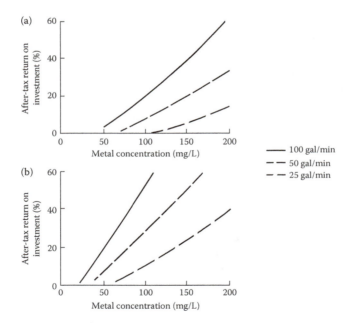

FIGURE 6.33 Return on investment of additional capital required for insoluble sulfide polishing system: (a) treatment requiring 2 times the stoichiometric FeS requirement and (b) treatment requiring 4 times the stoichiometric FeS requirement. Based on operating 4000 h/year: Return on investment calculated using same basis as Table 6.14; chemical and sludge disposal savings from Figure 6.32, and equipment cost from Table 6.11. (From USEPA, *Control and Treatment Technology for the Metal Finishing Industry: Sulfide Precipitation, Summary Report,* EPA 625/8-80-003, U.S. Environmental Protection Agency, The Industrial Environmental Research Laboratory, Cincinnati, OH, April 1980.)

dewater to the same level. The FeS reagent cost for the polishing system was derived from a required FeS dosage rate of 40 mg/L of wastewater. Figure 6.32 presents the difference in cost, rather than total treatment costs, of sulfide reagent and solid waste disposal for sulfide precipitation and sulfide polishing systems. Other costs associated with treatment should be similar for both systems.

A polishing system can achieve significant savings at higher wastewater metal concentrations. As an example, Figure 6.32 reveals that a system treating 3,000 gal/h (11,340 L/h) with a metal concentration of 100 mg/L and requiring twice the stoichiometric amount of FeS would save USD 7.00/h—(B minus A) × 3000 gal/h. At the same flow rate and metal concentration, the savings would be USD 14.00/h if the wastewater required 4 times the stoichiometric amount of FeS.

Using the savings shown in Figure 6.32. Figure 6.33 presents the return on investment for installing the additional treatment hardware needed for a polishing system over a range of metal concentrations and wastewater flow rates.

REFERENCES

1. Federal Register, *Resource Conservation and Recovery Act (RCRA),* 42 U.S. Code s/s 6901 et seq. 1976, U.S. Government, Public Laws, www.federalregister.gov, U.S. Environmental Protection Agency, Washington, DC, 2015.
2. USEPA, *Resource Conservation and Recovery Act (RCRA)—Orientation Manual,* U.S. Environmental Protection Agency, Report # EPA530-R-02-016, Washington, DC, January 2003.
3. USEPA, *Federal Hazardous and Solid Wastes Amendments (HSWA),* U.S. Environmental Protection Agency, Washington, DC, November 1984, http://www.epa.gov/, 2015.
4. Federal Register, *Clean Water Act (CWA),* 33 U.S.C. ss/1251 et seq. (1977), U.S. Government, Public Laws, Available at: www.federalregister.gov, 2015.
5. Suthee, J., *Biogenic Sulfide Production and Selective Metal Precipitation at Low pH for Semiconductor Wastewater Treatment,* http://www.sense.nl/docs/8232, 2015.
6. Shammas, N. K. and Wang, L. K., Treatment of nonferrous metals manufacturing wastes, in: *Handbook of Advanced Industrial and Hazardous Wastes Treatment,* Wang, L. K., Hung, Y. T., and Shammas, N. K. (Eds.), CRC Press, Taylor & Francis, Boca Raton, FL, pp. 71–150, 2010.
7. Shammas, N. K., Wang, L. K., Aulenbach, D. B., and Selke, W. A., Treatment of nickel-chromium plating waste, in: *Advances in Hazardous Industrial Waste Treatment,* Wang, L. K., Shammas, N. K., and Hung, Y. T.(Eds.), CRC Press, Taylor & Francis Group, Boca Raton, FL, pp. 517–544, 2009.
8. Shammas, N. K. and Wang, L. K., Treatment and management of metal finishing industry waste, in: *Heavy Metals in the Environment,* Wang, L. K., Chen, J., Hung, Y. T., and Shammas, N. K., (Eds.), CRC Press, Taylor & Francis Group, Boca Raton, FL, pp. 315–360, 2009.
9. Shammas, N. K. and Wang, L. K., Treatment of metal finishing industry wastes, in: *Waste Treatment in the Metal Manufacturing, Forming, Coating, and Finishing Industries,* Wang, L. K., Shammas, N. K., and Hung, Y. T. (Eds.), CRC Press, Taylor & Francis Group, Boca Raton, FL, pp. 345–380, 2009.
10. Kosińska, K. and Miśkiewicz, T., Precipitation of heavy metals from industrial wastewater by desulfovibrio desulfuricans, *Environment Protection Engineering,* 38, 2, 2012.
11. Wang, L. K., Yung, Y. T., Lo, H. H., and Yapijakis, C. (Eds.), *Hazardous Industrial Waste Treatment,* CRC Press, New York, pp. 289–360, 2007.
12. Wang, L. K., Vaccari, D. A., Li, Y., and Shammas, N. K., Chemical precipitation, in: *Physicochemical Treatment Processes,* Wang, L. K., Hung, Y. T. and Shammas, N. K. (Eds.), Humana Press, Totowa, NJ, pp. 141–198, 2005.
13. Shammas, N. K., Coagulation and flocculation, in: *Physicochemical Treatment Processes,* Wang, L. K., Hung, Y. T., and Shammas, N. K., (Eds.), Humana Press, Totowa, NJ, pp. 103–139, 2005.
14. Shammas, N. K., Kumar, I., Chang, S., and Hung, Y. T, Sedimentation, in: *Physicochemical Treatment Processes,* Wang, L. K., Hung, Y. T., and Shammas, N. K., (Eds.), Humana Press, Inc., Totowa, NJ, pp. 379–429, 2005.
15. Wang, L. K, Fahey, E. M., and Wu, Z., Dissolved air flotation. In: *Physicochemical Treatment Processes,* Wang, L. K., Hung, Y. T., and Shammas, N. K. (Eds.), Humana Press, Inc., Totowa, NJ, pp. 431–494, 2005.
16. Shammas, N. K., Wang, L. T., and Hahn, H., Fundamentals of wastewater flotation, in: *Flotation Technology,* Wang, L. K., Shammas, N. K., Selke, W., and Aulenbach, D. (Eds.), Humana Press, Inc., Totowa, NJ, 2010.

17. Chen, J., Shang, S.-U. Huang, J. Baumann, R., and Hung Y. T., Gravity filtration, in *Physicochemical Treatment Processes,* Wang, L. K., Hung, Y. T., and Shammas, N. K. (Eds.), Humana Press, Inc., Totowa, NJ, pp. 501–541, 2005.

18. USEPA, *Control and Treatment Technology for the Metal Finishing Industry: Sulfide Precipitation, Summary Report,* EPA 625/8-80-003, Environmental Protection Agency, The Industrial Environmental Research Laboratory, Cincinnati, OH, April 1980.

19. USEPA, *Waste Treatment: Upgrading Metal-Finishing Facilities to Reduce Pollution*, EPA 625/3-73-002, U.S. Environmental Protection Agency, Washington, DC. July 1973.

20. Metal Finishers Foundation, *Treatment of Metal Finishing Wastes by Sulfide Precipitation*, NTIS No. 267-284, EPA 600/2-77-049, National Technical Information Service, Springfield, VA, February 1977.

21. Sampaio, R.M.M., Timmers, R.A., Xu, Y., Keesman, K.J., and Lens, P.N.L., Selective precipitation of Cu from Zn in a pH controlled continuously stirred tank reactor, *Journal of Hazardous Materials*, 165, 256–265, 2009.

22. Soya, K., Mihara, N., Kuchar, D., Kubota, M., Matsuda, H., and Fukuta, T., Selective sulfidation of copper, zinc and nickel in plating wastewater using calcium sulfide, *International Journal of Civil and Environmental Engineering*, 2, 2, 2010.

23. Talbot, R.S., Co-precipitation of heavy metals with soluble sulfides using statistics for process control, In: *Proceedings of the 16th Mid-Atlantic Industrial Waste Conference*, Vol. 16, Technomic Publishing Co., Chicago, IL, pp. 279–288, 1984.

24. Peters, R. W. and Ku, Y., The effect of tartrate, a weak complexing agent, on the removal of heavy metals by sulfide and hydroxide precipitation, *Particulate Science and Technology*, 6, 4, 421–439, 1988.

25. Peters, R.W. and Ku, Y., The effect of citrate, a weak complexing agent, on the removal of heavy metals by sulfide precipitation, in: *Metals Speciation, Separation, and Recovery*, J.W. Patterson and R. Passino (Eds.), Lewis Publishers, Inc., Chelsea, MI, pp. 147–169, 1987.

26. Peters, R.W. and Ku, Y., Batch precipitation studies for heavy metal removal by sulfide precipitation, in: *AIChE Symposium Series, Separation of Heavy Metals and Other Contaminants*, Vol. 81, American Institute of Chemical Engineers, New York, pp. 9–27, 1985.

27. Ku, Y., Removal of heavy metals by sulfide precipitation in the presence of complexing agents, Ph.D. Dissertation, Purdue University, West Lafayette, IN, 1986.

28. Blais, J.F., Djedidi, Z., Ben Cheikh, R., Tyagi, R. D., and Mercier, G., Metals precipitation from effluents: Review, *Practice Periodical of Hazardous, Toxic, and Radioactive Waste Management*, ASCE, 12 (3), 135, 2008.

29. Zhuang, J.-M., Acidic rock drainage treatment: A review, *Recent Patents on Chemical Engineering*, 2 (3), 238–252 (Bentham Science Publishers Ltd.), 2009.

30. Naim, R., Kisay, L., Park, J., Qaisar, M., Zulfiqar, A. B., Noshin, M., and Jamil, K., Precipitation chelation of cyanide complexes in electroplating industry wastewater, *International Journal of Environmental Research*, 4 (4) 735–740, 2010.

31. Akpor, O. B. and Muchie M., Remediation of heavy metals in drinking water and wastewater treatment systems: Processes and applications, *International Journal of the Physical Sciences*, 5 (12), 1807–1817, 2010.

32. BioteQ Environmental Technologies, *Sulphide Precipitation Processes for Metal Removal and Recovery*, Vancouver, B.C., Canada, 2015.

33. Lantz, J. B., *Evaluation of a Developmental Heavy Metal Waste Treatment System*. Technical report prepared for Civil Engineering Laboratory, Naval Construction Battalion Center and U.S. Army Medical Research and Development Command, May, 1979.

34. Wang, L. K., Diatomaceous earth precoat filtration, in: *Advanced Physicochemical Treatment Processes*, Wang, L. K., Hung, Y. T., and Shammas, N. K. (Eds.), Humana Press, Totowa, NJ, pp. 155–190, 2006.

35. USEPA., *Environmental Pollution Control Alternatives: Economics of Wastewater Treatment Alternatives for the Electroplating Industry*, EPA 625/5-79-01 6 U.S. Environmental Protection Agency, Washington, DC, June 1979.

36. US ACE, Yearly average cost index for utilities. In: *Civil Works Construction Cost Index System Manual*, 110-2-1304, U.S. Army Corps of Engineers, Washington, DC, 44pp., 2015.

37. Shammas, N. K. and Wang, L. K. *Water Engineering: Hydraulics, Distribution and Treatment*. John Wiley, Hoboken, NJ, p. 775, 2016.

7 Stabilization of Cadmium in Waste Incineration Residues by Aluminum/Iron-Rich Materials

Kaimin Shih and Minhua Su

CONTENTS

ABSTRACT

Toxic metals enriched in the incineration residues of municipal solid waste (MSW) and sewage sludge are a substantial threat to ecosystems and human health. One example is cadmium, a toxic metal reported at concentrations ranging from 10 to 2100 mg/kg in fly ash. Various sorbents (e.g., bauxite, alumina, and calcium oxide) are usually injected into thermal treatment processes to immobilize toxic metals. However, this method has certain disadvantages including the agglomeration of sorbents, the clogging of sorption sites, and the need for additional ash stabilization. Solidification/stabilization (S/S) technologies are an alternative which aim to use physical and/or chemical mechanisms to prevent metal leaching from waste incineration residues. Common S/S technologies use sorption or cementation to immobilize metals but may not reliably control metal leaching in a variety of acidic environments. The development of a novel, economical, and reliable technology to stabilize toxic metals, such as cadmium, in waste incineration residues is a timely and important need. Gamma-alumina (γ-Al_2O_3) and hematite (α-Fe_2O_3) are common, low-cost industrial materials. It has been reported that γ-Al_2O_3 and α-Fe_2O_3 reacts with cadmium under thermal conditions to form crystal structures which immobilize and stabilize cadmium. However, for a reaction mechanism to become a feasible treatment technique, the optimal conditions for effectively incorporating cadmium into crystal structures using γ-Al_2O_3 and α-Fe_2O_3 precursors must be investigated in detail. In this study, γ-Al_2O_3 and α-Fe_2O_3 were employed to stabilize cadmium, and the operational parameters, such as treatment temperature and treatment time, were systematically evaluated. The chemical durability of reaction products was also evaluated using an acid leaching test to assess their metal stabilization effects. It was found that γ-Al_2O_3 and α-Fe_2O_3 were capable of incorporating cadmium into stable crystal structures under attainable thermal conditions, and the product phases, particularly ferrite spinel, showed a remarkably high acidic resistance.

7.1 INTRODUCTION

7.1.1 Cadmium Pollution: Sources, Toxicities, and Control Technologies

Global economic growth and the associated improvements in living standards result in a significant increase in the generation of solid waste such as municipal solid waste (MSW) and sewage sludge (1–3). More than 150 million tons of MSW are produced in China each year, and MSW generation is increasing at an annual rate of 8%–10% (2). In Hong Kong, over 5 million tons of MSW are generated annually from domestic, commercial, and industrial sources (3,4). Of that total, about 2.05 million tons of MSW were recovered in 2014, and the others were disposed of in landfills (4). Statistical data from 2014 have shown that approximately 9782 tons of MSW were daily disposed of in municipal landfills in Hong Kong (5). The disposal of solid waste in municipal landfills poses a serious environmental and sustainability issue due to the potential release of contaminants and the long stabilization period (6–8).

Incineration is an effective waste management technique that could effectively reduce waste mass by 70% and volume by 90% (2,9). Waste residues, such as bottom ash, grate sifting, heat recovery ash, fly ash, and air pollution control (APC) residues, are produced in the incineration process (9), and most of the hazardous heavy metals are thus concentrated in these waste residues (10). Low-boiling metals in particular are often released in volatile forms at high temperature and accumulate in fly ash (10–12). Cadmium (Cd) is a heavy metal present in MSW and has been reported at ranges from 10 to 2100 mg/kg in fly ash after incineration (13).

Cadmium in MSW originates from a variety of sources and its use is not dissipating. During the period 1950–1990, the production of cadmium increased annually. In 2014, the global production and consumption of cadmium yielded was 22,200 tons (14). Cadmium is usually discharged from manufacturing and municipal waste sources in the forms of the metal (e.g., from cadmium electroplating), salts (e.g., cadmium chloride ($CdCl_2$), cadmium sulfate ($CdSO_4$), and cadmium sulfide (CdS)) and alloys (e.g., solders, pigments, stabilizers, semiconductors, and batteries) (15,16). It has been reported that Ni–Cd batteries are the largest source of cadmium pollutants in MSW, accounting for 60%–70% of the input, and the second largest source is waste plastics (14).

When toxic metals are leached into the natural environment, they often lead to substantial risks to the ecosystem and public health (8). Cadmium has exhibited toxicity to humans and biota but has been reported to be highly mobile and soluble when discharged into the environment (air, soil, and water) by industrial activities (15,16). Cadmium can easily accumulate in the human body via food chains (Figure 7.1) and cause severe toxicological effects in living organisms, such as kidney damage, skeleton deformation, cancer, and so on (17,18).

To immobilize toxic metals, different sorbents and matrix materials (e.g., bauxite (19), alumina oxide (20,21), calcium oxide (22), montmorillonite [MMT], and silica (16), etc.) are usually injected into the thermal treatment process. However, such methods have certain drawbacks, including the agglomeration of sorbents, the clogging of sorption sites, and the need for additional ash residues stabilization (9,23–25). In order to stabilize ash residues, S/S technologies are frequently employed

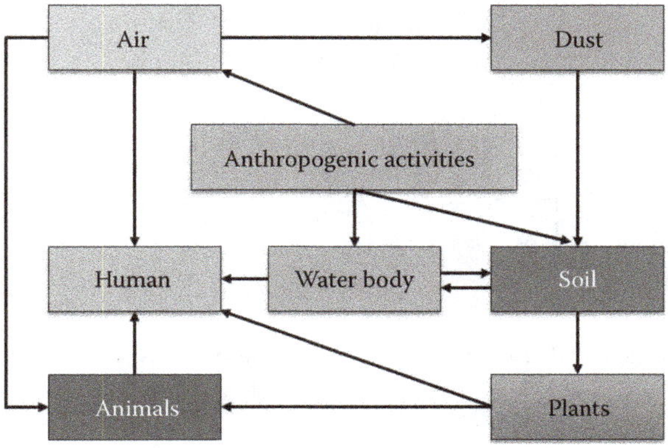

FIGURE 7.1 Pathways for cadmium entering into food chains and being up-taken by humans.

to incorporate and immobilize toxic metals with cementitious materials in a solid matrix prior to landfilling (10,26). However, common S/S technologies use sorption or cementation mechanisms to immobilize metals that may not reliably prevent metals from leaching in acidic environments (26,27). In addition, the need for additives in the S/S process will increase the treatment cost and product volume. Therefore, the development of a novel, economical, and reliable method to stabilize toxic metals in waste incineration residues is highly desirable.

7.1.2 POTENTIAL REACTIONS WITH GAMMA-ALUMINA (γ-AL$_2$O$_3$) AND HEMATITE (α-FE$_2$O$_3$)

Aluminum (Al) makes up 8 wt.% and iron (Fe) is 5 wt.% of the Earth's solid surface, the third and fourth most abundant elements (after oxygen and silicon) in the Earth's crust, respectively (28,29). Both metals thus have widespread applications in the construction and manufacturing industries (e.g., motor vehicles, ships, trucks, pipelines, and trains and railway tracks). Gamma-alumina (γ-Al$_2$O$_3$) and hematite (α-Fe$_2$O$_3$) are the most common oxide forms of aluminum and iron in the nature. Extensive studies have been conducted on γ-Al$_2$O$_3$ and α-Fe$_2$O$_3$ for chemical and environmental applications, because they are easily obtained and for their nontoxic, low cost, and highly stable properties (30,31).

It has been reported that γ-Al$_2$O$_3$ and α-Fe$_2$O$_3$ could react with cadmium under thermal conditions to immobilize and stabilize cadmium (32,33). Bauxite is a major natural source of aluminum and it mainly consists of hydrated aluminum oxide (Al$_2$O$_3 \cdot$ H$_2$O), together with different levels of hydrated iron oxide (Fe$_2$O$_3 \cdot$ H$_2$O) (34). The composition of bauxite provides an opportunity to stabilize cadmium in waste residues through reactions with aluminum oxide and iron oxide. However, to more reliably utilize this treatment strategy the optimal conditions for effectively initiating the mechanisms of cadmium incorporation by aluminum oxide and iron oxide precursors must be determined.

7.1.3 X-RAY DIFFRACTION TECHNIQUE FOR MONITORING PHASE TRANSFORMATIONS

X-ray diffraction (XRD) is an effective analytical technique to identify crystalline phases in solids (e.g., minerals and inorganic compounds) and to provide structural information on these phases (35–37). To obtain accurate and reliable data from XRD, the analyzed sample is usually finely ground and completely homogenized (35,38). The analysis of XRD data utilizes Bragg's law (Equation 7.1) (39) to indicate the peaks of crystal lattice scattering based on the conditions that: (a) the angle of incidence is equal to the angle of scattering and (b) the path length difference is equal to an integer number of wavelengths.

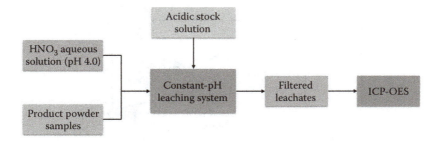

FIGURE 7.2 Design of the CPLT used for evaluating cadmium stabilization effects of the products.

$$d = \frac{n \cdot \lambda}{2 \cdot \sin \theta} \tag{7.1}$$

where n is an integer, λ is the used x-ray wavelength (nm), θ is the angle between the incident ray and the scattering planes (°), and d is the spacing between the lattice planes of the corresponding phase (nm).

The XRD technique has been widely employed to characterize natural and industrial materials with the support of extensive database information (36,38). XRD has been proven to be a reliable, precise, and reproducible method to identify phase compositions in samples (36,37). Each crystalline phase has a distinctive XRD pattern (37,38). Based on the peak positions (corresponding to d values) and peak intensities in the observed diffraction pattern the crystalline phases in the sample can be identified (38,39). The solid-state reaction route is one of the most important methods to incorporate metals into crystal structures, by driving atoms to their most energetically favorable positions under thermal conditions (37). The phase transformations that occur in the solid-state reactions of $CdO + \gamma\text{-}Al_2O_3$ and $CdO + \alpha\text{-}Fe_2O_3$ systems can be monitored with XRD.

7.1.4 EVALUATING THE CADMIUM STABILIZATION EFFECT WITH LEACHING TEST

Toxicity characteristic leaching procedure (TCLP) is the most commonly used method to assess the leachability of toxic metals from waste (40). During the leaching test, the pH value has a great impact on mineral dissolution, proton competition on surface binding sites, and surface potential (41,42). However, the pH value of leaching fluid is not maintained during TCLP, and this may influence the leaching process and/or lead to the reprecipitation of metal compound(s) (42,43). The constant-pH leaching test (CPLT) largely overcomes the disadvantages of TCLP, and can provide a measure of chemical durability for the tested materials or products in the leaching process (44). It can be used to compare the dissolution or leaching behavior of the test samples exposed to the diluted acid solutions, and the test is conducted with a constant pH value. The leachability of the tested sample is determined through the amounts of components released from the materials over the testing duration.

In this chapter, $\gamma\text{-}Al_2O_3$ and $\alpha\text{-}Fe_2O_3$ were used to stabilize cadmium, and the influences of operational parameters such as treatment temperature and time were systematically investigated. The CPLT was employed to evaluate the cadmium stabilization effects of the products, and the design of the CPLT procedure is shown in Figure 7.2.

7.2 EXPERIMENTAL SECTION

7.2.1 MATERIALS AND SAMPLE PREPARATION

Cadmium oxide (CdO), gamma-alumina ($\gamma\text{-}Al_2O_3$), and hematite ($\alpha\text{-}Fe_2O_3$) were used as the raw materials. CdO powder was purchased from Fisher Scientific (New Hampshire). The surface area of the CdO powder was determined to be 2.63 ± 0.05 m^2/g by the Brunauer–Emmett–Teller (BET) method on a Beckman Coulter SA3100 surface area and pore size analyzer at liquid nitrogen temperature (77 K). The γ-Al$_2$O$_3$ was thermally prepared at 650°C for 3 h from Pural SB alumina powder (Sasol) which was identified as boehmite (AlOOH) by XRD. To investigate the incorporation capability of cadmium by γ-Al$_2$O$_3$ and α-Fe$_2$O$_3$ precursors under thermal conditions, samples were prepared by mixing CdO powder with γ-Al$_2$O$_3$ or α-Fe$_2$O$_3$ precursor to a total dry weight of 60 g at Cd/Al and Cd/Fe molar ratios of 0.25 and 0.50, respectively. The mixtures were wet with distilled water, ball-milled for 18 h, and dried at 105°C for 24 h. Further powder homogenization was conducted by mortar grinding. Then the homogenized mixture was pelletized into Φ 20 mm pellets at 250 MPa and subjected to a thermal treatment scheme with a dwell time of 3 h at the targeted temperature.

7.2.2 XRD Analysis

The heated samples were air-cooled and ground into powder form for powder XRD analysis. Phase transformation was monitored by the analysis of XRD patterns of treated samples. The step-scanned XRD pattern of each powder sample was recorded by a Bruker D8 Advance x-ray powder diffractometer equipped with Cu K$\alpha_{1,2}$ x-ray radiation and a LynxEye detector. The 2θ scanning range was from 10° to 80°, and the step size was 0.02° with a scan speed of 0.5 s per step. Qualitative phase identification was executed by matching powder XRD patterns with those retrieved from the standard powder diffraction database of the International Centre for Diffraction Data (ICDD PDF-2 Release 2008).

7.2.3 Leaching Experiment

The leachability of single-phase cadmium-hosting samples (CdO, CdAl$_4$O$_7$, and CdFe$_2$O$_4$) was examined by a constant-pH leaching procedure in an acidic environment. The leaching fluid was a pH 4.0 nitric acid (HNO$_3$) aqueous solution, and the pH value was maintained at 4.0 ± 0.2 by the addition of 1 M HNO$_3$ aqueous solution of almost negligible volume (approximately 20 μL for each adjustment). The leaching test was carried out in a jar filled with 500 mL of leaching fluid and 0.5 g of tested powders, and magnetically stirred (200 rpm) throughout the leaching process. At 10-min intervals, 5 mL of leachate was withdrawn and filtered through a 0.2 μm syringe filter for subsequent Cd concentration determination by ICP-OES (inductively coupled plasma optical emission spectrometry) (Perkin Elmer 800). A set of cadmium standards showed a satisfactory calibration curve ($R^2 = 0.9999$ with the detection wavelength of 214.44 nm) in the measurement range (1–2000 ppb) of this study.

7.3 RESULTS AND DISCUSSION

7.3.1 CdAl$_4$O$_7$ Formation via Thermally Reacting with γ-Al$_2$O$_3$

7.3.1.1 Effect of Treatment Temperature

Gamma-alumina (γ-Al$_2$O$_3$) is an important metastable transition alumina phase derived from the thermal dehydration of boehmite or oxyhydroxides at temperatures ranging from 400°C to 700°C (45). Its crystal structure is generally considered to be a defect spinel structure. In the crystal structure of γ-Al$_2$O$_3$, Al atoms occupy both tetrahedral and octahedral positions, and nine cationic sites generate one vacancy for the disordered γ-Al$_2$O$_3$ phase. Phase transformations from transition phases ($\gamma \rightarrow \delta \rightarrow \eta \rightarrow \theta$) to the stable alumina phase (α-Al$_2$O$_3$) occur during the continuous heating of γ-Al$_2$O$_3$ (46). However, due to its unique and outstanding properties (large surface area, pore volume, and pore size) γ-Al$_2$O$_3$ exhibits much more application potential than the other alumina phases (47,48).

By thermally treating a CdO + γ-Al$_2$O$_3$ mixture, cadmium incorporation is expected to proceed via a crystallochemical reaction as follows:

$$CdO + 2\gamma - Al_2O_3 \rightarrow CdAl_4O_7 \tag{7.2}$$

The thermal incorporation capability of γ-Al$_2$O$_3$ for heavy metals, such as nickel (Ni), copper (Cu), and lead (Pb), is influenced by temperature (49–51). To investigate the effective temperature for γ-Al$_2$O$_3$ to incorporate cadmium into the CdAl$_4$O$_7$ crystal phase, a 3-h short-heating scheme at temperatures ranging from 850°C to 950°C was conducted. Figure 7.3 shows the evolution of CdAl$_4$O$_7$ during the stabilization of cadmium via heating a CdO + γ-Al$_2$O$_3$ sample with a Cd/Al molar ratio of 0.25. After thermal treatment at 850°C for 3 h, residual reactants still largely dominated in the treated sample (Figure 7.3). Besides the signals of reactants, a low-intensity diffraction peak was observed at 2θ between 25.8° and 26.1° and this indicated the initial formation of a new phase at this temperature. With an elevated treatment temperature of 900°C, this new phase was clearly observed and identified as the CdAl$_4$O$_7$ phase as shown in Figure 7.3. A previous study in equilibrium experiments (32) had reported that the formation of CdAl$_4$O$_7$ started at 800°C. However, our study found that the heating temperature for a 3-h treatment period should be above 900°C to effectively initiate the incorporation of Cd into the CdAl$_4$O$_7$ phase. Thermodynamic conditions and the diffusion process are the most important factors in solid-state reactions. The difference between our work and the result reported from equilibrium experiments may indicate that the formation of CdAl$_4$O$_7$ at temperatures below 900°C is mostly restricted by the slow diffusion process. Although it is thermodynamically feasible at temperatures above 800°C, the CdAl$_4$O$_7$ phase generated by the short-heating scheme at temperatures below 900°C may only occur at the grain boundary of reactants. The very small amount of the new surface phase is usually not able to be reflected in the XRD result.

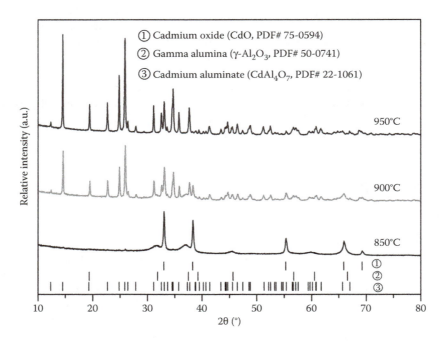

FIGURE 7.3 XRD patterns of the CdO + γ-Al$_2$O$_3$ system show the formation of the CdAl$_4$O$_7$ monoclinic phase at 850–950°C for 3 h. The standard patterns were derived from the ICDD database, including CdO (PDF# 75-0594), γ-Al$_2$O$_3$ (PDF# 50-0741), and CdAl$_4$O$_7$ (PDF# 22-1061).

Figure 7.4a provides a detailed XRD pattern comparison in a narrower 2θ range to monitor the formation process of the $CdAl_4O_7$ phase at elevated temperatures. Two major peaks of $CdAl_4O_7$ (PDF# 22-1061) are located at $2\theta = 24.86°$ and $25.93°$, corresponding to the diffraction planes of (220) and (−311), respectively. Both diffraction planes show substantial crystal growth in the 900°C treated sample and the peak intensities of the $CdAl_4O_7$ phase increased with the increase in temperature. The XRD patterns within the 2θ ranges of 37.2–39.0° were selected to further observe the development of peak intensity as an indication of the efficiency of cadmium incorporation at different heating temperatures (Figure 7.4b). Although only a small amount of new phase in the system was detected by XRD at 850°C, the peak intensity of CdO (200) showed a significant decrease at 850°C and this result may be an additional indication that a small amount of $CdAl_4O_7$ formed on the reactant surface as mentioned above. With an increase in treatment temperature, the signals of the CdO peaks gradually decreased and almost disappeared at 950°C.

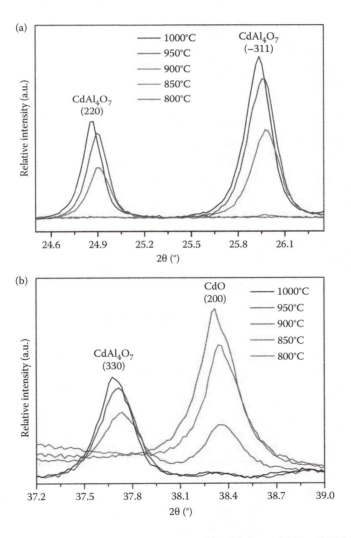

FIGURE 7.4 The comparison of XRD patterns between (a) $2\theta = 24.5°$ and $26.3°$ and (b) $2\theta = 37.2°$ and $39.0°$ for the CdO + γ-Al_2O_3 samples (with a Cd/Al molar ratio of 0.25) with treatment temperatures ranging from 800°C to 1000°C for 3 h.

7.3.1.2 Effect of Treatment Time

Treatment time is also another important factor for promoting solid-state reactions, particularly for diffusion-dominant processes. To minimize the treatment temperature and further reduce the treatment time to achieve sufficient transformation results, this study used 950°C to observe the influence of treatment time on the formation efficiency of $CdAl_4O_7$. Figure 7.5a shows the XRD results of heating the $CdO + \gamma\text{-}Al_2O_3$ samples at 950°C for 0.25–9 h. Even when the samples were treated for a very short heating time, as little as 0.25 h, a large amount of $CdAl_4O_7$ was observed in the product. This result indicates a strong cadmium incorporation capability of $CdAl_4O_7$ at 950°C. XRD patterns over a narrower 2θ range from 37.69° to 38.34° verify the development of $CdAl_4O_7$ and the decrease of CdO (Figure 7.5b). The curves confirm the continuous signal increase of the (330) diffraction plane of $CdAl_4O_7$ as the treatment time increased between 0.25 and 9 h. The (200) diffraction plane of CdO was found to disappear after 9 h of thermal treatment, while the formation of $CdAl_4O_7$ was found to reach its maximum after the same treatment period.

Figure 7.5 provides XRD patterns for the (a) heated $CdO + \gamma\text{-}Al_2O_3$ samples (with a Cd/Al molar ratio of 0.25) showing the thermal incorporation of cadmium at 950°C for different treatment times, and (b) the peak growth and decrease at 2θ between 36.2° and 39.0°. The inset illustrates the relative intensities for the (330) diffraction plane of $CdAl_4O_7$ and the (220) diffraction plane of CdO located at 2θ = 37.69° and 38.34°, respectively. The standard patterns were derived from the ICDD database, including CdO (PDF# 75-0594), $\gamma\text{-}Al_2O_3$ (PDF# 50-0741), and $CdAl_4O_7$ (PDF# 22-1061).

7.3.2 FERRITE SPINEL FORMATION BY THERMALLY REACTING WITH α-FE₂O₃

7.3.2.1 Effect of Treatment Temperature

Hematite ($\alpha\text{-}Fe_2O_3$) is the most thermodynamically stable iron oxide form, displaying a rhombohedral centered hexagonal structure in which two-thirds of the octahedral sites are occupied by Fe^{3+} ions (31). Due to its distinguished physical–chemical properties and environment-friendly and low-cost features, $\alpha\text{-}Fe_2O_3$ has attracted significant attention for applications such as catalysis, sorbents, pigments, anticorrosive agents, sensors, electrode materials, magnetic materials, etc. (52–54).

This study investigated the feasibility of transforming cadmium into the $CdFe_2O_4$ spinel phase (PDF# 22-1063) by heating a mixture of CdO and $\alpha\text{-}Fe_2O_3$ with a Cd/Fe molar ratio of 0.5 at 650–850°C for 3 h. The formation pathway of the $CdFe_2O_4$ spinel structure using $\alpha\text{-}Fe_2O_3$ and CdO as the precursors can be expressed as follows:

$$CdO + \alpha - Fe_2O_3(\text{hematite}) \rightarrow CdFe_2O_4 \qquad (7.3)$$

Figure 7.6a provides the XRD patterns for 650–850°C heated samples showing the formation of cadmium ferrite spinel. The growth of $CdFe_2O_4$ was first detected in the sample heated to 650°C although the residual reactants still dominated the system. The peak intensities of $CdFe_2O_4$ substantially increased at 700°C which can be considered an effective temperature for promoting Cd incorporation into the cadmium ferrite spinel structure. With an increase in temperature to 750°C, both CdO and $\alpha\text{-}Fe_2O_3$ diffraction peak signals nearly disappeared in the system (Figure 7.6b). This result indicates that the higher temperature had enabled a more intensive interaction between reactants. The only product phase observed at the highest temperatures (800°C and 850°C) was $CdFe_2O_4$.

7.3.2.2 Effect of Treatment Time

For a feasible environmental strategy, treatment time should be minimized to encourage adoption by industry. As noted in the previous section, treatment at 800°C for 3 h achieved satisfactory $CdFe_2O_4$ formation. Investigation to observe the influence of treatment time may be beneficial in revealing more energy saving and efficient treatment routes for effective cadmium incorporation. Therefore, hematite was used as a precursor to mix with CdO (with a Cd/Fe molar ratio of 0.5) for

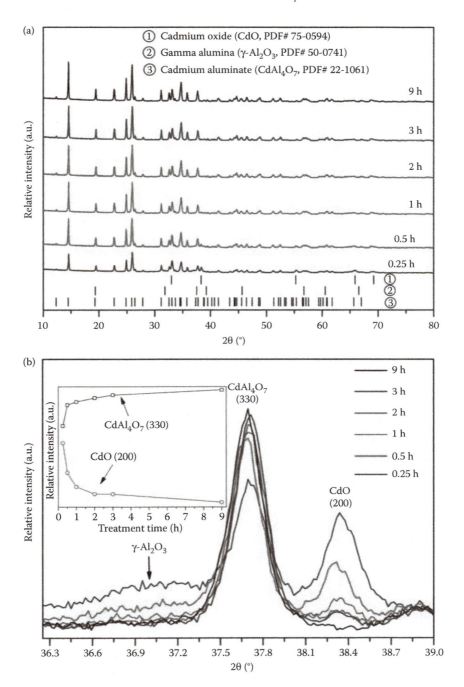

FIGURE 7.5 XRD patterns for the (a) heated CdO + γ-Al$_2$O$_3$ samples (with a Cd/Al molar ratio of 0.25) showing the thermal incorporation of cadmium at 950°C for different treatment times and (b) the peak growth and decrease at 2θ between 36.2° and 39.0°. The inset illustrates the relative intensities for the (330) diffraction plane of CdAl$_4$O$_7$ and the (220) diffraction plane of CdO located at 2θ = 37.69° and 38.34°, respectively. The standard patterns were derived from the ICDD database, including CdO (PDF# 75-0594), γ-Al$_2$O$_3$ (PDF# 50-0741), and CdAl$_4$O$_7$ (PDF# 22-1061).

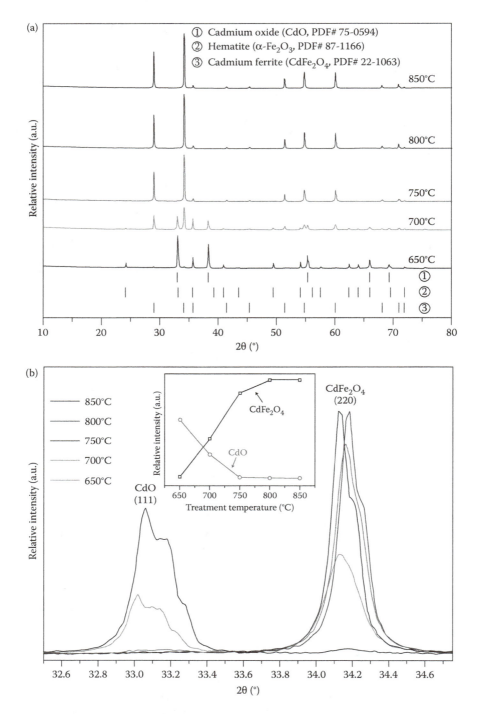

FIGURE 7.6 XRD patterns of (a) the CdO + α-Fe$_2$O$_3$ (hematite) mixture (with Cd/Fe molar ratio of 0.5) heated at 600–850°C for 3 h and (b) the peak growth and decrease at 2θ between 32.50° and 34.68°. The inset illustrates the relative intensities for the (220) diffraction plane of CdFe$_2$O$_4$ and the (110) diffraction plane of CdO, respectively. The standard patterns were derived from the ICDD database, including CdO (PDF# 75-0594), α-Fe$_2$O$_3$ (PDF# 87-1166), and CdFe$_2$O$_4$ (PDF# 22-1063).

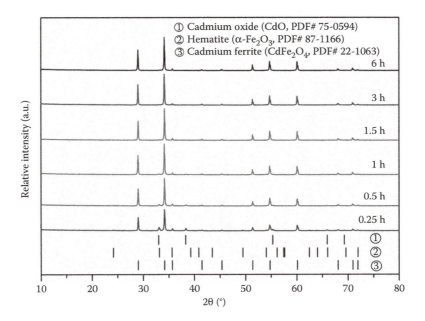

FIGURE 7.7 XRD patterns of the heated CdO + hematite (α-Fe_2O_3) samples (with a Cd/Fe molar ratio of 0.5) at 800°C for 0.25–6 h. The standard patterns were obtained from the ICDD database, including CdO (PDF# 75-0594), α-Fe_2O_3 (PDF# 87-1166), and $CdFe_2O_4$ (PDF# 22-1063).

treatment for times varying from 0.25 to 6 h at 800°C. The decrease of reactants and the increase of $CdFe_2O_4$ diffraction peak intensity indicate that the cadmium was rapidly incorporated into the crystal structure of the $CdFe_2O_4$ phase even with the shortest treatment time of 0.25 h (Figure 7.7). When prolonging the treatment time to 1.5 h, the signals of reactant diffraction peaks were found to disappear and this indicates the feasibility of using 800°C for cadmium incorporation.

7.4 EVALUATION OF CADMIUM STABILIZATION EFFECT

Since $CdAl_4O_7$ and $CdFe_2O_4$ are the potential cadmium-containing product phases when using γ-Al_2O_3 and α-Fe_2O_3 precursors, the capability of these phases to stabilize cadmium should be further evaluated. Utilizing constant-pH leaching tests (CPLTs), the intrinsic leachability of three single-phase cadmium-bearing samples, that is, CdO, $CdAl_4O_7$, and $CdFe_2O_4$, were quantitatively evaluated. The single-phase $CdAl_4O_7$ was obtained by heating a pelletized mixture with a Cd/Al molar ratio of 0.25 at 950°C for 36 h. Similarly, the single-phase $CdFe_2O_4$ was prepared from firing the pelletized mixture with a Cd/Fe molar ratio of 0.5 at 800°C for 48 h. The extended sintering time was to further ensure a complete transformation and homogeneity of the products. Before the leaching test, the single-phase products were ball milled into powdered form to maximize the surface area for leaching reactions. The leaching processes for CdO, $CdAl_4O_7$, and $CdFe_2O_4$ in an acidic environment can be described by the following reactions:

$$CdO_{(s)} + 2H^+_{(eq)} \rightarrow Cd^{2+}_{(eq)} + H_2O \tag{7.4}$$

$$CdAl_4O_{7(s)} + 14H^+_{(eq)} \rightarrow Cd^{2+}_{(eq)} + 4Al^{3+}_{(eq)} + 7H_2O \tag{7.5}$$

$$CdFe_2O_{4(s)} + 8H^+_{(eq)} \rightarrow Cd^{2+}_{(eq)} + 2Fe^{3+}_{(eq)} + 4H_2O \tag{7.6}$$

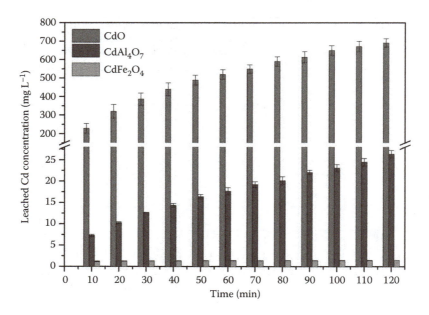

FIGURE 7.8 Comparison of leached cadmium from CdO, monoclinic $CdAl_4O_7$, and $CdFe_2O_4$ spinel phase powder samples.

Figure 7.8 reflects the leachabilities of cadmium from the samples in terms of the observed cadmium concentrations in leachates. At pH 4.0, the amounts of cadmium leached from both CdO and $CdAl_4O_7$ gradually increased as the leaching period was prolonged. However, at the end of the 120-min leaching period, the cadmium concentration observed in the CdO leachate was 691.2 mg/L, more than 20 times higher than from the $CdAl_4O_7$ monoclinic phase leachate (24.4 mg/L). The level of cadmium was much lower for the $CdFe_2O_4$ leachate (1.20–1.47 mg/L) and remained steady throughout the entire leaching process. The concentration of the leached cadmium from the $CdFe_2O_4$ phase was remarkably lower than for CdO and $CdAl_4O_7$ at the end of the leaching experiment. This indicates the superior stabilization achieved through the incorporation of cadmium into the $CdFe_2O_4$ spinel structure. The results of the CPLT procedure suggest that the phase transformations to monoclinic $CdAl_4O_7$ structure and $CdFe_2O_4$ spinel structure can largely enhance the intrinsic acid resistances of cadmium-bearing products as compared to CdO.

7.5 CONCLUSION

This chapter presents the successful incorporation of cadmium into a monoclinic $CdAl_4O_7$ structure and the $CdFe_2O_4$ spinel phase using γ-Al_2O_3 and α-Fe_2O_3 precursors at attainable temperatures in the ceramic industry. With the efficient transformations of cadmium into those two crystalline phases, the products were evaluated by the CPLT and demonstrated substantial decreases in cadmium leachability, particularly when cadmium was stabilized in the $CdFe_2O_4$ spinel structure. Because resistance to acidic environments is often the limiting factor in efforts to immobilize hazardous metals in the environment, the successful stabilization of cadmium reported in this study supports the reduction of the hazards from cadmium in waste incineration residues through the safe and reliable incorporation of these residues into the ceramic matrix.

ACRONYMS

APC	Air pollution control
BET	Brunauer–Emmett–Teller
CPLT	Constant-pH leaching test
ICDD	International Centre for Diffraction Data
ICP-OES	Inductively coupled plasma optical emission spectrometry
MMT	Montmorillonite
MSW	Municipal solid waste
PDF	Powder diffraction file
TCLP	Toxicity characteristic leaching procedure
XRD	X-ray diffraction

REFERENCES

1. Woon, K. S. and Lo, I. M. C. 2016. An integrated life cycle costing and human health impact analysis of municipal solid waste management options in Hong Kong using modified eco-efficiency indicator. *Resources, Conservation and Recycling.* Vol. **107**, 104–114.
2. Chi, Y., Dong, J., Tang, Y., Huang, Q., and Ni, M. 2015. Life cycle assessment of municipal solid waste source-separated collection and integrated waste management systems in Hangzhou, China. *Journal of Material Cycles and Waste Management.* Vol. **17**, No. 4, 695–706.
3. Ko, P. S. and Poon, C. S. 2009. Domestic waste management and recovery in Hong Kong. *Journal of Material Cycles and Waste Management.* Vol. **11**, No. 2, 104–109.
4. Environmental Protection Department (EPD). 2016. *Waste recycling statistics*, Hong Kong, available at https://www.wastereduction.gov.hk/en/quickaccess/stat_recycle.htm.
5. Environmental Protection Department (EPD). 2015. *Monitoring of Solid Waste in Hong Kong—Waste Statistics for 2014.* Environmental Protection Department, Hong Kong.
6. Gao, J. L., Oloibiri, V., Chys, M., Audenaert, W., Decostere, B., He, Y. L., van Langenhove, H., Demeestere, K., and van Hulle, S. W. H. 2015. The present status of landfill leachate treatment and its development trend from a technological point of view. *Reviews in Environmental Science and Bio/Technology.* Vol. **14**, No. 1, 93–122.
7. Laner, D., Crest, M., Scharff, H., Morris, J. W., and Barlaz, M. A. 2012. A review of approaches for the long-term management of municipal solid waste landfills. *Waste Management.* Vol. **32**, No. 3, 498–512.
8. Mojiri, A., Ziyang, L., Tajuddin, R. M., Farraji, H., and Alifar, N. 2016. Co-treatment of landfill leachate and municipal wastewater using the ZELIAC/zeolite constructed wetland system. *Journal of Environmental Management.* Vol. **166**, 124–130.
9. Lam, C. H. K., Ip, A. W. M., Barfordemail, J. P., and McKay, G. 2010. Use of incineration MSW ash: A review. *Sustainability.* Vol. **2**, No. 7, 1943–1968.
10. Quina, M. J., Bordado, J. M., and Quinta-Ferreira, R. M. 2014. Recycling of air pollution control residues from municipal solid waste incineration into lightweight aggregates. *Waste Management.* Vol. **34**, No. 2, 430–438.
11. Amutha Rani, D., Boccaccini, A. R., Deeganc, D., and Cheeseman, C. R. 2008. Air pollution control residues from waste incineration: Current UK situation and assessment of alternative technologies. *Waste Management.* Vol. **28**, No. 11, 2279–2292.
12. Nowak, B., Aschenbrenner, P., and Winter, F. 2013. Heavy metal removal from sewage sludge ash and municipal solid waste fly ash—A comparison. *Fuel Processing Technology*, Vol. **105**, 195–201.
13. Chandler, A. J., Eighmy, T. T., Hjelmar, O., Kosson, D. S., Sawell, S. E., Vehlow, J., van der Sloot, H. A., and Hartlen, J. (eds.). 1997. *Municipal Solid Waste Incinerator Residues.* Elsevier, Amsterdam, The Netherlands, pp. 305–307.
14. United States Geological Survey (USGS). 2015. USGS Minerals information: Cadmium, mineral commodity summaries. USGS, Reston, VA, USA.
15. Pinzani, M. C. C., Somogyi, A., Simionovici, A. S., Ansell, S., Steenari, B. M., and Lindqvist, O. 2002. Direct determination of cadmium speciation in municipal solid waste fly ashes by synchrotron radiation induced μ-X-ray fluorescence and μ-X-ray absorption spectroscopy. *Environmental Science and Technology.* Vol. **36**, No, 14. 3165–3169.

16. Lee, M.-H., Cho, K., Shah, A. P., and Biswas, P. 2005. Nanostructured sorbents for capture of cadmium species in combustion environments. *Environmental Science and Technology*. Vol. **39**, No. 21, 8481–8489.

17. Fowler, B. A. 2009. Monitoring of human populations for early markers of cadmium toxicity: A review. *Toxicology and Applied Pharmacology*. Vol. **238**, No. 3, 294–300.

18. Vromman, V., Saegerman, C., Pussemier, L., Huyghebaert, A., De Temmerman, L., Pizzolon, J. C., and Waegeneers, N. 2008. Cadmium in the food chain near non-ferrous metal production sites. *Food Additives and Contaminants: Part A*. Vol. **25**, No. 3, 293–301.

19. Scotto, M. U., Uberoi, M., Peterson, T. W., Shadman, F., and Wendt, J. O. L. 1994. Metal capture by sorbents in combustion processes. *Fuel Processing Technology*. Vol. **39**, No. 1–3, 357–372.

20. Kuo, J.-H., Lin, C.-L., and Wey, M. Y. 2009. Effects of agglomeration processes on the emission characteristics of heavy metals under different waste compositions and the addition of Al and Ca inhibitors in fluidized bed incineration. *Energy and Fuels*. Vol. **23**, No. 9, 4325–4336.

21. Liu, J., Abanades, S., Gauthier, D., Flamant, G., Zheng, C. G., and Lu, J. D. 2005. Determination of kinetic law for toxic metals release during thermal treatment of model waste in a fluid-bed reactor. *Environmental Science and Technology*. Vol. **39**, No. 23, 9331–9336.

22. Shi, C. and Fernández-Jiménez, A. 2006. Stabilization/solidification of hazardous and radioactive wastes with alkali-activated cements. *Journal of Hazardous Materials*. Vol. **137**, No. 3, 1656–1663.

23. Manovic, V., Stewart, M. C., Macchi, A., and Anthony, E. J. 2010. Agglomeration of sorbent particles during sulfation of lime in the presence of steam. *Energy and Fuels*. Vol. **24**, No. 12, 6442–6448.

24. Lee, E. M. and Clack, H. L. 2010. In situ detection of altered particle size distributions during simulated powdered sorbent injection for mercury emissions control. *Energy and Fuels*. Vol. **24**, No. 10, 5410–5417.

25. Ahmaruzzaman, M. 2010. A review on the utilization of fly ash. *Progress in Energy and Combustion Science*. Vol. **36**, No. 3, 327–363.

26. Billen, P., Verbinnen, B., De Smet, M., Dockx, G., Ronsse, S., Villani, K., and Vandecasteele, C. 2015. Comparison of solidification/stabilization of fly ash and air pollution control residues from municipal solid waste incinerators with and without cement addition. *Journal of Material Cycles and Waste Management*. Vol. **17**, No. 2, 229–236.

27. Lampris, C., Stegemann, J. A., Pellizon-Birelli, M., Fowler, G. D., and Cheeseman, C. R. 2011. Metal leaching from monolithic stabilised/solidified air pollution control residues. *Journal of Hazardous Materials*. Vol. **185**, No. 2–3. 1115–1123.

28. Zuckerman, B., Koester, D., Dufour, P., Melis, C., Klein, B., and Jura, M. 2011. An aluminum/calcium-rich, iron-poor, white dwarf star: Evidence for an extrasolar planetary lithosphere? *The Astrophysical Journal*. Vol. **739**, No. 2, 1–24.

29. Expert, D. and O'Brian, M. R. 2012. Iron, an element essential to life. In: Expert, D. (eds.) *Molecular Aspects of Iron Metabolism in Pathogenic and Symbiotic Plant-Microbe Associations*, Chapter 1. Springer, The Netherlands, pp. 1–6.

30. Losic, D. and Santos, A. (eds.). 2015. *Nanoporous Alumina: Fabrication, Structure, Properties and Applications*, Vol. **219**, Springer International Publishing, Switzerland, p. 31.

31. Cornell, R. M. and Schwertmann, U. (eds.). 2003. *The Iron Oxides: Structure, Properties, Reactions, Occurrences and Uses*. Wiley-VCH Verlag GmbH & Co. KGaA, Weinheim, Germany, pp. 509–551.

32. Colin F. 1968. Des phases formees au cours de la reduction de certains oxydes mixtes $nAl_2O_3 \cdot MO$ (Phases obtained during the reduction of some $nAl_2O_3 \cdot MO$ oxides). *Revue Internationale Des Huates Temperatures Et Des Refractaires*. Vol. **5**, No. 4, 267–283.

33. Karanjkar, M. M., Tarwal, N. L., Vaigankar, A. S., and Patil, P. S. 2013. Structural, Mössbauer and electrical properties of nickel cadmium ferrites. *Ceramics International*. Vol. **39**, No. 2, 1757–1764.

34. Belyaev, V. V. 2011. Mineralogy, spread, and use of bauxites. *Russian Journal of General Chemistry*. Vol. **81**, No. 6, 1277–1287.

35. Suryanarayana, C. and Norton, M. G. 2013. *X-Ray Diffraction: A Practical Approach*. Springer Science and Business Media, New York, USA, pp. 3–96.

36. Shih, K. (ed.). 2013. *X-Ray Diffraction: Structure, Principles and Applications*. Nova Science Publishers, New York, USA, pp. 117–134.

37. West, A. R. (ed.). 2013. *Solid State Chemistry and Its Applications*, 2nd edition (Student edition). John Wiley & Sons Inc., Chichester, UK, pp. 229–265.

38. Pecharsky, V. and Zavalij, P. (eds.). 2008. *Fundamentals of Powder Diffraction and Structural Characterization of Materials*. Springer Science and Business Media, LLC, New York, USA, pp. 301–346.

39. Bragg, W. H. and Bragg, W. L. 1913. The reflection of X-rays by crystals. *Proceedings of the Royal Society of London A*. Vol. **88**, No. 605, 428–438.
40. United States Environmental Protection Agency (USEPA). 1997. *Test Methods for Evaluating Solid Waste, Physical Chemical Methods*, SW-846. United States Environmental Protection Agency, Washington, DC, USA.
41. Potysz, A., Kierczak, J., Fuchs, Y., Grybos, M., Guibaud, G., Lens, P. N., and van Hullebusch, E. D. 2016. Characterization and pH-dependent leaching behaviour of historical and modern copper slags. *Journal of Geochemical Exploration*. Vol. **160**, 1–15.
42. Su, M. H., Liao, C. Z., Chuang, K.-H., Wey, M.-Y., and Shih, K. 2015. Cadmium stabilization efficiency and leachability by $CdAl_4O_7$ monoclinic structure. *Environmental Science and Technology*. Vol. **49**, No. 24, 14452–14459.
43. Hooper, K., Iskander, M., Sivia, G., Hussein, F., Hsu, J., DeGuzman, M., Odion, Z. et al. 1998. Toxicity characteristic leaching procedure fails to extract oxoanion-forming elements that are extracted by municipal solid waste leachates. *Environmental Science and Technology*. Vol. **32**, No. 23, 3825–3830.
44. Islam, M. Z., Catalan, L. J. J., and Yanful, E. K. 2004. Effect of remineralization on heavy-metal leaching from cement-stabilized/solidified waste. *Environmental Science and Technology*. Vol. **38**, No. 5, 1561–1568.
45. Paglia, G., Buckley, C. E., Rohl, A. L., Hart, R. D., Winter, K., Studer, A. J., Hunter, B. A., and Hanna, J. V. 2004. Boehmite derived γ-alumina system. 1. Structural evolution with temperature, with the identification and structural determination of a new transition phase, γ-alumina. *Chemistry of Materials*. Vol. **16**, No. 2, 220–236.
46. Wolverton, C. and Hass, K. 2000. Phase stability and structure of spinel-based transition aluminas. *Physical Review B*. Vol. **63**, No. 2, 024102.
47. Rahmanpour, O., Shariati, A., Khosravi Nikou, M. R., and Rohani, M. 2016. The effect of porosity and morphology of nano γ-Al_2O_3 and commercial γ-Al_2O_3 catalysts on the conversions of methanol to DME. *Synthesis and Reactivity in Inorganic, Metal-Organic, and Nano-Metal Chemistry*. Vol. **46**, No. 2, 171–176.
48. Wang, S. F., Zhuang, C. F., Yuan, Y. G., Xiang, X., Sun, G. Z., Ding, Q. P., and Zu, X. T. 2014. Synthesis and photoluminescence of γ-Al_2O_3 and C-doped γ-Al_2O_3 powders. *Transactions of the Indian Ceramic Society*. Vol. **73**, No. 1, 37–42.
49. Shih, K., White, T., and Leckie, J. O. 2006. Spinel formation for stabilizing simulated nickel-laden sludge with aluminum-rich ceramic precursors. *Environmental Science and Technology*. Vol. **40**, No. 16, 5077–5083.
50. Tang, Y., Chui, S. S.-Y., Shih, K., and Zhang, L. 2011. Copper stabilization via spinel formation during the sintering of simulated copper-laden sludge with aluminum-rich ceramic precursors. *Environmental Science and Technology*. Vol. **45**, No. 8, 3598–3604.
51. Lu, X., Shih, K., and Cheng, H. 2013. Lead glass-ceramics produced from the beneficial use of waterworks sludge. *Water Research*. Vol. **47**, No. 3, 1353–1360.
52. Tong, Q., Jiao, T., Guo, H., Zhou, J., Wu, Y., Zhang, Q., and Peng, Q. 2016. Facile synthesis of highly crystalline α-Fe_2O_3 nanostructures with different shapes as photocatalysts for waste dye treatment. *Science of Advanced Materials*. Vol. **8**, No. 5, 1005–1009.
53. Segev, G., Dotan, H., Malviya, K. D., Kay, A., Mayer, M. T., Grätzel, M., and Rothschild, A. 2016. High solar flux concentration water splitting with hematite (α-Fe_2O_3) photoanodes. *Advanced Energy Materials*. Vol. **6**, No. 1, 1500817–1500824.
54. Pouran, S. R., Raman, A. A. A., and Daud, W. M. A. W. 2014. Review on the application of modified iron oxides as heterogeneous catalysts in Fenton reactions. *Journal of Cleaner Production*, Vol. **64**, 24–35.

8 Arsenic in the Environment
Source, Characteristics, and Technologies for Pollution Elimination

Yu-Ming Zheng, Jiaping Paul Chen, and Liu Qing

CONTENTS

8.1　INTRODUCTION

Arsenic, a widely distributed metalloid element, ranks the 20th in abundance in the Earth's crust, the 14th in the seawater, and the 12th in the human body (Mandal and Suzuki, 2002). Due to its carcinogenesis and mutagenesis, arsenic is a notorious contaminant, which poses serious health risks to humans and animals. Health effects including cancers of the skin and internal organs have been linked to chronic exposure to arsenic in drinking water (Azcue and Nriagu, 1994; Frankenberger, 2002).

Arsenic has been widely used in various fields, such as medicine, agriculture, livestock, electronics, semiconductor, metallurgy, and so on. Furthermore, a large amount of arsenic in the ore mining and smelting process was never recovered and discharged directly to the environment. Consequently, arsenic contamination of natural resources, such as groundwater, surface water, and soil, has becoming one of the major public health problems in many countries. Water is one of the most important media through which arsenic enters the human body. Arsenic pollution in natural water is a worldwide problem, and has become an important issue and challenge for world engineers, researchers, and even for the policy maker. Humans may encounter arsenic in water from wells drilled into arsenic-rich ground strata or in water contaminated by industrial or agrochemical waste.

There is therefore an increasing worldwide concern for arsenic contamination in natural water. The World Health Organisation (WHO) in 1993 had recommended the maximum contaminant level (MCL) of arsenic in drinking water as 10 μg/L. Thereafter, the MCL of arsenic in drinking water has also been reduced from 50 to 10 μg/L by the European Commission and U.S. Environmental Protection Agency.

The topics of this chapter include source of arsenic, characteristics of arsenic, application and pollution of arsenic, technologies, and recent research and development for arsenic pollution elimination.

8.2　SOURCE OF ARSENIC

Source of arsenic in the environment is derived from both natural occurring and anthropogenic activities.

8.2.1 NATURAL SOURCES

Arsenic is widely distributed in the environment. The total amount of arsenic in the upper earth crust is estimated to be 4.01×10^{16} kg with an average of 6 mg/kg (Taylor and McLennan, 1985; Matschullat, 2000). In the global arsenic cycle, 3.7×10^6 kt occurs in the oceans, another 9.97×10^5 kt on land, 25×10^9 kt in sediments, and 8.12 kt in the atmosphere (Bissen and Frimmel, 2003).

Arsenic occurs as a major constituent in more than 200 minerals, including elemental arsenic, arsenides, sulfides, oxides, arsenates, and arsenites. Most of it are ore minerals or their alteration products. However, these minerals are relatively rare in the natural environment (Smedley and Kinniburgh, 2002). Arsenic is mainly associated with sulfide minerals in the natural environment. Orpiment (As_2S_3), realgar (AsS), mispickel (FeAsS), loellingite ($FeAs_2$), niccolite (NiAs), cobaltite (CoAsS), tennantite ($Cu_{12}As_4S_{13}$), and enargite (Cu_3AsS_4) are the most important arsenic bearing minerals. The most abundant arsenic ore mineral is arsenopyrite (FeAsS) (Smedley and Kinniburgh, 2002).

Baseline concentrations of arsenic in soils are generally of the order of 5–10 mg/kg. The concentration of arsenic in the atmosphere is normally low but increased by inputs from smelting and other industrial operations, fossil fuel combustion, and volcanic activity (Smedley and Kinniburgh, 2002). Concentrations around 10^{-5}–10^{-3} mg/m^3 have been recorded in unpolluted areas, increasing to 0.003–0.18 mg/m^3 in urban areas, and are greater than 1 mg/m^3 close to industrial plants (WHO, 2001).

The concentration of arsenic varies between 0.09 and 24 µg/L (with an average of 1.5 µg/L) in seawater, and between 0.15 and 0.45 µg/L (with a maximum of 1000 µg/L) in freshwater (Bissen and Frimmel, 2003). Arsenic concentration is also found up to 300 times of the mean concentration of arsenic in groundwater in mineral and thermal waters.

According to the recommended standard of MCL for arsenic by the WHO, which is 10 µg/L, natural sources of arsenic in groundwater used for drinking water purposes are a significant problem particularly in Bangladesh. The problem of groundwater contaminated by arsenic is also found in West Bengal in India, Vietnam, Taiwan, Mexico, Argentina, Chile, Hungary, Romania, and many parts of the United States (Mohan and Pittman, 2007).

8.2.2 ANTHROPOGENIC SOURCES

Human beings in the utilization of natural resources release arsenic into the water, air, and soil, which can ultimately affect the residue arsenic level in plants and animals. The major anthropogenic arsenic sources can be summarized as follows (Mandal and Suzuki, 2002).

8.2.2.1 Man-Made Sources

China, Russia, France, Mexico, Germany, Peru, Namibia, Sweden, and United States are the main arsenic producers, which produce 90% of the world production. About 80% of arsenic consumption was for agricultural purposes during the 1970s. At present, the utilization of arsenic for agriculture is declining. Approximately, only 3% of arsenic final products are metal for metallurgic additives, while 97% are in the form of white arsenic.

8.2.2.2 Pesticides and Insecticides

In the preparation of pesticides and insecticides, arsenic was widely used earlier. In 1955, the world production of white arsenic was 37,000 tons, while 10,800 tons were produced and more than 18,000 tons were used in the United States, respectively. Most of it are in the forms of pesticides, like lead arsenate, Ca_3AsO_4, copper acetoarsenite (Paris Green), H_3AsO_4, monosodium methanearsonate (MSMA), disodium methanearsonate (DSMA), and cacodylic.

8.2.2.3 Herbicides

Since 1890, inorganic arsenicals, such as sodium arsenite, have been widely used as weed killers, particularly as nonselective soil sterilants.

8.2.2.4 Desiccants and Wood Preservatives

Arsenic acid is used widely as a cotton desiccant for long time. It was reported, in 1964, 2500 tons of H_3AsO_4 was used as desiccant on 1,222,000 acres of U.S. cotton. Fluor-chrome-arsenate-phenol (FCAP) was the first wood preservative and used in early 1918 in United States. Earlier, ammoniacal copper arsenate (ACA) and chromated copper arsenate (CCA) were used in 99% of arsenical wood preservatives.

8.2.2.5 Drugs

The medicinal values of arsenic have been acclaimed for nearly 2500 years. In Austria, a large quantity of arsenic was used by peasants for the softness and cleanliness of the skin.

Other arsenic contained medicine include Fowler's solution (potassium arsenite), Donovan's solution (arsenic and mercuric iodides), Asiatic pills (arsenic trioxide and black pepper), de Valagin's solution (liquor arsenii chloridi), sodium cacodylate, arsphenamine (Salvarsan), neoarsphenamine, oxophenarsine hydrochloride (Mapharsen), arsthinol (Balarsen), acetarsone, tryparsamide, and carbarsone.

8.2.2.6 Feed Additives

Many arsenic compounds, such as H_3AsO_4, 3-nitro-4-hydroxy phenylarsonic acid, and 4-nitrophenylarsonic acid are used for feed additives. Under the Food Additives Law of 1958, all substituted phenylarsonic acids were used for feed additives.

8.3 CHARACTERISTICS OF ARSENIC

A large number of researchers with very diverse backgrounds, dealing with miscellaneous problems, have been concerned about the properties of arsenic. Some of these are summarized in the following sections.

8.3.1 Oxidation State and Mobility of Arsenic

Arsenic (As), a metalloid element, is situated in Group 15 or Main Group V of the periodic table directly below phosphorous with the atomic number of 33. It has only one natural isotope with the atomic weight of 74.9. In the natural environment, it exists in four oxidation states (−3, 0, +3, and +5), but the two predominated oxidation states in water are oxyanions of pentavalent arsenic (arsenate, As(V)) and trivalent arsenic (arsenite, (As(III)). Though not significant in most natural groundwater, organically bound arsenic, such as monomethylarsonic acid (MMA) and dimethylarsinic acid (DMA), may be contained where organic contaminants and arsenic interact with each other.

Under the pH range of 6.5–8.5, arsenic is particularly mobile, which is commonly found in groundwater. It can be present under both oxidizing and reducing conditions. The species of arsenic occurring at a certain location are primarily controlled by the water pH, redox potential, and possibly microbiological activity.

8.3.2 Arsenic Allotrope

Arsenic allotropes have noticeably different properties. Metallic gray, yellow, and black arsenic are the three most common allotropes.

Gray arsenic is the most common allotrope. Like black phosphorus (β-metallic phosphorus), it has a layered crystal structure and comprises many six-membered rings which are interlinked. The structure of gray arsenic is somewhat resembling that of graphite, that is, each arsenic atom is bound to three other ones in the layer and is coordinated by three arsenic atoms each in the lower and upper layer. This relatively close packing of gray arsenic results in a high density of 5.73 g/cm³.

Yellow arsenic (As4) has four atoms arranged in tetrahedral structure, in which each atom is bound to the other three by a single bond. Due to the structure, yellow arsenic is very instable and reactive. It can be rapidly transformed into the gray arsenic by light. It is more volatile and more toxic than the other allotropes. It is soft, waxy, and with a low density of 1.97 g/cm³. Yellow arsenic is produced by rapid cooling of arsenic vapor with liquid nitrogen.

Black arsenic is similar in structure to red phosphorus.

8.3.3 TOXICITY OF ARSENIC

The toxicology of arsenic is a complex phenomenon because arsenic is also considered to be an essential element (Jain and Ali, 2000). Inorganic arsenic has been recognized as a human poison since ancient times. As mentioned above, arsenite and arsenate are two major oxidation states of arsenic. Arsenite is more toxic than arsenate and other forms of arsenic.

Two types of arsenic toxicity, that is, acute and chronic poisoning, have been known for a long time.

8.3.3.1 Acute Arsenic Poisoning

Acute arsenic poisoning normally occurs through ingestion of contaminated food or drink. It has been estimated that the acute lethal dose of ingested inorganic arsenic in humans is about 1–3 mg/kg day⁻¹, whereas inhalation and dermal exposures to inorganic arsenic have not been associated with acute lethality (ATSDR, 2005). Symptoms of acute intoxication normally happen within 30 min of ingestion of arsenic, but the display of symptoms may be delayed if arsenic is taken with food. The most common and earliest manifestation of acute arsenic poisoning is the acute gastrointestinal syndrome, which starts with a garlic-like or metallic taste associated with burning lips, dry mouth, and dysphagia. Following the gastrointestinal phase, the damage of multisystem organs may occur.

8.3.3.2 Chronic Arsenic Poisoning

Chronic effects of prolonged low-level exposure to arsenic have recently shown up, but chronic arsenic poisoning is much more insidious in nature. The most obvious manifestations of chronic arsenic poisoning involve the skin, lungs, liver, and blood systems. Normally, only after several years of low-level arsenic exposure, various skin lesions appear. Due to the nonspecific characteristic of arsenical dermatosis, it was difficult to diagnose. In 1966, skin pigmentation, keratosis, and skin cancers were found by Tseng in Taiwan among people who drank arsenic-contaminated water. A very high incidence of lung, bladder, and other cancers was found by Dr. Chien-Jen Chen in Taiwan in 1986 and by Dr. Allan Smith and collaborators in Chile in 1993 (Wilson, 2009).

8.3.3.3 Stages of Clinical Features of Chronic Arsenic Poisoning

Chronic arsenic poisoning (arseniasis) develops insidiously after exposure to low levels of arsenic for 6 months to 2 years or more, depending on the amount of intake of arsenic laden water and arsenic concentration in the water. The higher the amount of daily arsenic laden water intake or the higher the concentration above the MCL, the earlier the appearance of symptoms. The features of arseniasis have been classified by Dr. Saha, which are now known as Saha's classification of stages. In general, there are four recognized stages of chronic arsenic poisoning (Saha et al., 1999; Choong et al., 2007):

1. Preclinical stage (asymptomatic): In this stage, no symptom is shown in the patient, but an amount of arsenic can be detected in urine or body tissue samples. Urine and body tissue show arsenic metabolites, such as dimethylarsinic acid (DMAA) and trimethylarsenic acid (TMAA).
2. Clinical stage (symptomatic): In this stage, the presence of clinical symptoms is confirmed by detection of higher arsenic concentration in nail, hair, and skin scales. Various effects

on the skin of the patient can be seen. The most common symptom is darkening of the skin, often observed on the palms. Dark spots on the chest, back, limbs, or gums may also be shown. Swelling of hands and feet (edema) is also often seen. A more serious symptom is keratosis, or hardening of skin into nodules, often on palms and soles.

3. Stage of internal complications: In this stage, clinical symptoms become more prominent, and internal organs are affected. Symptoms, such as the enlargement of the liver, kidneys, and spleen, have been reported. Some studies reported that conjunctivitis, bronchitis, and diabetes may be linked to arsenic exposure.

4. Stage of malignancy: Usually after 15–20 years from the onset of first symptoms, cancer develops. In this stage, the affected person may develop skin, lung, or bladder cancer. Tumors or cancers can be diagnosed on the affected skin or other organs.

8.3.4 Arsenic Compounds

From both the toxicological and the biological points of view, arsenic compounds can be classified into three major groups: (1) inorganic arsenic compounds, (2) organic arsenic compounds, and (3) arsine gas.

8.3.4.1 Inorganic Arsenic Compounds

As mentioned above, trivalent and pentavalent are the predominated oxidation states of arsenic. Arsenic trioxide, sodium arsenite, and arsenic trichloride are the most common inorganic trivalent arsenic compounds. Arsenic pentoxide, arsenic acid, and arsenate (such as lead arsenate and calcium arsenate) are pentavalent inorganic arsenic compounds. Arsenic trioxide with the formula of As_2O_3 is also termed as white arsenic. It is an important oxide of arsenic, which is the main precursor to other arsenic compounds, including elemental arsenic, arsenic alloys, arsenide semiconductors, organoarsenic compounds, sodium arsenite, and sodium cacodylate.

8.3.4.2 Organic Arsenic Compounds

Arsanilic acid, MSMA, methylarsonic acid, dimethylarsinic acid (cacodylic acid), and arsenobetaine are the most common forms of organic arsenic. Arsanilic acid is also called p-aminophenylarsenic acid. It is a colorless solid and used as a drug in the late nineteenth and early twentieth centuries but is now considered prohibitively toxic. Monosodium methyl arsenate (MSMA) is an arsenic-based herbicide and fungicide, which is a less toxic organic form of arsenic and has replaced the role of lead hydrogen arsenate in agriculture. It is one of the most widely used herbicides on golf courses. Dimethylarsinic acid is also called cacodylic acid with the formula $(CH_3)_2AsO_2H$. Its derivatives were frequently used as herbicides. For example, "Agent Blue," a mixture of cacodylic acid and sodium cacodylate, is one of the chemicals used during the Vietnam War. Arsenobetaine is the main source of arsenic found in fish.

8.3.4.3 Arsine

Arsine (AsH_3) also termed as hydrogen arsenide, is a colorless, inflammable gas with a slight garlic odor. Arsine is used in doping the silicon-based chips and in producing semiconductors, such as GaAs and InAs. It can be generated whenever nascent hydrogen is liberated in material comprising arsenic. As arsenic is usually present as an impurity in many metal ores, arsine may be generated in metal industries, nonferrous metal refineries, and in the manufacture of silicon steel. The toxicological mechanism of arsine is quite different from those of other organic or inorganic arsenic compounds. Arsine acts as a powerful hemolytic poison in cases of both acute and chronic exposure. Patients typically presented with decreased hematocrit values and red "port wine"—colored urine because of the presence of hemoglobin. Arsine poisoning is characterized by nausea, abdominal colic, vomiting, backache, and shortness of breath, followed by dark blood urine and jaundice.

The chief clinical effects observed in persons with acute occupational arsine exposure are massive hemolysis followed by death from renal failure. Renal dialysis can be effectively used to treat the acute arsine poisoning and greatly reduced the mortality.

8.3.5 ANALYSIS AND MONITORING OF ARSENIC

As mentioned above, arsenic is a relative common toxic element and also a known carcinogen. In the environment, arsenic can only be transformed into a form that is less toxic to organisms and cannot be transformed into a nontoxic substance like usual organic pollutants. As a permanent part of the environment, there is a long-term need for regular monitoring arsenic at sites where it occurs naturally at elevated concentrations and at sites where arsenic-containing waste is present. This section presents a summary of existing technologies that are available for detecting arsenic in liquid and solid media. The existing technologies for arsenic analysis in the field include colorimetric test, portable X-ray fluorescence, anodic stripping voltammetry, biological assays, electrophoresis techniques, laser-induced breakdown spectroscopy, microcantilever sensors, and surface-enhanced Raman spectroscopy (Melamed, 2005). Some of these methods are listed as below.

8.3.5.1 Colorimetric Test Kit

The field colorimetric test kit has been used extensively to test for arsenic in groundwater. The assay has been applied almost exclusively to water samples. For solid wastes and soils testing, either an acidic extraction or an acidic oxidation digestion of the sample must be done prior to the analysis.

8.3.5.2 Portable X-Ray Fluorescence

In the field, X-ray fluorescence is an effective technology for determining arsenic. It is one of the few techniques that can be directly used to measure arsenic in solid samples, such as soil, without aqueous extractions. Portable X-ray fluorescence has recently been accepted as a field technique to measure arsenic in dry solid samples.

8.3.5.3 Anodic Stripping Voltammetry

Electrochemical assays for the detection of arsenic have shown promising for detecting arsenic in the field. These methods work best for liquid samples, such as groundwater while solid samples must be digested or extracted before testing.

8.4 APPLICATION AND POLLUTION OF ARSENIC

8.4.1 WOOD PRESERVATION

Arsenic is an ideal component for the preservation of wood due to its toxicity to insects, bacteria, and fungi. Chromated copper arsenate, also known as Tanalith or CCA, is a widely used around the world as a heavy preservative for timber treatment. It is a mix of chromium, copper, and arsenic formulated as oxide or salts. It is one of the largest consumers of arsenic. The use of CCA on consumer products is banned by most countries due to the environmental problems caused by arsenic. In 2004, the ban took effect firstly in the European Union and the United Stated (Mandal and Suzuki, 2002).

CCA treated lumber was heavily used during the latter half of the twentieth century as structural and outdoor building materials, and are still widely used in many countries. Although the use of CCA as preservative has been banned by some countries after studies showed that arsenic could leach out from the wood into the surrounding soil, its application is still one of the most concerns to the general public. One of the risks is presented by the burning of older CCA timber. The direct or indirect ingestion of an amount of wood ash from burnt CCA treated lumber could cause serious poisonings in humans and fatalities in animals. There is also a concern about the widespread landfill disposal of CCA treated timber.

8.4.2 Medicine

The medicinal values of arsenic have been acclaimed for nearly 2500 years. In Austria, arsenic was used by peasants to soften and clean the skin, to give plumpness to the figure, to beautify and freshen the complexion, and also to improve breathing problems (Mandal and Suzuki, 2002). During the past centuries, arsphenamine and neosalvarsan were used to treat syphilis and trypanosomiasis, but now have been superseded by modern antibiotics. Arsenic trioxide has been widely used in a variety of ways over the past 500 years, but most commonly in the treatment of cancer. The U.S. Food and Drug Administration in 2000 approved this compound for the treatment of patients with acute promyelocytic leukemia that is resistant to ATRA (all-trans retinoic acid) (Rahman et al., 2004). Recently new studies have been done in locating tumors using arsenic-74. The advantages of using this isotope instead of the previously used iodine-124 is that the signal in the PET (positron emission tomography) scan is clearer as the iodine tends to transport iodine to the thyroid gland producing a lot of noise.

8.4.3 Military

Lewisite and Agent Blue are two infamous arsenic chemical weapons. Lewisite is an organoarsenic compound, specifically an arsine. As a chemical weapon, it acts as a vesicant (blister agent) and lung irritant. The compound is prepared by the addition of arsenic trichloride to acetylene: $AsCl_3 + C_2H_2 \rightarrow ClCHCHAsCl_2$. It can easily penetrate ordinary clothing and even rubber. In skin contacts it can cause severe chemical burns, resulting in immediate pain and itching with rash and swelling. Sufficient absorption can cause systemic poisoning leading to liver necrosis or death. Ingestion results in severe pain, nausea, vomiting, and tissue damage. Generalized symptoms of lewisite poisoning also include restlessness, weakness, subnormal temperature, and low blood pressure. After World War I, the United States built up a stockpile of 20,000 tons of lewisite, and the stockpile were neutralized with bleach and dumped into the Gulf of Mexico after the 1950s.

Agent Blue, one of the rainbow herbicides, is a mixture of two arsenic-containing compounds: sodium cacodylate and cacodylic acid, and is known for its use by the United States during the Vietnam War to deprive the Vietnamese of valuable crops. It is difficult to destroy rice with conventional explosives and rice does not burn, so the weapons of choice were herbicides. As a herbicide, Agent Blue destroys plants by causing them to dry out. As rice is highly dependent on water to live, Agent Blue is sprayed on paddy fields to ruin entire fields and leave them unsuitable for further planting. Today, arsenical herbicides comprising cadosylic acid as an active component are still used as weed killers. They are used widely in the United States, from backyards to golf courses. Before cotton harvesting, they are also used to dry out the cotton plants.

8.4.4 Pigments

Copper acetoarsenite, an extremely toxic blue green chemical, was used as a green pigment known under many different names, including Emerald Green and Paris Green, Schweinfurt Green, Imperial Green, Vienna Green, and Mitis Green. It may be prepared from copper(II) acetate and arsenic trioxide and was involved in four main uses: pigment, animal poison (rodenticide), insecticide, and blue colorant for fireworks. It is reportedly very difficult to obtain a good blue in fireworks with any other chemicals. It was once used to kill rats in Parisian sewers, hence the common name as Paris Green.

8.4.5 Other Usages

Besides the abovementioned applications, arsenic is involved in various agricultural insecticides, animal feeds, semiconductors, bronzing, pyrotechnics, lead alloys, brass, and so on. For examples, lead hydrogen arsenate is employed as an insecticide on fruiters; gallium arsenide is an important semiconductor material used in integrated circuits; up to 2% of arsenic is used in lead alloys for lead shots and bullets.

8.5 TECHNOLOGIES FOR ELIMINATION OF ARSENIC POLLUTION IN WATER

Arsenic, a highly toxic contaminant, is ubiquitous in the water environment as a result of both anthropogenic and natural activities, which poses severe risks to human health. Since 1993, an MCL of 10 µg/L arsenic in drinking water has been recommended by the WHO, which has later been adopted by the European Commission, the United States, China, and so on (Zheng et al., 2012).

The implementation of the new MCL of arsenic in drinking water has prompted a series of research and development activities in order to obtain cost-effective arsenic pollution elimination technologies. Generally, the technologies to remove arsenic from water can be categorized into five groups, including oxidation, coagulation, precipitation, membrane filtration, and adsorption. The techniques are discussed in detail as follows.

8.5.1 OXIDATION

Arsenic normally occurs in the oxidation states +3 (arsenite, As(III)) and +5 (arsenate, As(V)) in natural waters. Arsenite is the most toxic form of inorganic arsenic and its removal from drinking water is less effective as compared to arsenate. Therefore, As(III) usually has to be oxidized to As(V) prior to its removal. The redox reaction of As(III)/As(V) can be described as follows:

$$H_3AsO_4 + 2\,H^+ + 2\,e^- \rightarrow H_3AsO_3 + H_2O \quad E_o = +0.56\,V$$

The standard potential for the oxidation of As(III) to As(V) is lower than that for the oxidation of Fe(II) to Fe(III). It is known that, in the presence of air the oxidation of Fe(II) happens rapidly, however, the oxidation process of As(III) to As(V) is very slow.

The oxidation rate can be increased by using ozone, chlorine, hypochlorite, chlorine dioxide, or H_2O_2 as oxidants. The presence of manganese oxide or advanced oxidation processes is also possibly used to oxidize As(III) to As(V).

8.5.1.1 Chemical Oxidation

A lot of chemicals can be used for the oxidation of arsenite to arsenate, including air, ozone, chlorine, iron and manganese compounds, H_2O_2, Fenton's reagent, and so on.

The oxidation of arsenite with manganese oxides in water treatment was investigated by Driehaus et al. (1995). The obtained results showed, though arsenite persisted in aerated solutions even at high pH, it can be easily oxidized by δ-modification of manganese dioxide. The kinetics study demonstrated that the oxidation rate from As(III) to As(V) followed a second order kinetics with respect to the concentration of As(III). The oxidation rate depended strongly on the initial molar ratio of MnO_2 to As(III). Calcium had only a minor influence on the oxidation, however, pH had no effect at pH range from 5 to 10 with an initial molar ratio of MnO_2/As(III) of 14. No desorption of reduced manganese was observed in the batch tests at high initial molar ratios. In the study, the oxidation technique was successfully used in a preloaded filter. After 60 h, the increase of As(III) oxidation and the decrease of manganese concentration was observed, which could not be explained by an inorganic reaction mechanism. The reason may be due to the contribution of bacteria in this redox reaction with manganese oxides. The arsenite may be directly oxidized to arsenate by bacteria or react with biologically precipitated manganese oxides.

Chiu and Hering (2000) reported that arsenic occurs in the +3 oxidation state as a metastable species in oxic waters, and As(III) was both more mobile in natural waters and less efficiently removed by water treatment processes than As(V). The As(III) could be oxidized by manganite in hours. The overall conversion of As(III) to As(V) was slower at pH 6.3 than at pH 4. The presence of 200 μM phosphate (at pH 4) decreased the overall rate of conversion of As(III) to As(V), however, the presence of boric acid at 95 or 3 μM did not influence the conversion rate of As(III) to As(V).

Kim and Nriagu (2000) investigated the oxidation of As(III) into As(V) using oxygen or ozone in groundwater samples, which contained 46–62 μg/L total dissolved arsenic (more than 70% were As(III)), 100–1130 μg/L Fe, and 9–16 μg/L Mn. The obtained results demonstrated that the conversion of As(III) into As(V) was fast by using ozone, but the process was sluggish by using oxygen or air. In the study, the kinetics of As(III) oxidation were interpreted using modified pseudo-first-order reaction. The half-lives of As(III) in the experimental solutions, which were saturated with ozone were very short (approximately 4 min), however, the half-lives of As(III) in the solutions saturated with oxygen and air were much longer and depending on the Fe concentration, were 2–5 days and 4–9 days, respectively. The results also showed iron and manganese were also oxidized during the process, and played an important role in removing the resultant As(V). The sorption capacity of freshly precipitated $Fe(OH)_3$ was determined to be about 15.3 mg As/g.

The oxidation kinetics of As(III) with natural or technical oxidant is important for understanding the behavior of arsenic removal procedures. Hug and Leupin (2003) studied the oxidation of As(III) by dissolved oxygen and hydrogen peroxide at pH 3.5–7.5 in the presence of Fe(II, III) on a time scale of hours. In the time scale, no oxidation of arsenite was observed by using O_2, 20–100 μM H_2O_2, dissolved Fe(III), or iron(III) hydroxides as single oxidants, respectively. However, partial or complete oxidation of arsenite was observed in parallel to the oxidation of 20–90 μM Fe(II) by oxygen and by 20 μM H_2O_2 in aerated solutions. At low pH, the addition of ·OH radical scavenger, 2-propanol, quenched the As(III) oxidation. At neural pH, the addition of 2-propanol had little influence on the oxidation of arsenite. The oxidant formed at neutral pH oxidizes As(III) and Fe(II) but does not react competitively with 2-propanol. It was observed that high concentration of bicarbonate resulted in the increasing oxidation of arsenite. These obtained results indicated H_2O_2 and Fe(II) may form ·OH radicals at low pH, but a different oxidant, possibly an Fe(IV) species, at higher pH. In the presence of bicarbonate, carbonate radicals might also be produced.

Lee et al. (2003) reported that the arsenite could be oxidized to arsenate by Fe(VI) with a stoichiometry of 3:2 (As(III):Fe(VI)). The study showed that the reaction of As(III) with Fe(VI) was first order with respect to both reactant.

A manganese-loaded polystyrene matrix namely $R-MnO_2$ was developed and employed for the oxidation and removal of As(III) by Lenoble et al. (2004). The developed $R-MnO_2$ allowed the complete oxidation of As(III) in the solution, even at high concentration. Oxidation and adsorption of arsenite onto the MnO_2 were involved during the removal process. The mechanism study showed the oxidation of H_3AsO_3 by MnO_2 resulting in the formation of $HAsO_4^{2-}$ and Mn^{2+}. A novel Fe–Mn binary oxide adsorbent was developed for effective As(III) removal (Zhang et al., 2007b). The results showed the As(III) were oxidized and adsorbed onto the binary oxide. The oxidation ability of the manganese oxide played an important role during the process.

Leupin and Hug (2005) found that repeated contact of aerated water with zerovalent iron (ZVI) lead to continued release of Fe(II), and simultaneous oxidation of As(III) and Fe(II) with dissolved oxygen and without added oxidant.

The oxidation of arsenite with potassium permanganate ($KMnO_4$) was investigated under various conditions, including pH, initial As(III) concentration and dosage of Mn(VII) (Li et al., 2007). The results demonstrated that potassium permanganate was an effective oxidant for the oxidation of As(III) into As(V) in a wide pH range. The performance of Mn(VII) in the oxidation of As(III) is not significantly influenced by the solution pH. The main ending reduction products of Mn(VII) are Mn(II) and $Mn(OH)_2$ under acidic and basic conditions, respectively. The ratio of Mn(VII)/As(III) is about 2/5 for the oxidation of As(III) to As(V).

Jang and Dempsey (2008) investigated the single solute adsorption and coadsorption of As(III) and As(V) onto hydrous ferric oxide (HFO), oxidation of As(III). The results showed oxidation was negligible for single-adsorbate experiments, but significant oxidation was observed in the presence of As(V) and HFO.

A pilot study on the potential of enhancing As(III) removal was carried out for the process of oxidation, precipitation, and direct sand filtration as pretreatment before ultrafiltration (UF) by Sun et al. (2009). The obtained result showed that the pretreatment effectively facilitated the As(III) removal and residual arsenic concentration was below 10 µg/L for membrane effluent. Compared to chlorine, besides oxidation ability, the permanganate had positive seeding effects of *in situ* formed hydrous MnO_2 and the formation of larger floc, hence permanganate was more promising than chlorine in this process.

8.5.1.2 Catalytic Oxidation

Thermal and photochemical oxidation of As(III) were investigated in the lab on a time scale of hours by Hug et al. (2001). The water used contained 500 µg/L As(III), 0.06–5 mg/L Fe(II,III), and 4–6 mM bicarbonate at pH 6.5–8.0. It had been found that dissolved oxygen and micromolar hydrogen peroxide did not oxidize As(III) on a time scale of hours. In the dark, As(III) was partly oxidized by the addition of Fe(II) to aerated water, which may be due to the formation of reactive intermediates in the reduction of oxygen by Fe(II). It was observed that, under illumination with 90 W/m^2 UV-A light, over 90% of As(III) in solutions containing 0.06–5 mg/L Fe(II, III) could be oxidized photochemically within 2–3 h. The oxidation of As(III) could be strongly accelerated by the presence of citrate by forming Fe(III) citrate complexes.

The oxidation of As(III) by oxygen in the absence and in the presence of dissolved Fe(III) and illumination with near ultraviolet light was studied by Emett and Khoe (2001). The obtained results demonstrated that the oxidation rate of As(III) to As(V) by oxygen is increased by several orders of magnitude by the presence of dissolved Fe(III) and irradiation with near ultraviolet light. The study indicated that the free radicals mechanism could be well used to described the process, in which the rate of the initiation reaction is determined by the rate of photon absorption by the dissolved Fe(III)-hydroxo and Fe(III)-chloro species. The addition of arsenate or sulfate leads to lower quantum efficiencies for the As(III) photooxidation process. In the absence of dissolved oxygen, two moles of Fe(III) could oxidize 1 mole of As(III), and the dissolved Fe(II) significantly hindered the oxidation of As(III). However, under oxic conditions, both Fe(II) and As(III) can be oxidized simultaneously, and the presence of Fe(II) and reducing solution pH increased the photon efficiency. The results demonstrate that iron compounds were a good photooxidant due to the fact that ferric hydroxide is an excellent adsorbent for the resultant arsenate. The addition of ferrous salt in the presence of sunlight can be a practical method for the oxidation of As(III) in contaminated waters.

The photocatalytic oxidation of MMA and DMA using TiO_2 was studied by Xu et al. (2007). The study demonstrated that MMA and DMA were readily degraded upon TiO_2 photocatalysis. DMA is oxidized to MMA as the primary oxidation product followed by oxidizing to inorganic arsenate, As(V). The obtained results showed that the pH of the solution affects the adsorption and photocatalytic degradation process, due to the fact that the speciation of the arsenic substrates and surface charge of TiO_2 are pH dependent. The kinetics study indicated the mineralization of MMA and DMA by the TiO_2 photocatalysis follows the Langmuir–Hinshelwood kinetic model. During the photocatalysis, an addition of a hydroxyl radical scavenger, *tert*-butyl alcohol, obviously reduces the rate of degradation, which indicates that ·OH radical is the primary oxidant.

Advance oxidation methods, which utilize ultraviolet light and a photo absorber (iron salts or sulfite), had been developed (Zaw and Emett, 2002). The study demonstrated the application of the iron-based photooxidation process to oxidize and remove arsenic from mine water draining from a hard rock gold, silver, and lead mine in Montana. The results showed that the water treatment residues with and without cement were shown to be stable when subjected to leach testing using aerated water for 3 months. The application of a sunlight-assisted process to oxidize and remove arsenic from tubewell water was also studied in a village in Bangladesh. The obtained results showed the process simple to use for villagers in rural areas without electricity. The UV/sulfite process is preferred for use with UV as no solids are generated which may lead to the fouling of the lamps.

To understand the impact and fate of arsenic in the environment and for optimizing arsenic removal from drinking water, it is crucial to obtain the knowledge of arsenic redox kinetics. Voegelin and Hug (2003) reported that, in the presence of hydrogen peroxide (H_2O_2), rapid oxidation of arsenite adsorbed onto ferrihydrite (FH) might be an alternative technology due to two reasons. First, the adsorbed arsenite is supposed to be oxidized more readily than that of the species in solution. Second, decomposition of H_2O_2 on the surface of FH might also result in the oxidizing of arsenite. In the study, attenuated total reflection-Fourier transform infrared (ATR-FTIR) spectroscopy was employed to monitor the oxidation of the adsorbed As(III) on the FH surface *in situ*. The obtained results demonstrated that no oxidation of As(III) was observed within minutes to hours in the absence of H_2O_2. In the presence of H_2O_2, the oxidation rate coefficients for adsorbed As(III) increased. The solution pH did not significantly affect the As(III) oxidation. The experimental results also shown that Fe was necessary to induce As(III) oxidation by catalytic H_2O_2 decomposition.

The efficiency and mechanism of TiO_2-photocatalyzed oxidation of As(III) to As(V) at neutral pH and over a range of As(III) concentration were explored (Ferguson et al., 2005). The results showed that the complete oxidation of As(III) to As(V) was observed within 10–60 min of irradiation at 365 nm. The influence of addition of phosphate at 0.5–10 μM on the photooxidation rate was negligible. The mechanism study demonstrated the superoxide, $O_2^{\cdot-}$, play an important role during the photooxidation process.

Dutta et al. (2005) investigated the effects of As(III) concentration, pH, catalyst loading, light intensity, dissolved oxygen concentration, type of TiO_2 surfaces, and ferric ions on the performance and mechanism of photocatalytic oxidation of As(III) to As(V). The kinetics showed the photocatalytic oxidation of As(III) to As(V) occurs in minutes and follows zero-order kinetics. It had been found that the OH free radicals were involved and play an important role in the oxidation process.

Utilization of reactive intermediates, which were produced by the corrosion of ZVI in oxygen-containing water, to oxidize the arsenite to arsenate was explored by Katsoyiannis et al. (2008). The kinetics and mechanism of Fenton reagent generation, As(III) oxidation, and removal from aerated water by ZVI at pH 3–11 were investigated. The results showed, at pH 3–9, the observed half-lives for the oxidation of 500 μg/L As(III) with 150 mg/L ZVI were 26–80 min. However, at pH 11, no oxidation of As(III) was observed in the first 2 h. At pH 3, 5, and 7, the dissolved Fe(II) was determined as 325, 140, and 6 μM, and the peak concentration of H_2O_2 within 10 min was 1.2, 0.4, and less than 0.1 μM, respectively. The obtained experimental results suggested that the oxidation of As(III) mainly occurred in solution by Fenton reaction, and subsequently removed by sorption on freshly formed hydrous ferric oxides. During the oxidation process, OH· radials were identified as the main oxidant at low pH.

The photocatalytic oxidation of arsenite and simultaneous removal of the resultant arsenate from aqueous solution using a municipal solid waste melted slag containing iron oxide and TiO_2 in the presence of UV light were investigated (Zhang and Itoh, 2006). The results showed the oxidation of arsenite was rapid (within 3 h), whereas the adsorption of the generated arsenate was slow (within 10 h). The results indicated arsenite could also be oxidized to arsenate only by UV light at a slow rate of approximately one-third of that of the photocatalyzed reaction. Both alkaline and acidic conditions facilitated the oxidation reaction, and the optimum pH for the oxidation and adsorption was proposed to be around 3.

The photocatalyzed oxidation efficiency in an upflow-through, fixed-bed reactor, which was irradiated on above and used TiO_2 coated glass beads as packing materials, was examined by Ferguson and Hering (2006). In the study, the effects of reactor residence time, initial As(III) concentration, quantity of TiO_2 coatings on the beads, solution matrix, and light source on the performance were investigated. The results showed the beads could be repeatedly used for As(III) oxidation.

The competitively adsorbing anions, $NaNO_3$, did not significantly affect the catalyst activity. The designed TiO_2 fixed-bed reactor is expected to be an environmental benign method for As(III) oxidation.

Peroxydisulfate ions ($S_2O_8^{2-}$, KPS) was used as an oxidizing agent for the photochemical oxidation of As(III) to As(V) under UV light irradiation by Neppolian et al. (2008). It had been found that the rate of photochemical oxidation for As(III) by KPS was exceptionally high, and the oxidation of As(III) to As(V) using KPS was a simple and efficient method. In the study, the UV light intensity was proven to be primary importance for the dissociation of the KPS in producing sulfate anion radicals (SO_4^-), which favors a higher reaction rate. The variation of pH from 3 to 9 did not influence the reaction. However, the reaction rate was reduced (20%) by the continuous purging of nitrogen, which indicated the dissolved oxygen plays a role in the reaction. The presence of humic acid, even at 20 ppm, was found to have no detrimental effect on the oxidation reaction.

Yoon et al. (2008) investigated the usage of the vacuum–UV (VUV) lamp, which emits both 185 and 254 nm lights, as a new oxidation method for As(III). In the study, it was found that the employed VUV lamp showed a higher performance for As(III) oxidation compared to other photochemical oxidation methods (UV-C/H_2O_2, UV-A/Fe(III)/H_2O_2, and UV-A/TiO_2). The obtained results also showed that the presence of Fe(III) and H_2O_2 increased the As(III) oxidation efficiency, and humic acid did not cause a significant effect on the reaction.

In the presence of potassium iodide (typically 100 μM), the photooxidation of As(III) under 254 nm irradiation was investigated by Yeo and Choi (2009). The results showed that the presence of iodide dramatically enhanced the oxidation rate, and the quantum yields of As(III) photooxidation ranged from 0.08 to 0.6, which depends on the concentration of As(III) and iodide. The air- or N_2O-satuated solution enhanced the photooxidation of As(III), however N_2-saturated markedly reduced the photooxidation rate. The mechanisms study suggest that the excitation of iodides under 254 nm irradiation result in the generation of iodine atoms and triiodides, which seem to be involved in the oxidation process of As(III). It has been found that the UV254/KI/As(III) photooxidation process is essentially an iodide-mediated photocatalysis.

8.5.1.3 Biological Oxidation

A new heterotrophic bacterial strain, ULPAs1, was isolated from arsenic-contaminated water by Weeger et al. (1999). The isolated ULPAs1 shows rapid and extensive oxidation of As(III) into As(V). The study showed that the growth characteristics of the arsenite-oxidizing bacterium, ULPAs1, were independent of the presence of arsenic (1.33 mM as As(III)) in minimum medium containing lactate as the sole organic carbon source. However, no growth took place in the absence of organic carbon source or in a rich medium (i.e., Luria–Bertani). The doubling time of the ULPAs1 was 1.5 h. The minimum inhibitory concentration of arsenic for the strain was found to be 6.65 mM. The strain was demonstrated to be very effective for the oxidation of arsenic in a batch reactor. 16SrDNA sequence analysis showed that the strain belongs to the β-proteobacteria. The results demonstrated that this strain could represent a good candidate for arsenic remediation in heavily polluted water.

The cultivation and application of arsenic-oxidizing bacteria (ULPAs1) for arsenic oxidation were investigated (Lievremont et al., 2003). In the study, the strain was cultivated in batch reactors in the presence of two solid phases, chabazite and kutnahorite, which were used as microorganisms immobilizing materials. The results showed the arsenite oxidative properties of ULPAs1 were conserved when cultivated in the presence of quartz or chabazite. The experiments were carried out with induced (As+) or noninduced (As−) bacteria. It was found that the induced ULPAs1 oxidized As(III) in 2 days in the presence of chabazite, and As(V) was observed in the aqueous phase after 4 h. However, the oxidation rate of arsenite with the noninduced ULPAs1 was slower.

Two bacteria strains were isolated from acid waters originating from Carnoules mine tailings, and identified as Thomonas sp. (Lenoble et al., 2003). The acid water contained high dissolved concentration of arsenic and iron. It was found that the arsenic is precipitated very fast with Fe(III) during the flow of the acid water stream. The precipitation rate is related to the oxidation of iron and enhanced with iron-oxidizing bacteria. Rapid arsenic oxidation was observed in the acid water ascribed to the activity of arsenic-oxidizing bacteria.

Katsoyiannis et al. (2004) reported that indigenous iron- and manganese-oxidizing bacteria in groundwater are able to catalyze the oxidation of dissolved manganese [Mn(II)] to insoluble hydrous manganese oxide, which can subsequently be removed by filtration. The process leads to the formation of a natural coating on the surface of the filter medium. If arsenic is simultaneously present in the groundwater, it can subsequently be removed by sorption onto the manganese oxide. In the study, rapid oxidation of As(III) to As(V) was observed prior to removal by sorption onto the biogenic manganese oxide surfaces. The rates of As(III) oxidation were found to be significantly higher than the rates reported for abiotic As(III) oxidation by manganese oxides, indicating that bacteria play an important role in both the oxidation of As(III) and the generation of reactive manganese oxide surfaces for the removal of As(III) and As(V) from the solution. The obtained results also demonstrated the presence of phosphates at concentrations of around 600 μg/L did not affect the oxidation of As(III), however, it had an adverse influence on the As(III) removal, which decreased the overall removal efficiency by 50%, although it did not affect the oxidation of As(III).

A bacterial, strain B2, was isolated from the biofilm growing in a biological groundwater treatment system used for Fe removal by Casiot et al. (2006). The bacteria strain was proven to be able to oxidize arsenite into arsenate. This strain was found to be different from the genus Leptothrix commonly encountered in biological iron oxidation processes. The study revealed that this isolated strain B2 was the major population of the bacterial community in the biofilm. Therefore, it is probably one of the major contributors to arsenic oxidation in the treatment process.

8.5.2 Chemical Coagulation and Electrocoagulation

All waters, particularly surface waters, usually contain both dissolved and suspended particles. The suspended particles are stabilized (kept in suspension) by the action of physical forces on the particles themselves, and surface electrostatic repulsion plays a key role. Most suspended solids in water possess a negative charge and repel each other when they come close together. Coagulation and flocculation processes are commonly employed to separate the suspended solids from the water.

Though the terms coagulation and flocculation are often used interchangeably, in fact, they are two distinct processes. Coagulation and flocculation occur in consecutive steps to destabilize the suspended solids and facilitate the growth of the floc. Coagulation means the destabilization of colloids by neutralizing the forces, which keeps them apart. Cationic coagulants provide positive electric charges to neutralize or reduce the negative charge of the colloids, which leads the particles to collide to form larger particles. Flocculation is the action of polymers to form bridges between the larger mass particles or flocs, by which the particles were bound into large agglomerates or clumps. The bridging occurs when segments of the polymer chain are adsorbed onto different particles and help particles aggregate. Coagulation and flocculation are among the most common methods used for arsenic removal from aquatic systems.

8.5.2.1 Chemical Coagulation

The aluminum-based and iron-based chemical coagulation method is one of the most commonly used methods for arsenic removal from water. The removal efficiency of arsenic from source water and artificial freshwaters during chemical coagulation with alum and ferric chloride as coagulants were examined in a bench-scale reactor by Hering et al. (1997). The results showed that chemical coagulation by using ferric chloride or alum is capable of reducing final dissolved As(V) concentrations to no more than 2 μg/L for the range of influent As(V) concentration found in U.S. source waters. The suitable pH ranges and minimum dosage of coagulant needed are governed by the solubility of amorphous metal hydroxide solids. The range of pH for efficient As(V) removal with alum was more restricted that that with ferric chloride. At pH < 8, arsenate removal by either ferric chloride or alum was relatively insensitive to variation in source water compositions, however, at pH 8 and 9, the arsenate removal efficiency by ferric chloride was decreased in the presence of natural organic matter. Removal of arsenite from source waters by using ferric chloride as coagulant was both less efficient and more strongly affected by source composition than the removal of arsenate. The presence of natural organic matter (at pH 4–9) and sulfate (at pH 4 and 5) adversely affected the removal efficiency of arsenite by ferric chloride. Arsenite could not be removed from source waters by chemical coagulation with alum as coagulant.

In 2002, a modified conventional coagulation–flocculation process for arsenic removal from contaminated water was investigated (Zouboulis and Katsoyiannis, 2002). The modification referred to the introduction of the "pipe flocculation" process. Ferric chloride or alum was used as coagulant, cationic or anionic polyelectrolytes (organic polymers) were used as coagulant aids to enhance the arsenic removal efficiency. The results showed the modification of the conventional coagulation–flocculation technology was found to be very efficient for the removal of arsenic anions from wastewater and can also find applications in potable water treatment. The method was efficient with both iron and alum coagulants, and both types of coagulant aids (cationic or anionic polymers) were found to increase the overall removal efficiency of the method—reaching in some cases arsenic removals up to 99%. Compared with conventional coagulation processes, the modified technique presents several advantages: the overall flocculation process time was decreased during pipe flocculation, and there were less space requirements and capital costs.

Lee et al. (2003) investigated the arsenic removal using Fe(VI) as both an oxidant and a coagulant. The results showed that with minimum 2.0 mg/L Fe(VI), the arsenic concentration can be lowered from an initial 517 to below 50 μg/L, which is the regulation level for As in Bangladesh. Fe(VI) was demonstrated to be very effective in the removal of arsenic species from water at a relatively low dose level (2.0 mg/L). A combined use of a small amount of Fe(VI) (below 0.5 mg/L) and Fe(III) as a major coagulant was demonstrated to be a cost-effective method for arsenic removal. Ferric chloride and ferric sulfate were used as coagulants for arsenic removal from groundwater (Wickramasinghe et al., 2004).

The obtained results suggest that both coagulants can be well used for arsenic removal. However, coagulation with ferric sulfate results in a lower residual turbidity. The arsenic removal efficiency is highly dependent on the quality of raw water. An appropriate amount of coarse calcite with particle size 38–78 mm was added to enhance the arsenic removal efficiency from high-arsenic water using ferric ions as coagulant by Song et al. (2006). The enhancement of arsenic removal efficiency may be due to the coating of small arsenic-borne coagulates on the surface of calcite, which leads to greatly improve the gravitational sedimentation of the coagulates. The coating of small arsenic-borne coagulates on coarse calcite may be ascribed to the electrostatic attraction between coagulate and coarse calcite because of the reverse surface charge of the two particles. A very high arsenic removal (over 99%) from high-arsenic water in mine drainage systems can be obtained by the enhanced coagulation. Laboratory and field experiments were carried out to investigate the efficiency of a treatment process combining the biooxidation of As(III) and the subsequent removal of As(V) using coagulation with FeCl$_3$ for As removal from As-contaminated wastewater (Andrianisa et al., 2008). The obtained results suggests a high As removal efficiency (>95%) can be achieved by the combined treatment process, the residual As concentration of less than 10 μg/L in the supernatant can be obtained by addition of 24 or 85 mg/L FeCl$_3$ to the effluent of the mixed liquor.

8.5.2.2 Electrocoagulation

Electrocoagulation is one of the most promising methods for wastewater treatment where the coagulants are generated by *in situ* electrooxidation of a sacrificial anode, which generally is made up of iron or aluminum. In the electrocoagulation process, treatment is done without addition of any chemical coagulant or flocculant, thus reducing the amount of sludge produced during the process. Electrocoagulation technology has become an alternative to conventional chemical coagulation by using Al or Fe salts. In electrocoagulation, the coagulants (Al or Fe) are generated by electrolytic oxidation of anodes (Al or Fe). Compared with conventional chemical coagulation, the advantages of electrocoagulation include (Bagga et al., 2008): (1) no alkalinity consumption, (2) no change in bulk pH, (3) the direct handling of corrosive chemicals is nearly eliminated, and (4) can be easily adapted for use in potable water treatment units especially during emergencies.

A batch electrochemical reactor was used to study arsenic removal from smelter industry wastewater over a wide range of operating conditions (Balasubramanian and Madhavan, 2001). Stainless steel and mild steel plates were used as the cathode and anode, respectively. It had been observed that arsenic can be removed effectively by the electrocoagulation process. The production of coagulant (ferric ion) can easily be controlled during the electrocoagulation process by adjusting the operating conditions. The generation of solid sludge can be reduced significantly.

Electrocoagulation had been evaluated as a treatment technology for arsenite and arsenate removal from water by Kumar et al. (2004). In the study, laboratory scale experiments were conducted with three different electrode materials namely, iron, aluminum, and titanium to assess their efficiency. The obtained results demonstrated, in the electrocoagulation process, that arsenic removal efficiencies with different electrode materials follow the sequence: iron > titanium > aluminum. The electrocoagulation treatment process with iron as electrodes was able to remove more than 99% of arsenic and bring down the arsenic concentration to less than 10 µg/l, which could meet the drinking water standard. It was observed that arsenic removal is rapid at higher current densities, but when the results of different current density were converted into charge density, arsenic removal correlated well with charge density. Therefore, charge density was suggested as a design parameter for the process by the researcher. The results also showed the solution pH did not have significant effect on both As(III) and As(V) removal in the pH range 6–8. In the study, comparative evaluation of As(III) and As(V) removal using chemical coagulation (with ferric chloride as coagulant) and electrocoagulation has been carried out. The obtained results implied that electrocoagulation had better removal efficiency for As(III) than chemical coagulation, whereas As(V) removal performance of both electrocoagulation and chemical coagulation processes were nearly the same. In addition, the study indicated that the As(III) removal mechanism in electrocoagulation seems to be the oxidation of As(III) to As(V) and followed by surface complexation with iron hydroxides.

Parga et al. (2005) used a modified electrocoagulation process for arsenic removal from water. The arsenic-contaminated water was passed through a porous tube medium where air was injected before passing through the vertical electrodes in the EC (enterochromaffin) cell. The results showed that As(III) and As(V) can be effectively removed by the modified process. The study demonstrated 99% arsenic removal in the experimental electrocoagulation reactor was usually completed within 90 s or less for most experiments with approximately 100% current efficiency. The pilot plant study showed 99% total arsenic removal from well water. The solid products formed at iron electrodes during the EC process were analyzed by powder x-ray diffraction (PXRD), scanning electron microscopy, transmission Mossbauer spectroscopy, and Fourier transform infrared (FT-IR) spectroscopy. The results suggest that magnetite particles and amorphous iron oxyhydroxides present in the EC products remove arsenic(III) and arsenic(V) with an efficiency of more than 99% from groundwater in a field pilot-scale study. The obtained results indicated the electrocoagulation generated magnetic particles of magnetite and amorphous iron oxyhydroxides.

Electrocoagulation of As(V) solutions in a continuous flow reactor was studied by Hansen et al. (2006). The results demonstrated more than 98% As(V) could be removed from a 100 mg/L As(V) solution by using a current density of 1.2 A/dm^2 and a hydraulic retention time of approximately 9.4 min. However, less than 10% of As(III) was removed in the same operational conditions where around 80% of As(V) was removed. The Fe^{3+} and OH$^-$ dosage was increased with the increasing of current density, which facilitated the removal of As. On the other hand, it seems that the electrocoagulation process would not improve further by increasing the current density beyond a maximum value. This may be due to the passivation of the anode. A higher current reversal frequency was suggested to deal with this problem. Hansen et al. (2007) also investigated the effect of design of the electrocoagulation reactor and operation parameters on the efficiency of arsenic removal from a copper smelter wastewater stream. In the study, three types of electrocoagulation reactors, modified flow continuous reactor, turbulent flow reactor, and airlift reactor, were tested and compared. Iron was used as sacrificial anode in all the reactors. Comparing the different designs, all the electrocoagulation setup showed an efficient As removal. The results demonstrated all arsenic can be eliminated from a 100 mg-As(V)/L solution by using both modified continuous flow reactor and airlift reactor with current densities of around 120 A/m². The arsenic removal with the turbulent flow reactor did not reach the same level, but the ratio of Fe-to-As (mol/mol) achieved in the coagulation process was in this case lower than that with the other two reactors. Another important factor for the removal efficiency is the necessity to avoid anode passivation, which can be done either by optimization of the current reversal frequency or salt concentration.

Electrocoagulation with aluminum or iron or their combination as electrodes for arsenic removal from water with a wide range of arsenic concentration (1–1000 ppm) at different pH (4–10) was investigated by Gomes et al. (2007). The results showed that more than 99.6% of arsenic was removed at initial arsenic concentration of 13.4 ppm by using Fe–Fe electrode pair. When Al–Fe electrode pair was used, the removal efficiency varied from 78.9% to more than 99.6% at different initial arsenic concentrations (1.42–1230 ppm). A frequent change of electrode polarity was used during the electrocoagulation to provide an efficient way for removal of both organic and metallic pollutants from water. Electrochemically generated byproducts were analyzed by PXRD, X-ray photoelectron spectroscopy (XPS), scanning electron microscopy/energy dispersive spectroscopy (SEM-EDS), FT-IR, and Mossbauer spectroscopy. The spectral analysis revealed the expected crystalline iron oxides (magnetite (Fe_3O_4), lepidocrocite (FeO(OH)), iron oxide (FeO)) and aluminum oxides (bayerite ($Al(OH))_3$, diaspore (AlO(OH)), mansfieldite ($AlAsO_4 \cdot 2H_2O$)), as well as some interaction between the two phases. The results also indicated the presence of amorphous or ultrafine particular phase in the floc.

Different electrode materials, including zinc (Zn), brass (Cu–Zn), copper (Cu), and iron (Fe), were used as anodes in a lab scale electrocoagulation reactor for arsenic removal from a solution containing 70–130 mg/L arsenic at current density of 1.5, 3, and 12 mA/cm² for 60 min (Maldonado-Reyes et al., 2007). The obtained results demonstrated, at higher current density (12 mA/cm²), that rapid arsenic removal was achieved. The arsenic removal efficiencies followed the tendency given below (at 1.5 mA/cm²): Fe (>93%) ≈ Zn (>93%) > Cu–Zn (>73%) > Cu (>67%), and these efficiencies were relatively independent of the removal rate for all the initial arsenic concentrations investigated. However, at the early stages of the electro-removal process, the As removal rate with Fe is more rapid than that with Zn at low current densities, and Fe is considered as the most attractive material for practical applications. In addition, comparing with the addition of chemicals for arsenic removal, the As electro-removal process by iron as electrode has the advantage of producing a very low quantity of sludge. The proposal mechanism responsible for arsenate removal is the complexation of arsenate with the products from the sacrificial electrode materials. The formed products were determined as $(FeO)_2HAsO_4$, $Zn_{0.7}Al_{0.3}HAsO_4(CO_3)_{0.15} \times H_2O$, $CuHAsO_4$, and $ZnHAsO_4$ with Fe, Zn, Cu, and Cu–Zn alloy electrodes, respectively.

Basha et al. (2008) investigated the removal of arsenic from copper smelting industrial wastewater by electrodialysis or electrochemical ion-exchange technique, followed by electrocoagulation. The wastewater contained varying amounts of As(III) and As(V), oxyanion, arsenite, and arsenate with a very low pH. The results showed arsenic can be removed up to 91.4% and sulfate up to 37.1% using electrodialysis at a current density of 200 A/m², and arsenic can be removed up to 58.2% and sulfate up to 72.7% using electrochemical ion exchange at a current density of 300 A/m². The arsenic can be further removed up to below the detectable limit of an atomic absorption spectrometer by using electrocoagulation at a current density of 150 A/m². The results also demonstrated that the consumption of alkali needed to raise the pH can be effectively minimized by combining both the electrochemical ion-exchange and electrocoagulation processes.

Electrocoagulation is a promising remediation tool for the treatment of water containing As(V). Experiments showed the possibility of removing arsenic as adsorbed to or coprecipitated with iron(III)hydroxide. Increasing the current density from 0.5 to 1.5 A/dm^2 showed significant improvement in arsenic removal. However, beyond the current density of 1.5 A/dm^2 did not show any significant improvement. More than 98% of arsenic removal has been recorded in the present investigation. The electrocoagulation has been modeled using adsorption isotherm models and observed Langmuir isotherm models match satisfactorily with the experimental observations.

Batch experimental and modeling studies on arsenic removal using electrocoagulation with aluminum and mild steel as sacrificial anode were carried out by Balasubramanian et al. (2009). The obtained results showed that the efficiency of arsenic removal was significantly influenced with applied charged and solution pH. The arsenic removal efficiency was significantly enhanced by increasing the current density from 0.5 to 150 A/m^2. The maximum arsenic removal efficiency was observed as 94% under optimum condition. Adsorption isotherm kinetics was used to modeling the electrocoagulation mechanism, the results indicated that the Langmuir isotherm models match satisfactorily with the experimental observations.

8.5.3 Precipitation

Precipitation process is often used together with other physicochemical process to effectively remove arsenic from aqueous solutions. Nishimura and Umetsu (2001) investigated the removal performance of arsenic and manganese from an aqueous solution with pH range of 0.4–5.0 by oxidation–precipitation using ozone. In the study, the following results were obtained: (1) The oxidation of arsenic (III) to arsenic (V) takes place prior to the oxidation of manganese (II) when an O_3–O_2 gas mixture is supplied to solutions containing manganese (II) and arsenic (III). The resultant arsenic (V) reacts with the manganese to form a precipitate, which is believed to be $MnAsO_4 \cdot nH_2O$. (2) The residual arsenic concentration can be brought below the regulatory limit of 0.1 mg/L. (3) The performance of the oxidation–precipitation process is affected by the temperature. The remained arsenic (V) concentration in the Mn/As solutions is less than 0.1 mg/L at 25°C but rises to 2 mg/L with increasing temperature to 60°C before decreasing again to about 0.4 mg/L at 80°C. (4) In the pH 1–2, precipitation of arsenic with manganese by ozonation is effective for removing arsenic selectively where ferric arsenate and ferric hydroxide are not precipitated. (5) The removal of arsenic can be enhanced by an appropriate amount of ferrous ion coexisting with arsenic and manganese in the solution, particularly at pH 1–3.

A household coprecipitation and filtration system was developed and tested in the laboratory and field for arsenic removal from Bangladesh groundwater by Meng et al. (2001). The processes included coprecipitation of arsenic by adding a packet of about 2 g of ferric and hypochlorite salts to 20 L of well water and subsequent filtration of the water through a bucket sand filter. The obtained results showed that the household system could effectively remove arsenic from the Bangladesh well water. The experimental results indicate that elevated phosphate and silicate concentrations in Bangladesh well water dramatically decreased the adsorption of arsenic by ferric hydroxides. To reduce arsenic concentration from 300 to less than 50 μg/L in the Bangladesh well water, the Fe/As mass ratio should be greater than 40. It is estimated that the costs of chemical are less than US$4 annually for a family, based on a daily consumption of 50 L of filtered water.

A two-stage precipitation process was investigated to remove iron and arsenic from a wastewater stream produced in the leaching process for base metal recovery by Bolin and Sundkvist (2008). The presented method allows for selective disposal of iron and arsenic in a form that will easily settle and filtrate. The obtained product shows to have good sedimentation and filtration properties, which makes it easy to recover the iron–arsenic depleted solution by filtration and washing of the precipitate. The process also gives a possibility to optimize the pH profile for different temperatures and metal concentrations in the feed in a flexible way.

The reduction, precipitation, and transport of arsenic species by *Shewanella* sp., a facultative and versatile iron-reducing bacterium, were investigated through batch and column tests by Lim et al. (2008). The obtained results indicated that *Shewanella* sp. reduced As(V) to As(III), and reduced sulfate to sulfide, which resulted in the precipitation of arsenic precipitation with sulfides. As(V) was subject to both microbial reduction and precipitation. Due to microbial reduction of As(V), the As(III) concentration increased in early times, but was removed from the solution by precipitation in later times.

8.5.4 Membrane Filtration

In the past two decades, a lot of researches have been focused on membrane technologies for arsenic removal from water. Membrane filtration has been approved to be a viable method, which can be used to remove a wide range of pollutants from water. It will likely be increasingly applied for water treatment including arsenic removal, due to its being reliable, easy to produce, obtain, operate, and maintain.

The concept of multiple separation by chemisorptive filters was applied and investigated for arsenic removal from water by Jubinka et al. (1992). Multistage ion-exchange, adsorption, and chemical reaction inside the filter were involved in the process. The experimental results showed the chemisorptive filters exhibited remarkable efficiency in the removal of arsenic from water. When the initial arsenic concentration was 6.65×10^{-4} mol/L, a high degree of separation and a decrease in the arsenic concentration of more than 1000-fold could be obtained. The initial concentration, pH, and pollutants in anionic forms could significantly affect the process.

Brandhuber and Amy (1998) investigated the suitability of reverse osmosis, nanofiltration (NF), ultrafiltration, and microfiltration (MF) as an arsenic treatment method. Several conclusions were drawn from these pilot studies. These include (1) Combination of coagulation with MF is a technically feasible method for removal of arsenic from water to meet a 5 ppb or stricter MCL in the source waters. (2) Under the optimization conditions ($FeCl_3$ dosage of 7.0 mg/L, permeate flux of 102 gfd, and 90% recovery), averaged arsenic rejection of 84% and turbidity reduction of 64% were achieved. Air backwashes of the filter at 15 min intervals successfully controlled the fouling of the filter. (3) A low doses of $FeCl_3$ (2 mg/L) also could obtain significant (50%) arsenic rejection.

In 2001, Meng et al. successfully developed a household coprecipitation and filtration system for arsenic removal from Bangladesh groundwater. Brandhuber and Amy (2001) explored the influences of water quality and membrane operating conditions on the rejection of arsenic by a negatively charged UF membrane in the laboratory. The obtained results showed arsenic removal by the charged UF membrane is sensitive to the feed water composition and the membrane's hydraulic operating conditions, including permeate flux, membrane recovery, and cross flow velocity. The trends in arsenic rejection are qualitatively consistent with the Donnan theory. In particular, the existence of co-occurring divalent ions was demonstrated to have a negative influence on arsenate rejection. The presence of natural organic matter may play an intriguing role in the rejection of As(V) by charged membranes. The high concentrations of organic matter may improve arsenic rejection through the complexation of divalent ions.

Conventional coagulation–flocculation technology was modified and employed to remove arsenic from water by Zouboulis and Katsoyiannis (2002). The modifications refer to the introduction of a "pipe flocculation" process in the first stage of the technique, while direct sand filtration was used in the second step instead of separation by sedimentation. It was found that the modification process is very efficient for the removal of arsenic anions from wastewater and drinking water. The presence of cationic or anionic polyelectrolytes enhanced the coagulation efficiency of alum or ferric chloride in certain cases. It was found that the arsenic concentration can be reduced to 10 µg/L from initial concentrations of over 400 µg/L in almost all cases.

A ZW-1000 (Zenon) membrane module was used for the removal of arsenic from deep well water (Judit and Hideg, 2004). Before membrane filtration, pretreatments were done, including oxidation with potassium permanganate ($KMnO_4$), coagulation with ferrous(III) sulfate ($Fe_2(SO_4)_3$), fast mixing of chemicals with a mixer, coagulation with slow mixing and settlement. The process was shown to be able to reduce arsenic concentration to lower than 10 µg/L from an initial concentration of 200–300 µg/L. The obtained results showed that the technology was successful and is suitable to produce drinking water at the required quality from raw water with a high arsenic content in a pilot plant.

A combined coprecipitation and active filtration process was used to remove arsenic from drinking water (Newcombe et al., 2006). The combined process referred to a serpentine prereactor for ferric chloride reagent mixing which was combined with a moving bed active filter, followed by separation of waste residuals from clean water discharge. The pilot-scale testing showed that the arsenic concentration could be reduced to 3.3 ± 1.4 µg/L from initial concentration of 40.2 ± 1.0 µg/L under optimized experimental conditions. The optimized Fe/As molar ratio was found to be 133:1. The obtained research results demonstrated the formation and renewal of iron oxide-coated sand in the active filter is a viable mechanism for high efficiency arsenic removal (Ferella et al., 2007) examined the influences of cationic and anionic surfactants on the performance of surfactant-enhanced UF process for arsenic and lead removal from wastewater. In the study, dodecylbenzenesulfonic acid (DSA) was used as anionic surfactant, and dodecylamine was applied as cationic surfactant. The UF process was carried out by means of a monotubular ceramic membrane of nominal pore size 20 nm (molecular weight cutoff: 210 kDa). The results showed Pb and As ions are removed from the water flow one at a time using both DSA and dodecylamine.

The arsenic removal from water sources down to the residual concentration below 10 µg/L using chemisorption filtration was reported by Solozhenkin et al. (2007a,b). A layer of modified polystyrene granules was used in the chemisorption filtration process. The effects of basic physicochemical parameters on the performance of the process were investigated. The adsorption filtration was proved to be efficient for arsenic removal from water. It can reduce the arsenic concentration to less than 10 µg/L. Compared to other available technologies, it provides a number of advantages: it allows the reduction of the toxic slimes produced, expanding the area of surface for adsorption, and is applicable for arsenic removal from underground waters with low arsenic concentration.

Xia et al. (2007) investigated the removal of arsenic from synthetic waters by NF membrane. In the study, the influences of arsenic feed concentration, pH, existence of other ionic compounds, and natural organic matter on the performances were evaluated. The obtained results showed that there was a large difference in the removal of arsenate and arsenite. Arsenate was almost fully removed, while arsenite was removed about 5%. The existence of additional salts was demonstrated to have an impact on the rejection of arsenate. Increasing pH enhanced the arsenic rejection by the membrane. The study showed that the NF was particularly suitable to treat arsenic-rich groundwater in suburban China.

Hsieh et al. (2008) examined the removal of arsenic from groundwater by a laboratory scale electro-ultrafiltration (EUF) system. In the study, two groundwater samples taken from the northeastern part of Taiwan were studied. The As(III) to As(V) ratios of the well water were 1.8 and 0.4 for well-1 and well-2, respectively. The obtained results showed the presence of 25 V voltage in the UF system can increase the total arsenic removal efficiencies from 1% to 79% and 14% to 79% for well-1 and well-2 samples, respectively. The result also suggested the possible association between As(III) species and dissolved organic matter which enhanced the As removal.

NF and reverse osmosis were used on laboratory scale to concentrate arsenic-containing wastewater (Fogarassy et al., 2009). In the study, cross flow membrane filtration apparatus was applied in batch mode with recycling the retentate, while the retentate of membrane filtration was treated with lime ($Ca(OH)_2$) and sulfur hydrogen (H_2S) to help the precipitation for producing clean water and a low volume, highly concentrated As waste. The results showed 94%–99% arsenic rejection was reached by addition of $Ca(OH)_2$ to the high arsenic content model solution. The arsenic concentration of the clear liquid decreased from about 1300 µg/L to the drinking water level, 10 mg/L.

The application of MF and NF for arsenic removal was explored by Nguyen et al. (2009). The obtained results showed about 81% of As(V) and 57% of As(III) were removed from 500 µg/L arsenic solutions by NF (NTR729HF, Nitto Denko Corp., Japan) of 700 molecular weight (MW) cutoff, which indicated the performance of the nanofilter is better for removing As(V) than As(III).

The performance of MF for arsenic removal was much lower than that of NF due to its larger pore size. By comparison, only 40% of As(V) and 37% of As(III) were removed by MF (PVA membrane, Pure-Envitech, Korea). However, addition of 0.1 g/L nanoscale zerovalent iron (nZVI) significantly increased the removal efficiencies up to 90% with As(V) and 84% with As(III) by MF.

Pokhrel and Viraraghavan (2009) investigated the addition of iron to a biological sand filtration column for effective arsenic removal. The obtained results showed the addition of iron with Fe/As ration of 40:1 could reduce the arsenic to below 5 mg/L in a biological sand filtration column. At low Fe/As ration (10:1 and 20:1), the depth of filter was found to effect the arsenic removal efficiency, however, less influence of the filter depth on arsenic removal was observed at high Fe/As ration (30:1 and 40:1). The iron in the effluent was below 0.1 mg/L at all times.

As(III) and As(V) removal by direct contact membrane distillation (DCMD) were investigated with self-made polyvinylidene fluoride (PVDF) membranes by Qu et al. (2009). The results showed the maximum permeate flux of the membrane was 20.90 kg/m^2 h, and the PVDF membrane had high rejection of inorganic anions and cations which was independent of the solution pH and the temperature. The experimental results indicated that DCMD process had higher arsenic removal efficiency than pressure-driven membrane processes. The experimental results also implied that the permeate of As(III) and As(V) were lower than 10 µg/L until the feed As(III) and As(V) achieved 40 and 2000 mg/L, respectively.

8.5.5 ADSORPTION

Adsorption means the attachment of molecules or particles to a surface. In recent years, a tremendous amount of studies have been conducted to develop efficient adsorbents for arsenic removal from aqueous solutions. Adsorption becomes one of the most extensively methods used for arsenic removal because of its ease of operation and cost effectiveness.

In an adsorption process, arsenic is attached on the surface of the adsorbent by physical as well as chemical forces. There are several parameters influencing the adsorption efficiency significantly, including the active surface area of sorbent, the species of functional groups on the sorbent surface, pH of the solution, etc. In this section, adsorbents for arsenic removal are divided into a few classes based on the materials of the sorbent.

8.5.5.1 Activated Carbon

Modern activated carbon industrial production was established in 1900–1901 to replace bone char in sugar refining (Bansal et al., 1988). The commercial powdered activated carbon was first produced from wood and was widely used in the sugar industry in Europe in the early nineteenth century. In 1930, activated carbon was first reported for water treatment in the United States (Mantell, 1968). Due to its high porosity, large surface area, and high catalytic activity, activated carbon is generally recognized as an effective adsorbent and widely used for the removal of organic compounds in drinking water (Li et al., 2002). Compared with the uptake of organic compounds, the adsorption of metal ions on carbon is more complex due to the ionic charges affect. A lot of activated carbons, including commercial and synthetic, have been tested for their As(III) and As(V) adsorption capacity from water.

Fifteen different brands of commercial activated carbons were tested for their As(V) adsorption capacities over a wide pH region by Huang and Fu (1984). The obtained results showed that carbon type, total As(V) concentration, and pH were the major factors influencing the As(V) removal. Treatment of As(V)-loaded activated carbon with strong acid or base can effectively desorb As(V) but not restore As(V) adsorption capacity.

Eguez and Cho (1987) investigated the effects of pH and temperature on the adsorption of As(III) and As(V) on activated charcoal. The adsorption capacity of As(III) on the activated carbon was constant at pH 0.16–3.5. However, the carbon exhibits a maximum As(V) adsorption at pH 2.35 over the pH range of 0.86–6.33. The isosteric heat of As(III) adsorption varied from 4 to 0.75 kcal/mol, while the heat for As(V) adsorption is from 4 to 2 kcal/mol, which indicate that physisorption occurred due to weak Van der Waals forces. 2.5% As(V) and 1.2% As(III) (based on the weight of carbon) were observed to be adsorbed onto the activated carbon at an equilibrium concentration of 2.2×10^{-2} M of both As(V) and As(III).

The possibilities of arsenic, antimony, and bismuth impurities removal from copper electrolytes with activated carbon were investigated (Navarro and Alguacil, 2002). In the study, various variables which affect the metal adsorption/desorption operations are studied. The obtained results showed antimony and arsenic adsorption onto activated carbon can be used for separating these impurities from copper electrolytes and recycling the electrolytes to the electrorefining cells. A greater carbon/solution ratio and/or using a countercurrent device can enhance the extent of impurities removal from the copper electrolytes.

The efficiency of self-manufactured activated carbon produced from oat hulls for arsenate adsorption was tested in a batch reactor by Chuang et al. (2005). The experimental results indicated that the adsorptive capacity of activated carbon was significantly affected by initial pH value, with adsorption capacity decreasing from 3.09 to 1.57 mg As/g activated carbon when the initial pH values increased from 5 to 8. A modified linear driving force model conjugated with a Langmuir isotherm was developed to describe the arsenic adsorption kinetics on to the activated carbon. The obtained results demonstrate that rapid adsorption and slow adsorption take place simultaneously when the activated carbon is used to remove arsenate from the water solution.

A granular activated carbon was modified by polyaniline for arsenate adsorption (Yang et al., 2007). The obtained results showed that the modification does not change the specific surface area, however, the content of the aromatic ring structures and nitrogen-containing functional groups on the modified granular activated carbon is increased. It was found, in acidic solutions, the surface positive charge density is dramatically increased. The arsenate adsorption onto both granular activated carbons is highly pH dependent. The optimal pH range of the modified carbon for arsenate adsorption are 3.0–6.8 and 4.0–6.6 at initial arsenic concentrations of 0.15 and 8.0 mg/L, which are much broader than that of unmodified activated carbon. The maximum adsorption capacity of granular activated carbon is enhanced by 84% by the modification. The presence of humic acid does not significantly impact on the arsenic adsorption dynamics. XPS analysis indicates that the arsenate is reduced to arsenite during the adsorption process.

Natale et al. (2008) investigated the arsenate adsorption behaviors onto a granular activated carbon. In the study, the influences of initial arsenic concentration, solution pH, temperature, and salinity on equilibrium adsorption capacity had been studied. The obtained results showed the optimal experimental conditions for the arsenate adsorption are neutral solution pH, low salinity levels, and high temperatures. A model, based on the multicomponent Langmuir adsorption theory, was developed to describe the arsenic adsorption mechanism. The model demonstrates that the adsorption capacity is proportional to the concentration of arsenic anions in solution and decreases by increasing the concentration of competitive ions such as hydroxides and chlorides. It also can be used to interpret the pH and salinity effects on the adsorption capacity.

8.5.5.2　Metal Oxides

All kinds of metal oxides are widespread and abundant in the natural environment. Much research on the arsenic adsorption onto metal oxides have appeared. In general, the adsorption of arsenic on metal oxide can be classified as iron oxide, zirconium oxide, manganese oxide, and other metal oxides.

8.5.5.2.1　Iron Oxide

Porous iron oxides are being evaluated and selected for arsenic removal in potable water systems by Badruzzaman et al. (2004). In the study, granular ferric hydroxide (GFH), a typical and commercially available iron adsorbent, was used. In general, GFH is a highly porous adsorbent with micropore volume of ~0.0394 ± 0.0056 cm^3/g and mesopore volume of ~0.0995 ± 0.0096 cm^3/g.

The BET (Brunauer–Emmett–Teller) specific surface area of the adsorbent is 235 ± 8 m^2/g. The obtained from bottle-point isotherm and differential column batch reactor (DCBR) experiments were used to estimate Freundlich isotherm parameters (K and 1/n) as well as kinetic parameters (film diffusion (k_f) and intraparticle surface diffusion (D_s)). The obtained pseudo-equilibrium (18 days of contact time) arsenate adsorption density at pH 7 was 8 µg As/mg dry GFH at a liquid phase arsenate concentration of 10 µg As/L. A nonlinear relationship ($D_s = 3.0^{-9} \times R_p^{1.4}$) was observed between D_s and R_p (GFH particle radius) with D_s values ranging from $2.98^{-10} \times 10^{-12}$ cm^2/s for the smallest GFH mesh size (100×140) to 64×10^{-11} cm^2/s for the largest GFH mesh size (10×30).

The effectiveness of iron oxide-coated cement (IOCC) for As(III) adsorption from aqueous solutions was investigated by Kundu and Gupta (2007). The effects of adsorbent dose, pH, contact time, initial arsenic concentration, and temperature on the arsenic adsorption of the IOCC were studied. The experimental results showed the uptake of As(III) ion is very rapid and most of fixation occurs within the first 20 min of contact. The pseudo-second-order rate equation can successfully describe the adsorption kinetics. To describe the adsorption isotherms at different initial arsenite concentration at 30 g/L fixed adsorbent dose, the Langmuir, Freundlich, Redlich–Peterson (R–P), and Dubinin–Radushkevich (D–R) models were used. According to the Langmuir isotherm, the maximum adsorption capacity of IOCC for As(III) was determined as 0.69 mg/g. Based on the D–R isotherm, the mean free energy of adsorption (E) was calculated to be 2.86 kJ/moL, which implies that the process is a physisorption. The obtained thermodynamic parameters indicate the adsorption process is exothermic and spontaneous. The abovementioned results suggest that IOCC can be suitably used for As(III) removal from aqueous solutions.

The optimal operating conditions of the flow through column experiments and the influence of water composition on arsenate removal from water using an iron oxide-based sorbent were investigated by Zeng et al. (2008). The following results were obtained: (1) Both phosphate and silica influence arsenic adsorption to the iron-based sorbent. Silica has a much stronger inhibiting effect than phosphate at pH 7.5 due to its higher concentration in the test synthetic groundwater. (2) The arsenic removal efficiency decreases as empty bed contact time decreases and flow rate increases. (3) A pore and surface diffusion model can be used to predict the arsenate breakthrough curves at different empty bed contact times. The dominant intraparticle mass transfer process is surface diffusion.

Munoz et al. (2008) investigated the kinetics of absorption of As(V) on a Fe(III)-loaded sponge. The following results were obtained from the study: (1) The Fe(III)-loaded sponge is shown to be effective as an As(V) adsorbent, even in the existence of interfering anions, such as Cl$^-$, for the continuous column-type operation. (2) The adsorbent can be regenerated if a suitable desorbent is used. (3) The Fe(III)-loaded sponge has superior dynamic parameters than that of corresponding Fe(III)-loaded resin. (4) The Clark model can be used to predict the whole breakthrough curve for the sponge. (5) Due to the inferior dynamic properties of the resin, the Clark model has a poorer fit for the resin.

An iron oxide-based calcium alginate magnetic sorbent was developed and employed for the removal of inorganic and organic arsenic by Chen's group (Lim and Chen, 2007, Lim et al., 2009a,b). The schematic diagram of the magnetic sorbent is shown in Figure 8.1. The magnetic sorbent was prepared by an electro-syringing extrusion method as shown in Figure 8.2. The adsorption performance

FIGURE 8.1   Schematic diagram of the magnetic sorbent.

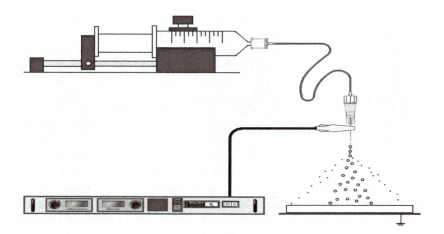

FIGURE 8.2   Schematic diagram of electro-syringing extrusion method.

and adsorption chemistry of inorganic and organic arsenic uptake onto the magnetic sorbent were studied. The obtained results show that the equilibrium sorption for both inorganic and organic arsenate can be attained within 25 h. The solution pH plays a key role in the removal of inorganic and organic arsenate from the solution, lower pH results in larger arsenate adsorption capacity. The maximum sorption capacity of the inorganic arsenate and organic arsenate were 11 and 8.57 mg As/g, respectively. The spectroscopy analysis indicates the –COOH and Fe–O groups in the sorbent are involved and play an important role in the adsorption process. It was observed that both inorganic and organic arsenates were partially reduced to arsenite during the adsorption process.

8.5.5.2.2   Zirconium Oxide

In the past two decades, zirconium has been received increasing attention for arsenic removal from aqueous solutions, and has been shown to have a good sorption capacity for arsenic.

Peräiniemi et al. (1994) reported that a zirconium-loaded activated charcoal can be suitably acted as a promising adsorbent for arsenic removal from aqueous solutions. A porous resin loaded with hydrous zirconium oxide was prepared and employed for the decontamination of arsenic wastewater by Suzuki et al. (2000). The hydrous zirconium oxide-loaded resin (Zr-resin) showed strong adsorption for both arsenate and arsenite. The adsorption of arsenate onto the Zr-resin was more favorable at a slightly acidic condition, while the arsenite was better adsorbed at pH 9–10. The Zr-resin revealed a remarkable selectivity toward the adsorption of arsenate, and common anions did not interfere with the adsorption of arsenate.

A zirconium(IV)-loaded chelating resin (Zr-LDA) with lysine-N^α, N^α diacetic acid functional groups was synthesized, and the adsorption performance of As(V) and As(III) onto the resin was evaluated by Balaji et al. (2005). The results showed that the Zr-LDA chelating resin can effectively remove arsenate and arsenite with a high adsorption capacity of 0.656 and 1.1843 mmoL/g, respectively. The adsorption mechanism is an additional complexation between arsenate or arsenite and Zr-LDA chelating resin. A type of activated carbon impregnated with zirconyl nitrate (Zr-AC) was prepared, and arsenate adsorption properties and mechanisms were investigated by Schmidt et al. (2008). The results suggested that Zr-AC is an effective adsorbent for arsenic removal due to its high surface area and the presence of high affinity surface hydroxyl groups. Biswas et al. (2008) examined the adsorption behavior of arsenate and arsenite onto zirconium-loaded orange waste. The results indicated this efficient and abundant biowaste could be employed for the remediation of aquatic environments polluted with arsenic.

A zirconium-based magnetic sorbent is developed using a coprecipitation technology and applied for arsenate removal by Zheng et al. (2009). In the study, the characterization of the sorbent and its adsorption behavior are systematically investigated. It is shown that the sorbent has a small mean diameter of 543.7 nm, a high specific surface area of 151 m^2/g, and a pH_{zpc} of 7. The sorption equilibrium can be obtained within 25 h. Better adsorption can be obtained at lower pH. The maximum adsorption capacity of the sorbent is 45.6 mg-As/g. FT-IR spectra analysis indicates –OH groups play an important role in the uptake. During the adsorption process, some of the arsenate is reduced to arsenite after its adsorption onto the magnetic sorbent; and the divalent iron in the sorbent may provide electrons for the reduction.

8.5.5.2.3 Manganese Oxide

A natural oxide sample, consisting basically of Mn-minerals and Fe-oxides, were tested for arsenic removal by Deschamps et al. (2005). In the experiments, As-spiked tap water and an As-rich mining effluent with As concentrations from 100 µg/L to 100 mg/L were used. The batch and column experimental results demonstrated the high adsorption capacity of the material, with the sorption of As(III) being higher than that of As(V). It is found, at pH 3.0, the maximum uptake for As(V) and for As(III)-treated materials were 8.5 and 14.7 mg/g, respectively. The oxidation of As(III) to As(V) was observed for both sorbed and dissolved As-species by the Mn minerals. Column experiments with the sample for an initial As concentration of 100 µg/L demonstrated a very efficient elimination of As(III), since the drinking water limit of 10 µg/L was exceeded only after 7400 bed volume.

A Mn-substituted iron oxyhydroxide ($Mn_{0.13}Fe_{0.87}OOH$) was prepared and used for arsenic removal by Lakshmipathiraj et al. (2006). X-ray diffraction analysis indicated that the sample was basically iron manganese hydroxide with bixbyite structure. The sorbent has a surface area of 101 m^2/g and a pore volume of 0.35 cm^3/g. Batch experiments were conducted to study the adsorption isotherm and kinetics of arsenite and arsenate species onto the sorbent. The obtained results showed the maximum uptake of arsenite and arsenate was found to be 4.58 and 5.72 mg/g, respectively. The Langmuir isotherm can be well used to describe the adsorption isotherm for both cases. It was found that the activation energies are on the order of 15–24 and 45–67 kJ/moL for arsenate and arsenite adsorption, respectively.

A novel Fe–Mn binary oxide adsorbent was developed and evaluated for arsenic removal, and the removal mechanism was also investigated by Zhang et al. (2007a,b). The sorbent was prepared by a simultaneous oxidation and coprecipitation method. The synthetic adsorbent showed a significantly higher As(III) uptake than As(V), the maximal adsorption capacities for arsenate and arsenite were found to be 0.93 and 1.77 mmol/g, respectively. Phosphate has a negative effect on arsenic adsorption. However, ionic strength, the presence of sulfate and humic acid had no significant effect on arsenic removal. The mechanism studied indicated that the manganese dioxide play an important role during the As(III) adsorption. The As(III) removal by the binary sorbent is an oxidation and adsorption process. The high uptake capability of the Fe–Mn binary oxide makes it a promising adsorbent for the removal of As(III) from aqueous solutions.

8.5.5.2.4 Other Metal Oxides/Hydroxides

Other metal oxides, including aluminum oxide, titanium oxide, and lanthanum hydroxide, are studied for arsenic removal from water solutions.

Due to the high surface area and a distribution of both macro- and micropores, the activated aluminum adsorption has been classified among the best available technologies for arsenic removal from aqueous solutions by the United Nations Environmental Program Agency (Mohan and Pittman, 2007). Activated alumina is normally prepared by the thermal dehydration of aluminum hydroxide. The application of activated aluminum for arsenic removal has received substantial attention. It is found that the optimal pH for arsenate sorption is between 6.0 and 8.0, where activated aluminum surfaces are positively charged. Singh and Pant (2004) reported that As(III) adsorption onto activated aluminum is strongly pH dependent and it exhibits a high affinity toward activated aluminum at pH 7.6.

The effectiveness of nanocrystalline titanium dioxide (TiO_2) for arsenate and arsenite removal, as well as the photocatalytic oxidation of arsenite were evaluated by Pena et al. (2005). In the study, batch adsorption and oxidation experiments were carried out with TiO_2 suspensions in a 0.04 M NaCl solution, in which the competing anions phosphate, silicate, and carbonate were present. The experimental results demonstrate the nanocrystalline TiO_2 is an effective adsorbent for As(V) and As(III) and an efficient photocatalyst. Kinetics study showed the adsorption of As(V) and As(III) reached equilibrium within 4 h, and a pseudo-second-order equation can be used to described the adsorption kinetics. The optimum pH form As(V) and As(III) were 8 and 7.5, respectively. At an equilibrium arsenic concentration of 0.6 mM, it was found that more than 0.5 mmoL/g of As(V) and As(III) was adsorbed by the TiO_2. At a neutral pH range, the presence of phosphate, silicate, and carbonate had a moderate effect on the adsorption capacities of the TiO_2 for As(III) and As(V). In the presence of sunlight and dissolved oxygen, 2 mg/L As(III) was completely converted to As(V) by 0.2 g/L TiO_2 through photocatalytic oxidation within 25 min.

Lanthanum hydroxide, lanthanum carbonate, and basic lanthanum carbonate can also be used to remove As(V) from aqueous solutions (Tokunaga et al., 1997).

8.5.5.3 Low-Cost Adsorbent

8.5.5.3.1 Biosorbent

Biosorbent is cost effective for removing traces of heavy metals from dilute aqueous solutions. Chitin, chitosan, cellulose, water hyacinth, and various biomasses have been evaluated for the arsenic removal from aqueous solutions.

Chitin is the most widely occurring natural carbohydrate polymer next to cellulose (Mohan and Pittman, 2007). It is a long, unbranched polysaccharide derivative of cellulose, where the C2 hydroxyl group has been replaced by the acetyl amino group $–NHCOCH_3$. Chitosan is derived from chitin by deacetylation of chitin using concentrated alkali at high temperature. Elson et al. (1980) studied a chitosan/chitin mixture for arsenic removal from contaminated water. The capacity of the mixture at pH 7 was found to be 0.13 μ-equiv. As/g mixture with a distribution coefficient of 65. The sorption of As(V) on molybdate-impregnated chitosan gel beads was investigated (Dambies et al., 2000). The impregnation of molybdate enhanced the sorption capacity of the raw chitosan for arsenic(V). It was found that the optimum pH for arsenic uptake was around 3. The pretreatment of the sorbent with phosphoric acid can remove the labile part of the molybdenum and decrease the release of molybdenum during the adsorption process. The As sorption capacity, over molybdenum loading, was almost 200 mg As/g Mo. The exhausted sorbent can be regenerated by phosphoric acid. Iron-loaded cellulose was investigated for arsenic removal (Munoz et al., 2002; Guo and Chen, 2005). Both studies showed the iron-loaded cellulose could be used for effective arsenic removal. Besides this, water hyacinth and other biomasses were investigated for arsenic removal (Mohan and Pittman, 2007).

8.5.5.3.2 Agricultural and Industrial Wastes

Agricultural and industrial wastes, such as rice husk, chars, coals, blast furnace slag, Fe(III)/Cr(III) hydroxide waste, fly ash, etc. have been widely investigated for their application for arsenic removal from aqueous solutions. Ocinski et al. (2016) utilized a type of water treatment residuals (WTRs), generated as a byproduct during the deironing and demanganization process of infiltration water, to adsorb arsenate and arsenite sorbent. WTRs were highly porous (120 $m^2/$ g) and mainly composed of iron and manganese oxides, which favored high arsenic removals, maximum Langmuir adsorption capacities of 132 mg As(III)/g and 77 mg As(V)/g, respectively. The presence of manganese oxide admixture played a key role for As(III) removal by As(III) oxidation and simultaneous creation of new adsorption sites on the adsorbent surface, contributing to a significant higher efficiency in arsenite removal. Moreover, this mechanism enables removal of As(III) with high efficiency also from acidic solutions, which is not possible when the only constituent of the sorbent is iron oxide. The kinetic studies indicated that As(V) adsorption on WTRs was mainly controlled by external and intraparticle diffusion, whereas the two-step chemisorptions mechanism, oxidation, and inner-sphere complexation, contributed more to the rate of As(III) adsorption. The regeneration of the spent sorbent was through NaOH/NaCl elution. Refer to the review paper by Mohan and Pittman (2007) for more examples of agricultural and industrial wastes applied for arsenic removal.

8.6 RECENT RESEARCH AND DEVELOPMENT

In the past decade, there was great development of nanotechnology applications in advanced water and wastewater treatment to improve treatment efficiency (Qu et al., 2013). Nanomaterials, such as nanoparticles (NPs), nanofibers (NFs), owing to their high specific surface area, abundant active binding sites, and fast reaction kinetics, are potential candidates for arsenic removal. Here, some advanced nanomaterials that demonstrated their effectiveness for arsenic removal were presented.

8.6.1 METAL OXIDE NANOPARTICLES

Iron NPs were chosen as anodes in an electrochemical peroxidation process (ECP) to remove high concentrations of arsenic (1300–3000 mg/L) in synthetic and real wastewater from a copper pyrometallurgical industry (Gutiérrez et al., 2015). Operating parameters, including initial pH and treatment time, were varied between 2.0–6.5 and 30–180 min, respectively, to treat both As(III) and As(V) synthetic wastewater and real copper smelter wastewater. A great dependency of the oxidation state and pH of arsenic present in the solution on the removal efficiency was observed. A maximum removal rate of 62.4% and 99.7% was found for As(III) and As(V) synthetic wastewater at pH of 6.5 and 5.0, respectively. Whereas, real copper smelter wastewater, which was treated using iron NPs and carbon electrodes for the first time, achieved As removal rates of 89–96% in the pH range of 3.5–6.5, with the maximum removal obtained at pH 6.5. Similar removal trending observed for As(III) synthetic and real copper smelter wastewater suggested that majority arsenic was present in the As(III) oxidation state.

Besides pure metal oxide NPs, binary oxide NPs combine the merits of dual elements which play different roles in the removal of arsenic. Zhang et al. (2013) reported a novel efficient and low-cost adsorbent for arsenic removal, nanostructured Fe–Cu binary oxide, synthesized by a facile coprecipitation method. Surface characterization indicated that the two-line ferrihydrite-like binary oxide was poorly crystalline and aggregated with lots of nanograins (around 50 nm). The Cu:Fe molar ratio was varied and that of 1:2 was found to be most effective in removing both As(V) and As(III), with the maximum adsorption capacities of 82.7 and 122.3 mg/g at pH 7.0, respectively. The superior performance over most reported adsorbents was mainly a result of the high specific surface area (282 m^2/g) and a combination effect of the copper and iron oxides. XPS analysis suggested there is no transformation of As(III) to As(V) during the adsorption. The presence of phosphate, instead of sulfate and carbonate, significantly affects the arsenic removal especially at high concentrations. The binary oxide NPs can be readily regenerated by simple washing by NaOH solution and drying.

Different from Fe–Cu binary oxide NPs, Fe–Mn binary oxide participated in the transformation of As valence state during adsorption (Zhang et al., 2014). The respective role of Fe and Mn contents in arsenic removal was investigated via direct *in situ* arsenic speciation determination by x-ray absorption spectroscopy (XAS). X-ray absorption near edge structure results revealed that Mn existed in both +3 and +4 valence states, and oxidizing As(III) to As(V) was mainly contributed by MnO$_x$ ($1.5 < x < 2$) via a two-step pathway, that is, reduction of Mn(IV) to Mn(III) and subsequent Mn(III) to Mn(II). Whereas, the FeOOH content was responsible for adsorbing the formed As(V), but made little contribution to As(III) oxidation when the system was exposed to air. Inner-sphere bidentate binuclear corner-sharing complex with an As–M (M = Fe or Mn) interatomic distance of 3.22–3.24 Å was formed between As and binary oxide according to the extended x-ray absorption fine structure result. The high adsorption effectiveness, low cost, and environmental friendly nature of Fe–Mn binary made it suitable as an efficient oxidant of As(III) and a sorbent for As(V) in environmental remediation and water treatment.

8.6.2 NANOFIBERS

One-dimensional (1D) NFs produced by electrospinning have been a research hot spot for the past two decades due to their high specific area, porosity, and interconnecting pore structures, and have found themselves in many applications, such as environmental engineering, tissue engineering, energy storage, etc. (Feng et al., 2013). Electrospinning is a simple and versatile method, where a high electric field was applied on a polymer jet resulting in elongated and stretched NFs with controllable surface morphology and chemical composition (Huang et al., 2003). A typical electrospinning setup is shown in Figure 8.3.

Min et al. (2015) reported the successful fabrication of a chitosan-based electrospun nanofiber membrane (CS-ENM) with average fiber diameter of 129 nm. CS-ENM was examined for As(V) removal, and the effect of contact time, initial As(V) concentration, solution pH, and ionic strength were investigated. A fast adsorption kinetics with equilibrium time about 0.5 h was observed, while the maximum adsorption capacity of As(V) on CS-ENM reached 30.8 mg/g, which was higher than most of the reported chitosan adsorbent. The high adsorption capacity was attributed to the high surface area, large pore volume, interconnecting pore structure, and the presence of high affinity surface hydroxyl and amine groups, which was also evidenced by the XPS analysis. Solution pH also played a key role in As(V) adsorption onto CS-ENM, with higher adsorption capacities obtained at lower pH. The adsorbed As(V) formed outer-sphere surface complexes with CS-ENM as suggested by the ionic strength effect study.

Incorporation of metals, which provides active site for As binding, can also be easily achieved by electrospinning. Fe^{3+} immobilized poly(vinyl alcohol) (PVA) NFs with smooth morphology and diameter ranging from 600 to 800 nm were electrospun from a Fe^{3+}/PVA mixture and cross-linked under ammonia vapor (Mahanta and Valiyaveettil, 2013). Fe^{3+} ions coordinated with the hydroxyl groups of PVA and served as cationic binding sites for negatively charged arsenic anions as demonstrated by FTIR and XPS. With the increase in Fe^{3+} ions content, the glass transition temperature was also enhanced. The main advantage of the nanofiber (NF) composite compared over Fe^{3+} incorporated carbon particles was the absence of leachable materials, easy handling, and storing of fibrous mats. The maximum adsorption capacity for As(III) and As(V) on the NFs was found

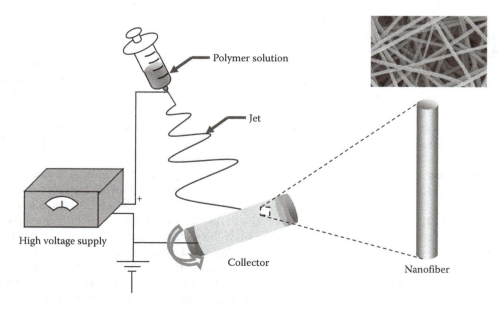

FIGURE 8.3 Schematic diagram of electropinning.

to be 67 and 36 mg/g, respectively. The presence of silicate anion reduced the extraction efficiency whereas humic acid had no significant interference on the adsorption.

Inorganic NFs prepared by electrospinning and calcination/acidic dissolution have also received great interest in water treatment applications. Vu et al. (2013) studied the phase effect of crystalline TiO_2 NFs on the adsorption of As(III). Different phase structures of TiO_2 NFs such as amorphous, anatase, etc. had significant impact on the As(III) adsorption rates and capacities. Among the various samples, amorphous TiO_2 NFs exhibited the highest As(III) adsorption capacity and take-up rate, which was mainly contributed by its higher surface area and porous volume. Phase-controlled fabrication of crystalline NFs by electrospinning demonstrated an effective way for arsenic removal from aqueous solutions.

Carbonaceous nanofibers (CNFs) produced by template-directed hydrothermal carbonization were also studied for the competitive sorption of As(V) and Cr(VI) (Cheng et al., 2016). The results showed that the maximum Langmuir sorption capacities of Cr(VI) and As(V) on CNFs in single-metal systems were 2.36 and 0.67 mmol/g, respectively. A greater affinity of CNFs to Cr(VI) than to As(V) in the binary As–Cr system, which was likely contributed by both the inner-sphere and outer-sphere surface complexation between Cr(VI) and CNFs, was in contrast to the electrostatic outer-sphere sorption of As(V) on CNFs.

8.6.3 Organics/Metal Oxide Nanocomposites

Organics/metal oxide nanocomposites have extended their usage in water treatment as photocatalysts, disinfectants, or adsorbents (Upadhyay et al., 2014). Recently, a novel zirconium-based NPs doped activated carbon fiber (ACF) prepared by the impregnation method was tested for the simultaneous removal of arsenic and natural organic matters (NOMs) (Zhao et al., 2016). ACF with high mechanical strength and specific surface area was chosen as a supporting matrix to avoid NPs aggregation in aqueous media, as well as to remove NOMs and various synthetic organic contaminants. The adsorption equilibrium was established within 30 h, while the optimal pH for As(V) adsorption was 3.0. The adsorption data were better described by the Langmuir isotherm with a maximum adsorption capacity of 21.7 mg As/g (pH 3.0). The presence of HA inhibited the uptake of As(V) to some extent, which was likely due to the blockages of the active adsorption sites on the sorbent by HA. The fixed-bed column filtration experiment demonstrated that the composite material could successfully produce 570.4 bed volumes meeting the MCL requirement of 10 μg/L when treating simulated arsenic-contaminated water with an initial concentration of 106 μg/L. The XPS analysis revealed that the As(V) adsorption was mainly through the ion-exchange reaction between hydrogen sulfate and arsenate ions.

Cross-linked anion exchangers (NS) with different pore size distributions were used as the hosts for confined growth of HFO NPs and to investigate the effect on the adsorption of As(V) (Li et al., 2016). As observed by TEM (transmission electron microscopy), the mean diameter of the confined HFO NPs reduced from 31.4 to 11.6 nm with the decrease in the average pore size of the NS hosts from 38.7 to 9.2 nm, whereas the density of active surface sites was increased as a result of the size-dependent effect. Via tailoring the pore size of the NS hosts, the adsorption capacity of As(V) could be improved from 24.2 to 31.6 mg/g, with the smallest pore size giving highest adsorption capacity. The adsorption kinetics were also slightly accelerated when pore size decreased. Besides, the enhanced adsorption of As(V) was observed over pH 3–10 for NS with smallest pore size, also in the presence of competing anions including chloride, sulfate, bicarbonate, nitrate, and phosphate. In addition, the fixed-bed working capacity increased from 2200 to 2950 BV (bed volumes) due to the size confinement effect, however, no adverse effect on As(V) desorption was observed.

Carboxylic graphene oxide decorated with akaganeite, β-FeOOH@GO-COOH nanocomposite, exhihibited more outstanding adsorption capability for both As(III) and As(V), compared to bare GO or iron oxides (Chen et al., 2015). The high adsorption capacities were mainly attributed to the surface complexation between arsenic and β-FeOOH@GO-COOH and electrostatic interaction as evidented by pH analysis. The nanocomposite was also effective to remove trace level of As(III) (100 μg/L) and As(V) (100 μg/L), with 100% and 97% removal rate after five succesive adsorption/ desorption cycles, respectively, and above 80% removal rate after 20 operation cycles. Furthermore, it is a promised candidate medium for preconcentration of ultra-trace inorganic arsenic with a detection limit of 29 ng/L for arsenate.

Although NPs can provide high specific area and reactivity, they usually suffered aggregation and difficulty in separation from aqueous solutions. Thus, extensive studies were focused on the development of magnetic NPs which can be easily isolated with external magnets, while the combination of low-cost biochar and magnetic materials seemed to be an attractive option for As removal (Wang et al., 2015). The magnetic biochar was produced by pyrolyzing hematite modified pinewood biomass. XRD examination confirmed the transformation form hematite to γ-Fe$_2$O$_3$, which possessed strong magnetic properties, during the pyrolysis. Adsorption ability was greatly improved due to the electrostatic interaction between As and the many sorption sites provided by γ-Fe$_2$O$_3$ particles on the carbon surface. This low-cost magnetic biochar can serve as an alternative remediation agent to mitigate the risk of As contamination.

REFERENCES

Andrianisa H. A., Ito A., Sasaki A., Aizawa J., Umita T. Biotransformation of arsenic species by activated sludge and removal of bio-oxidised arsenate from wastewater by coagulation with ferric chloride, *Water Res.*, 42, 2008, 4809–4817.

Azcue J. M., Nriagu, J. O. Arsenic: Historical perspectives. In: J.O. Nriagu (ed.), *Arsenic in the Environment. Part 1: Cycling and Characterization*, John Wiley & Sons, Inc: New York, pp. 1–15, 1994.

ATSDR. *Toxicological Profile for Arsenic, Draft for Public Comment*, Agency for Toxic Substances and Disease Registry: Atlanta, GA, 2005, 29–182.

Badruzzaman M., Westerhoff P., Knappe D. R. U. Intraparticle diffusion and adsorption of arsenate onto granular ferric hydroxide (GFH), *Water Res.*, 38, 2004, 4002–4012.

Bagga A., Chellam S., Clifford D. A. Evaluation of iron chemical coagulation and electrocoagulation pretreatment for surface water microfiltration, *J. Membr. Sci.*, 309, 2008, 82–93.

Balaji T., Yokoyama T., Matsunaga H. Adsorption and removal of As(V) and As(III) using Zr-loaded lysine diacetic acid chelating resin, *Chemosphere*, 59, 2005, 1169–1174.

Balasubramanian N., Kojima T., Basha C. A. Srinivasakannan C. Removal of arsenic from aqueous solution using electrocoagulation, *J. Hazard. Mater.*, 167, 2009, 966–969.

Balasubramanian N., Madhavan K. Arsenic removal from industrial effluent through electrocoagulation, *Chem. Eng. Technol.*, 24, 2001, 519–521.

Bansal R. P., Donnet J.-P., Stoeckli F. *Active Carbon.*, Marcel Dekker: New York, 1988.

Basha C. A., Selvi S. J., Ramasamy E., Chellammal S. Removal of arsenic and sulphate from the copper smelting industrial effluent, *Chem. Eng. J.*, 141, 2008, 89–98.

Bissen, M., Frimmel F. H. Arsenic—A review. Part I: Occurrence, toxicity, speciation, mobility, *Acta Hydrochim. Hydrobiol.*, 31, 2003, 9–18.

Biswas B. K., Inoue J. I., Inoue K., Ghimire K. N., Harada H., Ohto K., Kawakita H. Adsorptive removal of As(V) and As(III) from water by a Zr(IV)-loaded orange waste gel, *J. Hazard. Mater.*, 154, 2008, 1066–1074.

Bolin N. J., Sundkvist J. E. Two-stage precipitation process of iron and arsenic from acid leaching solutions, *Trans. Nonferrous Met. Soc. China*, 18, 2008, 1513–1517.

Brandhuber P., Amy G. Alternative methods for membrane filtration of arsenic from drinking water, *Desalination*, 117, 1998, 1–10.

Brandhuber P., Amy G. Arsenic removal by a charged ultrafiltration membrane—Influences of membrane operating conditions and water quality on arsenic rejection, *Desalination*, 140, 2001, 1–14.

Casiot C., Pedron V., Bruneel O., Duran R., Personne J. C., Grapin G., Drakides C., Elbaz-Poulichet F. A new bacterial strain mediating As oxidation in the Fe-rich biofilm naturally growing in a groundwater Fe treatment pilot unit, *Chemosphere*, 64, 2006, 492–496.

Chen M.-L., Sun Y., Huo C.-B., Liu C., Wang J.-H. Akaganeite decorated graphene oxide composite for arsenic adsorption/removal and its proconcentration at ultra-trace level, *Chemosphere*, 130, 2015, 52–58.

Cheng W., Ding C., Wang X., Wu Z., Sun Y., Yu S., Hayat T. Wang X. Competitive sorption of As(V) and Cr(VI) on carbonaceous nanofibers, *Chem. Eng. J.*, 293, 2016, 311–318.

Chiu V. Q., Hering J. G. Arsenic adsorption and oxidation at manganite surfaces, 1, Method for simultaneous determination of adsorbed and dissolved arsenic species, *Environ. Sci. Technol.*, 34, 2000, 2029–2034.

Choong T. S. Y., Chuah T. G., Robiah Y., Koay F. L. G., Azni I. Arsenic toxicity, health hazards and removal techniques from water: An overview, *Desalination*, 217, 2007, 139–166.

Chuang C. L., Fan M., Xu M., Brown R. C., Sung S., Saha B., Huang C. P. Adsorption of arsenic(V) by activated carbon prepared from oat hulls, *Chemosphere*, 61, 2005, 478–483.

Dambies L., Roze A., Guibal E. As(V) sorption on molybdate impregnated chitosan gel beads (MICB), *Adv. Chitin Sci.*, 4, 2000, 302–309.

Deschamps E., Ciminelli V. S. T., Holl W. H. Removal of As(III) and As(V) from water using a natural Fe and Mn enriched sample, *Water Res.*, 39, 2005, 5212–5220.

Driehaus W., Seith R., Jekel M. Oxidation of arsenate(III) with manganese oxides in water treatment, *Water Res.*, 29, 1995, 297–305.

Dutta P. K., Pehkonen S. O., Sharma V. K., Ray A. K. Photocatalytic oxidation of arsenic(III): Evidence of hydroxyl radicals, *Environ. Sci. Technol.*, 39, 2005, 1827–1834.

Eguez H. E., Cho E. H. Adsorption of arsenic on activated charcoal, *J. Met.*, 39, 1987, 38–41.

Elson C. M., Davies D. H., Hayes E. R. Removal of arsenic from contaminated drinking water by a chitosan/chitin mixture, *Water Res.*, 14, 1980, 1307–1311.

Emett M. T., Khoe G. H. Photochemical oxidation of arsenic by oxygen and iron in acidic solutions, *Water Res.*, 35, 2001, 649–656.

Feng C., Khulbe K. C., Matsuura T., Tabe S., Ismail A. F. Preparation and characterization of electro-spun nanofiber membranes and their possible applications in water treatment, *Sep. Purif. Technol.*, 102, 2013, 118–135.

Ferella F., Prisciandaro M., Michelis I. D., Veglio F. Removal of heavy metals by surfactant-enhanced ultrafiltration from wastewaters, *Desalination*, 207, 2007, 125–133.

Ferguson M. A., Hoffmann M. R., Hering J. G. TiO_2-photocatalyzed As(III) oxidation in aqueous suspensions: Reaction kinetics and effects of adsorption, *Environ. Sci. Technol.*, 39, 2005, 1880–1886.

Ferguson M. A., Hering J. G. TiO_2-photocatalyzed As(III) oxidation in a fixed-bed, flow-through reactor, *Environ. Sci. Technol.*, 40, 2006, 4261–4267.

Fogarassy E., Galambos I., Bekassy-Molnar E., Vatai G. Treatment of high arsenic content wastewater by membrane filtration, *Desalination*, 240, 2009, 270–273.

Frankenberger W. T. *Environmental Chemistry of Arsenic*, Marcel Dekker, Inc.: New York, 2002.

Gomes J. A. G., Daida P., Kesmez M. et al. Arsenic removal by electrocoagulation using combined Al–Fe electrode system and characterization of products, *J. Hazard. Mater.*, 139, 2007, 220–231.

Guo X., Chen F. Removal of arsenic by bead cellulose loaded with iron oxyhydroxide from groundwater, *Environ. Sci. Technol.*, 39, 2005, 6808–6818.

Gutiérrez C., Hansen H. K., Núñez P., Valdés E. Electrochemical peroxidation using iron nanoparticles to remove arsenic from copper smelter wastewater, *Electrochim. Acta*, 181, 2015, 228–232.

Hansen H. K., Nuñez P., Raboy D., Schippacasse I., Grandon R. Electrocoagulation in wastewater containing arsenic: Comparing different process designs, *Electrochim. Acta*, 52, 2007, 3464–3470.

Hansen H. K., Nunez P., Grandon R. Electrocoagulation as a remediation tool for wastewaters containing arsenic, *Miner. Eng.*, 19, 2006, 521–524.

Hering J. G., Chen P. Y., Wilkie J. A., Elimelech M. Arsenic removal from drinking water during coagulation, *J. Environ. Eng.*, 123, 1997, 800–807.

Hsieh L. H. C., Weng Y. H., Huang C. P., Li K. C. Removal of arsenic from groundwater by electro-ultrafiltration, *Desalination*, 234, 2008, 402–408.

Huang C. P., Fu P. L. K. Treatment of arsenic(V)-containing water by the activated carbon process, *J. Water Pollut. Control Fed.*, 56, 1984, 233–242.

Huang Z.-M., Zhang Y. Z., Kotaki M., Ramakrishna S. A review on polymer nanofibers by electrospinning and their applications in nanocomposites, *Compos. Sci. Technol.*, 63, 2003, 2223–2253.

Hug S. J., Canonica L., Wegelin M., Gechter D., Von Gunten U. Solar oxidation and removal of arsenic at circumneutral pH in iron containing waters, *Environ. Sci. Technol.*, 35, 2001, 2114–2121.

Hug S. J., Leupin O. Iron-catalyzed oxidation of arsenic(III) by oxygen and by hydrogen peroxide: pH-dependent formation of oxidants in the Fenton reaction. *Environ. Sci. Technol.*, 37, 2003, 2734–2742.

Jain C. K., Ali I. Arsenic: Occurrence, toxicity and speciation techniques. *Water Res.*, 34, 2000, 4304–4312.

Jang J. H., Dempsey B. A. Coadsorption of arsenic(III) and arsenic(V) onto hydrous ferric oxide: Effects on abiotic oxidation of arsenic(III), extraction efficiency, and model accuracy. *Environ. Sci. Technol.*, 42, 2008, 2893–2898.

Jubinka L., Rajakovic V., Mitrovic M. M. Arsenic removal from water by chemisorption filters, *Environ. Pollut.*, 75, 1992, 279–287.

Judit F., Hideg M. Application of ZW-1000 membranes for arsenic removal from water sources, *Desalination*, 162, 2004, 75–83.

Katsoyiannis I. A., Ruettimann T., Hug S. J. pH dependence of Fenton reagent generation and As(III) oxidation and removal by corrosion of zero valent iron in aerated water. *Environ. Sci. Technol.*, 42, 2008, 7424–7430.

Katsoyiannis I. A., Zouboulis A. I., Jekel M. Kinetics of bacterial As(III) oxidation and subsequent As(V) removal by sorption onto biogenic manganese oxides during groundwater treatment, *Ind. Eng. Chem. Res.*, 43, 2004, 486–493.

Kim M. J., Nriagu J. Oxidation of arsenite in groundwater using ozone and oxygen, *Sci. Total Environ.*, 247, 2000, 71–79.

Kumar P. R., Chaudhari S., Khilar K. C., Mahajan S. P. Removal of arsenic from water by electrocoagulation, *Chemosphere* 55, 2004, 1245–1252.

Kundu S., Gupta A. K. Adsorption characteristics of As(III) from aqueous solution on iron oxide coated cement (IOCC), *J. Hazard. Mater.*, 142, 2007, 97–104.

Lakshmipathiraj P., Narasimhan B. R. V., Prabhakar S., Bhaskar Raju G. Adsorption studies of arsenic on Mn-substituted iron oxyhydroxide, *J. Colloid Interface Sci.*, 304, 2006, 317–322.

Lee Y., Um I. H., Yoon J. Arsenic(III) oxidation by iron(VI) (ferrate) and subsequent removal of arsenic(V) by iron(III) coagulation, *Environ. Sci. Technol.*, 37, 2003, 5750–5756.

Lee Y. H., Um I. H., Yoon J. Y. Arsenic(III) oxidation by iron(VI) (ferrate) and subsequent removal of arsenic(V) by iron(III) coagulation, *Environ. Sci. Technol.*, 37, 2003, 5750–5756.

Lenoble V., Deluchat V., Serpaud B., Bollinger J. C. Arsenite oxidation and arsenate determination by the molybdene blue method, *Talanta*, 61, 2003, 267–276.

Lenoble W., Laclautre C., Serpaud B., Deluchat V., Bollinger J. C. Bacterial immobilization and oxidation of arsenic in acid mine drainage, *Water Res.*, 37, 2003, 2929–2936.

Lenoble W., Laclautre C., Serpaud B., Deluchat V., Bollinger J. C. As(V) retention and As(III) simultaneous oxidation and removal on a MnO_2-loaded polystyrene resin, *Sci. Total Environ.*, 326, 2004, 197–207.

Leupin O. X., Hug S. J. Oxidation and removal of arsenic(III) from aerated groundwater by filtration through sand and zero-valent iron, *Water Res.*, 39, 2005, 1729–1740.

Li H., Shan C., Zhang Y., Cai J., Zhang W., Pan B. Arsenate adsorption by hydrous ferric oxide nanoparticles embedded in cross-linked anion exchanger: Effect of the host pore structure, *ACS Appl. Mater. Interfaces*, 8, 2016, 3012–3020.

Li L., Quinlivan P. A., Knappe D. R. U. Effects of activated carbon surface chemistry and pore structure on the adsorption of organic contaminants from aqueous solution, *Carbon*, 40, 2002, 2085–2100.

Li N., Fan M. H., van Leeuwen J., Saha B., Yang H. Q., Huang C. P. Oxidation of As(III) by potassium permanganate, *J. Environ. Sci.*, 19, 2007, 783–786.

Lievremont D., N'negue M. A., Behra P., Lett M.-C. Biological oxidation of arsenite: Batch reactor experiments in presence of kutnahorite and chabazite, *Chemosphere*, 51, 2003, 419–428.

Lim M. S., Yeo I. W., Roh Y., Lee K. K., Jung M. C. Arsenic reduction and precipitation by *Shewanella* sp.: Batch and column tests, *Geosci. J.*, 12, 2008, 151–157.

Lim S. F., Chen J. P. Synthesis of an innovative calcium-alginate magnetic sorbent for removal of multiple contaminants, *Appl. Surf. Sci.*, 253, 2007, 5772–5775.

Lim S. F., Zheng Y. M., Chen J. P. Organic arsenic adsorption onto a magnetic sorbent, *Langmuir*, 25, 2009a, 4973–4978.

Lim S. F., Zheng Y. M., Chen J. P. Uptake of arsenate by an alginate-encapsulated magnetic sorbent: Process performance and characterization of adsorption chemistry, *J. Colloid Interface Sci.*, 333, 2009b, 33–39.

Mahanta N., Valiyaveettil S. Functionalized poly(vinyl alcohol) based nanofibers for the removal of arsenic from water, *RSC Adv.*, 3, 2013, 2776–2783.

Maldonado-Reyes A., Monetro-Ocampo C., Solorza-Feria O. Remediation of drinking water contaminated with arsenic by the electro-removal process using different metal electrodes, *J. Environ. Monit.*, 9, 2007, 1241–1247.

Mandal B. K., Suzuki K. T. Arsenic round the world: A review, *Talanta*, 58, 2002, 201–235.

Mantell C. L. *Carbon and Graphite Handbook*, John Wiley & Sons, Inc.: New York, 1968.

Matschullat J. Arsenic in the geosphere—A review, *Sci. Total Environ.*, 249, 2000, 297–312.

Melamed D. Monitoring arsenic in the environment: A review of science and technologies with the potential for field measurements, *Anal. Chim. Acta*, 532, 2005, 1–13.

Meng X. G., Korfiatis G. P., Christodoulatos C., Bang S. Treatment of arsenic in Bangladesh well water using a household co-precipitation and filtration system, *Water Res.*, 35, 2001, 2805–2810.

Min L.-L., Yuan Z.-H., Zhong L.-B., Liu Q., Wu R.-X., Zheng Y.-M. Preparation of chitosan based electrospun nanofiber membrane and its adsorptive removal of arsenate from aqueous solution, *Chem. Eng. J.*, 267, 2015, 132–141.

Mohan, D., Pittman C. U. Arsenic removal from water/wastewater using adsorbents—A critical review, *J. Hazard. Mater.*, 142, 2007, 1–53.

Munoz J. A., Gonzalo A. Valiente M., Arsenic adsorption by Fe(III)-loaded open-celled cellulose sponge. Thermodynamic and selectivity aspects, *Environ. Sci. Technol.*, 36, 2002, 3405–3411.

Munoz J. A., Gonzalo A., Valiente M. Kinetic and dynamic aspects of arsenic adsorption by Fe(III)-loaded sponge, *J Solution Chem.*, 37, 2008, 553–565.

Natale F. D., Erto A., Lancia A., Musmarra D. Experimental and modelling analysis of As(V) ions adsorption on granular activated carbon, *Water Res.*, 42, 2008, 2007–2016.

Navarro P., Alguacil F. J. Adsorption of antimony and arsenic from a copper electrorefining solution onto activated carbon, *Hydrometallurgy*, 66, 2002, 101–105

Neppolian B., Celik E., Choi H., Photochemical oxidation of arsenic(III) to arsenic(V) using peroxydisulfate ions as an oxidizing agent, *Environ. Sci. Technol.*, 42, 2008, 6179–6184.

Newcombe R. L., Hart B. K., Moller G., Arsenic removal from water by moving bed active filtration, *J. Environ. Eng.*, 132, 2006, 5–12.

Nguyen V. T., Vigneswaran S., Ngo H. H., Shon H. K., Kandasamy J., Arsenic removal by a membrane hydrid filtration system, *Desalination*, 236, 2009, 363–369.

Nishimura T., Umetsu Y. Oxidative precipitation of arsenic(III) with manganese(II) and iron(II) in dilute acidic solution by ozone, *Hydrometallurgy*, 62, 2001, 83–92.

Ocinski D., Jacukowicz-Sobala I., Mazur P., Raczyk J., Kocio.ek-Balawejder E. Water treatment residuals containing iron and manganese oxides for arsenic removal from water—Characterization of physicochemical properties and adsorption studies, *Chem. Eng. J.*, 294, 2016, 210–221.

Parga J. R., Cocke D. L., Valverde V., Gomes J. A. G., Kesmez M., Moreno H., Weir M., Mencer D. Characterization of electrocoagulation for removal of chromium and arsenic, *Chem. Eng. Technol.*, 28, 2005, 605–612.

Pena M. E., Korfiatis G. P., Patel M., Lippincott L., Meng X. Adsorption of As(V) and As(III) by nanocrystalline titanium dioxide, *Water Res.*, 11, 2005, 2327–2337.

Peräiniemi S., Hannonen S., Mustalahti H., Ahlgrén M. Zirconium-loaded activated charcoal as an adsorbent for arsenic, selenium and mercury, *Fresenius' J. Anal. Chem.*, 349, 1994, 510–515.

Pokhrel D., Viraraghavan T. Biological filtration for removal of arsenic from drinking water, *J. Environ. Manage.*, 90, 2009, 1956–1961.

Qu D., Wang J., Hou D. Y., Luan Z. K., Fan B., Zhao C. W. Experimental study of arsenic removal by direct contact membrane distillation, *J. Hazard. Mater.*, 163, 2009, 874–879.

Qu X., Alvarez P. J. J., Li Q. Applications of nanotechnology in water and wastewater treatment, *Water Res.*, 47, 2013, 3931–3946.

Rahman F. A., Allan D. L., Rosen C. J., Sadowsky M. J. Arsenic availability from chromated copper arsenate (CCA)-treated wood. *J. Environ. Qual.*, 33, 2004, 173–180.

Saha J. C., Dikshit A. K., Bandyopadhyay M., Saha, K. C. A review of arsenic poisoning and its effects on human health, *Crit. Rev. Environ. Sci. Technol.*, 29, 1999, 281–313.

Schmidt G. T., Vlasova N., Zuzaan D., Kersten M., Daus B. Adsorption mechanism of arsenate by zirconyl-functionalized activated carbon, *J. Colloid Interface Sci.*, 317, 2008, 228–234.

Singh T. S., Pant K. K. Equilibrium kinetics and thermodynamic studies for adsorption of As(III) on activated alumina, *Sep. Purif. Technol.*, 36, 2004, 139–147.

Smedley P. L., Kinniburgh D. G. A review of the source, behaviour and distribution of arsenic in natural waters, *Appl. Geochem.*, 17, 2002, 517–568.

Solozhenkin P. M., Zouboulis A. I., Katsoyiannis I. A. Removal of arsenic compounds by chemisorption filtration, *J. Min. Sci.*, 43, 2007a, 212–220.

Solozhenkin P. M., Zouboulis A. I., Katsoyiannis I. A. Removal of arsenic compounds from waste water by chemisorption filtration, *Theor. Found. Chem. Eng.*, 41, 2007b, 772–779.

Song S., Lopez-Valdivieso A., Hernandez-Campos D. J., Peng C., Monroy-Fernandez M. G. Razo-Soto I. Arsenic removal from high-arsenic water by enhanced coagulation with ferric ions and coarse calcite, *Water Res.*, 40, 2006, 364–372.

Sun L. H., Liu R. P., Xia S. J., Yang Y. L., Li G. B. Enhanced As(III) removal with permanganate oxidation, ferric chloride precipitation and sand filtration as pretreatment of ultrafiltration, *Desalination*, 243, 2009, 122–131.

Suzuki T. M., Bomani J. O., Matsunaga H., Yokoyama T. Preparation of porous resin loaded with crystalline hydrous zirconium oxide and its application to the removal of arsenic, *React. Funct. Polym.*, 43, 2000, 165–172.

Taylor, S. R., McLennan, S. M. *The Continental Crust: Its Composition and Evolution*, Blackwell Scientific: Oxford, 1985.

Tokunaga S., Wasay S. A., Park S. W. Removal of arsenic(V) ion from aqueous solutions by lanthanum compounds, *Water Sci. Technol.*, 35, 1997, 71–78.

Upadhyay R. K., Soin N., Roy S. S. Role of graphene/metal oxide composites as photocatalysts, adsorbents and disinfectants in water treatment: A review, *RSC Adv.*, 4, 2014, 3823–3851.

Voegelin A, Hug S. J. Catalyzed oxidation of arsenic(III) by hydrogen peroxide on the surface of ferrihydrite: An in situ ATR-FTIR study, *Environ. Sci. Technol.*, 37, 2003, 972–978.

Vu D., Li X., Li Z., Wang C. Phase-structure effects of electrospun TiO_2 nanofiber membranes on as(III) adsorption, *J. Chem. Eng. Data*, 58, 2013, 71–77.

Wang S., Gao B., Zimmerman A. R., Li Y., Ma L., Harris W. G., Migliaccio K. W. Removal of arsenic by magnetic biochar prepared from pinewood and natural hematite, *Bioresour. Technol.*, 175, 2015, 391–395.

Weeger W., Lievremont D., Perret M., Lagarde F., Hubert J.-C., Leroy M., Lett M.-C. Oxidation of arsenite to arsenate by a bacterium isolated from an aquatic environment, *Biometals*, 12, 1999, 141–149.

WHO. 2001. *Environmental Health Criteria 224: Arsenic Compounds*, 2nd edition, World Health Organisation: Geneva.

Wickramasinghe S. R., Han B., Zimbron J., Shen Z., Karim M. N. Arsenic removal by coagulation and filtration: Comparison of groundwaters from the United States and Bangladesh, *Desalination*, 169, 2004, 231–244.

Wilson R. Summary of the acute and chronic effects of arsenic and the extent of the world arsenic catastrophe, 2009. http://users.physics.harvard.edu/~wilson/arsenic/references/arsenic_project_introduction.html

Xia S. J., Dong B. Z., Zhang Q. L., Xu B., Gao N. Y., Causseranda C. Study of arsenic removal by nanofiltration and its application in China, *Desalination*, 204, 2007, 374–379.

Xu, T. L., Cai Y., O'Shea K. Adsorption and photocatalyzed oxidation of methylated arsenic species in TiO_2 suspensions, *Environ. Sci. Technol.*, 41, 2007, 5471–5477.

Yang L., Wu S.N., Chen J. P. Modification of activated carbon by polyaniline for enhanced adsorption of aqueous arsenate, *Ind. Eng. Chem. Res.*, 46, 2007, 2133–2140.

Yeo J., Choi W. Iodide-mediated photooxidation of arsenite under 254 nm irradiation, *Environ. Sci. Technol.*, 43, 2009, 3784–3788.

Yoon S. H., Lee J. H., Oh S. E., Yang J. E. Photochemical oxidation of As(III) by vacuum-UV lamp irradiation, *Water Res.*, 42, 2008, 3455–3463.

Zaw, M., Emett M. T., Arsenic removal from water using advanced oxidation processes, *Toxicol. Lett.*, 133, 2002, 113–118.

Zeng H., Arashiro M., Giammar D. E. Effects of water chemistry and flow rate on arsenate removal by adsorption to an iron oxide-based sorbent, *Water Res.*, 42, 2008, 4629–4636.

Zhang F. S., Itoh. H. Photocatalytic oxidation and removal of arsenite from water using slag-iron oxide-TiO_2 adsorbent, *Chemosphere*, 65, 2006, 125–131.

Zhang G. S., Liu F. D., Liu H.J., Qu J. H., Liu R. P. Respective role of Fe and Mn oxide contents for arsenic sorption in iron and manganese binary oxide: An x-ray absorption spectroscopy investigation, *Environ. Sci. Technol.*, 48, 2014, 10316–10322.

Zhang G. S., Qu J. H., Liu H. J., Liu R. P., Li G. T. Removal mechanism of As(III) by a novel Fe–Mn binary oxide adsorbent: Oxidation and sorption, *Environ. Sci. Technol.*, 41, 2007a, 4613–4619.

Zhang G. S., Qu J. H., Liu H. J., Liu R. P., Wu R. C. Preparation and evaluation of a novel Fe–Mn binary oxide adsorbent for effective arsenite removal, *Water Res.*, 41, 2007b, 1921–1928.

Zhang G. S., Ren Z. M., Zhang X. W., Chen J. Nanostructured iron(III)–copper(II) binary oxide: A novel adsorbent for enhanced arsenic removal from aqueous solutions, *Water Res.*, 47, 2013, 4022–4031.

Zhao D., Yu Y., Chen J. P. Fabrication and testing of zirconium-based nanoparticle-doped activated carbon fiber for enhanced arsenic removal in water, *RSC Adv.*, 6, 2016, 27020–27030.

Zheng Y. M., Yu L., Chen J. P. Removal of methylated arsenic using a nanostructured zirconia-based sorbent: Process performance and adsorption chemistry, *J. Colloid Interface Sci.*, 367, 2012, 362–369.

Zheng Y.-M., Lim S.-F., Chen J. P. Preparation and characterization of zirconium-based magnetic sorbent for arsenate removal, *J. Colloid Interface Sci.*, 338, 2009, 22–29.

Zouboulis A., Katsoyiannis I. Removal of arsenates from contaminated water by coagulation-direct filtration, *Sep. Sci. Technol.*, 37, 2002, 2859–2873.

9 Simultaneous Removal of Chromium and Arsenate

A Case Study Using Ferrous Iron

Xiaohong Guan and Haoran Dong

CONTENTS

ABSTRACT

Chromium and arsenic have been identified as cocontaminants in wastes from wood preservative manufacture, paint and ink manufactures, petroleum refineries, as well as some municipal wastewaters. Inadequate storage and improper disposal practices of chromium and arsenic have caused many incidences of soil and groundwater contamination in industrialized areas. Both chromium and arsenic represent potential threats to the environment, human health, and animal health due to their carcinogenic and toxicological effects. Hexavalent chromium Cr(VI) and arsenic have been considered as important priority pollutants worldwide owing to numerous health problems arising from groundwater contaminated by these two pollutants. Therefore, the World Health Organization (WHO) has established a provisional guideline of 10 µg/L for arsenic and 50 µg/L for Cr(VI) in drinking water. Recent public concern regarding arsenic and Cr(VI) in drinking water has promoted the investigation of treatment technologies with the potential to remove them simultaneously to levels well below the drinking water maximum contaminant level.

In this chapter, a case study of simultaneous removal of chromium and arsenate using Fe(II) is illustrated in detail. The feasibility and mechanisms of simultaneous removal of Cr(VI) and As(V) by Fe(II) were investigated. The influence of various parameters (e.g., pH, Fe(II) dosages and initial Cr(VI)/As(V) ratios) and the individual and combined influences of various geochemical constituents (e.g., calcium, phosphate, silicate, and humic acid) on the simultaneous removal of chromium and arsenate were also studied. The results indicate that Fe(II) is very effective for simultaneous removal of chromium and As(V) under neutral conditions. Chromium removal by Fe(II) is controlled by both the rate of Cr(VI) reduction by Fe(II) and the solubility of $Fe_{0.75}Cr_{0.25}(OH)_3$ at pH 4.0–6.0, but by the extent of Cr(VI) reduction under alkaline conditions under oxic conditions. The presence of As(V) resulted in a decrease in chromium removal by Fe(II) under neutral and alkaline conditions as a result of the depression in the Cr(VI) reduction by Fe(II) and inhibition of the $Fe_{0.75}Cr_{0.25}(OH)_3$ and FeOOH precipitation by $HAsO_4^{2-}$. As(V) removal by Fe(II) alone was trivial but was improved significantly at pH 4.0–9.0 due to the presence of Cr(VI). It was the oxidative property of Cr(VI) that resulted in the oxidization of Fe(II) to Fe(III) concomitantly facilitating the removal of As(V). As(V) was removed by both adsorption and coprecipitation with $Fe_{0.75}Cr_{0.25}(OH)_3$ and FeOOH precipitates. The presence of PO_4^{3-}, humic acid (HA), or SiO_3^{2-} affects chromium removal by Fe(II) through the following three routes: increase Cr(VI) reduction by Fe(II) at pH < 5.0, inhibit the precipitation of newly formed Cr(III), and decrease the amount of Cr(VI) reduced by Fe(II) under neutral and alkaline conditions. They exert influences on arsenate removal via two ways: compete for adsorption sites and depress the precipitation of $Fe_{0.75}Cr_{0.25}(OH)_3$. Singly present Ca^{2+} ions show negligible effect on chromium removal throughout the pH range 4.0–10.0, yet notably increase arsenate removal at pH > 7.0. The presence of Ca^{2+} promotes the aggregation of colloidal Cr(III)/Fe(III)–anion complexes, attenuating the detrimental impacts of anions on chromium removal under alkaline conditions. As(V) removal is increased correspondingly, but the degree of enhancement varies with respect to the competitive capability of the respective anion.

9.1 INTRODUCTION

9.1.1 CO-OCCURRENCE OF CR(VI) AND AS(V)

Chromium and arsenic can coexist in groundwater due to dissolution of natural minerals, leakage from landfill sites, or discharge of improperly treated wastewater (Agrafioti et al., 2014; Mandal et al., 2015; Poguberović et al., 2016). One major source of subsurface contamination associated with copresent Cr(VI) and As(V) is wood preservation industry because of the wide usage of chromated copper arsenic (CCA), which has been extensively used as wood preservatives for more than a half century (Gress et al., 2015; Ohgami et al., 2015). Timbers are treated with CCA preservatives to prevent decay by wood-boring crustaceans, molluscs, and fungi. Three CCA formulations with different compositions of Cr, Cu, and As have been developed while type C is the most commercially popular one (Table 9.1) (Cooper, 1994; Ohgami et al., 2015). The percentages of Cr, Cu, and As are varied in the three types of CCA preservative. In the United States, approximately 65.3 million kg of CCA is consumed annually for wood preservation (Hingston et al., 2001). Leaching and contamination of CCA has raised the public health concern due to the toxicity of Cu, Cr, and As (Ferrarini et al., 2016; Gress et al., 2016). These three metals have been listed as priority pollutants by the USEPA (United States Environmental Protection Agency).

Numerous sites have been reported to be contaminated by CCA, which are usually caused by accidental spill and leaching from CCA-treated wood in the wood processing industry (Bhattacharya et al., 2002; Gress et al., 2015), and by the leachate from discarded CCA-treated wood (Townsend et al., 2005). Most Cu in the CCA solution would be retained in soil by adsorption and precipitation (Zagury et al., 2003; Greven et al., 2007). Cu(II) could be strongly adsorbed by soil minerals such as iron oxides by forming inner-sphere complexes. Moreover, Cu(II) would be precipitated at neutral pH. However, Cr(VI) and As(V) are relatively mobile in soil and hence contaminate the groundwater (Greven et al., 2007). Owing to their toxicity, carcinogenicity, and high mobility, Cr(VI) and As(V) in groundwater threaten the public health especially in those countries and regions relying on groundwater as a drinking water source, and remediation options are called for in coping with them (Robinson et al., 2004; Agrafioti et al., 2014).

9.1.2 TOXICITY OF CR(VI)

Chromium has two stable oxidation states in natural environments: hexavalent chromium, Cr(VI), and trivalent chromium, Cr(III). The toxicity, aqueous concentration, and mobility of chromium in different geological environments are dependent on its oxidation state (Rai et al., 1989; Kim and Kang, 2016). The Eh–pH diagram for chromium, shown in Figure 9.1, indicates the oxidation

TABLE 9.1
CCA Formulations (Oxide Basis)

Type	% by Mass		
	CuO	CrO$_3$	As$_2$O$_5$
CCA-A	18.1	65.5	16.4
CCA-B	19.6	35.5	45.1
CCA-C	18.5	47.5	34.0

Source: Cooper, P.A. Leaching of CCA: Is it a problem? In: *Anonymous Environmental Considerations in the Manufacture, Use and Disposal of Pressure-Treated Wood.* Forest Products Society, Madison, WI, 1994.

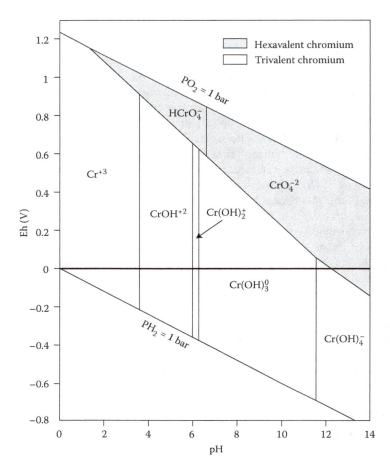

FIGURE 9.1 Eh–pH diagram for chromium. (Adapted from USEPA. *Technologies and Costs for Removal of Arsenic from Drinking Water.* EPA/815/R-00/028, USEPA; *In Situ Treatment of Soil and Ground Water Contaminated with Chromium, Technical Resource Guide.* EPA/625/R-00/005, USEPA, 2000; Palmer and Wittbrodt, 1991.)

states and chemical forms of the chromium species which exist within specific Eh and pH ranges (USEPA, 2000).

Chromium can be mobilized as stable Cr(VI) oxyanion species under oxidizing conditions, but forms cationic Cr(III) species in reducing environments and hence behaves like other trace cations (i.e., it is relatively immobile at near-neutral pH values). The main aqueous Cr(III) species are Cr^{3+}, $Cr(OH)_2^+$, $Cr(OH)_3^0$, and $Cr(OH)_4^-$. The presence, concentration, and forms of Cr(III) in a given environment depend on different chemical and physical processes, such as hydrolysis, complexation, redox reaction, and adsorption (Kotaś and Stasicka, 2000; Yirsaw et al., 2016). Cr(III) behaves as a typical "hard" Lewis acid and readily forms complexes with a variety of ligands including hydroxyl, sulfate, ammonium, cyanide, sulfocyanide, fluoride, chloride, as well as natural and synthetic organic ligands (Richard and Bourg, 1991; Chen et al., 2015). If the complexation with these ligands can be neglected, under normal redox and pH conditions, Cr(III) is removed from the solution as $Cr(OH)_3$ whose solubility is very low between pH 6 and 10.5 (Rai et al., 1987), or with the presence of Fe^{3+}, in the form of $(Cr_x, Fe_{1-x})(OH)_3$, (where x is the mole fraction of Cr) (Sass and Rai, 1987). The redox potential of the Cr(VI)/Cr(III) couple is high enough, so that only a few oxidants are present in the natural systems capable of oxidizing Cr(III) to Cr(VI). These oxidants include dissolved oxygen (DO) and manganese oxides (Rai et al., 1989). The oxidation of Cr(III) by manganese oxides is more rapid than by DO (Schroeder and Lee, 1975; Bartlett and James, 1979). Generally, Cr(III) would not be transported in natural systems as Cr(III) would precipitate as $Cr(OH)_3$ in neutral to alkaline pH range while Cr(III) tends to be adsorbed onto mineral surfaces in slightly acidic to neutral pH. Therefore, dissolved concentration of Cr(III) is maintained at low levels in natural water. The dissolved Cr(III) concentration can become high under acidic conditions (pH < 5) as it can be more mobile.

Compared with Cr(III), Cr(VI) is more toxic to bacteria, plants, and animals, and more soluble, mobile, and bioavailable (Chen et al., 2015; Kim and Kang, 2016). Cr(VI) mainly occurs in the form of $HCr_2O_7^-$, $Cr_2O_7^{2-}$, H_2CrO_4, $HCrO_4^-$, and CrO_4^{2-}. The relative proportions of species depend on both pH and total Cr(VI). $Cr_2O_7^{2-}$, H_2CrO_4, $HCrO_4^-$, and CrO_4^{2-} are predominant within the normal pH range in natural waters. Cr(VI) is a strongly oxidizing agent and readily reduced to Cr(III) (Du et al., 2012; Dong et al., 2016). It reacts with numerous reducing agents commonly found in the environment. Cr(VI) can be reduced in seconds by aqueous Fe^{2+} and in hours to days by Fe^{2+}-bearing material and sulfides (Eary and Rai, 1988; Patterson et al., 1997; Sedlak and Chan, 1997). Cr(VI) can also be reduced by organic matters (Goodgame et al., 1984; Stollenwerk and Grove, 1985; Wittbrodt and Palmer, 1995), which is favored in acidic conditions. Cr(VI), as an oxyanion, can be adsorbed by positively charged surfaces, such as Mn, Al, and Fe oxides and hydroxides, clay minerals, and natural soils and colloids (Bajda and Kłapyta, 2013; Nalbandian et al., 2016). The adsorption of Cr(VI) by these materials is favored in acidic solutions and increases with decreasing pH. Little or no adsorption occurs at a pH of 8.5 or above (Calder, 1988).

9.1.3 Toxicity of As(V)

Arsenic is considered to be an essential element but, at high concentration, a lot of arsenic compounds are toxic (Bissen and Frimmel, 2003; Ahoranta et al., 2016). The toxicity of arsenic depends on its forms. Organic arsenic compounds are less toxic than inorganic arsenic compounds. Acute and chronic poisoning of arsenic involves the respiratory, gastrointestinal, cardiovascular, nervous, and hematopoietic systems. Arsenic is carcinogenic and may cause cancers of the lungs, bladder, liver, kidneys, and skin (Azcue and Nriagu, 1994; Pontius et al., 1994; Guzmán et al., 2016). Therefore, the USEPA lowered the maximum contaminant level for arsenic in drinking water from 50 to 10 μg/L (USEPA, 2002).

Arsenic can occur in the environment in several oxidation states (−3, 0, +3, and +5) but is mostly found in inorganic forms as oxyanions of trivalent arsenite [As(III)] or pentavalent arsenate [As(V)] in natural waters (Guo et al., 2015; Guzmán et al., 2016). As(III) is commonly in the form of $As(OH)_3$, $As(OH)_4^-$, $HAsO_3^{2-}$, and AsO_3^{3-} while As(V) is found mainly in the form of AsO_4^{3-}, $HAsO_4^{2-}$, and $H_2AsO_4^-$. Like Cr, redox potential (Eh) and pH are the most essential factors controlling the speciation of As (Mishra and Mahato, 2016). The Eh–pH diagram for arsenic (Figure 9.2) indicates the oxidation states and chemical forms of the arsenic species which exist within specific Eh and pH ranges (Smedley and Kinniburgh, 2002). Under oxidizing conditions, $H_2AsO_4^-$ is dominant at low pH (less than about pH 6.9), while at higher pH, $HAsO_4^{2-}$ becomes dominant ($H_3AsO_4^0$ and AsO_4^{3-} may be present in extremely acidic and alkaline conditions, respectively). Under reducing conditions at a pH of less than 9.2, the uncharged $H_3AsO_3^0$ will predominate.

The mobility of arsenic in natural systems is mainly controlled by its adsorption onto metal oxide surfaces, involving surface complexation reactions in which the ligand exchange of arsenate or arsenite for a hydroxyl group on the metal oxide generates an inner-sphere complex (Waychunas et al., 1993; Manning et al., 1998; Jadhav et al., 2015). The oxides of Fe, Al, and Mn are the most important sorbents of arsenic in natural systems (Manning and Goldberg, 1997; Guo et al., 2015; Mishra and Mahato, 2016).

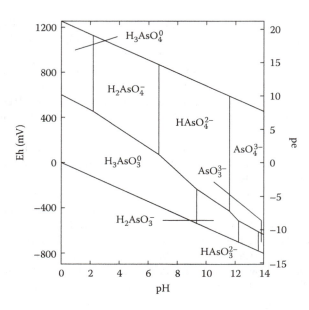

FIGURE 9.2 Eh–pH diagram for aqueous As species in the system As–O$_2$–H$_2$O at 25°C and at 1 bar total pressure. (Adapted from Smedley, P.L., Kinniburgh, D.G., *Appl. Geochem.*, 17, 517–568, 2002.)

9.2 COMMON METHODS OF SIMULTANEOUS REMOVAL OF CR(VI) AND AS(V)

9.2.1 Electrocoagulation

Electrocoagulation (EC) is an emerging water treatment technology that has been applied successfully to treat various wastewaters (Rincón and La Motta, 2014; Zewail and Yousef, 2014; Al-Shannag et al., 2015). It has been applied for the treatment of potable water, heavy metal-laden wastewater, restaurant wastewater, and pulp and paper mill wastewater (Thella et al., 2008). The advantages of EC over conventional technologies include high removal efficiency, compact treatment facility, and possibility of complete automation. EC also offers possibility of anodic oxidation, and *in situ* generation of adsorbents (such as hydrous ferric oxides and hydroxides of aluminum). EC operating conditions are highly dependent on the chemistry of the aqueous medium, especially conductivity and pH. Hansen et al. (2006) observed that arsenic can be removed effectively from smelter industrial wastewater through EC. Parga et al. (2005) demonstrated the removal of Cr(VI)/Cr(III) and As(III)/As(V) with an efficiency of more than 99% from both wastewater and wells. Balasubramanian and Madhavan (2001) reported that the efficient removal of arsenic takes about 7 h and the rate of arsenic removal by EC technique depends on initial arsenic concentration.

The mechanism for chromium removal from wastewater containing Cr(VI) and As(V) ions by the EC technique with iron sacrificial electrodes involves the ferrous iron generated by corrosion of the iron anode (Parga et al., 2005). The ferrous iron can reduce Cr(VI) to Cr(III) under alkaline conditions and is itself oxidized to ferric ion. The dissolved iron species immediately hydrolyze to iron hydroxides. The generated metal hydroxides are excellent coagulating material as they provide active surfaces for the adsorption of As(V). Coagulation occurs when these metal hydroxides combine with the negative particles carried toward the anode by electrophoretic motion. They are then removed by sedimentation or filtration (Thella et al., 2008). The Cr(III) species in aqueous solutions, however, may take the form of Cr^{3+} ion, Cr(OH)$^{2+}$ or Cr(OH)$_2$$^+$, depending on the solution pH values. *In situ* generated iron oxide/oxyhydroxide species will have a positive surface charge in acidic medium and a negative surface charge in basic medium. As these species carry positive electric charges, they readily adsorb onto the negatively charged iron particles. The precipitation and coagulation of species such as $Cr(OH)_3$ is aided by increasing the pH of the solution.

Although this process has the potential to eliminate the disadvantages of classical treatment techniques, a review of the literature shows that the potential of EC as an alternative to the conventional treatment process has not yet been adequately explored due to technical and economic reasons (Parga et al., 2005).

9.2.2 Zero-Valent Iron and/or Iron-Oxide-Coated Sand

9.2.2.1 Application of Zero Valent Iron in As and Cr(VI) Removal

Zerovalent iron (ZVI) has been examined as a reactive media in permeable reactive barriers (PRBs) for removing the copresent As(V) and Cr(VI) (Liu et al., 2009). Cr(VI) removal by ZVI is a chemical reduction process in which ZVI donates electrons and reduces Cr(VI) to Cr(III), and then Cr(III) precipitates as Cr(III) and mixed Fe/Cr (oxy)hydroxides. Different from that of Cr(VI), the principal removal mechanism of As(V) by ZVI is its adsorption onto or coprecipitation with iron corrosion products and the Cr precipitates on the surfaces of the ZVI in the form of $Cr_xFe_{1-x}(OH)_3$ that are formed on the surface of ZVI during iron corrosion (Lackovic et al., 2000; Su and Puls, 2001; Liu et al., 2009). Experimental results have shown that Cr(VI) removal was not affected by the presence of As(V). However, As(V) removal appeared to be inhibited by copresent Cr(VI), which probably resulted from competition between Cr(VI) and As(V) for the adsorption sites on the iron corrosion products (Liu et al., 2009).

Liu et al. (2009) investigated the influences of humic acid on Cr(VI) and As(V) removal by ZVI in laboratory batch settings and continuous flow column systems with the presence of various geochemical constituents, such as bicarbonate and Ca^{2+}. The results obtained in this study show that humic acid exerted different influences on Cr(VI) and As(V) removal. For Cr(VI) removal, the influences of humic acid varied significantly depending on the presence of Ca^{2+} in solutions. In the absence of Ca^{2+}, humic acid showed little inhibition to Cr(VI) removal. On the contrary, in the presence of Ca^{2+}, humic acid would greatly coaggregate with Fe (hydr)oxide colloids and progressively deposit on the ZVI surfaces, and hence inhibit transfer of electrons from the surface of ZVI to Cr(VI) and largely reduce the effective porosity of the ZVI matrix. As a result, the Cr(VI) removal capacity of ZVI has been significantly decreased. However, As(V) removal was observed to proceed differently facing the influences induced by humic acid. Humic acid significantly changed As(V) removal kinetics, by way of inhibiting Fe^{2+}/Fe^{3+} from forming hydroxides by binding with them and stabilizing the fine Fe hydroxides colloids (<0.45 μm) in solutions. These Fe hydroxides are the major adsorbents responsible for As(V) removal. As a result, the process of As(V) removal was retarded.

9.2.2.2 Application of Iron Oxides in As and Cr(VI) Removal

Khaodhiar et al. (2000) reported the use of iron-oxide-coated sand (IOCS) in removing CCA in groundwater. As(V) would inhibit Cr(VI) adsorption while Cr(VI) had no effect on As(V) adsorption. As(V) was strongly adsorbed by forming inner-sphere complexes with IOCS surface, while Cr(VI) was weakly adsorbed. The triple-layer model (TLM) was applied successfully in describing adsorption of As(V) and Cr(VI) in single-solute systems but the equilibrium constants determined from single-solute systems were unable to predict adsorption from multisolute systems. The researchers suggested that the heterogeneity of oxide surface sites and the formation of ternary complexes and/or solid phases that did not exist in single-solute systems may account for the failure of the model.

9.2.2.3 Application of ZVI/IOCS in As and Cr(VI) Removal

The feasibility of using ZVI and IOCS as a combination of reactive media in PRBs for removing Cr(VI) and As(V) from groundwater with various geochemical constituents such as hardness, alkalinity, and natural organic matter (NOM) was investigated (Mak et al., 2011). The results have shown that the Fe^0 and IOCS mixture performed better on the removal of both Cr(VI) and As(V), compared with using Fe^0 or IOCS alone. Compared with Fe^0 and quartz sand mixture for the column study, the Fe^0 and IOCS mixture achieved the highest removal of both Cr(VI) and As(V), while the effects of HA were marginal by using these reactive materials. A synergistic effect in these reactive materials occurred as Fe^{2+} was adsorbed onto the IOCS so that the iron oxides were transformed to magnetite, providing more reactive surface areas for Cr(VI) reduction and reducing the passivation on Fe^0. HA was adsorbed onto the IOCS so that the impact of the deposition of HA aggregates on the Fe^0 surface was reduced, thus enhancing the corrosion of Fe^0. The findings of this study suggest that the use of the combination of Fe^0 and IOCS can have a higher removal efficiency in Cr(VI) and As(V), and arouse a consideration in the design of a more environmentally sustainable PRB by using Fe^0 and IOCS together.

9.2.3 Nanoparticles

Arsenic and chromium in groundwater can be removed using nanomaterials, and a lot of research is being conducted in this field. Chowdhury and Yanful (2010) studied the application of magnetite–maghemite nanoparticles for arsenic and chromium removal. Electrostatic attraction between heavy metals and magnetite–maghemite is a key concept for the removal of arsenic and chromium from aqueous solutions. This study showed that the removal of arsenic and chromium from contaminated water depends on the pH, contact time, initial concentration of arsenic or chromium, PO_4^{3-} concentration in water, and on the adsorbent concentration. A comparison of the arsenic and chromium uptakes shows that the removal efficiency of arsenic was more than that of chromium in the groundwater pH range (6.5–8.5). Thus, arsenic removal by magnetite–maghemite particles from contaminated groundwater is more favorable than chromium in groundwater pH range.

Poguberović et al. (2016) investigated the removal of As(III) and Cr(VI) from aqueous solutions using "green" zero-valent iron nanoparticles (nZVI) produced by oak, mulberry, and cherry leaf extracts. Batch experiments showed that the adsorption kinetics followed pseudo-second-order rate equation and the obtained adsorption isotherm data could be well described by the Freundlich model. In addition, investigated pH effect showed that varying the initial pH value had a significant effect on As(III) and Cr(VI) removal. This study indicated that nZVI could potentially be used as a new green material for the remediation of water matrices contaminated with As(III) and Cr(VI).

Saikia et al. (2011) reported the efficient removal of chromate and arsenate from individual and mixed systems by malachite nanoparticles. In this study, malachite nanoparticles of 100–150 nm have been efficiently and for the first time used as an adsorbent for the removal of chromate and arsenate. A high adsorption capacity was reported for chromate and arsenate on malachite nanoparticle from both individual and mixed solution at pH ~4–5. However, the adsorption efficiency decreases with the increase in solution pH. Batch studies showed that initial pH, temperature, malachite nanoparticles dose, and initial concentration of chromate and arsenate were important parameters for the adsorption process. Thermodynamic analysis has shown that the adsorption of chromate and arsenate on malachite nanoparticles is endothermic and spontaneous. The adsorption data for both chromate and arsenate fitted well the Langmuir isotherm and preferentially followed the second-order kinetics. The binding affinity of chromate is found to be slightly higher than arsenate in a competitive adsorption process, which leads to the comparatively higher adsorption of chromate onto the surface of malachite nanoparticles.

Badruddoza et al. (2013) synthesized phosphonium silane-coated magnetic nanoparticles (PPhSi-MNPs) for the removal of both As(V) and Cr(VI) species. The solution pH plays a very important role upon the adsorption of both As(V) and Cr(VI) from an aqueous solution on PPhSi-MNPs and the optimal adsorption occurred at pH 3.0. The adsorption equilibrium data for both anions well fitted the Langmuir isotherm model. The modification of Fe_3O_4 MNPs by phosphonium silane greatly enhanced the adsorption capacities of both metal anions. The kinetic data closely fitted the pseudo-second-order model. From the mechanistic point of view, a synergy of electrostatic interaction and ion exchange between the positive ligand and negative pollutant anions is playing a significant role in enabling high adsorption.

9.3 THE PRINCIPLE AND PERFORMANCE OF SIMULTANEOUS REMOVAL OF CR(VI) AND AS(V) BY FE(II)

9.3.1 PRINCIPLES OF SIMULTANEOUS REMOVAL OF CR(VI) AND AS(V) BY FE(II)

As mentioned in the above section, chromium exists in natural waters in two main oxidation states, Cr(VI) and Cr(III) (Du et al., 2012; Dong et al., 2016). Cr(III) occurs primarily as a cation in solution and can be easily adsorbed onto the surface of iron oxides and oxyhydroxides at a pH higher than 4.0 by forming strongly bound inner-sphere complexes or precipitates (Charlet and Manceau, 1992; Pettine et al., 1998). Cr(III) hydroxide ($Cr(OH)_3$) exhibits a low solubility at a neutral pH range (Rai et al., 1987; Du et al., 2012). Furthermore, Cr(III) is generally considered to be benign and an essential trace nutrient for animals and humans (Qin et al., 2005). Therefore, Cr(VI) removal by reduction to Cr(III) with ferrous iron and subsequent precipitation, coprecipitation, or coagulation is well documented (Eary and Rai, 1988; Fendorf and Li, 1996; Buerge and Hug, 1997; Brown et al., 1998; Pettine et al., 1998; Schlautman and Han, 2001; Lee and Hering, 2003; Qin et al., 2005; Sharma et al., 2008; Palma et al., 2015; Dong et al., 2016). In the reaction between Cr(VI) and Fe(II), Cr(VI) is reduced to Cr(III) by Fe(II), while Fe(II) is oxidized to Fe(III), which forms ferric hydroxide rapidly. The reduced Cr(III) can be easily sorbed and/or coprecipitated with ferric hydroxide as in the following reaction (Lee and Hering, 2003; Palma et al., 2015):

$$CrO_4^{2-} + 3Fe^{2+} + 8H_2O \rightarrow 4Fe_{0.75}Cr_{0.25}(OH)_3(s) + 4H^+ \tag{9.1}$$

Equation 9.1 elucidates that the reduction of aqueous Cr(VI) by aqueous Fe(II) not only removes the toxic Cr(VI) species from solution but also results in the precipitation of $Fe_{0.75}Cr_{0.25}(OH)_3(s)$ (Eary and Rai, 1988). Our previous study had demonstrated that the Fe(III) formed *in situ* by oxidizing Fe(II) with permanganate was very powerful in the removal of As(V) (Guan et al., 2009). Moreover, Namasivayam and Senthilkumar (1998) showed that the Fe(III)/Cr(III) hydroxide could be effectively used for the removal of arsenate from solution. Thus, it is expected that the precipitates formed in the process of Cr(VI) reduction by Fe(II), $Fe_{0.75}Cr_{0.25}(OH)_3(s)$, have great capacity to entrap or coprecipitate As(V).

9.3.2 PERFORMANCE OF SIMULTANEOUS REMOVAL OF CR(VI) AND AS(V) BY FE(II)

9.3.2.1 The Kinetics of Chromium and Arsenate Removal by Fe(II)

The kinetics of chromium removal by Fe(II) in the absence or in the presence of arsenate was investigated at pH 6–8, as demonstrated in Figure 9.3. In the absence of arsenate, chromium removal by Fe(II) reached equilibrium in 120, 10, and 5 min, respectively, at pH 6, 7, and 8. At equilibrium, 99.1% and ~100% of chromium was removed at pH 6 and 7–8, respectively. The presence of 10 μmol L^{-1} arsenate had negligible effects on the removal rate of chromium at pH 7; however, it showed slight inhibitory effects on the removal rate of chromium at pH 6 and 8. The removal efficiency of chromium was reduced to 2.8% and 6.5%, respectively, at pH 6 and 8 due to the presence of arsenate.

Arsenic removal by Fe(II) was very slow at pH 6–8 and did not achieve equilibrium in 120 min in the absence of chromate. As shown in Figure 9.3d, only 6.4%–23.8% of arsenate was removed by 45 μmol L^{-1} Fe(II). The presence of 10 μmol L^{-1} chromate remarkably enhanced the removal rate of arsenate at pH 6–8. In the presence of chromate, arsenic removal increased rapidly in the first 45 min and then increased gradually. The removal rate of arsenic at pH 7 and 8 was greater than

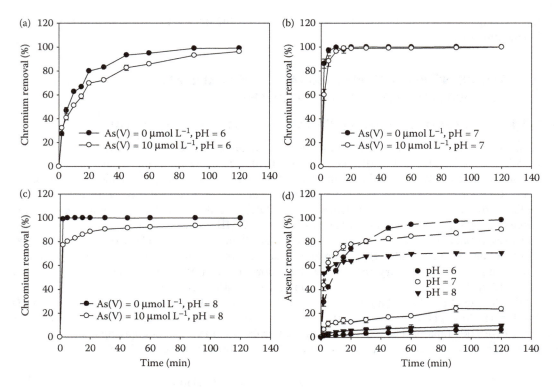

FIGURE 9.3 Kinetics of chromium removal and arsenic removal by Fe(II) under various conditions: (a) chromium removal at pH 6 in the presence of 0 or 10 μmol L^{-1} arsenate; (b) chromium removal at pH 7 in the presence of 0 or 10 μmol L^{-1} arsenate; (c) chromium removal at pH 8 in the presence of 0 or 10 μmol L^{-1} arsenate; and (d) arsenic removal at pH 6–8 in the presence of 0 μmol L^{-1} chromate (the solid lines) or 10 μmol L^{-1} chromate (the dashed lines) (As(V) = 0 or 10 μmol L^{-1}, Cr(VI) = 0 or 10 μmol L^{-1}, Fe(II) = 45 μmol L^{-1}).

that at pH 6 in the first 20 and 10 min, respectively, which may be attributable to the more rapid oxidation of Fe(II) by chromate at pH 7–8 than that at pH 6.

9.3.2.2 Chromium Removal by Fe(II) in the Absence and Presence of Arsenate

Chromium removal by Fe(II) in the absence of arsenate as a function of pH and Fe(II) dosages was illustrated in Figure 9.4. Chromium removal was strongly influenced by pH and Fe(II) dosages. At various Fe(II) dosages, chromium removal increased to a maximum with increasing pH and then decreased with further increase in pH. When Fe(II) was dosed at 20 μmol L^{-1}, the maximum removal of chromium was 68.6% which was achieved at pH 6. When the dosage of Fe(II) was increased to 30 μmol L^{-1}, 97.5% of chromium was removed at pH 7. Chromium removal of up to 99%–100% was observed over a pH range of 5.9–7.7 and 5.8–7.8, respectively, when Fe(II) was dosed at 45 and 60 μmol L^{-1}. Decrease in pH or increment in pH out of this range resulted in a sag in chromium removal. The increment in Fe(II) dosage resulted in an improvement in chromium removal and the improvement was more remarkable under alkaline conditions than that under acidic conditions. Chromium removal was enhanced over pH 5–10 by increasing the Fe(II) dosage from 20 to 45 μmol L^{-1} and a further rise in Fe(II) dosage from 45 to 60 μmol L^{-1} only resulted in an improvement in chromium removal at pH 8–10.

The effects of 10 μmol L^{-1} arsenate on chromium removal by Fe(II) were strongly dependent on pH and Fe(II) dosages, as illustrated in Figure 9.4. The presence of arsenate had more drastic effects on chromium removal by Fe(II) under alkaline conditions than that under acidic and neutral conditions. For instance, chromium removal was decreased by 5.4%–11.2% at pH 4–7 and by

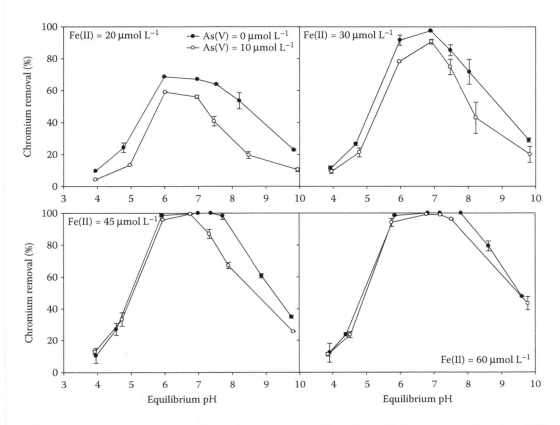

FIGURE 9.4 Chromium removal by Fe(II) in the presence of 0 or 10 μmol L⁻¹ arsenate as a function of pH and Fe(II) dosages (Cr(VI) = 10 μmol L⁻¹, As(V) = 0 or 10 μmol L⁻¹).

12.4%–33.9% at pH 7.5–10 due to the presence of 10 μmol L⁻¹ arsenate when Fe(II) was applied at 20 μmol L⁻¹. Increasing Fe(II) dosage could alleviate the inhibitive effects from arsenate for chromium removal by Fe(II). When Fe(II) was dosed at 45 μmol L⁻¹, the presence of 10 μmol L⁻¹ arsenate had negligible effects on chromium removal at pH 4–7 but reduced chromium removal under alkaline conditions. When Fe(II) was applied at 60 μmol L⁻¹, chromium removal was only slightly affected by the presence of 10 μmol L⁻¹ arsenate at pH 8–9. More than 99% of chromium was removed at pH 6.8 even in the presence of 10 μmol L⁻¹ arsenate when Fe(II) was dosed at 45 or 60 μmol L⁻¹.

9.3.2.3 Arsenate Removal by Fe(II) in the Absence and Presence of Chromate

It was expected that arsenic could be removed by ferric hydroxide derived from oxidation of Fe(II) by DO in the solution and accordingly, arsenic removal by Fe(II) at various pH levels and Fe(II) dosages was investigated, as shown in Figure 9.5. When Fe(II) was dosed at 20–45 μmol L⁻¹, arsenic removal varied from 2.2% to 14.7% at pH 4–6 and reached maximum at pH 6.7–6.9. With further increment in pH, arsenic removal experienced a reduction and then a slight increase. However, a different removal edge for arsenate was observed when Fe(II) was applied at 60 μmol L⁻¹. Arsenic removal rose slowly from pH 4.0 to 6.0 but increased sharply to 92.7% over the pH range of pH 6.0–7.1. Under alkaline conditions, arsenic removal was lowered sharply from pH 7.1 to 7.4 but decreased gradually at pH 7.4–9.7.

The presence of 10 μmol L⁻¹ chromate dramatically improved arsenic removal by Fe(II) under most conditions investigated in this study, as shown in Figure 9.5. At various Fe(II) dosages, optimal arsenic removal was achieved at pH 5.8–6.0 and the increase or reduction in pH resulted in a

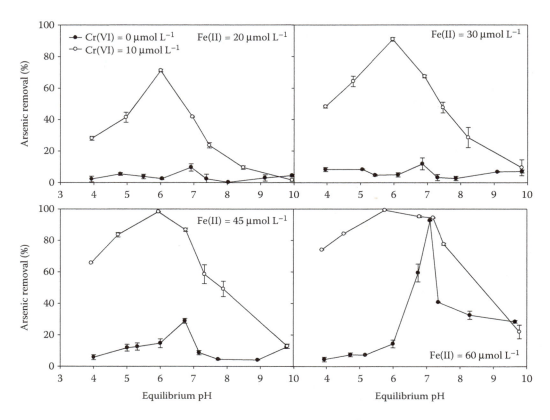

FIGURE 9.5 Arsenic removal by Fe(II) in the presence of 0 or 10 μmol L⁻¹ chromate as a function of pH and Fe(II) dosages (As(V) = 10 μmol L⁻¹ and Cr(VI) = 0 or 10 μmol L⁻¹).

decline in arsenic removal. When Fe(II) was dosed at 20 μmol L⁻¹, arsenic removal rose gradually from 28.1% to 71.1% as the pH increased from 4.0 to 6.0 and then reduced gradually from 71.1% to 9.5% as the pH varied from 6.0 to 8.5. The variation of arsenic removal with pH at an Fe(II) dosage of 30 or 45 μmol L⁻¹ was very similar to that at an Fe(II) dosage of 20 μmol L⁻¹, except that a higher Fe(II) dosage resulted in a higher arsenic removal at pH 3.9–9.8. In particular, when Fe(II) was applied at 60 μmol L⁻¹, arsenic removal over the pH range of 3.8–9.6 can be divided into three stages: a slow increase from 74.0% to 99.2% at pH 3.9–5.8, a very slow decline from 99.2% to 94.4% at pH 5.8–7.2, and a sharp decrease from 94.4% to 22.1% over the pH range of 7.2–9.8. Arsenic removal was improved by 9.3%–68.7%, 26.0%–86.0%, 44.7%–83.6%, and 1.7%–84.9%, respectively, at pH 4–9 due to the presence of 10 μmol L⁻¹ chromate at Fe(II) dosages of 20, 30, 45, and 60 μmol L⁻¹. Moreover, arsenic removal by Fe(II) in the presence of chromate was enhanced by the increase of Fe(II) dosages over the pH range of 4–10.

9.3.2.4 Effects of Initial Cr(VI)/As(V) μmolar Ratios on Chromium and Arsenate Removal by Fe(II)

The removal of chromium and arsenate by Fe(II) was examined when the initial concentrations of chromate and arsenate were 20 and 10 μmol L⁻¹ (initial Cr(VI)/As(V) μmolar ratio = 2:1), respectively, and the results were illustrated in Figure 9.6. In the presence of 10 μmol L⁻¹ arsenate, chromium removal by Fe(II) increased with increasing pH from 3.9 to 5.9 and remained almost constant over a pH range of 5.9–7.4 before a decline with further increase in pH. The maximum

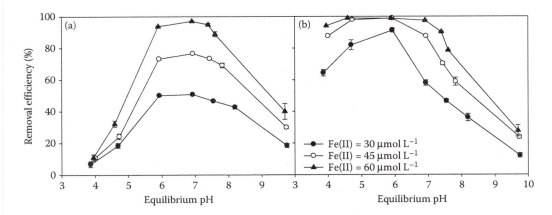

FIGURE 9.6 Simultaneous removal of chromium and arsenic as a function of pH and Fe(II) dosages at Cr(VI)/As(V) µmolar ratio of 2:1 (a) chromium removal and (b) arsenic removal (Cr(VI) = 20 µmol L^{-1} and As(V) = 10 µmol L^{-1}).

chromium removal was achieved at pH 6.9 at various Fe(II) dosages. The increment in Fe(II) dosage from 30 to 60 µmol L^{-1} resulted in a drastic enhancement in optimum chromium removal from 50.7% to 97.1%.

In the presence of 20 µmol L^{-1} chromate, arsenic removal rose gradually from 64.3% to 91.3% as the pH increased from 3.9 to 5.9 and then reduced gradually from 91.3% to 12.0% as the pH varied from 6.0 to 9.8, when Fe(II) was dosed at 30 µmol L^{-1}. As the Fe(II) dosage was applied at 45 or 60 µmol L^{-1}, a broad removal maximum was achieved for arsenate, with 97.4%–98.9% arsenate uptake at pH 4.7–5.9 and 4.6–6.9, respectively. Interestingly, it was found that increasing chromate concentration from 10 to 20 µmol L^{-1} resulted in an improvement in arsenic removal by 16.1%–20% and 14.3%–17.7%, respectively, at pH 4.0 and 4.6 when Fe(II) was dosed at 30–60 µmol L^{-1}, as illustrated in Figure 9.7.

This study also examined chromium and arsenate uptake by Fe(II) dosed at 30–60 µmol L^{-1} when the initial concentrations of chromate and arsenate were 10 and 20 µmol L^{-1} (initial Cr(VI)/As(V) µmolar ratio = 1:2), respectively, as shown in Figure 9.8. In the presence of 20 µmol L^{-1} arsenate, optimum chromium removal was achieved at pH 5.8 at various Fe(II) dosages and an increase or decrease in pH resulted in a sharp decrease in chromium removal. Figure 9.8a shows that chromium removal in the presence of 20 µmol L^{-1} arsenate was not greatly affected by Fe(II) dosages, especially under neutral and alkaline conditions. The increase in Fe(II) dosage from 30 to 60 µmol L^{-1} only led to a slight improvement in chromate uptake at pH 5.8 from 84.0% to 97.7%. Arsenic removal by Fe(II) in the presence of chromate was strongly dependent on pH but moderately dependent on Fe(II) dosage when the initial Cr(VI)/As(V) µmolar ratio was 1:2, as illustrated in Figure 9.8b. Arsenic removal improved gradually as the pH increased from 3.9 to 5.8 but reduced sharply with further increase in pH. Arsenic removal was enhanced by only 12.1%–26.7% at pH 3.9–6.8 when the Fe(II) dosage was increased from 30 to 60 µmol L^{-1}; however, there was almost no improvement under neutral and alkaline conditions. It was found that chromium removal by Fe(II) was increased by 1.2%–19.6% at pH 3.9–5.8 due to the presence of 20 µmol L^{-1} arsenate, compared with the case where arsenate was 10 µmol L^{-1}, as shown in Figure 9.9. On the other hand, the presence of 20 µmol L^{-1} arsenate dramatically decreased chromium removal by Fe(II) over the pH range of 6.7–9.8, as illustrated in Figures 9.8a and 9.9. Furthermore, under neutral and alkaline conditions, increasing Fe(II) dosage could not mediate the detrimental effects from arsenate of 20 µmol L^{-1} on chromium removal by Fe(II).

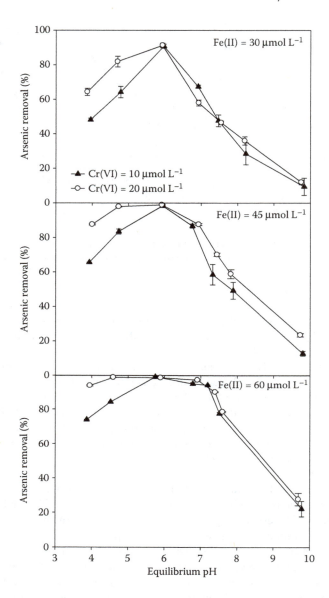

FIGURE 9.7 Effects of concentration of coexisting chromate on arsenic removal as a function of pH and Fe(II) dosages (As(V) = 10 μmol/L and Cr(VI) = 10 or 20 μmol/L).

9.4 MECHANISMS OF SIMULTANEOUS REMOVAL OF CR(VI) AND AS(V) BY FE(II)

9.4.1 CHROMIUM REMOVAL BY FE(II) AT VARIOUS pH LEVELS

Chromium removal by Fe(II) as a function of pH in the absence of arsenate (i.e., in the presence of 0 μmol L^{-1} arsenate) is presented in Figure 9.10a. Chromium removal increased from 11.3% to 97.5% as the pH increased from 3.9 to 6.9 and then decreased gradually to 29.1% with a further increase in pH to 9.8. The species of residual chromium and iron in the process of Cr(VI) removal by Fe(II) were determined and are shown in Figure 9.11. The concentration of residual Cr(III) in the supernatant was in the range of 0.83–1.11 μmol L^{-1} at pH 3.9–4.7 and below 0.43 μmol L^{-1} at

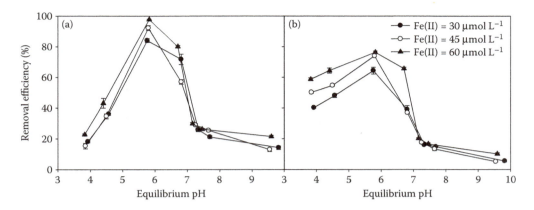

FIGURE 9.8 Simultaneous removal of chromium and arsenic as a function of pH and Fe(II) dosages at Cr(VI)/As(V) μmolar ratio of 1:2: (a) chromium removal and (b) arsenic removal (Cr(VI) = 10 μmol L^{-1} and As(V) = 20 μmol L^{-1}).

pH > 4.7. The predominant residual chromium species present in solution was Cr(VI) over the pH range of 3.9–9.8. The concentration of residual Cr(VI) at pH 3.9 was as high as 7.43 μmol L^{-1} but dropped to 0.22 μmol L^{-1} at pH 7. However, the concentration of soluble Cr(VI) rose markedly with further increase in pH and was 5.9 μmol L^{-1} at pH 9.8. The concentration of residual Fe(II) decreased from 23.8 to 4.1 μmol L^{-1} as the pH increased from 3.9 to 6.0 while it was negligible at pH > 6.0, indicating the nearly complete oxidation of Fe(II) under neutral and alkaline conditions. The concentration of soluble Fe(III) varied from 0.18 to 1.53 μmol L^{-1} throughout the pH range of 3.9–9.8.

The Cr 2p and Fe 2p line x-ray photoelectron spectroscopy (XPS) spectra of the precipitates collected in the process of Cr(VI) removal by Fe(II) at various pH levels are illustrated in Figures 9.12 and 9.13, respectively. The Cr 2p1/2 and Cr 2p3/2 lines appear at 587.0 ± 0.2 and 577.0 ± 0.3 eV, respectively, indicating that chromium in the precipitates is present as Cr(III), confirming that the reduction of Cr(VI) to Cr(III) occurs during reaction with Fe(II) (Lee and Hering, 2003). The single and smooth Gaussian-shaped peak indicates the formation of Cr(OH)$_3$ precipitate, Fe$_x$Cr$_{1-x}$(OH)$_3$ coprecipitate, or possibly a hydrous Cr(III) oxide (CrOOH) instead of Cr$_2$O$_3$ (Manning et al., 2007). The Fe 2p1/2 and Fe 2p3/2 lines appear at 724.8 ± 0.1 and 711.8 ± 0.1 eV, respectively, suggesting that the iron in the precipitate is present as Fe(III) and is typical of iron oxyhydroxides (FeOOH) (Zhang et al., 2007). The XPS results have also shown that the Fe/Cr μmolar ratios of the precipitates collected in the process of Cr(VI) reduction by Fe(II) was 2.74–3.23 at pH < 7.0. In addition, speciation analysis of chromium and iron in the solution has shown that the μmolar ratio of residual Fe(II) and Cr(VI) in solution at pH 3.9–6.0 is 3.0 ± 0.20. Thus, it was concluded that the reaction between Cr(VI) and Fe(II) at pH < 7.0 follows the equation proposed by other investigators (Schlautman and Han, 2001) with 3.0 μmol of aqueous Fe(II) consumed in the reduction of 1.0 μmol of aqueous Cr(VI) with the formation of solid Fe$_{0.75}$Cr$_{0.25}$(OH)$_3$; that is,

$$\frac{1}{4}Cr(VI) + \frac{3}{4}Fe(II) + 3H_2O \rightarrow Fe_{0.75}Cr_{0.25}(OH)_3 + 3H^+ \tag{9.2}$$

At pH 3.9–4.7, less than 40% of the original chromate was reduced to Cr(III) by Fe(II). The presence of high concentrations of both Cr(VI) and Fe(II) in the solution under acidic conditions after 2 h indicated that the reaction of Cr(VI) with Fe(II) under these conditions was slow. The reduction rates of Cr(VI) by Fe(II) at pH 4–5 were examined and are presented in Figure 9.14, which confirmed the slow reaction rate of Cr(VI) with Fe(II) under acidic conditions. Many other researchers,

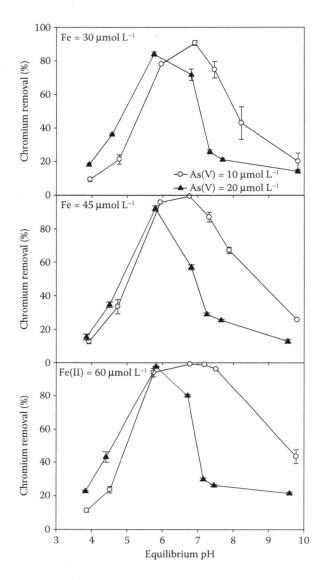

FIGURE 9.9 Effects of concentration of coexisting arsenate on chromium removal as a function of pH and Fe(II) dosages (Cr(VI) = 10 μmol L^{-1} and As(V) = 10 or 20 μmol L^{-1}).

including Buerge and Hug (1997) and Pettine et al. (1998) also reported that the reduction rate of Cr(VI) by Fe(II) was very slow under acidic conditions. Moreover, Fe$_{0.75}$Cr$_{0.25}$(OH)$_3$ precipitates are more soluble at pH 3.9–4.7 than those at higher pH, which also contributed to the low removal rate of chromium under acidic conditions. Therefore, the variation in removal efficiency of chromium by Fe(II) over the pH range 3.9–6.9 was mainly ascribed to the different rates of Cr(VI) reduction by Fe(II) and solubility of Fe$_{0.75}$Cr$_{0.25}$(OH)$_3$ precipitates occurring at various pH levels.

Figure 9.10b shows that, in the presence of 0 μmol L^{-1} arsenate, the amount of iron in the precipitates rose gradually with increase in pH and then reached a plateau. Figures 9.10b and 9.13 indicate that at pH above 7, more than 98% of Fe(II) was oxidized to Fe(III) and precipitated. The molar ratio of Fe(II) oxidized to Cr(VI) reduced was larger than the expected stoichiometric value of 3 under alkaline condition and this ratio increased significantly with increasing pH, implying that the competition from oxygen resulted in a more significant decrease in the amount of chromate that could be

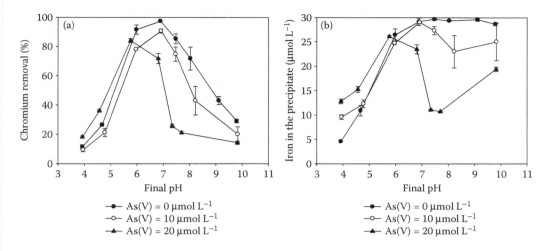

FIGURE 9.10 (a) Chromium removal by Fe(II) in the presence of arsenate of various concentrations at different pH levels and (b) the amount of iron entrapped in the precipitate generated in this process ($Cr(VI) = 10\ \mu mol\ L^{-1}$ and $Fe(II) = 30\ \mu mol\ L^{-1}$).

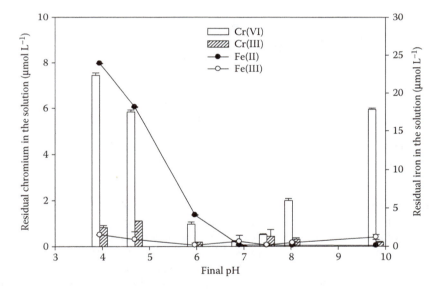

FIGURE 9.11 Speciation of residual chromium and iron in the process of chromate removal by Fe(II) ($Cr(VI) = 10\ \mu mol\ L^{-1}$ and $Fe(II) = 30\ \mu mol\ L^{-1}$).

reduced by Fe(II) at higher pH (Lee and Hering, 2003). The reduction rate of Cr(VI) by Fe(II) in the presence of oxygen can be expressed by the following equation, adapted from Pettine et al. (1998):

$$-\frac{d[Cr(VI)]}{dt} = K_{Cr}[Cr(VI)]\left[Fe(II)\cdot exp^{-K_{O_2}[O_2][OH^-]t} \right] \tag{9.3}$$

where K_{Cr} and K_{O_2} are the overall constants for the reaction of Fe(II) with Cr(VI) and O_2, respectively. Equation 9.3 includes a decaying term for Fe(II) due to the presence of oxygen and clearly shows that the competitive effect from oxygen is more obvious at higher pH. Figure 9.15 showed

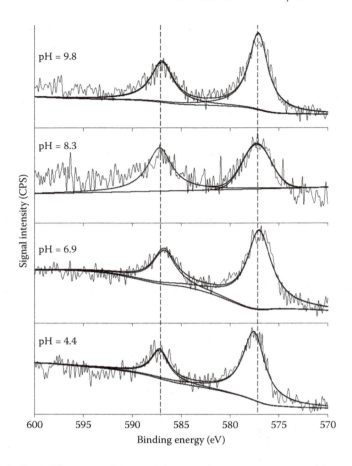

FIGURE 9.12 Cr 2p line XPS spectra of the precipitates collected in the process of Cr(VI) removal by Fe(II) at various equilibrium pH levels. The smooth lines are the results of quantitative Gaussian–Lorentzian curve fitting (Cr(VI) = 10 μmol L^{-1} and Fe(II) = 30 μmol L^{-1}).

that chromate in the deoxygenated solution could be removed almost completely by Fe(II) under alkaline conditions, suggesting that Cr(VI) reduction by Fe(II) can complete in 2 h and the incomplete reduction of Cr(VI) by Fe(II) under oxic condition under alkaline conditions be not resulted from the slow rate of reduction of Cr(VI) by Fe(II). Therefore, Figure 9.15 confirmed that a fraction of Fe(II) was oxidized by oxygen instead of chromate in the oxic systems under alkaline conditions and formed FeOOH according to the information provided in Figure 9.13. Some previous studies also reported that DO competed with Cr(VI) in the oxidation of Fe(II) and that chromate reacted very rapidly with Fe(II) under alkaline conditions (Buerge and Hug, 1997; Pettine et al., 1998; Singh and Singh, 2002). However, the DO in our system exhibited stronger capability in competing Fe(II) with Cr(VI) compared with those reported in the literature, which might have resulted from the enhanced oxidation rate of Fe(II) by oxygen caused by the presence of HCO$_3^-$ (King, 1998). Thus, the minute concentration of Cr(III), the high concentration of Cr(VI) remaining in the solution at pH 7.5–9.8, and the XPS spectra collected at pH > 7 suggest that chromium removal under alkaline conditions was mainly controlled by the magnitude of chromate reduction by Fe(II).

9.4.2 Effect of Arsenate on Chromium Removal

The effects of arsenate of 10 or 20 μmol L^{-1} on chromium removal by Fe(II) was strongly dependent on pH, as shown in Figure 9.10. The presence of arsenate had a minor influence on chromium

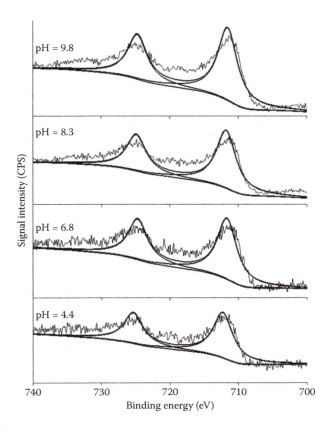

FIGURE 9.13 Fe 2p line XPS spectra of the precipitates collected in the process of Cr(VI) removal by Fe(II) at various equilibrium pH levels. The smooth lines are the results of quantitative Gaussian–Lorentzian curve fitting (Cr(VI) = 10 μmol L^{-1} and Fe(II) = 30 μmol L^{-1}).

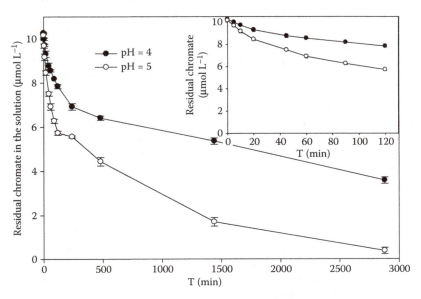

FIGURE 9.14 Kinetics of chromate reduction by Fe(II) at initial pH 4 and 5 (Cr(VI) = 10 μmol L^{-1} and Fe(II) = 30 μmol L^{-1}). (The inset illustrates kinetics of chromate reduction by Fe(II) in 2 h.)

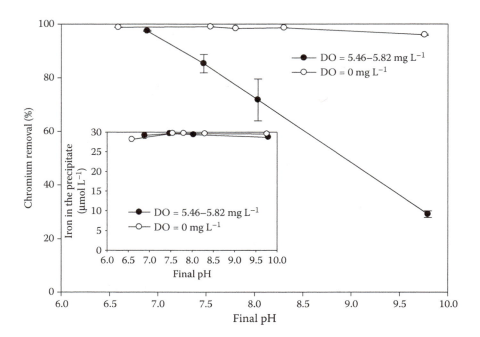

FIGURE 9.15 Effect of DO on chromium removal by Fe(II) under neutral and alkaline conditions (The inset shows the amount of iron entrapped in the precipitate collected in this process.) (Cr(VI) = 10 μmol L^{-1} and Fe(II) = 30 μmol L^{-1}).

removal under acidic conditions while it decreased the chromium removal to different extents, depending on the concentration of arsenate. Arsenate of 10 μmol L^{-1} decreased chromium removal by 9.0%–28.7% at pH 7.5–9.8 while the presence of 20 μmol L^{-1} arsenate had detrimental effects on chromium removal at pH 6.7–9.8. The influence of arsenate on Fe(II)-mediated chromium removal is presumably associated with the effects of arsenate on the reduction of Cr(VI) by Fe(II) and the precipitation of Cr(III). As shown in Figure 9.16a, the presence of 10 μmol L^{-1} arsenate decreased the concentration of residual Cr(VI) in the solution at pH 3.9–4.7 and a higher arsenate concentration resulted in a lower concentration of residual Cr(VI). However, the presence of 10 μmol L^{-1} arsenate decreased the reduction of Cr(VI) with Fe(II) by approximately 3.4%–20.0% at pH > 6.9 and arsenate of elevated concentrations resulted in stronger inhibition in the reduction of Cr(VI) by Fe(II), which was very similar to the influence of phosphate on Cr(VI) reduction by Fe(II) in the presence of trace amounts of oxygen and under alkaline conditions. The rate of the reaction between aqueous Fe(II) and DO relative to the rate of the reaction between aqueous Fe(II) and Cr(VI) increased with increasing pH. The presence of arsenate caused a more significant enhancement in the rate of Fe(II) oxidation by oxygen than that by Cr(VI) (Tamura et al., 1976). Therefore, the amount of Fe(II) that reacted with aqueous Cr(VI) decreased with increasing pH and arsenate concentration, resulting in an elevation in unreacted Cr(VI) concentration at pH > 6.9.

Figure 9.16b demonstrates that the presence of 10 μmol L^{-1} arsenate elevated the concentration of soluble Cr(III) from 0.02–1.11 to 0.37–2.07 μmol L^{-1} over the pH range of 4–10. The concentration of soluble Cr(III) was up to 5.00 μmol L^{-1} at pH 7.3 in the presence of 20 μmol L^{-1} arsenate and decreased sharply with increase or decrease in pH. The speciation of soluble iron at the end of reaction was also analyzed, as shown in Figure 9.17. The high soluble Cr(III) concentration commonly accompanies elevated concentrations of soluble Fe(III) at pH 6.9–9.8. The increase in the concentration of both soluble Cr(III) and Fe(III) at pH 6.9–9.8 caused by the presence of arsenate should be ascribed to the formation of soluble complexes between arsenate and Fe(III)/Cr(III) and inhibition of the precipitation of Fe$_{0.75}$Cr$_{0.25}$(OH)$_3$ and FeOOH, as indicated in Figure 9.10b. Rai et al. (2004)

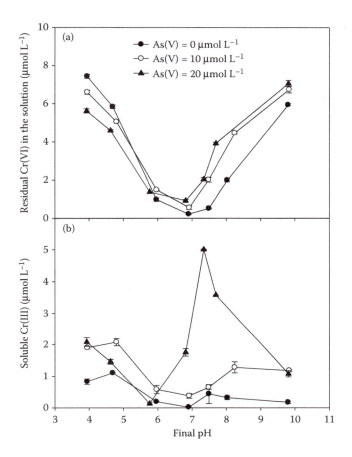

FIGURE 9.16 The concentration of (a) residual Cr(VI) and (b) soluble Cr(III) in the process of chromate removal by Fe(II) in the absence or presence of arsenate (Cr(VI) = 10 μmol L^{-1} and Fe(II) = 30 μmol L^{-1}).

observed an obvious increase in Cr(OH)$_3$ solubility in the presence of phosphate at pH 4.5–12 and concluded that the increase in solubility resulted from the formation of soluble complexes between Cr(III) and phosphate species. Guan et al. (2009) reported that the presence of competing anions decreased the removal of arsenic by reducing the formation of ferric hydroxide precipitate derived from the oxidation of Fe(II). Stumm and Morgan (1996) quantified the solubility of (hydr)oxides considering the possibility of complex formation with ligand L by

$$Me_T = [Me]_{free} + \sum Me[OH]_n + m \sum_i [Me_m H_k L_n [OH]]_i \qquad (9.4)$$

where L stands for the ligand other than OH$^-$. Equation 9.4 shows that the solubility of (hydr)oxides is determined by both OH$^-$ and L while OH$^-$ should play a more important role with increasing pH. HAsO$_4^{2-}$ is the dominate arsenate species at pH 7.3 when the maximum concentration of soluble Cr(III) and Fe(III) is observed, indicating that the complexation of HAsO$_4^{2-}$ with Fe$_{0.75}$Cr$_{0.25}$(OH)$_3$ and FeOOH resulted in the high concentration of soluble Cr(III) and Fe(III). The significant reduction in the concentration of soluble Cr(III) from pH 7.3 to 9.8 should be associated with the much higher concentration and stronger complexation ability of OH$^-$ at pH 9.8.

Arsenate decreased the concentration of soluble Fe(III) but had dual influences on the concentration of soluble Cr(III) at pH < 6.0. Arsenate of lower concentration coordinated with Fe(III) species preferentially under acidic conditions compared with Cr(III); therefore, arsenate of lower

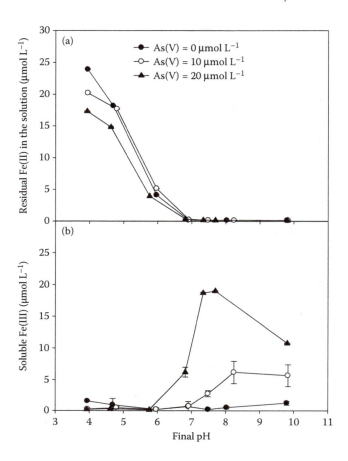

FIGURE 9.17 The concentration of (a) residual Fe(II) and (b) soluble Fe(III) in the process of chromate removal by Fe(II) in the absence or presence of arsenate (Cr(VI) = 10 μmol L^{-1}, Fe(II) = 30 μmol L^{-1}, and As(V) = 0–20 μmol L^{-1}).

concentration resulted in more soluble Cr(III). As shown in Figure 9.18, Cr^{3+} and Fe^{3+} are present in solution as $CrOH^{2+}$ and $Fe(OH)_2^+$ at pH 3.9–5.8, respectively, while arsenate exists as the anion $H_2AsO_4^-$ (Sass and Rai, 1987). Thus, the presence of 20 μmol L^{-1} arsenate may facilitate the precipitation of $Fe_{0.75}Cr_{0.25}(OH)_3$ or the adsorption of $CrOH^{2+}$ and $Fe(OH)_2^+$ on the $Fe_{0.75}Cr_{0.25}(OH)_3$ precipitates at pH 4.6–6.0 through ternary surface complex formation in a manner similar to that proposed to account for the arsenate enhancement of uranium sorption on aluminum oxide (Tang and Reeder, 2009).

9.4.3 Effect of Chromate on Arsenate Removal by Fe(II)

Fe(II) alone was not effective for arsenate removal, as demonstrated in Figure 9.19a, and only 2.6%–8.2% of arsenate was removed by Fe(II) at pH 3.9–9.8 in the absence of chromate. Roberts et al. (2004) reported a much higher arsenate removal efficiency by Fe(II) than that obtained in this study, which should be due to the high concentration of Ca^{2+} and CO_3^{2-} contained in their synthetic groundwater. In the presence of 10 μmol L^{-1} chromate, arsenate removal increased from 48.2% to 90.8% as the pH increased from 3.9 to 6.0 but it decreased gradually with further increase in pH. Increasing chromate concentration from 10 to 30 μmol L^{-1} resulted in an improvement in arsenate removal by 30.9% and 35.4%, respectively, at pH 3.9 and 4.8. However, the increase in chromate concentration had little effect on arsenate removal in the pH range of 6.0–9.8. In the system with

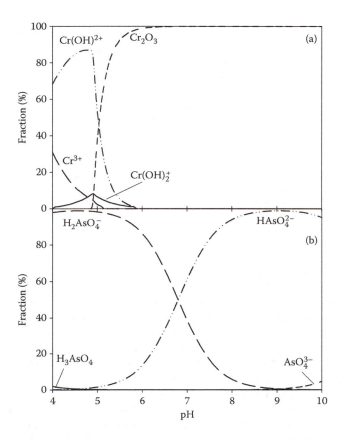

FIGURE 9.18 Species distribution diagrams of (a) Cr(III) and (b) As(V) at various pH levels.

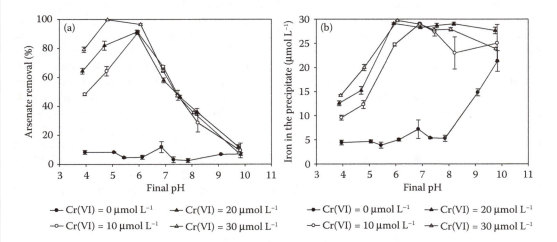

FIGURE 9.19 (a) Arsenate removal by Fe(II) in the presence of chromate of various concentrations at different pH values and (b) the amount of iron entrapped in the precipitate generated in this process $(As(V) = 10 \ \mu mol \ L^{-1}$ and $Fe(II) = 30 \ \mu mol \ L^{-1})$.

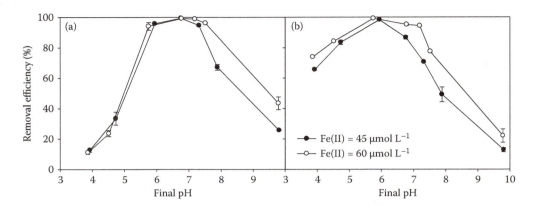

FIGURE 9.20 Simultaneous removal of (a) chromium and (b) arsenate by Fe(II) as a function of pH and Fe(II) dosages (Cr(VI) = 10 μmol L^{-1}, As(V) = 10 μmol L^{-1}, and Fe(II) = 45–60 μmol L^{-1}).

initial Cr(VI) and As(V) concentration of 10 μmol L^{-1} each, Fe(II) dosed at 30 μmol L^{-1} was impossible to reduce Cr(VI) and As(V) simultaneously to satisfy the drinking water standard. However, the residual concentration of both Cr(VI) and As(V) at pH 5.9 could meet the drinking water standard when Fe(II) was applied at 60 μmol L^{-1}, as illustrated in Figure 9.20.

Arsenate adsorption on Cr(OH)$_3$ solid was examined and it was found that arsenate could not be removed by Cr(OH)$_3$ (data not shown). Thus, arsenate removal in the process of simultaneous removal of Cr(VI) and As(V) by Fe(II) should be associated with the precipitated iron. The remarkable influence of chromate on arsenate removal by Fe(II) may be correlated with the oxidative property of chromate, which resulted in the oxidization of Fe(II) to Fe(III) thus facilitating the removal of arsenate. The amount of iron in the precipitates as a function of pH in the presence of chromate was examined with results shown in Figure 9.19b. The presence of 10 μmol L^{-1} chromate greatly increased the amount of iron in the precipitates over the pH range of 3.9–9.0, which should contribute to the enhancement in arsenate removal as a result of the presence of chromate. The great improvement in arsenate removal at pH 3.9–4.8 caused by increasing chromate concentration from 10 to 30 μmol L^{-1} should be associated with the increase in the amount of precipitated iron in this pH range.

As only the precipitated iron could mediate arsenate removal, the amount of arsenate removed per unit of precipitated iron was calculated and presented in Figure 9.21. In the presence of 10 μmol L^{-1} chromate, the amount of arsenate removed per unit of precipitated iron elevated slightly as the pH increased from 3.9 to 4.8 and then decreased gradually with increasing pH. The maximum arsenate removal per unit of precipitated iron was observed at pH 4.8, which may be associated with the appearance of the highest concentration of CrOH^{2+} at this pH level, as demonstrated in Figure 9.18. The decline in arsenate removal per unit of precipitated iron on increasing pH from 4.8 to 9.8 may be ascribed to the gradual shift of H$_2$AsO$_4^-$ species to HAsO$_4^{2-}$ species and the increased competition with hydroxide ions at higher pH (Guan et al., 2008, 2009). Figure 9.21 also demonstrates that the presence of chromate drastically enhances the amount of arsenate removed per unit of precipitated iron under acidic conditions. For freshly precipitated ferric hydroxide, the total concentration of surface sites available for sorption is approximately 0.2 mol/mol Fe (Dzombak and Morel, 1990). However, the amounts of arsenate removed per unit of precipitated iron at pH 3.9–6.9 varied from 0.23 to 0.52 μmol As(V)/μmol Fe, suggesting that arsenate was removed by both adsorption and coprecipitation with the precipitated Fe$_{0.75}$Cr$_{0.25}$(OH)$_3$ and FeOOH.

9.4.4 ARSENIC K-EDGE EXTENDED X-RAY ABSORPTION FINE STRUCTURE ANALYSIS

Extended x-ray absorption fine structure (EXAFS) spectra were used to determine the local coordination environments of arsenate entrapped in the precipitates, as demonstrated in Figure 9.22.

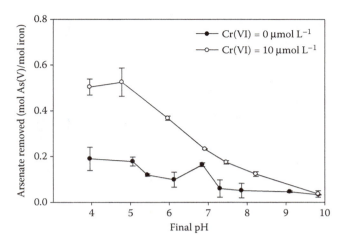

FIGURE 9.21 The amount of arsenate removed by per unit precipitated iron in the presence of 0 or 10 μmol L^{-1} chromate as a function of pH (As(V) = 10 μmol L^{-1}, Cr(VI) = 0 or 10 μmol L^{-1}, and Fe(II) = 30 μmol L^{-1}).

Figure 9.23a and b shows the k^3 weighted As K-edge EXAFS spectra and the corresponding radial structure functions (RSFs) as Fourier transform (FT) versus radial distance obtained for the arsenate entrapped in the precipitates at various pH levels, respectively. The resolved structural parameters obtained by fitting the theoretical paths to the experimental spectra are shown in Table 9.2. The FT of the EXAFS spectra isolates the contributions of different coordination shells, in which the peak positions correspond to the interatomic distances. However, these peak positions in Figure 9.23b are uncorrected for the phase shift, so they deviate from the true distance by 0.3–0.5 Å (Guan et al., 2008). As shown in Figure 9.23b and Table 9.2, the first peak in the RSF (radial structure function) was the result of backscattering from the nearest neighbor As(V)–O shell. The As–O interatomic distances display a narrow range of variation from 1.68 to 1.69 Å and the coordination number varies from 3.5 to 3.8, which were in agreement with the values previously reported for the tetrahedral arsenate geometry and were diagnostic for the arsenate species (Guo et al., 2007). The theoretical

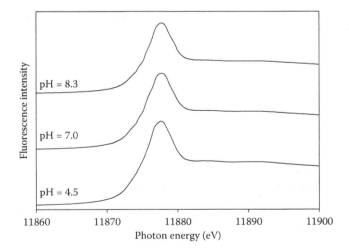

FIGURE 9.22 As K-edge EXAFS spectra of the precipitates collected in the process of simultaneous removal of chromium and arsenate by Fe(II) at various final pH levels (Cr(VI) = 10 μmol L^{-1}, As(V) = 10 μmol L^{-1}, and Fe(II) = 30 μmol L^{-1}).

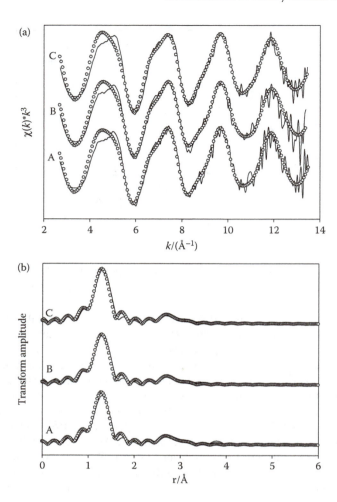

FIGURE 9.23 (a) Raw (solid line) and fitted (dotted line) of k^3 weighted $\chi(k)$ spectra and (b) the corresponding RSFs for arsenate entrapped in the precipitates (Cr(VI) = 10 µmol L^{-1}, As(V) = 10 µmol L^{-1}, and Fe(II) = 30 µmol L^{-1}): (A) pH = 4.5, (B) pH = 7.0, and (C) pH = 8.3. The peak positions are uncorrected for phase shift.

TABLE 9.2

Structural Parameters for As(V) Entrapped in the Precipitates Collected in the Process of Simultaneous Removal of Chromium and Arsenate by Fe(II) at Various Final pH Levels (Cr(VI) = 10 µmol L^{-1}, As(V) = 10 µmol L^{-1}, and Fe(II) = 30 µmol L^{-1})

	As–O				As–Fe			
	N	R	σ^2	E_0 (eV)	N	R	σ^2	E_0 (eV)
pH = 4.5	3.5	1.68	0.0012	9.23	2.6	3.26	0.0084	1.85
pH = 7	3.5	1.69	0.0016	9.95	1.9	3.25	0.0070	2.23
pH = 8.3	3.8	1.69	0.0015	9.90	2.4	3.25	0.0081	3.78

paths of As–Fe, As–Cr, or a combination of As–Fe and As–Cr were attempted when fitting the raw k^3 weighted $\chi(k)$ function in the data reduction process, and the fits were not successful for As–Cr or a combination of As–Fe and As–Cr. Consequently, As–Fe was finally used and the best fit results were shown in Table 9.2. On the basis of these findings, it was inferred that the second shell was primarily attributed to As(V)–Fe bonding and that arsenate mainly coordinated with Fe(III), consistent with the results of arsenate adsorption on $Cr(OH)_3$ solid. Fitting the As–Fe peak was completed in both k-space and R-space using a single As–Fe shell, resulting in a coordination number (CN) of 1.92–2.60. The As–Fe interatomic distances are relatively uniform from 3.25 to 3.26 Å and iron coordination numbers range from 1.9 to 2.6 for the precipitates collected at different pH levels. These results are consistent with the local structural data of ferric arsenate and the bidentate-binuclear attachment of arsenate to FeO(OH) octahydra (Sherman and Randall., 2003; Paktunc et al., 2008).

9.5 INFLUENCE OF COEXISTING IONS ON SIMULTANEOUS REMOVAL OF CR(VI) AND AS(V) BY FE(II)

9.5.1 INFLUENCE OF PHOSPHATE, HUMIC ACID, AND SILICATE ON THE TRANSFORMATION OF CHROMATE BY FE(II)

9.5.1.1 Effect of Phosphate

Chromium removal by Fe(II) in the absence of phosphate (i.e., in the presence of 0 mg L^{-1} P) and iron retained in the precipitate were investigated under suboxic conditions at pH 4–10 within 2 h and the results are shown in Figure 9.24a and b. Chromium removal was as low as 9.9% at pH 4.0 but increased markedly to 63.1% as the pH increased to 5.0, and further rose to 96.0%–99.0% at pH ≥ 6. The trend of iron retained in the precipitate in the absence of phosphate almost coincided with that of chromium removal, implying the synchronous removal of chromium and iron. The species of

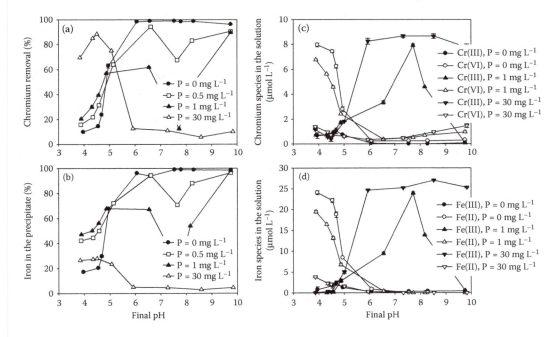

FIGURE 9.24 (a) Chromium removal by Fe(II) and (b) iron retained in the precipitate in the presence of phosphate of various concentrations at different pH levels; (c) speciation of residual chromium; and (d) speciation of residual iron in the solution in this process (Cr(VI) = 10 μmol L^{-1}, Fe(II) = 30 μmol L^{-1}, and P = 0–30 mg L^{-1}).

residual chromium and iron in the process of Cr(VI) removal by Fe(II) were also determined and are demonstrated in Figure 9.24c and d. High concentrations of Cr(VI) and Fe(II) existed in the solution at pH < 6, indicating the slow reduction rate of Cr(VI) by Fe(II), in agreement with the results reported in the literature (Buerge and Hug, 1997; Pettine et al., 1998). The molar ratio of residual Fe(II) and Cr(VI) in the solution at pH 3.9–6.0 was approximately 3.0, same as the initial molar ratio of Fe(II) and Cr(VI) used in this experiment, which was in conformity with the findings that 3.0 mol of aqueous Fe(II) was consumed in the reduction of 1.0 mol of aqueous Cr(VI) (Eary and Rai, 1988; Buerge and Hug, 1997). It was not surprising since Fe(II) was oxidized to Fe(III) by donating one electron and Cr(VI) was reduced to Cr(III) by capturing three electrons in the reaction of Cr(VI) reduction by Fe(II) (Eary and Rai, 1988). XPS analysis was also carried out to determine the molar ratios of Fe/Cr in the precipitates collected at different pH values, as listed in Table 9.3, and the results show that they were in the range of 2.74–3.23, approximate to the theoretical value of 3.0. Thus, it was concluded that $Fe_{0.75}Cr_{0.25}(OH)_3$ precipitates were formed in the process of Cr(VI) reduction by Fe(II), consistent with the results reported in the literature (Fendorf and Li, 1996). The adsorption of Cr(VI) on the precipitates generated in the process of Cr(VI) reduction by Fe(II) was found to be negligible in our study (the results are not shown). Thus, it was believed that Cr(VI) removal was mainly attributable to the reduction of Cr(VI) to Cr(III) by Fe(II) and the following precipitation of Cr(III) as $Fe_{0.75}Cr_{0.25}(OH)_3$. The concentration of Cr(III) in the solution at pH < 6 was slightly higher than that under neutral and alkaline conditions. Thus, the low chromium removal at pH < 6 should be ascribed to both the slow reduction rate of Cr(VI) by Fe(II) and the dissolution of newly formed Cr(III). Figure 9.24c and d shows that the concentrations of residual Cr(VI) and Fe(II) are in the range of 0.24–0.31 μmol L^{-1} and not detectable, respectively, over the pH range of 6.0–9.8, implying the nearly complete reduction of Cr(VI) by Fe(II) in 2 h. Therefore, chromium removal at pH ≥ 6 resulted from the fast reaction rate of Cr(VI) with Fe(II) and the negligible solubility of Cr(III).

The influences of phosphate on chromium removal by Fe(II) at pH 4–10 are also illustrated in Figure 9.24. The presence of 0.5 mg L^{-1} phosphate slightly increased chromium removal at pH 3.9–4.6 but decreased it to different extents with the pH varying from 5.1 to 9.8. An increase in phosphate concentration from 0.5 to 1.0 mg L^{-1} resulted in a more considerable increase and a greater drop in chromium removal at pH 3.9–4.6 and pH 5.1–9.8, respectively. Chromium removal experienced a sharp decline as the pH increased from 6.6 to 7.7 and a significant elevation with further increase in pH in the presence of 0.5 or 1 mg L^{-1} phosphate. Phosphate dosed at 30 mg L^{-1} resulted in a significant improvement in chromium removal at pH < 5.1; however, there was depressed chromium removal from 75.1% to 12.6% as the pH increased from 5.1 to 5.9. Chromium removal was as low as 5.7%–11.0% in the pH range of 6.0–9.8. Interestingly, it was observed that the amount of iron retained in the precipitate deviated from the trend of chromium removal in the presence of phosphate at pH < 5.1, as demonstrated in Figure 9.24a and b, which may be ascribed to the different complexing constants of Fe(III) and Cr(III) with phosphate.

The presence of phosphate decreased the concentration of residual Cr(VI) and Fe(II) in the solution at pH 3.9–4.9 and phosphate of higher concentrations triggered greater decrease, as illustrated

TABLE 9.3

μmolar Ratios of Fe/Cr in the Precipitates Collected at Different pH levels in the Process of Cr(VI) Reduction by Fe(II) (Cr(VI) = 10 μmol L^{-1} and Fe(II) = 30 μmol L^{-1})

Final pH	4.3	6.8	8.3	9.8
Content of Fe (%)	16.05	18.8	18.85	19.10
Content of Cr (%)	5.75	6.1	6.36	5.80
Molar ratio of Fe/Cr	2.74	3.03	2.91	3.23

in Figure 9.24c and d, implying an increase in the reduction rate of Cr(VI) by Fe(II) in this pH range (Sedlak and Chan, 1997). As phosphate forms stronger complexes with Fe(III) than with Fe(II), the redox potential of the Fe(III)/Fe(II) redox couple decreases in the presence of phosphate and Fe(II)–phosphate complexes are more redox reactive than free Fe(II) ions (Stumm and Morgan, 1996). Therefore, the enhancement in the reduction rate of Cr(VI) by Fe(II) at pH 3.9–4.9 in the presence of phosphate should be ascribed to the formation of Fe(II)–phosphate complexes, as illustrated in Figure 9.25.

The kinetics of chromate reduction by Fe(II) in the absence or presence of phosphate at initial pH 4 and 5 were examined and shown in Figure 9.26. In the absence of ligands, Cr(VI) reduction by Fe(II) is not an elementary reaction but can be simply formulated in terms of three successive elementary steps (Buerge and Hug, 1997):

$$Cr(VI) + Fe(II) \Leftrightarrow Cr(V) + Fe(III) \tag{9.5}$$

$$Cr(V) + Fe(II) \Leftrightarrow Cr(IV) + Fe(III) \tag{9.6}$$

$$Cr(IV) + Fe(II) \Leftrightarrow Cr(III) + Fe(III) \tag{9.7}$$

The Cr(VI) to Cr(V) electron transfer is taken to be the rate-limiting step when the dissolved Fe(III) concentrations are small as compared with Fe(II) (Buerge and Hug, 1997). The subsequent reductions to Cr(IV) and Cr(III) are fast and Fe(III) formed *in situ* rapidly precipitated, and therefore the backward reactions are negligible. Thus, the reduction of Cr(VI) by Fe(II) exhibited overall second-order kinetics in the absence of other ligands apart from OH⁻, according to the following rate law (Buerge and Hug, 1997):

$$-\frac{d[Cr(VI)]}{dt} = k_{obs}[Fe(II)][Cr(VI)] \tag{9.8}$$

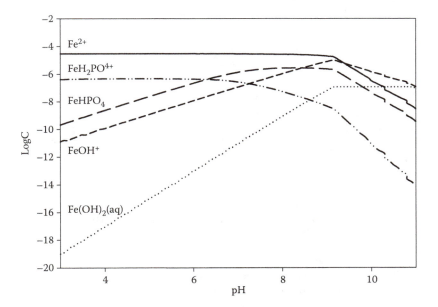

FIGURE 9.25 Complex formation of Fe(II) by phosphate: $TOT_{Fe(II)} = 30\,\mu mol\,L^{-1}$ and phosphate = $32.2\,\mu mol\,L^{-1}$ (1 mg L^{-1}).

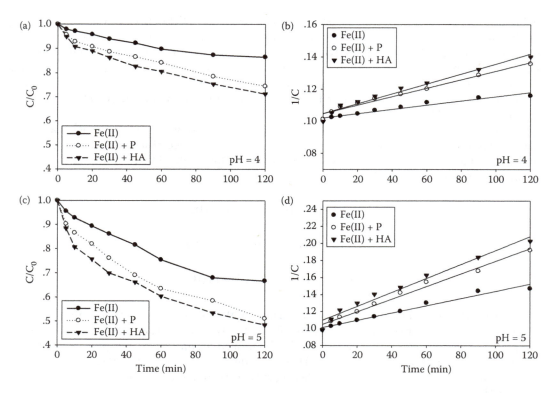

FIGURE 9.26 The kinetics of chromate reduction by Fe(II) in the presence of phosphate or humic acid at initial pH 4 and 5 (a, c). Straight lines are fitted according to the rate law $-d[Cr(VI)]/dt = k_{obs}(pH)[Cr(VI)]$ $[Fe(II)]$ (b, d) (Cr(VI) = 10 μmol L^{-1}, Fe(II) = 30 μmol L^{-1}, P = 1 mg L^{-1}, and HA = 3.81 mg L^{-1}).

As suggested by Buerge and Hug (1998), the overall reduction rate of Cr(VI) by Fe(II) in the presence of Fe(III)-stabilizing ligands could be expressed as follows:

$$-\frac{d[Cr(VI)]}{dt} = \left(\sum k_L[Fe(II)L]\right)[Cr(VI)] \tag{9.9}$$

where Fe(II)L represents individual known iron–ligand complexes and k_L represents the respective pH-dependent rate constants of chromate reduction by Fe(II)L. As illustrated in Figure 9.25, the predominant ferrous species are free Fe(II) and Fe(H$_2$PO$_4$)$^+$ in the presence of phosphate and the amount of Fe(H$_2$PO$_4$)$^+$ is much smaller than that of free Fe(II). Thus, Equation 9.9 can be modified to (the detailed process of deriving this equation is shown in the supporting information)

$$-\frac{d[Cr(VI)]}{dt} = -(k_f + k_L K[L])[Fe(II)_T][Cr(VI)] \tag{9.10}$$

where k_f is the reaction constants of chromate reduction by free Fe(II); K is the stability constant of Fe(H$_2$PO$_4$)$^+$; [L] is the concentration of H$_2$PO$_4^-$ in the system; and [Fe(II)]$_T$ is the total concentration of Fe(II). Equation 9.10 reveals that reduction of Cr(VI) by Fe(II) still follows pseudo-second-order kinetics in case that the concentration of complexed Fe(II) is much smaller than that of free Fe(II). The solid lines in Figure 9.26b and d represent the least squares fit to the experimental data set, obtained with Equation 9.10. The solid lines fit closely to the experimental

TABLE 9.4

Second-Order Rate Coefficients k_{obs}(pH) and k_L of Cr(VI) Reductions

System	k_{obs}(pH) (M^{-1} S^{-1})		k_L (M^{-1} S^{-1})	
	Final pH 3.9 ± 0.03	Final pH 4.87 ± 0.04	Final pH 3.9 ± 0.03	Final pH 4.87 ± 0.04
Cr + Fe	0.73 M^{-1} s^{-1}	2.38 M^{-1} s^{-1}	–	–
Cr + Fe + P	1.47 M^{-1} s^{-1}	4.12 M^{-1} s^{-1}	40.79	95.92
Cr + Fe + HA	1.72 M^{-1} s^{-1}	4.54 M^{-1} s^{-1}	–	–

Note: Cr(VI) = 10 μmol L^{-1}, Fe(II) = 30 μmol L^{-1}, P = 1 mg L^{-1}, and HA = 3.81 mg L^{-1}.

data set, providing strong support for the rate law in Equation 9.10. The pH-dependent overall rate coefficients (k_{obs}(pH)) and k_L are summarized in Table 9.4. The kinetics study confirmed that the formation of Fe(II)–phosphate complexes considerably enhanced the reduction rate of Cr(VI) by Fe(II) at pH 4–5.

Figure 9.24c demonstrates that the presence of phosphate resulted in an elevation in residual Cr(VI) concentration at pH 6.0–9.8 and the increase was greater at higher pH and higher phosphate concentration. However, the residual Fe(II) was nondetectable in the solution at pH 6.0–9.8 in the presence of phosphate, as illustrated in Figure 9.24d, indicating the nonstoichiometric Cr(VI) reduction under alkaline conditions. The solutions were deoxygenated by purging with N$_2$ for 30 min but still contained ~0.12 mg L^{-1} DO. Eary and Rai (1988) examined the reduction of Cr(VI) with Fe(II) in oxygenated solutions containing phosphate of various concentrations and found an increase in the molar ratios of Fe(II) oxidized by Cr(VI) reduced with the increase in phosphate concentration. The rate of the reaction between aqueous Fe(II) and DO relative to the rate of the reaction between aqueous Fe(II) and Cr(VI) increases with increasing pH (Eary and Rai, 1988). The presence of phosphate causes a more significant enhancement in the rate of Fe(II) oxidation by oxygen than that by Cr(VI) (Tamura et al., 1976). Therefore, the amount of Fe(II) that reacts with aqueous Cr(VI) decreased with the increase in pH and phosphate concentration, resulting in an elevation in unreacted Cr(VI) concentration at pH 6.0–9.8. The influence of oxygen on Cr(VI) reduction by Fe(II) in the presence of phosphate observed in this study was much more pronounced than that reported by Eary and Rai (1988) but the reason was unknown at present.

Figure 9.24c and d shows that the presence of 1 mg L^{-1} phosphate had significant influence on the concentrations of Cr(III) and Fe(III) in the solution at pH > 4.7. As the pH was raised from 4.9 to 7.7, the concentrations of soluble Cr(III) and Fe(III) increased to 7.91 and 23.9 μmol L^{-1}, respectively. However, they decreased sharply to 0.07 and 0.53 μmol L^{-1}, respectively, as the pH varied from 7.7 to 9.8. The high concentrations of Cr(III) and Fe(III) in the solution at pH 4.9–9.0 should be ascribed to the formation of soluble complexes between phosphate and the newly formed Cr(III)/Fe(III). Rai et al. (2004) reported an obvious increase in Cr(OH)$_{3(am)}$ solubility in the presence of phosphate at pH 4.5–12. They concluded that the increase in Cr(OH)$_{3(am)}$ solubility resulted from the formation of soluble complexes between Cr(III) and phosphate species, that is, Cr(OH)$_3$H$_2$PO$_4^-$, Cr(OH)$_3$(H$_2$PO$_4$)$^{2-}$, and Cr(OH)$_3$HPO$_4^{2-}$.

The total solubility of (hydr)oxides considering the possibility of complex formation with ligands can be expressed by the following equation (Stumm and Morgan, 1996):

$$Me_T = [Me]_{free} + \sum Me[OH]_n + m \sum [Me_m H_k L_n [OH]_i]$$ (9.11)

where L stands for the ligands other than OH$^-$. As the presence of 1 mg L^{-1} phosphate had significant effect on the concentrations of Cr(III) and Fe(III) in the solution at pH > 4.7 and the maximum solubility of Cr(III) and Fe(III) was observed at pH 7.7, where HPO_4^{2-} was the predominant phosphate species in the solution, the solubility of Cr(III) and Fe(III) can be expressed by Equations 9.12 and 9.13, respectively:

$$Cr_T = [Cr^{3+}]_{free} + [CrOH^{2+}] + [Cr(OH)_2^+] + [Cr(OH)_3]_{(aq)} + [Cr(OH)_4^-] + [Cr(HPO_4)^+] \\ + [Cr(OH)(HPO_4)]_{(aq)} \quad (9.12)$$

$$Fe_T = [Fe^{3+}]_{free} + [Fe\,OH^{2+}] + [Fe(OH)_2^+] + [Fe(OH)_3]_{(aq)} + [Fe(OH)_4^-] + [Fe(HPO_4)^+] \\ + [Fe(OH)(HPO_4)]_{(aq)} \quad (9.13)$$

Equations 9.12 and 9.13 show that the solubility of (hydr)oxides is determined by both OH^- and HPO_4^{2-}, depending on the complexing ability of OH^- and HPO_4^{2-} as well as their relative concentrations. Therefore, it was the complexation of Cr(III)/Fe(III) with HPO_4^{2-} controlling the solubility of the $Fe_{0.75}Cr_{0.25}(OH)_3$ at pH 7.7. On the other hand, the significant reduction in the concentration of soluble Cr(III) and Fe(III) from pH 7.7 to 9.8 should be ascribed to the much higher concentrations of OH^- at pH 9.8 and the resulted $Cr(OH)_3$ and/or $Fe_{0.75}Cr_{0.25}(OH)_3$ precipitation. Equations 9.12 and 9.13 also indicate that the complexation of Cr(III)/Fe(III) with HPO_4^{2-} may outweigh the complexation of Cr(III)/Fe(III) with OH^- at high pH when phosphate is present at extremely high concentrations. It should be noted that the effect of 30 mg L^{-1} phosphate on the reduction products of chromate by Fe(II) was examined just to prove this assumption. As shown in Figure 9.24c and d, the concentrations of soluble Cr(III) and Fe(III) at pH > 5.9 were kept at 7.74–8.63 and 24.7–27.0 μmol L^{-1}, respectively, in the presence of 30 mg L^{-1} phosphate. The results proved that the solubility of Cr(III) and Fe(III) was controlled by the complexation of Cr(III)/Fe(III) with HPO_4^{2-} at pH 5.9–9.8 in this case.

9.5.1.2 Effect of HA

The influence of HA dosed at two concentrations, 1.27 and 3.81 mg L^{-1}, on chromium removal over a wide pH range was demonstrated in Figure 9.27a. HA applied at both concentrations improved chromium removal at pH < 5.0, but the improvement were not closely proportional to the HA concentration. The presence of 1.27 mg L^{-1} HA had little effect on chromium removal at pH 6.4 but decreased it from ~98.4% to 84.9% and 7.4% at pH 7.7 and 8.2, respectively. The removal rate of chromium was decreased to 3.7%–15.4% at pH 6.5–8.2 in the presence of 3.81 mg L^{-1} HA. As the pH was raised from 8.2 to 9.8, a similar phenomenon of elevation in chromium removal was observed as in the case of phosphate.

In many environmental compartments, HA is one of the dominant reductants. However, blank experiments carried out in this study indicated that chromate reduction by HA was extremely slow and chromium removal by HA alone was negligible in 2 h (as illustrated in Figure 9.28). However, HA exerted influence on chromate reduction by Fe(II). As shown in Figure 9.27b and c, the concentrations of both Cr(VI) and Fe(II) in the solution were reduced in the presence of 3.81 mg L^{-1} HA at pH 3.9–5.0, indicating that the presence of HA resulted in faster reaction rate of Cr(VI) with Fe(II) at pH 3.9–5.0, which was confirmed by the results of kinetics study, as demonstrated in Figure 9.26. Buerge and Hug (1998) also reported the accelerated Cr(VI) reduction by Fe(II) complexes with various Fe(III)-stabilizing organic ligands (e.g., carboxylates and phenolates) at pH 4.0–5.5. The Fe(II)–SRFA (Suwannee River fulvic acid) system also exhibited significantly higher chromate reduction rate than the Fe(II)-only system under certain conditions (Agrawal et al., 2009). Agrawal et al. (2009) showed that the kinetics of Cr(VI) reduction by Fe(II) in the presence of organic ligands becomes complex because of side reactions and the pH dependence of complex formation. However, overall stoichiometric relationship of 3 moles of Fe(II) oxidized for every mole of Cr(VI) reduced was observed by Buerge and Hug (1998) even in the presence of organic ligands. Our observation in

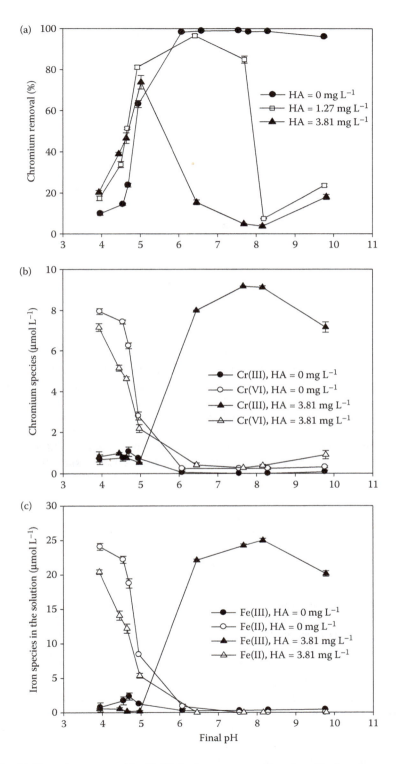

FIGURE 9.27 (a) Chromium removal by Fe(II) in the presence of humic acid of various concentrations at different pH levels, (b) speciation of residual chromium, and (c) speciation of residual iron in the solution in the process (Cr(VI) = 10 μmol L^{-1}, Fe(II) = 30 μmol L^{-1}, and TOC = 0–3.81 mg L^{-1}).

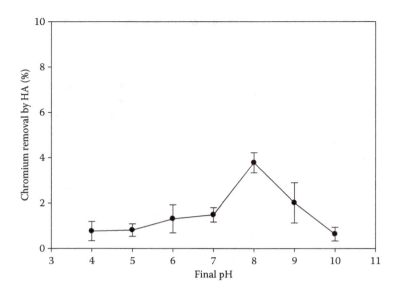

FIGURE 9.28 Chromium removal at various pH levels by HA alone (Cr(VI) = 10 μmol L⁻¹ and HA = 3.81 mg L⁻¹).

the presence of humic acid agreed with that reported by Buerge and Hug (1998). The side reaction of Fe(III) re-reduction to Fe(II) found by Agrawal et al. (2009) might be due to high dissolved organic carbon (DOM) concentrations, long time scale, and due to the possible photochemical effects of light on Fe(III) reduction if the experiments were not performed in the dark. HA's compositional heterogeneity precludes us from constructing a kinetic model as shown in Equation 9.10 because the intrinsic stability constants of HA and Fe(II) are not known. By ignoring the side reaction and assuming that $K[L]$ was much smaller than 1, Equation 9.14 was used to simulate the data of kinetics study, as shown in Figure 9.26 and Table 9.4:

$$\frac{1}{[Cr(VI)]_t} = 3k'_{obs}(pH)t + \frac{1}{[Cr(VI)]_0} \tag{9.14}$$

$$k'_{obs}(pH) = k_f + k_L K[L] \tag{9.15}$$

Similar to the case of chromium removal in the presence of phosphate, the presence of HA also resulted in a slight increase in residual Cr(VI) concentration at pH 6.0–9.8 and the increase was greater at higher pH, as illustrated in Figure 9.27b. This elevation in residual Cr(VI) concentration may be associated with the enhancement in the rate of the reaction between aqueous Fe(II) and DO caused by the presence of HA. As shown in Figure 9.27b, the Cr(III) concentration in the solution increased sharply from 0.53 μmol L⁻¹ at pH 5.0 to 8.00 μmol L⁻¹ at pH 6.5, and was up to 9.18 μmol L⁻¹ with a further pH increase to 7.7. A similar trend was found regarding the concentration of Fe(III) in the solution (Figure 9.27c). Guan et al. (2009) reported that the presence of HA strongly inhibited the precipitation of ferric hydroxide by forming soluble complexes with Fe(III) under alkaline conditions. Liu et al. (2009) found that humic acid suppressed iron precipitation due to the formation of soluble Fe–humate complexes and stably dispersed fine Fe (oxy)hydroxide colloids. Therefore, the significant influence of HA on the solubility of Cr(III) and Fe(III) at pH 6.5–9.8 should be associated with the formation of soluble Cr(III)–humate and Fe(III)–humate complexes.

9.5.1.3 Effect of Silicate

Different from phosphate and HA, silicate had little impact on chromium removal at pH < 5.0 as shown in Figure 9.29. Figure 9.29b and c illustrated that the presence of 9.3 mg L^{-1} silicate had little effect on both residual Cr(VI) and Fe(II) at pH < 5.0 and thus little effect on the reduction rate of Cr(VI) by Fe(II) at pH < 5.0. However, chromium removal was depressed drastically from 96.9% at pH 7.6 to 25.9% at pH 8.5 in the presence of 9.3 mg L^{-1} silicate and a further increase in pH from 8.5 to 9.8 resulted in a minor elevation in chromium removal. Silicate of higher concentration had adverse impact on chromium removal over a wider pH range. Figure 9.29b shows that there was only a slight increase in the residual Cr(VI) concentration but a considerable rise in the concentration of residual Cr(III) and Fe(III) at pH > 7.6. The different influences of silicate on the transformation of chromium by Fe(II) at different pH values were associated with the species distribution of silicic acid with pH. Silicic acid exists as a neutral molecule at pH below 7 which does not complex with Fe(II)/Cr(III)/Fe(III). Therefore, silicate has little effect on the reductive capacity of Fe(II) and the reduction of Cr(VI) by Fe(II). However, when silicic acid starts to dissociate under alkaline conditions, its complex power with Cr(III)/Fe(III) increases significantly, resulting in the formation of soluble complexes with Cr(III)/Fe(III) and thus sequesters the precipitation of Fe$_{0.75}$Cr$_{0.25}$(OH)$_3$. Similar to phosphate, complexation of silicate with Fe(II) gave rise to an enhancement in the rate of aqueous Fe(II) oxidation by DO and the amount of Fe(II) that reacted with aqueous Cr(VI) decreased with increasing pH, resulting in a slight increase in the residual Cr(VI) concentration at pH > 7.6. Therefore, the significant impact of silicate on chromium removal under alkaline conditions should be largely ascribed to the inhibition of Fe$_{0.75}$Cr$_{0.25}$(OH)$_3$ precipitation. Previous researchers also reported that silicate could interact with Fe(III) to form soluble polymers and highly dispersed colloids that were not removable by filtration (Davis et al., 2001; Liu et al., 2007). The small reduction in the concentration of Cr(III) and Fe(III) in the solution at pH > 8.5 (Figure 9.29b and c), corresponding to the slight increase in chromium removal (Figure 9.29a), should be ascribed to the stronger complexation ability of OH$^-$ at higher pH, as mentioned above.

9.5.1.4 Influence of Ionic Strength on the Precipitation of Cr(III)

As mentioned before, the filter paper with a pore size of 0.45 μm was used to separate particulate from dissolved matter in this study. However, the "dissolved matter" could be further divided into two kinds of species, that is, truly dissolved complexes and colloidally dispersed particles (d < 0.45 μm) that cannot be retained by the filter paper. The colloidal stability can be decreased by increasing the ionic strength (addition of salts), which can compress the electric double layer of the colloids (Stumm and Morgan, 1996). Thus, the effect of ionic strength on the chromium removal by Fe(II) in the presence of various ligands at pH 7.7 was examined to verify the existence of colloidally dispersed particulate Cr(III) species that are sufficiently small to pass through a membrane filter.

As shown in Figure 9.30, chromium removal was markedly enhanced as ionic strength increased from 10 to 200 mmol L^{-1} in the presence of various ligands. Since chromium removal was mainly controlled by the precipitation of newly formed Cr(III) species at pH ~7.6, increasing ionic strength facilitated the aggregation and precipitation of Cr(III) species. However, the degree of increase in chromium removal caused by increasing ionic strength varied with different ligands, ascribed to the different stabilities of different Cr(III) ligands. It is difficult to differentiate the amount of truly dissolved species and the colloidal species as the chromium removal keeps a rising tendency at ionic strengths of up to 200 mmol L^{-1}, indicating that more Cr(III) would precipitate with further increase in ionic strength. Nonetheless, a conclusion could be drawn that fine colloidal particles (d < 0.45 μm) were formed in the system containing phosphate, silicate or HA, and the stability of these colloids followed this order: Cr(III)-HA (3.81 mg L^{-1})> Cr(III)-silicate (18.6 mg L^{-1}) > Cr(III)-phosphate (1 mg L^{-1}).

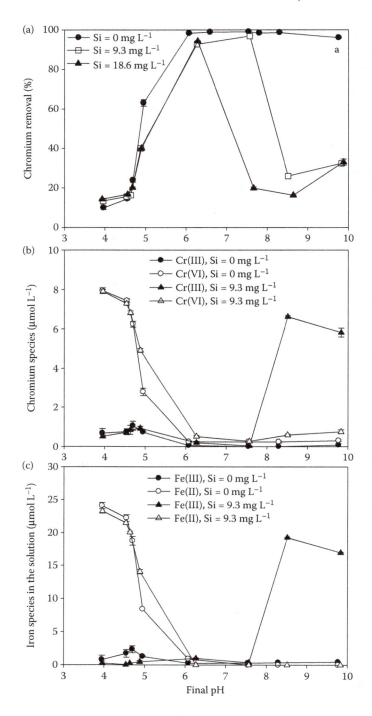

FIGURE 9.29 (a) Chromium removal by Fe(II) in the presence of silicate of various concentrations at different pH levels, (b) speciation of residual chromium, and (c) speciation of residual iron in the solution in the process (Cr(VI) = 10 μmol L^{-1}, Fe(II) = 30 μmol L^{-1}, and Si = 0–18.6 mg L^{-1}).

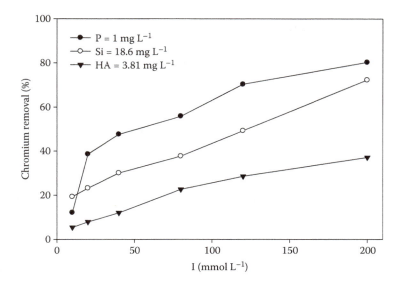

FIGURE 9.30 Effect of ionic strength on chromium removal by Fe(II) at pH 7.7: (Cr(VI) = 10 μmol L^{-1}, Fe(II) = 30 μmol L^{-1}, P = 1 mg L^{-1}, Si = 18.6 mg L^{-1}, and HA = 3.81 mg L^{-1}).

9.5.2 INDIVIDUAL AND COMBINED INFLUENCE OF CALCIUM AND ANIONS ON SIMULTANEOUS REMOVAL OF CHROMATE AND ARSENATE BY FE(II)

9.5.2.1 Simultaneous Removal of Chromium and Arsenate by Fe(II)

The performance of simultaneous removal of chromium and arsenate by Fe(II) was examined under suboxic conditions (Figure 9.31). Chromium could be removed efficiently at pH > 6 with the removal efficiency varying from 87.0% to 94.7%, as demonstrated in Figure 9.31a. Comparing with the cases in the absence of arsenate as discussed in the previous part, chromium removal was increased at pH < 5.0 and declined at pH ≥ 5.0 due to the presence of 10 μmol L^{-1} arsenate, which should be ascribed to the complexing ability of arsenate with Fe(II) and the newly formed Cr(III)/Fe(III). As only the precipitated Fe(III) formed *in situ* could mediate arsenic removal (McNeill and Edwards, 1997), the concentration of iron entrapped in the precipitate was also examined and shown in Figure 9.31c. Arsenate removal improved significantly from 45.5% to 90.4% as the pH increased from 4.0 to 6.3 (as shown in Figure 9.31b), which was mainly ascribed to the increase in the amount of precipitated iron. As the pH increased from 6.3 to 9.8, arsenate removal decreased with increasing pH although the amount of precipitated iron kept almost constant in this pH range. It should be associated with the shift of arsenate species from $H_2AsO_4^-$ to $HAsO_4^{2-}$ (Guan et al., 2008) and the decrease in the surface charge of the precipitates with increasing pH, as shown in Figure 9.31d. Considering the concomitant removal of chromium and arsenate, pH 6.3 was obviously the optimum pH.

9.5.2.2 Effect of Calcium Ions

The influence of calcium on the coremoval of chromium and arsenate by Fe(II) is demonstrated in Figure 9.31. Calcium dosed at 20–50 mg L^{-1} showed negligible effect on chromium removal throughout the pH range of 4–10. Although the presence of Ca^{2+} had no significant effect on arsenate removal at pH < 6.7, notable enhancement in arsenate removal was observed at pH > 7 in the presence of calcium and a higher concentration of Ca^{2+} resulted in greater arsenate removal. The results of previous research (Guan et al., 2009) indicated that the improvement of arsenic removal at pH 7–9 in the $KMnO_4$–Fe^{2+} process caused by the introduction of calcium was associated with the following three possible reasons: (1) the specific adsorption of Ca^{2+} increased the surface charge; (2) the formation of

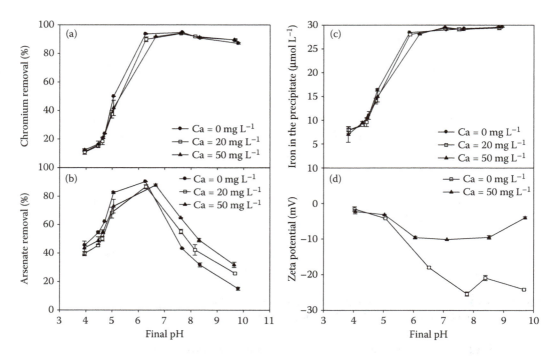

FIGURE 9.31 Influence of coexisting Ca^{2+} on simultaneous removal of chromium and arsenate by Fe(II): (a) chromium removal, (b) arsenate removal, (c) iron retained in the precipitate, and (d) zeta potential of the flocs formed in this process (Cr(VI) = 10 μmol L^{-1}, As(V) = 10 μmol L^{-1}, Fe(II) = 30 μmol L^{-1}, and Ca^{2+} = 0–50 mg L^{-1}).

calcium carbonate precipitate which can coprecipitate arsenate; and (3) the introduction of calcium results in more precipitated ferrous hydroxide or ferric hydroxide. As shown in Figure 9.31c, Ca^{2+} had no influence on the amount of iron retained in the precipitate under alkaline conditions, the third reason mentioned above, contributing to arsenic removal in the $KMnO_4$–Fe^{2+} process, was not applicable to this process under given conditions. Figure 9.31d illustrated that the presence of 50 mg L^{-1} Ca^{2+} had a minor effect on the zeta potential at pH 4 and 5 but it increased the surface charge of the precipitates at pH > 5 and the enhancement was greater at higher pH, contributing to the enhanced retention of arsenate under neutral and alkaline conditions. Since bicarbonate was used in this system to provide alkalinity, the concentrations of CO_3^{2-} were calculated, according to the equilibrium of dissolved carbonate (closed system) (Stumm and Morgan, 1996), and the results obtained were 5.48×10^{-6}, 5.31×10^{-5}, and 0.36×10^{-3} mol L^{-1}, respectively, at pH 8, 9, and 10, respectively. The concentrations of Ca^{2+} used in this study were $(0.5–1.25) \times 10^{-3}$ mol L^{-1} (20–50 mg L^{-1}) and the K_{sp} of $CaCO_3$ at 22°C is about $10^{-8.4}$ (Stumm and Morgan, 1996). Therefore, the solution was undersaturated with respect to calcium carbonate at pH 8 and saturated at pH 9 and 10.

The FTIR spectra were collected for the precipitates collected at pH 4.3, 7, and 8.9 in the presence of 50 mg/L Ca^{2+} (Figure 9.32). The FTIR spectra of the precipitates collected at pH 4.3 and 7 resembled each other and no evident bands corresponding to calcium carbonate were found under both conditions. However, strong and well-resolved bands appearing at 1415, 874, and 712 cm^{-1} were observed in the FTIR spectrum of precipitates collected at pH 8.9. The peak at 1415 cm^{-1} could be assigned to the formation of amorphous calcium carbonate while those at 874 and 712 cm^{-1} might be assigned to the characteristic vibration of a calcite phase (Shen et al., 2006). Accordingly, the enhancement of arsenate removal in the presence of calcium ions at pH > 7 might be ascribed to the following two reasons: (1) the decrease of negative surface charge on the precipitates due to the adsorption of calcium at pH > 7 and (2) due to the coprecipitation of calcium carbonate and calcite at pH > 8.

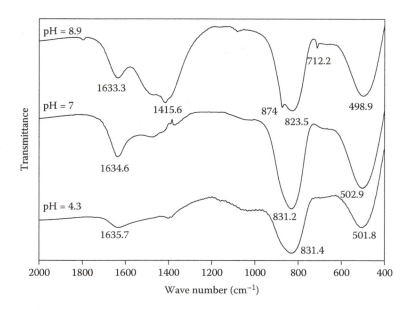

FIGURE 9.32 FTIR spectra of the precipitates collected in the process of simultaneous removal of chromium and arsenate by Fe(II) in the presence of calcium at various final pH levels (Cr(VI) = 10 μmol L^{-1}, As(V) = 10 μmol L^{-1}, Fe(II) = 30 μmol L^{-1}, and Ca^{2+} = 50 mg L^{-1}).

9.5.2.3 Effect of Phosphate Ions

The influences of phosphate dosed at two concentrations, 0.5 and 1 mg L^{-1} (as P), on the coremoval of chromium and arsenate over a wide pH range were demonstrated in Figure 9.33. The influence of phosphate on the transformation of Cr(VI) by Fe(II) (in the absence of arsenate) was systematically investigated in the previous part. By comparing the influence of phosphate on chromium removal in the absence of arsenate with that in the presence of arsenate (as demonstrated in Figure 9.33a), it was found that the influences of phosphate on chromium removal under these two cases were very similar. Accordingly, the mechanism of the influence of phosphate should be the same in both the systems as follows: the elevation in chromium removal at pH < 5.0 in the presence of phosphate was due to the formation of Fe–phosphate complexes, which are more redox active than the free Fe(II) species, accelerating the reduction of chromate; the sharp drop in chromium removal at pH 6.0–7.7 was due to the formation of soluble Cr(III)/Fe(III)–phosphate complexes, sequestering the precipitation of Fe(III)/Cr(III) species; the significant reduction in the concentration of soluble Cr(III) from pH 7.7 to 9.8 should be ascribed to the much higher concentration of OH$^-$ at pH 9.8 and the resultant Fe$_{0.75}$Cr$_{0.25}$(OH)$_3$ precipitation.

The decrease in arsenate removal caused by the presence of phosphate was dependent on pH and phosphate concentrations, as shown in Figure 9.33b. The most significant influence of phosphate on arsenate removal occurred at pH 6.3. The increase in phosphate concentration resulted in a more pronounced decrease in arsenate removal over the pH range of 4–9. The presence of phosphate increased the amount of iron retained in the precipitate at pH < 5.0 while significantly inhibited the precipitation of Fe(III) formed *in situ* at pH 7–9 and only 26.4% of the total iron was precipitated at pH 7.7 in the presence of 1 mg L^{-1} phosphate (shown in Figure 9.33c). Phosphate removal in this process was also determined and illustrated in Figure 9.33d. The trend of phosphate removal as a function of pH was very similar to that of arsenate, resulting from the similar structure and deprotonation constants of phosphate and arsenate in solution (Luengo et al., 2007). Although more iron was precipitated at pH < 5, that is, formation of more adsorption sites, caused by the presence of phosphate, arsenate removal was still decreased considerably, indicating the strong competition

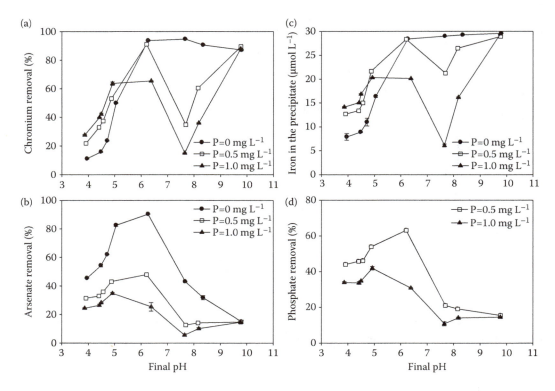

FIGURE 9.33 Influence of coexisting phosphate on simultaneous removal of chromium and arsenate by Fe(II): (a) chromium removal, (b) arsenate removal, (c) iron retained in the precipitate, and (d) phosphate removed in this process $(Cr(VI) = 10\ \mu mol\ L^{-1}$, $As(V) = 10\ \mu mol\ L^{-1}$, $Fe(II) = 30\ \mu mol\ L^{-1}$, and $P = 0-1.0\ mg\ L^{-1})$.

from phosphate (Jain and Loeppert, 2000; Zhao and Stanforth, 2001; Guan et al., 2009). On the other hand, the great drop in arsenic removal caused by the presence of phosphate at pH 7–9 was associated with not only the competition of phosphate for the adsorption sites but also the reduction in the surface sites. As the pH increased from pH 7.7 to 9.8, the amount of precipitated iron rose significantly, being associated with the high concentration and strong complexation ability of OH⁻ as discussed previously, but arsenate removal did not rise correspondingly. This should be attributable to the accumulation of negative charge on the precipitates which increased the repulsion between arsenate species and the precipitates.

9.5.2.4 Effect of Silicate

The influence of silicate on simultaneous removal of chromium and arsenate was investigated, as shown in Figure 9.34. Both chromium and arsenate removal were almost unaffected in the presence of 9.3–18.6 mg L⁻¹ silicate (as Si) at pH < 6.3; however, they were significantly inhibited with further increase in pH. The pH-dependent influence of silicate on the removal efficiency of chromium and arsenate should be associated with species distribution of silicic acid as a function of pH. Silicic acid is a weak acid, which is not dissociated and exist in the solution as neutral molecules at pH < 7. Thereby silicate generally has a weak affinity for iron hydroxide surface and is difficult to coordinate with Fe(III) formed *in situ* at pH < 7, consequently exerting little effect on arsenic removal under acidic conditions. When silicate is dissociated under alkaline conditions, its affinity for iron hydroxide surface increases significantly, resulting in a stronger competition with arsenate (Guan et al., 2009). Smith and Edwards (2005) observed a significant reduction in arsenic adsorption on amorphous Fe(OH)₃ or activated alumina at pH 8.5 by the coexisting

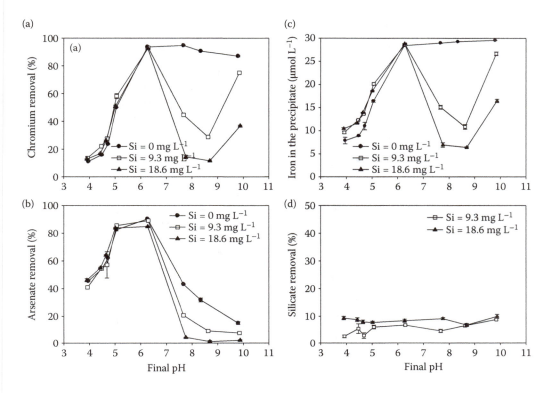

FIGURE 9.34 Influence of coexisting silicate on simultaneous removal of chromium and arsenate by Fe(II): (a) chromium removal, (b) arsenate removal, (c) iron retained in the precipitate, and (d) silicate removed in this process (Cr(VI) = 10 μmol L^{-1}, As(V) = 10 μmol L^{-1}, Fe(II) = 30 μmol L^{-1}, and Si = 0–18.6 mg L^{-1}).

40 mg/L silicate. Figure 9.35c shows that silicate seriously inhibited the precipitation of newly formed Cr(III)/Fe(III) and reduced the amount of precipitated iron at pH 7–8.6. Only less than 10% silicate was entrapped in the precipitate throughout the pH range of 4–10, as demonstrated in Figure 9.34d, showing that the adsorption of silicate on the precipitates was trivial. Therefore, the marked decrease in arsenate removal induced by silicate at pH 7.0–8.6 should be mainly ascribed to the formation of soluble Cr(III)/Fe(III)–silicate species. The formation of Fe$_{0.75}$Cr$_{0.25}$(OH)$_3$ precipitates significantly increased with the rising of pH from 8.6 to 9.8, due to the increasing concentration of OH$^-$, as discussed above. However, it did not enhance arsenate removal, which should be associated with the competition of OH$^-$ for the adsorption sites on the surface of Fe$_{0.75}$Cr$_{0.25}$(OH)$_3$ precipitates.

9.5.2.5 Effect of Humic Acid

The influences of HA dosed at 1.27 or 3.81 mg/L on simultaneous removal of chromium and arsenate are demonstrated in Figure 9.35. Considering the reductive characteristics of HA, blank experiments were carried out to examine chromate reduction by HA in the presence of arsenate and results indicated that the reaction rate was extremely slow and both chromium removal and arsenate removal by HA alone was negligible in 2 h (date not shown). As shown in Figure 9.35a, the presence of HA enhanced chromium removal at pH < 5.0 and the enhancement was greater at higher HA concentrations. Nonetheless, chromium removal was decreased radically from 95.3%–98.3% to 4.9%–21.2% at pH 7.5–9.8 and to 7.0%–18.5% at pH 6.3–9.8, respectively, in the presence of 1.27 and 3.81 mg L^{-1} HA, respectively. The variation in chromium removal induced by the presence of HA was consistent with that observed in the Fe(II)-mediated chromium removal process. The

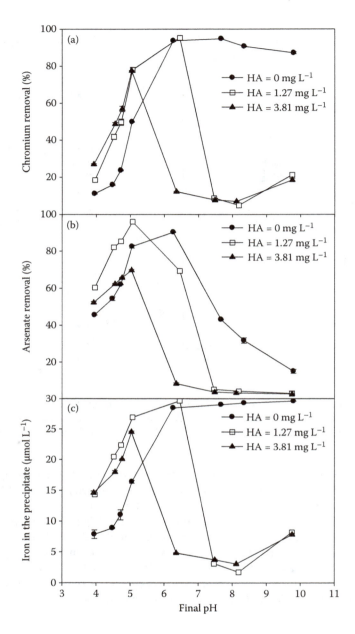

FIGURE 9.35 Influence of coexisting humic acid on simultaneous removal of chromium and arsenate by Fe(II): (a) chromium removal, (b) arsenate removal, and (c) iron entrapped in the precipitate (Cr(VI) = 10 μmol L^{-1}, As(V) = 10 μmol L^{-1}, Fe(II) = 30 μmol L^{-1}, and HA = 0–3.81 mg L^{-1} as DOC).

presence of HA improved the chromium removal at pH < 5 by accelerating the rate of chromate reduction and intensively depressed the chromium removal at pH > 7 as a result of the formation of soluble Cr(III)/Fe(III)–humate complexes.

The presence of 1.27 mg L^{-1} of HA improved arsenate removal by ~10.4% at pH 4.0–5.1, as illustrated in Figure 9.35b. However, arsenate removal dropped to 70.0% at pH 5.1 as HA was applied at 3.81 mg L^{-1}. The amount of iron entrapped in the precipitate was examined and shown in Figure 9.35c. The amount of precipitated iron was increased by 64.1%–130.1% at pH 4.0–5.1 due to the presence of 1.27 mg L^{-1} HA and it decreased slightly by increasing HA concentration from 1.27 to 3.81 mg L^{-1}. Thus, arsenate removal in the presence of HA at pH ≤ 5.1 was determined not only by the amount of precipitated iron but also by the competition from HA, which becomes more remarkable with increasing HA concentration.

HA had a detrimental effect on arsenate removal in the pH range of 5.1–9.8. With the presence of 1.27 mg L^{-1} HA, arsenate removal was decreased significantly from 96.1% at pH 5.1%–69.3% at pH 6.5, and to only 2.9%–5.2% at pH 7.5–9.8. Raising HA concentration from 1.27 to 3.81 mg L^{-1} resulted in a detrimental decrease in arsenate removal over a wider pH range (6.3–9.8), which should be mainly associated with the inhibition of precipitation of newly formed Fe(III) caused by complexation of HA (shown in Figure 9.35c). Our previous study noted that HA sequestered the precipitation of *in situ* formed Fe(III) derived from the oxidation of Fe(II) by permanganate under alkaline conditions and resulted in low arsenic removal (Guan et al., 2009). Some other researchers also reported that HA, oxalate, and salicylic acid could form soluble complexes with Fe and Al to dissolve clay minerals and adsorption was reduced accordingly (Rahni and Legube, 1996; Luo et al., 2006).

9.5.2.6 Combined Effects of Coexisting Calcium and Anions

The combined impacts of phosphate and Ca^{2+}, silicate and Ca^{2+}, HA and Ca^{2+} on coremoval of chromium and arsenate at pH 7.6 were studied, respectively, and the results are shown in Figure 9.36. Figure 9.36a illustrates that in the presence of 1 mg L^{-1} phosphate, the addition of 5 mg L^{-1} calcium caused a sharp elevation in chromium removal from 13.3% to 76.0% and chromium removal rose slowly to 95.8% as the concentration of calcium increased to 80 mg L^{-1}. However, arsenate removal was enhanced slowly with the increase in Ca^{2+} concentration. Arsenate was increased from 5.6% in the absence of calcium to 19.4% at 10 mg L^{-1} Ca^{2+} and mounted up to only 30.2% even in the presence of 80 mg L^{-1} Ca^{2+}. In the presence of HA, as compared with the case with phosphate, a similar variation trend in the removal of chromium and arsenate was observed with the increasing concentration of Ca^{2+}, as illustrated in Figure 9.36c. The application of 20 mg L^{-1} Ca^{2+} resulted in a dramatic increase from 7.7% to 91.8% in chromium removal, while arsenate removal was increased from 3.7% to 36.5% accordingly. Chromium and arsenate removal increased slowly to 95.6% and 45.7%, respectively, as Ca^{2+} concentration was further increased from 20 to 80 mg L^{-1}. However, the application of Ca^{2+} resulted in a different changing trend in chromium and arsenate removal in the presence of silicate, as demonstrated in Figure 9.36b. Ca^{2+} applied at 20 mg L^{-1} increased chromium and arsenate removal by only 13.2% and 8.4%, respectively. Then chromium and arsenate removal experienced a rapid rise to 91.4% and 49.4%, respectively, as Ca^{2+} concentration increased to 40 mg L^{-1}. Chromium and arsenate removal increased slowly to 97.5% and 58.6%, respectively, as Ca^{2+} concentration was further increased from 20 to 80 mg L^{-1} in the presence of 18.6 mg L^{-1} silicate.

As discussed previously, the trivial removal of both chromium and arsenate in the presence of phosphate, silicate, or HA at pH 7.6~7.7 should be ascribed to the formation of soluble or colloidal complexes that are not filterable. The amount of precipitated iron increased significantly with increasing concentrations of calcium and more than 95% of added iron was entrapped in the precipitates at 80 mg L^{-1} Ca^{2+}, in a trend almost as identical as that of chromium removal, as shown in Figure 9.36. Therefore, the remarkable improvement in chromium removal caused by the presence of Ca^{2+} should be ascribed to the promoted aggregation of the colloidal complexes. The introduction of Ca^{2+} decreased the negative surface charge of precipitates, facilitated the formation of larger Cr(III)/Fe(III) hydroxide flocs, increased the amount of precipitated solids, and thereby improved chromium removal significantly. Figure 9.36 shows that, in the presence of phosphate, silicate, or HA, arsenate removal was increased rapidly at lower Ca^{2+} concentrations when the amount of precipitated iron increased significantly and it was enhanced slowly even when the amount of precipitated iron maintained almost constant at higher Ca^{2+} concentration. Thus, the increase in arsenate removal with increasing Ca^{2+} concentration was mainly ascribed to both the

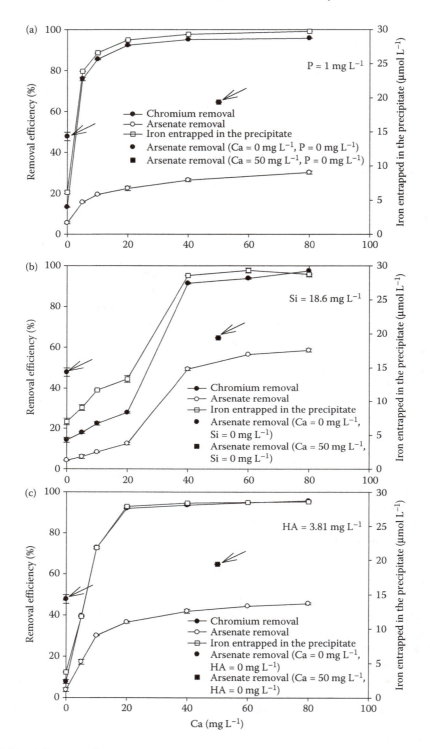

FIGURE 9.36 Effect of Ca^{2+} concentration on simultaneous removal of chromium and arsenate by Fe(II) in the presence of (a) phosphate, (b) silicate, (c) humic acid (final pH = 7.6–7.7, Cr(VI) = 10 μmol L^{-1}, As(V) = 10 μmol L^{-1}, Fe(II) = 30 μmol L^{-1}, P = 1.0 mg L^{-1}, Si = 18.6 mg L^{-1}, HA = 3.81 mg L^{-1}, and Ca^{2+} = 0–80 mg L^{-1}). (The black spot marked by arrow stands for arsenate removal in the absence of the four ions.)

increase in the amount of precipitated iron and the decrease in the negative surface charge of precipitates caused by Ca^{2+}.

At pH 7.7, arsenate was removed by 43.1% in the absence of Ca^{2+} or anions and by 64.7% in the presence of 50 mg L^{-1} Ca^{2+}, which are marked in Figure 9.36 by arrows. In the presence of 80 mg L^{-1} Ca^{2+}, arsenate was removed by 30.2%, 58.6%, and 45.7%, respectively, in the presence of 1 mg L^{-1} phosphate, 18.6 mg L^{-1} silicate, and 3.81 mg L^{-1} humic acid. The different arsenate removal efficiencies in the presence of 80 mg L^{-1} Ca^{2+} and anions of various concentrations should be ascribed to the different competitive capability and respective concentration of anions because more than 95% of iron was precipitated in all cases. The presence of 80 mg L^{-1} Ca^{2+} could not balance the negative influence of 1 mg L^{-1} phosphate on arsenate removal at pH 7.7, indicating the strong competition of phosphate with arsenate for adsorption sites. The combined influence of Ca^{2+} and HA or silicate resulted in an arsenate removal efficiency greater than that in the absence of Ca^{2+} or anions but lower than that in the presence of Ca^{2+} alone. On the basis of the discussion above, the following conclusions could be made: (i) HA or silicate exerted negative effect on the arsenate removal mainly by inhibiting the formation of precipitates and thus decreasing the number of adsorption sites, while phosphate could not only hinder the formation of precipitates but also strongly compete for the available adsorption sites with arsenate; (ii) the degree of competitive effects of these three anions on arsenate removal decreased in the following order: 1 mg L^{-1} phosphate > 3.81 mg L^{-1} humic acid > 18.6 mg L^{-1} silicate; and (iii) the presence of calcium could only balance the negative effect of anions on the precipitation of Cr(III)/Fe(III) but failed to alleviate the competition between the anions and arsenate for adsorption sites.

9.6 CONCLUSIONS AND ENVIRONMENTAL IMPLICATIONS

This chapter gives a detailed introduction to a case study of simultaneous removal of chromium and arsenate by using Fe(II). The principle and performance of the process were studied. It was demonstrated that Fe(II) was effective for simultaneous removal of chromium and arsenate from contaminated groundwater under neutral conditions. Arsenate had limited effects on Cr(VI) reduction by Fe(II) but markedly inhibited the precipitation of Cr(III) species in the solution under neutral and alkaline conditions, leading to the decrease in chromium removal at pH > 6. However, the presence of chromate significantly increased arsenate removal by Fe(II) at pH 4–9. The remarkable increase was associated with the oxidative property of chromate, which could oxidize Fe(II) to Fe(III) formed *in situ* and thus facilitated the removal of arsenate. Arsenate was removed by both adsorption and coprecipitation with precipitated $Fe_{0.75}Cr_{0.25}(OH)_3$ and FeOOH.

Compared with other technologies (e.g., single-adsorption process and Fe^0) reported in the literature, Fe(II) is much more efficient for simultaneous removal of chromate and arsenate. Thus, Fe(II) is believed to be a feasible reagent for above-ground treatment of groundwater contaminated by both chromate and arsenate. However, the removal performance could be influenced by the presence of cations and anions (e.g., calcium, phosphate, silicate, and humic acid) in the water system. Taking into account the common groundwater pH of 6–8, the *in situ* remediation of chromate by Fe(II) could possibly result in the formation of soluble or colloidal Cr(III)/Fe(III) complexes in the presence of the above anions of high concentration. This fact would pose hazardous effect to water security which results from the potential conversion of Cr(III) to toxic Cr(VI) in the presence of the naturally occurring manganese oxides and also from the toxicity of Cr(III) itself when at high concentrations. Moreover, the retention of arsenate would be seriously inhibited. Nevertheless, the copresence of Ca^{2+} could completely offset the detrimental effects of anions on Cr(III) precipitation and could also improve the arsenate removal to different extents depending on the competitive capacity of anions. The concentration range for calcium in the groundwater is 0.04–5.28 mM. With regard to groundwater with low hardness, combination of Ca^{2+} and Fe^{2+} would be a good choice for both chromium and arsenate removal under alkaline conditions.

REFERENCES

Agrafioti, E., Kalderis, D., Diamadopoulos, E. Arsenic and chromium removal from water using biochars derived from rice husk, organic solid wastes and sewage sludge. *Journal of Environmental Management*, 133, 309–314, 2014.

Agrawal, S.G., Fimmen, R.L., Chin, Y.P. Reduction of Cr(VI) to Cr(III) by Fe(II) in the presence of fulvic acids and in lacustrine pore water. *Chemical Geology*, 262, 328–335, 2009.

Ahoranta, S.H., Kokko, M.E., Papirio, S., Özkaya, B., Puhakka, J.A. Arsenic removal from acidic solutions with biogenic ferric precipitates. *Journal of Hazardous Materials*, 306, 124–132, 2016.

Al-Shannag, M., Al-Qodah, Z., Bani-Melhem, K., Qtaishat, M.R., Alkasrawi, M. Heavy metal ions removal from metal plating wastewater using electrocoagulation: Kinetic study and process performance. *Chemical Engineering Journal*, 260, 749–756, 2015.

Azcue, J.M., Nriagu, J.O. Arsenic: Historical perspectives In: *Arsenic in the Environment, Part I: Cycling and Characterization*, edited by Nriagu, J.O., pp. 1–15, John Wiley & Sons, New York, 1994.

Badruddoza, A.Z., Shawon, Z.B., Wei, J.D.T., Hidajat, K., Uddin, M.S. Fe_3O_4/cyclodextrin polymer nanocomposites for selective heavy metals removal from industrial wastewater. *Carbohydrate Polymers*, 91(1), 322–332, 2013.

Bajda, T., Kłapyta, Z. Adsorption of chromate from aqueous solutions by HDTMA-modified clinoptilolite, glauconite and montmorillonite. *Applied Clay Science*, 86, 169–173, 2013.

Balasubramanian, N., Madhavan, K. Arsenic removal from industrial effluent through electrocoagulation. *Chemical Engineering and Technology*, 24(5), 519–521, 2001.

Bartlett, R.J., James, B.R. Behavior of chromium in soils: III. Oxidation. *Journal of Environmental Quality*, 8, 31–35, 1979.

Bhattacharya, P., Mukherjee, A.B., Jacks, G., Nordqvist, S. Metal contamination at a wood preservation site: Characterisation and experimental studies on remediation. *Science of the Total Environment*, 290(1–3), 165–180, 2002.

Bissen, M., Frimmel, F.H. Arsenic—A review. Part I: Occurrence, toxicity, speciation, mobility. *Acta Hydrochimica et Hydrobiologica*, 31, 9–18, 2003.

Brown, R.A., Leahy, M.C., Pyrih, R.Z. In situ remediation of metals comes of age. *Remediation*, 8, 81–96, 1998.

Buerge, I.J., Hug, S.J. Kinetics and pH dependence of chromium(VI) reduction by iron(II). *Environmental Science and Technology*, 31, 1426–1432, 1997.

Buerge, I.J., Hug, S.J. Influence of organic ligands on chromium(VI) reduction by Fe(II). *Environmental Science and Technology*, 32, 2092–2099, 1998.

Calder, L.M. Chromium contamination of groundwater. In: *Chromium in Natural and Human Environments*, edited by Nriagu, J. O. and Nieboer, E., pp. 215–229, John Wiley & Sons, Inc., New York, 1988.

Charlet, L., Manceau, A.A. X-ray absorption spectroscopic study of the sorption of Cr(III) at the oxide-water interface, II. Adsorption, coprecipitation, and surface precipitation on hydrous ferric oxide. *Journal of Colloid and Interface Science*, 148, 443–458, 1992.

Chen, T., Zhou, Z., Xu, S., Hongtao Wang, Wenjing Lu. Adsorption behavior comparison of trivalent and hexavalent chromium on biochar derived from municipal sludge. *Bioresource Technology*, 190, 388–394, 2015.

Chowdhury, S.R. and Yanful, E.K. Arsenic and chromium removal by mixed magnetite-maghemite nanoparticles and the effect of phosphate on removal. *Journal of Environmental Management*, 91, 2238–2247, 2010.

Cooper, P.A. Leaching of CCA: Is it a problem? In: *Environmental Considerations in the Manufacture, Use and Disposal of Pressure-Treated Wood*. Forest Products Society, Madison, WI, 1994.

Davis, C.C., Knocke, W.R., Edwards, M. Implications of aqueous silica sorption to iron hydroxide: Mobilization of iron colloids and interference with sorption of arsenate and humic substances. *Environmental Science and Technology*, 35, 3158–3162, 2001.

Dong, H.R., He, Q., Zeng, G.M., Tang, L., Zhang, C., Xie, Y.K., Zeng, Y.L., Zhao, F., Wu, Y.N. Chromate removal by surface-modified nanoscale zero-valent iron: Effect of different surface coatings and water chemistry. *Journal of Colloid and Interface Science*, 471, 7–13, 2016.

Du, J.J., Lu, J.S., Wu, Q., Jing, C.Y. Reduction and immobilization of chromate in chromite ore processing residue with nanoscale zero-valent iron. *Journal of Hazardous Materials*, 215–216, 152–158, 2012.

Dzombak, D.A., Morel, F.M.M. *Surface Complexation Modeling: Hydrous Ferric Oxide*. Wiley-Interscience, New York, 1990.

Eary, L.E., Rai, D. Chromium removal from aqueous wastes by reduction with ferrous ion. *Environmental Science and Technology*, 22(8), 972–977, 1988.

Fendorf, S.E., Li, G.C. Kinetics of chromate reduction by ferrous ion. *Environmental Science and Technology*, 30, 1614–1617, 1996.

Ferrarini, S.F., dos Santos, H.S., Miranda, L.G., Azevedo, C.M.N., Maia, S.M., Pires, M. Decontamination of CCA-treated eucalyptus wood waste by acid leaching. *Waste Management*, 49, 253–262, 2016.

Goodgame, D.M.L., Hayman, P.B., Hathway, D.E. Formation of water soluble chromium(V) by the interaction of humic acid and the carcinogen chromium(VI). *Inorganica Chimica Acta*, 91, 113–115, 1984.

Gress, J., da Silva, E.B., de Oliveira, L.M., Di Zhao, Anderson, G., Heard, D., Stuchal, L.D., Ma, L.Q. Potential arsenic exposures in 25 species of zoo animals living in CCA-wood enclosures. *Science of the Total Environment*, 551–552, 614–621, 2016.

Gress, J., de Oliveira, L.M., da Silva, E.B., Lessl, J.M., Wilson, P.C., Townsend, T., Ma, L.Q. Cleaning-induced arsenic mobilization and chromium oxidation from CCA-wood deck: Potential risk to children. *Environment International*, 82, 35–40, 2015.

Greven, M., Green, S., Robinson, B., Clothier, B., Vogeler, I., Agnew, R., Neal, S., Sivakumaran, S. The impact of CCA-treated posts in vineyards on soil and ground water. *Water Science and Technology*, 56(2), 161–168, 2007.

Guan, X.H., Ma, J., Dong, H.R., Jiang, L. Removal of arsenic from water: Effect of Ca^{2+} on As(III) removal in $KMnO_4$–Fe(II) process. *Water Research*, 43, 5119–5128, 2009.

Guan, X.H., Su, T.Z., Wang, J.M. Removal of arsenic from water using granular ferric hydroxide: Macroscopic and microscopic studies. *Journal of Hazardous Materials*, 156, 178–185, 2008.

Guo, L.Q., Ye, P., Wang, J., Fu, F.F., Wu, Z.J. Three-dimensional Fe_3O_4-graphene macroscopic composites for arsenic and arsenate removal. *Journal of Hazardous Materials*, 298, 28–35, 2015.

Guo, X.J., Du, Y.H., Chen, F.H., Park, H.S., Xie, Y.N. Mechanism of removal of arsenic by bead cellulose loaded with iron oxyhydroxide (β-FeOOH): EXAFS study. *Journal of Colloid and Interface Science*, 314, 427–433, 2007.

Guzmán, A., Nava, J.L., Coreño, O., Rodríguez, I., Gutiérrez, S. Arsenic and fluoride removal from groundwater by electrocoagulation using a continuous filter-press reactor. *Chemosphere*, 144, 2113–2120, 2016.

Hansen, H.K., N'ūnez, P., Grandon, R. Electrocoagulation as a remediation tool for wastewaters containing arsenic. *Minerals Engineering*, 19, 521–524, 2006.

Hingston, J.A., Collins, C.D., Murphy, R.J., Lester, J.N. Leaching of chromated copper arsenate wood preservatives: A review. *Environmental Pollution*, 111(1), 53–66, 2001.

Jadhav, S.V., Bringas, E., Yadav, G.D., Rathod, V.K., Ortiz, I., Marathe, K.V. Arsenic and fluoride contaminated groundwaters: A review of current technologies for contaminants removal. *Journal of Environmental Management*, 162, 306–325, 2015.

Jain, A., Loeppert, R.H. Effect of competing anions on the adsorption of arsenate and arsenite by ferrihydrite. *Journal of Environmental Quality*, 29, 1423–1430, 2000.

Khaodhiar, S., Azizian, M.F., Osathaphan, K., Nelson, P.O. Copper, chromium, and arsenic adsorption and equilibrium modeling in an iron-oxide-coated sand, background electrolyte system. *Water, Air, and Soil Pollution*, 119(1), 105–120, 2000.

Kim, J.H., Kang, J.C. The toxic effects on the stress and immune responses in juvenile rockfish, *Sebastes schlegelii* exposed to hexavalent chromium. *Environmental Toxicology and Pharmacology*, 43, 128–133, 2016.

King, D.W. Role of carbonate speciation on the oxidation rate of Fe(II) in aquatic systems. *Environmental Science and Technology*, 32, 2997–3003, 1998.

Kotaś, J., Stasicka, Z. Chromium occurrence in the environment and methods of its speciation. *Environmental Pollution*, 107, 263–283, 2000.

Lackovic, J.A., Nikolaidis, N.P., Dobbs, G.M. Inorganic arsenic removal by zero-valent iron. *Environmental Engineering Science*, 17(1), 29–39, 2000.

Lee, G., Hering, J.G. Removal of chromium(VI) from drinking water by redox-assisted coagulation with iron(II). *Journal of Water Supply: Research and Technology-Aqua*, 52(5), 319–332, 2003.

Liu, R.P., Li, X., Xia, S.J., Yang, Y.L., Wu, R.C., Li, G.B. Calcium-enhanced ferric hydroxide co-precipitation of arsenic in the presence of silicate. *Water Environment Research*, 79, 2260–2264, 2007.

Liu, T.Z., Rao, P.H., Mak, M.S.H., Wang, P., Lo, I.M.C. Removal of co-present chromate and arsenate by zero-valent iron in groundwater with humic acid and bicarbonate. *Water Research*, 43(9), 2540–2548, 2009.

Luengo, C., Brigante, M., Avena, M. Adsorption kinetics of phosphate and arsenate on goethite: A comparative study. *Journal of Colloid and Interface Science*, 311, 354–360, 2007.

Luo, L., Zhang, S.Z., Shan, X.Q., Zhu, Y.G. Effects of oxalate and humic acid on arsenate sorption by and desorption from a Chinese red soil. *Water Air and Soil Pollution*, 176, 269–283, 2006.

Mak, M.S.H., Rao, P., Lo, I.M.C. Zero-valent iron and iron oxide-coated sand as a combination for removal of co-present chromate and arsenate from groundwater with humic acid. *Environmental Pollution*, 159, 377–382, 2011.

Mandal, S., Mahapatra, S.S., Patel, R.K. Neuro fuzzy approach for arsenic(III) and chromium(VI) removal from water. *Journal of Water Process Engineering*, 5, 58–75, 2015.

Manning, B.A., Fendorf, S.E., Goldberg, S. Surface structures and stability of arsenic(III) on goethite: Spectroscopic evidence for inner-sphere complexes. *Environmental Science and Technology*, 32, 2383–2388, 1998.

Manning, B.A., Goldberg, S. Adsorption and stability of arsenic(III) at the clay mineral–water interface. *Environmental Science and Technology*, 31, 2005–2011, 1997.

Manning, B.A., Kiser, J.R., Kwon, H., Kanel, S.R. Spectroscopic investigation of Cr(III)- and Cr(VI)-treated nanoscale zerovalent iron. *Environmental Science and Technology*, 41, 586–592, 2007.

McNeill, L.S., Edwards, M. Predicting As removal during metal hydroxide precipitation. *Journal of the American Water Works Association*, 89, 75–86, 1997.

Mishra, T., Mahato, D.K. A comparative study on enhanced arsenic(V) and arsenic(III) removal by iron oxide and manganese oxide pillared clays from ground water. *Journal of Environmental Chemical Engineering*, 4, 1224–1230, 2016.

Nalbandian, M.J., Miluo Zhang, Sanchez, J., Yong-Ho Choa, Jin Nam, Cwiertny, D.M., Nosang V. Myung. Synthesis and optimization of Fe_2O_3 nanofibers for chromate adsorption from contaminated water sources. *Chemosphere*, 144, 975–981, 2016.

Namasivayam, C., Senthilkumar, S. Removal of arsenic(V) from aqueous solution using industrial solid waste: Adsorption rates and equilibrium studies. *Industrial and Engineering Chemistry Research*, 37, 4816–4822, 1998.

Ohgami, N., Yamanoshita, O., Thang, N., Yajima, I., Nakano, C., Wenting, W., Ohnuma, S., Kato, M. Carcinogenic risk of chromium, copper and arsenic in CCA-treated wood. *Environmental Pollution*, 206, 456–460, 2015.

Paktunc, D., Dutrizac, J., Gertsman, V. Synthesis and phase transformations involving scorodite, ferric arsenate and arsenical ferrihydrite: Implications for arsenic mobility. *Geochimica Cosmochimica Acta*, 72, 2649–2672, 2008.

Palma, L.D., Gueye, M.T., Petrucci, E. Hexavalent chromium reduction in contaminated soil: A comparison between ferrous sulphate and nanoscale zero-valent iron. *Journal of Hazardous Materials*, 281, 70–76, 2015.

Parga, J.R., Cocke, D.L., Valverde, V., Gomes, J.A.G., Kesmez, M., Moreno, H., Weir, M., Mencer, D. Characterization of electrocoagulation for removal of chromium and arsenic. *Chemical Engineering and Technology*, 28(5), 605–612, 2005.

Patterson, R.R., Fendorf, S., Fendorf, M. Reduction of hexavalent chromium by amorphous iron sulfide. *Environmental Science and Technology*, 31(7), 2039, 1997.

Pettine, M., D'Ottone, L., Campanella, L., Millero, F.J., Passino, R. The reduction of chromium(VI) by iron(II) in aqueous solutions. *Geochimica et Cosmochimica Acta*, 62(9), 1509–1519, 1998.

Poguberović, S., Krčmar, D.M., Maletić, S.P., Zoltán Kónya, Tomašević Pilipović, D.D., Kerkez, D.V., Rončević, S.V. Removal of As(III) and Cr(VI) from aqueous solutions using "green" zero-valent iron nanoparticles produced by oak, mulberry and cherry leaf extracts. *Ecological Engineering*, 90, 42–49, 2016.

Pontius, F.W., Puls, R.W., Chen, C.J. Health implications of arsenic in drinking water. *Journal of the American Water Works Association*, 86, 52–63, 1994.

Qin, G., Mcguire, M.J., Blute, N.K., Seidel, C., Fong, L. Hexavalent chromium removal by reduction with ferrous sulfate, coagulation, and filtration: A pilot-scale study. *Environmental Science and Technology*, 39, 6321–6327, 2005.

Rahni, M., Legube, B. Mechanism of salicylic acid precipitation by Fe(III) coagulation. *Water Research*, 30, 1149–1160, 1996.

Rai, D., Eary, L.E., Zachara, J.M. Environmental chemistry of chromium. *Science of the Total Environment*, 86(1–2), 15–23, 1989.

Rai, D., Moore, D.A., Hess, N.J., Rao, L., Clark, S.B. Chromium(III) hydroxide solubility in the aqueous Na^+-OH^--$H_2PO_4^-$-HPO_4^{2-}-PO_4^{3-}-H_2O system: A thermodynamic model. *Journal of Solution Chemistry*, 33, 1213–1242, 2004.

Rai, D., Sass, B.M., Moore, D.A. Chromium(III) hydrolysis constants and solubility of chromium(III) hydroxide. *Inorganic Chemistry*, 26(3), 345–349, 1987.

Richard, F.C., Bourg, A.C.M. Aqueous geochemistry of chromium: A review. *Water Research*, 25, 807–816, 1991.

Rincón, G.J., La Motta, E.J. Simultaneous removal of oil and grease, and heavy metals from artificial bilge water using electro-coagulation/flotation. *Journal of Environmental Management*, 144, 42–50, 2014.

Roberts, L.C., Hug, S.J., Ruettimann, T., Billah, M.M., Khan, A.W., Rahman, M.T. Arsenic removal with iron(II) and iron(III) in waters with high silicate and phosphate concentrations. *Environmental Science and Technology* 38, 307–315, 2004.

Robinson, B., Clothier, B., Bolan, N.S., Mahimairaja, S., Greven, M., Moni, C., Marchett, M., van den Dijssel, C., Milne, G. Arsenic in the New Zealand environment. In: *SuperSoil: 3rd Australian New Zealand Soils Conference*, December 5–9, pp. 1–8, University of Sydney, Australia, 2004.

Saikia, J., Saha, B., Das, G. Efficient removal of chromate and arsenate from individual and mixed system by malachite nanoparticles. *Journal of Hazardous Materials*, 186, 575–582, 2011.

Sass, B.M., Rai, D. Solubility of amorphous chromium(III)–iron(III) hydroxide solid solutions. *Inorganic Chemistry*, 26, 2228–2232, 1987.

Schlautman, M.A., Han, I. Effect of pH and dissolved oxygen on the reduction of hexavalent chromium by dissolved ferrous iron in poorly buffered aqueous systems. *Water Research*, 35, 1534–1546, 2001.

Schroeder, D.C., Lee, G.F. Potential transformation of chromium in natural waters. *Water, Air, and Soil Pollution*, 4, 355–365, 1975.

Sedlak, D.L., Chan, P.G. Reduction of hexavalent chromium by ferrous iron. *Geochimica et Cosmochimica Acta,-* 61, 2185–2192, 1997.

Sharma, S.K., Petrusevski, B., Amy, G. Chromium removal from water: A review. *Journal of Water Supply Research and Technology*, 57(8), 541–553, 2008.

Shen, Q., Wei, H., Zhou, Y., Huang, Y.P., Yang, H.R., Wang, D.J., Xu, D.F. Properties of amorphous calcium carbonate and the template action of vaterite spheres. *Journal of Physical Chemistry B*, 110, 2994–3000, 2006.

Sherman, D.M., Randall, S.R. Surface complexation of arsenic(V) to iron(III) (hydr)oxides: Structural mechanism from ab initio molecular geometries and EXAFS spectroscopy. *Geochimica et Cosmochimica Acta* 67, 4223–4230, 2003.

Singh, L.B., Singh, D.R. Influence of dissolved oxygen on the aqueous Cr(VI) removal by ferrous ion. *Environmental Technology* 23, 1347–1353, 2002.

Smedley, P.L., Kinniburgh, D.G. A review of the source, behavior and distribution of arsenic in natural waters. *Applied Geochemistry*, 17, 517–568, 2002.

Smith, S.D., Edwards, M. The influence of silica and calcium on arsenate sorption to oxide surfaces. *Journal of Water Supply Research and Technology*, 54, 201–211, 2005.

Stollenwerk, K.G., Grove, D.B. Adsorption and desorption of hexavalent chromium in an alluvial aquifer. *Journal of Environmental Quality*, 14, 150–155, 1985.

Stumm, W., Morgan, J.J. *Aquatic Chemistry—Chemical Equilibria and Rates in Natural Waters*. John Wiley & Sons Inc., New York, 1996.

Su, C., Puls, R.W. Arsenate and arsenite removal by zerovalent iron: Effects of phosphate, silicate, carbonate, borate, sulfate, chromate, molybdate, and nitrate, relative to chloride. *Environmental Science and Technology*, 35(22), 4562–4568, 2001.

Tamura, H., Goto, K., Nagayama, M. Effect of anions on oxygenation of ferrous ion in neutral solutions. *Journal of Inorganic and Nuclear Chemistry*, 38, 113–117, 1976.

Tang, Y., Reeder, R.J. Enhanced uranium sorption on aluminum oxide pretreated with arsenate. Part 1: Batch uptake behavior. *Environmental Science and Technology*, 43, 4452–4458, 2009.

Thella, K., Verma, B. et al. Electrocoagulation study for the removal of arsenic and chromium from aqueous solution. *Journal of Environmental Science and Health, Part A: Toxic/Hazardous Substances and Environmental Engineering*, 43(5), 554–562, 2008.

Townsend, T., Dubey, B., Tolaymat, T., Solo-Gabriele, H. Preservative leaching from weathered CCA-treated wood. *Journal of Environmental Management* 75, 105–113, 2005.

USEPA. *Technologies and Costs for Removal of Arsenic from Drinking Water*. EPA/815/R-00/028, USEPA; *In Situ Treatment of Soil and Ground Water Contaminated With Chromium, Technical Resource Guide*. EPA/625/R-00/005, USEPA, Washington, DC, 2000.

USEPA. *Implementation Guidance for the Arsenic Rule-Drinking Water Regulations for Arsenic and Clarifications to Compliance and New Source Contaminants Monitoring*. EPA/816/K-02/018, USEPA, Washington, DC, 2002.

Waychunas, G.A., Rea, B.A., Fuller, C.C., Davis, J.A. Surface chemistry of ferrihydrite: Part 1. EXAFS studies of the geometry of coprecipitated and adsorbed arsenate. *Geochimica et Cosmochimica Acta*, 57, 2251–2269, 1993.

Wittbrodt, P.R., Palmer, C.D. Reduction of Cr(VI) in the presence of excess soil fulvic acid. *Environmental Science and Technology*, 29, 255–263, 1995.

Yirsaw, B.D., Megharaj, M., Zuliang Chen, Naidu, R. Reduction of hexavalent chromium by green synthesized nano zero valent iron and process optimization using response surface methodology. *Environmental Technology and Innovation*, 5, 136–147, 2016.

Zagury, G.J., Samson, R., Deschenes, L. Occurrence of metals in soil and ground water near chromated copper arsenate-treated utility poles. *Journal of Environmental Quality*, 32(2), 507–514, 2003.

Zewail, T.M., Yousef, N.S. Chromium ions (Cr6 + &Cr3+) removal from synthetic wastewater by electrocoagulation using vertical expanded Fe anode. *Journal of Electroanalytical Chemistry*, 735, 123–128, 2014.

Zhang, G.S., Qu, J.H., Liu, H.J., Liu, R.P., Li, G.T. Removal mechanism of As(III) by a novel Fe–Mn binary oxide adsorbent: Oxidation and sorption. *Environmental Science and Technology*, 41, 4613–4619, 2007.

Zhao, H.S., Stanforth, R. Competitive adsorption of phosphate and arsenate on goethite. *Environmental Science and Technology*, 35, 4753–4757, 2001.

10 E-Waste Recycling
Environmental and Health Impacts

Rajasekhar Balasubramanian and
Obulisamy Parthiba Karthikeyan

CONTENTS

ABSTRACT

Waste electrical and electronic equipment (WEEE or E-waste) generation is identified as one of the fastest growing waste streams. Developing countries in Asia are currently importing more than 80% of the E-waste generated worldwide (20–50 million tons/year) and recovering valuable components through informal waste recycling approaches. These recycling centers are neither using appropriate recycling technologies, nor handling the E-waste properly taking into consideration environmental and health impacts. In many cases, precious metals are recovered from the electrical components (including electrical wires) by simple burning, which releases deadly toxic airborne pollutants such as phthalates and dioxins into the environment. Also, acid leaching, wet chemical processing, and melting treatment are in place for metal recovery purposes. These processes, when carried out under unsafe and environmentally risky conditions, pose great risks to health and create impacts to the surrounding environment. These aspects of E-waste recycling are given prime focus in this chapter. Further, the need to comply with the existing rules and regulations for the safe treatment and disposal of E-waste, especially in developing Asian countries, is also highlighted. The absence of a management infrastructure, the refusal of extended producer responsibilities, and the lack of institutional capacities are identified and discussed as the major issues in managing E-waste in this chapter.

10.1 OVERVIEW

Worldwide, waste electrical and electronic equipment (WEEE or E-waste) disposals are escalating at a faster rate due to rapid production of new electronic items, new technological innovations, replacement of old electronic goods, and globalized market expansions associated with population growth (1–4). According to the organization for economic co-operation and development (OECD) definition "Any appliance using an electric power supply that has reached its end-of-life called E-waste." These OECD groups of countries generate huge amount of waste, which is highly saturated with the markets for electric and electronic items (2,5). About 10% of the E-waste is illegally transported from the OECD countries to developing Asian countries (1). Despite the implementation of the Basel Convention of 1992 (6) to prevent illegal transport of E-waste, its export still continues through clandestine operations, legal loopholes, and by countries that have not ratified the convention. Figure 10.1 shows the known and suspected routes of E-waste movement from developed to developing Asian countries for recycling or disposal practices. There are, however, no confirmed figures available on how substantial these transboundary E-waste streams.

In recent years, the environmental and health effects of E-waste disposal are drawing a great deal of attention, given the volume of this waste being generated and the hazardous components present in them. It is estimated that 20–50 million tons of E-waste are discarded in a year worldwide and about 12 million tons are disposed, especially in Asian developing countries (2,7,8). However, countries like China, India, and others have adjusted their laws to fight WEEE imports very recently (9,10). No infrastructure and protocols exist for safe recycling and disposal practices, nor is there legislation dealing specifically with E-waste in developing Asian countries (4,7,9,11,12). Moreover, the lack of extended producer responsibilities and lack of institutional capacities have also been identified as the major issues in managing E-waste in these regions. Impacts resulting from such

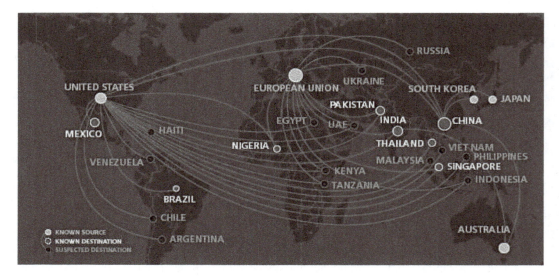

FIGURE 10.1 Major routes of E-waste transport for disposal/recycling practices. (Silicon Valley Toxic Coalition.)

improper disposal/recycling practices have been documented, especially from Asian regions by various researchers (13–18). The United States Environmental Protection Agency (USEPA) regards improper recycling of E-waste as a greater risk than disposal in landfills. An unsafe, unskillful, and environmentally risky practice adopted by the informal sectors in Asian countries poses great risks to health and the surrounding environment (Table 10.1).

E-waste contains more than hundreds of toxic substances, including beryllium in computer motherboards, cadmium in semiconductors, chromium in floppy disks, mercury in thermostats, position sensors, relays, and switches, lead in batteries and computer monitors, and mercury in alkaline batteries and fluorescent lamps, etc. To recover copper and precious metals, the electrical components (including electrical wires) are burnt, releasing deadly toxic airborne pollutants such as phthalates, which are used as plasticizers in flexible plastics, and dioxins. Also, acid leaching, wet chemical processing, and melting treatment are employed for metal recovery purposes. Other E-waste items are dismantled and sorted manually to recover fractions, including printed circuit boards (PCBs), cathode ray tubes (CRTs), cables, plastics, metals, condensers, and batteries. Plastics treated with brominated flame retardants (BFRs) make them harder to recycle. Lead, mercury, cadmium, and polybrominated flame retardants are all persistent, bioaccumulative toxins (PBTs). The PBTs are released into the bio-chain when the computers are incinerated, landfilled, or melted and are known to cause cancer, nerve damage, and reproductive disorders in exposed human beings.

TABLE 10.1
Health Impacts Associated with E-Waste Component Recycling

Pollutant	Associated Health/Environmental Impacts
	Human Health-Related Issues
Nickel (Ni)	Asthma, skin damages, lung diseases, and cancer
Arsenic (As)	Skin and lung cancer
Barium (Ba)	Gastro intestinal disorders, muscle weakness, paralysis, altered heartbeat, and death
Beryllium (Be)	Pneumonia, respiratory problems, and lung cancer
Cadmium (Cd)	Lung and kidney damage, and death
Lead (Pb)	Anorexia, muscle pain, headache, weakness, brain damage, affects reproductive system, and causes death
Mercury (Hg)	Lung damage, nausea, vomiting, diarrhea, eye and skin irritation, kidney and brain damage
Antimony	Skin and eye irritations, hair loss, fertility, lung, and heart damage
Polybrominated diphenyl ethers (PBDE)	Anemia, thyroid, skin, liver, and stomach damage, thyroxin, disturbs endocrine system and thyroid damage
Tetrabromobisphenol-A (TBBP-A)	Endocrine damage, mutation, or carcinogen effects
Polybrominated biphenyls (PBB)	Kidney, liver, and thyroid damage
Polyvinyl chloride (PVC)	Damages animal kidneys
	Environment-Related Issues
Nickel (Ni)	Contained by air particulates
Cadmium (Cd)	Absorbed by plants
Anitmony (Sb)	Absorbed in soil containing steel, magnesium, and aluminum
Chlorofluro carbon (CFC)	Damages ozone layer
Polybrominated biphenyls (PBB)	Passes into food chain
Tetrabromobisphenol-A (TBBP-A)	Toxic to aquatic organisms

Source: Kiddee, P., Naidu, R., Wong, M.H., *Waste Management*, 33(5), 1237–1250, 2013.

However, for an effective management system, we need to understand the generation and characteristics of E-waste, waste generators, and assess the risks involved which is lacking especially in developing regions.

In this chapter, basic information related to E-waste quantities, estimation, composition, basal conventions for E-waste import/export, recycling approaches, and health and environmental impacts associated with informal recycling methods and recent developments in E-waste recycling methods are provided.

10.2   E-WASTE CATEGORIES AND QUANTITIES

E-waste is generally generated from four major sectors: (i) individual households; (ii) small business sectors; (iii) original manufacturing sectors; and (iv) large corporations, institutions, and governmental sectors. E-waste can broadly be categorized under profitable and nonprofitable E-waste types. The profitable E-waste (e.g., PCBs and cell phones) means that the output value generated is more than the input logistics and processing costs for the recycling. The nonprofitable E-waste refers to bulky products such as televisions, refrigerators, and computer monitors. In general, plastics, nonferrous metals, and precious metals are distributed as the largest components in both profitable and nonprofitable E-wastes for recycling.

10.2.1   STATISTICS OF E-WASTE GENERATION

As mentioned earlier, the world's E-waste production was estimated to be around 20–50 million tons per year (19), representing 1%–3% of the global municipal solid waste production of 1636 million tons per year (20). Generally, the contribution of a particular electrical/electronic item to the annual E-waste production depends on the mass of the item, the number of units in service, and its average lifespan, and calculated based on the equation given below (Equation 10.1).

$$E = \frac{MN}{L} \tag{10.1}$$

where E = E-waste production in kilograms per year; M = mass of the item in kilograms; N = number of units; and L = life span of the item in years.

In Equation 1, the number of units and the life span for the electronic equipment can be estimated based on following methods:

1. Market supply method: Uses past domestic sales data coupled with the average life span in particular regions
2. Consumption and use method: Extrapolation from the average amount of electronic equipment in a typical household
3. Saturated market method: Assume that for each new appliance bought, the old one reaches its end of life

Typical weights and life time of specific E-waste items are listed in Table 10.2.

The global E-waste production is likely to increase with changes in its composition as economies grow and new technologies are developed. For a country, the potential E-waste items are strongly correlated with the country's gross domestic product, because electrical and electronic items are essential for the functioning of all but the most primitive economies (22). For example, the per capita E-waste generation of Japan and the United States was around 7.5 and 6.7 kg/citizen/year, respectively. In the former 15 European member countries (EU15), the amount of WEEE generated varied between 3.3 and 3.6 kg per capita for the period 1990–1999, and was projected to rise to 3.9–4.3 kg

TABLE 10.2
List of Common and Noncommon WEEE Items and Their Lifespan

Item	Weight of Item (kg)	Typical Life (Year)
WEEE Normally Considered as E-Waste		
Computer	25	3
Facsimile machine	3	5
High-fidelity system	10	10
Mobile telephone	0.1	2
Electronic game	3	5
Photocopier	60	8
Radio	2	10
Television	30	5
Video recorder and DVD player	5	5
WEEE Not Normally Considered as E-Waste		
Air conditioning unit	55	12
Dish washer	50	10
Electric cooker	60	10
Electric heater	5	20
Food mixer	1	5
Freezer	35	10
Hair dryer	1	10
Iron	1	10
Kettle	1	3
Microwave	15	7
Refrigerator	35	10
Telephone	1	5
Toaster	1	5
Tumble dryer	35	10
Vacuum cleaner	10	10
Washing machine	65	8

Source: Robinson, H.B., *Science of the Total Environment*, 408, 183–191, 2009.

per capita for the period 2000–2010 (23). Developing countries such as India and China generate 0.4 and 1.7 kg/citizen/year of E-waste, respectively (Table 10.3). Although the percapita waste production is still relatively small in these populous countries, the total absolute volume of E-waste generated is relatively higher than that in developed countries. Additionally, some developing and industrializing countries import considerable quantities of E-waste, even though the Basel Convention restricts its transboundary trade. For instance, China receives some 70% of all exported E-waste (24), while significant quantities are also exported to India, Pakistan, Vietnam, Philippines, Malaysia, Nigeria, and Ghana (25), and possibly to Brazil and Mexico (22).

Apart from national economic growth, the technological advancement/innovations in developed countries will also affect overall E-waste generation. In 1994, it was estimated that approximately 20 million personal computers (PCs) (about 7 million tons) became obsolete. By the end of year 2004, this figure increased to over 100 million PCs. Also, 130 million mobile phones were obsolete in the year 2005 (5). Similar quantities of electronic waste are also expected for all kinds of portable electronic devices such as PDAs (personal digital assistants), MP3 (portable media players) players, computer games, and peripherals. Also, short innovation cycles of hardware have led to a high turnover of devices such as PCs, mobile phones, and televisions. Cobbing (26) calculated that

TABLE 10.3
WEEE Generation Statistics in Selected Countries and Their Management Schemes

Country	Generation (tons/year)	Per Capita Generation (kg/person)	Collection and Treatment Routes
Germany	1,100,000 (2005)	13.3	Public waste management authorities and retailers take back
UK	940,000 (2003)	15.8	Distributor take back scheme and producer compliance scheme
Switzerland	66,042 (2003)	9	Swiss association for information, communication, and organization technology, Swiss foundation for waste management, and Swiss light recycling foundations
China	2,212,000 (2007)	1.7	Informal collection and recycling
India	439,000 (2007)	0.4	Informal and formal recycling
Japan	860,000 (2005)	6.7	Collection via retailers
Nigeria	12,500 (2001–2006)	NA	Informal collection and recycling
Kenya	7,350 (2007)	0.2	Informal collection and recycling
South Africa	59,650 (2007)	1.2	Informal and formal recycling
Argentina	100,000 (NA)	2.5	Small number of take back schemes and municipal waste services
Brazil	679,000 (NA)	3.5	Municipalities and recyclable waste collectors
USA	2,250,000 (2007)	7.5	Municipal waste services and a number of voluntary schemes
Canada	86,000 (2002)	2.7	A number of voluntary schemes
Australia	NA	NA	Proposed national recycling scheme from 2011; voluntary take back scheme

Source: Ongondo, F.O., Williams, I.D., and Cherrett, T.J., *Waste Management*, 31, 714–730, 2011.
Note: NA, not available.

computers, mobile telephones, and television sets would contribute 5.5 million tons to the E-waste stream. For emerging economies, these material flows from waste imports not only offer a business opportunity, but also satisfy the demand for cheap second-hand electrical and electronic equipment.

10.2.2 MATERIAL COMPOSITION OF E-WASTE

E-waste contains a diverse range of materials, and it is difficult to provide a generalized material composition (10). However, E-waste can be divided into five major categories of materials: ferrous metals, nonferrous metals, glass, plastics, and others as shown in Figure 10.2. From the figure it is very clear that iron and steel are the most commonly found materials and account for almost 50% of the total weight. Plastics and nonferrous metals contribute approximately 21% and 13% to the total weight, respectively (2,5). The association of plastics manufacturers in Europe stated that 12% of all plastics used in television housing and computer monitors contained BFRs. This is higher in percentage, that is, 55% from consumer electronic equipment specifically, and BFRs emit dioxins when burned (1). Two primary families of BFRs have been used in electronic components/products, that is, polybrominated diphenyl ethers (PBDEs) in cabinets and phenolic component (tetrabromobisphenol-A [TBBP-A]) in PCBs (27). Also, polyvinylchloride (PVC) is widely used in electrical and electronic equipment polymers, often as insulation coating on wires and cables. Uncontrolled burning emits polychlorinated dibenzo-p-dioxins and furans (PCDDs/Fs) from PVCs.

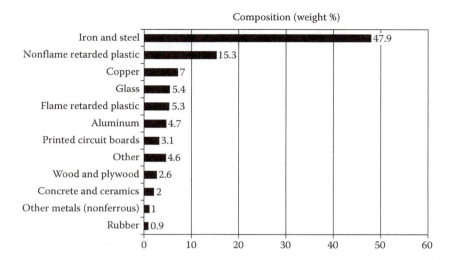

FIGURE 10.2 Individual E-waste components and their distribution percentage. (From Widmer, R. et al., *Environmental Impact Assessment Review*, 25, 436–458, 2005.)

Nonferrous metals contain considerable quantities of precious metals such as copper, aluminum, gold, and platinum along with other toxic metal components such as lead, mercury, arsenic, cadmium, selenium, and hexavalent chromium (Cr (VI)). Approximately 1 ton of E-waste contains up to 0.2 tons of copper (which is 7% in nonferrous metal). Similarly, early generation personal computers contained 4 g of gold in each; however this has decreased to about 1 g today. The platinum group of metals are also used in switching contacts (relays and switches) or as sensors to ascertain the electrical measure. Table 10.4 provides a compilation of the metal composition of individual E-waste scraps (28). Cui and Roven (1) reported that there is no average scrap composition, even the values given as typical averages actually only represent scrap of a certain age and manufacturer. In addition, material composition varies with the technological developments and pressure on manufacturers due to environmental conservation initiatives and developments taken by various regulatory authorities/NGOs worldwide. For instance, replacement of CRTs in televisions and computer monitors with liquid crystal display (LCD) irreversibly reduced the usage of 2 kg of lead, however, LCD displays contain mercury, zinc, and tin (21,29). Similarly, fiber optics, which replaced some copper wires, instead contains fluorine, lead, yttrium, and zirconium. Rechargeable battery composition has also changed dramatically, from old nickel–cadmium, to nickel metal hydrides, to lithium ion batteries (21). Therefore, the metal content has remained the dominant fraction, well over 50%, as compared to hazardous components which have seen a steady decline.

10.3 BASEL CONVENTIONS AND EXTENDED PRODUCER RESPONSIBILITIES FOR E-WASTE MANAGEMENT

A special problem related with E-waste is the international trade from OECD countries to the developing regions. In 1970s and early 1980s, hazardous waste exported from developed countries to developing Asian countries caused serious environmental issues. Therefore, the "Basel Convention on the Control of Transboundary Movements of Hazardous Wastes and Their Disposal" was formulated in the year 1989 and implemented effectively in 1992. The Basel Convention is the foremost global initiative, aimed at tackling the issues related to E-waste transboundary movements. According to the Basel Convention, the trading of E-waste for final disposal has been replaced by the trading of waste for reuse and recycling. End-of-life (EoL) electrical and electronic equipment

TABLE 10.4

Metal Composition in WEEE and Their Marker Prices

WEEE Items	Weight (%)					Weight (ppm)		
	Iron (Fe)	Copper (Cu)	Aluminum (Al)	Lead (Pb)	Nickel (Ni)	Silver (Ag)	Gold (Au)	Palladium (Pd)
TV board scrap	28	10	10	1	0.3	280	20	10
PC board scrap	7–20	7–20	5–14	1.5–6	1–0.85	189–1,000	16–250	3–110
Mobile phone scrap	5	13	1	0.3	0.1	1,380	350	210
Portable audio scrap	23	21	1	0.14	0.03	150	10	4
DVD player scrap	62	5	2	0.3	0.05	115	15	4
Calculator scrap	4	3	5	0.1	0.5	260	50	5
PC mainboard scrap	4.5	14.3	2.8	2.2	1.1	639	566	124
PCB and scrap	5.3–12	10–26.8	1.9–7	0–1.2	0.47–0.85	280–3,300	80–110	NA
TV scrap (CRTs removed)	NA	3.4	1.2	0.2	0.038	20	<10	<10
Electronic scrap	8.3–37.4	8.5–18.2	0.71–19	1.6–3.2	0–2.0	6–210	12–150	0–20
Typical E-waste mixture	8–36	4.1–20	4.9–2	0.29–2	1.0–2	0–2,000	0–1,000	0–50
E-scrap (1972 sample)	26.2	18.6	NA	NA	NA	1,800	220	30
Prices[a] (\$/tons)	500	7,175	2,164	2,129	25,650	601,500	38,100,000	17,000,000

Source: Cui, J. and Roven, H.R., 2011. *Waste: A Handbook for Management*, Elsevier's Science and Technology Rights Department, Oxford, UK, Chapter 20, pp. 281–296, doi: 10.1016/B978-0-12-381475-3.10020-8.

Note: NA, not available.

[a] The metal price data are from the London Metal Exchange (LME) official prices for cash seller and settlement (base metals) or London Fix Prices (precious metals) on the 4 May 2010: Ag 18.71 \$ per troy oz; Au 1185 \$ per troy oz; Pd 529 \$ per troy oz. Here ton refers to a US short tons which is equivalent to 0.9072 metric tons.

produced in developed countries has to a large extent been exported to developing countries as second-hand products. For emerging economies, these material flows from waste imports not only offer a business opportunity, but also satisfy the demand for cheap second-hand electrical and electronic equipment.

"Basel Action Network" called Basel BAN, an amendment in the Basel Convention, was introduced in the year 1995 to prohibit any export of hazardous E-waste from developed to developing regions. However, it has not yet been translated into practical actions, since it requires the signature of three quarters of the countries included in Basel Conventions. In early March 2004, a public–private initiative called "Solving the E-waste Problem (StEP)" was coinitiated and coordinated by the various UN organizations (United Nations University—UNU, United Nations Environment Programme—UNEP, and United Nations Conference on Trade and Development—UNCTAD) together with industry, governments, donors, and academic institutions. The main goal of StEP initiative is to standardize the recycling processes, extend the product life, harmonize world legislative and policy approaches for effective E-waste management, capacity building, and knowledge management (30,31). During the 8th Conference of the Parties of the Basel Convention in Nairobi declared in 2006, E-waste was a priority issue and emphasized the need for creative and innovative solutions for an environmentally sound management system.

In addition, different countries have established their own national policies to solve the problems related with E-waste disposal and management issues (Table 10.5). In OECD member countries and other nations, extended producer responsibility (EPR) principles were developed and legislated for E-waste disposal/recycling practices. The definition of EPR is that "policy principle to promote total life cycle environmental improvements of product systems by extending the responsibilities of the manufacturers of the product to various parts of the entire life cycle of the product, and especially to take back, recycling and final disposal of the product." Hence, the effective system developed by the producers requires establishing an adequate infrastructure for collection and treatment. Consequently, high shares of the discarded items are diverted from landfills and incinerators to recycling facilities (33). In the United States, implementation of EPR mechanism is voluntary by the industries whereas in European Union, the impetus comes from the local governments. In the United States, EPR is known as extended product responsibility to emphasize that the responsibility is shared—the producer is not the only responsible party but also the packaging manufacturer, the consumer, and the retailer (30).

However, in developing Asian countries it may be difficult to apply EPR because of the following reasons:

1. In countries with rural communities that have low home appliance rates, used household appliances flow from cities into the countryside. In addition, reuse is the norm and even with appliances that are beyond repair, parts are replaced, and the appliances continue to be used, which makes it difficult to collect the EoL equipment as is.
2. Recycling is undertaken by the informal sector, meaning that even when responsibility for this task is assigned to producers and importers, collecting used home appliances would not be an easy task.
3. It is also difficult to establish where the responsibility lies for used products that have been repaired or modified and smuggled, with the producer or with the importer? This also applies to "cloned" computer systems in which components and modules made by different manufacturers are used.
4. Where used products are being imported, there are no figures on the number of importing agents that handle such products as "new" products, and there are believed to be innumerable importers of the products of just one brand.
5. There are also products that have been brought in as private imports, and it is thus difficult to identify which product was imported by whom.

TABLE 10.5
Legislative Policies in E-Waste Collection and Recycling from Selected Countries

Country	E-Waste Specific Legislation (year)	Approach	Disposal/Recycling Approaches
Germany	ElektroG (2005)	EPR	Illegal exports
UK	UK WEEE regulations (2007)	EPR	Illegal exports
Switzerland	NA	EPR	NA
China	China RoHS (2007) and China WEEE (2011)	NA	Illegal imports
India	NA	NA	Illegal imports
Japan	HARL	EPR, postpaid recycling fee	Illegal dumping
Nigeria	NA	NA	Open burning and open dumping
South Africa	NA	NA	NA
Argentina	NA	NA	Landfilling
Brazil	NA	NA	Disposed in dumps
USA	State regulations and no federal level	EPR, ARF, voluntary	Illegal exports and landfilling
Canada	No federal level	Voluntary	Landfilling
Australia	NA	EPR, Voluntary	Landfilling
Thailand	Thai WEEE strategy	EPR, DRS	Dumping, open burning

Source: Ongondo, F.O., Williams, I.D., Cherrett, T.J., *Waste Management*, 31, 714–730, 2011; Manomaivibool, P., Vassanadumrongdee, S., *Journal of Industrial Ecology*, 15(2), 185–205, 2011.

Note: Elekro, G., Legal act governing the sale, return, and environmentally sound disposal of electrical and electronic equipment in Germany; RoHS, restriction of the use of certain hazardous substances in electrical and electronic equipment; HARL, home appliance recycling law; EPR, extended producer responsibility; ARF, advanced recycling fee; NA, not available; and DRS, deposit-refund system.

Despite the existence of agreements, conventions and policy regulations, the United States, Canada, Australia, Europe, Japan, and Korea transfer E-waste to Asian countries such as China, India, and Pakistan (26). Moreover, emerging economies such as China and India are themselves large generators of E-waste, facing a problem with rapidly increasing amount of E-waste, both, from domestic generation and illegal imports (5). In addition, the lack of national regulations and/or strict enforcement of existing laws are promoting the growth of a semi-formal or informal economy in these developing countries.

10.4 E-WASTE MANAGEMENT PRACTICES IN DEVELOPING ASIAN COUNTRIES

Generally through trading in the second-hand market and through donations, E-waste migrates from developed to developing regions. A study compiled that the total supply of E-waste in the United States amounted to 6.6×10^6 tons, of which 20% was estimated to be exported to Asia. Also, about 10% of E-waste was shipped from OECD countries to non-OECD countries in Asia, Africa, and Eastern Europe every year. These E-waste imports from developed to developing countries include both illegal and legal shipments; and distinguishing between the two is not so straightforward (2,34). Therefore, accurate data regarding how much of E-waste is generated, how much is imported, how it is managed, and where it is processed (either domestically or abroad) was largely not available from developing Asian countries.

Worldwide, China is the largest exporter as well as importer of electronic items. It is reported that the 35 million tons of E-waste were imported to China from developed countries (24,35). Further,

China itself produced 160 million tons (including TV, PCs, air conditioners, refrigerators, and washing machines) of E-waste in the year 2010 which is increasing at a rate of 13%–15% annually (2,36). However, there is no any actual official estimation of E-waste generation in China according to Yang et al. (34). Similarly, domestic E-waste generation is also significant in addition to the illegal imports in China (37). India has been one of the main E-waste disposal destinations for the OECD countries and it was estimated that around 50 kilo tons are imported every year (38). In 2007, it was estimated that 823.6 kilo tons of electronic items were put into the market and 439 kilo tons of E-waste disposed of in India (39). Despite the decline in the price of new computers, Pakistan consumes over 500,000 second-hand PCs every year (www.ewasteguide.info), but no reliable data or inventory are available with the total E-waste generated or imported within the country. Other developing Asian countries like Sri Lanka and Bangladesh are yet to determine the extent of E-waste quantities generated in their respective nations and associated problems. It was reported that, the total number of PCs, TVs, mobile phones, and refrigerators were 600,000, 1,252,000, 58,000,000 and 2,200,000 numbers, respectively in Bangladesh during the year 2006. Apart from computers, imported second-hand electronic equipment has not been regulated in Cambodia. In Vietnam, though the import of second-hand electronic equipment (including home appliances and computers) was banned in 2001, in reality, the practice still in place due to the lack of control and management on the part of the government. In Thailand, 536 tons of E-waste were estimated as obsolete during the year 2003. Every year 5%, 20%, and 15% of the refrigerator, mobile phones, and PCs are obsolete in Thailand (40). In Malaysia, the E-waste generation escalated from 40,275 to 134,035 tons between the years 2006 and 2009. A study also projected a total of 761.507 million units of E-waste would be generated between the years 2008 and 2020 in Malaysia (41). Over a 10-year period from 1995 to 2005, approximately 25 million units became obsolete. An additional 14 million units were projected to become obsolete at the end of 2010 and more than a million units go into landfill and storage every year (42). Further, it is highlighted that there is an illegal transport of E-waste, that is, around 500 shipping containers, through Lao every month. Now, there are NGOs such as Greenpeace that campaign against this hidden flow of E-waste from developed to developing countries (26).

10.4.1 Role of Informal Sectors in E-Waste Recycling

More than 95% of E-waste is being handled by the informal sectors in developing Asian regions. These informal economies were considered as "shadow economies," circulating the money immediately into the official economy, resulting in a stimulating effect within the country. Most of the E-waste recycling practices involved a group of small enterprises, which are widespread. It is very difficult to identify/regulate this kind of operation in developing Asian countries. They take advantage of poor and marginalized social groups, who are mainly dependent on scavenging of electronic scraps for their livelihood. Around 2% of the population mainly depends on this waste scavenging for their survival in Asian countries (43). The informal group of people risk their health and environment because of the crude recycling methods adopted for E-waste disposal. They never use any sophisticated instruments or personal protective measures during waste handling. All the dismantling works are done by bare hands and using hammers, screw drivers, etc. The valuable fractions, namely, directly reusable components and secondary raw materials, were shifted to formal sectors for the further processing (43). The tasks are mainly undertaken by young and old people of both genders with the combination of urban dwellers and rural migrants. In many cases, children were also involved in this informal recycling of E-waste. For metal recovery purposes, they use strong acids without using any protective measures. Also, they work in poorly ventilated enclosed areas without any masks. However, the formal and informal waste recycling professions vary from country to country and they also have peculiar local appellations (44). Generally, small-scale E-waste recycling operations are largely invisible to state scrutiny because they border on the informal economy and are therefore not included in the official statistics (45). Figure 10.3 depicts the functions of informal and formal sectors in the recycling of E-waste from China as an example. The major

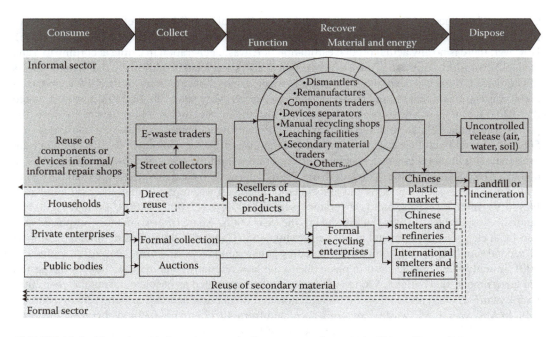

FIGURE 10.3   Formal and informal sectors in E-waste recycling within China. (From Chi, X. et al., *Waste Management*, 31, 731–742, 2011.)

reasons for existence of such low-end E-waste processing technologies in developing Asian regions are given below (44,46–48):

1. Lack of public awareness and unwillingness of consumers to return their old E-waste
2. High level of import of E-waste as second-hand devices
3. Lack of strict regulations for E-waste and lack of funds and investment to finance improvements in E-waste recycling
4. Absence of recycling infrastructure or appropriate management of E-waste
5. Lack of awareness of the potential hazards of E-waste and lack of interest/incentives in E-waste management

10.4.2   Recycling, Open Burning, and Landfilling Scenarios in Developing Countries

Currently, the main options that exist for E-waste management in developing Asian countries are recycling, incineration, and dumping/landfilling (1,49). In the general E-waste management hierarchy, reuse of electronic items comes first, then remanufacturing, recycling, material recovery, incineration, and landfilling/dumping appear at the final stage (1). However, the recycling industries in China, India, Pakistan, Vietnam, and the Philippines are often use crude methods and do not have the appropriate facilities to safeguard environmental and human health. For example, India and China use more complex processes for E-waste recycling as shown in Figure 10.4.

10.4.2.1   Recycling of E-Waste

E-waste recycling includes disassembly and recovery of materials. Most of the disassembly processes are done manually using uneducated laborers in developing regions by the informal economy. They collect, sort, and manually separate electrical and electronic items in a crude way to segregate substances from the bulk items (Figure 10.5). Crude dismantling activities release dust particles loaded with heavy metals and flame retardants into the atmosphere (37).

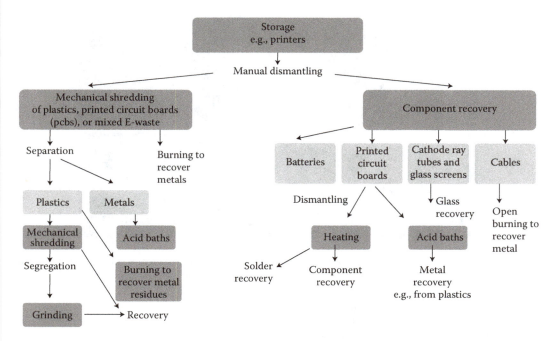

FIGURE 10.4 E-waste recycling approaches in India and China. Different processes are used to recover materials from the same components. Not all the processes shown here are used in all cases and there is no set sequence which applies to all locations or all types of electronic equipment. (From Tsydenova, O., Bengtsson, M., *Waste Management*, 31, 45–58, 2011.)

FIGURE 10.5 Dismantling of E-waste for recycling purposes in developing Asian countries. (From Widmer, R. et al., *Environmental Impact Assessment Review*, 25, 436–458, 2005.)

These low-technology recycling practices separate component parts such as plastic cases, metal chassis, CRTs, circuit boards, wires, and printer toner cartridges. These cases and chassis are sold for their scrap value while CRTs and wires are further processed for recovering recyclable materials. Toner cartridges are opened for the remaining toner, and electronic chips are separated from the circuit boards and dissolved in strong acidic solutions to recover the valuable metals (Figure 10.6). CRTs are broken to recover copper contained in electronic guns (50).

10.4.2.2 Open Burning and Incineration of E-Waste

In order to extract precious metals such as copper from wire scraps and PCBs, they were simply burnt in the open environment for the removal of plastic insulations (Figure 10.7). Subsequently, the primary copper metal and tin are separated from the ash by simple water floatation. Other miscellaneous component parts are also burnt as supplementary fuel for the metal recovery purposes,

FIGURE 10.6 Crude recycling of E-waste for metal recovery. (From Widmer, R. et al., *Environmental Impact Assessment Review*, 25, 436–458, 2005.)

FIGURE 10.7 Open burning of E-waste components emitting atmospheric air pollutants. (a) From Basel Action Network, (b) From Swiss Federal Laboratories for Materials Science and Technology, and (c) From http://www.atterobay.com/blogs/.

or disposed of in dumping grounds (50). During the open burning process, due to less oxygen and low temperatures, many more toxic air pollutants are released into the ambient environment. The residual burnt ash also contains numerous toxic elements, which are dumped in landfills or open dumps, posing threat to the soil and aquatic biota.

Because of the variety of substances found in E-waste, incineration is associated with a major risk of generating and dispersing contaminants and toxic substances. The gases released during the burning and the residue ash are often toxic. This is especially true for incineration or coincineration of E-waste with no pretreatment or sophisticated flue gas purification. Incineration also leads to the loss of valuable trace elements, which could have been recovered had they been sorted and processed separately.

10.4.2.3 Open Dumping and Landfilling of E-Waste

Nonrecyclable E-waste and dismantled unsalvageable waste components are finally dumped into open areas or to the river side (Figure 10.8). In addition, it is a common practice that burnt waste components and ash contents are also disposed of in open dumpsites or in landfill facilities along with municipal solid waste. However, it is common knowledge that all landfills leak. The leachate often contains heavy metals and other toxic substances which can contaminate ground and water

FIGURE 10.8 Open dumping of E-waste near river side and in open ground. (a) From Greenpeace, (b) From http://blogs.ubc.ca/sabrinatkk/, and (c) From http://blogs.whattheythink.com/going-green/2010/11.

resources (8,51). Older landfill sites and uncontrolled dumps pose a much greater danger of releasing hazardous emissions. Besides leaching, vaporization is also of concern in landfills. For example, volatile compounds such as mercury or a frequent modification of it, dimethyl mercury can be released. In addition, landfills are also prone to uncontrolled fires which can release toxic fumes. Disposal of acids after the crude extraction of metals from acid baths contributes to soil, air, and aquatic pollution. Through depositions, rain water runoff and dissolution, the contaminants finally get into aquatic environments.

10.5 ENVIRONMENTAL AND HEALTH IMPACTS ASSOCIATED WITH INFORMAL E-WASTE RECYCLING APPROACHES

E-waste, which is chemically and physically distinct from other waste categories, mainly contains valuable as well as hazardous materials (4,9,43). It contains a large amount of hazardous substances such as heavy metals (e.g., mercury, cadmium, lead, etc.), flame retardants (e.g., pentabromophenol (PBP), PBDEs, TBBP-A, etc.), which are all considered to be PBTs. Other than PBTs, rare earth elements (REEs), precious metals, polychlorinated biphenyls (PCBs), chlorofluorocarbons (CFCs), and polycyclic aromatic hydrocarbons (PAHs) are reported to be found in E-waste scrap in defined concentrations (Table 10.6). Therefore, special handling and recycling methods are required to avoid environmental contamination and the detrimental effects of these harmful chemicals on human health. However, E-waste recycling is largely unregulated in Asian regions; virtually no data are available to track its fate at a global level. Therefore, it is very difficult to understand the pollution load and risk factors associated with the current E-waste disposal practices in these regions.

10.5.1 ENVIRONMENTAL FATE OF PBTs

PBTs refer to broad categories of chemicals that do not degrade very easily in the environment, and are commonly present in E-waste. The PBTs are lipophilic and bioaccumulative substances.

TABLE 10.6

Potential Environmental Pollutants from WEEE Recycling or Disposal

Contaminant	Relationship with E-Waste	Typical Concentration (mg/kg)	Annual Global Emission from E-Waste (tons)[a]
Polybrominated diphenyl ethers (PBDEs); Polybrominated biphenyls (PBB); Tetrabromo bisphenol-A (TBBP-A)	Flame retardants		
Polychlorinated biphenyls (PCBs)	Condensers, transformers	14	280
Chlorofluorocarbon (CFC)	Cooling units, insulation foam		
Polycyclic aromatic hydrocarbons (PAHs)	Product of combustion		
Poly halogenated aromatic hydrocarbons (PHAHs)	Product of low temperature combustion		
Polychloronated dibenzo-p-dioxins (PCDDs); polychlorinated dibenzofurans (PCDFs)	Product of low temperature combustion of PVCs and other plastics		
Americium (Am)	Smoke detectors		
Antimony	Flame retardants, plastics	1,700	34,000
Arsenic (As)	Doping material for Si		
Barium (Ba)	Getters in CRTs		
Beryllium (Be)	Silicon controlled rectifiers		
Cadmium (Cd)	Batteries, toners, plastics	180	3,600
Chromium (Cr)	Data tapes and floppy disk	9,900	198,000
Copper (Cu)	Wiring	41,000	820,000
Gallium (Ga)	Semiconductors		
Indium (In)	LCD displays		
Lead (Pb)	Solder, CRTs, batteries	2,900	58,000
Lithium (Li)	Batteries		
Mercury (Hg)	Fluorescent lamps, batteries, switches	0.68	13.6
Nickle (Ni)	Batteries	10,300	206,000
Selenium (Se)	Rectifiers		
Silver (Ag)	Wiring, switches		
Tin (Sn)	Solder, LCD screens	2,400	48,000
Zinc (Zn)		5,400	102,000
REEs	CRT screens		

Source: Robinson, H.B., *Science of the Total Environment*, 408, 183–191, 2009.

[a]   Assuming a global WEEE production of 20 million tons per year.

They typically accumulate in fatty tissues, and are slowly metabolized, often increasing in concentration within the food chain, and biomagnified (Figure 10.8). PBTs can easily move through the environment, which is a major concern as they make their way into remote regions. They can travel through long distances via the "grasshopper effect," involving a complex cycle of long-range atmospheric transport, deposition, and revolatilization, and accumulate in cold regions through "global distillation." Because of its volatile nature, PBTs can enter anywhere into the atmosphere and be carried with the wind over long distances. Through atmospheric processes they are deposited onto land, or into water ecosystems, where they accumulate and may cause damage. Through these processes, some PBTs can move thousands of kilometers from their sources of emission and accumulate in polar latitudes. Aquatic organisms are very efficient at accumulating these chemicals through their diet and ambient environment sources, resulting in an extremely high concentration in their body. These harmful chemicals eventually get into the human beings through the food chain. Certain PBTs such as lead, mercury, cadmium, beryllium, hexavalent chromium, arsenic, selenium, antimony, BFRs, PCBs, phthalates, and PAHs have been linked to adverse health effects in both humans and animals as discussed (adapted from http://ewasteguide.info/hazardous-substances) in Figure 10.9 and also summarized below.

1. **Lead** is most widely used metal after iron, aluminum, copper, and zinc. It is used in solder, lead acid batteries, electronic components, cable sheathing, in the glass of CRTs, etc. Short-term exposure to high levels of lead can cause vomiting, diarrhea, convulsions, coma, or even death. Other symptoms are appetite loss, abdominal pain, constipation, fatigue, sleeplessness, irritability, and headache. Continued excessive exposure can affect the kidneys. It is particularly dangerous for young children because it can damage nervous connections and cause blood and brain disorders. Lead is also a possible carcinogen and has very high chronic and acute effects on microorganisms, plants, and animals.

2. **Mercury** is used in relays (used in telecommunication circuit boards, commercial/industrial electric ranges, and other equipment) and switches (used in a variety of consumer, commercial, and industrial products, including appliances, space heaters, ovens, air handling units, security systems, leveling devices, and pumps), batteries, and gas discharge lamps (used for backlighting in LCDs in a wide range of electronic equipment, including computers, flat screen TVs, cameras, camcorders, cash registers, digital projectors, copiers, and fax machines). Mercury is most toxic yet widely used metals in the production of electrical and electronic applications. It can cause brain and liver damage if ingested or inhaled.

3. **Cadmium** is used in some contacts, switches, and solder joints. Many devices contain rechargeable nickel–cadmium (Ni–Cd) batteries which contain cadmium oxide. Cadmium compounds have also been used as stabilizers within PVC formulations, including those used as wire insulation. Cadmium sulfide has been also used in CRTs as a phosphor on the interior surface of the screen to produce light. Cadmium, when released to aquatic environments, is more mobile than most other metals. Cadmium is highly toxic to plants, animals, and humans, having no known biochemical or nutritional function. Many animals and plants, including those consumed by humans, can also accumulate cadmium, providing an additional route of dietary exposure for humans. Cadmium exposure can occur occupationally through the inhalation of fumes, or dusts containing cadmium and its compounds, or through environmental exposures, primarily diet. Cadmium is a cumulative toxicant and long-term exposure can result in damage to the kidneys and bone toxicity (52).

4. **Beryllium** is most commonly used in beryllium-copper alloys to increase flexibility and strength in components that need to be capable of flexing, such as contacts and springs. Some of the greatest risks from beryllium occur in manufacturing and recycling facilities, where dust or fumes expose workers to one of the most toxic metals if inhaled. Beryllicosis can cause permanent scarring of the lungs, sometimes years after initial exposure, and can be fatal (53). Beryllium has recently been classified as a human carcinogen because exposure to it can cause lung cancer.

5. **Hexavalent chromium** and its oxides are widely used because of their high conductivity and anticorrosive properties. While some form of chromium are nontoxic, Chromium (VI) is easily absorbed in the human body and can produce various toxic effects within cells. Most chromium (VI) compounds are irritating to eyes, skin, and mucous membranes. Chronic exposure to chromium (VI) compounds can cause permanent eye injury, unless properly treated. Chromium (VI) may also cause DNA damage.

6. **Arsenic** in the form of gallium arsenide is present in light-emitting diodes. Arsenic is a poisonous metallic element which is present in dust and soluble substances from E-waste recycling centers. Chronic exposure to arsenic can lead to various diseases of the skin and

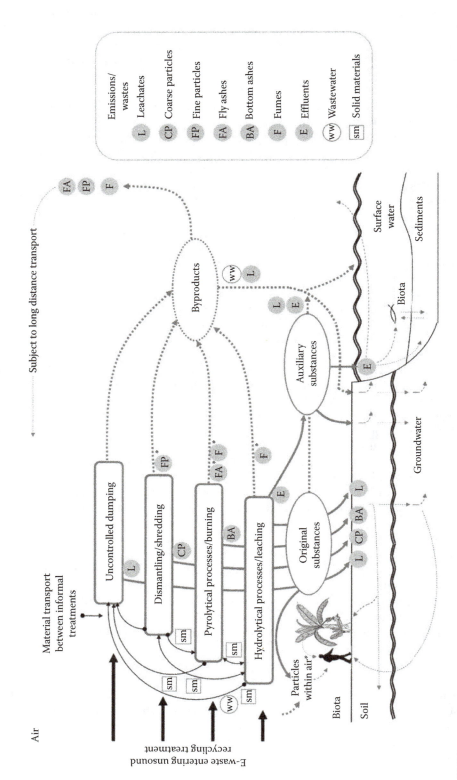

FIGURE 10.9　Transformation of persistent pollutants in food chain from E-waste disposal practices.

decrease nerve conduction velocity. Chronic exposure to arsenic can also cause lung cancer and often be fatal.

7. **Selenium** exposure to high concentrations causes selenosis. The major signs of selenosis are hair loss, nail brittleness, and neurological abnormalities.

8. **Antimony** compounds are used in semiconductor manufacture (antimony trihydride) and in flame retardant formulations in plastics (antimony trioxide), normally in combination with BFRs, especially PBDEs, though there are also reports of their use in combination with phosphorus-based flame retardants. Antimony is also used in the manufacture of lead acid starter batteries and can occur as a component of electrical solders. Antimony shows many chemical similarities to arsenic. Like arsenic, it can undergo methylation as a result of microbiological activity (i.e., to form its trimethyl derivative, often called trimethylstibine), albeit at slower rates than for arsenic. It also shows some similarities in its toxic effects, especially to skin cells. Antimony compounds have been associated with dermatitis and irritation of the respiratory tract, as well as interfering with normal function of the immune system. Antimony trioxide and antimony trisulfide have been listed by the International Agency for Research on Cancer (IARC) as "possibly carcinogenic to humans," with inhalation of dusts and vapors being the critical route of exposure (52,54–58).

9. **Brominated flame retardant** are used in PCBs, components such as connectors and plastic covers, and in cables. BFRs are also used in a multitude of products, including, but not exclusively, plastic covers of television sets, carpets, paints, upholstery, and domestic kitchen appliances. Also, a number of different types of BFRs are currently used in electronics, some of which are known to be damaging to human health and the environment, while others are still being tested. The following are BFRs in use:
   a. Hexabromocyclododecane (HBCD)
   b. Polybrominated biphenyls (PBBs)
   c. Polybrominated diphenyl ethers (PBDEs)
      - Decabromodiphenyl ether (Deca-BDE)
      - Octabromodiphenyl ether (Octa-a-BDE)
   d. Tetrabromobisphenol (TBBP-A)

   Among these compounds, PBDEs have been used extensively over the past two decades as additive BFRs in most types of polymers applied to computer monitors, television sets, computer cases, wire and cable insulation, and electrical/electronic connectors, at levels ranging between 5% and 30%. These PDBEs are associated with cancer, liver damage, neurological and immune system problems, thyroid dysfunction, and endocrine disruption. In worst case, in the presence of copper, they emit dioxins and furans during the incineration of E-waste. Furthermore, if this incineration occurs at low temperatures, as is commonly found in E-waste recycling operations in developing nations, the incomplete combustion generates even higher amounts of dioxins and furans (53).

10. **Polychlorinated biphenyls** were historically used as coolants and lubricants in transformers and capacitors, and as hydraulic and heat exchange fluids in electrical/electronic equipment. PCBs can also be produced during the combustion of chlorinated organic materials, including PVC. Once released to the environment from whatever source, PCBs are highly persistent. PCBs can be absorbed through the skin as well as through ingestion and inhalation. For the general population today, food is undoubtedly the primary route of exposure to PCBs although dermal exposure may be dominant among those directly handling PCBs or PCB-contaminated materials. PCBs exhibit a wide range of toxic effects in animals, including immunosuppression, liver damage, tumor promotion, neurotoxicity, behavioral changes, and damage to both male and female reproductive systems. PCBs may also affect many endocrine systems (52). The use of PCBs is prohibited in OECD countries. However, these compounds can still be found in WEEE as well as in some other wastes.

11. **Phthalates** are nonhalogenated chemicals with a diversity of uses, dominated by use as plasticizers (or softeners) in plastics, especially PVC (e.g., in coated wires and cables and other flexible components). Other applications include their use as components of inks, adhesives, sealants, and surface coatings. Phthalates are commonly found in human tissues, including in blood and, as metabolites, in urine. In humans and other animals, they are relatively rapidly metabolized to their monoester forms, but these are frequently more toxic than the parent compounds. For example, DEHP, one of the most widely used to date, is a known reproductive toxin, capable (in its monoester form MEHP) of interfering with development of the testes in early life. Butylbenzyl phthalate and dibutyl phthalate have also been reported to exert reproductive toxicity (52).

Airborne particulate matter emitted during open burning contains all these major toxic pollutants along with the other elements and cause major occupational health issues to workers. Moreover, their excessive release from the improper E-waste recycling is one of the important sources from which these chemicals enter the global environment (27,37,59).

10.5.2 E-Waste Recycling and Health Risk Assessment

Using the health risk assessment calculation (detailed below), one can measure the risks associated with exposure to pollution emissions from E-waste recycling processes. This involves four important steps

1. Hazard identification: The elements which have known toxicity values are considered. For example, elements, aluminum, chromium, and manganese induce noncarcinogenic effects, while arsenic, cadmium, chromium, nickel, and cobalt induce carcinogenic health effects.
2. Exposure assessment: This involves estimation of chronic daily intake (CDI) of these elements calculated from the following equations:

$$CDI = \frac{TD\,(mg/m^3) \times IR(m^3/day)}{Body\,weight\,(kg)} \qquad (10.2)$$

where total dose (TD) = C × E, where C is concentration of pollutant and E is deposition fraction of particles by size given Volckens and Leith (60).

$$E = -0.081 + 0.23\,lm(Dp)^2 + 0.23\,sqrt(Dp),$$

where Dp is the diameter of airborne particles

Inhalation rate (IR) = differs with the different age and gender groups along with their respective body weights. These factors are highly variable and need to be recorded during the sampling.
3. Dose response assessment: It is the probability of health effects to the given dose of pollutants. For the different routes (inhalation, ingestion, and dermal), the dose response assessment needs to be studied. Assuming only inhalation as the major exposure route, the reference dose (RfD, mg/kg/day) for toxic elements that are noncarcinogenic can be calculated from reference concentrations (Rfc, mg/m^3) provided by the United States Environmental Protection Agency (USEPA). Likewise, for carcinogenic elements the inhalation slope factor (SF, mg/kg/day) can be calculated from the inhalation unit risk value (IR, mg/m^3) provided by the USEPA.
4. Risk characterization or estimation of health risk: This is calculated based on the exposure and dose response assessments. For noncarcinogenic metals, it is indicated by hazard quotient (HQ) = CDI/Rfd. For carcinogenic metals, the total carcinogenic risk is estimated in

terms of excess life time cancer risk given by (ELCR) = CDI × SF (by the United States Department of Energy).

10.6 SUSTAINABLE E-WASTE RECYCLING APPROACHES

The complexity of the composition of E-waste imposes significant challenges in disposal and recycling practices. Most commonly, mechanical shredding, acid-bath leaching, and incineration (burning) are considered for the treatment of the E-waste. Alternatively, cryogenic decomposition can also be considered for PCB recycling. In addition, a few other approaches can be effectively used for the recycling of E-waste in developing Asian countries and they are discussed below.

10.6.1 CEMENTATION TECHNOLOGY

Cementation technology is one of the solidification technologies, involving the use of a solidifying agent (i.e., cement, in this case) to trap hazardous E-wastes (such as mercury-containing batteries or equipment). Conventional cementation technology has two major drawbacks: (a) the hazardous substances may leak out through the cement pores and (b) the solidified cement or concrete blocks are not strong enough. An improved solidification (cementation) technology is proposed by Dr. Lawrence K. Wang, Lenox Institute of Water Technology, Massachusetts, USA. The improved solidification (cementation) technology involves: (a) use of special dry powder mixture to improve the crystalline structure that permanently seals the concrete against the penetration or movement of water and other hazardous liquids from any direction; (b) use of special nonmetal reinforcement to provide better tensile and compressive strengths to the concrete blocks; and (c) use of special chemical crystallization treatment for the waterproofing and protection of the concrete blocks' surface. By means of diffusion, the reactive chemicals in the agent use water as a migrating medium to enter and travel through the capillary tracts in the concrete. This process precipitates the natural chemical byproducts of cement hydration such as calcium hydroxide, mineral salts, mineral oxides, and unhydrated and partially hydrated cement particles. It forms crystallization at the end and, ultimately, a nonsoluble crystalline structure that plugs the pores and capillary tracts of the concrete, thereby rendering it impenetrable by water and other liquids from any direction. The chemical treatment is permanent. It is unique, and crystalline growth will not deteriorate under wide conditions. The treated concrete block is structurally strong, and is not affected by a wide range of aggressive chemicals including acids, solvents, chlorides, and caustic materials in the pH range of 3–11(61).

10.6.2 NANOTECHNOLOGY

Nanotechnology is still in its infancy stage and poses great applications in environmental pollutant remediation. In electronics, a number of different nanomaterials are already being used commercially, or are being used for research and development purposes. At the same time nanotechnology can be effectively used to treat pollutants, especially the volatile organic components and persistant organic pollutants (POPs) in E-waste.

10.6.3 BIO-METALLURGY

Bio-metallurgy or bio-hydrometallurgy is a biotechnological approach to solubilize metals from E-waste components and is considered as one of the promising technologies. It offers a number of advantages over pyro- and hydro-metallurgical techniques in terms of (a) low operating costs, (b) use of less hazardous chemicals, (c) eco-friendliness, and (d) low energy requirement. Generally, iron and sulfur-oxidizing microbes (Table 10.7) are used to support direct or indirect bioleaching of metals from E-waste.

TABLE 10.7

List of Iron and Sulfur Oxidizers Involved in Bioleaching of Metals

Substrate/Temperature	Mesophiles	Moderate Thermophiles	Thermophiles
Iron oxidizers	• *Leptospirillum ferrooxidans* • *Ferroplasma* spp. • *Ferrimicrobium acidophilum*	• *Acidimicrobium ferrooxidans* • *Leptospirillum thermoferrooxidans*	
Iron/sulfur oxidizers	• *Acidithiobacillus ferrooxidans* • *Thiobacillus prosperus* • *Sulfobaciulls montserratensis*	• *Sulfobacillus thermosulfidooxidans* • *Leptospirillum ferriphilum* • *Sulfobacillus acidophilus*	• *Acidianus* spp. • *Sulfolobus metallicus* • *Sulfurococcus yellowstonensis*
Sulfur oxidizers	• *Acidithiobacillus thiooxidans* • *Thiomonas cuprina*	• *Acidithiobacillus caldus*	• *Metallosphaera* spp.

Source: Plumb, J.J., Hawkes, R.B., and Franzmann, P.D., 2007. *Biomining*, Springer-Verlag, Berlin, Heidelberg, 217–235.

These microbes gain energy by oxidizing ferrous iron (Fe^{2+}) into ferric ion (Fe^{3+}) and elemental sulfur (S^0) to sulfuric acid (H_2SO_4), in turn, contributing to the indirect leaching of metals from E-waste (shown in Equations R.1 and R.2). The regenerated Fe^{3+} provides oxidative attack and H_2SO_4 provides proton attack to leach out the metals in bioleaching solutions. However, the presence of a high concentration of metal and organic solvent residues in E-waste components is reported to be inhibiting to the sulfur oxidation processes (62,63):

$$4Fe^{2+} + O_2 + 4H^+ \xrightarrow{\text{biological}} 4Fe^{3+} + 2H_2O \tag{R.1}$$

$$2S^0 + 3O_2 + 2H_2O \xrightarrow{\text{biological}} 2H_2SO_4 \tag{R.2}$$

A feasibility study for using fungi (*Aspergillus niger* and *Penicillium simplicissimum*) to leach metals from electronic scrap by a two-step process has also been investigated by Brandl et al. (64). However, biological extraction of metals is severely affected by a number of factors such as pulp density (weight of substrate in bioleaching tank), solution pH, O_2/CO_2 distribution levels, inoculum size, oxidation reduction potential, temperature, etc., (65,67). All these hindering factors are closely interlinked with each other. Heap leaching is the easiest approach to handle E-waste. At the same time, it is very difficult to manipulate the number of environmental factors that affect overall recovery. Therefore, the use of appropriate bioreactors and operating conditions allow a good control over these variables, resulting in a better bioleaching of metals from E-waste.

10.6.4 Fast Pyrolysis

Fast pyrolysis is a technically advanced technology that is capable of converting WEEE into liquid products, particularly oil (8). Furthermore, emissions of toxic gases are far less with the use of fast pyrolysis as compared to other technologies when it is cotreated with other organic wastes. Fast pyrolysis has thus advantages over landfilling or incineration approaches and its energy consumption is found to be less than 10%. During the use of fast pyrolysis, WEEEs are thermally decomposed under 700–900 K in an oxygen-free system into byproducts such as oil, gas, and char products. Based on the WEEE composition and process conditions, the quality of product yields could vary widely (8,67). The technical terminologies of the subjects presented in this chapter can be found in the literature (68).

10.7 SUMMARY

Overall, the current E-waste recycling practices in developing Asian countries are found to be rudimentary and have not addressed environmental and health concerns sufficiently. The common problems identified with these practices are (i) no national level policies or guidelines are made available to handle hazardous E-waste; (ii) technologically advanced methods are not used to recycle E-waste; (iii) unwanted components in the E-waste are open burned or landfilled without considering their environmental impacts; and (iv) health risks associated with E-waste recycling remain poorly quantified and understood. Although E-waste has received increased attention from governments, NGOs, manufacturing companies, and consumers worldwide, the progress toward an environmentally sound management system has been slow in developing countries. Also, the available data on environmental and occupational health impacts associated with E-waste recycling practices in developing countries are very limited. Considering the complexity and uncertainty in E-waste recycling, proper inventorization and management approaches need to be developed and implemented. It is also very important for the developing countries to come up with environmentally benign technologies that are suitable for their specific regions.

REFERENCES

1. Cui, J. and Roven, H.R., 2011. Electronic waste. In: *Waste: A Handbook for Management*, Letcher, T.M. and Vallero, D.A. (Eds.), Elsevier's Science and Technology Rights Department, Oxford, UK, Chapter 20, pp. 281–296, doi: 10.1016/B978-0-12-381475-3.10020-8.
2. Ongondo, F.O., Williams, I.D., and Cherrett, T.J., 2011. How are WEE doing? A global review of the management of electrical and electronic wastes. *Waste Management*, 31, 714–730.
3. Lau, W.K.Y., Chung, S.S., and Zhang, C., 2013. A material flow analysis on current electrical and electronic waste disposal from Hong Kong households. *Waste Management*, 33 (3), 714–721.
4. Kiddee, P., Naidu, R., and Wong, M.H., 2013. Electronic waste management approaches: An overview. *Waste Management*, 33 (5), 1237–1250.
5. Widmer, R., Oswald-Krapf, H., Sinha-Khetriwal, D., Schnellmann, M., and Boni, H., 2005. Global perspectives on e-waste. *Environmental Impact Assessment Review*, 25, 436–458.
6. UNEP—United Nations Environment Programme, 2009. *Basel Convention on the Control of Transboundary Movements of Hazardous Wastes and Their Disposal*. United Nations Environment Programme. http://www.basel.int/.
7. Herat, S. and Agamuthu, P., 2012. E-waste: A problem or an opportunity? Review of issues, challenges and solutions in Asian countries. *Waste Management and Research*, 30 (11), 1113–1129.
8. Liu, W.J., Tian, K., Jiang, H., Zhang, X.S., and Yang, G.X., 2013. Preparation of liquid chemical feedstocks by co-pyrolysis of electronic waste and biomass without formation of polybrominated dibenzo-p-dioxins. *Bioresource Technology*, 128, 1–7.
9. Townsend, T.G., 2011. Environmental issues and management strategies for waste electronic and electrical equipment. *Journal of the Air and Waste Management Association*, 61 (6), 587–610.
10. Veit, H.M. and Bernardes, A.M., 2015. Electronic waste: Generation and management. In: *Electronic Waste Recycling Techniques-Topics in Mining, Metallurgy and Materials Engineering*, Veit, H.M. and Bernardes, A.M., (Eds.), Springer International Publishing, Switzerland, doi: 10.1007/978-3-319-15714-6, pp. 3–12.
11. Visvanathan, C., Yin, N.H., and Karthikeyan, O.P., 2010. Co-disposal of electronic waste with municipal solid waste in bioreactor landfills. *Waste Management*, 30, 2608–2614.
12. Frazzoli, C., Orisakwe, O.E., Dragone, R., and Mantovani, A., 2010. Diagnostic health risk assessment of electronic waste on the general population in developing countries' scenarios. *Environmental Impact Assessment Review*, 30, 388–399.

13. Brigden, K., Labunska, I., Santillo, D., and Allsopp, D., 2005. *Recycling of Electronic Wastes in China and India: Workplace and Environmental Contamination*. Report, Greenpeace International. University of Exeter, UK.

14. Leung, A.O., Luksemburg, W.J., Wong, A.S., and Wong, M.H., 2007. Spatial distribution of polybrominated diphenyl ethers and polychlorinated dibenzo-p-dioxins and dibenzofurans in soil and combusted residue at Guiyu, an electronic waste recycling site in southeast China. *Environmental Science and Technology*, 41, 2730–2737.

15. Zhao, G., Wang, Z., Dong, M.H., Rao, K., Luo, J., Wang, D., Zha, J., Huang, S., Xu, Y., and Ma, M., 2008. PBBs, PBDEs, and PCBs levels in hair of residents around e-waste disassembly sites in Zhejiang Province, China and their potential sources. *Science of the Total Environment*, 397, 46–57.

16. Zheng, L., Wu, K., Li, Y., Qi, Z., Han, D., Zhang, B., Gu, C. et al., 2008. Blood lead and cadmium levels and relevant factors among children from an e-waste recycling town in China. *Environmental Research*, 108, 15–20.

17. Chen, D., Bi, X., Zhao, J., Chen, L., Tan, J., Mai, B., Sheng, G., Fu, J., and Wong, M., 2009. Pollution characterization and diurnal variation of PDBEs in the atmosphere of an e-waste dismantling region. *Environmental Pollution*, 157, 1051–1057.

18. Xing, G.H., Chan, J.K.Y., Leung, A.O.W., Wu, S.C., and Wong, M.H., 2009. Environmental impact and human exposure to PCBs in Guiyu: An electronic waste recycling site in China. *Environment International*, 35, 76–82.

19. UNEP—United Nations Environment Programme 2006. Call for Global Action on E-waste. http://www.unep.org/Documents.Multilingual/Default.asp?ArticleID=5447&DocumentID=496&1=en (retrieved on July 10, 2011)

20. OECD—Organisation for Economic Cooperation and Development, 2008. Environmental Outlook to 2030. http://213.253.134.43/oecd/pdfs/browseit/9708011E.PDF.

21. Robinson, H.B., 2009. E-waste: An assessment of global production and environmental impacts. *Science of the Total Environment*, 408, 183–191.

22. Balde, C.P., Wang, F., Kuehr, R., and Huisman, J., 2015. *The Global E-waste Monitor-2014, United Nations University*. IAS-SCYCLE, Bonn, Germany (ISBN: 978-92-808-4555-6).

23. EEA—European Environmental Agency 2003. *Waste Electrical and Electronic Equipment (WEEE)*, EEA, Copenhagen, Denmark.

24. Liu, X., Tanaka, M., and Matusi, Y., 2006. Generation amount prediction and material flow analysis of electronic waste: A case study in Beijing, China. *Waste Management and Research*, 24, 434–445.

25. Puckett, J., Westervelt, S., Gutierrez, R., and Takamiya, Y., 2005. *The Digital Dump, Exporting Re-Used and Abuse to Africa*. The Basel Action Network (BAN), Seattle, USA.

26. Cobbing, M., 2008. *Toxic Tech: Not in Our Backyard, Uncovering the Hidden Flows of E-waste*. Greenpeace International, Amsterdam.

27. Tsydenova, O. and Bengtsson, M., 2011. Chemical hazards associated with treatment of waste electrical and electronic equipment. *Waste Management*, 31, 45–58.

28. Cui, J. and Zhang, L., 2008. Metallurgical recovery of metals from electronic waste: A review. *Journal of Hazardous Materials*, 158, 228–256.

29. Li, J., Duan, H., and Yuan, W., 2009. Case study of a Suzhou pilot project on the suitable treatment technology for scrap computers in China. In: *ISSST 2009, IEEE International Symposium on Sustainable Systems and Technology*, Phoenix, AZ, USA, May 18–20, 2009, pp. 1–5.

30. Nnorom, I.C. and Osibanjo, O., 2008. Overview of electronic waste (e-waste) management practices and legislations, and their poor applications in the developing countries. *Resources, Conservation and Recycling*, 52, 843–858.

31. Boeni, H., Silva, U., and Ott, D., 2008. E-waste recycling in Latin America: Overview, challenges and potential. In: *Proceedings of the 2008 Global Symposium on Recycling*, Waste Treatment and Clean Technology, Rewas, pp. 665–673.

32. Manomaivibool, P. and Vassanadumrongdee, S., 2011. Extended producer responsibility in Thailand: Prospects for policies on waste electrical and electronic equipment. *Journal of Industrial Ecology*, 15 (2), 185–205.

33. Bengtsson, M., Hayashi, S., and Hotta, Y., 2009. Conclusions: Toward an extended producer responsibility policy with international considerations. In: *Extended Producer Responsibility Policy in East Asia—In Consideration of International Resource Circulation*, Hotta, Y., Hayashi, S., Bengtsson, M., Mori, H. (Eds.), IGES, Hayama, pp. 169–175.

34. Yang, J., Lu, B., and Xu, C., 2008. WEEE flow and mitigating measures in China. *Waste Management*, 28 (9), 1589–1597.

35. Yu, J., Ju, M., and Williams, E., 2009. Waste electrical and electronic equipment recycling in China: Practices and strategies. In: *2009 IEEE International Symposium on Sustainable Systems and Technology*, ISSST, Tempe, Arizona, USA, May 18–20.

36. He, H., Li, L., and Ding, W., 2008. Research on recovery logistics network of waste electronic and electrical equipment in China. In: *Industrial Electronics and Applications, 2008, ICIEA 2008, 3rd IEEE Conference*, Singapore, June 3–5, pp. 1797–1802.

37. Sepulveda, A., Schluep, M., Renaud, F.G., Streicher, M., Kuehr, R., Hageluken, C., and Gerecke, A.C., 2010. A review of the environmental fate and effects of hazardous substances released from electronic equipments during recycling: Examples from China and India. *Environmental Impact Assessment Review*, 30 (1), 28–41.

38. Manomaivibool, P., 2009. Extended producer responsibility in a non-OECD context the management of waste electrical and electronic equipment in India. *Resources, Conservation and Recycling*, 53 (3), 136–144.

39. Schluep, M., Hageluken, C., Kuehr, R., Magalini, F., Maurer, C., Meskers, C.E., Mueller, E., and Wang, F., 2009. *Recycling—From E-waste to Resources*. United Nations Environmental Programme and United Nations University, Germany, http://isp.unu.edu/news/2010/files/UNEP_eW2R_publication.pdf.

40. EEI—Electrical and Electronics Institute 2007. *Development of E-Waste Inventory in Thailand*. http://archive.basel.int/techmatters/e_wastes/E-waste%20Inventory%20in%20Thailand.pdf

41. Babington, J.C., Siwar, C., Fariz, A., and Ara, R.B., 2010. Bridging the gaps: An E-waste management and recycling assessment of material recycling facilities in Selangor and Penang. *International Journal of Environmental Sciences*, 1 (3), 383–391.

42. Peralta, G.L. and Fontanos, P.M., 2006. E-waste issues and measures in the Philippines. *Journal of Material Cycles and Waste Management*, 8, 34–39.

43. Perkins, D.N., Drisse, M.N.N., Nxele, T., and Sly, P.D., 2014. E-waste: A global hazard. *Annals of Global Health*, 80, 286–295.

44. Chi, X., Streicher-Porte, M., Wang, M.Y.L., and Reuter, M.A., 2011. Informal electronic waste recycling: A sector review with special focus on China. *Waste Management*, 31, 731–742.

45. Kurian, J., 2007. Electronic waste management in India—Issues and strategies. In: *Proceedings Sardinia 2007, Eleventh International Waste Management and Landfill Symposium*, Cagliari, Italy, 1–5 October, 2007.

46. Finlay, A. 2005. E-waste challenges in developing countries: South Africa case study. APC Issue Papers. Association for Progressive Communications. http://www.apc.org.

47. Hicks, C., Dietmar, R., and Eugster, M., 2005. The recycling and disposal of electrical and electronic waste in China—Legislative and market responses. *Environmental Impact Assessment Review*, 25, 459–471.

48. Osibanjo, O. and Nnorom, I.C., 2007. The challenge of electronic waste (e-waste) management in developing countries. *Waste Management and Research*, 25, 489–501.

49. Barba-Gutierrez, Y., Adenso-Diaz, B., and Hopp, M., 2008. An analysis of some environmental consequences of European electric and electronic waste regulation. *Resources, Conservation and Recycling*, 52, 481–495.

50. Gullett, B.K., Linak, W.P., Touati, A., Wasson, S.J., Gatica, S., and Kingm C.J., 2007. Characterization of air emissions and residual ash from open burning of electronic wastes during simulated rudimentary recycling operations. *Journal of Material Cycles and Waste Management*, 9, 69–79.

51. Kasassi, A., Rakimbei, P., Karagiannidis, A., Zabaniotou, A., Tsiouvaras, K., Nastis, A., and Tzafeiropoulou, K., 2008. Soil contamination by heavy metals: Measurements from a closed unlined landfill. *Bioresource Technology*, 99, 8578–8584.

52. Brigden, K., Labunska, I., Santillo, D., and Johnston, P., 2008. *Chemical Contamination at E-waste Recycling and Disposal Sites in Accra and Korforidua, Ghana*. Greenpeace Research Laboratories, University of Exeter, UK.

53. BAN—Basal Action Network 2004. *Mobile Toxic Waste: Recent Findings on the Toxicity of End-of-Life Cell Phones*. A Report—http://archive.ban.org/library/mobilephonetoxicityrep.pdf, pp. 1–7.

54. IARC—International Agency for Research on Cancer, 1989. *IARC Monographs Programme on the Evaluation of Carcinogenic Risks to Humans: Some Organic Solvents, Resin Monomers and Related Compounds, Pigments and Occupational Exposures in Paint Manufacture and Painting*, France, October 18–25, 1988, vol. 47, pp. 291–306. http://monographs.iarc.fr/ENG/Monographs/vol47/

55. Jenkins, R.O., Morris, T.A., Craig, P.J., Goessler, W., Ostah, N., and Wills, K.M., 2000. Evaluation of cot mattress inner foam as a potential site for microbial generation of toxic gases. *Human and Experimental Toxicology*, 19 (12), 693–702.

56. Patterson, T.J., Ngo, M., Aronoy, P.A., Reznikova, T.V., Green, P.G., and Rice, R.H., 2003. Biological activity of inorganic arsenic and antimony reflects oxidation state in cultured human keratinocytes. *Chemical Research in Toxicology*, 16 (12), 1624–1631.

57. Lau, J.H., Wong, C.P., Lee, N.C., and Ricky Lee, S.W., 2003. *Electronics Manufacturing with Lead-Free, Halogen-Free and Conductive-Adhesive Materials*. McGraw-Hill, New York, USA, ISBN 0071386246.

58. Andrewes, P., KitChendian, K.T., and Wallace, K., 2004. Plasmid DNA damage caused by stibine and trimethylstibine. *Toxicology and Applied Pharmacology*, 194, 41–48.

59. Wong, M.H., Wu, S., Deng, W., Yu, X., Luo, Q., Leung, A., Wong, C., Luksemburg, W.J., and Wong, A.S., 2007. Export of toxic chemicals—A review of the case of uncontrolled electronic-waste recycling. *Environmental Pollution*, 149 (2), 131–140.

60. Volckens, J. and Leith, D. 2003. Effects of sampling bias on gas-particle partitioning of semi-volatile compounds, *Atmos. Environ.* 37, 3385–3393.

61. Wang, L.K., 2009. Recycling and disposal of hazardous solid waste containing heavy metals and other toxic substances. In: *Heavy Metals in the Environment,* Wang, L. K. , Chen, J. P., Hung, Y.-T., Shammas, N. K. (Eds.), CRC Press, Florida, US, 361–380. ISBN 9781420073164.

62. Ilyas, S., Anwar, M., Niazi, S., and Ghauri, M., 2007. Bioleaching of metals from electronic scrap by moderately thermophilic acidophilic bacteria. *Hydrometallurgy*, 88, 180–188.

63. Vestola, E.A., Kuusenaho, M.K., Narhi, H.M., Tuovinen, O.H., Puhakka, J.A., Plumb, J.J., and Kaksonen, A.H., 2010. Acid bioleaching of solid waste materials from copper, steel and recycling industries. *Hydrometallurgy*, 103, 74–79.

64. Brandl, H., Bosshard, R., and Wegmann, M., 2001. Computer-munching microbes: Metal leaching from electronic scrap by bacteria and fungi. *Hydrometallurgy* 59: 319–326.

65. Cordoba, E.M., Munoz, J.A., Blazquez, M.L., Gonzalez, F., and Ballester, A., 2008. Leaching of chalcopyrite with ferric ion (Part I): General aspects. *Hydrometallurgy*, 93 (3–4), 81–87.

66. Xia, L., Dai, S., Yin, C., Hu, Y., Liu, J., and Qiu, G., 2009. Comparison of bioleaching behaviors of different compositional sphalerite using *Leptospirillum ferriphilum, Acidithiobacillus ferrooxidans* and *Acidithiobacillus caldus. Journal of Industrial Microbiology Biotechnology*, 36, 845–851.

67. Alston, S.M., Clark, A.D., Arnold, J.C., and Stein, B.K., 2011. Environmental impact of pyrolysis of mixed WEEE plastics. Part 1: Experimental pyrolysis data. *Environmental Science and Technology*, 45, 9380–9385.

68. Wang, M.H.S. and Wang, L.K., 2016. Glossary of land and energy resources engineering. In: *Natural Resources and Control Processes*, Wang, L.K., Wang, M.H.S., Hung, Y.T., and Shammas, N.K. (Eds). Springer, New York, USA.

11 Site Assessment and Cleanup Technologies of Metal Finishing Industry

CONTENTS

ABSTRACT

Preparing brownfields sites for productive reuse requires integration of many elements: financial issues, community involvement, liability considerations, environmental assessment, and cleanup and regulatory requirements. The challenge to any brownfields program is to cleanup sites in accordance with redevelopment goals. Such goals may include cost-effectiveness, timeliness, avoidance of adverse effects to site structures and neighboring communities, and redevelopment of land in a way that benefits communities and local economies. The Triad approach focuses on management of decision uncertainty by incorporating systematic project planning, dynamic work planning strategies, and use of real-time measurement technologies, including innovative technologies, to accelerate and improve the cleanup process.

Specifically, the objective of this chapter is to provide decision makers with: An understanding of common industrial processes at metal finishing facilities and the relationship between such processes and potential releases of contaminants to the environment; information on the types of contaminants likely to be present at a metal finishing site; a discussion of site assessment and cleanup technologies that can be used to assess and cleanup the types of contaminants likely to be present at metal finishing sites; a conceptual framework for identifying potential contaminants at the site, pathways by which contaminants may migrate offsite, and environmental and human health concerns; information on developing an appropriate cleanup plan for metal finishing sites where contamination levels must be reduced to allow a site's reuse; a discussion of pertinent issues and factors that should be considered when developing a site assessment and cleanup plan and selecting appropriate technologies for brownfields, given time and budget constraints.

11.1 INTRODUCTION

11.1.1 BACKGROUND

The Comprehensive Environmental Response, Compensation, and Liability Act (CERCLA) or Superfund (1) defines brownfields sites as "real property, the expansion, redevelopment, or reuse of which may be complicated by the presence or potential presence of a hazardous substance, pollutant, or contaminant." According to the U.S. Environmental Protection Agency (USEPA), brownfields sites are abandoned, idled, or under-used industrial and commercial facilities where expansion or redevelopment is complicated by real or perceived environmental contamination (2). Concerns about liability, cost, and potential health risks associated with brownfields sites often prompt businesses to migrate to "greenfields" outside the city. Left behind are communities burdened with environmental contamination, declining property values, and increased unemployment.

USEPA's Brownfields Economic Redevelopment Initiative was established to enable states, site planners, and other community stakeholders to work together in a timely manner to prevent, assess, safely cleanup, and sustainably reuse brownfields sites (3). With the enactment of the Small Business Liability Relief and Brownfields Revitalization Act in 2002, USEPA assistance was expanded to provide greater support for brownfields cleanup and reuse. Many states and local jurisdictions also help businesses and communities adapt environmental cleanup programs to the special needs of brownfields sites.

Preparing brownfields sites for productive reuse requires integration of many elements: financial issues, community involvement, liability considerations, environmental assessment, and cleanup and regulatory requirements, and more, the coordination among many groups of stakeholders (4). The assessment and cleanup of a site must be carried out in a way that integrates all these factors into the overall redevelopment process. In addition, the cleanup strategy will vary from site to site. At some sites, the cleanup will be completed before the properties are transferred to new owners. At other sites, the cleanup may take place simultaneously with construction and redevelopment activities.

Regardless of when and how cleanups are accomplished, the challenge to any brownfields program is to cleanup sites in accordance with redevelopment goals. Such goals may include cost-effectiveness, timeliness, avoidance of adverse effects to site structures and neighboring communities, and redevelopment of land in a way that benefits communities and local economies. Regulators and site managers are increasingly recognizing the value of implementing a more dynamic approach to streamline assessment and cleanup activities at brownfields sites. This approach, referred to as the Triad, is flexible and recognizes site-specific decisions and data needs (4).

The Triad approach focuses on management of decision uncertainty by incorporating (a) systematic project planning; (b) dynamic work planning strategies; and (c) use of real-time measurement technologies, including innovative technologies, to accelerate and improve the cleanup process. The Triad approach can reduce costs, improve decision certainty, expedite site closeout, and positively affect regulatory and community acceptance. This approach is well aligned with brownfields site priorities, which are affected by the economics of redevelopment, community involvement, and liability considerations.

Numerous technology options are available to assist those involved in brownfields cleanup. USEPA's Office of Superfund Remediation and Technology Innovation (OSRTI) encourages the use of smarter solutions for characterizing and cleaning up contaminated sites by advocating more effective, less costly technological approaches. Use of innovative technologies to characterize and cleanup brownfields sites provides opportunities for stakeholders to reduce cleanup costs and accelerate cleanup schedules. Often, innovative approaches are also more acceptable to communities.

The cornerstone of USEPA's Brownfields Initiative is the pilot program. Under this program, USEPA is funding more than 200 brownfields assessment pilot projects in states, cities, towns, counties, and tribes across the country (2). The pilots, each funded at up to USD 200,000 over 2 years, are bringing together community groups, investors, lenders, developers, and other affected parties to address the issues associated with assessing and cleaning up contaminated brownfields sites and returning them to appropriate, productive use. USEPA's regional brownfields coordinators can provide communities with technical assistance such as targeted brownfields assessments. In addition to the hundreds of brownfields sites being addressed by these pilots, over 40 states have established brownfields or voluntary cleanup programs (VCPs) to encourage municipalities and private sector organizations to assess, cleanup, and redevelop brownfields sites.

11.1.2 METALS AND METALLOIDS

Metals are one of the three groups of elements distinguished by their ionization and bonding properties, along with metalloids and nonmetals. Metals have certain characteristic physical properties: they are usually shiny, have a high density, are ductile and malleable, usually have a high melting point, are usually hard, and conduct electricity and heat well. Metalloids have properties that are intermediate between those of metals and nonmetals. There is no unique way of distinguishing a metalloid from a true metal, but the most common way is that metalloids are usually semiconductors rather than conductors (4).

Locations where metals and metalloids may be found include artillery and small arms impact areas, battery disposal areas, burn pits, chemical disposal areas, contaminated marine sediments, disposal wells and leach fields, electroplating and metal finishing shops, firefighting training areas, landfills and burial pits, leaking storage tanks, radioactive and mixed waste disposal areas, oxidation ponds and lagoons, paint stripping and spray booth areas, sand blasting areas, surface impoundments, and vehicle maintenance areas. Typical metals and metalloids encountered at many sites include those listed in Table 11.1.

11.1.3 PURPOSE

USEPA has developed a set of technical guides to assist communities, states, municipalities, and the private sector to more effectively address brownfields sites. Each guide in this series contains

---

TABLE 11.1

Typical Metals and Metalloids at Brownfields Sites

Metals and Metalloids

Aluminum	Calcium	Mercury	Tin
Antimony	Chromium	Molybdenum	Titanium
Arsenic	Cobalt	Nickel	Vanadium
Barium	Copper	Potassium	Zinc
Beryllium	Iron	Selenium	Zirconium
Bismuth	Lead	Silver	
Boron	Magnesium	Sodium	
Cadmium	Manganese	Thallium	

Source: USEPA. *Road Map to Understanding Innovative Technology Options for Brownfields Investigation and Cleanup*, 4th Edition. EPA 542-B-05-001, U.S. Environmental Protection Agency, Washington, DC, September 2005.

---

information on a different type of brownfields site (classified according to former industrial use). In addition, a supplementary guide contains information on cost-estimating tools and resources for brownfields sites (4–6).

The overview of the technical process involved in assessing and cleaning up brownfields sites can assist planners in making decisions at various stages of the project. An understanding of land use and industrial processes conducted in the past at a site can help the planner to conceptualize the site and identify likely areas of contamination that may require cleanup. Numerous resources are suggested to facilitate the characterization of the site and consideration of cleanup technologies (2–6).

Specifically, the objective of this chapter is to provide decision makers with

1. An understanding of common industrial processes at metal finishing facilities and the relationship between such processes and potential releases of contaminants to the environment.
2. Information on the types of contaminants likely to be present at a metal finishing site.
3. A discussion of site assessment (also known as site characterization), screening and cleanup levels, and cleanup technologies that can be used to assess and cleanup the types of contaminants likely to be present at metal finishing sites.
4. A conceptual framework for identifying potential contaminants at the site, pathways by which contaminants may migrate offsite, and environmental and human health concerns.
5. Information on developing an appropriate cleanup plan for metal finishing sites where contamination levels must be reduced to allow a site's reuse.
6. A discussion of pertinent issues and factors should be considered when developing a site assessment and cleanup plan and selecting appropriate technologies for brownfields, given time and budget constraints.

11.2   INDUSTRIAL PROCESSES AND CONTAMINANTS AT METAL FINISHING SITES

Understanding the industrial processes used during a metal finishing facility's active life and the types of contaminants that may be present provides important information to guide planners in the assessment, cleanup, and restoration of the site to an acceptable condition for sale or reuse. This section provides a general overview of the processes, chemicals, and contaminants used or found at metal finishing sites. Specific metal finishing brownfields sites may have had a different combination of these processes, chemicals, and contaminants. Therefore, this information can be used only to develop a framework of likely past activities. Planners should obtain facility-specific information on industrial processes at their site whenever possible. Site-specific information is also important to obtain because the site may have been used for other industrial purposes at other times in the past.

This section describes waste-generating surface preparation operations; metal finishing operations and the types of waste streams and specific contaminants associated with each process; auxiliary areas at metal finishing sites that may produce contaminants and nonprocess-related contamination problems associated with metal finishing sites. Figure 11.1 presents typical metal finishing processes and land areas, along with the types of waste streams associated with each area (7). Table 11.2 lists the specific contaminants associated with each waste stream (2).

11.2.1   Surface Preparation Operations

Metal finishing processes are typically housed within one structure. The surface of metal products generally requires preparation (i.e., cleaning) prior to applying a finish. An initial set of degreasing tanks ([A] in Figure 11.1) are used to remove oils, grease, and other foreign matter from the surface of the metal so that a coating can be applied. Metal finishing facilities may use solvents or emulsion

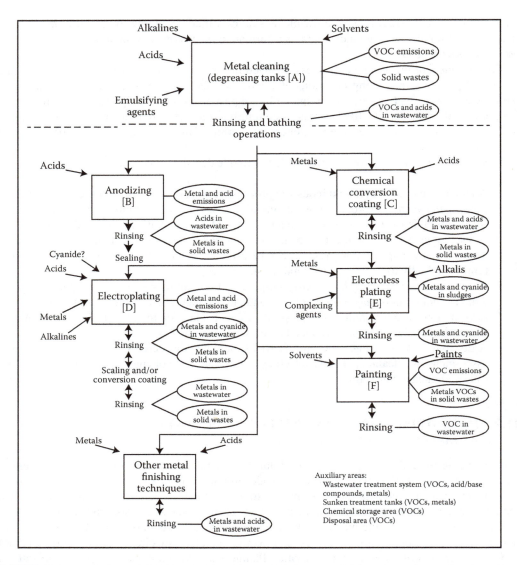

FIGURE 11.1 Typical metal finishing facility. (From USEPA. *Brownfields and Land Revitalization—Tools and Technical Information.* U.S. Environmental Protection Agency, Washington, DC, http://www.epa.gov/ brownfields/toolsandtech.htm, 2015.)

solutions (i.e., solvents dispersed in an aqueous medium with the aid of an emulsifying agent) in the degreasing tanks to clean and prepare the surfaces of metal parts. Wastewaters generated from cleaning operations are primarily rinse waters, which are usually combined with other metal finishing wastewaters and treated onsite by conventional chemical precipitation. These wastewaters may contain solvents, as listed in Table 11.2. Solid wastes such as wastewater treatment sludges, still bottoms, and cleaning tank residues may also be generated.

11.2.2 METAL FINISHING OPERATIONS

Metal finishing operations are typically performed in a series of tanks (baths) followed by rinsing cycles. Acid or alkaline baths "pickle" the surface of the steel to improve the adherence of the

TABLE 11.2

Common Contaminants at Metal Finishing Sites

Contaminant Group	Contaminant Name
Volatile organic compounds (VOCs)	Acetone, benzene, isopropyl alcohol, 2-dichlorobenzene, 4-trimethylbenzene, dichloromethane, ethyl benzene, freon 113, methanol, methyl isobutyl ketone, methyl ethyl ketone, phenol, tetrachloroethylene, toluene, trichloroethylene (TCE), xylene (mixed isomers)
Metals/inorganics	Aluminum, antimony, arsenic, asbestos (friable), barium, cadmium, chromium, cobalt, copper, lead, cyanide, manganese, mercury, nickel, silver, zinc
Acids	Hydrochloric acid, nitric acid, phosphoric acid, sulfuric acid

Source: USEPA. *Technical Approaches to Characterizing and Cleaning Up Metal Finishing Sites under the Brownfields Initiative.* EPA/625/R-98/006, U.S. Environmental Protection Agency, Cincinnati, OH, March 1999.

coating. After the pickling baths, the metal products are moved to plating tanks, where the final coat is applied. Wastes generated during finishing operations derive from the solvents and cleansers applied to the surface and the metal-ion-bearing aqueous solutions used in acid/alkaline rinsing and bathing operations. Common metal finishing operations include anodizing, chemical conversion coating, electroplating, electroless plating, and painting. Common waste streams include metals and acids in the wastewater; metals in sludges and solid waste; and solvents from painting operations, as listed in Table 11.2. If these wastes were managed or disposed of onsite, it is possible that pollutants were released into the environment. Even at facilities where wastes were not stored onsite, releases may have occurred during the handling and use of chemicals. Metal finishing operations are described below (2).

11.2.2.1 Anodizing Operations

Anodizing is an electrolytic process that uses acids from the combined electrolytic solution/acid bath tank to convert the metal surface into an insoluble oxide coating ([B] in Figure 11.1). After anodizing, metal parts are typically rinsed and then sealed. Anodizing operations produce contaminated wastewaters and solid wastes.

11.2.2.2 Chemical Conversion Coating

Chemical conversion coating ([C] in Figure 11.1) includes the following processes:

Chromating. Chromate conversion coatings are produced on various metals by chemical or electrochemical treatment. Acid solutions react with the metal surface to form a layer of a complex mixture of the constituent compounds, including chromium and the base metal.

Phosphating. Phosphate conversion coating involves the immersion of steel, iron, or zinc-plated steel into a dilute solution of phosphate salts, phosphoric acid, and other reagents to condition the surfaces for further processing.

Metal coloring. Metal coloring involves chemically converting the metal surface into an oxide or similar metallic compound to produce a decorative finish.

Passivating. Passivating is the process of forming a protective film on metals by immersing them in an acid solution (usually nitric acid or nitric acid with sodium dichromate.)

Pollutants associated with chemical conversion processes enter the wastestream through rinsing and batch dumping of process baths. Wastewater containing chromium is usually pretreated; this process generates a sludge that is sent offsite for metals reclamation and/or disposal.

11.2.2.3   Electroplating

Electroplating is the production of a surface coating of one metal upon another by electrodeposition ([D] in Figure 11.1). In electroplating, metal ions (in either acid, alkaline, or neutral solutions) are reduced on the cathodic surfaces of the work pieces being plated. Electroplating operations produce contaminated wastewater and solid wastes. Contaminated wastewater results from work piece rinsing and process cleanup water. Rinse water from electroplating is usually combined with other metal finishing wastewater and treated onsite by conventional chemical precipitation, which results in wastewater treatment sludges. Other wastes generated from electroplating include spent process solutions and quench baths that may be discarded periodically when the concentrations of contaminants inhibit their proper functions.

11.2.2.4   Electroless and Immersion Plating

Electroless plating involves chemically depositing a metal coating onto a plastic object by immersing the object in a plating solution ([E] in Figure 11.1). Immersion plating produces a thin metal deposit, commonly zinc or silver, by chemical displacement. Both produce contaminated wastewater and solid wastes. Facilities generally treat spent plating solutions and rinse water chemically to precipitate the toxic metals; however, some plating solutions can be difficult to treat because of the presence of chelates. Most waste sludges resulting from electroless and immersion plating contain significant concentrations of toxic metals.

11.2.2.5   Painting

Painting is the application of predominantly organic coatings for protective and/or decorative purposes ([F] in Figure 11.1). Paint is applied in various forms, including dry powder, solvent diluted formulations, and waterborne formulations, most commonly via spray painting and electrodeposition. Painting operations may result in solvent-containing waste and the direct release of solvents, paint sludge wastes, and paint-bearing wastewater. Paint cleanup operations also may contribute to the release of chlorinated solvents. Discharge from water curtain booths generates the most wastewater. Onsite wastewater treatment processes generate a sludge that is taken offsite for disposal. Other sources of wastes include emission control devices (e.g., paint booth collection systems and ventilation filters) and discarded paints. Sandblasting may be performed to remove paint and to clean metal surfaces for painting or resurfacing; this practice may be of particular concern if the paint being removed contains lead.

11.2.2.6   Other Metal Finishing Techniques

Polishing, hot dip coating, and etching are other processes used to finish metal. Wastewater is often generated during these processes. For example, after polishing operations, area cleaning and wash down can produce metal-bearing wastewaters. Hot-dip coating techniques, such as galvanizing, use water for rinses following precleaning and for quenching after coating. Hot dip coatings also generate a solid waste, oxide dross that is periodically skimmed off the heated tank. Etching solutions are composed of strong acids or bases which may result in etching solution wastes that contain metals and acids.

11.2.3   Auxiliary Activity Areas and Potential Contaminants

11.2.3.1   Wastewater Treatment

Many of the operations involved in metal finishing produce wastewaters, which usually are combined and treated onsite, often by conventional chemical precipitation. Even though the facility would have been required to meet state wastewater discharge standards before releasing wastes, spills of process wastewater may have occurred in the area. At abandoned sites, any remaining wastewaters left in tanks or floor drains could contain solvents, metals, and acids, such as those listed in Table 11.2. In addition, it is possible that wastewater sludges, which can contain metals, were left at the site in baths or tanks.

11.2.3.2 Sunken Wastewater Treatment Tank

Some metal finishing facilities have wastewater treatment tanks sunk into the concrete slab to rest on the underlying soils. This is done by design to aid facility operators in accessing the tanks. If these tanks develop leaks, the lost material, which may contain VOCs (volatile organic compounds) and metals, may be released directly to the soils beneath the building.

11.2.3.3 Chemical Storage Area

At most metal finishing sites an area for storing chemicals used in the various operations was designated. Bulk containers stored in these areas may have leaked or spilled, resulting in discharges to floor drains or cracks in the floor. VOCs such as those listed in Table 11.2 may be found in such areas. Acids and alkaline reagents may also be found in this area.

11.2.3.4 Disposal Area

Materials, both liquid and solid, from process baths may have been disposed of at a designated area at the site. Such areas may be identified by stained soils or a lack of vegetation. These areas may contain VOCs, such as those listed in Table 11.2.

11.2.3.5 Other Considerations

Not all releases are related to the industrial processes described above. Some releases result from the associated services required to maintain the industrial processes. For example, electroplating facilities are large consumers of electricity, which requires a number of transformers. At older facilities, these transformers may have been disposed of in unmarked areas of the facility, which makes it difficult to know where leaks of polychlorinated biphenyl (PCB)-laden oils used as coolants may have occurred. Similarly, large machinery used to move metal pieces requires periodic maintenance. In the past, chemicals used for maintenance operations, such as solvents, oils, and grease, may have been flushed down drain and sumps after use. Stormwater runoff from paved areas such as parking lots may contain petroleum hydrocarbons and oils, which can contaminate areas located downgradient. When conducting initial site evaluations, planners should expand their investigations to include these types of activities.

In addition, metal finishing facilities may have been located in older buildings that contain lead paint and asbestos insulation and tiling. Any structure built before 1970 should be assessed for the presence of these materials. They can cause significant problems during demolition or renovation of the structures for reuse. Special handling and disposal requirements under state and federal laws can significantly increase the cost of construction.

11.3 SITE ASSESSMENT

The site investigation phase focuses on confirming whether any contamination exists at a site, locating any contamination, and characterizing the nature and extent of that contamination (8). It is essential that an appropriately detailed study of the site be performed to identify the cause, nature, and extent of contamination and the possible threats to the environment or to any people living or working nearby. For brownfields sites, the results of such a study can be used in determining goals for cleanup, quantifying risks, determining acceptable and unacceptable risk, and developing effective cleanup plans that minimize delays or costs in the redevelopment and reuse of property. To ensure that sufficient information is obtained to support future decisions, the proposed cleanup measures and the proposed end use of the site should be considered when identifying data needs during the site investigation (4).

The elements of a site assessment are designed to help planners build a conceptual framework of the facility, which will aid site characterization efforts (9). The conceptual framework should identify (2)

1. Potential contaminants that remain in and around the facility.
2. Pathways along which contaminants may move.
3. Potential risks to the environment and human health that exist along the migration pathways.

This section highlights the key role that state environmental agencies usually play in brownfields projects. The types of information that planners should attempt to collect to characterize the site in a Phase I site assessment (i.e., the facility's history) are discussed. Information is presented about where to find and how to use this information to determine whether or not contamination is likely. Additionally this section provides information to assist planners in conducting a Phase II site assessment, including sampling the site and determining the magnitude of contamination. Other considerations in assessing iron and steel sites are also discussed, and general sampling costs are included. The linking of the decision to be taken to the collected data and technologies is illustrated in Figure 11.2.

11.3.1 THE CENTRAL ROLE OF THE STATE AGENCIES

A brownfields redevelopment project involves partnerships among site planners (whether private or public sector), state and local officials, and the local community. State environmental agencies often are key decision makers and a primary source of information for brownfields projects. Brownfields sites are generally cleaned up under state programs, particularly state voluntary cleanup or brownfields programs; thus, planners will need to work closely with state program managers to determine their particular state's requirements for brownfields development. Planners may also need to meet additional federal requirements. Key state functions include (2)

1. Overseeing brownfields site assessment and cleanup processes, including the management of VCPs.
2. Providing guidance on contaminant screening levels.
3. Serving as a source of site information, as well as legal and technical guidance.

11.3.1.1 State VCPs

State VCPs are designed to streamline brownfields redevelopment, reduce transaction costs, and provide state liability protection for past contamination. Planners should be aware that state cleanup requirements vary significantly and should contact the state brownfields manager; brownfields managers from state agencies will be able to identify their state requirements for planners and will clarify how their state requirements relate to federal requirements.

11.3.1.2 Levels of Contaminant Screening and Cleanup

Identifying the level of site contamination and determining the risk, if any, associated with that contamination level is a crucial step in determining whether cleanup is needed. Some state environmental agencies, as well as federal and regional USEPA offices, have developed screening levels for certain contaminants, which are incorporated into some brownfields programs. Screening levels represent breakpoints in risk-based concentrations of chemicals in soil, air, or water. If contaminant concentrations are below the screening level, no action is required; above the level, further investigation is needed.

In addition to screening levels, USEPA regional offices and some states have developed cleanup standards; if contaminant concentrations are above cleanup standards, cleanup must be pursued. Section 11.3.7 provides more information on screening levels and Section 11.4 provides more information on cleanup standards.

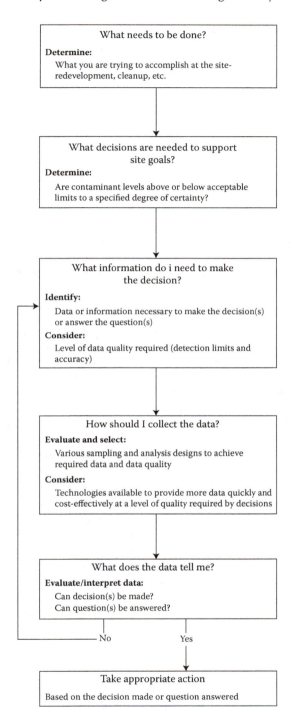

FIGURE 11.2 Linking the decision, data, and technology. (From USEPA. *Road Map to Understanding Innovative Technology Options for Brownfields Investigation and Cleanup*, 4th Edition. EPA 542-B-05-001, U.S. Environmental Protection Agency, Washington, DC, September 2005.)

11.3.2   PERFORMING A PHASE I SITE ASSESSMENT: OBTAINING FACILITY BACKGROUND INFORMATION FROM EXISTING DATA

Planners should compile a history of the iron and steel manufacturing facilities to identify likely site contaminants and their probable locations. Financial institutions typically require a Phase I site assessment prior to lending money to potential property buyers to protect the institution's role as mortgage holder (10). In addition, parties involved in the transfer, foreclosure, leasing, or marketing of properties recommend some form of site evaluation. The site history should include

1. A review of readily available records (e.g., former site use, building plans, and records of any prior contamination events).
2. A site visit to observe the areas used for various industrial processes and the condition of the property.
3. Interviews with knowledgeable people (e.g., site owners, operators, and occupants; neighbors; and local government officials).
4. A report that includes an assessment of the likelihood that contaminants are present at the site.

The Phase I site assessment should be conducted by an environmental professional, and may take 3–4 weeks to complete. Site evaluations are required in part as a response to concerns over environmental liabilities associated with property ownership. A property owner needs to perform "due diligence," that is, fully inquire into the previous ownership and uses of a property to demonstrate that all reasonable efforts to find site contamination have been made. Because brownfields sites often contain low levels of contamination and pose low risks, due diligence through a Phase I site assessment will help to answer key questions about the levels of contamination. Several federal and state programs exist to minimize owner liability at brownfields sites and facilitate cleanup and redevelopment; planners should contact the state environmental or regional USEPA office for further formation.

Information on how to review records, conduct site visits and interviews, and develop a report during a Phase I site assessment is provided below.

11.3.2.1   Facility Records

Facility records are often the best source of info on former site activities. If past owners are not initially known, a local records office should have deed books that contain ownership history. Generally, records pertaining specifically to the site in question are adequate for review purposes. In some cases, however, records of adjacent properties may also need to be reviewed to assess the possibility of contaminants migrating from or to the site, based on geologic or hydrogeologic conditions. If the brownfields property resides in a low-lying area, in close proximity to other industrial facilities or formerly industrialized sites, or downgradient from current or former industrialized sites, an investigation of adjacent properties is warranted.

11.3.2.2   Other Sources of Recorded Information

Planners may need to use other sources in addition to facility records to develop a complete history. American Society for Testing and Materials (ASTM) Standard 1527 identifies standard sources such as historical aerial photographs, fire insurance maps, property tax files, recorded land title records, topographic maps, local street directories, building department records, zoning/land use records, and newspaper archives (10).

Some metal finishing site managers may have worked with state environmental regulators; these offices may be key sources of information. Federal (e.g., USEPA) records may also be useful. The types of information provided by regulators may include facility maps that identify activities and disposal areas, lists of stored pollutants, and the types and levels of pollutants released. State offices and other sources where planners can search for site-specific information are presented below.

1. The state offices responsible for industrial waste management and hazardous waste should have a record of any emergency removal actions at the site (e.g., the removal of leaking drums that posed an "imminent threat" to local residents); any Resource Conservation and Recovery Act (RCRA) (11) permits issued at the site; notices of violations issued; and any environmental investigations.
2. The state office responsible for discharges of wastewater to water bodies under the National Pollutant Discharge Elimination System (NPDES) (12) program will have a record of any permits issued for discharges into surface water at or near the site. The local publicly owned treatment works (POTW) will have records for permits issued for indirect discharges into sewers (e.g., floor drain discharges to a sanitary sewer).
3. The state office responsible for underground storage tanks (USTs) may also have records of tanks located at the site, as well as records of any past releases.
4. The state office responsible for air emissions may be able to provide information on air pollutants associated with particular types of onsite contamination.
5. USEPA's Comprehensive Environmental Response, Compensation, and Liability Information System (CERCLIS) (13) of potentially contaminated sites should have a record of any previously reported contamination at or near the site.
6. USEPA regional offices can provide records of sites that have hazardous substances. Information is available from the Federal National Priorities List (NPL) and lists of treatment, storage, and disposal (TSD) facilities subject to corrective action under RCRA. RCRA non-TSD facilities, RCRA generators, and Emergency Response Notification System (ERNS) information on contaminated or potentially contaminated sites can help to determine if neighboring facilities are recorded as having released hazardous substances into the immediate environment.
7. State and local records may indicate any permit violations or significant contaminant releases from or near the site.
8. Residents and former employees may be able to provide useful information on waste management practices, but these reports should be substantiated.
9. Local fire departments may have responded to emergency events at the facility. Fire departments or city halls may have fire insurance maps or other historical maps or data that indicate the location of hazardous waste storage areas at the site.
10. Local waste haulers may have records of the facility's disposal of hazardous or other waste materials.
11. Utility records.
12. Local building permits.

11.3.2.3   Identifying Migration Path Ways and Potentially Exposed Populations

Offsite migration of contaminants may pose a risk to human health and the environment; planners should gather as much readily available information on the physical characteristics of the site as possible. Migration pathways, that is, soil, groundwater, and air, will depend onsite-specific characteristics such as geology and the physical characteristics of the individual contaminants (e.g., mobility). Information on the physical characteristics of the general area can play an important role in identifying potential migration pathways and focusing environmental sampling activities, if needed. Planners should collect three types of information to obtain a better understanding of migration pathways, including topographic, soil and subsurface, and groundwater data, as described below (14–17).

11.3.3 Gathering Topographic Information

In this preliminary investigation, topographic information will be helpful in determining whether the site may be subject to contamination by adjoining properties or may be the source of contamination of other properties. Topographic information will help planners identify low-lying areas of the facility where rain and snowmelt (and any contaminants in them) may collect and contribute both water and contaminants to the underlying aquifer or surface runoff to nearby areas. The U.S. Geological Survey (USGS) of the Department of the Interior has topographic maps for nearly every part of the country.

11.3.4 Gathering Soil and Subsurface Information

Planners should know about the types of soils at the site from the ground surface extending down to the water table because soil characteristics play a large role in how contaminants move in the environment. For example, clay soils limit downward movement of pollutants into underlying groundwater but facilitate surface runoff. Sandy soils, on the other hand, can promote rapid infiltration into the water table while inhibiting surface runoff. Soil information can be obtained through a number of sources (2):

1. Local planning agencies should have soil maps to support land use planning activities. These maps provide a general description of the soil types present within a county (or sometimes a smaller administrative unit, such as a township).
2. The Natural Resource Conservation Service and Co-operative Extension Service offices of the U.S. Department of Agriculture (USDA) are also likely to have soil maps.
3. Well-water companies are likely to be familiar with local subsurface conditions, and local water districts and state water divisions may have well-logging information.
4. Local health departments may be familiar with subsurface conditions because of their interest in septic drain fields.
5. Local construction contractors are likely to be familiar with subsurface conditions from their work with foundations.

Soil characteristics can vary widely within a relatively small area, and it is common to find that the top layer of soil in urban areas is composed of fill materials, not native soils. While local soil maps and other general soil information can be used for screening purposes such as in a Phase I assessment, site-specific information will be needed in the event that cleanup is necessary.

11.3.5 Gathering Groundwater Information

Planners should obtain general groundwater information about the site area, including

1. State classifications of underlying aquifers
2. Depth to the groundwater tables
3. Groundwater flow direction and rate

This information can be obtained by contacting state environmental agencies or from several local sources, including water authorities, well drilling companies, health departments, and Agricultural Extension and Natural Resource Conservation Service offices.

11.3.5.1 Identifying Potential Environmental and Human Health Concerns

Identifying possible environmental and human health risks early in the process can influence decisions regarding the viability of a site for cleanup and the choice of cleanup methods used. A visual inspection of the area will usually suffice to identify onsite or nearby wetlands and water bodies that may be particularly sensitive to releases of contaminants during characterization or cleanup activities. Planners should also review available information (e.g., from state and local environmental agencies) to ascertain the proximity of residential dwellings, nearby industrial/commercial activities, and wetlands/water bodies, and to identify people, animals, or plants that might receive migrating contamination; any particularly sensitive populations in the area (e.g., children and endangered species); and whether any major contamination events have occurred previously in the area (e.g., drinking water problems and groundwater contamination).

For environmental information, planners can contact the U.S. Army Corps of Engineers, state environmental agencies, local planning and conservation authorities, the USGS, and the USDA Natural Resource Conservation Service. State and local agencies and organizations can usually provide information on local fauna and the habitats of any sensitive and/or endangered species.

For human health information, planners can contact

1. *State and local health assessment organizations.* Organizations such as health departments should have data on the quality of local well water used as a drinking water source, as well as any human health risk studies that have been conducted. In addition, these groups may have other relevant information, such as how certain types of contaminants (e.g., volatile organics, such as benzene and phenols) might pose a health risk (e.g., dermal exposure to volatile organics during site characterization); information on exposures to particular contaminants and potential associated health risks can also be found in health profile documents developed by the Agency for Toxic Substances and Disease Registry (ATSDR). In addition, ATSDR may have conducted a health consultation or health assessment in the area if an environmental contamination event that may have posed a health risk occurred in the past; such an event and assessment should have been identified in the Phase I records review of prior contamination incidents at the site if any occurred.

2. *Local water and health departments.* During the site visit (described below), when visually inspecting the area around the facility, planners should identify any residential dwellings or commercial activities near the facility and evaluate whether people there may come into contact with contamination along one of the migration pathways. Where groundwater contamination may pose a problem, planners should identify any nearby waterways or aquifers that may be impacted by groundwater discharge of contaminated water, including any drinking water wells that may be downgradient of the site, such as a municipal well field. Local water departments will have a count of well connections to the public water supply. Planners should also pay particular attention to information on private wells in the area downgradient of the facility, since, depending on their location, they may be vulnerable to contaminants migrating offsite even when the public municipal drinking water supply is not vulnerable. Local health departments often have information on the locations of private wells.

In addition to groundwater sources and migration pathways, surface water sources and pathways should be evaluated since groundwater and surface waters can interface at some (or several) point(s) in the region. Contaminants in groundwater can eventually migrate to surface water, and contaminants in surface water can migrate to groundwater.

11.3.5.2 Community Involvement

It is important that brownfields decision makers encourage acceptance of redevelopment plans and cleanup alternatives by involving members of the community early in the decision-making process through community meetings, newsletters, or other outreach activities. For an individual site, the community should be informed about how the use of a proposed technology might affect redevelopment plans or the adjacent neighborhood (4). For example, the planting of trees for the use of phytoremediation may create esthetic or visual improvements; on the other hand, the use of phytoremediation may bring about issues related to site security or long-term maintenance that could affect access to the site.

Community-based organizations represent a wide range of issues, from environmental concerns to housing issues to economic development. These groups can often be helpful in educating planners and others in the community about local brownfields sites, which can contribute to successful brownfields site assessment and cleanup activities. In addition, most state VCPs require that local communities be adequately informed about brownfields cleanup activities. Planners can contact the local Chamber of Commerce, local philanthropic organizations, local service organizations, and neighborhood committees for community input. State and local environmental groups may be able to supply relevant information and identify other appropriate community organizations. Local community involvement in brownfields projects is a key component in the success of such projects (2).

USEPA can assist members of the brownfields community by directing its members to appropriate resources and providing opportunities to network and participate in the sharing of information. A number of Internet sites, databases, newsletters, and reports provide opportunities for brownfields stakeholders to network with other stakeholders to identify information about cleanup and technology options. USEPA's Brownfields and Land Revitalization Technology Support Center is a valuable resource for brownfields decision makers.

11.3.5.3　Conducting a Site Visit

In addition to collecting and reviewing available records, planners need to conduct a site visit to visually and physically observe the uses and conditions of the property, including both outdoor areas and the interior of any structure or property. Current and past uses involving the use, treatment, storage, disposal, or generation of hazardous substances or petroleum products should be noted. Current or past uses of abutting properties that can be observed readily while conducting the site visit also should be noted. In addition, readily observable geologic, hydrologic, and topographic conditions should be identified, including any possibility of hazardous substances migrating on or offsite.

Roads, water supplies, and wastewater systems should be identified, as well as any storage tanks, whether above or below ground. If any hazardous substances or petroleum products are found, their type, quantity, and storage conditions should be noted. Any odors, pools of liquids, drums or other containers, and equipment likely to contain PCBs should be noted. Additionally, indoors, heating and cooling systems should be noted, as well as any stains, corrosion, drains, or sumps. Outdoors, any pits, ponds, lagoons, stained soil or pavement, stressed vegetation, solid waste, wastewater, and wells should be noted (10).

11.3.5.4　Conducting Interviews

In addition to reviewing available records and visiting the site, conducting interviews with the site owner and/or site manager, site occupants, and local officials is highly recommended to obtain information about the prior and/or current uses and conditions of the property, and to inquire about any useful documents that exist regarding the property. Such documents include environmental audit reports, environmental permits, registrations for storage tanks, material safety data sheets, community right-to-know plans, safety plans, government agency notices or correspondence, hazardous waste generator reports or notices, geotechnical studies, or any proceedings involving the property (10). Interviews with at least one staff person from the following local government agencies are recommended: the fire department, health agency, and the agency with authority for hazardous waste disposal or other environmental matters. Interviews can be conducted in person, by telephone, or in writing.

ASTM standard 1528 (18) provides a questionnaire that may be appropriate for use in interviews for certain sites. ASTM suggests that this questionnaire be posed to the current property owner, any major occupant of the property (or at least 10% of the occupants of the property if no major occupant exists), or "any occupant likely to be using, treating, generating, storing, or disposing of hazardous substances or petroleum products on or from the property." A user's guide accompanies the ASTM questionnaire to assist the investigator in conducting interviews, as well as researching records and making site visits.

11.3.5.5 Developing a Report

Toward the end of the Phase I assessment, planners should develop a report that includes all of the important information obtained during record reviews, the site visit, and interviews. Documentation, such as references and important exhibits, should be included, as well as the credentials of the environmental professional that conducted the Phase I environmental site assessment. The report should include all information regarding the presence or likely presence of hazardous substances or petroleum products on the property and any conditions that indicate an existing, past, or potential release of such substances into property structures or into the ground, groundwater, or surface water of the property (10). The report should include the environmental professional's opinion of the impact of the presence or likely presence of any contaminants, and a findings and conclusion section that either indicates that the Phase I environmental site assessment revealed no evidence of contaminants in connection with the property, or discusses what evidence of contamination was found.

Additional sections of the report might include a recommendations section (e.g., for a Phase II site assessment, if appropriate); and sections on the presence or absence of asbestos, lead paint, lead, and radon in drinking water and wetlands. Some states or financial institutions may require information on these substances.

If the Phase I site assessment adequately informs state and local officials, planners, community representatives, and other stakeholders that no contamination exists at the site, or that contamination is so minimal that it does not pose a health or environmental risk, then those involved may decide that adequate site assessment has been accomplished and the process of redevelopment may proceed. In some cases where evidence of contamination exists, stakeholders may decide that enough information is available from the Phase I site assessment to characterize the site and determine an appropriate approach for site cleanup of the contamination. In other cases, stakeholders may decide that additional site assessment is warranted, and a Phase II site assessment would be conducted.

11.3.6 THE TRIAD APPROACH: STREAMLINING SITE INVESTIGATIONS AND CLEANUP DECISIONS

The modernization of the collection, analysis, interpretation, and management of data to support decisions about hazardous waste sites rests on USEPA's three-pronged or Triad approach (19–21). The introduction of new technologies in a dynamic framework allows project managers to meet clearly defined objectives. Such an approach incorporates the elements described below (4,19–21).

Systematic planning is a commonsense approach to assuring that the level of detail in project planning matches the intended use of the data being collected. Once cleanup goals have been defined, systematic planning is undertaken to chart a course for the project that is resource effective, as well as technically sound and defensible to reach these project-critical goals. A team of multidisciplinary, experienced technical staff works to translate the project's goals into realistic technical objectives. The Conceptual Site Model (CSM) is the planning tool that organizes the information that already is known about the site; the CSM helps the team identify the additional information that must be obtained. The systematic planning process ties project goals to individual activities necessary to reach these goals by identifying data gaps in the CSM. The team then uses the CSM to direct the gathering of needed information, allowing the CSM to evolve and mature as work progresses at the site.

A *dynamic working strategy* approach relies on real-time data to reach decision points. The logic for decision-making is identified and responsibilities, authority, and lines of communication are established. Dynamic work strategy implementation relies on and is driven by critical project decisions needed to reach closure. It uses a decision-tree and real-time uncertainty management practices to reach critical decision points in as few mobilizations as possible. Success of a dynamic approach depends on the presence of experienced staff in the field empowered to make decisions based on the decision logic and their capability to deal with new data and any unexpected issues, as they arise. Field staff maintains close communication with regulators or others overseeing the project during implementation of dynamic work plans.

The use of onsite analytical tools, rapid sampling platforms, and onsite interpretation and management of data makes dynamic work strategies possible. Such *real-time measurement tools* are among the key streamlined site investigation tools because they provide the data that are used for onsite decision-making. The tools are a broad category of analytical methods and equipment that can be applied at the sample collection site. They include methods that can be used outdoors with handheld, portable equipment, as well as more rigorous methods that require the controlled environments of a mobile laboratory (transportable). During the planning process, the team identifies the type, rigor, and quantity of data needed to answer the questions raised by the CSM. Those decisions then guide the design sampling modifications and the selection of analytical tools.

The Triad approach enables project managers to minimize uncertainty while expediting site cleanup and reducing project costs. For example, USEPA collaborated with the town of Greenwich, Connecticut to implement the Triad approach to characterize a former power plant site scheduled for redevelopment as a waterfront park. The Triad approach yielded an estimated cost savings of 50%–60% when compared with a traditional approach involving two mobilizations and comprehensive analytical methods at a fixed laboratory. The city of Trenton, New Jersey began implementing the Triad approach in 2001 as part of its program to redevelop a large number of abandoned industrial sites. Overall, the Triad approach eliminated costs associated with follow-on investigation activities while accelerating the redevelopment schedule and reducing decision uncertainty. Additional details about these and other examples are available in the USEPA's Technology News and Trends newsletter (22).

11.3.7 PERFORMING A PHASE II SITE ASSESSMENT: SAMPLING THE SITE

A Phase II site assessment (23) typically involves taking soil, water, and air samples to identify the types, quantity, and extent of contamination in these various environmental media. The types of data used in a Phase II site assessment can vary from existing site data (if adequate), to limited sampling of the site, to more extensive contaminant-specific or site-specific sampling data. Planners should use knowledge of past facility operations whenever possible to focus the site evaluation on those process areas where pollutants were stored, handled, used, or disposed. These will be the areas where potential contamination will be most readily identified. Generally, to minimize costs, a Phase II site assessment will begin with limited sampling (assuming readily available data do not exist that adequately characterize the type and extent of contamination on the site) and will proceed to more comprehensive sampling if needed (e.g., if the initial sampling could not identify the geographical limits of contamination).

This section explains the importance of setting data quality objectives (DQOs) and provides brief guidance for doing so; describes screening levels to which sampling results can be compared; and provides an overview of environmental sampling and data analysis, including sampling methods and ways to increase data certainty.

11.3.7.1 Setting Data Quality Objectives

USEPA has developed a guidance document that describes key principles and best practices for brownfields site assessment quality assurance and quality control based on program experience (16,24,25).

USEPA has adopted the DQOs process (25) as a framework for making decisions. The DQO process is commonsense, systematic planning tool based on the scientific method. Using a systematic planning approach, such as the DQO process, ensures that the data collected to support defensible site decision-making will be of sufficient quality and quantity, as well as be generated through the most cost-effective means possible. DQOs, themselves, are statements that unambiguously communicate the following:

1. The study objective
2. The most appropriate type of data to collect
3. The most appropriate conditions under which to collect the data
4. The amount of uncertainty that will be tolerated when making decisions

It is important to understand the concept of uncertainty and its relationship to site decision-making (26–28). Regulatory agencies, and the public they represent, want to be as confident as possible about the safety of reusing brownfields sites. Public acceptance of site decisions may depend on the site manager's being able to scientifically document the adequacy of site decisions. During negotiations with stakeholders, effective communication about the tradeoffs between project costs and confidence in the site decision can help set the stage for a project's successful completion. When the limits on uncertainty (e.g., only a 5%, 10%, or 20% chance of a particular decision error is permitted) are clearly defined in the project, subsequent activities can be planned so that data collection efforts will be able to support those confidence goals in a resource-effective manner. On the one hand, a manager would like to reduce the chance of making a decision error as much as possible, but on the other hand, reducing the chance of making that decision error requires collecting more data, which is, in itself, a costly process.

Striking a balance between these two competing goals—more scientific certainty versus less cost—requires careful thought and planning, as well as the application of professional expertise (26–28).

The following steps are involved in systematic planning:

1. *Agree on intended land reuse.* All parties should agree early in the process on the intended reuse for the property because the type of use may strongly influence the choice of assessment and cleanup approaches. For example, if the area is to be a park, removal of all contamination will most likely be needed. If the land will be used for a shopping center, with most of the land covered by buildings and parking lots, it may be appropriate to reduce, rather than totally remove, contaminants to specified levels (e.g., state cleanup levels; see Section 11.4).

2. *Clarify the objective of the site assessment.* What is the overall decision(s) that must be made for the site? Parties should agree on the purpose of the assessment. Is the objective to confirm that no contamination is present? Or is the goal to identify the type, level, and distribution of contamination above the levels, which are specified, based on the intended land use. These are two fundamentally different goals that suggest different strategies. The costs associated with each approach will also vary.

   As noted above, parties should also agree on the total amount of uncertainty allowable in the overall decision(s). Conducting a risk assessment involves identifying the levels of uncertainty associated with characterization and cleanup decisions. A risk assessment involves identifying potential contaminants and analyzing the pathways through which people, other species of concern, or the environment can become exposed to those contaminants. Such an assessment can help identify the risks associated with varying the levels of acceptable uncertainty in the site decision and can provide decision makers with greater confidence about their choice of land use decisions and the objective of the site assessment. If cleanup is required, a risk assessment can also help determine how clean the site needs to be, based on expected reuse (e.g., residential or industrial), to safeguard people from exposure to contaminants.

3. *Define the appropriate type(s) of data that will be needed to make an informed decision at the desired confidence level.* Parties should agree on the type of data to be collected by

defining a preliminary list of suspected analytes, media, and analyte-specific action levels (screening levels). Define how the data will be used to make site decisions. For example, data values for a particular analyte may or may not be averaged across the site for the purposes of reaching a decision to proceed with work. Are there maximum values, which a contaminant(s) cannot exceed? If found, will concentrations of contaminants above a certain action level (hotspots) be characterized and treated separately? These discussions should also address the types of analyses to be performed at different stages of the project. Planners and regulators can reach an agreement to focus initial characterization efforts in those areas where the preliminary information indicates potential sources of contamination may be located. It may be appropriate to analyze for a broad class of contaminants by less expensive screening methods in the early stages of the project in order to limit the number of samples needing analysis by higher quality, more expensive methods later. Different types of data may be used at different stages of the project to support interim decisions that efficiently direct the course of the project as it moves forward.

4. *Determine the most appropriate conditions under which to collect the data.* Parties should agree on the timing of sampling activities, since weather conditions can influence how representative the samples are of actual conditions.

5. *Identify appropriate contingency plans/actions.* Certain aspects of the project may not develop as planned. Early recognition of this possibility can be a useful part of the DQO process. For example, planners, regulators, and other stakeholders can acknowledge that screening-level sampling may lead to the discovery of other contaminants on the site than were originally anticipated. During the DQO process, stakeholders may specify appropriate contingency actions to be taken in the event that contamination is found. Identifying contingency actions early in the project can help ensure that the project will proceed even in the light of new developments. The use of a dynamic workplan combined with the use of rapid turnaround field analytical methods can enable the project to move forward with a minimum of time delay and wasted effort.

6. Develop a sampling and analysis plan that can meet the goals and permissible uncertainties described in the proceeding steps. The overall uncertainty in a site decision is a function of several factors: the number of samples across the site (the density of sample coverage), the heterogeneity of analytes from sample to sample (spatial variability of contaminant concentrations), and the accuracy of the analytical method(s). Studies have demonstrated that analytical variability tends to contribute much less to the uncertainty of site decisions than does sample variability due to matrix heterogeneity. Therefore, spending money to increase the sample density across the site will usually (for most contaminants) make a larger contribution to confidence in the site decision, and thus be more cost-effective, than will spending money to achieve the highest data quality possible, but a lower sampling density.

    Examples of important consideration for developing a sampling and analysis plan include

    a. Determine the sampling location placement that can provide an estimate of the matrix heterogeneity and thus address the desired certainty. Is locating hotspots of a certain size important? Can composite sampling be used to increase coverage of the site (and decrease overall uncertainty due to sample heterogeneity) while lowering analytical costs?

    b. Evaluate the available pool of analytical technologies/methods (both field methods and laboratory methods, which might be implemented in either a fixed or mobile laboratory) for those methods that can address the desired action levels (the analytical methods quantification limit should be well below the action level). Account for possible or expected matrix interferences when considering appropriate methods. Can field analytical methods produce data that will meet all of the desired goals when sampling uncertainty is also taken into account? Evaluate whether a combination of screening and definitive methods may produce a more cost-effective means to generate data. Can economy of scale be used? For example, the expense of a mobile laboratory is seldom cost-effective for a single small site, but might be cost-effective if several sites can be characterized sequentially by a single mobile laboratory.

c. When the sampling procedures, sample preparation and analytical methods have been selected, design a quality control protocol for each procedure and method that ensures that the data generated will be of known, defensible quality.

7. Through a number of iterations, refine the sampling and analysis plan to one that can most cost-effectively address the decision-making needs of the site planner.

8. Review agreements often. As more information becomes available, some decisions that were based on earlier, limited information should be reviewed to see if they are still valid. If they are not, the parties can again use the DQO framework to revise and refine site assessment and cleanup goals and activities.

The data needed to support decision-making for brownfields sites generally are not complicated and are less extensive than those required for more heavily contaminated, higher-risk sites (e.g., superfund sites). But data uncertainty may still be a concern at brownfields sites because knowledge of past activities at a site may be less than comprehensive, resulting in limited site characterization. Establishing DQOs can help address the issue of data uncertainty in such cases. Examples of DQOs include verifying the presence of soil contaminants, and assessing whether contaminant concentrations exceed screening levels.

11.3.7.2   Screening Levels

In the initial stages of a Phase II site assessment an appropriate set of screening levels for contaminants in soil, water, and/or air should be established. Screening levels are risk-based benchmarks which represent concentrations of chemicals in environmental media that do not pose an unacceptable risk. Sample analyses of soils, water, and air at the facility can be compared with these benchmarks. If onsite contaminant levels exceed the screening levels, further investigation will be needed to determine if and to what extent cleanup is appropriate.

Some states have developed generic screening levels (e.g., for industrial and residential use). These levels may not account for site-specific factors that affect the concentration or migration of contaminants. Alternatively, screening levels can be developed using site-specific factors. While site-specific screening levels can more effectively incorporate elements unique to the site, developing site-specific standards is a time- and resource-intensive process. Planners should contact their state environmental offices and/or USEPA regional offices for assistance in using screening levels and in developing site-specific screening levels.

Risk-based screening levels are based on calculations/models that determine the likelihood that exposure of a particular organism or plant to a particular level of a contaminant would result in a certain adverse effect. Risk-based screening levels have been developed for tap water, ambient air, fish, and soil. Some states or USEPA regions also use regional background levels (or ranges) of contaminants in soil and maximum contaminant levels (MCLs) in water established under the Safe Drinking Water Act (29) as screening levels for some chemicals. In addition, some states and/or USEPA regional offices have developed equations for converting soil screening levels to comparative levels for the analysis of air and groundwater.

When a contaminant concentration exceeds a screening level, further site assessment (such as sampling the site at strategic locations and/or performing more detailed analysis) is needed to determine that: (a) the concentration of the contaminant is relatively low and/or the extent of contamination is small and does not warrant cleanup for that particular chemical or (b) the concentration or extent of contamination is high, and that site cleanup is needed.

Using state cleanup standards for an initial brownfields assessment may be beneficial if no industrial screening levels are available or if the site may be used for residential purposes.

USEPA's soil screening guidance is a tool developed by USEPA to help standardize and accelerate the evaluation and cleanup of contaminated soils at sites on the NPL where future residential land use is anticipated. This guidance may be useful at corrective action or VCP (voluntary cleanup program) sites where site conditions are similar. However, use of this guidance for sites where residential land use assumptions do not apply could result in overly conservative screening levels.

11.3.7.3 Environmental Sampling and Data Analysis

Environmental sampling and data analysis are integral parts of a Phase II site assessment process. Many different technologies are available to perform these activities, as discussed below.

11.3.7.4 Levels of Sampling and Analysis

There are two levels of sampling and analysis: screening and contaminant-specific. Planners are likely to use both at different stages of the site assessment.

11.3.7.4.1 Screening

Screening sampling and analysis use relatively low-cost technologies to take a limited number of samples at the most likely points of contamination and analyze them for a limited number of parameters. Screening analyses often test only for broad classes of contaminants, such as total petroleum hydrocarbons (TPHs), rather than for specific contaminants, such as benzene or toluene. Screening is used to narrow the range of areas of potential contamination and reduce the number of samples requiring further, more costly, analysis. Screening is generally performed onsite, with a small percentage of samples (e.g., generally 10%) submitted to a state-approved laboratory for a full organic and inorganic screening analysis to validate or clarify the results obtained.

Some geophysical methods are used in site assessments because they are noninvasive (i.e., do not disturb environmental media as sampling does). Geophysical methods are commonly used to detect underground objects that might exist at a site, such as USTs, dry wells, and drums. The two most common and cost-effective technologies used in geophysical surveys are ground-penetrating radar and electromagnetics (30).

11.3.7.4.2 Contaminant-Specific

For a more in-depth understanding of contamination at a site (e.g., when screening data are not detailed enough), it may be necessary to analyze samples for specific contaminants. With contaminant-specific sampling and analysis, the number of parameters analyzed is much greater than for screening level sampling, and analysis includes more accurate, higher-cost field and laboratory methods. Such analyses may take several weeks.

Computerization, microfabrication, and biotechnology have permitted the recent development of analytical equipment that can be generated in the field, onsite in a mobile laboratory, and offsite in a laboratory. The same kind of equipment might be used in two or more locations.

11.3.8 Increasing the Certainty of Sampling Results

One approach to reducing the level of uncertainty associated with site data is to implement a statistical sampling plan. Statistical sampling plans use statistical principles to determine the number of samples needed to accurately represent the contamination present. With the statistical sampling method, samples are usually analyzed with highly accurate laboratory or field technologies, which increase costs and take additional time. Using this approach, planners can negotiate with regulators and determine in advance specific measures of allowable uncertainty (e.g., an 80% level of confidence with a 20% allowable error).

Another approach to increasing the certainty of sampling results is to use lower-cost technologies with higher detection limits to collect a greater number of samples. This approach would provide a more comprehensive picture of contamination at the site, but with less detail regarding the specific contamination. Such an approach would not be recommended to identify the extent of contamination by a specific contaminant, such as benzene, but may be an excellent approach for defining the extent of contamination by total organic compounds with a strong degree of certainty. Planners will find that there is a tradeoff between scope and detail. Performing a limited number of detailed analyses provides good detail but less certainty about overall contamination, while performing a larger number of general analyses provides less detail but improves the understanding and certainty of the scope of contamination.

11.3.9 Site Assessment Technologies

This section discusses the differences between using field and laboratory technologies and provides an overview of applicable site assessment technologies (31,32). In recent years, several innovative technologies that have been field-tested and applied to hazardous waste problems have emerged. In many cases, innovative technologies may cost less than conventional techniques and can successfully provide the needed data. Operating conditions may affect the cost and effectiveness of individual technologies.

11.3.9.1 Field versus Laboratory Analysis

The principal advantages of performing field sampling and field analysis are that results are immediately available and more samples can be taken during the same sampling event; also, sampling locations can be adjusted immediately to clarify the first round of sampling results if warranted. This approach may reduce costs associated with conducting additional sampling events after receipt of laboratory analysis. Field assessment methods have improved significantly over recent years; however, while many field technologies may be comparable to laboratory technologies, some field technologies may not detect contamination at levels as low as laboratory methods, and may not be contaminant-specific. To validate the field results or to gain more information on specific contaminants, a small percentage of the samples can be sent for laboratory analysis. The choice of sampling and analytical procedures should be based on DQOs established earlier in the process, which determine the quality (e.g., precision, level of detection) of the data needed to adequately evaluate site conditions and identify appropriate cleanup technologies.

11.3.9.2 Sample Collection and Analysis Technologies

Tables 11.3 and 11.4 list sample collection technologies for oil in subsurface and groundwater that may be appropriate for metal finishing brownfields sites. Technology selection depends on the medium being sampled and the type of analysis required, based on DQOs. Soil samples are generally collected using spoons, scoops, and shovels. The selection of a subsurface sample collection technology depends on the subsurface conditions (e.g., consolidated materials, bedrock), the required sampling depth and level of analysis, and the extent of sampling anticipated. For example, if subsequent sampling efforts are likely, then installing semipermanent well casings with a well drilling rig may be appropriate. If limited sampling is expected, direct push methods, such as cone penetrometers, may be more cost-effective. The types of contaminants will also play a key role in the selection of sampling methods, devices, containers, and preservation techniques.

Table 11.5 lists analytical technologies that may be appropriate for assessing metal finishing sites, the types of contamination they can measure, applicable environmental media, and the relative cost of each. The final two columns of the table contain the applicability (e.g., field and or laboratory) of analytical methods and the technology's ability to generate quantitative versus qualitative results. Less expensive technologies that have rapid turnaround times and produce only qualitative results generally should be sufficient for many brownfields sites.

TABLE 11.3
Soil and Subsurface Sampling Tools

Technique/Instrumentation	Media Soil	Media Groundwater	Relative Cost Per Sample	Sample Quality
Drilling Methods				
Cable	X	X	Mid-range expensive	Soil properties will most likely be altered
Casing advancement	X	X	Most expensive	Soil properties will likely be altered
Direct air rotary with rotary hammer	X	X	Mid-range expensive	Soil properties will most likely be altered
Direct mud rotary	X	X	Mid-range expensive	Soil properties may be altered
Directional drilling	X	X	Most expensive	Soil properties may be altered
Hollow-stem auger	X	X	Mid-range expensive	Soil properties may be altered
Jetting methods	X	X	Least expensive	Soil properties may be altered
Rotary diamond drilling	X	X	Most expensive	Soil properties may be altered
Rotating core	X		Mid-range expensive	Soil properties may be altered
Solid flight and bucket augers	X	X	Mid-range expensive	Soil properties will likely be altered
Sonic drilling	X	X	Most expensive	Soil properties will most likely not be altered
Split and solid barrel	X		Least expensive	Soil properties may be altered
Thin-wall open tube	X		Mid-range expensive	Soil properties will most likely not be altered
Thin-wall piston/specialized thin wall	X		Mid-range expensive	Soil properties will most likely not be altered
Direct Push Methods				
Cone penetrometer	X	X	Mid-range expensive	Soil properties may be altered
Driven wells		X	Mid-range expensive	Soil properties may be altered
Handheld Methods				
Augers	X	X	Least expensive	Soil properties may be altered
Rotating core	X		Mid-range expensive	Soil properties may be altered
Scoop, spoons, and shovels	X		Least expensive	Soil properties may be altered
Split and solid barrel	X		Least expensive	Soil properties may be altered
Thin-wall open tube	X		Mid-range expensive	Soil properties will most likely not be altered
Thin-wall piston/specialized thin wall	X		Mid-range expensive	Soil properties will most likely not be altered
Tubes	X		Least expensive	Soil properties will most likely not be altered

Source: USEPA. *Technical Approaches to Characterizing and Cleaning Up Metal Finishing Sites under the Brownfields Initiative.* EPA/625/R-98/006, U.S. Environmental Protection Agency, Cincinnati, OH, March 1999.

11.3.10 Additional Considerations for Assessing Metal Finishing Sites

When assessing a metal finishing brownfields site, planners should focus on the most likely areas of contamination. Although the specific locations vary from site to site, this section provides some general guidelines.

11.3.10.1 Where to Sample

Most metal finishing facilities perform all operations indoors. Consequently, most site assessment activities should focus on contamination inside and underneath the facility. Outdoor assessment

TABLE 11.4

Groundwater Sampling Tools

Technique/ Instrumentation	Contaminants	Relative Cost Per Sample	Sample Quality
Portable Grab Samplers			
Bailers	Metals, VOCs	Least expensive	Liquid properties may be altered
Pneumatic depth-specific samplers	Metals, VOCs	Mid-range expensive	Liquid properties will most likely not be altered
Portable In Situ Groundwater Samplers/Sensors			
Cone penetrometer samplers	Metals, VOCs	Least expensive	Liquid properties will most likely not be altered
Direct drive samplers	Metals, VOCs	Least expensive	Liquid properties will most likely not be altered
Hydropunch	Metals, VOCs	Mid-range expensive	Liquid properties will most likely not be altered
Fixed Situ Samplers			
Multilevel capsule samplers	Metals, VOCs	Mid-range expensive	Liquid properties will most likely not be altered
Multiple-port casings	Metals, VOCs	Least expensive	Liquid properties will most likely not be altered
Passive multilayer samplers	VOCs	Least expensive	Liquid properties will most likely not be altered

Source: USEPA. *Technical Approaches to Characterizing and Cleaning Up Metal Finishing Sites under the Brownfields Initiative.* EPA/625/R-98/006, U.S. Environmental Protection Agency, Cincinnati, OH, March 1999.

activities should evaluate points where drain pipes may have carried contaminated wastewater or spilled materials.

The typical metal finishing facility is comprised of one or more large, warehouse-type buildings that contain the bath tanks, chemical storage areas, and wastewater treatment system. The floors are likely to be a continuous concrete slab containing several drains leading to a central storm drain or sewer access. In older facilities, the feed lines from bath to wastewater tanks are underneath the floor slab. In newer facilities, the bath tanks and/or the wastewater tanks will likely be partially submerged in the floor slab and positioned directly on the ground.

A visual inspection of the site should identify the most likely points of potential contaminant releases. These include the areas surrounding

1. Floor drains in chemical storage and process bath areas
2. Sludges left in process bath and wastewater treatment tanks
3. Pipes underneath the floor slab
4. Tanks set through the floor slab
5. Cracks in floor or stains in low spots in the floor

Solvents can be highly mobile on release, and can seep into and through the concrete flooring, which is porous. The inspection of the facility floor should look not only for cracks through which solvents could migrate, but also for stained areas where spilled solvents may have pooled. Wipe samples should be taken along the walls of the facility, as solvent vapors may have penetrated wall materials.

Since metal finishing operations are typically conducted inside the facility, outside points of potential release are likely to be limited to

1. Points of discharge from effluent pipes
2. Waterways, canals, and ditches at points of pipe discharge
3. Areas where process bath materials may have been dumped

TABLE 11.5

Sample Analysis Technologies

Technique/Instrumentation	Analytes	Media Soil	Media Groundwater	Media Gas	Relative Cost Per Analysis	Application	Produces Quantitative Data
Laser-induced breakdown spectrometry	Metals	X			Least expensive	Usually used in field	Additional effort required
Titrimetry kits	Metals	X	X		Least expensive	Usually used in laboratory	Additional effort required
Particle-induced x-ray emissions	Metals	X	X		Mid-range expensive	Usually used in laboratory	Additional effort required
Atomic adsorption spectrometry	Metals	X*	X	X	Most expensive	Usually used in laboratory	Yes
Inductively coupled plasma-atomic emission spectroscopy	Metals	X	X	X	Most expensive	Usually used in laboratory	Yes
Field bioassessment	Metals	X	X		Most expensive	Usually used in field	No
X-ray fluorescence	Metals	X	X	X	Least expensive	Laboratory and field	Yes (limited)
Chemical colorimetric kits	VOCs	X	X		Least expensive	Can be used in field, usually used in laboratory	Additional effort required
Flame ionization detector (hand held)	VOCs	X	X	X	Least expensive	Immediate, can be used in field	No
Explosimeter	VOCs	X	X*	X	Least expensive	Immediate, can be used in field	No
Photoionization detector (hand held)	VOCs,	X	X	X	Least expensive	Immediate, can be used in field	No
Catalytic surface oxidation	VOCs	X*	X	X	Least expensive	Usually used in laboratory	No
Near IR reflectance/trans spectroscopy	VOCs	X			Mid-range expensive	Usually used In laboratory	Additional effort required
Ion mobility spectrometer	VOCs	X*	X	X*	Mid-range expensive	Usually used in laboratory	Yes
Raman spectroscopy/SERS	VOCs	X	X	X	Mid-range expensive	Usually used in laboratory	Additional effort required
Infrared spectroscopy	VOCs	X	X	X	Mid-range expensive	Usually used in laboratory	Additional effort required
Scattering/absorption Lidar	VOCs	X*	X	X	Mid-range expensive	Usually used in laboratory	Additional effort required
FTIR spectroscopy	VOCs	X	X	X	Mid-range expensive	Laboratory and field	Additional effort required
Synchronous luminescence/fluorescence	VOCs	X	X		Mid-range expensive	Usually used in laboratory, can be used in field	Additional effort required
Gas chromatography (GC) (can be used with numerous detectors)	VOCs	X*	X	X	Mid-range expensive	Usually used in laboratory, can be used in field	Yes
UV–visible spectrophotometry	VOCs	X	X	X	Mid-range expensive	Usually used in laboratory	Additional effort required
UV fluorescence	VOCs	X	X	X	Mid-range expensive	Usually used in laboratory	Additional effort required
Ion trap	VOCs	X	X*	X	Most expensive	Laboratory and field	Yes
Other chemical reaction-based test papers	VOCs, metals	X	X		Least expensive	Usually used in field	Yes
Immunoassay and colorimetric kits	VOCs, metals	X	X		Least expensive	Usually used in laboratory, can be used in field	Additional effort required

Source: USEPA. *Technical Approaches to Characterizing and Cleaning Up Metal Finishing Sites under the Brownfields Initiative.* EPA/625/R-98/006, U.S. Environmental Protection Agency, Cincinnati, OH, March 1999.

VOCs Volatile organic compounds.

X* Indicates there must be extraction of the sample to gas or liquid phase.

While discharge points may be visually obvious, areas of dumping may be less apparent. Often these areas are marked by stained soils and a lack of vegetation. Low-lying areas should also be investigated, as they make natural dumping areas and contaminants may drain to these points.

11.3.10.2   How Many Samples to Collect

Samples should be taken in and around the areas of potential release mentioned above (33). Planners should expect that two to three samples will be required in each area, depending on DQOs. A cost-effective approach is to perform screening analyses using field methods on all samples and then to submit one sample to a laboratory for analysis by an accepted USEPA method. Although the screening analyses can be conducted for broad contaminant groups, such as total organics, a contaminant-specific analysis should be conducted as a full screen for organic and inorganic contaminants and to validate the screening analyses. Contaminant-specific analyses may be conducted either in the field using appropriate technologies and protocols or in a laboratory.

11.3.10.3   What Types of Analysis to Perform

The selection of analytical procedures will be based on the DQOs established. Generally, the following analyses may be appropriate at metal finishing sites:

1. Residuals taken from drain sumps in storage areas should be screened for total organics and acids. Screening analyses for these contaminants can be performed inexpensively using a photoionization detector (PID) or flame ionization detector (FID) for total organics.
2. Residuals taken from drains in the process and wastewater treatment areas should be screened for a similar range of organic contaminants, but additional analyses should be performed to screen for the presence of inorganic contaminants, such as the metals used in the metal finishing process. Immunoassays are an inexpensive field technology that can be used to perform the screening analyses for organic contaminants and mercury. X-ray fluorescence (XRF) is another innovative technology that can be used to perform either field or laboratory analyses.
3. Soil gas should be collected at points underneath the floor slab, particularly near any tanks that are set through the floor slab, to detect the presence of solvents and other organic contaminants. These samples can be analyzed with the PID/FID technology described above. Corings of the floor slab may need to be taken and sent to a laboratory to determine if contaminants have penetrated floor slabs.
4. Wipe samples taken from walls should be analyzed for organic compounds. These analyses can be performed using the same technologies that are used to analyze residuals samples.
5. Soils and sediments at points of pipe discharge should be screened for both organic and inorganic contaminants using PID/FID technology. XRF can be used for field or laboratory analyses.
6. Water samples collected in swales, canals, and ditches should be screened for organics. Inorganic contamination can sometimes be detected in water samples, but conditions do not always allow it.

In addition, as discussed earlier, many older structures contain lead paint and asbestos insulation and tiling. Numerous kits are readily available to test for lead paint. Experienced professionals may be able to visually identify asbestos insulation, but specialized equipment may be needed to confirm the presence of asbestos in other areas. Core or wipe samples can be analyzed for asbestos using polarized light microscopy (PLM). Local and state laws regarding lead and asbestos should be consulted to determine how they may affect the selection of DQOs, sampling, and analysis.

11.3.11   General Sampling Costs

Site assessment costs vary widely, depending on the nature and extent of the contamination and the size of the sampling area. The sample collection costs discussed below are based on an assumed labor rate of USD 48/h plus USD 14 per sample for shipping and handling. All costs have been updated to 2016 USD using USACE Yearly Average Cost Index for Utilities (34).

11.3.11.1 Soil Collection Costs

Surface soil samples can be collected with tools as simple as a stainless steel spoon, shovel, or hand auger. Samples can be collected using hand tools in soft soil for as low as USD 14 per sample (assuming that a field technician can collect 10 samples/h). When soils are hard, or deeper samples are required, a hammer-driven split spoon sampler or a direct push rig is needed. Using a drill rig equipped with a split spoon sampler or a direct push rig typically costs more than USD 840/day for rig operation (35), with the cost per sample exceeding USD 42 (assuming that a field technician can collect 2 samples/h). Labor costs generally increase when heavy machinery is needed.

11.3.11.2 Groundwater Sampling Costs

Groundwater samples can be extracted through conventional drilling of a permanent monitoring well or using the direct push methods listed in Table 11.3. The conventional, hollow stem auger-drilled monitoring well is more widely accepted but generally takes more time than direct push methods. Typical quality assurance protocols for the conventional monitoring well require the well to be drilled, developed, and allowed to achieve equilibrium for 24–48 h. After the development period, a groundwater sample is extracted. With the direct push sampling method, a probe is either hydraulically pressed or vibrated into the ground, and groundwater percolates into a sampling container attached to the probe. The direct push method costs are contingent upon the hardness of the subsurface, depth to the water table, and permeability of the aquifer. Costs for both conventional and direct push techniques are generally more than USD 56 per sample (assuming that a field technician can collect 1 sample/h); well installation costs must be added to that number.

11.3.11.3 Surface Water and Sediment Sampling Costs

Surface water and sediment sampling costs depend on the location and depth of the required samples. Obtaining surface water and sediment samples can cost as little as USD 42 per sample (assuming that a field technician cam collect 2 samples/h). Sampling sediment in deep water or sampling a deep level of surface water, however, requires the use of larger equipment, which drives up the cost. Also, if surface water presents a hazard during sampling and protective measures are required, costs will increase greatly.

11.3.11.4 Sample Analysis Costs

Costs for analyzing samples in any medium can range from as little as USD 38 per sample for a relatively simple test (e.g., an immunoassay test for metals) to greater than USD 564 per sample for a more extensive analysis (e.g., for semivolatiles) and up to USD 1,680 per sample for dioxins (31). Major factors that affect the cost of sample analysis include the type of analytical technology used, the level of expertise needed to interpret the results, and the number of samples to be analyzed. Planners should make sure that laboratories that have been certified by state programs are used.

For information on costs for brownfields cleanup, the reader is referred to USEPA document (36), guide (37), and remediation cost compendium (38).

11.4 SITE CLEANUP

The purpose of this section is to guide planners in the selection of appropriate cleanup technologies. The principal factors that will influence the selection of a cleanup technology include (2)

1. Types of contamination present
2. Cleanup and reuse goals
3. Length of time required to reach cleanup goals

4. Posttreatment care needed
5. Budget

The selection of appropriate cleanup technologies often involves a tradeoff between time and cost. The USEPA document on cost-estimating tools and resources (36) provides information on cost factors and developing cost estimates. In general, the more intensive the cleanup approach, the more quickly the contamination will be mitigated and the more costly the effort. In the case of brownfields cleanup, this can be a major point of concern, considering the planner's desire to return the facility to the point of reuse as quickly as possible. Thus, the planner may wish to explore a number of options and weigh carefully the costs and benefits of each. One effective method of comparison is the cleanup plan, as discussed below. Planners should involve stakeholders in the community in the development of the cleanup plan.

The intended future use of a brownfields site will drive the level of cleanup needed to make the site safe for redevelopment and reuse. Brownfields sites are by definition not Superfund NPL sites; that is, brownfields sites usually have lower levels of contamination present and therefore generally require less extensive cleanup efforts than Superfund NPL sites. Nevertheless, all potential pathways of exposure, based on the intended reuse of the site, must be addressed in the site assessment and cleanup; if no pathways of exposure exist, less cleanup (or possibly none) may be required.

Some regional USEPA and state offices have developed cleanup standards for different chemicals, which may serve as guidelines or legal requirements for cleanups. It is important to understand that screening levels are different from cleanup levels. Screening levels indicate whether further site investigation is warranted for a particular contaminant. Cleanup levels indicate whether cleanup action is needed and how extensive it needs to be. Planners should check with their state environmental office for guidance and/or requirements for cleanup standards.

This section contains information on developing a cleanup plan; various alternatives for addressing contamination at the site (i.e., institutional controls and containment and cleanup technologies); using different technologies for cleaning up metal finishing sites, and postconstruction issues that planners need to consider when considering alternatives.

11.4.1 Developing a Cleanup

If the results of the site evaluation indicate the presence of contamination above acceptable levels, planners will need to have a cleanup plan developed by a professional environmental engineer that describes the approach that will be used to contain and possibly cleanup the contamination present at the site. In developing this plan, planners and their engineers should consider a range of possible options, with the intent of identifying the most cost-effective approaches for cleaning up the site, given time and cost concerns. The cleanup plan can include the following elements (2,4,39,40):

1. A clear delineation of environmental concerns at the site. Areas should be discussed separately if the cleanup approach for an area is different than that for other areas of the site. Clear documentation of existing conditions at the site and a summarized assessment of the nature and scope of contamination should be included.
2. A recommended cleanup approach for each environmental concern that takes into account expected land reuse plans and the adequacy of the technology selected.
3. A cost estimate that reflects both expected capital and operating/maintenance costs.
4. Postconstruction maintenance requirements for the recommended approach.
5. A discussion of the assumptions made to support the recommended cleanup approach, as well as the limitations of the approach.

Planners can use the framework developed during the initial site evaluation and the controls and technologies described below to compare the effectiveness of the least costly approaches for meeting the required cleanup goals established in the DQOs. These goals should be established at levels that are consistent with the expected reuse plans. A final cleanup plan may include a combination of actions, such as institutional controls, containment technologies, and cleanup technologies, as discussed below.

11.4.1.1 Institutional Controls

Institutional controls may play an important role in returning a metal finishing brownfields site to marketable condition. Institutional controls are mechanisms that control the current and future use of, and access to, a site. They are established, in the case of brownfields, to protect people from possible contamination. Institutional controls can range from a security fence prohibiting access to a certain portion of the site to deed restrictions imposed on the future use of the site. If the overall cleanup approach does not include the complete cleanup of the facility (i.e., the complete removal or destruction of onsite contamination), a deed restriction will likely be required that clearly states that hazardous waste is being left in place within the site boundaries. Many state brownfields programs include institutional controls.

11.4.1.2 Containment Technologies

Containment technologies, in many instances, will be the likely cleanup approach for landfilled waste and wastewater lagoons (after contaminated wastewater has been removed) at metal finishing facilities. The purpose of containment is to reduce the potential for offsite migration of contaminants and, possible subsequent exposure. Containment technologies include engineered barriers such as caps (41) for contaminated soils, slurry walls (42), and hydraulic containment. Often, soils contaminated with metals can be solidified (43,44) by mixing them with cement-like materials, and the resulting stabilized material can be stored onsite in a landfill. Like institutional controls, containment technologies do not remove or destroy contamination, but mitigate potential risk by limiting access to it.

If contamination is found underneath the floor slab at metal finishing facilities, leaving the contaminated materials in place and repairing any damage to the floor slab may be justified. The likelihood that such an approach will be acceptable to regulators will depend on whether potential risk can be mitigated and managed effectively over the long term. In determining whether containment is feasible, planners should consider (2,4)

1. *Depth to groundwater.* Planners should be prepared to prove to regulators that groundwater levels will not rise, due to seasonal conditions, and come into contact with contaminated soils.
2. *Soil types.* If contaminants are left in place, the native soils should not be highly porous, as are sandy or gravelly soils, which enable contaminants to migrate easily. Clay and fine silty soils provide a much better barrier.
3. *Surface water control.* Planners should be prepared to prove to regulators that rainwater and snowmelt cannot infiltrate under the floor slab and flush the contaminants downward.
4. *Volatilization of organic contaminants.* Regulators are likely to require that air monitors be placed inside the building to monitor the level of organics that may be escaping upward through the floor and drains.

11.4.1.3 Types of Cleanup Technologies

Cleanup may be required to remove or destroy onsite contamination if regulators are unwilling to accept the level of contamination present or if the types of contamination are not conducive to the use of institutional controls or containment technologies. Cleanup technologies fall broadly into two categories: ex situ and in situ, as described below.

1. *Ex situ.* An ex situ technology treats contaminated materials after they have been removed and transported to another location. After treatment, if the remaining materials,

or residuals, meet cleanup goals, they can be returned to the site. If the residuals still do not meet cleanup goals, they can be subjected to further treatment, contained onsite, or moved to another location for storage or further treatment. A cost-effective approach to cleaning up a metal finishing brownfields site may be the partial treatment of contaminated soils or groundwater, followed by containment, storage, or further treatment offsite (2). For example, it is common practice for operating metal finishing facilities to treat wastewaters to an intermediate level and then send the treated water to the local POTW.

2. *In situ.* The use of in situ technologies has increased dramatically in recent years. In situ technologies treat contamination in place and are often innovative technologies. Examples of in situ technologies include bioremediation (45), soil flushing (46), oxygen releasing compounds (47), air sparging (48), and treatment walls (49). In some cases, in situ technologies are feasible, cost-effective choices for the types of contamination that are likely at metal finishing sites. Planners, however, do need to be aware that cleanup with in situ technologies is likely to take longer than with ex situ technologies.

Maintenance requirements associated with in situ technologies depend on the technology used and vary widely in both effort and cost. For example, containment technologies such as caps and liners will require regular maintenance, such as maintaining the vegetative cover and performing periodic inspections to ensure the long-term integrity of the cover system. Groundwater treatment systems will require varying levels of postcleanup care. If an ex situ system is in use at the site, it will require regular operations support and periodic maintenance to ensure that the system is operating as designed.

11.4.2 Keys to Technology Selection and Acceptance

Innovative technologies and technology approaches offer many advantages in the cleanup of brownfields sites (50–56). Stakeholders in such sites, however, first must accept the technology. Brownfields decision makers should consider the following elements to increase the likelihood that the technology will be accepted, thereby facilitating the cleanup of the site (4).

1. Focus on the decisions that support site goals: The Triad approach systematic planning is an important element of all cleanup activities. Clear and specific planning to meet explicit decision objectives is essential in managing the process of cleaning up contaminated sites: site assessment, site investigation, site monitoring, and remedy selection. With good planning, brownfields decision makers can establish the cleanup goals for the site, identify the decisions necessary to achieve those goals, and develop and implement a strategy for addressing the decision needs. Technology decisions are made in the context of the requirements for such decisions. All cleanup activities are driven by the project goals. An explicit statement of the decisions to be made and the way in which the planned approach supports the decisions should be included in the work plan.

2. Build consensus: Investing time, before the site work begins, in developing decisions that are acceptable to all decision makers will foster more efficient site activities and make successful cleanup more likely. Conversely, allowing work to begin at a site before a common understanding and acceptance of the decisions have been established increases the likelihood that the cleanup process will be inefficient, resulting in delays and inefficient use of time and money. Further, decision makers must understand that there is uncertainty in all scientific and technical decisions. Clearly defining and accepting uncertainty thresholds before making decisions about the site remedy will build consensus. Decisions also should be made in the context of applicable regulatory requirements, political considerations, budget available for the project, and time constraints.

3. Understand the technology: A thorough knowledge of a technology's capabilities and limitations is necessary to secure its acceptance. All technologies are subject to limitations in performance. Planning for the strengths and weaknesses of a technology maximizes understanding of its benefits and its acceptance. "Technology approvers," typically regulators, community groups, and financial service providers are likely to be more receptive of a new approach if the proposer provides a clear explanation of the rationale for its use and demonstrates confidence in its applicability to specific site conditions and needs. This latter point underscores the importance of carefully selecting an experienced, multidimensional team of professionals who have the expertise necessary to plan, present, and implement the chosen approach.

4. Allow flexibility: Streamlining site activities, whether site assessment, site investigation, removal, treatment, or monitoring, requires a flexible approach. Site-specific conditions, including various physical conditions, contamination issues, stakeholder needs, uses of the site, and supporting decisions, require that all decision makers understand the need for flexibility. Although presumptive remedies, standard methods, applications at other sites, and program guidance can serve as the basis for designing a site-specific cleanup plan and can help decision makers avoid "starting from scratch" at each site, decision makers should be wary of depending too heavily on "boilerplate language" and prescriptive methodologies, as well as standard operating procedures and "accepted" methods. While such tools provide excellent starting points, they lack the flexibility to meet site-specific goals. To ensure an efficient and effective cleanup, the actual technology approach, whether established or innovative, must focus on decisions specific to the site.

5. Narrow the list of potential technologies that are most appropriate for addressing the contamination identified at the site and that are compatible with the specific conditions of the site and the proposed reuse of the property:
   a. Network with other brownfields stakeholders and environmental professionals to learn about their experiences and to tap their expertise
   b. Determine whether sufficient data are available to support identification and evaluation of cleanup alternatives
   c. Evaluate the options against a number of factors, including toxicity levels, exposure pathways, associated risks, future land use, and economic considerations
   d. Analyze the applicability of a particular technology to the contamination identified at a site
   e. Determine the effects of various technology alternatives on redevelopment objectives

6. Continue to work with appropriate regulatory agencies to ensure that regulatory requirements are addressed properly:
   a. Consult with the appropriate federal, state, local, and tribal regulatory agencies to include them in the decision-making process as early as possible
   b. Contact the USEPA regional brownfields coordinator to identify and determine the availability of USEPA support programs

7. Integrate cleanup alternatives with reuse alternatives to identify potential constraints on reuse and time schedules and to assess cost and risk factors

8. To provide a measure of certainty and stability to the project, investigate environmental insurance policies, such as protection against cost overruns, undiscovered contamination, and third-party litigation, and integrate their cost into the project financial package

9. Select an acceptable remedy that not only achieves cleanup goals and addresses the risk of contamination, but also best meets the objectives for redevelopment and reuse of the property and is compatible with the needs of the community

10. Communicate information about the proposed cleanup option to brownfields stakeholders, including the affected community

11.4.3 Summary of Technologies for Treating Metals/Metalloids at Brownfield Sites

Chemical treatment, also known as chemical reduction/oxidation (redox) (47), typically involves redox reactions that chemically convert hazardous contaminants into compounds that are nonhazardous, less toxic, more stable, less mobile, or inert. Redox reactions involve the transfer of electrons from one compound to another. Specifically, one reactant is oxidized (loses electrons) and one reactant is reduced (gains electrons). The oxidizing agents used for treatment of hazardous contaminants in soil include ozone, hydrogen peroxide, hypochlorites, potassium permanganate, Fenton's reagent (hydrogen peroxide and iron), chlorine, and chlorine dioxide. This method may be applied in situ or ex situ to soils, sludges, sediments, and other solids and may also be applied to groundwater in situ or ex situ chemical treatment using pump and treat technology. Chemical treatment may also include use of ultraviolet (UV) light in a process known as UV oxidation.

Electrokinetics is based on the theory that a low-density current will mobilize contaminants in the form of charged species. A current passed between electrodes is intended to cause aqueous media, ions, and particulates to move through soil, waste, and water. Contaminants arriving at the electrodes can be removed by means of electroplating or electrodeposition, precipitation or coprecipitation, adsorption, complexing with ion exchange resins, or pumping of water (or other fluid) near the electrodes.

Flushing. For flushing, a solution of water, surfactants, or co-solvents is applied to soil or injected into the subsurface to treat contaminated soil or groundwater (46). When soil is being treated, injection is often designed to raise the water table into the contaminated soil zone. Injected water and treatment agents are recovered together with flushed contaminants.

Permeable reactive barriers, also known as passive treatment walls, are installed across the flow path of a contaminated groundwater plume, allowing the water portion of the plume to flow through the wall (49). These barriers allow passage of water while prohibiting movement of contaminants by means of treatment agents within the wall such as zerovalent metals (usually zerovalent iron), chelators, sorbents, compost, and microbes. The contaminants are either degraded or retained in a concentrated form by the barrier material, which may need to be replaced periodically.

Physical separation processes use physical properties to separate contaminated and uncontaminated media or to separate different types of media (57–59). For example, different-sized sieves and screens can be used to separate contaminated soil from relatively uncontaminated debris. Another application of physical separation is dewatering of sediments or sludge.

Phytoremediation is a process in which plants are used to remove, transfer, stabilize, or destroy contaminants in soil, sediment, or groundwater. The mechanisms of phytoremediation include enhanced rhizosphere biodegradation (which takes place in soil or groundwater immediately around plant roots), phytoextraction (also known as phytoaccumulation, the uptake of contaminants by plant roots and the translocation and accumulation of contaminants into plant shoots and leaves), phytodegradation (metabolism of contaminants within plant tissues), and phytostabilization (production of chemical compounds by plants to immobilize contaminants at the interface of roots and soil). The term phytoremediation applies to all biological, chemical, and physical processes that are influenced by plants (including the rhizosphere) and that aid in the cleanup of contaminated substances (60–63). Phytoremediation may be applied in situ or ex situ to soils, sludges, sediments, other solids, or groundwater.

Environment Canada (63) studied the effectiveness of phytoremediation in Quebec's climate using herbaceous plants (Indian mustard and fescue) and shrubs (willow) to absorb heavy metals (lead, copper, and zinc). They reported that metal concentration levels in the leaves reached 1500–2300 mg/kg that resulted in total extraction of between 2 and 13 kg of metal per ha, per growth period.

Pump and treat involves extraction of groundwater from an aquifer and treatment of the water above the ground. The extraction step is usually conducted by pumping groundwater from a well or trench (64). The treatment step can involve a variety of technologies such as adsorption, air stripping, bioremediation, chemical treatment, filtration, ion exchange, metal precipitation, and membrane filtration (57–59).

Soil washing. For soil washing, contaminants sorbed onto fine soil particles are separated from bulk soil in a water-based system based on particle size (65). The wash water may be augmented with a basic leaching agent, surfactant, or chelating agent or by adjustment of pH to help remove contaminants. Soils and wash water are mixed ex situ in a tank or other treatment unit. The wash water and various soil fractions are usually separated by means of gravity settling (57).

Solidification/stabilization (S/S) reduces the mobility of hazardous substances and contaminants in the environment through both physical and chemical means (43,44). The S/S process physically binds or encloses contaminants within a stabilized mass. S/S can be performed both ex situ and in situ. Ex situ S/S requires excavation of the material to be treated, and the treated material must be disposed of. In situ S/S involves use of auger or caisson systems and injector head systems to add binders to contaminated soil or waste without excavation, and the treated material is left in place (66,67).

Solvent extraction involves use of an organic solvent as an extractant to separate contaminants from soil. The organic solvent is mixed with contaminated soil in an extraction unit. The extracted solution is then passed through a separator, where the contaminants and extractant are separated from the soil (68).

Vitrification involves use of an electric current to melt contaminated soil at elevated temperatures (1600–2000°C or 2900–3650°F). Upon cooling, the vitrification product is a chemically stable, leach-resistant, glass and crystalline material similar to obsidian or basalt rock. The high-temperature component of the process destroys or removes organic materials. Radionuclides and heavy metals are retained within the vitrified product. Vitrification may be conducted in situ or ex situ (69).

11.4.4 Cleanup Technologies Options for Metal Finishing Sites

Table 11.6 presents the technologies that may be appropriate for use at metal finishing sites. In addition to more conventional technologies, a number of innovative technology options are listed. Many possible cleanup approaches use institutional controls and one or a combination of the technologies described in Table 11.6. Whatever cleanup approach is ultimately chosen, planners should explore a number of cost-effective options.

Cleanup at metal finishing facilities will most likely entail removing a complex mix of contaminants, primarily organic solvents and metals. The cleanup will usually require more than one technology, or treatment train, because single technologies tend not to address both metal and organic contaminants. S/S can address metal contamination by limiting mobility (solubility) and thereby limit risk. Approaches at metal finishing sites depend on local conditions. At larger metal finishing sites, one approach may be to excavate and stabilize the contaminated material with either onsite or offsite disposal or treatment of material (70–81). Access to contaminated soils may be limited at smaller sites requiring excavation and offsite treatment or disposal. The stabilized material can be placed onsite or sent to a USEPA-approved landfill.

11.4.5 Postconstruction Care

Many of the cleanup technologies that leave contamination onsite, either in containment systems or because of the long periods required to reach cleanup goals, will require long-term maintenance and possibly operation. If waste is left onsite, regulators will likely require long-term monitoring of applicable media (i.e., soil, water, and/or air) to ensure that the cleanup approach selected is continuing to function as planned (e.g., residual contamination, if any, remains at acceptable levels and is not migrating). If long-term monitoring is required (e.g., by the state), periodic sampling, analysis, and reporting requirements will also be involved. Planners should be aware of these requirements and provide for them in cleanup budgets. Postconstruction sampling, analysis, and reporting costs in their cleanup budgets can be a significant problem as these costs can be substantial.

TABLE 11.6

Cleanup Technologies for Metal Finishing Brownfields Sites Sample Analysis Technologies

Applicable Technology	Description	Examples of Applicable Land/ Process Areas	Contaminants Treated by This Technology	Limitations
		Containment Technologies		
Sheet piling	• Steel or iron sheets are driven into the ground to form a subsurface barrier • Low-cost containment method • Used primarily for shallow aquifers	• Metal cleaning, rinsing and bathing operations, chemical, storage, wastewater treatment	• Not contaminant-specific	• Not effective in the absence of a continuous aquitard • Can leak at the intersection of the sheets and the aquitard or through pile wall joints
Grout curtain	• Grout curtains are injected into subsurface soils and bedrock • Forms an impermeable barrier in the subsurface	• Metal cleaning, rinsing and bathing operations, chemical storage, wastewater treatment	• Not contaminant-specific	• Difficult to ensure a complete curtain without gaps through which the plume can escape; however, new techniques have improved continuity of curtain
Slurry walls	• Consist of a vertically excavated slurry-filled trench • The slurry hydraulically shores the trench to prevent collapse and forms a filtercake to reduce groundwater flow • Often used where the waste mass is too large for treatment and where soluble and mobile constituents pose an imminent threat to a source of drinking water • Often constructed of a soil, bentonite, and water mixture	• Metal cleaning, rinsing and bathing operations, chemical storage, wastewater treatment	• Not contaminant-specific	• Contains contaminants only within a specified area • Soil-bentonite backfills are not able to withstand attack by strong acids, bases, salt solutions, and some organic chemicals • Potential for the slurry walls to degrade or deteriorate over time

(Continued)

TABLE 11.6 (*Continued*)
Cleanup Technologies for Metal Finishing Brownfields Sites Sample Analysis Technologies

Applicable Technology	Description	Examples of Applicable Land/ Process Areas	Contaminants Treated by This Technology	Limitations
Capping	• Used to cover buried waste materials to prevent migration • Made of a relatively impermeable material that will minimize rainwater infiltration • Waste materials can be left in place • Requires periodic inspections and routine monitoring • Contaminant migration must be monitored periodically	• Anodizing, solid wastes from anodizing, electroplating, electroplating wastewaters and solid wastes, finishing wastewaters, chemical conversion coating wastewaters and solid wastes, electroless plating, electroless plating wastewaters, solid wastes from painting, wastewater treatment system, sunken treatment tank	• Metals	• Costs associated with routine sampling and analysis may be high • Long-term maintenance may be required to ensure impermeability • May have to be replaced after 20 to 30 years of operation • May not be effective if groundwater table is high
Excavation/ offsite disposal	• Removes contaminated material to an EPA-approved landfill	**Ex situ technologies** • Wastes from painting, wastewater treatment system, sunken treatment tanks, chemical storage, disposal	• Not contaminant-specific	• Generation of fugitive emissions may be a problem during operations • The distance from the contaminated site to the nearest disposal facility will affect cost • Depth and composition of the media requiring excavation must be considered • Transportation of the soil through populated areas may affect community acceptability • Disposal options for certain waste (e.g., mixed waste or transuranic waste) may be limited. There is currently only one licensed disposal facility for radioactive and mixed waste in the United States

(Continued)

TABLE 11.6 (Continued)
Cleanup Technologies for Metal Finishing Brownfields Sites Sample Analysis Technologies

Applicable Technology	Description	Examples of Applicable Land/ Process Areas	Contaminants Treated by This Technology	Limitations
Chemical oxidation/ reduction	• Reduction/oxidation (redox) reactions chemically convert hazardous contaminants to nonhazardous or less toxic compounds that are more stable, less mobile, or inert • Redox reactions involve the transfer of electrons from one compound to another • The oxidizing agents commonly used are ozone, hydrogen peroxide, hypochlorite, chlorine, and chlorine dioxide	• Wastes from anodizing, electroplating, finishing, chemical conversion coating, electroless plating, painting, rinsing operations, wastewater treatment system, sunken treatment tank	• Metals • Cyanide	• Not cost-effective for high contaminant concentrations because of the large amounts of oxidizing agent required • Oil and grease in the media should be minimized to optimize process efficiency
UV oxidation	• Destruction process that oxidizes constituents in wastewater by the addition of strong oxidizers and irradiation with UV light • Practically any organic contaminant that is reactive with the hydroxyl radical can potentially be treated • The oxidation reactions are achieved through the synergistic action of UV light in combination with ozone or hydrogen peroxide • Can be configured in batch or continuous flow models, depending on the throughput rate under consideration	• Wastes from metal cleaning, painting, rinsing operations, wastewater treatment system, sunken treatment tank, chemical storage area, disposal area	• VOCs	• The aqueous stream being treated must provide for good transmission of UV light (high turbidity causes interference) • Metal ions in the wastewater may limit effectiveness • VOCs may volatilize before oxidation can occur • Off-gas may require treatment • Costs may be higher than competing technologies because of energy needs • Handling and storage of oxidizers require special safety precautions

(Continued)

TABLE 11.6 (Continued)
Cleanup Technologies for Metal Finishing Brownfields Sites Sample Analysis Technologies

Applicable Technology	Description	Examples of Applicable Land/Process Areas	Contaminants Treated by This Technology	Limitations
Precipitation	• Involves the conversion of soluble heavy metal salts to insoluble salts that will precipitate • Precipitate can be removed from the treated water by physical methods such as clarification or filtration • Often used as a pretreatment for other treatment technologies where the presence of metals would interfere with the treatment processes • Primary method for treating metal-laden industrial wastewater	• Wastes from anodizing, electroplating, finishing, chemical conversion coating, electroless plating, painting, rinsing operations, wastewater treatment system, sunken treatment tank	• Metals	• Contamination source is not removed • The presence of multiple metal species may lead to removal difficulties • Discharge standard may necessitate further treatment of effluent • Metal hydroxide sludges must pass TCLP (toxicity characteristic leaching procedure) criteria prior to land disposal • Treated water will often require pH adjustment
Liquid phase carbon adsorption	• Groundwater is pumped through a series of vessels containing activated carbon, to which dissolved contaminants adsorb • Effective for polishing water discharges from other remedial technologies to attain regulatory compliance • Can be quickly installed • High contaminant-removal efficiencies	• Wastes from metal cleaning, painting, rinsing operations, wastewater treatment system, sunken treatment tank, chemical storage area, disposal area	• VOCs	• The presence of multiple contaminants can affect process performance • Metals can foul the system • Costs are high if used as the primary treatment on waste streams with high contaminant concentration levels • Type and pore size of the carbon and operating temperature will impact process performance • Transport and disposal of spent carbon can be expensive • Water soluble compounds and small molecules are not adsorbed well

(Continued)

TABLE 11.6 (Continued)
Cleanup Technologies for Metal Finishing Brownfields Sites Sample Analysis Technologies

Applicable Technology	Description	Examples of Applicable Land/Process Areas	Contaminants Treated by This Technology	Limitations
Air stripping	• Contaminants are partitioned from groundwater by greatly increasing the surface area of the contaminated water exposed to air • Aeration methods include packed towers, diffused aeration, tray aeration, and spray aeration • Can be operated continuously or in a batch mode, where the air stripper is intermittently fed from a collection tank • The batch mode ensures consistent air stripper performance and greater efficiency than continuously operated units because mixing in the storage tank eliminates any inconsistencies in feed water composition	• Wastes from metal cleaning, painting, rinsing operations, wastewater treatment system, sunken treatment tank, chemical storage area, disposal area	• VOCs	• Potential for inorganic (iron greater than 5 ppm, hardness greater than 800 ppm) or biological fouling of the equipment, requiring pretreatment of groundwater or periodic column cleaning • Consideration should be given to Henry's law constant of the VOCs in the water stream and the type and amount of packing used in the tower • Compounds with low volatility at ambient temperature may require preheating of the groundwater • Off-gases may require treatment based on mass emission rate and state and federal air pollution laws
Natural attenuation	**In Situ Technologies** • Natural subsurface processes such as dilution, volatilization, biodegradation, adsorption, and chemical reactions with subsurface media can reduce contaminant concentrations to acceptable levels • Consideration of this option requires modeling and evaluation of contaminant degradation rates and pathways • Sampling and analyses must be conducted throughout the process to confirm that degradation is proceeding at sufficient rates to meet cleanup objectives	**In Situ Technologies** • Metal cleaning, metal cleaning wastewater, painting, painting wastewater and solid SwastTes, wastewater treatment system, sunken treatment tank, chemical storage area, disposal area	• VOCs	• Intermediate degradation products may be more mobile and more toxic than original contaminants • Contaminants may migrate before they degrade • The site may have to be fenced and may not be available for reuse until hazard levels are reduced • Source areas may require removal for natural attenuation to be effective • Modeling contaminant degradation rates, and sampling and analysis to confirm modeled predictions extremely expensive

(Continued)

TABLE 11.6 (*Continued*)
Cleanup Technologies for Metal Finishing Brownfields Sites Sample Analysis Technologies

Applicable Technology	Description	Examples of Applicable Land/Process Areas	Contaminants Treated by This Technology	Limitations
Soil vapor extraction	• A vacuum is applied to the soil to induce controlled air flow and remove contaminants from the unsaturated (vadose) zone of the soil • The gas leaving the soil may be treated to recover or destroy the contaminants • The continuous air flow promotes in situ biodegradation of low-volatility organic compounds that may be present	• Metal cleaning, metal cleaning wastewaters, painting, painting wastewaters and solid wastes, wastewater treatment system, sunken treatment tank, chemical storage area, disposal area	• VOCs	• Tight or extremely moist content (>50%) has a reduced permeability to air, requiring higher vacuums • Large screened intervals are required in extraction wells for soil with highly variable permeabilities • Air emissions may require treatment to eliminate possible harm to the public or environment • Off-gas treatment residual liquids and spent activated carbon may require treatment or disposal • Not effective in the saturated zone
Soil flushing	• Extraction of contaminants from the soil with water or other aqueous solutions • Accomplished by passing the extraction fluid through in-place soils using injection or infiltration processes • Extraction fluids must be recovered with the underlying aquifer and recycled when possiblei	• Anodizing, solid wastes from anodizing, electroplating, electroplating wastewater and solid wastes, finishing waste-water, chemical conversion coating wastewater and solid wastes, electroless plating, electroless plating wastewater, solid wastes from painting, wastewater treatment system, sunken treatment tank	• Metals	• Low-permeability soils are difficult to treat • Surfactants can adhere to soil and reduce effective soil porosity • Reactions of flushing fluids with soil can reduce contaminant mobility • Potential of washing the contaminant beyond the capture zone and the introduction of surfactants to the subsurface

(*Continued*)

TABLE 11.6 (Continued)
Cleanup Technologies for Metal Finishing Brownfields Sites Sample Analysis Technologies

Applicable Technology	Description	Examples of Applicable Land/Process Areas	Contaminants Treated by This Technology	Limitations
Air sparging	• In situ technology in which air is injected under pressure below the water table to increase groundwater oxygen concentrations and enhance the rate of biological degradation of contaminants by naturally occurring microbes • Increases the mixing in the saturated zone, which increases the contact between groundwater and soil • Air bubbles traverse horizontally and vertically through the soil column, creating an underground stripper that volatilizes contaminants • Air bubbles travel to a soil vapor extraction (SVE) system • Air sparging is effective for facilitating extraction of deep contamination, contamination in low-permeability soils, and contamination in the saturated zone	• Metal cleaning, metal cleaning wastewater, painting, painting wastewater and solid wastes, wastewater treatment system, sunken treatment tank, chemical storage area, disposal area	• VOCs	• Depth of contaminants and specific site geology must be considered • Air flow through the saturated zone may not be uniform • A permeability differential such as a clay layer above the air injection zone can reduce the effectiveness • Vapors may rise through the vadose zone and be released into the atmosphere • Increased pressure in the vadose zone can build up vapors in basements, which are generally low-pressure areas
Passive treatment walls	• A permeable reaction wall is installed inground, across the flow path of a contaminant plume, allowing the water portion of the plume to passively move through the wall • Allows the passage of water while prohibiting the movement of contaminants by employing such agents as iron, chelators (ligands selected for their specificity for a given metal), sorbents, microbes, and others • Contaminants are typically completely degraded by the treatment wall	• Appropriately selected location for wall	• VOCs • Metals	• The system requires control of pH levels. When pH levels within the passive treatment wall rise, it reduces the reaction rate and can inhibit the effectiveness of the wall • Depth and width of the plume. For large-scale plumes, installation cost may be high • Cost of treatment medium (iron) • Biological activity may reduce the permeability of the wall • Walls may lose their reactive capacity, requiring replacement of the reactive medium

(Continued)

TABLE 11.6 (*Continued*)
Cleanup Technologies for Metal Finishing Brownfields Sites Sample Analysis Technologies

Applicable Technology	Description	Examples of Applicable Land/Process Areas	Contaminants Treated by This Technology	Limitations
Biodegradation	• Indigenous or introduced microorganisms degrade organic contaminants found in soil and groundwater • Used successfully to remediate soils, sludges, and groundwater • Especially effective for remediating low-level residual contamination in conjunction with source removal	• Metal cleaning, metal cleaning wastewater, painting, painting wastewater and solid wastes, wastewater treatment system, sunken treatment tank, chemical storage area, disposal area	• VOCs	• Cleanup goals may not be attained if the soil matrix prevents sufficient mixing • Circulation of water-based solutions through the soil may increase contaminant mobility and necessitate treatment of underlying groundwater • Injection wells may clog and prevent adequate flow rates • Preferential flow paths may result in nonuniform distribution of injected fluids • Should not be used for clay, highly layered, or heterogeneous subsurface environments • High concentrations of heavy metals, highly chlorinated organics, long chain hydrocarbons, or inorganic salts are likely to be toxic to microorganisms • Low temperatures slow bioremediation • Chlorinated solvents may not degrade fully under certain subsurface conditions

Source: USEPA. *Technical Approaches to Characterizing and Cleaning Up Metal Finishing Sites under the Brownfields Initiative.* EPA/625/R-98/006, U.S. Environmental Protection Agency, Cincinnati, OH, March 1999.

11.5 CONCLUSION

Brownfields redevelopment contributes to the revitalization of communities across the United States. Reuse of these abandoned, contaminated sites spurs economic growth, builds community pride, protects public health, and helps maintain our nation's "greenfields," often at a relatively low cost. This chapter provides brownfields planners with the technical methods that can be used to achieve successful site assessment and cleanup, which are two key components in the brownfields redevelopment process.

While the general guidance provided in this chapter will be applicable to many brownfields projects, it is important to recognize the heterogeneous nature of brownfields work. That is, no two brownfields sites will be identical, and planners will need to base site assessment and cleanup activities on the conditions at their particular site. Some of the conditions that may vary by site include the type of contaminants present, the geographic location and extent of contamination, the availability of site records, hydrogeological conditions, and state and local regulatory requirements. Based on these factors, as well as financial resources and desired timeframes, planners will find different assessment and cleanup approaches appropriately.

Consultation with state and local environmental officials and community leaders, as well as careful planning early in the project, will assist planners in developing the most appropriate site assessment and cleanup approaches. Planners should also determine early on if they are likely to require the assistance of environmental engineers. A site assessment strategy should be agreeable to all stakeholders and should address

1. The type and extent of contamination, if any, present at the site (82–85)
2. The types of data needed to adequately assess the site
3. Appropriate sampling and analytical methods for characterizing contamination
4. An acceptable level of data uncertainty

When used appropriately, the site assessment methods described in this chapter will help to ensure that a good strategy is developed and implemented effectively.

Once the site has been assessed and stakeholders agree that cleanup is needed, planners will need to consider cleanup options. Many different types of cleanup technologies are available. The guidance provided in this chapter on selecting appropriate methods directs planners to base cleanup initiatives onsite- and project-specific conditions. The type and extent of cleanup will depend in large part on the type and level of contamination present, reuse goals, and the budget available. Certain cleanup technologies are used onsite, while others require offsite treatment. Also, in certain circumstances, containment of contamination onsite and the use of institutional controls may be important components of the cleanup effort. Finally, planners will need to include budgetary provisions and plans for postcleanup and postconstruction care if it is required at the brownfields site. By developing a technically sound site assessment and cleanup approach that is based on site-specific conditions and addresses the concerns of all project stakeholders, planners can achieve brownfields redevelopment and reuse goals effectively and safely.

REFERENCES

1. Federal Register. *Comprehensive Environmental Response, Compensation, and Liability Act (CERCLA or Superfund)* 42 U.S.C. s/s 9601 et seq. (1980), United States Government, Public Laws. Available at: www.epa.gov/laws-regulations/summary-comprehensive-environmental-response-compensation-and-liability-act, 2016.
2. USEPA. *Technical Approaches to Characterizing and Cleaning Up Metal Finishing Sites under the Brownfields Initiative.* EPA/625/R-98/006, U.S. Environmental Protection Agency, Cincinnati, OH, March 1999.
3. USEPA. *Brownfields Home Page.* U.S. Environmental Agency, http://www.epa.gov/brownfields, 2015.

4. USEPA. *Road Map to Understanding Innovative Technology Options for Brownfields Investigation and Cleanup*, 4th Edition. EPA 542-B-05-001, U.S. Environmental Protection Agency, Washington, DC, September 2005.
5. USEPA. *Brownfields Tool Kit*. U.S. Environmental Protection Agency, Cincinnati, OH, https://archive.epa.gop/brownfields/policy/web/html/initiatives_sb.html, 2016.
6. USEPA. *Brownfields and Land Revitalization—Tools and Technical Information*. U.S. Environmental Protection Agency, Washington, DC, http://www.epa.gov/brownfields/toolsandtech.htm, 2015.
7. USEPA. *Profile of the Fabricated Metal Products Industry*. EPA 3 10-R-95-007, U.S. Environmental Protection Agency, Washington, DC, 1995.
8. Brebbia, C. A. (Ed.). *Brownfields III: Prevention, Assessment, Rehabilitation and Development of Brownfield Sites. WIT Transactions on Ecology and the Environment Series*, Vol. 94, 228pp, Wessex Institute of Technology (WIT), UK, 2006.
9. CERP. *Brownfields Identification*. The Community Environmental Resource Program (CERP), St Louis, MO, http://stlcin.missouri.org/cerp/brownfields/identification.cfm, 2015.
10. ASTM. *Standard Practice for Environmental Site Assessments: Phase I Environmental Site Assessment Process*. E 1527-00, American Society for Testing and Materials, West Conshohocken, PA, 2003.
11. Federal Register. *Resource Conservation and Recovery Act (RCRA)*. 42 US Code s/s 6901 et seq. (1976), U.S. Government, Public Laws, www.federalregister.gov/resource-conservation-and-recovery-act-rcra, 2016.
12. USEPA. *National Pollutant Discharge Elimination System (NPDES)*. U.S. Environmental Protection Agency, Washington, DC, NPDES Web Site is at: http://cfpub.epa.gov/npdes, 2015.
13. USEPA. *Comprehensive Environmental Response, Compensation, and Liability Information System (CERCLIS)*. U.S. Environmental Protection Agency, Washington, DC, CERCLIS Web Site is at: http://cumulis.epa.gov/supercpad/cursites/srchsites.cfm, 2016.
14. ASTM. *Standard Guide for Process of Sustainable Brownfields Development*. E 1984-03, American Society for Testing and Materials, West Conshohocken, PA, 2003.
15. NEWMOA. Improving decision quality: Making the case for adopting next-generation site characterization practices. Northeast Waste Management Officials' Association. *Remediation*, p. 91, Spring, 2003.
16. USEPA. *Quality Assurance Guidance for Conducting Brownfields Site Assessments*. EPA 540-R-98-038, U.S. Environmental Protection Agency, Washington, DC, 1998.
17. Pediaditi, K., Wehrmeyer, W., and Chenoweth, J. Sustainability indicators for brownfield redevelopment projects. In: *Proceedings of Sustainable Urban Environments: EPSRC Conference*, University of Birmingham, UK. February 2005.
18. ASTM. *ASTM Standard Practice for Environmental Site Assessments: Transaction Screen Process*. E 1528-00, American Society for Testing and Materials, West Conshohocken, PA, 2000.
19. USEPA. *Brownfields Technology Primer: Using the Triad Approach to Streamline Brownfields Site Assessment and Cleanup*. EPA 542-B-03-002, U.S. Environmental Protection Agency, Washington, DC, 2003.
20. USEPA. *Improving Sampling, Analysis, and Data Management for Site Investigation and Cleanup*. EPA 542-F-04-001a, U.S. Environmental Protection Agency, Washington, DC, 2004.
21. USEPA. *The Triad Resource Center*, www.triadcentral.org, 2015.
22. USEPA. *Technology News and Trends*, www.epa.gov/tio/download/newsltrs/tnandt0704.pdf, 2015.
23. ASTM. *ASTM Standard Guide for Environmental Site Assessments: Phase II Environmental Site Assessment Process*. E1903-97, American Society for Testing and Materials, West Conshohocken, PA, 2002.
24. USEPA. *Clarifying DQO Terminology Usage to Support Modernization of Site Cleanup Practices*. EPA 542-R-01-014, U.S. Environmental Protection Agency, Washington, DC, 2001.
25. USEPA. *Data Quality Objective Process for Hazardous Waste Site Investigations*. EPA 600-R-00-007, U.S. Environmental Protection Agency, Washington, DC, 2000.
26. USEPA and USACE. Managing uncertainty in environmental decisions. *Environmental Science and Technology*, American Chemical Society, p. 405, October 2001.
27. USACE. *Engineering and Design: Requirements for the Preparation of Sampling and Analysis Plans*. EM 200-1-3, U.S. Army Corps of Engineers, Washington, DC, February 2001.
28. USEPA-OSRTI. In search of representativeness: Evolving the environmental data quality model. *Quality Assurance,* 9, 179–190, 2002.
29. Federal Register. *Safe Drinking Water Act (SDWA)*. 42 U.S.C. s/s 300f et seq. 1974, United States Government, Public Laws. Available at: http://www.epa.gop/sdwa, 2016.

30. USEPA. *Subsurface Characterization and Monitoring Techniques: A Desk Reference Guide.* EPA/ 625/R-93-003a, U.S. Environmental Protection Agency, Washington, DC, 1993.
31. Robbat, A., Jr. *Dynamic Workplans and Field Analytics: The Keys to Cost Effective Site Characterization and Cleanup.* Tufts University under Cooperative Agreement with the U.S. Environmental Protection Agency, Boston, MA, October 1997.
32. USEPA. *Field Analytical and Site Characterization Technologies: Summary of Applications.* EPA 542-R-97-011, U.S. Environmental Protection Agency, Washington, DC, 1997.
33. USEPA. *Electroplating.* U.S. Environmental Protection Agency, Mid-Atlantic Brownfields, http://www. epa.gov/reg3hscd/bfs/regional/industry/electroplating.htm, 2015.
34. USACE. Yearly average cost index for utilities. In: *Civil Works Construction Cost Index System Manual.* 110-2-1304, U.S. Army Corps of Engineers, Washington, DC, 44pp. PDF file is available at www. publications.usac.army.mil/USAC-publications/Engineer-manuals/udt_43544_param_page/5/, 2016.
35. Geo-Environmental Solutions. *Rental Rate Sheet.* Geoprobe Systems, Inc., http://www.gesolutions.com/ assess.htm, September 15, 1998.
36. USEPA. *Cost Estimating Tools and Resources for Addressing the Brownfields Initiatives.* EPA 625-R-99-001, U.S. Environmental Protection Agency, Washington, DC, 1999.
37. USEPA. *Guide to Documenting and Managing Cost and Performance Information for Remediation Projects.* EPA 542-B-98-007, U.S. Environmental Protection Agency, Washington, DC, 1998.
38. USEPA. *Remediation Technology Cost Compendium—Year 2000.* EPA 542-R-01-009, U.S. Environmental Protection Agency, Washington, DC, September 2001.
39. Al-Tabbaa, A. Impact of and response to climate change in UK brownfield remediation. In: *Paper Presented to the Chartered Institute of Water and Environmental Management Hong Kong*, The Hong Kong Institution of Engineers, Hong Kong, May 2007.
40. Catney, P., Yount, K., Henneberry, J., and Meyer, P. Can we really compare brownfield regulation and redevelopment in the United States and European Union? In: *Revit and Cabernet 2nd International Conference on Managing Urban Land*, Theaterhaus Stuttgart, Germany, April 2007.
41. USEPA. *Capping.* EPA 542-F-01-022, U.S. Environmental Protection Agency, Washington, DC, 2001.
42. USEPA. *Evaluation of Subsurface Engineered Barriers at Waste Sites.* EPA 542-R-98-005, U.S. Environmental Protection Agency, Washington, DC, 1998.
43. USEPA. *Solidification/Stabilization Use at Superfund Sites.* EPA 542-R-00-010, U.S. Environmental Protection Agency, Washington, DC, 2000.
44. USEPA. *Solidification/Stabilization.* EPA 542-F-01-024, U.S. Environmental Protection Agency, Washington, DC, 2001.
45. USEPA. *Bioremediation.* EPA 542-F-01-001, U.S. Environmental Protection Agency, Washington, DC, 2001.
46. USEPA. *In Situ Flushing.* EPA 542-F-01-011, U.S. Environmental Protection Agency, Washington, DC, 2001.
47. USEPA. *Chemical Oxidation.* EPA 542-F-01-013, U.S. Environmental Protection Agency, Washington, DC, 2001.
48. USEPA. *Soil Vapor Extraction (SVE) and Air Sparging.* EPA 542-F-01-006, U.S. Environmental Protection Agency, Washington, DC, 2001.
49. USEPA. *Permeable Reactive Barriers.* EPA 542-F-01-00, U.S. Environmental Protection Agency, Washington, DC, 2001.
50. USEPA. *Site Remediation Technology InfoBase: A Guide to Federal Programs, Information Resources, and Publications on Contaminated Site Cleanup Technologies*, 2nd Edition. EPA 542-B-00-005, U.S. Environmental Protection Agency, Washington, DC, 2000.
51. USEPA. *Innovative Remediation Technologies: Field-Scale Demonstration Projects in North America*, 2nd Edition. EPA 542-B-00-004, U.S. Environmental Protection Agency, Washington, DC, 2000.
52. USEPA. *Brownfields Technology Primer: Requesting and Evaluating Proposals that Encourage Innovative Technologies for Investigation and Cleanup.* EPA 542-R-01-005, U.S. Environmental Protection Agency, Washington, DC, 2001.
53. Xia, B., Shen, S., and Xue, F., Phytoextraction of heavy metals from highly contaminated soils using *Sauropus androgynus. Soil and Sediment Contamination Journal*, 22, 6, 631–640, 2013.
54. Laidlaw, W. S., Arndt, S. K., Huynh, T. T., Gregory, D., and Baker, A. J. M., Phytoextraction of heavy metals by Willows growing in biosolids under field conditions. *Journal of Environmental Quality,* Soil Science Society of America, https://www.researchgate.net/publication/43528532_phytoextraction_ of_heavy_metals_from_an_aged_biosolids_stockpile_by_willow_Salix_species_after_one_year, 2016.

55. USEPA. *Innovative Remediation and Site Characterization Technologies Resources.* EPA 542-C-04-002, U.S. Environmental Protection Agency, Washington, DC, 2004.
56. Adejumo, S.A., Togun, A.O., Adediran, J.A., and Ogundiran, M.B. In-situ remediation of heavy metal contaminated soil using Mexican sunflower (*Tithonia diversifolia*) and cassava waste composts. *World Journal of Agricultural Sciences*, 7, 2, 224–233, 2011.
57. Wang, L. K., Hung, Y. T., and Shammas, N. K. (Eds.). *Physicochemical Treatment Processes.* Humana Press, Totowa, NJ, 2005.
58. Wang, L. K., Hung, Y. T., and Shammas, N. K. (Eds.). *Advanced Physicochemical Treatment Processes.* Humana Press, Totowa, NJ, 2006.
59. Wang, L. K., Hung, Y. T., and Shammas, N. K. (Eds.). *Advanced Physicochemical Treatment Technologies.* Humana Press, Totowa, NJ, 2007.
60. *USEPA. Phytoremediation.* EPA 542-F-01-002, U.S. Environmental Protection Agency, Washington, DC, 2001.
61. Zhang, X., Wang, H., He. L., Lu, K., Sarmah, A., Li, J., Bolan, N.S., Pei, J., and Huang, H. Using biochar for remediation of soils contaminated with heavy metals and organic pollutants. *Environmental Science and Pollution Research International*, p. 8472, April 2015.
62. USEPA. *Use of Field-Scale Phytotechnology for Chlorinated Solvents, Metals, Explosives and Propellants, and Pesticides—Status Report.* EPA 542-R-05-002, U.S. Environmental Protection Agency, Washington, DC, April 2005.
63. Environment Canada. *Phytoremediation of Soil Containing heavy metals and Hydrocarbons.* Environmental Protection, Quebec Region, Government of Canada Publication, 2004.
64. USEPA. *Pump and Treat.* EPA 542-F-01-025, U.S. Environmental Protection Agency, Washington, DC, 2001.
65. USEPA. *Soil Washing.* EPA 542-F-01-008, U.S. Environmental Protection Agency, Washington, DC, 2001.
66. Harbottle, M. J. and Al-Tabbaa, A. Combining stabilization/solidification with biodegradation to enhance long-term remediation performance. In: *Proceedings of the 2nd IASTED International Conference on Advanced Technology in the Environmental Field*, Lanzarote, Spain, pp. 222–227, 2006.
67. Harbottle, M. J., Al-Tabbaa, A., and Evans, C. W. The technical sustainability of in-situ stabilization/solidification. In: *Proceedings of the International Conference on Stabilization/Solidification Treatment and Remediation*, Al-Tabbaa, A. and Stegemann, J. (Eds.). Cambridge, UK, pp. 159–170, April 2005.
68. USEPA. *Solvent Extraction.* EPA 542-F-01-009, U.S. Environmental Protection Agency, Washington, DC, 2001.
69. USEPA. *Vitrification.* EPA 542-F-01-017, U.S. Environmental Protection Agency, Washington, DC, 2001.
70. Wang, L. K., Chen, J. P., Hung, Y. T., and Shammas, N. K. (Eds.). *Heavy Metals in the Environment.* CRC Press, Taylor & Francis, Boca Raton, FL, 2009.
71. Wang, L. K, Hung, Y. T., and Shammas, N. K. (Eds.). *Handbook of Advanced Industrial and Hazardous Wastes Treatment.* CRC Press, Taylor & Francis, Boca Raton, FL, 2010.
72. Wang, L. K., Chen, J. P., Hung, Y. T., and Shammas, N. K. (Eds.). *Membrane and Desalination Technologies.* Humana Press, Totowa, NJ, 2011.
73. Wang, L. K., Shammas, N. K., Selke, W. A., and Aulenbach, D. B. (Eds.). *Flotation Technology.* Humana Press, Totowa, NJ, 2010.
74. Carbtrol[R]. *Heavy Metal Removal System*, http://www.carbtrol.com/heavy_metal.html, 2015.
75. Wastech Control and Engineering. *Heavy Metal Contamination Removal*, http://www.wastechengineering.com/heavy-metal-removal-systems.html, 2015.
76. Christian, D., Wong, E., Crawford, R. L., Cheng, I. F., and Hess. T. F. Heavy metals removal from mine runoff using compost bioreactors. *Environmental Technology*, 31, 14, 1533–1546, 2010.
77. Veolia. *Heavy Metal Removal, Ceramic Membranes*, Available at: http://www.wateronline.com/doc/heavy_metals_removal_with_ceramem_ceramc_0001, 2016.
78. Wuana, R. A, Okieimen, F. E., and Imborvungu, J. A., Removal of heavy metals from a contaminated soil using organic chelating acids, *International Journal of Environmental Science and Technology*, 7, 3, 485–496, 2010.
79. Mohammadi, M., Fotovat, A., and Haghnia, G. Sand–soil–organic matter filter column for removal of heavy metals from industrial waste water, *International Conference on the Biogeochemistry of Trace Elements*, Chihuahua, Mexico, July 14–16, 2009.
80. Lee, K.-Y. and Kim, K.-W. Heavy metal removal from shooting range soil by hybrid electrokinetics with bacteria and enhancing agents. *Environmental Science and Technology*, 44, 24, 9482–9487, 2010.

81. Mohanty, B. and Mahindrakar, A. B. Removal of heavy metal by screening followed by soil washing from contaminated soil. *International Journal of Technology and Engineering System*, 2, 3, 290–203, 2011.

82. Deka, J. and Sarma, H.P. Heavy metal contamination in soil in an industrial zone and its relation with some soil properties. *Archives of Applied Science Research*, 4, 2, 831–836, 2012.

83. Liang, J., Chen, C., Song, X., Han, Y., and Liang, Z. Assessment of heavy metal pollution in soil and plants from Dunhua sewage irrigation area. *International Journal of Electrochemical Science*, 6, 5314–5324, 2011.

84. Gaddis, M. *Heavy Metal Contamination in Agricultural Soils.* Global Issues Seminar, Colorado Mountain College, CO, http://www.google.com/search?sourceid=navclient&ie=UTF-8&rlz=1T4SKPT_enUS446US446&q=Soil+contaminated+with+Heavy+Metals, 2015.

85. Wuana, R.A. and Okieimen, F.E. Heavy metals in contaminated soils: A review of sources, chemistry, risks and best available strategies for remediation. *ISRN Ecology Journal*, 2011, 20, 2011, Article ID 402647.

12 Adsorptive Removal of Arsenic from Water Sources Using Novel Nanocomposite Mixed Matrix Membranes

R. Jamshidi Gohari, Woei Jye Lau, Takeshi Matsuura, and Ahmad F. Ismail

CONTENTS

ABSTRACT

Arsenic is viewed as being synonymous with toxicity. Dangerous arsenic concentrations in natural waters are now a worldwide problem particularly in the countries like Bangladesh, India, the United States, Canada, etc. and is often referred to as a twenty-first century calamity. In order to remove arsenic from polluted water to a safe level, adsorption based on nanosized metal oxide adsorbents is found to be very promising, as recent studies have reported that many metal oxide adsorbents could exhibit very favorable sorption to arsenic, resulting in excellent removal of toxic metals to meet increasingly strict regulations. However, the use of metal oxide adsorbents alone is prone to agglomeration due to van der Waals forces or other interactions, causing their high capacity and selectivity of heavy metal to decrease or even be lost. Furthermore, the small size of metal oxides has made them unable to be used in fixed beds or any other flow-through systems. In view of this, integrating porous host media with metal oxide adsorbents has become a hot topic in the development of applicable and reliable treatment technology. Among the many porous supports ever investigated, the use of polymeric microporous membrane as host media for nanoparticles is reported to be advantageous and unique in arsenic removal. This chapter is intended to highlight the significant advantages of using novel mixed matrix membranes (MMMs) in the adsorptive removal of arsenic in comparison to other treatment methods and to provide in-depth discussion on the important factors influencing the performance of MMMs during the treatment process as well as the regeneration process of the membrane.

12.1   INTRODUCTION

Arsenic is a toxic heavy metal with a name derived from the Greek word *arsenikon*, meaning potent (Choong et al., 2007). Nowadays, the toxicity of arsenic has been well known and many organizations around the world adjusted the maximum acceptable concentration of arsenic in contaminated water to a very low level. For instance, since 2006, the USEPA (United States Environmental Protection Agency) and WHO (World Health Organization) decided to reduce the maximum contaminant level (MCL) of arsenic concentration in drinking water from 50 part per billion (ppb) to 10 ppb (Mohan and Pittman, 2007). The stiffening of regulations generates strong demands to improve methods for removing arsenic from water and controlling water treatment residuals.

With respect to worldwide research and development activities, the growth of the research publications related to arsenic removal have increased tremendously since 1995 as shown in Figure 12.1. Statistics from the Scopus database reveal that the total number of relevant articles published in year 2013 is more than 500 in comparison to only 32 documented in 1995. Conventionally, many treatment methods such as chemical precipitation (Harper and Kingham, 1992), coagulation and flocculation (Bilici Baskan and Pala, 2010), and ion exchange (Kartinen Jr and Martin, 1995) could be employed for arsenic decontamination, but they are found to have inconsistent and/or incomplete elimination of arsenic. In order to meet the MCL required by law, additional posttreatment process is always required to complete the treatment process, which indirectly would increase

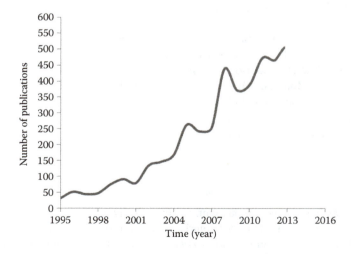

FIGURE 12.1   Number of publications related to arsenic removal for periods between 1995 and 2013 according to the Scopus database.

the overall cost of treatment. Although membrane technology is reported to be potential for arsenic removal when it is operated in reverse osmosis (RO) or nanofiltration (NF) mode (Shih, 2005), the relatively high energy consumption resulted from high operating pressure remains as a major concern to many. Low-pressure-driven membranes like microfiltration (MF) and ultrafiltration (UF) on the other hand are not effective in removing arsenic, mainly due to their porous structure which offers minimal/no resistance against arsenic (Brandhuber and Amy, 1998).

Adsorption is now recognized as an effective and economic method for arsenic removal. The adsorption process offers flexibility in design and operation and in many cases will produce high-quality treated water. In addition, because adsorption is sometimes reversible, the adsorbents used can be regenerated by suitable desorption process (Harper and Kingham, 1992; Fu and Wang, 2011). Recent investigations show many of the metal oxide nanoparticles to exhibit high adsorption capacity and selectivity for removing arsenic from contaminated water (Deliyanni et al., 2007; Hua et al., 2012). This can be mainly attributed to their high surface areas and activities due to size-qualification effect (Henglein, 1989). However, as the size of metal oxides could be down to nanometer levels, the increased surface energy inevitably leads to their poor stability. Consequently, nanosized metal oxides are prone to agglomeration due to van der Waals forces or other interactions (Pradeep and Anshup, 2009; Hua et al., 2012). Because of this, their high adsorption capacity and selectivity would be significantly reduced or even lost. Furthermore, metal oxide nanoparticles are not suitable for use in fixed beds or any other flow-through system due to excessive pressure drops or difficult particles separation from aqueous solutions after treatment process. According to Li et al. (2012), it is extremely difficult to remove completely nanosized absorbents from aqueous solutions, even though they have unique properties. An effective approach to overcome these technical bottlenecks is to fabricate hybrid adsorbents by impregnating or coating particles into/onto porous supports of larger size (Hua et al., 2012).

12.2 CHARACTERISTICS OF ARSENIC

Arsenic, symbolized As, is a member of group 15 (IVA) of the periodic table with an atomic number of 32 and a relative atomic weight of 74.92 g/mol. It has a melting point of 817°C and density of 5.73 g/cm³ at 25°C. Arsenic is often attributed to as a heavy metal in the literature, but in fact it is semimetallic and is usually white in color. Typically, arsenic in the environment can exist in four valence states: −3, 0, +3, and +5. However, rarely, it occurs in its elemental state, that is, As(0). In aqueous solution, arsenic tends to form two classes of colorless compounds; namely arsenate, As(V) and arsenite, As(III) (Smedley and Kinniburgh, 2002). In surface waters, it exists in the form of As(V). Examples of arsenic compounds of relevance to drinking water systems include $H_2AsO_4^-$, $HAsO_4^{2-}$, H_3AsO_3, and H_2As_3 in which the predominant form of arsenic present in the water is highly a function of the pH and the redox potential (Mohan and Pittman, 2007). Figure 12.2 demonstrates the forms that arsenic can take in water as a function of pH. Note that Eh is the oxidation/reduction potential (ORP) of the water (Kartinen and Martin, 1995).

As(III) is generally found in water as arsenious acid form which ionizes according to the following equations:

$$H_3AsO_3 \rightarrow H^+ + H_2AsO_3^- \ pK_a = 9.22 \tag{12.1}$$

$$H_2AsO_3^- \rightarrow H^+ + HAsO_3^{2-} \ pK_a = 12.3 \tag{12.2}$$

where pK_a is the pH at which the disassociation of the reactant is 50% complete.

Unlike surface water, arsenic in ground water mostly exists in the form of As(III). Compared to As(V), removal of As(III) is rather difficult. Furthermore, As(III) is considerably more toxic, soluble, and mobile than As(V) (Singh and Pant, 2004; Hossain, 2006). Therefore, it is usually

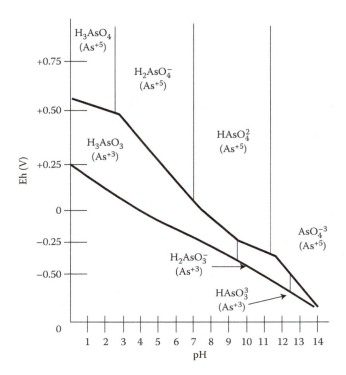

FIGURE 12.2 Species of arsenic in water. (Adapted from Kartinen Jr, E. O. and C. J. Martin. 1995a. *Desalination* 103(1): 79–88.)

necessary to change As(III) to As(V) form by adding an oxidant (generally chlorine) to reduce its toxicity level. Compared to As(III), As(V) which is normally found in water as arsenic acid, could be ionized according to the following equations:

$$H_3AsO_4 \rightarrow H^+ + H_2AsO_4^- \; pK_a = 2.2 \tag{12.3}$$

$$H_2AsO_4^- \rightarrow H^+ + HAsO_4^{2-} \; pK_a = 7.08 \tag{12.4}$$

$$HAsO_4^{2-} \rightarrow H^+ + AsO_4^{3-} \; pK_a = 11.5 \tag{12.5}$$

A review of Figure 12.2 and the equations listed above show that in the pH range normally found in municipal water supplies (say pH 6–9) trivalent arsenic is found primarily as H_3AsO_3, which is not ionized. On the other hand, in this same pH range, pentavalent arsenic is found primarily as $H_2AsO_4^-$ and $HAsO_4^{2-}$. This pentavalent form of arsenic is more easily removed from water than trivalent arsenic (Kartinen and Martin, 1995).

12.2.1 CHEMISTRY AND OCCURRENCE OF ARSENIC IN WATER

Arsenic is the 20th most abundant naturally occurring element in the Earth's crust with the average content of 2 mg/kg and is a component of more than 245 minerals. Table 12.1 shows the contents of arsenic in the soils of various countries (Mandal and Suzuki, 2002). In general, the sources of arsenic that enter the environment can be divided into two main categories: natural source and anthropogenic source. Relatively high concentrations of naturally occurring As can occur in some

TABLE 12.1

Arsenic Contents in the Soils of Various Countries

Country	Types of Soil/ Sediment	Number of Samples	Range (mg/kg)	Mean (mg/kg)
West Bengal, India	Sediments	2,235	10–196	–
Bangladesh	Sediments	10	9.0–28	22.1
Argentina	All types	20	0.8–22	5
China	All types	4,095	0.01–626	11.2
France	All types	–	0.1–5	2
Germany	Berlin region	2	2.5–4.6	3.5
Italy	All types	20	1.8–60	20
Japan	All types	358	0.4–70	11
Mexico	All types	18	2–40	14
South Africa	All types	2	3.2–3.7	3
Switzerland	All types	2	2–2.4	2.2
United States	Various states	52	1–20	7.5
United States	Tiller	1,215	1.6–72	7.5

Source: Mandal, B. K. and K. T. Suzuki. 2002. *Talanta* 58(1): 201–235.

areas as a result of inputs from geothermal sources or high-As groundwater sources. Arsenic concentrations in river waters from geothermal areas have been typically reported at around 10–70 ppb, although higher concentrations have been found. For instance, As concentrations up to 370 ppb have been reported in Madison River water (Wyoming and Montana) as a result of geothermal inputs from the Yellowstone geothermal system (Smedley and Kinniburgh, 2002). Uncontrolled anthropogenic activities, including manufacturing of metals and alloys, petroleum refining, pharmaceutical manufacturing, pesticide manufacturing and application, chemical manufacturing, burning of fossil fuels, and waste incineration may also release arsenic directly to the environment (Ning, 2002). A review conducted by Mukherjee et al. (2007) showed that Bangladesh and India are the two major countries in the world with major incidences of arsenic contamination. Figure 12.3 further shows that besides these two countries, many other countries in the world also face the similar problem.

12.2.2 ARSENIC TOXICITY AND HUMAN HEALTH EFFECTS

Historically, arsenic is the first chemical for which carcinogenic properties were understood. As early as 1879, it is reported that a high rate of lung cancer was detected in miners which was caused by inhaled arsenic (Smith et al., 2002). Far before this, the toxicity of arsenic has been well known at least since Roman times, when arsenolite (As_2O_3) was often used as a poison (Reimann et al., 2009; Bilici Baskan and Pala, 2010). The two most common valence states of inorganic arsenic to which humans might be environmentally exposed are the forms of As(III) and As(V). The soluble forms of inorganic arsenic are absorbed from the gastrointestinal tract and distributed in the tissues. Inorganic arsenic is cleared from the blood within a few hours after it is absorbed via drinking water. The inorganic arsenicals are known to be taken up by the liver, transformed to monomethylated (MMA) and dimethylated arsenicals (DMA) through consecutive reductive methylation, and then excreted into urine as pentavalent methylated arsenic form (Mandal and Suzuki, 2002; Naranmandura et al., 2006; Marchiset-Ferlay et al., 2012).

It must also be noted that the presence of arsenic in drinking water could cause other chronic diseases and health problems such as gastrointestinal symptoms, respiratory tract issues, abnormalities of the cardiovascular and nervous system, hematopoietic system, etc. Long-term exposure to potable

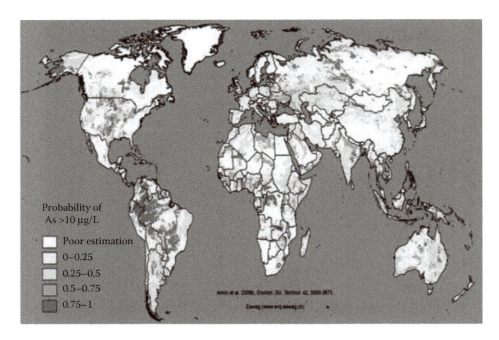

FIGURE 12.3 Worldwide arsenic contamination scenario. (From Winkel, L., Berg, M., Amini, M., Hug, S.J., and Johnson, C.A. *Nat. Geosci.*, 1(8), 536–542, 2008.)

water contaminated with arsenic is strongly associated to various kinds of cancers such as skin, lung, bladder, and kidney (Jain and Ali, 2000; Mandal and Suzuki, 2002; Mohan and Pittman, 2007). The adverse effects of arsenic in groundwater used for irrigation water on crops and aquatic ecosystems are also of major concern. The fate of arsenic in agricultural soils is often less well studied compared to groundwater, and in general has been studied in the context of arsenic uptake by different plants. Crop quality and the effect of arsenic on crop quality and yield is becoming a major worldwide concern, particularly for rice, which forms the staple for many South-Asian countries where groundwater is widely used for irrigation (Meharg and Rahman, 2003; Bhattacharya et al., 2007).

12.2.3 REGULATIONS AND GUIDELINES APPLICABLE TO ARSENIC

Generally, human exposure to arsenic compounds comes not only from polluted water but also from food and air contaminated by industrial and agricultural activities (DeSesso et al., 1998; Santra et al., 2013). This is of special concern for the reason that in most naturally occurring incidents, the liquid form of arsenic is odorless, colorless, and tasteless, making it almost impossible to detect by sight only (Atkinson, 2006). Some studies show that long-term drinking of arsenic contaminated groundwater can lead to cancer of the bladder, lungs, skin, kidney, and liver (Santra et al., 2013). In the United States, the Safe Drinking Water Act (SDWA) of 1974 called for the establishment of MCL as national drinking water standards, and required the Environmental Protection Agency (EPA) to periodically revise the standard. Based on a Public Health Service standard established in 1942, the EPA established a standard of 50 ppb as the maximum arsenic level in drinking water in 1975. In 1984, the WHO followed with the same 50 ppb recommendation. Since that time, rapidly accumulated toxicity information prompted a revision of the standard and a provisional guideline of 10 ppb was further recommended by the WHO in 1993. In January 2001, EPA published a revised standard that would require public water supplies to reduce arsenic to 10 ppb by 2006 (Ning, 2002). Furthermore, about 130 million people have access to only drinking water containing more than 10 mg As/L (Marchiset-Ferlay et al., 2012). A lower limit of 2 ppb was suggested for a better safety margin but was rejected due to financial implications (Smith et al., 2002). However, it must be pointed out that many countries with high population including China and Bangladesh have retained the earlier WHO guideline of 50 ppb as their standard or as an interim target.

12.3 TECHNOLOGIES IN ARSENIC REMOVAL

Conventionally, there are several technologies for arsenic elimination. These technologies include coagulation and flocculation, ion exchange, adsorption, and membrane filtration. Each technology will be explained briefly together with the related works in the following sections.

12.3.1 COAGULATION AND FLOCCULATION

In arsenic decontamination methods, coagulation and flocculation are among the most common methods employed. Coagulation is a process to destabilize colloids by neutralizing the forces that keep them apart. Cationic coagulants provide positive electric charges to reduce the negative charge (zeta potential) of the colloids. As a result, the particles collide to form larger particles. Rapid mixing is required to disperse the coagulant throughout the liquid. Flocculation on the other hand is the process of forming bridges between the larger mass particles or flocs and binds the particles into large agglomerates or clumps. Bridging occurs when segments of the polymer chain adsorb on different particles and help particles aggregate. An anionic flocculant will react against a positively charged suspension, adsorbing on the particles and causing destabilization either by bridging or charge neutralization (Choong et al., 2007).

The most commonly used coagulants are metal salts. These are generally classified into two groups; those based on aluminum (e.g., $Al_2(SO_4)_3 \cdot 18H_2O$) and those based on iron (e.g., $Fe_2(SO_4)_3 \cdot 7H_2O$). Aluminum- and iron-based coagulants are popular due to their relatively low cost and their abundance (Fu and Wang, 2011). Arsenic in the arsenate form can be readily removed by adding ferric salts. However, it must be pointed out that As(III) is generally less efficiently removed by ferric salts in comparison to As(V). Results showed that >95% of As(V) could be effectively removed using $FeSO_4$ within pH range of 5–7.5 (EPA, 2000). Besides $FeSO_4$, $FeCl_3$ was also reported to be efficient to remove as high as 80% As(V) at the pH ranging between 4 and 8 (Hering et al., 1997). Similar like ferric salts, aluminum sulfate is only effective to treat water contaminated with As(V), not As(III) (Ali et al., 2011).

12.3.2 ION EXCHANGE

Ion exchange is a reversible physical/chemical reaction. Exchange of arsenic ions generally takes place between the resin and the feed water. The resins are normally elastic three dimensional hydrocarbon compounds enclosing a huge amount of ionizable groups electrostatically bound to the resin. Ion exchange has been widely applied for the removal of arsenic from wastewater (Kartinen and Martin, 1995; Vaaramaa and Lehto, 2003; Kim and Benjamin, 2004). The process is normally used for the removal of specific undesirable cations or anions from water. As the resin becomes exhausted, it needs to be regenerated. The following reactions show the arsenic exchange with resin and how the resin is to be practically regenerated by salt solution (Ahmed, 2001).

Arsenic exchange with resin

$$2R - Cl + HAsO_4^- \rightarrow R_2HAsO_4 + 2Cl^- \tag{12.6}$$

Resin regeneration using NaCl solution

$$R_2HAsO_4 + 2Na^+ + 2Cl^- \rightarrow 2R - Cl + HAsO_4^- + 2Na^+ \tag{12.7}$$

where R stands for ion exchange resin.

Since chemical reagents must be used to regenerate ion exchange resins when they are exhausted, it is unavoidable that secondary pollution occurs. Furthermore, the regeneration process is rather expensive, especially when a large amount of water containing metal ions in low concentration is needed for the treatment (Fu and Wang, 2011). During the ion exchange process, arsenic decontamination occurs by continuously passing contaminated water under pressure through one or more columns packed with ion exchange resins. Ion exchange resin can decontaminate As(V) efficiently, but hardly remove As(III) (EPA, 2000; Mohan and Pittman, 2007). Thus, it limits the wide applications of this technology for arsenic removal. Furthermore, ion exchange resin systems are limited mostly to small- and medium-scale operations and have relatively higher treatment cost than conventional treatment technologies. Moreover, the presence of some anions such as sulfate, fluoride, and nitrate in the solution could further compete with arsenic ions and affect the efficiency of arsenic removal. Other factors affecting the use of the ion exchange process include contact time and spent regenerate disposal (Amin et al., 2006; Chiban et al., 2012).

12.3.3 Adsorption Process

Adsorption is a process that uses solids (act as adsorbents) for removing arsenic from liquid solutions. Industrially, adsorption operations employing solids such as activated carbon (AC) and metal hydrides are used widely in purification of water and wastewater. The process of adsorption involves separation of a substance from one phase accompanied by its accumulation or concentration at the surface of another. Mainly, van der Waals forces and electrostatic forces between adsorbate molecules and the atoms, which compose the adsorbent surface, cause physical adsorption. Thus, adsorbents are characterized first by surface properties such as surface area and polarity (Choong et al., 2007). In the following sections, a brief of different types of most commonly used adsorbents for arsenic removal will be explained.

12.3.3.1 AC Adsorbents

AC also known as activated coal, carbon active, or activated charcoal is a form of carbon that is produced as a permeable product with a very high surface area. For as much as only 1 g of AC, a surface area of >500 m² could be obtained, allowing it to be very effective in removing specific impurities. Hindus in ancient India used charcoal to filter drinking water, while in ancient Egypt (1500 BC) used carbonized wood as a medical adsorbent and purifying agent. AC in the powder form was first made commercially from wood in Europe in the early nineteenth century and was widely applied in the sugar industry. While in the United States, AC was first reported for water treatment in 1930 (Mohan and Pittman, 2007).

Many have reported the use of AC for arsenic elimination from water sources (Eguez and Cho, 1987; Navarro and Alguacil, 2002; Jahan et al., 2008; Vitela-Rodriguez and Rangel-Mendez, 2013). A huge arsenic sorption capacity of 2860 mg/g was able to obtain on this coal-derived commercial carbon. Some ACs impregnated with metallic silver and copper was also used for arsenic remediation (Rajaković, 1992; Mohan and Pittman, 2007). In general, the adsorption capacity of AC strongly depends on the characteristics of solution such as chemical properties, temperature, and pH. Many ACs are available but only few are selective. Therefore, the research thrust over the years is leading to find improved and tailor-made materials, which will meet several requirements such as regeneration capability, easy availability, cost effectiveness, etc. (Chiban et al., 2012).

12.3.3.2 Iron Oxide-Coated Sand Adsorbents

Iron oxide-coated sand (IOCS) is a rare adsorbent, which is mostly applied in fixed-bed columns to eliminate various dissolved metal ions. IOCS, which has exhibited some affinity for arsenic elimination is prepared by treating river sand with acidic solution, then mixed with iron (III) nitrate monohydrate with a ratio of 10:1 and heating to 110°C for at least 20 h (Yuan et al., 2002). The metal ions are exchanged with the surface hydroxides on the IOCS. Like other adsorption processes, when the bed is exhausted it needs to be regenerated by a sequence of operations consisting of rinsing and flushing with water, and neutralizing with strong acid. Mostly, sodium hydroxide (NaOH) is used as the regenerating agent while sulfuric acid (H_2SO_4) is used as neutralizer (EPA, 2000). Several studies have exhibited that IOCS is effective for arsenic removal. For instance, batch experiments using IOCS exhibited that an arsenic level under 10 ppb could be reached with adsorption capacity of 136 µg/g (Thirunavukkarasu et al., 2005). Another study also reported that IOCS showed a removal efficiency of 68% and 83% for As(III) and As(V), respectively. Nevertheless, a very strong hardness of water due to the existence of high concentration of calcium carbonate (612.5 mg/L $CaCo_3$) could affect the removal efficiency of arsenic (Yuan et al., 2002). Other factors such as pH, arsenic oxidation state, competing ions, and regeneration process also have significant effects on arsenic removal achieved with IOCS.

12.3.3.3 Metal Oxide Adsorbents

Nanotechnology is a promising novel solution to environmental engineering problems such as water and wastewater treatment, pollution control, surface and groundwater remediation (Zhang, 2003; Tratnyek and Johnson, 2006; Ramos et al., 2016). Many different types of metal oxide nanoparticles with high surface area could exhibit specific affinity toward arsenic for the purpose of water decontamination. These include titanium dioxide (TiO_2), hydrous TiO_2, titanium nanotube (TNT), Fe–Mn binary oxide (FMBO), zirconia nanoparticle, etc. (Nabi et al., 2009; Niu et al., 2009; Xu et al., 2010; Zhang et al., 2012; Zheng et al., 2012). To date, it has become a hot topic to develop new technologies to synthesize nanosized metal oxides. The shape, size, and surface area of metal oxides are important factors affecting their adsorption capacity. Generally, these nanomaterials can be presented in different forms, like particles and tubes. Table 12.2 lists down some of the famous metal oxide adsorbents that have been studied for arsenic removal.

12.3.3.3.1 Importance of Using Metal Oxide Adsorbents with Porous Host Medias

Previous research work has shown that metal oxide nanomaterials are highly efficient adsorbents for removing selective heavy metals, which also include arsenic from water/wastewater. They in general exhibit various advantages such as fast kinetics, high capacity, and preferable sorption toward metal ions in water and wastewater. Nevertheless, it must be pointed out that the practical

TABLE 12.2
Arsenic Removal by Metal Oxide Adsorbents

Metal Oxide Adsorbent	As Species	Adsorption Capacity (mg/g)	Reference
Activated aluminum	As(V)	11–24	Ghosh and Yuan (1987)
Hematite	As(V)	0.2	Singh et al. (1996)
Titanium dioxide	As(V)	4.65	Pena et al. (2005)
	As(III)	59.93	
Akaganeite	As(V)	1.8	Deliyanni et al. (2003)
Hydrous iron oxide	As(V)	7.0	Lenoble et al. (2002)
	As(III)	28.0	
Copper oxide	As(V)	26.9	Martinson and Reddy (2009)
Nano-iron–titanium mixed oxide	As(V)	14.3	Gupta and Ghosh (2009)
	As(III)	85	
Iron–manganese binary oxide	As(V)	69.8	Zhang et al. (2012)
Zirconium oxide	As(V)	45.6	Zheng et al. (2009)
Iron–zirconium binary oxide	As(V)	46.1	Ren et al. (2011)
	As(III)	120.0	

TABLE 12.3
Host-Supported Metal Oxides for Arsenic Decontamination

Metal Oxide Adsorbent	Host Media	Arsenic Species	Adsorption Capacity (mg/g)	Reference
Iron(III) oxide	Gel resin	As(III)	31.56	Styles et al. (1996)
Titanium dioxide	Amberlite	As(V)	4.72	Driehaus et al. (1998)
Iron(III) oxide	Loaded chelating resin	A(III)	9.74	Rau et al. (2000)
		As(V)	60.0	
Iron oxide	Coated sand	As(III)	0.14	Vaishya and Gupta (2003)
Iron oxide	Coated sand	As(III)	0.14	Thirunavukkarasu et al. (2005)
Zirconium oxide	Membrane	As(V)	21.5	Zheng et al. (2011)
Manganese oxide	Ion exchanger	As(III)	47.6	Li et al. (2012b)
Iron–manganese binary oxide	Membrane	As(III)	73.5	Jamshidi Gohari et al. (2013)
Zirconium nanoparticle	Membrane	As(V)	131.8	He et al. (2014b)

application of nanosized metal oxide adsorbents alone in the abatement of arsenic pollution is still very challenging due to some technical bottlenecks. For instance, when metal oxide adsorbents are applied in aqueous solution, they tend to aggregate into large size particles due to van der Walls force or other interactions, causing them to lose their adsorption capacity. Moreover, metal oxide adsorbents are unusable in fixed beds or any other flow-through systems because of the excessive pressure drops (or the difficult separation from aqueous systems) and poor mechanical strength (Hua et al., 2012). Fortunately, fabrication of new nanosized metal oxide-based composite adsorbents seems to be an effective approach to respond to all the above-mentioned technical problems (Cumbal and SenGupta, 2005; Deliyanni et al., 2007; Hua et al., 2012). Table 12.3 presents some host-supported metal oxides for arsenic decontamination.

12.3.4 ARSENIC REMOVAL BASED ON MEMBRANE TECHNOLOGY

Membrane filtration is a viable technology which also can remove a wide range of contaminants from water. Membranes are reliable, easy to produce, obtain, operate, and maintain and will likely be increasingly applied to water treatment as tougher regulations are promulgated. The presence of billions of pores (or microscopic holes) on a membrane surface tends to act as selective barriers, allowing some constituents to pass through, while at the same rejecting/excluding others. The movements of molecules across the membrane need a driving force, such as a potential difference between the two sides of the membrane (Shih, 2005). Pressure-driven membrane processes are often classified by pore size into four categories: MF, UF, NF, and RO. Typical pore size classification ranges are given in Figure 12.4. High-pressure processes (i.e., NF and RO) have a relatively small pore size compared to low-pressure processes (i.e., MF and UF) with typical pressure ranges for each membrane process shown in Table 12.4. In the following sections, a brief on different types of membrane technologies used for arsenic decontamination is provided.

12.3.4.1 Reverse Osmosis

Reverse osmosis as a semi-permeable membrane is the oldest of membrane technologies, and also has been identified as a likely best available technology to help small water treatment systems to remove arsenic from water and meet the MCL regulation. The removal efficiency for As(V) was reported over 90% when RO membranes were first evaluated for arsenic removal in the 1980s by using the cellulose-acetate RO membrane with operating pressure set at around 400 psi (approximately 27.3 bar). However, removal efficiency for As(III) is not promising, recording at <70%.

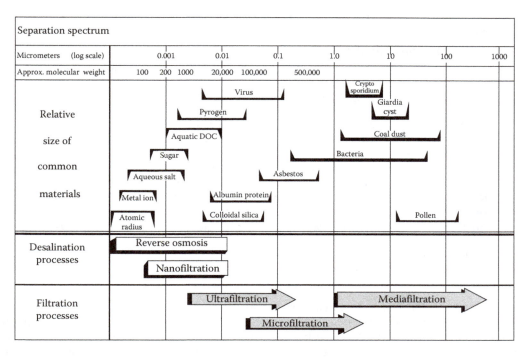

FIGURE 12.4 Pressure-driven membrane process classification. (From EPA. 2000. *Technologies and Costs for Removal of Arsenic from Drinking Water.* United States Environmental Protection Agency. http://www.epa/gov/.)

TABLE 12.4
Typical Pressure Ranges for Membrane Processes

Membrane Process	Typical Pressure Range (psi)
MF	5–45
UF	7–50
NF	100–220
RO	>220

Although the oxidation process could improve removal efficiency, the oxidant used could damage the RO membrane (Shih, 2005). A number of earlier research results for arsenic decontamination by RO are summarized in Table 12.5. It shows that this technology can be an operative method for arsenic elimination; however, membrane property and operating conditions are very important factors for arsenic decontamination. The major problem of RO may be the high-energy consumption as a result of pumping pressure and restoration in the membrane.

12.3.4.2 Nanofiltration

Similar to RO membrane, NF is also able to eliminate heavy metal ions from solutions, but with a relatively lower removal rate. Its degree of separation in general depends on membrane surface pores and charge property. Previously, Brandhuber and Amy (1998) used three different types of NF membranes to remove arsenic ions from aqueous solutions. The results showed that as high as

TABLE 12.5
Arsenic Decontamination by RO Process

| Model | Suppliers | Water | Rejection% | |
			As(III)	As(V)
TFC 4921	Fluid systems	Ground water	63	95
TFC 4820-ULPT	Fluid systems	Ground water	77	99
AG 4040	Desal	Ground water	70	99
4040 LSA CPA2	Hydranautics	Ground water	85	99
TFC ULP RO	Koch Membrane System	Ground water	99	100
ES 10	Nitto Electric Industrial Co.	Distilled water	75	95
NTR-729HF	Nitto Electric Industrial Co.	Distilled water	20–43	80–95
DK2540F	Desal	Lake water	5	96
HR3155	Toyobo Co. Ltd.	Ground water	55	95

Source: Trina Dutta, C.B. and S. Bhattacherjee. *Int. J. Eng. Res. Technol.*, 1(9), 1–23, 2012.

90% As(V) was possible to remove in comparison to 20%–53% reported for As(III). A similar rate of As(V) removal was also reported by Harisha et al. (2010) with the use of thin film composite NF membrane. Saitua et al. (2011) further studied arsenic decontamination from naturally contaminated ground water by using a NF pilot plant. But, a slight decrease in As(V) rejection (<90%) was reported. Figoli et al. (2010) on the other hand related the performance of a commercial NF membrane in removing As(V) from synthetic water to pH and operating temperature. It is found that with decreasing operating temperature and increasing pH range, As(V) removal efficiency was increased. Other results of arsenic removal using commercial NF membrane are summarized in Table 12.6. Although NF membranes show good results in removing arsenic in particular As(V), the fouling problem is still unavoidable. The small membrane pore size has made NF prone to fouling compared to other microporous membrane like UF and MF. Furthermore, this technology is also associated with a very high capital and running cost, although high rejection is generally possible to achieve (Mohan and Pittman, 2007).

TABLE 12.6
Arsenic and Removal by Different Types of NF Process

| Model | Supplier | Water Origin | Rejection % | |
			As(III)	As(V)
NF70 4040-B	Film Tec	Colorado River	53	99
HL-4040F1550	Hydranautics	Idem	21	99
4040-UHA-ESNA	Film Tec	Idem	30	97
NF-45	Nitto-Denko Co. Ltd.	Synthetic water	10	90
ES-10	Nitto Electric Industrial Co.	Ground water	50–89	87–93
ES10	Nitto Electric Industrial Co.	Synthetic water	80	97
NTR-729HF	Nitto Electric Industrial Co.	Synthetic water	21	94
NTR-7250	Nitto Electric Industrial Co.	Ground water	10	86
NF70	Film Tec	Fresh water	99	99
NF270	Film Tec	Ground water (Osijek)	–	99

Source: Trina Dutta, C.B. and S. Bhattacherjee. *Int. J. Eng. Res. Technol.*, 1(9), 1–23, 2012.

12.3.4.3 Ultrafiltration/Microfiltration

Today many UF processes are often being used in industries and water purification systems, and more are being investigated in order to find the best efficiency in use. However, UF which operates at low pressure is only suitable for the removal of large particles. Since the pore sizes of UF membranes are significantly larger than dissolved metal ions in the form of hydrated ions or as low molecular weight complexes, these ions would easily pass through UF membranes. To gain high elimination efficiency of arsenic, the micellar-enhanced ultrafiltration (MEUF) and polymer-enhanced ultrafiltration (PEUF) are proposed (Pookrod et al., 2005; Beolchini et al., 2006; Fu and Wang, 2011). The removal characteristics of As(V) using MEUF were studied by Iqbal et al. (2007) using four different cationic surfactants. Of the surfactants studied, hexadecylpyridinium chloride (CPC) showed the highest removal efficiency of arsenic (96%) followed by hexadecyltrimethylammonium bromide (CTAB) (94%), octadecylamine acetate (ODA) (80%), and benzalkonium chloride (BC) (57%). Without surfactant micelles, the control PES (polyethersulfone) membrane was found to be ineffective for arsenic elimination. Brandhuber and Amy (1998) in their earlier work conducted a series of bench-scale tests to study the effect of membrane charge on arsenic elimination efficiency. Negatively charged GM2540F UF membrane and uncharged FV2450F UF membrane were applied in the investigation. From the results, it was found that GM2540F membrane achieved better rejection of As(V) at neutral pH compared to acidic pH whereas FV2540F membrane gave poor rejection for both As(V) and As(III) species. The high rejection by charged membrane could be attributed to the electrostatic interaction between arsenic ions and membrane surface negative charge (Trina Dutta and Bhattacherjee, 2012). It must be pointed out that it is almost impossible to achieve any elimination of dissolved As(V) and As(III) species from contaminated water by the use of macrostructured MF alone. Its arsenic removal capability however is achievable by increasing the particle size of arsenic bearing species through coagulation and flocculation.

12.4 ADSORPTIVE REMOVAL OF ARSENIC BY MIXED MATRIX MEMBRANES

12.4.1 Advantages of Membrane as Host-Supported Media

Recent studies suggested that many metal oxide adsorbents exhibit very favorable sorption to arsenic in terms of high capacity and selectivity, making them increasingly important to meet strict regulations (Gupta and Ghosh, 2009; Xu et al., 2010; Hua et al., 2012). However, the main disadvantage of the metal oxide adsorbents in batch processes is the difficulty in the particle separation after the treatment. To promote the applicability of metal oxide adsorbent nanoparticles in real treatment processes, many researchers in recent years have focused on impregnating nanoparticles into porous host media such as bentonite (Ranđelović et al., 2012), zeolite (Li et al., 2011), diatomite (Jang et al., 2006), cellulose (Guo et al., 2007), and porous polymer (Pan et al., 2009; Su et al., 2009).

Compared to other host materials, porous polymeric hosts are a particularly attractive option mainly because of their controllable pore size and surface chemistry, in addition to their excellent mechanical strength (Hua et al., 2012). If the nanoparticles are impregnated in a MF or UF membrane, they can still act as an adsorbent for the pollutant elimination. More significantly, the nanoparticles can stay in the membrane matrix and no addition separation device is required (Singh et al., 1996).

To overcome the shortages of standalone adsorption and membrane technology, there has been a great interest in recent years in the fabrication of adsorptive mixed matrix membranes (MMMs) for various applications (Arcibar-Orozco et al., 2014; Chatterjee and De, 2014; Mukherjee and De, 2014a,b; Jamshidi Gohari et al., 2015; Hao et al., 2016). Besides being able to effectively remove pollutants from aqueous solutions, MMMs are also found to be attractive with their low operating pressure (Klein, 2000). So far, three attempts have been made to remove arsenic using MMMs and their important findings will be highlighted in the following sections.

12.4.2 Polyvinylidene Difluoride/Zirconia Flat Sheet Membrane

In this work, five different polyvinylidene difluoride (PVDF)/zirconia flat sheet membranes with different zirconia loadings (i.e., M0, M0.5, M1.0, M1.5, and M2.0) were successfully fabricated through phase inversion process by casting the solutions on a glass plate using a film applicator (Zheng et al., 2011). The research findings of this work opened the discussion on the potential of using MMMs in continuous filtration for improving the adsorption process compared to the batch adsorption process. Extensive theories and studies are given behind using of the PVDF/zirconia MMMs for arsenate removal, as well as an adsorption isotherm study and continuous filtration study.

Figure 12.5 shows the effect of pH on the As(V) adsorption onto the PVDF/zirconia blend membrane under the pH ranging from 3 to 11 with an initial arsenate concentration of 1.0 mg/L. The experimental results showed more than 95% of As(V) could be eliminated in the pH ranging from 3 to 8. However, when pH was higher than 9, the adsorption of As(V) onto the membrane was decreased, and declined to <40% at pH > 11.The negative pH effect at its higher alkali range however is useful for the regeneration of the membrane adsorption capacity through the desorption process. Figure 12.5 also presents that the leakage of Zr(IV) ions from the blend membrane is negligible during the filtration process, causing no harmful effects to human beings.

The kinetics of arsenate removal and change in concentrations of two batch adsorption experiments with different arsenate concentrations were also conducted to study the adsorption kinetics of arsenate onto the PVDF/zirconia blend membrane in aqueous phase and the results are shown in Figure 12.6. Clearly, the PVDF/zirconia blend membrane could effectively and quickly adsorb most of the arsenate in the first 10 h and achieved adsorption equilibrium after 25 h. The adsorption kinetic of this kind of MMM can be well described by the pseudo-second-order rate model.

Figure 12.7 shows the experimental data of adsorption isotherms for PVDF membrane (labeled as M0), zirconia particles, and PVDF/zirconia blend membranes (M0.5, M1.0, M1.5, and M2.0). The maximal adsorption capacity of the hydrous zirconia approximately was approximately 39 mg/g. No adsorption capacity of the PVDF membrane was recorded, mainly because of the absence of adsorbents. Comparing between the PVDF/zirconia membranes, it is found that the maximum

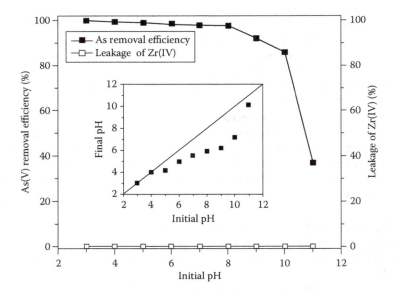

FIGURE 12.5 Effect of pH on the adsorption of arsenic on PVDF/zirconia membrane with highest zirconia loading (M2.0). Conditions: [As(V)]$_o$ = 1.0 mg/L, m = 1.0 g/L, T = 293 K, and contact time = 48 h.

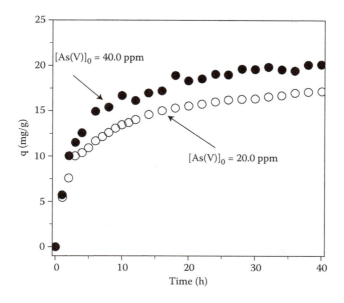

FIGURE 12.6 Adsorption kinetics of As(V) on the PVDF/zirconia blend membrane (M2.0). Conditions: m = 1.0 g/L, T = 293 K, and initial pH = 3–4.

adsorption capacity of M0.5, M1.0, M1.5, and M2.0 membrane was 4.5, 10.1, 15.1, and 21.5 mg/g, respectively. The results indicated that the higher the amount of zirconia in the PVDF membrane, the greater the adsorption capacity.

Figure 12.8 shows the performance of optimum PVDF/zirconia blend membrane (M2.0) for decontaminating As(V) before and after regeneration in the continuous filtration process. M2.0 was selected due to its highest adsorption capacity and the influent pH of 3–4 was chosen based on the pH effect study presented in Figure 12.6. Experimental results showed that the virgin membrane which operated at 1 psi was able to maintain the concentration of As(V) below the MCL of 10 μg/L

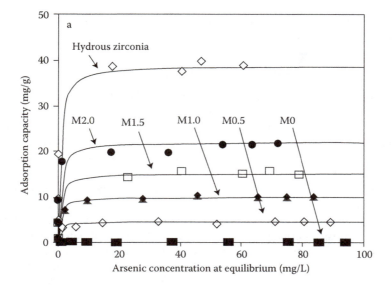

FIGURE 12.7 Adsorption isotherms of As(V) on PVDF/zirconia blend membrane.

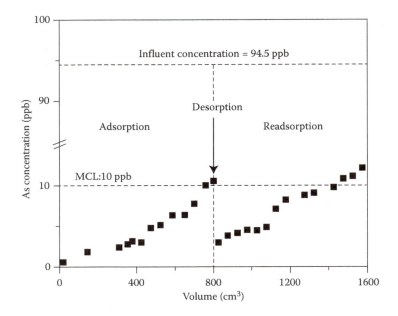

FIGURE 12.8 Arsenate removal from aqueous solution by M2.0 with continuous filtration mode.

for nearly 750 cm^3 of permeate collected, before failing to produce permeate of high quality. As the adsorption rate was relatively slow, a longer reaction time was needed.

The As(V)-adsorbed M2.0 membrane was then subjected to a desorption process to regenerate its adsorption capacity before it was tested again using the same concentration of feed solution. Based on the result obtained from the pH effect study, a diluted NaOH solution with pH of 11 was used to regenerate the arsenate-loaded membrane. As shown in the figure, the regenerated membrane was able to further treat the arsenic containing water by producing another 650 cm^3 permeate sample with As(V) concentration <10 μg/L. The high recovery rate of M2.0 membrane indicated that it can be applied for multiple treatment cycles in removing As(V) solution before losing its function as an adsorbent in the membrane matrix. It is anticipated that the blend membrane would provide a better engineering solution for treatment of arsenate contaminated groundwater treatment for production of drinking water.

12.4.3 Polyethersulfone/Fe–Mn Binary Oxide Flat Sheet Membrane

According to the literature, Chakravarty et al. (2002) and Deschamps et al. (2005) were the pioneers in using natural FMBO particles for removing arsenite from contaminated water samples (Jamshidi Gohari et al., 2013). In 2012, Zhang et al. (2012) successfully synthesized high capacity FMBO nanoparticles through a coprecipitation process for As(III) and As(V) removal. Other recent research works focused on the use of FMBO particles for As(III) removal can also be found elsewhere (Zhang et al., 2007; Szlachta et al., 2012).

In order to promote the applicability of FMBO particles in arsenic decontamination, researchers have made attempts by impregnating FMBO particles into porous host media such as diatomite (Chang et al., 2009) and anion exchanger resins (Li et al., 2012). In view of the significance of developing a treatment process for efficient and applicable As(III) removal, Jamshidi Gohari et al. (2013) recently proposed a novel method to remove As(III) from contaminated water samples by PES/FMBOMMM. Four different PES/FMBO flat sheet MMMs composed of different PES/FMBO weight ratio, that is, 0, 0.5, 1.0, and 1.5 (with membrane labeled as M0, M0.5, M1.0, and M1.5 according to the ratio) were successfully fabricated through phase inversion. Figure 12.9 shows

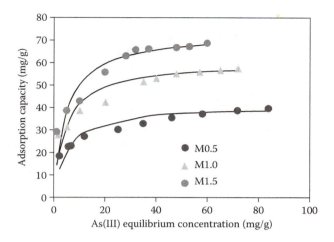

FIGURE 12.9 Adsorption isotherms for As(III) by MMMs with different FMBO:PES ratio: (●) M0.5, (▲) M1, and (●) M1.5 membranes. (Experimental conditions: membrane weight = 1.0 g/L, pH = 3–4, temperature = 298 K, and contact time = 48 h.)

As(III) adsorption capacity of MMMs prepared from different FMBO/PES weight ratio from 0.5 to 1.5. The adsorption capacity of the PES membrane was not conducted as it was prepared without addition of any FMBO particles and thus exhibited no adsorption against As(III).The results showed that with increasing FMBO/PES ratio from 0.5 to 1.5, the adsorption capacity of MMM significantly improved. The highest As(III) adsorption capacities that could be achieved by the M0.5, M1.0, and M1.5 membranes were 41.3, 60.6, and 73.5 mg/g, respectively. The increasing adsorption rate could be attributed to the presence of a greater quantity of adsorbent available to adsorb a higher amount of As(III).

Figure 12.10 shows the changes in As(III) removal efficiency and Fe–Mn leakage using the M1.5 membrane as a function of pH in the range of 2–11. Clearly, the M1.5 membrane could easily achieve

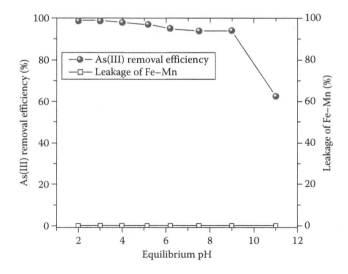

FIGURE 12.10 Effect of pH on the As(III) uptake of optimum M1.5 membrane. (Experimental conditions: initial As(III) concentration = 10 mg/L, membrane weight = 1.0 g/L, temperature = 298 K, and contact time = 48 h.)

at least 90% of As(III) removal in the pH ranging from 2 to 8 after 48 h contact time when it was tested with 10 mg/L initial As(III) concentration. Since the pH of ground water is between 6 and 8, this membrane can be used to treat the water without any pH adjustment. Results also revealed that the As(III) removal rate of membrane decreased sharply at pH > 9, which can be attributed to the electrostatic repulsion between the electronegative H_2AsO_3 species and the negatively charged adsorbent at pH higher than 9. At pH > 8, H_2AsO_3 is the predominant arsenic species compared to H_3AsO_3 species found at pHs between 2 and 8. On the other hand, no leakage of FMBO from the PES membrane matrix was detected within the pH studied, indicating excellent compatibility between the nanoparticles and membrane matrix. Overall, the results are in accordance with the study of As(III) adsorption by PVDF/zirconia membrane as highlighted in Section 12.4.2.

The adsorption kinetics for As(III) adsorption as a function of time was also investigated by the optimum M1.5 membrane for 25 h contact and the results are presented in Figure 12.11. As can be seen, the reaction between the As(III) and the FMBO particles was particularly fast in the first 2.5 h, during which period close to 75% of the initial As(III) content was adsorbed. This can be attributed to the fine particle size of the FMBO, offering many active sites for adsorbing As(III) from the bulk solution. In principle, As(III) is first transported to the solid–water interface by convection or diffusion from the bulk solution. It is followed by being adsorbed onto the surface of the nanoparticles. The adsorbed arsenite ions near the manganese atoms would then be oxidized to arsenate by the manganese oxide and the formed arsenate ions would be released into the aqueous solution. During this process, fresh active adsorption sites tend to form at the solid surface. Arsenate ions are later transported to the solid–water interface and adsorbed into the surface of the FMBO adsorbent, occupying empty adsorption sites or replacing sportive arsenite ions. The entire process can be represented by the following equations:

$$As(III)(aq) + (-S_{Fe-Mn}) \rightarrow As(III) - S_{Fe-Mn} \tag{12.8}$$

$$As(III) - S_{Fe-Mn} + MnO_2 + 2H^+ \rightarrow As(V)(aq) + Mn^{2+} + H_2O \tag{12.9}$$

$$As(V)(aq) + As(III)(aq) \rightarrow As(V) - S_{Fe-Mn} + As(III)(aq) \tag{12.10}$$

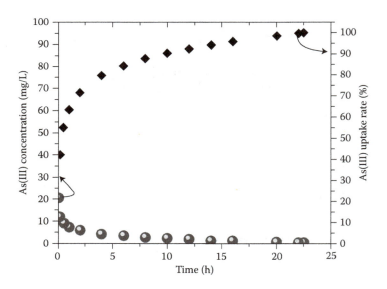

FIGURE 12.11 As(III) adsorption kinetics onto optimum M1.5 membrane and change of As(III) concentration in aqueous solution as a function of time. (Experimental conditions: initial As(III) concentration = 20 mg/L, membrane weight = 1.0 g/L, pH = 7.5, and temperature = 298 K.)

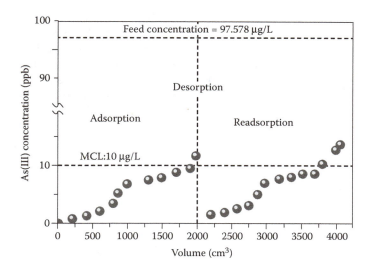

FIGURE 12.12 As(III) removal from aqueous solution using optimum M1.5 membrane with continuous UF process before and after membrane regeneration (Experimental conditions: initial As(III) concentration: 97.58 µg/L, pH: 7.5, pressure: 2 bar, and temperature: 298 K.)

where $(-S_{Fe-Mn})$ represents an adsorption site on the FMBO particles surface, $As(III) - S_{Fe-Mn}$ represents the As(III) surface species, and $As(V)—S_{Fe-Mn}$ represents the As(V) surface species. This process is continued until As(III) or the available manganese oxide is completely depleted.

Figure 12.12 shows the performance of the optimum M1.5 membrane in removing As(III) before and after the regeneration process in a continuous UF experiment. Using 97.58 ppb As(III) solution as feed, it was found that the M1.5 membrane was able to keep the As(III) concentration under the MCL of 10 ppb even after nearly 2000 cm³ of permeate was collected. Equations 12.8 through 12.10 confirm that As(III) can be oxidized to As(V) by Mn(IV) oxide during its sequestration by a large amount of FMBO nanoparticles impregnated in the M1.5, whereas Mn(IV) is reduced to Mn(II) simultaneously. Once the As(III) concentration in the permeate exceeded 10 ppb, the As-loaded M1.5 was subjected to a desorption process to regenerate its adsorption capacity. A solution mixture containing NaOCl and NaOH was used to regenerate the membrane and the desorption process lasted for 2 h before the membrane was reused in a new filtration. In principle, NaOH is used for desorption of the adsorbed As(III) of membrane while NaOCl is applied to oxidize Mn(II) to Mn(IV). XPS (x-ray photoelectron spectroscopy) study by Nesbitt et al. (1998) also showed all the Mn 2p2/3 peaks could be deconvolved to three components assigned to Mn(IV), Mn(III), and Mn(II), respectively. After adsorption, the amount of Mn(IV) was reduced while both Mn(III) and Mn(II) were increased.

After regeneration, the amount of Mn(IV) was recovered near to its original value, while those of Mn(III) and Mn(II) were reduced. The results demonstrated that the Mn(III) and Mn(II) species resulting from Mn(IV) reduction during As(III) adsorption could be effectively oxidized back to Mn(IV) (Nesbitt et al., 1998; Li et al., 2012b). After the regeneration process, it was experienced that 87.5% of the original adsorption capacity of the M1.5 membrane could be recovered, indicating that this membrane could be possibly used for multiple cycles in treating As(III) solution before losing its function as an adsorbent in the membrane matrix.

12.4.4 Polysulfone/Zirconia Hollow Fiber Membrane

In the most recent literature, He et al. (2014a) described the adsorptive removal of arsenic from aqueous solutions by polysulfone (PSF)/Zr blend membranes of different properties by varying the Zr/PSF weight ratio from 0.5 to 1.5. Unlike the PVDF/Zr membranes mentioned earlier, these

PSF/Zr blend membranes were fabricated in hollow fiber configuration through a spinning technique. Figure 12.13 demonstrates that the As(V) removal on the M1.5 is strongly pH dependent and maximum As(V) adsorption occurs at pH 3.5–4.5 with adsorption capacity of around 70 mg/g. As expected, the As(V) uptake tended to decrease with increasing the pH from acidic to alkali. When the initial pH increased to 11.5, As(V) uptake dramatically dropped to around 10 mg/g. The decrease in As(V) uptake at pH above 7 may be due to the competition between hydroxide ions and arsenic species for the exchange with sulfate. It should be also noted from this figure that no Zr(IV)

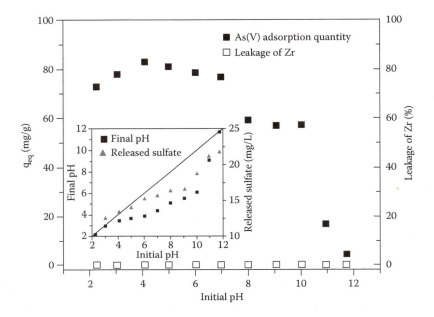

FIGURE 12.13 Effect of initial pH on As(V) adsorption on blend membrane M1.5 (membrane dose: 0.5 g/L, initial As(V) concentration: 53.30 mg/L, and T: 20°C).

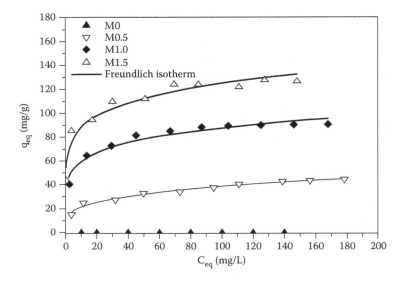

FIGURE 12.14 Adsorption isotherm of As(V) for the blend membranes and pure PSF membrane. Conditions: membrane dosage: 0.5 g/L, pH: 3.5, and T: 20°C.

ion is detected in the solution after the adsorption experiment at any initial pH values, indicating no occurrence of Zr leaching during the adsorption process.

Figure 12.14 shows the As(V) adsorption capacity for the PSF membranes incorporated with and without nanoparticles. PSF membrane made of not any zirconia nanoparticles (labeled as M0) exhibited no adsorption of As(V). However, with increasing Zr/PSF ratio from 0.5 to 1.5, the adsorption capacity of nanoparticles-loaded membranes was significantly increased as evidenced in the M0.5, M1.0, and M1.5 membranes. The highest As(V) adsorption capacity that could be achieved by each membrane was 44.60, 95.13, and 131.78 mg/g for the M0.5, M1.0, and M1.5 membrane, respectively. The increase in the adsorption capacity of the membranes could be attributed to the presence of a greater quantity of adsorbent available to a adsorb higher amount of As(V).

Figure 12.15a shows the performance of the optimum M1.5 membrane for decontaminating As(V) in a continuous filtration process. The experimental results presented that the M1.5 membrane

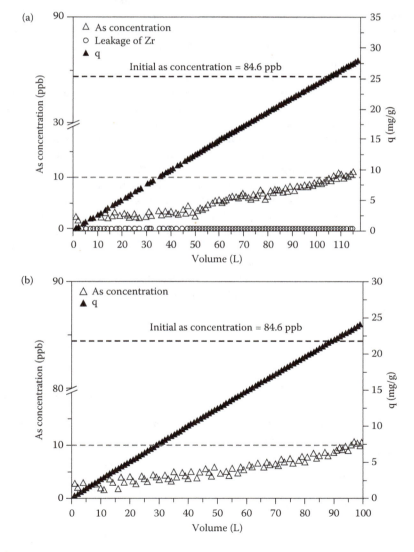

FIGURE 12.15 As(V) removal by cross-flow filtration from aqueous solution by the blend membrane M1.5: (a) fresh membrane and (b) regenerated membrane. Conditions: initial As(V) concentration: 84.6 mg/L, membrane weight: 0.3282 g, pH: 3.5–4.5, trans-membrane pressure: 0.4 bar, and T: 20°C.

was able to keep the concentration of As(V) in the 106 L of permeate below 10 ppb before failing to produce permeate of high quality, which meets the USEPA and WHO standard of 10 ppb of arsenic for drinking water. Figure 12.15b on the other hand shows the performance of the membrane after the regeneration process. The regeneration of the membrane was performed by back washing with 0.01 M NaOH solution followed by 0.01 M H_2SO_4 solution. It was experienced that 90.1% of the original adsorption capacity of M1.5 was able to recover by producing a 95 L permeate sample with the As(V) concentration <10 ppb. This finding indicates that the blend membrane can be regenerated and reused for the treatment of arsenic contaminated water for multiple cycles with high efficiency.

12.5 CONCLUSION

In recent years, a tremendous amount of research has been conducted to develop novel technologies for arsenic removal, specifically low-cost, low-tech, and environmental friendly systems that can be applied in rural and urban areas. It is apparent from the literature that many of metal oxide nanoparticle adsorbents exhibit high adsorption capacity coupled with selectivity for decontaminating selective hazardous metal ions such as arsenic from water sources. However, these nanosized adsorbents are not suitable to use in real treatment systems due to the difficulty of complete separation from aqueous solutions after the adsorption process. Hence, researchers in recent years have been searching for the ideal host media for nanoparticles impregnation. One of the promising candidates is microporous support made of polymeric material.

MMMs which were fabricated by impregnating nanoparticles into a polymeric membrane matrix were successfully evaluated by several research groups through a typical phase inversion technique. The results showed that nanoparticles such as Zr and FMBO could be practically embedded in either flat sheet or hollow fiber membranes without any sign of leakage during the treatment process. The new generation of MMMs is not only able to effectively remove the arsenic from polluted water sources but also be able to operate at low operating pressure with minimum maintenance cost. The membranes also have huge potential to meet the MCL of arsenic concentration in drinking water as required by law by producing permeate containing <10 ppm arsenic. With a simple desorption process using either NaOH solution or NaOH/NaOCl mixture, an excellent recovery rate on the adsorption capacity of arsenic-loaded membranes could be practically achieved. However, it has to be pointed out that MMMs in the hollow fiber configuration are more favored than flat sheet ones, mainly because of the larger membrane area per volume and better flexibility in the hollow fiber module.

REFERENCES

Ahmed, M. F. 2001. An overview of arsenic removal technologies in Bangladesh and India. In: *BUET-UNU International Workshop on Technologies for Arsenic Removal from Drinking Water*, Dhaka, May 5–7, pp. 251–269.

Ali, I., T. A. Khan, and M. Asim. 2011. Removal of arsenic from water by electrocoagulation and electrodialysis techniques. *Separation and Purification Reviews* 40(1): 25–42.

Amin, M. N., S. Kaneco, T. Kitagawa, A. Begum, H. Katsumata, T. Suzuki, and K. Ohta. 2006. Removal of arsenic in aqueous solutions by adsorption onto waste rice husk. *Industrial and Engineering Chemistry Research* 45(24): 8105–8110.

Arcibar-Orozco, J. A., D.-B. Josue, J. C. Rios-Hurtado, and J. R. Rangel-Mendez. 2014. Influence of iron content, surface area and charge distribution in the arsenic removal by activated carbons. *Chemical Engineering Journal* 249: 201–209.

Atkinson, S. 2006. Filtration technology verified to remove arsenic from drinking water. *Membrane Technology* 2006(3): 8–9.

Beolchini, F., F. Pagnanelli, I. De Michelis, and F. Vegliò. 2006. Micellar enhanced ultrafiltration for arsenic (V) removal: Effect of main operating conditions and dynamic modelling. *Environmental Science and Technology* 40(8): 2746–2752.

Bhattacharya, P., A. H. Welch, K. G. Stollenwerk, M. J. McLaughlin, J. Bundschuh, and G. Panaullah. 2007. Arsenic in the environment: Biology and chemistry. *Science of the Total Environment* 379(2): 109–120.

Bilici Baskan, M. and A. Pala. 2010. A statistical experiment design approach for arsenic removal by coagulation process using aluminum sulfate. *Desalination* 254(1–3): 42–48.

Brandhuber, P. and G. Amy. 1998. Alternative methods for membrane filtration of arsenic from drinking water. *Desalination* 117(1): 1–10.

Chang, F., J. Qu, H. Liu, R. Liu, and X. Zhao. 2009. Fe–Mn binary oxide incorporated into diatomite as an adsorbent for arsenite removal: Preparation and evaluation. *Journal of Colloid and Interface Science* 338(2): 353–358.

Chakravarty, S., V. Dureja, G. Bhattacharyya, S. Maity, and S. Bhattacharjee. 2002. Removal of arsenic from groundwater using low cost ferruginous manganese ore. *Water Research* 36(3): 625–632.

Chatterjee, S. and S. De. 2014. Adsorptive removal of fluoride by activated alumina doped cellulose acetate phthalate (CAP) mixed matrix membrane. *Separation and Purification Technology* 125: 223–238.

Chiban, M., M. Zerbet, G. Carja, and F. Sinan. 2012. Application of low-cost adsorbents for arsenic removal: A review. *Journal of Environmental Chemistry and Ecotoxicology* 4(5): 91–102.

Choong, T. S., T. Chuah, Y. Robiah, F. Gregory Koay, and I. Azni. 2007. Arsenic toxicity, health hazards and removal techniques from water: An overview. *Desalination* 217(1): 139–166.

Cumbal, L. and A. K. SenGupta. 2005. Arsenic removal using polymer-supported hydrated iron (III) oxide nanoparticles: Role of Donnan membrane effect. *Environmental Science and Technology* 39(17): 6508–6515.

Deliyanni, E. A., D. N. Bakoyannakis, A. I. Zouboulis, and K. A. Matis. 2003. Sorption of As(V) ions by akaganéite-type nanocrystals. *Chemosphere* 50(1): 155–163.

Deliyanni, E. A., E. N. Peleka, and K. A. Matis. 2007. Removal of zinc ion from water by sorption onto iron-based nanoadsorbent. *Journal of Hazardous Materials* 141(1): 176–184.

Deschamps, E., V. S. T. Ciminelli, and W. H. Höll. 2005. Removal of As(III) and As(V) from water using a natural Fe and Mn enriched sample. *Water Research* 39(20): 5212–5220.

DeSesso, J., C. Jacobson, A. Scialli, C. Farr, and J. Holson. 1998. An assessment of the developmental toxicity of inorganic arsenic. *Reproductive Toxicology* 12(4): 385–433.

Driehaus, W., M. Jekel, and U. Hildebrandt. 1998. Granular ferric hydroxide—A new adsorbent for the removal of arsenic from natural water. *Journal of Water Supply: Research and Technology—AQUA* 47(1): 30–35.

Eguez, H. E. and E. H. Cho. 1987. Adsorption of arsenic on activated charcoal. *Journal of Metals* 39(7): 38–41.

EPA. 2000. Technologies and costs for removal of arsenic from drinking water. United States Environmental Protection Agency. http://nepis.epa.gov/Exe/ZyPURL.cgi?Dockey=P1004WDI.txt

Figoli, A., A. Cassano, A. Criscuoli, M. S. I. Mozumder, M. T. Uddin, M. A. Islam, and E. Drioli. 2010. Influence of operating parameters on the arsenic removal by nanofiltration. *Water Research* 44(1): 97–104.

Fu, F. and Q. Wang. 2011. Removal of heavy metal ions from wastewaters: A review. *Journal of Environmental Management* 92(3): 407–418.

Ghosh, M. M. and J. R. Yuan. 1987. Adsorption of inorganic arsenic and organoarsenicals on hydrous oxides. *Environmental Progress* 6(3): 150–157.

Guo, X., Y. Du, F. Chen, H.-S. Park, and Y. Xie. 2007. Mechanism of removal of arsenic by bead cellulose loaded with iron oxyhydroxide (FeOOH): EXAFS study. *Journal of Colloid and Interface Science* 314(2): 427–433.

Gupta, K. and U. C. Ghosh. 2009. Arsenic removal using hydrous nanostructure iron(III)–titanium(IV) binary mixed oxide from aqueous solution. *Journal of Hazardous Materials* 161(2–3): 884–892.

Hao, L., T. Zheng, J. Jiang, G. Zhang, and P. Wang. 2016. Removal of As(III) and As(V) from water using iron doped amino functionalized sawdust: Characterization, adsorptive performance and UF membrane separation. *Chemical Engineering Journal* 292: 163–173.

Harisha, R. S., K. M. Hosamani, R. S. Keri, S. K. Nataraj, and T. M. Aminabhavi. 2010. Arsenic removal from drinking water using thin film composite nanofiltration membrane. *Desalination* 252(1–3): 75–80.

Harper, T. R. and N. W. Kingham. 1992. Removal of arsenic from wastewater using chemical precipitation methods. *Water Environment Research* 64(3): 200–203.

He, J., T. Matsuura, and J. P. Chen. 2014a. A novel Zr-based nanoparticle-embedded PSF blend hollow fiber membrane for treatment of arsenate contaminated water: Material development, adsorption and filtration studies, and characterization. *Journal of Membrane Science* 452: 433–445.

He, J., T.-S. Siah, and J. Paul Chen. 2014b. Performance of an optimized Zr-based nanoparticle-embedded PSF blend hollow fiber membrane in treatment of fluoride contaminated water. *Water Research* 56: 88–97.

Henglein, A. 1989. Small-particle research: Physicochemical properties of extremely small colloidal metal and semiconductor particles. *Chemical Reviews* 89(8): 1861–1873.

Hering, J., P. Chen, J. Wilkie, and M. Elimelech. 1997. Arsenic removal from drinking water during coagulation. *Journal of Environmental Engineering* 123(8): 800–807.

Hossain, M. 2006. Arsenic contamination in Bangladesh—An overview. *Agriculture, Ecosystems and Environment* 113(1): 1–16.

Hua, M., S. Zhang, B. Pan, W. Zhang, L. Lv, and Q. Zhang. 2012. Heavy metal removal from water/wastewater by nanosized metal oxides: A review. *Journal of Hazardous Materials* 211–212(0): 317–331.

Iqbal, J., H.-J. Kim, J.-S. Yang, K. Baek, and J.-W. Yang. 2007. Removal of arsenic from groundwater by micellar-enhanced ultrafiltration (MEUF). *Chemosphere* 66(5): 970–976.

Jahan, M. I., M. A. Motin, M. Moniuzzaman, and M. Asadullah. 2008. Arsenic removal from water using activated carbon obtained from chemical activation of jute stick. *Indian Journal of Chemical Technology* 15(4): 413–416.

Jain, C. and I. Ali. 2000. Arsenic: Occurrence, toxicity and speciation techniques. *Water Research* 34(17): 4304–4312.

Jamshidi Gohari, R., W. J. Lau, E. Halakoo, A.F. Ismail, F. Korminouri, T. Matsuura, M. S. Jamshidi Gohari, and Md. N. K. Chowdhury. 2015. Arsenate removal from contaminated water by a highly adsorptive nanocomposite ultrafiltration membrane. *New Journal of Chemistry* 39: 8263–8272.

Jamshidi Gohari, R., W.J. Lau, T. Matsuura, and A.F. Ismail. 2013. Fabrication and characterization of novel PES/Fe–Mn binary oxide UF mixed matrix membrane for adsorptive removal of As(III) from contaminated water solution. *Separation and Purification Technology* 118: 64–72.

Jang, M., S.-H. Min, T.-H. Kim, and J. K. Park. 2006. Removal of arsenite and arsenate using hydrous ferric oxide incorporated into naturally occurring porous diatomite. *Environmental Science and Technology* 40(5): 1636–1643.

Kartinen Jr, E. O. and C. J. Martin. 1995. An overview of arsenic removal processes. *Desalination* 103(1): 79–88.

Kim, J. and M. M. Benjamin. 2004. Modeling a novel ion exchange process for arsenic and nitrate removal. *Water Research* 38(8): 2053–2062.

Klein, E. 2000. Affinity membranes: A 10-year review. *Journal of Membrane Science* 179(1): 1–27.

Lenoble, V., O. Bouras, V. Deluchat, B. Serpaud, and J.-C. Bollinger. 2002. Arsenic adsorption onto pillared clays and iron oxides. *Journal of Colloid and Interface Science* 255(1): 52–58.

Li, J., H. Chang, L. Ma, J. Hao, and R. T. Yang. 2011. Low-temperature selective catalytic reduction of NO_x NH$_3$ over metal oxide and zeolite catalysts—A review. *Catalysis Today* 175(1): 147–156.

Li, X., K. He, B. Pan, S. Zhang, L. Lu, and W. Zhang. 2012. Efficient As(III) removal by macroporous anion exchanger-supported Fe–Mn binary oxide: Behavior and mechanism. *Chemical Engineering Journal* 193: 131–138.

Mandal, B. K. and K. T. Suzuki. 2002. Arsenic round the world: A review. *Talanta* 58(1): 201–235.

Marchiset-Ferlay, N., C. Savanovitch, and M.-P. Sauvant-Rochat. 2012. What is the best biomarker to assess arsenic exposure via drinking water? *Environment International* 39(1): 150–171.

Martinson, C. A. and K. J. Reddy. 2009. Adsorption of arsenic(III) and arsenic(V) by cupric oxide nanoparticles. *Journal of Colloid and Interface Science* 336(2): 406–411.

Meharg, A. A. and M. M. Rahman. 2003. Arsenic contamination of Bangladesh paddy field soils: Implications for rice contribution to arsenic consumption. *Environmental Science and Technology* 37(2): 229–234.

Mohan, D. and C. U. Pittman Jr. 2007. Arsenic removal from water/wastewater using adsorbents—A critical review. *Journal of Hazardous Materials* 142(1–2): 1–53.

Mukherjee, A., A. E. Fryar, and H. D. Rowe. 2007. Regional-scale stable isotopic signatures of recharge and deep groundwater in the arsenic affected areas of West Bengal, India. *Journal of Hydrology* 334(1): 151–161.

Mukherjee, R. and S. De. 2014a. Adsorptive removal of nitrate from aqueous solution by polyacrylonitrile-alumina nanoparticle mixed matrix hollow-fiber membrane. *Journal of Membrane Science* 466: 281–292.

Mukherjee, R. and S. De. 2014b. Adsorptive removal of phenolic compounds using cellulose acetate phthalate–alumina nanoparticle mixed matrix membrane. *Journal of Hazardous Materials* 265: 8–19.

Nabi, D., I. Aslam, and I. A. Qazi. 2009. Evaluation of the adsorption potential of titanium dioxide nanoparticles for arsenic removal. *Journal of Environmental Sciences* 21(3): 402–408.

Naranmandura, H., N. Suzuki, and K. T. Suzuki. 2006. Trivalent arsenicals are bound to proteins during reductive methylation. *Chemical Research in Toxicology* 19(8): 1010–1018.

Navarro, P. and F. J. Alguacil. 2002. Adsorption of antimony and arsenic from a copper electrorefining solution onto activated carbon. *Hydrometallurgy* 66(1): 101–105.

Nesbitt, H. W., G. W. Canning, and G. M. Bancroft. 1998. XPS study of reductive dissolution of 7Å-birnessite by H_31AsO_3, with constraints on reaction mechanism. *Geochimica et Cosmochimica Acta* 62(12): 2097–2110.

Ning, R. Y. 2002. Arsenic removal by reverse osmosis. *Desalination* 143(3): 237–241.

Niu, H., J. Wang, Y. Shi, Y. Cai, and F. Wei. 2009. Adsorption behavior of arsenic onto protonated titanate nanotubes prepared via hydrothermal method. *Microporous and Mesoporous Materials* 122(1): 28–35.

Pan, B., B. Pan, W. Zhang, L. Lv, Q. Zhang, and S. Zheng. 2009. Development of polymeric and polymer-based hybrid adsorbents for pollutants removal from waters. *Chemical Engineering Journal* 151(1): 19–29.

Pena, M. E., G. P. Korfiatis, M. Patel, L. Lippincott, and X. Meng. 2005. Adsorption of As(V) and As(III) by nanocrystalline titanium dioxide. *Water Research* 39(11): 2327–2337.

Pookrod, P., K. J. Haller, and J. F. Scamehorn. 2005. Removal of arsenic anions from water using polyelectrolyte-enhanced ultrafiltration. *Separation Science and Technology* 39(4): 811–831.

Pradeep, T. and Anshup. 2009. Noble metal nanoparticles for water purification: A critical review. *Thin Solid Films* 517(24): 6441–6478.

Rajaković, L. V. 1992. The sorption of arsenic onto activated carbon impregnated with metallic silver and copper. *Separation Science and Technology* 27(11): 1423–1433.

Ramos, M. L. P., J. A. González, S. G. Albornoz, C. J. Pérez, M. E. Villanueva, S. A. Giorgieri, and G. J. Copello. 2016. Chitin hydrogel reinforced with TiO_2 nanoparticles as an arsenic sorbent. *Chemical Engineering Journal* 285: 581–587.

Ranđelović, M., M. Purenović, A. Zarubica, J. Purenović, B. Matović, and M. Momčilović. 2012. Synthesis of composite by application of mixed Fe, Mg (hydr) oxides coatings onto bentonite—A use for the removal of Pb(II) from water. *Journal of Hazardous Materials* 199: 367–374.

Rau, I., A. Gonzalo, and M. Valiente. 2000. Arsenic(V) removal from aqueous solutions by iron(III) loaded chelating resin. *Journal of Radioanalytical and Nuclear Chemistry* 246(3): 597–600.

Reimann, C., J. Matschullat, M. Birke, and R. Salminen 2009. Arsenic distribution in the environment: The effects of scale. *Applied Geochemistry* 24(7): 1147–1167.

Ren, Z., G. Zhang, and J. Paul Chen. 2011. Adsorptive removal of arsenic from water by an iron–zirconium binary oxide adsorbent. *Journal of Colloid and Interface Science* 358(1): 230–237.

Saitua, H., R. Gil, and A. P. Padilla 2011. Experimental investigation on arsenic removal with a nanofiltration pilot plant from naturally contaminated groundwater. *Desalination* 274(1–3): 1–6.

Santra, S. C., A. C. Samal, P. Bhattacharya, S. Banerjee, A. Biswas, and J. Majumdar. 2013. Arsenic in food-chain and community health risk: A study in Gangetic West Bengal. *Procedia Environmental Sciences* 18(0): 2–13.

Shih, M.-C. 2005. An overview of arsenic removal by pressure-driven membrane processes. *Desalination* 172(1): 85–97.

Singh, D. B., G. Prasad, and D. C. Rupainwar. 1996. Adsorption technique for the treatment of As(V)-rich effluents. *Colloids and Surfaces A: Physicochemical and Engineering Aspects* 111(1–2): 49–56.

Singh, T. S. and K. Pant. 2004. Equilibrium, kinetics and thermodynamic studies for adsorption of As(III) on activated alumina. *Separation and Purification Technology* 36(2): 139–147.

Smedley, P. and D. Kinniburgh. 2002. A review of the source, behaviour and distribution of arsenic in natural waters. *Applied Geochemistry* 17(5): 517–568.

Smith, A. H., P. A. Lopipero, M. N. Bates, and C. M. Steinmaus. 2002. Arsenic epidemiology and drinking water standards. *Science* 2965576: 2145–2146.

Styles, P. M., M. Chanda, and G. L. Rempel. 1996. Sorption of arsenic anions onto poly(ethylene mercapto-acetimide). *Reactive and Functional Polymers* 31(2): 89–102.

Su, Q., B. Pan, B. Pan, Q. Zhang, W. Zhang, L. Lv, X. Wang, J. Wu, and Q. Zhang. 2009. Fabrication of polymer-supported nanosized hydrous manganese dioxide (HMO) for enhanced lead removal from waters. *Science of the Total Environment* 407(21): 5471–5477.

Szlachta, M., V. Gerda, and N. Chubar. 2012. Adsorption of arsenite and selenite using an inorganic ion exchanger based on Fe–Mn hydrous oxide. *Journal of Colloid and Interface Science* 365(1): 213–221.

Thirunavukkarasu, O. S., T. Viraraghavan, K. S. Subramanian, O. Chaalal, and M. R. Islam. 2005. Arsenic removal in drinking water—Impacts and novel removal technologies. *Energy Sources* 27(1–2): 209–219.

Tratnyek, P. G. and R. L. Johnson. 2006. Nanotechnologies for environmental cleanup. *Nano Today* 1(2): 44–48.

Trina Dutta, C. B. and S. Bhattacherjee. 2012. Removal of arsenic using membrane technology—A review. *International Journal of Engineering Research and Technology* 1(9): 1–23.

Vaaramaa, K. and J. Lehto. 2003. Removal of metals and anions from drinking water by ion exchange. *Desalination* 155(2): 157–170.

Vaishya, R. C. and S. K. Gupta. 2003. Modelling arsenic(III) adsorption from water by sulfate-modified iron oxide-coated sand (SMIOCS). *Journal of Chemical Technology and Biotechnology* 78(1): 73–80.

Vitela-Rodriguez, A. V. and J. R. Rangel-Mendez. 2013. Arsenic removal by modified activated carbons with iron hydro (oxide) nanoparticles. *Journal of Environmental Management* 114: 225–231.

Winkel, L., M. Berg, M. Amini, S. J. Hug, and C. A. Johnson. 2008. Predicting groundwater arsenic contamination in Southeast Asia from surface parameters. *Nature Geoscience* 1(8): 536–542.

Xu, Z., Q. Li, S. Gao, and J. K. Shang. 2010. As(III) removal by hydrous titanium dioxide prepared from one-step hydrolysis of aqueous $TiCl_4$ solution. *Water Research* 44(19): 5713–5721.

Yuan, T., J. Yong Hu, S. L. Ong, Q. F. Luo, and W. Jun Ng. 2002. Arsenic removal from household drinking water by adsorption. *Journal of Environmental Science and Health, Part A* 37(9): 1721–1736.

Zhang, G., H. Liu, J. Qu, and W. Jefferson. 2012. Arsenate uptake and arsenite simultaneous sorption and oxidation by Fe–Mn binary oxides: Influence of Mn/Fe ratio, pH, Ca^{2+}, and humic acid. *Journal of Colloid and Interface Science* 366(1): 141–146.

Zhang, G.-S., J.-H. Qu, H.-J. Liu, R.-P. Liu, and G.-T. Li. 2007. Removal mechanism of As(III) by a novel Fe–Mn binary oxide adsorbent: Oxidation and sorption. *Environmental Science and Technology* 41(13): 4613–4619.

Zhang, W.-X. 2003. Nanoscale iron particles for environmental remediation: An overview. *Journal of Nanoparticle Research* 5(3–4): 323–332.

Zheng, Y.-M., S.-F. Lim, and J. P. Chen. 2009. Preparation and characterization of zirconium-based magnetic sorbent for arsenate removal. *Journal of Colloid and Interface Science* 338(1): 22–29.

Zheng, Y.-M., L. Yu, D. Wu, and J. Paul Chen. 2012. Removal of arsenite from aqueous solution by a zirconia nanoparticle. *Chemical Engineering Journal* 188(0): 15–22.

Zheng, Y.-M., S.-W. Zou, K. G. N. Nanayakkara, T. Matsuura, and J. P. Chen 2011. Adsorptive removal of arsenic from aqueous solution by a PVDF/zirconia blend flat sheet membrane. *Journal of Membrane Science* 374(1–2): 1–11.

13 Treatment of Photographic Processing Waste

Irvan Dahlan, Hamidi Abdul Aziz,
Yung-Tse Hung, and Lawrence K. Wang

CONTENTS

ABSTRACT

The excellent characteristics of silver have made it a very valuable raw material in many industrial fields, especially in photographic processing industries due to its photosensitive properties. However, photographic processing industries have contributed a significant amount of silver waste to the environment. During the development of photographic processing, silver ions precipitate with many organic and inorganic compounds. The effect of silver and its fate in the environment are still controversial and remains a topic of scientific study. Nevertheless, as different degrees of toxicity are shown by different forms of silver, it is necessary to remove this metal from photographic processing wastewater. Most research and development for photographic processing waste treatment had generally emphasized on how to efficiently remove and recover the valuable silver from photographic processing waste. This chapter reviews information on the historical development of the photographic processing industry, process description of black and white paper/film, development of color processes, and sources and as well as characteristics of photographic processing waste. Silver criteria in the environment were also briefly discussed. This is followed by the main discussion of this chapter, namely treatment methods and recovery techniques of silver from photographic processing waste.

13.1 INTRODUCTION

Photography is one of the most noteworthy discoveries of the nineteenth century. Different practical works related to photography were discovered long before the first photographs were made. Subsequently, a major role has been played by photographic processing industries around the world. With the rapid development of photographic processing industries, especially before the digital revolution period, many innovations have been made particularly related to the amount of energy, silver, and water needed to process photographic materials. The photographic processing industry is dependent heavily on silver, since it was found that silver has unique properties and this light-sensitive compound has the capability of forming photographic images.

The main environmental concerns associated with photographic processing are silver and other organic chemicals. Purcell and Peters (1) reported that the major sources of silver in the environment are wastes from photographic and imaging materials and processing. Besides, nonphotographic industries also contribute in generating silver waste to the environment such as in production of electronic equipment/materials, bearings, silverware, jewelry, batteries and catalysts (silver oxide), alloys and solders (silver chloride), electroplating (silver cyanide), mirrors (silver nitrate), and in some medical and dental applications (excluding x-ray processing).

Silver is dissolved in various waste streams during the development of photographic processing. This photographic processing waste has been taken into consideration not only by the government or regulatory agencies, but also by the community due to its hazardous characteristics. Since a large amount of silver cannot be removed from wastewater by municipal sewage treatment plants, other methods are required. A number of different methods have been used to treat and recover silver from photographic processing facilities. However, it should be noted that the specific method used will ultimately depend on several factors such as environmental and ecological processes, the price of silver, capital and operating costs for its recovery from waste solutions, experience and knowledge with a specific removal system, and practical operational considerations (2).

This chapter seeks to provide information on the historical development of the photographic processing industry. Then, an overview of process description of black and white paper/film, development of color processes, and sources and characteristics of photographic processing waste is given. It also includes information on the silver criteria in the environment which is very important in verifying the silver form and predicting the undesirable effects of silver. This is then followed by the discussion of a number of methods used to treat and recover silver from photographic processing waste which is the core of this chapter.

13.2 HISTORICAL DEVELOPMENT OF THE PHOTOGRAPHIC PROCESSING INDUSTRY

Two components were essential in performing the work related to photography, that is, an optical and mechanical device for forming an image, and the other a chemical technique for creating a permanent record of the image (3). The optical concepts and devices had been known for thousands of years. Greek and Islamic scholars had developed a device based on those optical concepts, called a *camera obscura* ("dark chamber") to observe eclipses and other phenomena and to make illustrations (4). In 1720s, Johann Heinrich Schulze, a professor at the University of Altdorf, Germany had shown the existence of a chemical that was sensitive to light rather than to heat through unexpected observation. He discovered that silver nitrate ($AgNO_3$) in a bottle on a shelf of his lab became darkened in the presence of light. Schulze proved that light acts as a catalyst in this occurrence and his discovery became the principal aspect of all photography. Although his discovery did not provide the means of preserving an image, it provided the basis for further effort in fixing images (5).

The first photograph was created in 1827 by the French chemist, Joseph Nicephore Niépce. This first permanent photograph involved an 8-h contact to produce a positive image on a metal plate. Niépce called his invention heliography where he created an image using a substance, thinner asphalt. In this process, he covered a highly polished plate of pewter with thinner asphalt and positioned it inside the back of a camera facing out the window. The resulting image on a metal plate is small, rough but permanent and both are negative and positive depending on the lighting. This process was not practical and the procedure did not involve the use of light-sensitive silver compounds (which is the essential component in photography) (3,6).

In 1839, Louis Jacques Mandé Daguerre discovered the first "practical" photograph using a process called the daguerreotype. In this process, Daguerre used a highly polished, silver-plated piece of copper. The silver surface was treated with fumes from heated crystals of iodine to make it light sensitive and exposed to the focused image in a camera obscura. Then, the invisible latent image was formed on the plate and to develop a visible image Daguerre treated the silver plate with mercury vapor. To make the permanent image (so that the whole image did not turn black with time), the plate was immersed in a solution of hyposulfite of soda to remove the unexposed or undeveloped parts of the plate to be no longer light sensitive, and hence the image/photo would be preserved permanently. The daguerreotype was the earliest practical photographic process, and for his discovery of the first practical method of recording images in a camera, Daguerre is rightfully known as the father of photography (3).

The daguerreotype had some limitations such as the usage of toxic chemicals (e.g., mercury and iodine), a metallic (heavy) picture, and not allowing duplicate copies to be made. Therefore, in the 1840s, William Henry Fox Talbot developed the idea of coating a piece of high-quality writing paper with a light-sensitive silver compound (i.e., silver nitrate, potassium iodide, and gallic acid solutions) and placed it into a camera with the sensitized side facing the lens. The permanent image was made by immersing it in a solution of hyposulfite of soda, followed by washing. The developed image paper was not the end product of a photograph, but rather a negative. By this technique, Talbot was able to use and develop the negative print to produce multiple copies of positive prints, which could not be achieved in the daguerreotype process. The final image (positive print) was relatively different from a daguerreotype. The process for both stabilized negative and the following positive print was called photogenic drawing. Talbot's invention was called calotypes (from the Greek word *kalos*, meaning "beautiful") (3,6). Since then, many scientists continued the work to discover new photographic materials and methods to produce photographs that were viable commercially, and later created the new profession of "photographer" around the world.

In 1861, the British physicist James Clerk Maxwell invented the first durable color image by mixing red, blue, and green light in varying proportions. Maxwell's achievement gave inspiration to others in effort to capture an effective color image. Later, a significant discovery of a black and white emulsion that was sensitive to all colors (called *panchromatic* sensitivity) took place in 1904. When color separation was involved in photography, panchromatic or color-specific emulsions were important materials. More developments in color rapidly followed in the mid-1930s and 1940s. In this period, two companies, Agfa Company and Eastman Kodak, released new color negative films and advanced printing paper that completely improved the quality of color prints, which led to the development of a commercial photofinisher. Then Eastman Kodak led to the improvement of this modern color photography, reducing significantly the amount of energy, silver, and water needed to process photographic film and papers (3,7).

Three major evolutions were involved in the historical development of color photography. In the earliest period, the reproduction of spectral distributions of light observed in the original scene was involved. Then followed the synthesis of colors in pictures through the additive mixtures of separate red, blue, and green beams of light. The third development (chromogenic period), which was the greatest current interest, involved the invention of dye-forming couplers. In this evolution, the use of cyan, magenta, and yellow colorants in a single beam of white light, known as subtractive systems, became essential to obtain a color photographic image in a single processing step (8,9). Nevertheless, in the beginning of the 2000s, the digital revolution has started to lead modern photography to reduce the need for conventional processing methods and this remains until today.

Digital photography became the dominant system for various imaging applications due to simplicity, automation, and optimum results.

13.3 DESCRIPTION OF THE PHOTOGRAPHIC PROCESSING INDUSTRY

The photographic processing industry comprised of individuals and companies that offer photographic services which included developing and photo finishing, such as photographic processing laboratories, graphic arts film processors, and medical imaging (x-ray) processors. The manufacturing of photographic films and papers related to medical imaging (to diagnose medical problems, identify structural defects, document, record, and transfer information, and to preserve memories) can be found at only a few sites around the world, while the processing of these materials can be found at numerous facilities. Around 90% of the total number of photographic processing facilities is small and medium sized, which include small hospitals, doctors, dentists, veterinarians' and chiropractors' offices, neighborhood clinics, schools, portrait studios, minilabs, custom labs, professional processing labs, small microfilm facilities, printers, motion picture processors, and a large number of municipal, state, and federal facilities where some in-house photographic processing is prepared. Large photographic processors represent about 9% of all photographic processing facilities (10).

The photographic processing industry falls under Standard Industrial Classification (SIC) codes 7221, 7335, 7336, and 7384 (code developed by the U.S. Department of Labor) (11). All of the photographic processing companies use many types of films and paper products in the preparation of printed materials using their own specific photographic processing solutions. The developing agent in the photographic processing industry is a common material. For instance, numerous type of black and white films (which still exists commercially) are made based on Eastman Kodak technologies through a number of photographic processing solutions recognized as D-76, HC-110, DK-50, and D-19. Eastman Kodak Co. also formulated the photographic processing solutions for both color negative films and color paper known as C-22/C-41 and EP-2/RA-4, respectively (12,13). The photographic processing solutions varies between manufacturers depending on the particular quality of their product since all manufacturers want to market their own solutions as well as their films to make as much profit as possible. The historical development of the photographic processing industry showed continuous modernization with the aim of reducing energy and water consumption during film/photography processing to lower costs and consequently obtain greater revenues. By late 1970s, photographic processing industries (especially Eastman Kodak, Konica, Agfa, and Fuji) decided to use similar color processes and chromogenic methods (C-22/C-41 and EP-2/RA-4) developed by Eastman Kodak Co. In actual fact, currently most of the photographic processing industries worldwide use this process/method (13).

According to the 2002 Economic Census, there were about 4791 photofinishing facilities (*NAICS 81292*) in the United States, not including the establishment of photographic services (*NAICS 54192*). This number decreased to 1915 in 2007, which mean that the number of establishments in this industry decreased 61% between 2002 and 2007. Meanwhile, the annual revenue growth declined steadily from USD3.9 billion in 2002 to USD2.1 billion in 2007, mainly due to the rise of digital photography for various professional and popular imaging applications that remains so nowadays. And based on more recent available data from the U.S. Economic Census, the revenue growth for photofinishing continued to decrease to USD1.9 billion in 2009 (14–16).

A study conducted by the U.S. Environmental Protection Agency (U.S. EPA) of the U.S. photographic processing industry at its highest production rate (1994–1996) shows that about 296 million ft^2 or 700 million rolls of color film (C-22/C-41 method) and 4130 million ft^2 of color paper (EP-2/RA-4 method) were processed per year. These facts comprise the retail photographic processing stores (drugstores and independent retail facilities, attributed to 20%–30% of the total market) and wholesale laboratories (which were more industrialized and centralized facilities). In addition, numerous photographic processing facilities were available across the country and the majority of these commercial photographic film and paper facilities were considered small in size with less than

10 employees. Form this study, it was also shown that in order to process 716 million rolls of film, photographic processing industries were releasing approximately 2260 million gallons of wastewater in 1994 that declined to 1840 million gallons of wastewater in 2003 (17). Although the wastewater released from the photographic processing industry was declining, water usage has always been a major historical concern, especially in early generations of color film and color paper.

13.4 PROCESS DESCRIPTIONS, SOURCES, AND CHARACTERISTICS OF PHOTOGRAPHIC PROCESSING WASTE

13.4.1 PROCESS DESCRIPTIONS OF PHOTOGRAPHIC PROCESSING

During the history of the photographic processing industry, various methods and processes were used. The processing of photographic film and paper involves the usage of chemicals to develop and produce finished photographic materials which include acids, bases, and salts. Currently, commercial photographic films and papers are based on photoactive chemicals, that is, silver halide compounds, such as silver chloride, silver bromide, or silver iodide. These compounds consist of an emulsion of fine crystals of a light-sensitive material suspended in a gelatin on a substrate of film or paper (18). Principally, a silver halide can be prepared by combining either a halide or a halogen with elemental silver or silver nitrate. Halogens are the elements of chlorine (Cl), bromine (Br), iodine (I), etc. from the periodic table, while halides are a combination of one of these halogens with an elemental metal like cadmium (Cd), sodium (Na), or potassium (K) (8). When exposed to light, the ionic silver will reduce to elemental or metallic silver. This reduction depends on the intensity of light and exposure time. The reduced silver creates a latent image on the exposed film or paper which is then enhanced to a visible image and preserved in the processing steps (18).

Currently, photographic processing is dominated by color print film, prints, and slides. Only about 10% of the market involves black and white films and papers. An increasing portion of the color market is being taken by minilabs, which are automated machines that occupy little space. These machines are the ones used by the popular 1-h developing centers (19). Color photographic processing materials are comprised of three superimposed silver halide-gelatin layers sensitive to red, blue, and green spectral regions. With subsequent exposures to colored light different silver images are formed in each layer. Basically, photographic processing is the process of transforming the latent image on photographic emulsions into pictures. In order to accomplish this, there are three processing steps involved in the development of photographic films and papers: (1) developing the image, (2) fixing the image, and (3) stabilizing of the image by washing residual processing chemicals out of the emulsion layer with water or a stabilizing solution (12,18,20).

13.4.1.1 Developing the Image

The development of the image occurred near the center of the exposed areas, whereby the latent image in an exposed emulsion was converted into a visible image by converting the silver ions into black metallic silver. This is performed by bathing the emulsion in a solution of developer containing a reducing agent. The most popular agent used in developing solutions is hydroquinone, an organic compound which actually makes the latent image visible. The grains inside a latent image are blackened whereas other areas remain unaffected as silver bromine. The developing solution is usually has a pH between 8 and 12. Nevertheless, developing agents are imperfect solutions, therefore other components need to be added in the process of developing an image, such as

1. Activator, a strong base (alkaline) solution, used for increasing/maintaining the pH of agents in an alkaline condition since the agents do not work well in a neutral solution. Common activators are sodium hydroxide, sodium borate, and sodium carbonate.
2. Preservative used to prevent/reduce the oxidation process (developing agents lose their electrons when exposed to oxygen). Sodium sulfite is generally used as a preservative.

3. Restrainers/anti-foggants used to protect the unexposed compounds from the action (developmental fog or chemical fog) of the developing agent. Potassium bromide is generally added to the developer solution.
4. Stop bath used to terminate the developing action of the image, and the remaining silver removed using the fixer. A typical stop bath consists of water and acetic acid. Stop baths are used to prevent the image become darker when the emulsion is stored too long in a developer. Besides stopping the developing action, acetic acid helps to save the fixing bath by changing the pH of the solution from basic to acid (12,18).

Processing black and white films and papers is less complex than color materials. For color photographic processing, different developers are needed. The most popular color developers used in both color negative films and color papers are generally based on *p*-phenylenediamines compounds. Since different dyes are needed for color photographic processing (due to different applications), following the image development, a bleaching step is needed to oxidize and remove the developed metallic silver as soluble silver ions. In the past, potassium bromide and potassium ferrycyanate were used in the bleaching step, however, today a silver-dye bleach process (known as the Cibachrome process) is used to obtain sharper images (12,18).

13.4.1.2 Fixing the Image

The image after development must be "fixed." In the fixing process, some or all of the remaining silver halides (photosensitive material that could fade the image over time) are removed. Typically, the commercial fixers (chemicals) used are sodium thiosulfate (hypo) in black and white processes and ammonium thiosulfate in color processes. Besides, sodium hyposulfite is also being used but to a lesser extent. These chemicals are used selectively to dissolve silver halides without influencing the adjacent metallic image (18). The characteristic of fixer solutions can be neutral, acidic, or alkali. Neutral fixers have a short tray life and are low priced, while acidic fixers are the most popular since they can neutralize any residual alkali from the developer. On the other hand, alkali fixers are less common in commercial applications, and more so in specific applications. However at equivalent thiosulfate concentration, alkali fixers work slightly faster than acidic fixers and are also removed faster during the final print washing (21).

The thiosulfate salts in the fixing process will react with the silver halides to form a soluble silver thiosulfate complex $(Ag(S_2O_3)_2)^{3-}$. Consequently, fixer solutions contain silver in higher concentrations than the other spent processing chemicals. Fixers must be retained at a low pH to neutralize the alkalinity of any residual developing solution carried over with the photographic media and to stop any further developing action within the emulsion layers. Acetic acid is usually used to maintain low pH. In black and white processes, the image after development is metallic silver. The nonimage areas (the residual solid silver halides) are discarded in the fixing process as a soluble silver thiosulfate complex. On the average almost 40% of the silver will remain in black and white products as the metallic image (18). The film contains a negative image of the scene after being recorded by the camera, while a positive print is prepared by exposing a photosensitive sheet of paper to a light source passing through the negative film image. Subsequently, the paper is processed using developer, stop bath, fixer, and washing (22). Figure 13.1 shows the diagram for black and white processing for both film and paper.

Whereas in color processes metallic silver is formed during the development process where the image dyes absorb the complementary color of the white light based on the subtractive principle. It is silver-dye bleach materials which produce the coloration of the final image. The bleach contains an oxidizing agent, usually an iron-complexed ethylenediaminetetraacetic acid (EDTA). After the bleach bath process, the silver halides are discarded as a soluble silver thiosulfate complex in the subsequent fixer solution. In several paper processes, the bleach and the fixer baths are combined as a single solution called bleach–fix. Almost all of the silver is removed from color films and papers during photographic processing. In a common practical process, the film is introduced into a

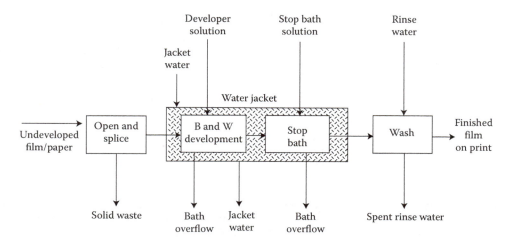

FIGURE 13.1 Development of black and white processes. (From Little, A. D. *Waste Minimization Audit Study of the Photographic Processing Industry.* Report to California Department of Health Services, Alternative Technology Section, Toxic Substances Control Division, April 1989.)

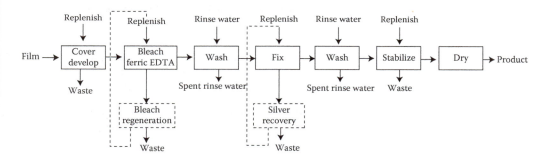

FIGURE 13.2 Process development of color negative film. (From Little, A. D. *Waste Minimization Audit Study of the Photographic Processing Industry.* Report to California Department of Health Services, Alternative Technology Section, Toxic Substances Control Division, April 1989.)

stabilizer bath (after the fixer solution) to counterbalance the emulsion and increase the stability of the dye image to light (18). A schematic color negative film processing is presented in Figure 13.2.

From the film negative recorded by the camera, the positive color print can be prepared by exposing color paper or other appropriate print media to light over the developed film. The print media contains a combination of color-sensitive emulsion layers similar to the film, which then processed through the same step of solutions in order to obtain the final color negative paper. Figure 13.3 shows the process for color paper development (22).

13.4.1.3 Stabilizing of the Image

After the fixing process, the silver image is washed with water or a stabilizing solution. The unwashed fixed image contains a significant amount of thiosulfate, which must be removed to improve the lastingness of the silver image. The main purpose of washing is to reduce the remaining thiosulfate to a concentration of 0.015 g/m^2 or less, including the small amount of concentration of soluble silver thiosulfate complexes, which otherwise remain in the paper. For an effective washing process, there are three components that need to be considered, that is, washing aid, water replenishment, and temperature. Commercial washing aids are also known as hypo-clearing agents. These agents help in removing the thiosulfate and improve washing efficiency. Hypo-clearing agents contain 2%

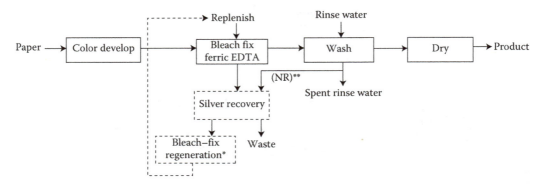

FIGURE 13.3 Process development of color negative paper. (From Little, A. D. *Waste Minimization Audit Study of the Photographic Processing Industry.* Report to California Department of Health Services, Alternative Technology Section, Toxic Substances Control Division, April 1989.)

TABLE 13.1
Recommended Washing Time for Various Photographic Media

Media	Without Hypo-Clearing Agent (min)	With Hypo-Clearing Agent (min)
Films	20–30	5
Resin-coated paper	4	Not recommended
Single-weight fiber-based paper	60	10
Double-weight fiber-based paper	120	20

Source: Reprinted from *Focal Encyclopedia of Photography*, 4th Edition, Williams, S., The chemistry of developers and the development processes, pp. 654–664, Copyright 2007, with permission from Elsevier.

of sodium sulfite and a low concentration of citrate and sulfate salts. Nevertheless, over-washing is a risk with some resin-coated papers, and therefore the use of a washing aid is not recommended for this type of paper. Water replenishment is important for even and thorough washing. The usage of a multi-slot vertical print washer is recommended when involve washing of many images at the same time, however the correct water flow rate must be controlled properly. The recommended water temperature is within a range of 20–27°C since washing efficiency increases with water temperature. Washing temperatures higher than 27°C will soften the emulsion beyond safe print handling. On the contrary, when it is impossible to heat the wash water (below 20°C), the washing time needs to be increased. Moreover, testing needs to be carried out to verify the washing efficiency. Washing temperature lower than 10°C must be avoided (21). In addition, washing time is also another important factor to be considered based on the need for archival products and the type of medium washed. The recommended length of washing times is shown in Table 13.1 for various photographic media with or without a hypo-clearing agent.

13.4.2 SOURCES AND CHARACTERISTICS OF PHOTOGRAPHIC PROCESSING WASTE

Photographic processing and its related activities generate a complex mixture of wastewater which contains organic and inorganic compounds. Liquid effluents are the major wastes generated by this industry. The waste streams generated vary widely depending on the type and volume of photographic processing. Color photographic processing industries generate a larger waste stream volume since they represent a larger production volume of the operations at a given location. This wastewater can be categorized as process bath wastes, color developer wastes, and bleach/fix/bleach–fix wastes (24). In addition, spent rinse water is also generated by photographic processing industries (19).

Silver and organic waste chemicals are the major component of wastewater generated from photographic processing. Silver, in different forms and different concentrations, has become one of the main environmental concerns. The wastewater generated from photographic processing, its constituents, and the associated environmental concerns are presented in Table 13.2. Based on the silver wastes generated, photographic processing industries are categorized into four groups. Table 13.3 shows the type of industrial photographic processing together with total wastewater generated as well as the recovery efficiency with which the industry needs to comply. According to Eisler (25) approximately 2.47 million kg of silver are discharged each year to the environment, mostly (82%) as a result of human activities. The photographic processing industry accounts for about 47% of all silver discharged into the environment from anthropogenic sources. It was estimated that about 150,000 kg of silver enter the aquatic environment every year from the photographic processing industry, mine tailings, and electroplaters.

The silver released from photographic processing facilities mainly in the form of silver-thiosulfate complexes as listed in Table 13.4. From used film-fixer solutions alone, it can contain significant amounts of silver higher than 3000 mg/L (1,27), and this concentration value can fetch up to 6000 mg/L (28). According to Huang et al. (28), a typical black and white photographic processing laboratory generates approximately 1 gal of used fixer/day. From the fixer bath process, the silver concentrations generated were in the range of 1000–10,000 mg/L, but the typical value was reported as 5000 mg/L. The residual silver was also found in the wash baths which was considerably more difficult to recover. Generally, the residual concentration of silver in the wash bath ranges from 1 to 10 mg/L (29). In addition, photographic processing wastewater contains a high chemical oxygen demand (COD) value. For example, developing baths from x-ray photographic plates have a COD value of 150,000 mg/L due to several reducing compounds and as well as various nonpolar and polar organic compounds (30,31). According to Vengris et al. (32) the cumulative photographic processing wastewater is characterized by a high COD value range from 80,000 to 120,000 mg/L due to various organic binders, photosensitive compounds, and dyes.

TABLE 13.2
Liquid Wastes Generated from Photographic Processing

Solution	Constituents	Environmental Concerns
Prehardeners, hardeners, and prebaths	Organic chemicals, chromium compounds	Oxygen demand, toxic metals
Developers	Organic chemicals	Oxygen demand
Stop baths	Organic chemicals	Oxygen demand
Ferricyanide bleaches	Ferricyanide	Toxic chemical
Dichromate bleaches	Organic chemicals, chromium compounds	Oxygen demand, toxic metals
Clearing baths	Organic chemicals	Oxygen demand
Fixing baths	Organic chemicals, silver, thiocyanate, ammonium compounds, sulfur compounds	Oxygen demand, toxic metals, toxic chemicals, ammonia, possible H_2S generation
Neutralizers	Organic chemicals	Oxygen demand
Stabilizers	Phosphate	Bio-nutrients
Soundtrack fixer or redeveloper	Organic chemical, ammonium compounds	Oxygen demand, ammonia
Monobaths	Organic chemicals	Oxygen demand

Source: U.S. EPA. *Guides to Pollution Prevention: The Photographic Processing Industry.* U.S. Environmental Protection Agency, October 1991. http://nepis.epa.gov/Adobe/PDF/200061U8.PDF (November 2011).

TABLE 13.3

Category of Industrial Photographic Processing

Type of Industrial Photographic Processing	Total Process Effluent (GPD)	Production of Silver Rich Solutions (GPD)	Silver Recovery (Percent Efficiency)	Example of Facilities
Small-sized	<1,000	<2	≥90	Small hospitals, doctors,
Medium-sized	<10,000	2–20	≥95	dentists, veterinarians' and chiropractors' offices, neighborhood clinics, schools, portrait studios, minilabs, custom lab, professional processing labs, small microfilm facilities, printers, motion picture processors, and some in-house photographic processing in a large number of municipal, state, and federal facilities
Large-sized	>10,000	>20	≥99	Hospital, printer/graphic arts professional lab, motion picture lab
Significant industrial user (SIU)	>25,000	Facilities within SIUs • May establish a silver-discharge limit based on mass loading or flow-based concentration limits • May follow same recommendations as small, medium, or large facilities depending on category definition		Major motion picture film processors, large hospitals, and a few diagnostic clinics, commercial printers, and photofinishers

Source: AMSA and The Silver Council. *Code of Management Practice for Silver Dischargers*, 1997. http://www1.honolulu.gov/env/wwm/envquality/cmpforsilverdischargers.pdf (November 2011); Stasch, P. *Pollution Prevention and Treatment Alternatives for Silver-Bearing Effluents with Special Emphasis on Photographic Processing.* Washington State Department of Ecology, January 1997. http://www.owr.ehnr.state.nc.us/ref/05/04841.pdf (November 2011); Eastman Kodak Company. *Using the Code of Management Practice to Manage Silver in Photographic Processing Facilities.* Eastman Kodak Company, Rochester, 1999. http://www.kodak.com/ek/uploadedFiles/J217ENG.pdf (January 2012).

13.4.3 THE FATE AND EFFECT OF SILVER IN THE ENVIRONMENT

The influence of silver in photographic processing wastewater is not well identified. Factors controlling the fate of silver in the environment (including silver transformations in water and soil and as well as the role of microorganisms) are also not well characterized. The environmental fate of silver comprises a series of changes in chemical and physical properties that govern the ultimate disposition of silver in the environment (25,34). As mention earlier, several forms of silver are essential in the manufacturing and processing of photographic materials. Different degrees of toxicity are shown by different forms of silver. For freshwater fish, the acute toxicity of silver is solely due to Ag^+, interacting at the gills, inhibiting basolateral Na^+, K^+-ATPase activity causing

TABLE 13.4

Complex Formation Equilibrium for Silver and Thiosulfate

Equilibrium	pK (0.1 mol/L Ionic Strength)
$Ag^+ + S_2O_3^{2-} \leftrightarrow AgS_2O_3^-$	6.93
$Ag^+ + 2S_2O_3^{2-} \leftrightarrow Ag(S_2O_3)_2^{3-}$	12.72
$Ag^+ + 3S_2O_3^{2-} \leftrightarrow Ag(S_2O_3)_3^{3-}$	14.78
$2Ag^+ + 4S_2O_3^{2-} \leftrightarrow Ag_2(S_2O_3)_4^{6-}$	28.23
$3Ag^+ + 5S_2O_3^{2-} \leftrightarrow Ag_3(S_2O_3)_5^{7-}$	42.58
$6Ag^+ + 8S_2O_3^{2-} \leftrightarrow Ag_6(S_2O_3)_8^{10-}$	85.23

Source: Smith, R.M. and Martell, A.E. *Critical Stability Constants*, Vol. IV. Plenum Press, New York, 1976.

the osmoregulation failure of fish. Silver nitrate in freshwater is more toxic than in seawater. This difference might be due to the fact that free Ag^+ concentration (the toxic moiety in freshwater) in seawater is low. Nevertheless, high concentrations of silver nitrate are toxic to marine invertebrates although in the absence of Ag^+ this related to the bioavailability of stable sliver-chloro complexes. Silver ions were 300 times more toxic than silver chloride to fathead minnow (a species of temperate freshwater fish), 15,000 times more toxic than silver sulfide, and exceeding 17,500 times more toxic than silver thiosulfate complex. However, under these conditions (i.e., increasing pH between 7.2 and 8.6, increasing water hardness between 50 and 250 mg/L as $CaCO_3$, and increasing humic acid and copper concentrations), silver was less toxic to fathead minnow (25).

Generally, under static test conditions (i.e., low concentration of dissolved Ag^+, increasing water pH, sulfides, hardness, and dissolved and particulate loadings), silver ion is less toxic to freshwater aquatic organisms. The ability of aquatic organisms to accumulate dissolved Ag^+ ranges widely between species. A number of studies have reported that bioconcentration factors (in unit mg Ag/kg FW organism or mg Ag/L of medium) are 210 in diatoms, 240 in brown algae, 330 in mussels, 2,300 in scallops, and 18,700 in oysters. Among all trace metals, silver is the most strongly accumulated by marine bivalve mollusks. However, less data were established on the effect of silver compounds on avian or mammalian wildlife. Meanwhile, acute toxic effects of silver in humans were only occasioned mainly from accidental or suicidal overdoses of medical forms of silver (25).

The silver released from photographic processing facilities could be in the form of soluble and free to any other atoms, while in solution is known by free silver, ionic silver, and hydrated silver ion. Generally, free silver is the most toxic form. This toxicity is the basis of regulations on the release of silver compounds. Certain silver compounds release ionic silver very slowly due to very low solubility (e.g., silver sulfide) or complexation of the silver (e.g., silver thiosulfate). These compounds are over 15,000 times less toxic than silver nitrate to aquatic organisms and this correlation is illustrated in Figure 13.4 (25,34).

13.4.4 SILVER CRITERIA IN THE ENVIRONMENT

Measuring the accurate silver concentration in the environment and verifying the silver form are very important in predicting the undesirable effects. Regrettably, most measurement of silver is very difficult since silver is present in such low concentrations (parts per trillion or ppt). These measurements usually rely on scientific advances of analytical equipment. Therefore, factors governing the environmental fate of silver are not well characterized, including silver transformations in water and soil and the role of microorganisms. For example, further research is needed to verify the toxic potential of silver chloride complexes in seawater and the role of sediments as

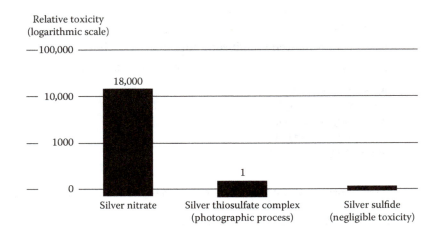

FIGURE 13.4 Toxicity of some silver compounds. (From Eastman Kodak Company. *The Fate and Effects of Silver in the Environment*. Eastman Kodak Company, Rochester, 2003. http://www.kodak.com/ek/uploadedFiles/J-216_ENG.pdf (January 2012).)

sources of silver contamination for the food web. The available data is also insufficient on correlations between tissue residues of silver with the health of aquatic organisms, therefore more exploration is needed on the inference of silver residues in tissues. In addition to that, more information is required in food chain transfer on sources and forms of silver as well as data on silver concentrations in field collections of flora and fauna, particularly close to hazardous waste sites. Correspondingly, the technology to recover silver from waste media before its escape to the environment must be improved (25).

The criteria of silver in aquatic ecosystems are now under constant amendment by regulatory agencies. The U.S. EPA has recommended silver criteria formulation by dissolved silver to replace total recoverable silver. Dissolved silver is closely related to the bioavailable fraction of silver in the water column than total recoverable silver. The recommended criteria for dissolved silver are approximately 0.85 times than total recoverable silver under specific conditions, however they may differ significantly depending on other compounds' availability. The recommended silver criteria in the environment are listed in Table 13.5. Since silver is one of the most hazardous pollutants released into the environment, a threshold limit or emission standard have been introduced, especially in the United States. The current air level of silver exposure in the United States is about 100.0 μg total Ag daily/person. The recommended threshold limit value for silver (in the air) is within a range of 0.01 to <0.1 mg total Ag/m³. While for human drinking water, the proposed silver criteria is within a range of 50.0 to <200.0 μg total Ag/L which does not appear to represent a hazard to human health. However, much lower concentrations were proposed to freshwater and marine organisms to avoid adverse effects.

As mentioned earlier, the presence of soluble silver in wastewater is a problem shared by numerous industrial processes, especially the photographic processing industry. Due to the harmful impact of silver, regulatory agencies have enacted strict regulations to limit discharge concentration of silver to 5 ppm for soluble silver (35). Also, in establishing the limitations of silver, U.S. EPA takes various factors into consideration by which silver may be discharged from a photographic processing point source. For example, the age and size of plant, raw materials, manufacturing processes, items produced, treatment technology, energy requirements, and costs. Based on the application of the best practicable control technology (BPT) currently available, the U.S. EPA has set the effluent limitations guidelines representing the degree of effluent reduction attainable. This silver concentration (effluent) from a photographic processing point source was calculated from the production normalized amounts and the average production normalized hydraulic load for the particular

TABLE 13.5
Recommended Silver Criteria

Resource, Standard, and Other Variables	Effective Silver Concentration
Agricultural Crops	
• Soils	<100.0 mg total Ag/kg dry weight soil for most species; <10.0 mg/kg for sensitive species
• Groundwater	<50.0 µg total Ag/L
Freshwater Aquatic Life	
(a) Acute exposure	
• Total recoverable silver	<1.32 µg total Ag/L
• Acid-soluble silver[a]	Four-day average not to exceed 0.12 µg/L more than once every 3 years; 1-h average not to exceed 0.92 µg/L more than once every 3 years
(b) Chronic exposure	<0.12 to <0.13 µg total recoverable Ag/L
(c) Tissue residues	
• Adverse effects on growth of the Asiatic clam, *Corbicula fluminea*	>1.65 mg total Ag/kg soft tissues, fresh weight basis
Marine Life	
(a) Acute exposure	
• Total recoverable silver	<2.3 µg/L at any time
• Acid-soluble silver[a]	Four-day average not to exceed 0.92 µg/L more than once every 3 years; 1-h average not to exceed 7.2 µg/L more than once every 3 years
(b) Tissue residues	
• Marine clams, soft part	
Normal	<1.0 mg total Ag/kg dry weight
Stressful or fatal	>100.0 mg total Ag/kg dry weight
Human Health	
(a) Air (USA)	
• Current level of exposure (nationwide)	100.0 µg total Ag daily/person
• Short-term exposure limit (15 min; up to 4 times daily with 60 min intervals at <0.01 mg Ag/m³ air)	<0.03 mg total Ag/m³
• Threshold limit value (8 h daily, 5 days weekly)	
Aerosol silver compounds	<0.01 mg total Ag/m³
Metallic silver dust	<0.1 mg total Ag/m³
(b) Diet (USA)	
• Current level of exposure	35.0–40.0 µg daily/person
(c) Drinking water	
• United States	
Long-term exposure (>10 days)	<50.0 µg total Ag/L
Proposed long-term exposure	<90.0 µg total Ag/L
Short-term exposure (1–10 days)	<1142.0 µg total Ag/L
California	<10.0 µg/L
• Germany	<100.0 µg/L

Source: Eisler, R., *Eisler's Encyclopedia of Environmentally Hazardous Priority Chemicals.* Elsevier, Amsterdam, 2007.

[a] Silver that passes through a 0.45 µm membrane after the sample has been acidified to a pH between 1.5 and 2.0 with nitric acid.

TABLE 13.6
Effluent Limitations Guidelines for Silver Based on BPT
Currently Available

	Effluent Limitations	
	Daily Maximum	30-Day Average
Parameter	(kg/1,000 m² Product)	
Silver	0.14	0.07
pH	Within the range 6.0–9.0	

Source: Code of Federal Regulations. *Title 40: Protection of Environment (Parts 425 to 699).* Part 459: Photographic point source category, Subpart A: Photographic processing subcategory (revised as of July 1, 2011).

industry (36). Table 13.6 shows the effluent limitations guidelines for silver representing the degree of effluent reduction attainable by the application of the BPT currently available.

13.5 PHOTOGRAPHIC PROCESSING WASTE TREATMENT METHODS

Various hazardous wastes are produced from photographic processing and its related activities and these wastes need to be managed appropriately to protect the environment, and as well as the safety and health of workers. Currently, the basic treatment methods used for photosensitive wastes can be categorized into several basic methods such as physical, chemical, biological, and thermal methods; including a combination between these methods. However, the most cost-effective and environmentally responsible methods is to minimize photographic processing waste using the 3Rs—reduce, reuse, and recycle—approach.

13.5.1 ADSORPTION AND ION EXCHANGE PROCESS

Adsorption is considered as one of the dominant processes in removing silver from water/wastewater. Various types of solid sorbent are being used to remove or recover silver from water/wastewater, such as clays and organic matter, kaolin, concrete particles, chitosan, and activated carbon. Begum (37) studied the adsorption of silver using concrete particles under pH and silver concentrations parameters on their chemical interactions. The adsorbent with a size fraction of 0.18–0.54 mm was prepared from concrete blocks obtained from ordinary Portland cements. It was found that at room temperature, silver removal is favored by low concentration and high pH. Silver nitrate used in these batch adsorption experiments is insoluble at pH <2 and it was assumed that the surface sites for silver are anionic and can act as weak brØnsted bases. Therefore, there is a greater concentration of H^+ to compete with Ag^+ for these anionic sites at lower pH resulting in lower silver removal. For maximum silver removal at low concentration, a greater amount of adsorbent (10 g/100 mL) was needed. The results of this study also indicated that the pH_{PZC} values (from zeta potential measurements) agree reasonably well with the calculated pH_{PZC} values. This indicates that the concrete particle adsorbent behaves amphoterically.

Resins with various functionalities have also been extensively used for removing and recovering noble metal ions including silver. In a research work carried out by Atia (38), the adsorption of silver (together with gold, Au) was studied using various resins. The resins were synthesized by polymerization of bisthiourea (BS) with formaldehyde (HCHO) at different molar ratios as shown in Table 13.7. The results showed that the uptake capacities of silver increases with increasing amount of BS

TABLE 13.7

Preparation of BS/HCHO Resins at Different Molar Ratios

Resin	Molar Ratio (BS/HCHO)	Weight (g) BS	Weight (g) HCHO	% BS in Resin	% HCHO in Resin
R1	2:3	1	0.3	87.57	12.43
R2	1:1	1	0.2	91.36	8.64
R3	3:2	1	0.133	94.07	5.91
R4	2:1	1	0.10	95.48	4.51

Source: Atia, A.A., *Hydrometallurgy*, 80, 98–106, 2005.

in the BS/HCHO resin. Among various resins prepared, resin R4 shows a maximum uptake capacity of 8.25 mmol/g. It was also shown that the adsorption of silver was controlled by the rate of the intraparticular diffusion. The results from equilibrium adsorption suggest that pore diffusion and adsorption behavior followed the monolayer Langmuir isotherm. Resin regeneration efficiency was obtained to be 95% over five cycles. Compared to commercial resins, resin R4 also showed a high efficiency toward the removal and recovery of silver from photographic processing waste.

Research has also been conducted on the use of biopolymers for removal of heavy metal ions from industrial wastewater. Chitosan is known as one of biopolymers that has the natural capability to bind various metal ions in aqueous solutions even at low metal ion concentrations. Lasko and Hurst (39) examined the effectiveness of chitosan in removing free (hydrated) silver ion as well as the ammonia, thiocyanate, thiosulfate, and cyanide complexes of silver in simulated wastewater at an initial concentration of 50 ppm and in a pH range of 2–10. A stirred-batch experiment and a column reactor were used separately to determine the ability of chitosan to bind the various forms of silver in simulated wastewater. From the results of the batch method experiments, generally positive ions were better bound at high pH where the chitosan amine groups are unprotonated and the electron pair on the amine nitrogen is available for donation to silver, while the anions were bound at low pH, where the amine group on chitosan is protonated. At pH range of 4–8, chitosan was found to be effective (80%–95%) to bind silver cations (i.e., Ag^+ and $Ag(NH_3)_2^+$) and at pH 2, 92% and 75% of $Ag(S_2O_3)_2^{3-}$ and $Ag(SCN)_3^{2-}$, respectively were bound to chitosan. Results for chitosan using the column experiment using 0.5 g of chitosan shows that 5 ppm of silver ion concentration in the effluent was obtained when treating the bed volumes (effluent volume/resin volume) of 50 ppm silver in simulated wastewater containing 160 bed volumes of Ag^+, 875 bed volumes of $Ag(NH_3)_2^+$, 715 bed volumes of $Ag(S_2O_3)_2^{3-}$, and 190 bed volumes of $Ag(SCN)_3^{2-}$. It was also found that chitosan did not successfully bind $Ag(CN)_2^-$ at any pH tested. This column study indicated that 42 mg of silver are bound per gram of chitosan used.

The use of activated carbon to selectively adsorb silver has been studied extensively based on the gold–silver cyanide complex adsorption principles. From the study conducted by Gallagher et al. (40), carbon adsorptions for Au/Ag complexes were obtained in the following order (from high to low affinity): Au halide > $Au(CN)_2^-$ > $Au[CS(NH_2)]_2^+$ > $Au_2(S_2O_3)_2^{3-}$. From various gold thiosulfates and cyanides adsorption studies, Gittins (41) and Adani (42) have made the analogy for silver adsorption in the following order of preference on carbon as $Ag(CN)_2^-$ > $AgSCN^-$ > $Ag[CS(NH_2)]_2^+$ > CH_3COOAg^+ > $Ag(NH_3)_2^+$ > $AgNO_3$ > Ag_2SO_4 > $Ag(S_2O_3)_2^{3-}$. To establish a better understanding on the effects of silver adsorption in photographic and medical x-ray process effluents, Adani et al. (43) studied the adsorption of silver from synthetic photographic and spent fix solutions on granulated activated carbon in a batch process. The photographic and medical x-ray wastes were obtained from local photography and hospital laboratories and the results of the analysis of these wastes are shown in Table 13.8. High silver adsorptions and recoveries were obtained when carbon was pretreated using 2 mol/L HNO_3 and H_2SO_4 at 25°C, respectively. High silver recoveries were also noticed in a

TABLE 13.8

Metal Ion Concentrations of Raw Medical X-Ray and Photographic Effluents

Metal Ion	Concentration (mg/L)	
	Medical X-Ray	Photographic
Ag^+	4,196	4,050
Na^+	3,809	3,050
K^+	6,471	3,500
Cu^{2+}	0.15	0.01
Ca^{2+}	29.0	18.9
Fe^{2+}	1.00	0.35
Al^{3+}	762.10	220.54
Cr^{2+}	0.46	0.031

Source: Adani, K.G., Barley, R.W., and Pascoe, R.D., *Miner. Eng.*, 18, 1269–1276, 2005.

narrow range of pH between 3 and 4. However, when the silver adsorptions were carried out under alkaline conditions very low silver recoveries were observed (less than 15%). This result shows that silver adsorptions in thiosulfate solutions are pH dependent.

From the silver recovery study, Adani et al. (43) suggested that there is a potential to purify and concentrate silver from large volumes of medical x-ray and photographic process effluents. This can be achieved through selective adsorption followed by stripping, where the results showed that the silver concentration in eluate was increased by threefold from 500 to 3250 mg/L. According to Adani et al. (43), the effective silver recovery using pretreated (HNO_3 and H_2SO_4) carbon might happened under a two-way mechanism where silver was probably adsorbed as the $H_3[Ag(S_2O_3)_2]$ complex ion. The $M_3[Ag(S_2O_3)_2]$ complex ion may probably then be reduced by substituting the M^+ for H^+ to form the $H_3[Ag(S_2O_3)_2]$ complex just before being adsorbed on the carbon. The M represents simple cations of Na^+ or NH_4^+ in photographic and medical x-ray wastes. This indicates that $3H^+$ may have substituted M^{3+} in synthetic solutions to generate $H_3[Ag(S_2O_3)_2]$ when in contact with excess HNO_3 and H_2SO_4. The NO_3^- and SO_4^{2-} ions are assumed to remain in solution and/or attracted toward the protonated carbon surface. Then $H_3[Ag(S_2O_3)_2]$ gradually moved to the protonated carbon surface and attached to the functional groups inside the carbon matrix through coulombic attraction. This was then followed by possible formation of $C_x H_3[Ag(S_2O_3)_2]$, where C_x represents one of the oxidized surface functional groups. In addition, silver also might be adsorbed through electrostatic attraction that existed between the protonated carbon surface and silver thiosulfate anions. Simplified mechanisms involving the silver thiosulfate complex anion adsorption on carbon under the alkaline and acidic conditions are shown in Figure 13.5. From illustrated mechanisms, it shows that high silver adsorptions and recoveries in thiosulfate solutions can be obtained when the carbon was pretreated in acidic condition (Figure 13.5a). Contrarily, very low silver adsorptions can be obtained under alkaline conditions (Figure 13.5b).

The use of biomass as an adsorbent for the removal of heavy metals has also been widely investigated over last three decades. In a very recent work, Saman et al. (44) reported the utilization of agricultural residue of coconut fiber (CF) as adsorbents for silver removal from aqueous and photographic waste solutions. The CF was prepared by treating the pure CF with NaOH solution. The maximum silver adsorption capacity of 0.502 and 0.612 mmol/g were obtained for pure CF and CF-NaOH, respectively. The selectivity of adsorbents was studied using liquid photographic waste. The results showed that the pure CF had high selectivity toward Fe, followed by Na, Ag, and K whereas the selectivity of CF-NaOH was high toward Na followed by Fe, Ag, and K. It was

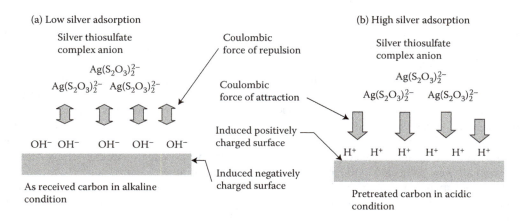

FIGURE 13.5 Mechanisms of silver adsorption on carbon under alkaline and acidic conditions. (a) Low silver adsorption and (b) high silver adsorption. (From Adani, K.G., Barley, R.W., and Pascoe, R.D., *Miner. Eng.*, 18, 1269–1276, 2005.)

observed that selectivity of CF-NaOH toward Ag and Na was almost twice higher compared with the pure CF. The results showed that the adsorption capacity and selectivity of CF-NaOH toward silver had improved compared to pure CF. In addition, numerous types of microorganisms have also been studied and were found to possess natural capabilities to remove heavy metals, including silver. Recently, Li et al. (45) studied the potential application of *Bacillus cereus* strain HQ-1 isolated from a lead and zinc mine as metal adsorbent to remove silver from aqueous solutions. The kinetics of biosorption of silver ions onto *B. cereus* biomass was studied with initial silver and biomass concentrations of 203.5 mg/L and 2 g/L, respectively. It was found that more than 90% of maximum silver biosorption uptake capacity was obtained during a 90-min experiment. To correlate the equilibrium data of the silver biosorption and to obtain the kinetic constants, pseudo first- and second-order kinetic models were used. The silver biosorption using *B. cereus* biomass is best described by pseudo second-order kinetic model. From scanning electron microscopy (SEM) and x-ray photoelectron spectroscopy (XPS) analysis, the complex chemical interaction between the silver ions and biopolymers from the cell wall and the entrapment of metals result in the crystal precipitation, which explained a possible mechanism of silver biosorption using *B. cereus* biomass.

The use of natural mineral materials, especially nanoscopic materials, has been investigated as an alternative for currently available adsorbent for removing heavy metals (including silver) from wastewater. Among these materials, halloysite nanotubes (HNTs) were found to have good adsorption capacity for the removal of silver ions from aqueous solutions. In a recent study, Kiani (46) reported the utilization of HNTs for silver ions adsorption in aqueous solutions in a batch system. It was found that the amounts of the silver ions adsorbed onto HNTs were influenced by various process variable conditions and silver ions adsorption increased with increasing initial silver ion concentration, initial pH, and temperature. The maximum adsorption capacity of 109.79 mg/g (99.8% removal) of silver ions was obtained for the initial silver concentration of 110 mg/L. This finding suggested that due to high adsorption capacity, relative low cost, and easy availability, HNTs could be used as an alternative and effective adsorbent for the removal of silver ions.

13.5.2 Biological Treatment Process

The application of a biological method for the removal of photographic processing wastewater has been studied over the last four decades. This method has been adapted to both aerobic and anaerobic treatments and does not unfavorably affect the treatment processes. However, the effect of silver on microorganisms was documented in the late nineteenth century when the processing of silver in water treatments started (47). From previous studies it was found that silver toxicity was related to free silver ion and not to the total silver concentration. Pavlostathis and Maeng (48) reported the aerobic biodegradation of a photographic processing wastewater and extractability of silver from the resulting waste using laboratory-scale activated sludge bioreactors at a volumetric loading of 40% (with an organic mixture). It was revealed that the aerobic biodegradation of the organic mixture was not affected by the photographic processing wastewater. No adverse effects on the activated sludge process were observed when the influent total silver concentration was 1.85 mg/L. On the other hand, all silver was mixed with the sludge solids, where effluent silver concentration was less than 0.01 mg/L. From this study, they also found that the resulting silver concentration was at least 40 times lower than the regulatory limit (5 mg Ag/L) when raw sludge and aerobically digested sludge solids were subjected to the toxicity characteristic leaching procedure.

In another study, Pavlostathis and Maeng (49) reported the anaerobic biodegradability of a silver compound from fixer-derived photographic processing wastewater on the anaerobic digestion process. All experiments were carried out at 35°C in the dark. It was found that the maximum biodegradability of a silver-bearing waste activated sludge (5.0 g silver/kg sludge dry solids) was 61% as compared to 59% for the control (i.e., silver-free) sludge. Both sludge samples also had a similar rate and extent of methane production. In addition to this, there was no effect on the rate and extent of methane production when silver nitrate or silver sulfide was added to mixed cultures up to 100 mg Ag/L. Hence, in this study the anaerobic digestion systems can accept a relatively high concentration of silver (at least up to 100 mg Ag/L). Based on these two studies (48,49), they concluded that biological treatment (i.e., aerobic and anaerobic) of a silver compound from photographic processing wastewater is feasible and effective without any operational problems. Consequently, this biological treatment has significant implications on the management of photographic processing wastewater.

13.5.3 ELECTROLYSIS PROCESS

Currently, the most commonly used method for silver treatment and recovery is electrolysis. Nevertheless, effluent silver concentrations from the electrolytic unit usually are in the range of 200–800 mg/L and a secondary or tailing recovery technique is required in order to meet stringent regulatory discharge limits. In addition, there are some constraints to most electrolytic equipment such as high cost (requires larger-than-normal cell size), and the fact that it may generate noxious by products (50). In the electrolysis method, it is important to achieve a suitable space–time–yield (amount of deposited metal per unit of time and cell volume) in the electrochemical reactor by providing high mass transfer conditions to the electrode or to increase the specific electrode area. Pollet et al. (51) studied how to improve the mass transfer conditions to the electrode using ultrasonic vibration applied to the electrochemical cell. They studied the effect of ultrasound on the electrochemical recovery of silver from photographic processing wastes using a newly designed electrochemical cell called SonoEcoCell. The result showed that the magnitude of the cathodic potential plays a major role in the removal of silver under silent conditions. The optimum cathodic potential of about −500 mV versus SCE (saturated calomel electrode) was obtained for the removal of silver from synthetic photographic processing solutions. It was also revealed that rotating the cylinder electrode improves the rate of silver removal and the appearance quality of the electrodeposit. In addition, combining high-power ultrasound (107 W/cm^2) with rotation also improved the rate constant for the removal of silver below 1500 rpm (face-on geometry) and 2000 rpm (side-on geometry) compared with that for rotation alone.

13.5.4 ADVANCED OXIDATION PROCESS TECHNOLOGIES

The advanced oxidation process is one of the alternative and very attractive treatment for photographic processing waste since this wastewater contains a number of reducing compounds which have strong oxygen demand. Previous investigations have shown that TiO_2/UV, O_3/UV, O_3/H_2O_2, and Fenton reactions can effectively be used for destroying organic constituents in wastewater. Few studies also have proposed advanced oxidation process for reducing the silver found in photographic processing wastewater.

The potential degradation of effluents from medical x-ray film processes by advanced oxidation process with both photo- and thermal-Fenton reactions was reported by Stalikas et al. (52). The optimization of this study was evaluated as a function of different variables, that is, $[H_2O_2]$, $[Fe^{2+}]$, pH, and temperature, where the effectiveness of this process was evaluated by observing the COD values and residual H_2O_2 concentration. The optimum pH obtained ranges from 2 and 3.5. Inhibition occurred at pH values lower than 2 due to prior complexation of Fe^{3+} with H_2O_2. In contrast, when pH increased above 3.5, the iron concentration in the solution decreased due to precipitation of Fe^{3+} to amorphous oxyhydroxides ($Fe_2O_3 \cdot nH_2O$) which then ultimately lead to ineffective incidental irradiation. The temperature also strongly affects the oxidation rate and the total conversion of the organic compounds in medical x-ray film wastewater. Complete destruction of hydrogen peroxide was detected after 24 h of treatment at temperature values above 40°C. Under optimal degradation conditions, about 97% of the COD was removed within 6 h of treatment. It is concluded from this study that the degradation of effluents from medical x-ray film processes are very effective at 60°C for both photo- and thermal-Fenton reactions and accordingly can be one of the alternatives for an *in situ* treatment.

In a recent effort, Chun-du et al. (53) investigated the degradation behavior of high-strength photographic processing wastewater using an ozone oxidation process. The average characteristics of photographic processing wastewater used in their study were as follows; 3874 mg/L COD, 464.8 mg/L BOD_5 (biochemical oxygen demand), 2.0 cyanide mg/L, 97.4 mg/L aniline, and 2.5 g/L silver. The results showed that ozone oxidation was affected by pH value and the alkaline condition is favorable compared to acid condition. From the pretreatment process, about 66% of the silver could be removed and during the ozone oxidation process about 80%, 92%, and 89% of the COD, cyanide, and aniline respectively were removed. In addition, the biodegradation of the photographic processing wastewater (BOD_5/COD) was enhanced significantly from 0.12 to 0.54.

On the other hand, TiO_2-based photocatalysis is another option used for removal of silver ions from photographic processing wastewater. However, only limited study has been done on the use of TiO_2 photocatalysis in photographic processing wastewater treatment. Herrmann et al. (54) reported the first study on the applications of UV-irradiated TiO_2 to reduce silver ions from $AgNO_3$ solution. From the results, they suggested that adsorption is required in the preliminary step of the photocatalytic process. It was also observed that there was no decrease in the reaction rate when a long-term experiment (100 h) was conducted with total silver concentrations of 0.1 M and that the silver did not saturate or block the surface of the catalyst. In addition, they successfully proposed kinetic models to explain the deposition of Ag on TiO_2.

While the study of Hermann et al. (54) only use synthetic photographic processing wastewater ($AgNO_3$ solution), Huang et al. (28) examined the UV/TiO_2-based photocatalysis for reduction of silver ions from actual commercial photographic processing wastewater. In their study, sunlight can be used directly to power the photo-electrochemical silver removal process. Their results showed that the silver ion was reduced to its metallic particles. These metallic particles were deposited on the TiO_2 catalyst and using transmission electron microscopy (TEM), their size is significantly larger than the 25 nm particle size of the TiO_2 catalyst. As their experimental results showed that particle size of silver in the spent catalyst was markedly increased with the silver loading, and the metallic silver can be separated from the TiO_2 by the physical process of sonication. The amount of silver recovered reached almost three times the TiO_2. The study also reported there was no difference in reaction rate for TiO_2 loadings of 0.1 and 0.2 wt.%. Apart from that, the role of sodium thiosulfate (the major component in the spent fixer) on the reduction of silver ions in solution was also studied. It was found that thiosulfate plays a complicated role in the reduction of silver ions; thiosulfate act as a hole scavenger, where it can increase the rate of silver removal, and also act as a stabilizer, and hinder photocatalysis when present at high concentration. The photocatalysis study incorporating a TiO_2-based catalyst could be a promising part of a photographic processing waste management in future.

13.6 RECOVERY OF SILVER FROM PHOTOGRAPHIC PROCESSING WASTE

To choose the method used to recover silver from photographic processing waste, many factors need to be considered including (a) silver-discharge limit base (regulatory), (b) economic (budget), and (c) volume of wastewater to be treated. Nevertheless, the economic and regulatory factors are the main reason for recovering the silver. Table 13.9 presents the overview of the most common types of silver recovery methods. Obviously, silver recovery processing may prove financially beneficial to the photographic processing industry, depending on the amount of waste generated.

As shown in Table 13.9, electrolysis and metallic replacement are the most common methods used for silver recovery from photographic processing wastewater. Efficiencies above 90% are easily reachable when recovering silver from black and white processing fixers. Recovering as much as 90% of silver is possible for bleach–fix and fixer solutions from color processing with conditions of higher current densities, pH adjustments, and longer contact times. In addition,

TABLE 13.9
Evaluation of Silver Recovery Methods

Method	Advantages	Disadvantages	Applications	Recovery Efficiency
Electrolysis	High silver recovery as pure metal (>90% pure silver)	Relatively high final silver concentration; may require secondary recovery; potential for sulfide formation	All photographic processing facilities except very small facilities	>90%
Chemical precipitation	Very high silver recovery (can attain 0.1 mg Ag^+/L); low investment; easy to monitor	Not available for all processes; complex operation; silver recovered as sludge; treated solution cannot be reused; potential H_2S release	Very small and large facilities	>99%
Metallic replacement	Low investment; low operating cost; simplest operation	Difficult to know when to replace; discharges iron; silver recovered as sludge; high silver concentration in effluent unless two units are in series; in some cases, not consistent; limited by some sewer codes	All photographic processing facilities	>95%
Ion exchange	High silver recovery (can attain 0.1–2.0 mg Ag^+/L)	Only for dilute influent; complex operation; high investment		~98%
Reverse osmosis	Also recovers other chemicals; purified water is recyclable	Concentrate requires further processing; high investment; high operating cost		~90%
Evaporation	Minimum aqueous effluent; water conservation	High energy requirement; silver recovered as a sludge; organic contaminant buildup; potential air emissions		~90%

Source: U.S. EPA. *Guides to Pollution Prevention: The Photographic Processing Industry.* U.S. Environmental Protection Agency, October 1991. http://nepis.epa.gov/Adobe/PDF/200061U8.PDF (November 2011); Eastman Kodak Company. *Recovering Silver from Photographic Processing Solutions.* Eastman Kodak Company, Rochester, 1999. http://www.kodak.com/ek/uploadedFiles/J215ENG.pdf (November 2011).

the secondary recovery process is required to reduce silver concentrations below 5 mg/L. As in the metallic replacement method, an oxidation–reduction reaction occurred between silver in solution with active solid metal. The active solid metal (such as iron particles, iron impregnated resin, or steel wool) is added to a commercially available unit known as a metallic replacement cartridge (MRC), silver recovery cartridge (SRC), or chemical recovery cartridge (CRC). A series arrangement of cartridges will be placed to recover more than 99% of silver from silver-rich solutions.

The other potential methods are chemical precipitation, ion exchange, reverse osmosis, and evaporation. Generally, these methods can meet stringent regulatory discharge limits, however capital and operating costs prevent these methods to be practiced by most small photographic processing facilities. Although above 99% of silver (in terms of sludge) can be recovered from silver-rich solutions, chemical precipitation is not a common method since it required chemicals and skilled employees. It is also not easy working with the ion exchange process since it is only effective at low silver concentrations as the resin is quickly saturated at high silver concentrations. Compared to most other methods, reverse osmosis involves high capital investment. Reverse osmosis requires high pressure/energy and faces some challenging issues including fouling of the membrane and biological growth. This is why the reverse osmosis process did not receive much attention in photographic processing waste treatment.

13.7 CONCLUSIONS

For decades photographic processing industries have played an important role in our daily life. Since photographic processing and its related activities generate a significant amount of waste, therefore photographic processing waste became an important issue not only from the point of waste treatment but also from the recovery of precious metals, especially silver. Many methods have been applied to treat and recover silver from photographic processing facilities. Although electrolysis and metallic replacement are among the most common and very efficient methods, other technologies used to treat and recover silver need to be improved so as to fulfill stringent regulatory discharge limits set by regulatory agencies (55,56).

ACKNOWLEDGMENTS

The authors wish to acknowledge the financial support from Universiti Sains Malaysia under Iconic Grant Scheme (A/C. 1001/CKT/870023) for research associated with the Solid Waste Management Cluster (Science and Engineering Research Centre, Engineering Campus, USM).

REFERENCES

1. Purcell, T.W. and Peters, J.J. Sources of silver in the environment. *Environmental Toxicology and Chemistry*, 17, 4, pp. 539–546, 1998.
2. Zouboulis, A.I. Silver recovery from aqueous streams using ion flotation. *Minerals Engineering*, 8, 12, pp. 1477–1488, 1995.
3. Sandler, M.W. *Photography: An Illustrated History*. Oxford University Press, New York, 2002.
4. Krebs, R.E. *Groundbreaking Scientific Experiments, Inventions and Discoveries of the Middle Ages and the Renaissance*. Greenwood Press, London, 2004.
5. Hodgson, A. Silver halide materials: General emulsion properties. In: *Focal Encyclopedia of Photography*, 4th Edition, Peres, M.R. (Ed.). Elsevier, Oxford, pp. 641–649, 2007.
6. Osterman, M. The technical evolution of photography in the 19th century. In: *Focal Encyclopedia of Photography*, 4th Edition, Peres, M.R. (Ed.). Elsevier, Oxford, pp. 27–36, 2007.
7. Coote, J.H. *The Illustrated History of Colour Photography*. Fountain Press, Surrey, 1993.
8. Osterman, M. Introduction to photographic equipment, processes, and definitions of the 19th century. In: *Focal Encyclopedia of Photography*, 4th Edition, Peres, M.R. (Ed.). Elsevier, Oxford, pp. 36–123, 2007.

9. Kapecki, J.A. Color photography. In: *Focal Encyclopedia of Photography*, 4th Edition, Peres, M.R. (Ed.). Elsevier, Oxford, pp. 692–700, 2007.

10. AMSA and The Silver Council. *Code of Management Practice for Silver Dischargers*, 1997. http://infohouse.p2ric.org/ref/02/01994.pdf (July 2016).

11. U.S. Department of Labor. *Standard Industrial Classification (SIC) Codes*. http://www.osha.gov/pls/imis/sic_manual.html (December 2011).

12. Rogers, D. *The Chemistry of Photography-From Classical to Digital Technologies*. RSC Publishing, Cambridge, 2007.

13. Delaveau, B. *Technology in Photography: A Case Study toward Simplicity and Environmental Responsibility*, 2009. http://www.benoitdelaveau.com/ENVS250_Final_paper.pdf (December 2011).

14. U.S. Census Bureau. *Service Annual Survey 2007*. U.S. Department of Commerce, Economics and Statistics Administration, March 2008. http://www2.census.gov/services/sas/data/Historical/sas-07.pdf (December 2011).

15. U.S. Census Bureau. *Service Annual Survey 2009*. U.S. Department of Commerce, Economics and Statistics Administration, February 2011. http://www2.census.gov/services/sas/data/Historical/sas-09.pdf (December 2011).

16. U.S. Census Bureau. *Industry Snapshots for Other Services—By Industry*. http://thedataweb.rm.census.gov/TheDataWeb_HotReport2/econsnapshot/2012/snapshot.hrml?NAICS=81292 (July 2016).

17. Matuszko, J. *Memorandum to Public Record for the 2006 Effluent Guidelines Program Plan. Re: Photographic Processing* (DCN 02096), EPA Docket Number OW-2004–0032 U.S. EPA, August 2005. http://www.regulations.gov/#!documentDetail;D=EPA-HQ-OW-2004-0032-0695;oldLink=false (February 2012).

18. Stasch, P. *Pollution Prevention and Treatment Alternatives for Silver-Bearing Effluents with Special Emphasis on Photoprocessing*. Washington State Department of Ecology, January 1997. http://infohouse.p2ric.org/ref/05/04841.pdf (July 2016).

19. U.S. EPA. *Guides to Pollution Prevention: The Photographic Processing Industry*. U.S. Environmental Protection Agency, October 1991. http://nepis.epa.gov/Adobe/PDF/200061U8.PDF (November 2011).

20. DOE. *DOE Industry Profile: Profile of Miscellaneous Industries, Incorporating Charcoal Works, Dry-Cleaners, Fibreglass and Fibreglass Resins Manufacturing Works, Glass Manufacturing Works, Photographic Processing Industry, Printing and Bookbinding Works*. Department of Environment, 1996. http://dclg.ptfs-europe.com/AWData/Library1/Departmental%20Publications/1996/Profile%20of%20Miscellaneous%20industries.pdf (July 2016).

21. Lambrecht, R.W. From exposure to print: The essentials of silver-halide photography required for long-lasting, high-quality prints. In: *Focal Encyclopedia of Photography*, 4th Edition, Peres, M.R. (Ed.). Elsevier, Oxford, pp. 664–692, 2007.

22. Little, A. D. *Waste Minimization Audit Study of the Photographic Processing Industry*. Report to California Department of Health Services, Alternative Technology Section, Toxic Substances Control Division, April 1989.

23. Williams, S. The chemistry of developers and the development processes. In: *Focal Encyclopedia of Photography*, 4th Edition, Peres, M.R. (Ed.). Elsevier, Oxford, pp. 654–664, 2007.

24. Freeman, H.M. *Hazardous Waste Minimization*. McGraw-Hill, New York, 1990.

25. Eisler, R. Silver. In: *Eisler's Encyclopedia of Environmentally Hazardous Priority Chemicals*, Chapter 29. Elsevier, Amsterdam, pp. 761–782, 2007.

26. Eastman Kodak Company. *Using the Code of Management Practice to Manage Silver in Photographic Processing Facilities*. Eastman Kodak Company, Rochester, 1999. http://www.kodak.com/ek/uploaded-Files/J217ENG.pdf (January 2012).

27. Pethkar, A.V. and Paknikar, K.M. Thiosulfate biodegradation-silver biosorption process for the treatment of photofilm processing wastewater. *Process Biochemistry*, 38, pp. 855–860, 2003.

28. Huang, M., Tso, E., Datye, A.K., Prairie, M.R., and Stange, B.M. Removal of silver in photographic processing waste by TiO_2-based photocatalysis. *Environmental Science and Technology*, 30, 10, pp. 3084–3088, 1996.

29. Anderson, P.R., Kim, B., and O'Connor, C. *Photocatalytic Process for Silver Recovery and Wash Water Reuse*. Desalination Research and Development Program, Report No. 50, June 2000. https://www.usbr.gov/research/AWT/reportpdfs/report050.pdf (July 2016).

30. Lunar, L., Sicilia, D., Rubio, S., Perez-Bendito, D., and Nickel, U. Degradation of photographic developers by Fenton's reagent: Condition optimization and kinetics for metol oxidation. *Water Research*, 34, 6, pp. 1791–1802, 2000.

31. Lunar, L., Rubio, S., and Perez-Bendito, D. Ion trap LC/MS characterisation of toxic polar organic pollutants in colour photographic wastewaters and monitoring of their chemical degradation. *Environmental Technology*, 25, 2, pp. 173–184, 2004.

32. Vengris, T., Binkiene, R., Butkiene, R., Nivinskiene, O., Melvydas, V., and Manusadzianas, L. Microbiological degradation of a spent offset-printing developer. *Journal of Hazardous Materials*, B113, pp. 181–187, 2004.

33. Smith, R.M. and Martell, A.E. *Critical Stability Constants, Vol. IV. Inorganic Complexes.* Plenum Press, New York, 1976.

34. Eastman Kodak Company. *The Fate and Effects of Silver in the Environment.* Eastman Kodak Company, Rochester, 2003. http://www.kodak.com/ek/uploadedFiles/J-216_ENG.pdf (January 2012).

35. Code of Federal Regulations. *Title 40: Protection of Environment (Parts 260 to 265).* Part 261: Identification and listing of hazardous waste, Subpart C: Characteristics of hazardous waste (revised as of July 1, 2010).

36. Code of Federal Regulations. *Title 40: Protection of Environment (Parts 425 to 699).* Part 459: Photographic point source category, Subpart A: Photographic processing subcategory (revised as of July 1, 2011).

37. Begum, S. Silver removal from aqueous solution by adsorption on concrete particles. *Turkish Journal of Chemistry*, 27, 5, pp. 609–617, 2003.

38. Atia, A.A. Adsorption of silver (I) and gold (III) on resins derived from bisthiourea and application to retrieval of silver ions from processed photo films. *Hydrometallurgy*, 80, pp. 98–106, 2005.

39. Lasko, C.L. and Hurst, M.P. An investigation into the use of chitosan for the removal of soluble silver from industrial wastewater. *Environmental Science and Technology*, 33, pp. 3622–3626, 1999.

40. Gallagher, N.P., Hendrix, J.L., Milosavljevic, E.B., Nelson, J.H., and Solujic, L. Affinity of activated carbon towards some gold(I) complexes. *Hydrometallurgy*, 25, 3, pp. 305–316, 1990.

41. Gittins, P.M. Sorption of phosphine, water vapour and Ag(I) and Cu(II) salts by activated carbon cloth. M.Phil. thesis, Chemistry Department, University of Exeter, UK, 1990.

42. Adani, K.G. Silver adsorption from photographic and medical x-ray process effluents by activated carbon. MSc. thesis, Camborne School of Mines, University of Exeter, Cornwall, UK, 2003.

43. Adani, K.G., Barley, R.W., and Pascoe, R.D. Silver recovery from synthetic photographic and medical x-ray process effluents using activated carbon. *Minerals Engineering*, 18, pp. 1269–1276, 2005.

44. Saman, N., Johari, K., Song, S.T., and Mat, H. Silver adsorption enhancement from aqueous and photographic waste solutions by mercerized coconut fiber. *Separation Science and Technology*, 50, pp. 937–946, 2015.

45. Li, L., Hu, Q., Zeng, J., Qi, H., and Zhuang, G. Resistance and biosorption mechanism of silver ions by *Bacillus cereus* biomass. *Journal of Environmental Sciences*, 23, 1, pp. 108–111, 2011.

46. Kiani, G. High removal capacity of silver ions from aqueous solution onto Halloysite nanotubes. *Applied Clay Science*, 90, pp. 159–164, 2014.

47. Tilton, R.C. and Rosenberg, B. Reversal of the silver inhibition of microorganisms by agar. *Applied and Environmental Microbiology*, 35, 6, pp. 1116–1120, 1978.

48. Pavlostathis, S.G. and Maeng, S.K. Aerobic biodegradation of a silver-bearing photographic processing wastewater. *Environmental Toxicology and Chemistry*, 17, 4, pp. 617–624, 1998.

49. Pavlostathis, S.G. and Maeng, S.K. Fate and effect of silver on the anaerobic digestion process. *Water Research*, 34, 16, pp. 3957–3966, 2000.

50. Eastman Kodak Company. *Recovering Silver from Photographic Processing Solutions.* Eastman Kodak Company, Rochester, 1999. http://www.kodak.com/ek/uploadedFiles/J215ENG.pdf (November 2011).

51. Pollet, B., Lorimer, J.P., Phull, S.S., and Hihn, J.Y. Sonoelectrochemical recovery of silver from photographic processing solutions. *Ultrasonics Sonochemistry*, 7, pp. 69–76, 2000.

52. Stalikas, C.D., Lunar, L., Rubio, S., and Perez-Bendito, D. Degradation of medical x-ray film developing wastewaters by advanced oxidation processes. *Water Research*, 35, 6, pp. 3845–3856, 2001.

53. Chun-du, W., Jin-yu, C., Ning, L., and Cheng-wu, Y. Ozone oxidation of photographic processing wastewater in a batch reactor. *International Journal of Plasma Environmental Science and Technology*, 1, 2, pp. 135–140, 2007.

54. Herrmann, J-M., Disdier, J., and Pichat, P. Photocatalytic deposition of silver on powder titania: Consequences for the recovery of silver. *Journal of Catalysis*, 113, pp. 72–81, 1988.

55. Wang, L.K. and Wang, M.H.S. (Eds.). *Handbook of Industrial Waste Treatment.* Marcel Dekker Inc., New York, pp. 173–228, 1992.

56. Wang, L.K., Hung, Y.T., Lo, H.H., and Yapijakis, C. (Eds.). *Handbook of Industrial and Hazardous Wastes Treatment.* Marcel Dekker Inc., New York, pp. 275–322, 515–584, 2004.

14 Toxicity, Source, and Control of Barium in the Environment

Hamidi Abdul Aziz, Miskiah Fadzilah Ghazali,
Yung-Tse Hung, and Lawrence K. Wang

CONTENTS

ABSTRACT

Barium (Ba) is a silver-white metal that makes up 0.05% of the Earth's crust. Very small amounts of naturally occurring barium are sometimes present in food and drinking water. Two barium compounds barium sulfate and barium carbonate are often found in underground deposits. Naturally occurring levels of barium are very low. Groundwater erosion of sedimentary rocks is the primary source of naturally occurring barium in drinking water. Natural soil erosion releases barium into the air. The air most people breathe contains less than 0.0015 parts of barium per billion parts (ppb) of air. Barium and barium compounds are used for many commercial processes. Barium sulfate is mined and used in oil and gas production, medical procedures, and the manufacture of paints, bricks, tiles, glass, and rubber. Other barium compounds are used in the manufacture of ceramics, pesticides, and oil and fuel additives. Barium can enter the body in three ways: through consumption of certain foods and/or drinking water; by inhalation of airborne barium compounds; and through direct skin contact with material containing barium. The latter is a rare occurrence, unless working in a chemical laboratory or similar occupation. Persons working in industries that manufacture or use barium compounds may also be exposed to barium in the air. Such exposure may be hazardous. The amount of barium in food and water supplies poses little or no health concern. In fact, the human body requires a certain level of barium in order to maintain good health. Barium is not a carcinogen, according to the most recent research. The Environmental Protection Agency has established a maximum level of 2 parts of barium per million parts (ppm) of water. Federal agencies like the Occupational Safety and Health Administration regulate barium releases in both water and workplace air in order to protect human health and the environment.

14.1   INTRODUCTION

14.1.1   BRIEF BACKGROUND

Barium was first isolated in 1808 by the English chemist Sir Humphry Davy (1778–1829). In 1807 and 1808, Davy also discovered five other new elements: sodium, potassium, strontium, calcium, and magnesium. All of these elements had been recognized much earlier as new substances, but Davy was the first to prepare them in pure form. Barium had first been identified as a new material in 1774 by the Swedish chemist Carl Wilhelm Scheele (1742–1786) (1).

Barium (Ba) is a silver-white metal that makes up 0.05% of the Earth's crust. It is a naturally occurring component of minerals that are found in small but widely distributed amounts in the Earth's crust, especially in igneous rocks, sandstone, shale, and coal. Barium enters the environment naturally through the weathering of rocks and minerals. Barium is present in the atmosphere, urban and rural surface water, soils, and many foods. In addition to its natural presence in the Earth's crust, and therefore its natural occurrence in most surface waters, barium is also released to the environment via industrial emissions. Anthropogenic releases are primarily associated with industrial processes. The residence time of barium in the atmosphere may be up to several days (2,3).

Barium is a member of the alkaline earth metals. The alkaline earth metals make up Group 2 (IIA) of the periodic table. The other elements in this group are beryllium, magnesium, calcium, strontium, and radium. These elements tend to be relatively active chemically and form a number of important and useful compounds. They also tend to occur abundantly in the Earth's crust in a number of familiar minerals such as aragonite, calcite, chalk, limestone, marble, travertine, magnesite, and dolomite. Alkaline earth compounds are widely used as building materials. Barium itself tends to have relatively few commercial uses. However, its compounds have a wide variety of applications in industry and medicine. Barium sulfate is used in x-ray studies of the gastrointestinal (GI) system. The GI system includes the stomach, intestines, and associated organs (1–3).

Barium exists in nature only in ores containing mixtures of elements. It also combines with other chemicals such as sulfur or carbon and oxygen to form barium compounds. The most important of these combinations are peroxide, chloride, sulfate, carbonate, nitrate, and chlorate. Barium compounds are solids and they do not burn well. Barium sulfate and barium carbonate are two forms of barium that are normally found in nature as underground ore deposits (1).

Barium sulfate exists as a white orthorhombic powder or crystals. Barite, the mineral from which barium sulfate is produced, is a moderately soft crystalline white opaque to transparent mineral. The most important impurities are iron (III) oxide, aluminum oxide, silica, and strontium sulfate. Barite is used primarily as a constituent in drilling muds in the oil industry. It is also used as filler in a range of industrial coatings, as dense filler in some plastics and rubber products, in brake linings, and in some sealants and adhesives. The use dictates the particle size to which barite is milled. For example, drilling muds are ground to an average particle diameter of 44 μm, with a maximum of 30% of particles less than 6 μm in diameter (2,3).

14.1.2 Chemical and Physical Information

Barium is an active metal. It combines easily with oxygen, the halogens, and other nonmetals. The halogens are Group 17 (VIIA) of the periodic table and include fluorine, chlorine, bromine, iodine, and astatine. Barium also reacts with water and with most acids. It is so reactive that it must be stored under kerosene, petroleum, or some other oily liquid to prevent it from reacting with oxygen and moisture in the air. Of the alkaline family, only radium is more reactive. Barium does not exist in nature in the elemental form but occurs as the divalent cation in combination with other elements. Two commonly found forms of barium are barium sulfate (CAS No. 7727-43-7) and barium carbonate (CAS No. 513-77-9), often found as underground ore deposits. These forms of barium are not very soluble in water: 0.020 g/L (at 20°C) for barium carbonate and 0.001 15 g/L (at 0°C) for barium sulfate (1–4).

Under natural conditions, barium is stable in the +2 valence state and is found primarily in the form of inorganic complexes. Conditions such as pH, Eh (oxidation–reduction potential), cation-exchange capacity, and the presence of sulfate, carbonate, and metal oxides will affect the partitioning of barium and its compounds in the environment, The major features of the biogeochemical cycle of barium include wet and dry deposition to land and surface water, leaching from geological formations to groundwater, adsorption to soil and sediment particulates, and biomagnifications in terrestrial and aquatic food chains.

Barium sulfate exists as a white orthorhombic powder or crystals. Barite, the mineral from which barium sulfate is produced, is a moderately soft crystalline white opaque to transparent mineral. The most important impurities are iron (III) oxide, aluminum oxide, silica, and strontium sulfate. Some of the more commonly used synonyms of barium sulfate include barite, barites, heavy spar, and blanc fixe (5).

Pure barium is a pale yellow, somewhat shiny, somewhat malleable metal. Malleable means capable of being hammered into thin sheets. It has a melting point of about 700°C (1300°F) and a boiling point of about 1500°C (2700°F). Its density is 3.6 g/cm³. When heated, barium compounds give off a pale yellow-green flame. This property is used as a test for barium. Table 14.1 shows the chemical and physical properties of barium and barium compounds.

TABLE 14.1

Chemical and Physical Properties of Barium and Barium Compounds

	Barium	Barium Acetate	Barium Carbonate	Barium Chloride	Barium Hydroxide	Barium Oxide	Barium Sulfate
CAS registry number	7440-39-3	543-80-6	513-77-9	10361-37-2	17194-00-2	1304-28-5	7727-43-7
Molecular formula	Ba	$Ba(C_2H_3O_2)_2$	$BaCO_3$	$BaCl_2$	$Ba(OH)_2@8H_2O$	BaO	$BaSO_4$
Molecular weight	137.34	255.43	197.35	208.25	315.48	153.34	233.4
Melting point (°C)	725	41[a]	1740 (at 90 atm)[a]	963	78	1923	1580 (decomposes)
Boiling point (°C)	1640	No data	1560	Decomposes	550[a]	2000	1149 (monoclinal transition point)[a]
Vapor pressure (mm Hg)	10 at 1049°C	No data	Essentially zero[a]	Essentially zero[a]	No data[a]	Essentially zero[a]	No data[a]
Water solubility (g/L)	Forms barium hydroxide	588 at 0°C, 750 at 100°C	0.02 at 20°C, 0.06 at 100°C	375 at 20°C[a],	56 at 15°C, 947 at 78°C	38 at 20°C, 908 at 100°C	0.00222 at 0°C, 0.00413 at 100°C
Specific gravity	3.5 at 20°C	2.468	4.43	3.856 at 24°C	2.18 at 16°C	5.72	4.50 at 15°C

Source: US Department of Health and Human Services Public Health Service Agency for Toxic Substances and Disease Registry. 2007. *Toxicological for Barium and Barium Compounds.* Atlanta, GA.

[a] All information obtained from Lide 2005 except where noted.

14.1.3 USES

Barium and barium compounds are used for many important purposes. Barium sulfate ore is mined and used in several industries. It is used primarily as a constituent in drilling muds in the oil industry. Drilling muds make it easier to drill through rock by keeping the drill bit lubricated. The use dictates the particle size to which barium sulfate is milled. For example, drilling muds are ground to an average particle diameter of 44 μm, with a maximum of 30% of particles less than 6 μm in diameter. Barium sulfate is also used to make paints, bricks, tiles, glass, rubber, and other barium compounds. It is also used as filler in a range of industrial coatings, as dense filler in some plastics and rubber products, in brake linings, and in some sealants and adhesives (5,6).

Some barium compounds, such as barium carbonate, barium chloride, and barium hydroxide, are used to make ceramics, insect and rat poisons, additives for oils and fuels, and many other useful products. Barium sulfate is sometimes used by doctors to perform medical tests and take x-ray photographs of the stomach and intestines (5,6).

Barium and its compounds have several important medical uses as well. Barium chloride was formerly used in treating complete heart block, because periods of marked bradycardia and asystole were prevented through its use. This use was abandoned, however, mainly due to barium chloride's toxicity. Characterized by extreme insolubility, chemically pure barium sulfate is nontoxic to humans. It is frequently utilized as a benign, radiopaque aid to x-ray diagnosis, because it is normally not absorbed by the body after oral intake. In addition to the extensive use of barium sulfate in studying GI motility and diagnosis of GI disease, barium sulfate may be chosen as the opaque medium for the x-ray examination of the respiratory and urinary systems as well (7).

14.2 SOURCE IN THE ENVIRONMENT

14.2.1 NATURAL OCCURRENCE

Barium is a naturally occurring component of minerals that is found in small but widely distributed amounts in the Earth's crust, especially in igneous rocks, sandstone, shale, and coal. Barium enters the environment naturally through the weathering of rocks and minerals. Anthropogenic releases are primarily associated with industrial processes. Barium is present in the atmosphere, urban and rural surface water, soils, and many foods (1).

Barium in water comes primarily from natural sources. The acetate, nitrate, and halides are soluble in water, but the carbonate, chromate, fluoride, oxalate, phosphate, and sulfate are quite insoluble. The primary source of barium in the atmosphere is industrial emissions. Barium concentrations ranging from 2×10^{-4} to 2.8×10^{-2} g/m^3 have been detected in urban areas of North America. Barium naturally occurs in most surface waters and in public drinking water supplies. Barium content in US drinking water supplies ranges from 1 to 20 g/L; in some areas barium concentrations as high as 10,000 g/L have been detected (8). Barium is ubiquitous in soils, with concentrations ranging from 15 to 3000 ppm (6).

The two most prevalent naturally occurring barium ores are barite (barium sulfate) and witherite (barium carbonate). Barite occurs largely in sedimentary formations, as residual nodules resulting from weathering of barite-containing sediments, and in beds along with fluorspar, metallic sulfides, and other minerals. Witherite is found in veins and is often associated with lead sulfide. Barium is found in coal at concentrations up to 3000 mg/kg, as well as in fuel oils (6). Estimates of terrestrial and marine concentrations of barium are 250 and 0.006 g/tonne, respectively.

Barite ore is the raw material from which nearly all other barium compounds are derived. Barite is mined in Morocco, China, India, and the United Kingdom. Crude barite ore is washed free of clay and other impurities, dried, and then ground before use. Barite is usually imported as crude ore or crushed ore for milling or as ready-milled ore. Barite can be 90%–98% barium sulfate. World production of barite in 1985 was estimated to be 5.7 million tons.

Anthropogenic sources of barium are primarily industrial. Emissions may result from mining, refining, or processing of barium minerals and manufacture of barium products. Barium is released to the atmosphere during the burning of coal, fossil fuels, and waste. Barium is also discharged in wastewater from metallurgical and industrial processes. Deposition on soil may result from human activities, including the disposal of fly ash and primary and secondary sludge in landfills. Estimated releases of barium and barium compounds to the air, water, and soil from manufacturing and processing facilities in the United States during 1998 were 900, 45, and 9300 tons, respectively (5–7).

14.2.2 MAN-MADE SOURCES

Barite ore is the raw material from which nearly all other barium compounds are derived. World production of barite in 1985 was estimated to be approximately 5.7 million tons. The major world producers of barite are China, the United States, USSR, India, Mexico, Morocco, Ireland, Germany, and Thailand. Other producers are Canada, France, Spain, Czechoslovakia, and England. China as the world's leading producer, accounted for about 1.0 million tons or 17% of world output in 1984. The Unites States, the second largest producer, accounted for 0.70 million tons in 1984 and also imported 1.6 million tons. Canada produced approximately 64,000 tons and consumed around 78,000 tons in 1984 (9).

Emissions of barium into the air from mining, refining, and processing barium ore can occur during loading and unloading, stockpiling, materials handling, and grinding and refining of the ore. Emission into water may occur during the purification of barite ore and subsequent discharge of the industrial water to the environment. Coal-fired power plants emit barium into the atmosphere via ash. Some barium escapes into the atmosphere as fly ash, while the rest is generally disposed of in landfills. Barium in coal ash ranges from 100 to 5000 mg/kg (6).

In the Unites States, the barium chemical industries released an estimated 1200 tons of particulates into the atmosphere in 1972. Waste water from barium chemical production processes is another potential source of barium emission.

Although most fugitive dust emissions and process effluents are reduced by control technologies, an area of concern is the emission of soluble barium into the atmosphere from dryers and calciners. Bag houses can reduce the uncontrolled emission factor to 0.25 g/kg. The release of soluble barium into the atmosphere around these plants was estimated at 56 tons for 1972, but it has decreased as barium chemical production has declined. The plastics industry is a relatively important source of barium emission to the atmosphere. It utilizes barium as a stabilizer to prevent discoloration during processing. Another source of barium emission is the manufacture of glass. Emissions of barium-containing particulates with an average size of 1 μm have been reported by various authors.

The detonation of nuclear devices in the atmosphere is a source of atmospheric radioactive barium. The radioactive isotopes ^{140}Ba and ^{143}Ba are products of the decay chains from thermal-neutron fission of ^{235}U. Among the isotopes of barium, ^{140}Ba has the longest half-life about 12.8 days and contributes 10% of the total fission products at 10 days after nuclear fission. At 60 days,

TABLE 14.2
Main Uses of Some Barium Compounds

Barium Compounds	Uses
Acetate	Catalyst for organic reactions; textile mordant; oil and grease lubricator; paint and varnish driers
Aluminate	In ceramics; in water treatment
Bromate	Analytical reagent; oxidizing agent; corrosion inhibitor in low carbon steel; in the preparation of rare earth bromates.
Bromide	In the manufacture of other bromides; in photographic compounds; in the preparation of phosphors
Carbonate	In the treatment of brines in chlorine alkali cells to remove sulfates; as a rodenticide; in ceramic flux, optical glass, case hardening baths, ferrites, radiation-resistant glass for color television tubes; in manufacturing paper.
Chloride	In the manufacture of pigments, color lakes, glass; as a mordant for acid dyes; in pesticides, lube oil additives, boiler compounds, and aluminum refining; as a flux in the manufacture of magnesium metal; in leather tanner and finisher, in photographic paper and textiles
Fluoride	In ceramics; in the manufacture of other fluorides; in crystals for spectroscopy; in electronics; in dry-film lubricants; in embalming; in glass manufacture; manufacture of carbon brushes for DC motors and generators
Hypophosphite	In medicine and nickel plating
Iodide	In the preparation of other iodides
Manganate (VI)	As a paint pigment
Metaphosphate	In glasses, porcelain, and enamels
Nitrate	In pyrotechnics (green light); in incendiaries; chemicals (barium peroxide); ceramic glazes; as a rodenticide; in the vacuum-tube industry
Oxide	As a dehydrating agent; in the manufacture of lubricating oil, detergents
Permanganate	As a strong disinfectant; in the manufacture of permanganates; as a dry cell depolarizer
Selenide	In photocells; in semiconductors
Thiosulfate	In explosives, luminous paints, matches, varnishes; as an iodometry standard; in photographic diffusion-transfer processes.
Tungstate	As a pigment; in x-ray photography for the manufacture of intensifying and phosphorescent screens
Zirconate	In the manufacture of silicone rubber compounds stable up to 246°C; in electronics

Source: US Department of Health and Human Services, Public Health Service, Agency for Toxic Substances and Disease Registry, ATSDR. 1992. *Toxicological Profile for Barium.* Atlanta, GA.

however, its contribution falls to 2% of total activity. The concentration of barium particles in the atmosphere due to this source, in terms of actual weight, is immeasurably small. Due to the short half-life and low concentrations of barium radionuclide, this source is not considered a significant source of barium in the environment (6). The main use of some barium compounds is shown in Table 14.2.

14.3 RELEASE INTO THE ENVIRONMENT

Barium is a highly reactive metal that occurs naturally only in a combined state. The element is released into environmental media by both natural processes and anthropogenic sources. Barium waste may be released into air, land, and water during industrial operations. Barium is released into the air during the mining and processing of ore and during manufacturing operations (1–5).

The length of time that barium will last in the environment following the release into air, land, and water depends on the form of barium released. Barium compounds that do not dissolve well in water, such as barium sulfate and barium carbonates, can last a long time in the environment. Barium compounds that dissolve easily in water usually do not last a long time in the environment. Barium that is dissolved in water quickly combines with sulfate or carbonates ions and becomes the longer lasting forms such as barium sulfate and barium carbonate. Barium sulfate and barium carbonate are the forms of barium most commonly found in the soil and water. If barium sulfate and barium carbonate are released onto land, they will combine with particles of soil (6,7,9).

14.3.1 Air

Barium is released primarily into the atmosphere as a result of industrial emissions during the mining, refining, and production of barium and barium chemicals, fossil fuel combustion (Miner 1969a), and entrainment of soil and rock dust into the air. In addition, coal ash, containing widely variable amounts of barium, is also a source of airborne barium particulates. In 1969, an estimated 18% of the total US barium emissions into the atmosphere resulted from the processing of barite ore, and more than 28% of the total was estimated to be from the production of barium chemicals. The manufactures of various end products such as drilling well muds, and glass, paint, and rubber products and the combustion of coal were estimated to account for an additional 23% and 26% of the total barium emissions for 1969, respectively.

Estimates of barium releases from individual industrial processes are available for particulate emissions from the drying and calcining of barium compounds and for fugitive dust emissions during the processing of barite ore. Soluble barium compounds are emitted as particulates from barium chemical dryers and calciners to the atmosphere during the processing of barium carbonate, barium chloride, and barium hydroxide. Uncontrolled particulate emissions of soluble barium compounds from chemical dryers and calciners during barium processing operations may range from 0.04 to 10 g/kg of the final product. Controlled particulate emissions are less than 0.25 g/kg of the final product. Fugitive dust emissions occur during processing which is the grinding and mixing of barite ore and may also occur during the loading of the bulk product of various barium compounds into railroad hopper cars. Based on an emission factor of 1 g/kg, total emissions of fugitive dust from the domestic barium chemicals industry during the grinding of barite ore have been estimated to be approximately 90 metric tons per year.

The use of barium in the form of organometallic compounds as a smoke suppressant in diesel fuels results in the release of solids to the atmosphere. The maximum concentration of soluble barium in exhaust gases containing barium-based smoke suppressants released from test diesel engines and operating diesel vehicles is estimated to be 12,000 mg/m^3, when the barium concentration in the diesel fuel is 0.075% by weight and 25% of the exhausted barium is soluble. Thus, 1 L of this exhaust gas contains an estimated 12 mg soluble barium or 48 mg total barium (6,7,9).

14.3.2 WATER

The primary source of naturally occurring barium in drinking water results from the leaching and eroding of sedimentary rocks into groundwater. Although barium occurs naturally in most surface water bodies, releases of barium to surface waters from natural sources are much lower than those to groundwater. About 80% of the barium produced is used as barite to make high-density oil and gas well drilling muds, and during offshore drilling operations there are periodic discharges of drilling wastes in the form of cuttings and muds into the ocean. For example, in the Santa Barbara Channel region, about 10% of the muds used are lost into the ocean. The use of barium in offshore drilling operations may increase barium pollution, especially in coastal sediments (6).

14.3.3 SOIL

The process of drilling for crude oil and natural gas generates waste drilling fluids or muds, which are often disposed of by land farming. Most of these fluids are water based and contain barite and other metal salts. Thus barium may be introduced into soils as the result of land farming these slurried reserve pit wastes. The use of barium fluorosilicate and carbonate as insecticides might also contribute to the presence of barium in agricultural soils (6).

14.4 ENVIRONMENTAL DISTRIBUTION AND TRANSFORMATION

Both specific and nonspecific adsorption of barium onto oxides and soils has been observed. Specific sorption occurs onto metal oxides and hydroxides. Adsorption onto metal oxides probably acts as a control over the concentration of barium in natural waters. Electrostatic forces account for a large fraction of the nonspecific sorption of barium on soil and subsoil. The retention of barium, like that of other alkaline earth cations, is largely controlled by the cation-exchange capacity of the sorbent.

Examination of dust falls and suspended particulates indicates that most contain barium. The presence of barium is mainly attributable to industrial emissions, especially the combustion of coal and diesel oil and waste incineration, and may also result from dusts blown from soils and mining processes. Barium sulfate and carbonate are the forms of barium most likely to occur in particulate matter in the air, although the presence of other insoluble compounds cannot be excluded. The residence time of barium in the atmosphere may be several days, depending on the particle size. Most of these particles, however, are much larger than 10 μm in size and rapidly settle back to earth. Particles can be removed from the atmosphere by rainout or washout wet deposition. Soluble barium and suspended particulates can be transported great distances in rivers, depending on the rates of flow and sedimentation.

Barium sulfate is present in soil through the natural process of soil formation; barium concentrations are high in soils formed from limestone, feldspar, and biotite micas of the schists and shales. When soluble barium-containing minerals weather and come into contact with solutions containing sulfates, barium sulfate is deposited in available geological faults. If there is insufficient sulfate to combine with barium, the soil material formed is partially saturated with barium. In soil, barium that replaces other sorbed alkaline barium sulfate in soils is not expected to be very mobile because of the formation of water-insoluble salts and its inability to form soluble complexes with humic and fulvic materials (10).

14.5 METABOLISM AND DISPOSITION

14.5.1 ABSORPTION

The soluble forms of barium salts are rapidly absorbed into the blood from the intestinal tract. The rates of absorption of a number of barium salts have been measured in rats following oral exposure to small quantities (30 mg/kg body weight). The relative absorption rates were found to be: barium chloride > barium sulfate > barium carbonate. Large doses of barium sulfate do not increase the uptake of this salt because of its low solubility. Systemic toxic effects have been observed following both oral and inhalation exposure. No absorption kinetics is available following inhalation exposure, although it is obvious that absorption does occur (10).

14.5.2 Distribution

Barium enters the human body when people breathe air, eat food, or drink water containing barium. It may also enter the body to a small extent when humans have direct skin contact with barium compounds. It seems to enter the bloodstream very easily after breathing the compounds. Barium does not seem to enter the bloodstream as well from the stomach or intestines. Barium absorbed into the bloodstream disappears in about 24 h. However, it is deposited in the muscles, lungs, and bone. Very little is stored in the kidneys, liver, spleen, brain, heart, or hair. It remains in the muscles about 30 h after which the concentration decreases slowly. The deposition of barium into bone is similar to calcium but occurs at a faster rate. The half-life of barium in bone is estimated to be about 50 days (10–12).

14.5.3 Metabolism

Some barium compounds such as barium chloride can enter the human body through the skin, but this is very rare and usually occurs in industrial accidents at factories where they make or use barium compounds. Barium at hazardous waste sites may enter the human body if people breathe dust, eat soil or plants, or drink water polluted with barium. Barium can also enter the body if polluted soil or water touches the skin. About 54% of the barium dose is protein bound. Barium is known to activate the secretion of catecholamine from the adrenal medulla without prior calcium deprivation. It may displace calcium from the cell membranes, thereby increasing permeability and providing stimulation to muscles. Eventual paralysis of the central nervous system can occur.

14.5.4 Excretion

Barium that enters the human by breathing, eating, or drinking is removed mainly in feces and urine. Most of the barium that enters their body is removed within a few days, and almost all of it is gone within 1–2 weeks. Most barium that stays in the body goes into the bones and teeth. A tracer study in rats using ^{140}Ba demonstrated that 7% and 20% of the barium dose was excreted in 24 h in the urine and feces, respectively. In contrast, calcium is primarily excreted in the urine. The clearance of barium is enhanced with saline infusion (2).

14.6 ENVIRONMENTAL TRANSPORT

14.6.1 Air

Examination of dust falls and suspended particulates indicates that most contain barium. The presence of barium is mainly attributable to industrial emissions, especially the combustion of coal and diesel oil and waste incineration, and may also result from dusts blown from soils and mining processes. Barium sulfate and carbonate are the forms of barium most likely to occur in particulate matter in the air, although the presence of other insoluble compounds cannot be excluded. The residence time of barium in the atmosphere may be several days, depending on particle size. Most of these particles, however, are much larger than 10 µm in size, and rapidly settle back to earth. Particles can be removed from the atmosphere by rainout or washout wet deposition. These two forms of deposition efficiently clear the atmosphere of pollutants, but they are not well understood. Without knowing the amount of barium in the atmosphere, it is difficult to evaluate the processes of deposition, transport, and distribution (2,6).

14.6.2 Water

Soluble barium and suspended particulates can be transported great distances in rivers, depending on the rates of flow and sedimentation. In the absence of any possible removal mechanisms, the residence time of barium in aquatic systems could be several hundred years. Unless it is removed by precipitation, exchange with soil, or other processes, barium in surface waters ultimately reaches the ocean. Once freshwater sources discharge into sea water, barium and the sulfate ions present in salt water form barium sulfate. Due to the relatively higher concentration of sulfate present in the oceans, only an estimated 0.006% of the total barium brought by freshwater sources remains in solution. This estimate is supported by evidence that outer shelf sediments have lower barium concentrations than those closer to the mainland.

Upon entering the ocean, barium is transported downward by the physical processes of mixing. It is depleted in the upper layers of the ocean by incorporation into biological matter, which settles toward the ocean floor. The higher concentration of barium in deep water relative to surface water probably reflects the deposition of barium onto suspended particles forming at the ocean surface and the subsequent release of barium to the deep water as the particles are destroyed in transit to the ocean floor. In the ocean, barium is in steady state; the amount entering the ocean through rivers is balanced by the amount falling to the bottom as particles forming a permanent part of the sediment (13).

14.6.3 Soil

Barium added to soils may either be taken up by vegetation or transported through soil with precipitation. Relative to the amount of barium found in soils, little is bio-concentrated by plants. However, this transport pathway has not been comprehensively studied. Barium is not very mobile in most soil systems. The rate of transportation of barium in soil is dependent on the characteristics of the soil material. Soil properties that influence the transportation of barium to groundwater are cation-exchange capacity and calcium carbonate ($CaCO_3$) content. In soil with a high cation-exchange capacity, barium mobility will be limited by adsorption. High $CaCO_3$ content limits mobility by precipitation of the element as $BaCO_3$.

Barium is more mobile and is more likely to be leached from soils in the presence of chloride due to the increased solubility of barium chloride as compared to other chemical forms of barium. Barium complexes with fatty acids such as in acidic landfill leachate will be much more mobile in the soil due to the lower charge of these complexes and subsequent reduction in adsorption capacity. Barium mobility in soil is reduced by the precipitation of barium carbonate and sulfate. Humic and fulvic acid have not been found to increase the mobility of barium (9).

14.6.4 Vegetation

Despite relatively high concentrations in soils, only a limited amount of barium accumulates in plants. Barium is actively taken up by legumes, grain stalks, forage plants, red ash leaves, and the black walnut, hickory, and Brazil nut trees. The Douglas fir tree and plants of the genus *Astragallus* also accumulate barium. No studies of barium particle uptake from the air have been reported, although vegetation is capable of removing significant amounts of contaminants from the atmosphere. Plant leaves act only as deposition sites for particulate matter. There is no evidence that barium is an essential element in plants.

14.7 TOXICITY

Barium and its entire compound are very toxic. The soluble salts of barium are toxic in mammalian systems. They are absorbed rapidly from the GI tract and are deposited in the muscles, lungs, and bone. At low doses, barium acts as a muscle stimulant and at higher doses affects the nervous system eventually leading to paralysis. The Department of Health and Human Services, the International Agency for Research on Cancer, and the Environmental Protection Agency (EPA) has not classified barium as to its carcinogenicity (11).

The different barium compounds have different solubility in water and body fluids and therefore they serve as variable sources of the Ba^{2+} ion. The Ba^{2+} ion and the soluble compounds of barium (notably chloride, nitrate, and hydroxide) are toxic to humans. The insoluble compounds of barium (notably sulfate and carbonate) are inefficient sources of Ba^{2+} ion because of limited solubility and are therefore generally nontoxic to humans (14).

The insoluble, nontoxic nature of barium sulfate has made it practical to use this particular barium compound in medical applications such as enema procedures and in x-ray photography of the GI tract. Barium provides an opaque contrasting medium when ingested or given by enema prior to x-ray examination. Under these routine medical situations, barium sulfate is generally safe. However, barium sulfate or other insoluble barium compounds may potentially be toxic when it is introduced into the GI tract under conditions where there is colon cancer or perforations of the GI tract and barium is allowed to enter the blood stream (12,15).

Barium has been associated with a number of adverse health effects in both humans and experimental animals. Both human and animal evidence suggests that the cardiovascular system may be one of the primary targets of barium toxicity. In addition to cardiovascular effects, exposure of humans and/or animals to barium has been associated with respiratory, GI, hematological, musculoskeletal, hepatic, renal, neurological, developmental, and reproductive effects. No data or insufficient data are available to draw conclusions regarding the immunological, genotoxic, or carcinogenic effects of barium. Death has been observed in some individuals following acute oral exposure to high concentrations of barium.

14.8 EXPOSURE TO BARIUM AND BARIUM COMPOUNDS

EPA identifies the most serious hazardous waste sites in the nation. These sites are then placed on the National Priorities List (NPL) and are targeted for long-term federal clean-up activities. Barium and barium compounds have been found in at least 798 of the 1684 current or former NPL sites; however, the total number of NPL sites evaluated for these substances is not known (7).

Background levels of barium in the environment are very low. They are in water, air, and soil. The maximum amount of barium that is recommended by the EPA is 2 ppm.

Literature reviews from 1983 to 2009 reported that the concentration of barium in both groundwater and drinking water in several regions around the world varied from 0.001 to 6.4 mg/L. Thus, the removal of barium from water is in some of these places was necessary (16). The air that most people breathe contains about 0.0015 parts of barium per billion parts of air (ppb). The air around factories that release barium compounds into the air has about 0.33 ppb or less of barium. The amount of barium found in soil ranges from about 15 to 3500 ppm. Some foods, such as Brazil nuts, seaweed, fish, and certain plants, may contain high amounts of barium. The amount of barium found in food and water usually is not high enough to be a health concern (7).

Barium is mainly found in many food groups. In most foods, the barium content is relatively low (<3 mg/100 g) except in Brazil nuts, which have a very high barium content (150–300 mg/100 g) (7,13).

14.8.1 HUMAN EXPOSURE

If somebody is exposed to barium and barium compounds, many factors will determine whether that person will be harmed. These factors include the dose, how much, the duration, how long, and how the individual come in contact with the barium and barium compounds. People must also consider any other chemicals they are exposed to and their age, sex, diet, family traits, lifestyle, and state of health (7).

The greatest group known to risk exposure to high levels of barium is those working in industries that make or use barium compounds. Most of these exposed persons breathe air that contains barium sulfate or barium carbonate. Sometimes they are exposed to one of the more harmful barium compounds; for example, barium chloride or barium hydroxide by breathing the dust from these compounds or by getting them on their skin.

Barium carbonate can be harmful if accidentally eaten because it will dissolve in the acids within the stomach unlike barium sulfate, which will not dissolve in the stomach. Many hazardous waste sites contain barium compounds, and these sites may be a source of exposure for people living and working near them. Exposure near hazardous waste sites may occur by breathing dust, eating soil or plants, or drinking water that is polluted with barium. People near these sites may also get soil or water that contains barium on their skin. Barium sulfate is the major barium compound used in medicinal diagnostics; it is applied as an opaque contrast medium for Roentgenographic studies of the GI tract, providing another possible source of human exposure to barium (5,7,13).

Barium enters the human body when people breathe air, eat food, or drink water containing barium. It may also enter the body to a small extent when they have direct skin contact with barium compounds. The amount of barium that enters the bloodstream after breathing, eating, or drinking depends on the barium compound. Some barium compounds that are soluble, such as barium chloride, can enter the bloodstream more easily than insoluble barium compounds such as barium sulfate. Some barium compounds such as barium chloride can enter the human body through the skin, but this is very rare and usually occurs in industrial accidents at factories where they make or use barium compounds. Barium at hazardous waste sites may enter the body if people breathe dust, eat soil or plants, or drink water polluted with barium from this area (6).

14.9 HEALTH EFFECTS

The health effects of the different barium compounds depend on how well the specific barium compound dissolves in water. For example, barium sulfate does not dissolve well in water and has few adverse health effects. Doctors sometimes give barium sulfate orally or by placing it directly in the rectum of patients for purposes of making x-rays of the stomach or intestines. The use of this particular barium compound in this type of medical test is not harmful to people. Barium compounds such as barium acetate, barium carbonate, barium chloride, barium hydroxide, barium nitrate, and barium sulfide that dissolve in water can cause adverse health effects (5,6,9).

Most results from studies indicated that a small number of individuals were exposed to fairly large amounts of barium for short periods. Eating or drinking very large amounts of barium compounds that dissolve in water may cause paralysis or death in a few individuals. The toxicity of barium compounds depends on the specific species, but lethal doses in humans usually range from 1 to 30 g (17). Barium is a dermal chemical irritant and may cause dermal lesions. When this element is ingested orally or inhaled it can cause tachycardia, hypertension, and benign granulomatous pneumoconiosis (18). Some people who eat or drink somewhat smaller amounts of barium for a short period may potentially have difficulties in breathing, increased blood pressure, changes in heart rhythm, stomach irritation, minor changes in blood, muscle weakness, changes in nerve reflexes, swelling of the brain, and damage to the liver, kidney, heart, and spleen.

One study showed that people who drank water containing as much as 10 ppm of barium for 4 weeks did not have increased blood pressure or abnormal heart rhythms. We have no reliable information about the possible health effects in humans who are exposed to barium by breathing or by direct skin contact. However, many of the health effects might be similar to those seen after eating or drinking barium.

The health effects of barium have been studied more often in experimental animals than in humans. Rats that ate or drank barium over short periods had buildup of fluid in the trachea (windpipe), swelling and irritation of the intestines, changes in organ weights, decreased body weight, and increased numbers of deaths. Rats that ate or drank barium over long periods had increased blood pressure and changes in the function and chemistry of the heart. Mice that ate or drank barium over a long period had a shorter life span.

At low doses, barium acts as a muscle stimulant and at higher doses affects the nervous system eventually leading to paralysis. Acute and subchronic oral doses of barium cause vomiting and diarrhea, followed by decreased heart rate and elevated blood pressure. Higher doses result in cardiac irregularities, weakness, tremors, anxiety, and dyspnea. A drop in serum potassium may account for some of the symptoms. Death can occur from cardiac and respiratory failure. Acute doses around 0.8 g can be fatal to humans (7,9).

Subchronic and chronic oral or inhalation exposure primarily affects the cardiovascular system resulting in elevated blood pressure. A lowest-observed-adverse-effect level (LOAEL) of 0.51 mg barium/kg/day based on increased blood pressure was observed in chronic oral rat studies whereas human studies identified a no-observed-adverse-effect level (NOAEL) of 0.21 mg barium/kg/day. Subchronic and chronic inhalation exposure of human populations to barium-containing dust can result in a benign pneumoconiosis called *baritosis*. This condition is often accompanied by an elevated blood pressure but does not result in a change in pulmonary function. Exposure to an air concentration of 5.2 mg barium carbonate/m^3 for 4 h/day for 6 months has been reported to result in elevated blood pressure and decreased body weight gain in rats (2).

The Department of Health and Human Services, the International Agency for Research on Cancer, and EPA has not classified barium as to its carcinogenicity.

14.10 ENVIRONMENTAL LEVELS

Environmental levels are generally reported as total barium ion rather than as specific barium compounds.

14.10.1 AIR

The levels of barium in air are not well documented, and in some cases the results are contradictory. No distinct pattern between ambient levels of barium in the air and the extent of industrialization was observed. In general, however, higher concentrations were observed in areas where metal smelting occurred. In the USA survey, ambient barium concentrations ranged from 0.0015 to 0.95 mg/m^3 (9–11). In three communities in New York City, barium was measured in dust fall and household dust. With standard methods, (USEPA, 1974), the dust fall was found to contain an average of 137 mg barium/g, while house dust contained 20 mg barium/g.

14.10.2 WATER

The presence of barium in sea water, river water, and well water has been well documented. It occurs in almost all surface waters that have been examined. The concentration present is extremely variable and depends on factors that affect aquifers and any water treatment that has occurred. The concentration of barium in water is related to the hardness of the water, which is defined as the sum of the polyvalent cations present, including the ions of calcium, magnesium, iron, manganese, copper, barium, and zinc. Barium concentrations of 7–15 000 µg/L have been measured in fresh water and 6 µg/L in sea water. In the United States, levels of barium in water vary greatly depending on local geochemical influences.

14.10.3 DRINKING WATER

Municipal water supplies depend upon the quality of surface water and groundwater, and these, in turn, depend upon local geochemical influences. Studies of the water quality in cities in the United States have revealed levels of barium ranging from a trace to 10,000 µg/L. Drinking water levels of at least 1000 µg barium/L have been reported when barium is present mainly in the form of insoluble salts. Levels of barium in Canadian water supplies have been reported to range from 5 to 600 μg/L and municipal water levels in Sweden ranging from 1 to 20 μg/L have been measured (6).

14.10.4 Sea Water

The concentration of barium in sea water varies greatly with factors such as latitude, depth, and the ocean in question. Several studies have shown that the barium content in the open ocean increases with the depth of water. The level is not same in different places.

14.10.5 Soil and Sediment

The presence of barium in soils has received attention since it was first documented in the muds of the River Nile and the soils of the United States. In Earth's crust the barium concentration is 400–500 mg/kg. Later works have verified the levels found in the early studies. The background level of barium in soils is considered to range from 100 to 3000 mg/kg, the average abundance being 500 mg/kg.

14.10.6 Food

The concentrations of barium in sediments of the Iowa River are 450–3000 mg/kg, suggesting that barium in the water is removed by precipitation and silting and may possibly affect the ecology of benthic organisms. Barium is also present in wheat, although most is concentrated in the stalks and leaves rather than in the grain. Tomatoes and soybeans also concentrate soil barium, the bio-concentration factor ranging from 2 to 20. In the beverages group, tea and cocoa had the highest barium content (2.7 and 1.2 mg/100 g, respectively) on a dry-weight basis. Among breads, cereal products, and cracker products, bran flakes (0.39 mg/100 g), and enriched instant cream of wheat (0.2 mg/100 g) had the highest levels. Eggs were found to have 0.76 mg/100 g, and Swiss cheese 0.22 mg/100 g. Fruits and fruit juice had low barium levels, the highest values being in raw, unpeeled apples (0.075 mg/100 g). These levels are similar to those found in grapes (0.05 mg/100 g) and cooked prunes (0.064 mg/100 g). All meats showed concentrations of 0.04 mg per 100 g or less. Vegetables had relatively low barium levels, with the exception of beets (0.26 mg/100 g) and sweet potatoes (0.22 mg/100 g). Among nuts, pecans had the highest barium content (0.67 mg/100 g) (6).

14.10.7 Nuclear Fallout

The principal potential source of radioactive isotopes of barium is nuclear weapons testing. Atmospheric testing suspends radioactive dusts in the upper troposphere where, depending on atmospheric conditions, dusts are carried around the world several times.

The lightest dust particles reach the stratosphere. Several years are required for the bulk of this radioactive material to be deposited on the ground. Since 1952, when tests began on nuclear weapons with high explosive yields, fallout from the stratosphere has been more or less continuous. Most of this nuclear fallout occurs in the temperate and polar regions of the earth. The total radiation from nuclear testing has added 10%–15% to the naturally occurring radiation throughout the world.

Because ^{140}Ba and ^{143}Ba are radioactive byproducts of the thermal nuclear fission of ^{235}U, their concentration in the environment increases after the detonation of a nuclear device in the atmosphere. Radioactive particles are normally cleared from the atmosphere by rain and snow.

14.11 REMOVAL TECHNIQUES

There are several removal techniques of barium compounds including ion exchange, reverse osmosis (RO), lime softening, or electrodialysis. Ion exchange for soluble Ba uses charged cation resin to exchange acceptable ions from the resin for undesirable forms of Ba in the water. It is a very effective and well-developed technique. RO for soluble Ba uses a semipermeable membrane, and the application of pressure to a concentrated solution causes water, but not suspended or dissolved solids (soluble Ba), to pass through the membrane. This technique can produce high quality water. Lime softening for soluble Ba uses $Ca(OH)_2$ in sufficient quantity to raise the pH to about 10 to precipitate carbonate hardness and heavy metals like Ba. Electrodialysis reversal uses semipermeable membranes in which ions migrate through the membrane from a less concentrated to a more concentrated solution as a result of the ions' representative attractions to direct electric current.

14.11.1 ION EXCHANGE

In solution, salts separate into positively charged cations and negatively charged anions. Deionization can reduce the amounts of these ions. Cation ion exchange is a reversible chemical process in which ions from an insoluble, permanent, solid resin bed are exchanged for ions in water. The process relies on the fact that water solutions must be electrically neutral. Therefore, ions in the resin bed are exchanged with ions of similar charge in the water.

As a result of the exchange process, no reduction in ions is obtained. In the case of Ba reduction, the operation begins with a fully recharged cation resin bed, having enough positively charged ions to carry out the cation exchange. Usually a polymer resin bed is composed of millions of medium sand grain size, spherical beads. As water passes through the resin bed, the positively charged ions are released into the water, being substituted or replaced with the Ba ions in the water (ion exchange). When the resin becomes exhausted of positively charged ions, the bed must be regenerated by passing a strong, usually NaCl (or KCl), solution over the resin bed, displacing the Ba ions with Na or K ions. Many different types of cation resins can be used to reduce dissolved Ba concentrations. The use of ion exchange to reduce concentrations of Ba will be dependent on the specific chemical characteristics of the raw water (4). Cation ion exchange, commonly termed water softening, can be used with low flows (up to 200 GPM) and when the ratio of hardness to Ba is greater than 1.

Advantages:

1. Acid addition, degasification, and repressurization are not required
2. Ease of operation and highly reliable
3. Lower initial cost; resins will not wear out with regular regeneration
4. Effective and widely used
5. Suitable for small and large installations
6. Variety of specific resins is available for removing specific contaminants

Disadvantages:

1. Pretreatment lime softening may be required
2. Requires salt storage, regular regeneration
3. Concentrate disposal
4. Usually not feasible with high levels of total dissolved solids (TDS)
5. Resins are sensitive to the presence of competing ions

14.11.2 REVERSE OSMOSIS

RO is a physical process in which contaminants are removed by applying pressure on the feed water to direct it through a semipermeable membrane. The process is the reverse of natural osmosis (water diffusion from dilute to concentrate through a semipermeable membrane to equalize ion concentration) as a result of the applied pressure to the concentrated side of the membrane, which overcomes the natural osmotic pressure. RO membranes reject ions based on size and electrical charge.

The raw water is typically called feed. The product water is called permeate and the concentrated reject is called concentrate.

Common RO membrane materials include asymmetric cellulose acetate or polyamide thin film composite. Common membrane construction includes spiral wound or hollow fine fiber. Each material and construction method has specific benefits and limitations depending upon the raw water characteristics and pretreatment. A typical large RO installation includes a high pressure feed pump, parallel first and second stage membrane elements (in pressure vessels), valving and feed, permeate, and concentrate piping. All materials and construction methods require regular maintenance. Factors influencing membrane selection are cost, recovery, rejection, raw water characteristics, and pretreatment. Factors influencing performance are raw water characteristics, pressure, temperature, and regular monitoring and maintenance (4).

Advantages:

1. Produces highest water quality
2. Can effectively treat wide range of dissolved salts and minerals, turbidity, health and esthetic contaminants, and certain organics; some highly maintained units are capable of treating biological contaminants
3. Low pressure (<100 psi), compact, self-contained, single membrane units are available for small installations

Disadvantages:

1. Relatively expensive to install and operate
2. Frequent membrane monitoring and maintenance; monitoring of rejection percentage for Ba removal
3. Pressure, temperature, and pH requirements to meet membrane tolerances. May be chemically sensitive

14.11.3 LIME SOFTENING

Lime softening uses a chemical addition followed by an up flow SCC to accomplish coagulation, flocculation, and clarification. Chemical addition includes adding $Ca(OH)_2$ in sufficient quantity to raise the pH while keeping the levels of alkalinity relatively low, to precipitate carbonate hardness. Heavy metals, like Ba, precipitate as $Ba(OH)_2$. In the up flow SCC, coagulation and flocculation and final clarification occur. In the up flow SCC, the clarified water flows up and over the weirs, while the settled particles are removed by pumping or other collection mechanisms (4).

Advantages:

1. Other heavy metals are also precipitated; reduces corrosion of pipes
2. Proven and reliable
3. Low pretreatment requirements

Disadvantages:

1. Excessive insoluble Ba may be formed requiring coagulation and filtration
2. Operator care required with chemical handling
3. Produces high sludge volume

14.11.4 ELECTRODIALYSIS REVERSAL

Electrodialysis reversal (EDR) is an electrochemical process in which ions migrate through ion-selective semipermeable membranes as a result of their attraction to two electrically charged electrodes. A typical EDR system includes a membrane stack with a number of cell pairs, each consisting of a cation transfer membrane, a demineralized flow spacer, an anion transfer membrane, and a concentrate flow spacer. Electrode compartments are at opposite ends of the stack. The influent feed water and concentrated reject flow in parallel across the membranes and through the demineralized and concentrates flow spacers, respectively. The electrodes are continually flushed to reduce fouling or scaling. Careful consideration of flush feed water is required.

Typically, the membranes are cation or anion exchange resins cast in sheet form; the spacers are HDPE; and the electrodes are inert metal. Electrodialysis reversal stacks are tank contained and often staged. Membrane selection is based on careful review of raw water characteristics. A single-stage EDR system usually removes 50% of the TDS; therefore, for water with more than 1000 mg/L TDS, blending with higher quality water or a second stage is required to meet 500 mg/L TDS. EDR uses the technique of regularly reversing the polarity of the electrodes, thereby freeing accumulated ions on the membrane surface. This process requires additional plumbing and electrical controls, but increases membrane life, does not require added chemicals, and eases cleaning (4).

Advantages:

1. EDR can operate with minimal fouling or scaling, or chemical addition
2. Low pressure requirements; typically quieter than RO
3. Long membrane life expectancy; EDR extends membrane life and reduces maintenance

Disadvantages:

1. Not suitable for high levels of Fe and Mn, H2S, chlorine, or hardness
2. Limited current density; current leakage; back diffusion
3. At 50% rejection of TDS per pass, process favors low TDS water

14.12   REGULATIONS

The USEPA has determined that barium is not classifiable as a human carcinogen and has assigned it the cancer classification, by using their 1986 guidelines. Using their recent guidelines, the USEPA determined that barium is considered not likely to be carcinogenic to humans following oral exposure and its carcinogenic potential cannot be determined following inhalation exposure.

The Agency for Toxic Substances and Disease Registry (ATSDR) has derived an intermediate-duration oral maximum residue limit (MRL) of 0.2 mg barium/kg/day for barium. This MRL is based on a NOAEL of 65 mg barium/kg/day and a LOAEL of 115 mg barium/kg/day for increased kidney weight in female rats and an uncertainty factor of 100 (10 to account for animal to human extrapolation, and 10 for human variability) and a modifying factor of 3 to account for the lack of an adequate developmental toxicity study.

The USEPA (15) has derived an oral reference dose (RfD) for barium of 0.2 mg/kg/day, based on a BMDL05 of 63 mg/kg/day for nephropathy in male mice and an uncertainty factor of 300 (10 to account for animal to human extrapolation, 10 for human variability, and 3 for database deficiencies, particularly the lack of a two-generation reproductive toxicity study and an adequate investigation of developmental toxicity). The USEPA (15) has not recommended an inhalation reference concentration (RfC) for barium at this time.

ATSDR has derived a chronic-duration oral MRL of 0.2 mg barium/kg/day for barium. The MRL is based on a BMDL05 of 61 mg barium/kg/day for nephropathy in male mice and an uncertainty factor of 100 (10 to account for animal to human extrapolation and 10 for human variability) and a modifying factor of 3 to account for the lack of an adequate developmental toxicity study.

Regulation of Ba contaminated soils has received little attention in general. Current regulatory guidance does not account for differences in speciation and soil properties controlling mobility. In Australia, the National Environmental Protection Council (NEPC) has not provided human health guidance although it does provide an ecological investigation level of 300 mg/kg (19).

The international and national regulations and guidelines regarding barium and barium compounds in air, water, and other media are summarized in Table 14.3.

TABLE 14.3
Regulations and Guidelines Applicable to Barium and Barium Compounds

Agency	Description	Information	References
INTERNATIONAL			
Guidelines:			
IARC	Carcinogenicity classification	No data	IARC (2004)
WHO	Air quality guidelines	No data	WHO (2000)
	Drinking water quality guidelines	0.7 mg/L	WHO (2004)
NATIONAL			
Regulations and			
Guidelines:			
a. Air			ACGIH (2004)
ACGIH	TLV (TWA)	0.5 mg/m^3	
	Barium and soluble compounds (as barium)		
	Barium sulfate	10 mg/m^3	
NIOSH	REL (TWA)		NIOSH (2005a,b)
	Barium chloride[a]	0.5 mg/m^3	
	Barium sulfate	10 mg/m^3 (total)	
		5.0 mg/m^3 (respiratory)	
	IDLH		
	Barium chloride	50 mg/m^3	
	Barium sulfate	No data	
OSHA	PEL(8-h TWA) for general industry		OSHA (2005c)
	Barium, soluble compounds (as Ba)	0.5 mg/m^3	29 CFR 1910.1000
	Barium sulfate	15 mg/m^3 (total dust)	
		5.0 mg/m^3 (respirable fraction)	
	PEL (8-h TWA) for construction industry		OSHA (2005b)
	Barium, soluble compounds (as Ba)	0.5 mg/m^3	29 CFR 1910.1000
	PEL (8-h TWA) for shipyard industry	15 mg/m^3 (total dust)	
	Barium, soluble compounds (as Ba)	5.0 mg/m^3 (respirable	
	Barium sulfate	fraction)	OSHA (2005a)
			29 CFR 1910.1000
b. Water			
EPA	Drinking water standards and health advisories		EPA (2004)
	1-day health advisory for a 10-kg child	0.7 mg/L	
	10-day health advisory for a 10-kg child	0.7 mg/L	
	National primary drinking water standards		EPA (2002a)
	MCLG	2.0 mg/L	EPA (2005b)
	MCL	2.0 mg/L	40 CFR 117.3
	Reportable quantities of hazardous substances (barium cyanide) designated pursuant to Section 311 of the Clean Water Act	10 pounds	
	Water quality criteria for human health consumption of:		EPA (2002b)
	Water + organism	1.0 mg/L	
	Organism only	No data	
c. Food			
FDA	Bottled drinking water	2.0 mg/L	FDA (2004)
			21 CFR 165.110

(Continued)

TABLE 14.3 (*Continued*)

Regulations and Guidelines Applicable to Barium and Barium Compounds

Agency	Description	Information	References
d. Other			
ACGIH	Carcinogenicity classification	A4[b]	ACGIH (2004)
EPA	Carcinogenicity classification	Group D[c]	IRIS (2006)
NTP	RfC	Not recommended at this time	NTP (2005)
	RfD	0.2 mg/kg/day	
	Carcinogenicity classification	No data	

Source: US Department of Health and Human Services Public Health Service Agency for Toxic Substances and Disease Registry. 2007. *Toxicological for Barium and Barium Compounds.* Atlanta, GA.

Note: ACGIH = American Conference of Governmental Industrial Hygienists; CFR = Code of Federal Regulations; DWEL = drinking water equivalent level; U.S. EPA = Environmental Protection Agency; FDA = Food and Drug Administration; IARC = International Agency for Research on Cancer; IDLH = immediately dangerous to life or health; MCL = maximum contaminant level; MCLG = maximum contaminant level goal; NIOSH = National Institute for Occupational Safety and Health; NTP = National Toxicology Program; OSHA = Occupational Safety and Health Administration; PEL = permissible exposure limit; REL = recommended exposure limit; RfC = inhalation reference concentration; RfD = oral reference dose; TLV = threshold limit values; TWA = time-weighted average; and WHO = World Health Organization.

[a] The REL also applies to other soluble barium compounds (as Ba) except barium sulfate.

[b] A4: Not classifiable as a human carcinogen.

[c] Group D: Not classifiable as to human carcinogenicity.

14.13 DISPOSAL

In case of a spill, it is suggested that persons not wearing protective equipment be restricted from the area. Furthermore, ventilation should be provided in the room and the spilled material collected in as safe a manner as possible. Barium compounds (particularly soluble ones) should be placed in sealed containers and reclaimed or disposed of in a secured sanitary landfill (12). It is also suggested that all federal, state, and local regulations concerning barium disposal should be followed. No other guidelines or regulations concerning disposal of barium and its compounds were found (20,21).

REFERENCES

1. Barium. http://www.chemistryexplained.com/elements/A-C/Barium.html (Accessed July 18, 2016).
2. Chemtrails and Barium Toxicity. www.Rense.com (Accessed September 16, 2008).
3. ToxFAQs™ for Barium. http://www.atsdr.cdc.gov/tfacts24.html (Accessed 17 September, 2008).
4. https://www.usbr.gov/research/AWT/reportpdfs/Ba.pdf (Accessed July 19, 2016).
5. US EPA. 1998. Toxicological review of barium and barium compounds. In: *Support of Summary Information on the Integrated Risk Information System (IRIS).* CAS No. 7440-39-3. US EPA: Washington, DC.
6. US Department of Health and Human Services, Public Health Service, Agency for Toxic Substances and Disease Registry, ATSDR. 1992. *Toxicological Profile for Barium.* ATSDR: Atlanta, GA.
7. US Department of Health and Human Services, Public Health Service, Agency for Toxic Substances and Disease Registry. 2007. *Toxicological for Barium and Barium Compounds.* ATSDR: Atlanta, GA.
8. World Health Organization. 2004. *Barium in Drinking Water.* Background document for development of WHO Guidelines for Drinking-Water Quality. WHO: Geneva, Switzerland.
9. US EPA. 1984. *Health Effects Assessment for Barium.* Prepared by the Office of Health and Environmental Assessment, Environmental Criteria and Assessment Office, Cincinnati, OH, for the Office of Emergency and Remedial Response: US EPA: Washington, DC.
10. US EPA. 1984. *Health Effects Assessment for Barium.* US EPA: Washington, DC.

11. US EPA. 1995b. *Integrated Risk Information System (IRIS). Health Risk Assessment for Barium.* Office of Health and Environmental Assessment, Environmental Criteria and Assessment Office, Cincinnati, OH. US EPA: Washington, DC.

12. National Institute for Occupational Safety and Health, Occupational Safety and Health Administration, NIOSH/OSHA. 1978. Occupational health guidelines for chemical hazards: Soluble barium compounds (as barium).

13. World Health Organization. 2001. *Barium and Barium Compounds.* Concise International Chemical Assessment Document 33. WHO: Geneva, Switzerland.

14. ILO. 1983. Barium and compounds. In: Parmeggiani L., ed. *International Labour Office Encyclopedia of Occupational Health and Safety.* Vols. I and II. International Labour Office, pp. 242–244: Geneva, Switzerland.

15. IRIS. 2006. *Barium: Integrated Risk Information System.* US EPA: Washington, DC.

16. Baeza-Alvarado, M.D. and M.T. Olguín. 2011. Surfactant-modified clinoptilolite-rich tuff to remove barium (Ba^{2+}) and fulvic acid from mono- and bi-component aqueous media. *Microporous Mesoporous Mater.*, 139, 81–86.

17. Lukasik-Glebocka, M., K. Sommerfeld, A. Hanc, A. Grzegorowski, D. Baralkiewicz, M. Gaca, B. Zielinska-Psuja. 2014. Barium determination in gastric contents, blood and urine by inductively coupled plasma mass spectrometry in the case of oral barium chloride poisoning. *J. Anal. Toxicol.*, 38, 380–382.

18. Richard, S. 2011. Pappas; toxic elements in tobacco and in cigarette smoke: Inflammation and sensitization. *Metallomics.*, 3, 1181–1198.

19. Abbasi, S., D.T. Lamb, T. Palanisami, M. Kader, V. Matanitobuaa, M. Megharaj, R. Naidu. 2016. Bioaccessibility of barium from barite contaminated soils based on gastric phase *in vitro* data and plant uptake. *Chemosphere*, 144, 1421–1427.

20. Wang, M.H.S. and L.K. Wang. 2015. Environmental water engineering glossary. In: *Advances in Water Resources Engineering.* Yang, C.T., and L.K. Wang (eds). Springer: New York, pp. 471–556.

21. Shammas, N.K. and L.K. Wang. 2016. *Water Engineering: Hydraulics, Distribution and Treatment.* John Wiley and Sons: Hoboken, NJ, 805.

22. Lide, D.R. 2005. CRC Handbook of Chemistry and Physics. CRC Press: New York, pp. 4–50, 4–51, 14–17.

15 Toxicity, Sources, and Control of Selenium, Nickel, and Beryllium in the Environment

Joseph F. Hawumba, Yung-Tse Hung, and Lawrence K. Wang

CONTENTS

ABSTRACT

Heavy metals, which are the stable metals or metalloids whose density is greater than 4.5 g/cm³, include, among others, lead, copper, nickel, cadmium, platinum, zinc, mercury, antimony, arsenic, beryllium, chromium, cobalt, molybdenum, selenium, silver, tellurium, thallium, tin, titanium, uranium, and vanadium. These metals are stable and cannot be degraded or destroyed, and therefore accumulate in the environment. This chapter explores the sources of three metals, namely selenium, nickel, and beryllium, including man-made sources such as industrial point sources and natural sources. Such anthropogenic activities release Se, Ni, and Be into the environment: air, soil, and water, where they not only affect water quality but also enter the food chain where they accumulate in organisms such as fish, making it unfit for human consumption and reducing the growth of some plants. For instance, in high concentrations, heavy metals cause adverse effects on health such as the deterioration of the immune system, nervous system, and metabolic activities. This chapter discusses the various toxic effects of selenium, nickel, and beryllium in man and other animals with specific reference to well-studied aquatic fish and birds.

In order to mitigate the environment of Se, Ni, and Be pollution, a number of processes and methods have been developed to either remove or reduce the levels to acceptable standards. This chapter explains some of the processes that are currently being used to mitigate the environment of the three metal pollutants. Finally, the standards and regulatory role of the U.S. Environmental Protection Agency (USEPA) are highlighted.

15.1   INTRODUCTION

Heavy metals are the stable metals or metalloids whose density is greater than 4.5 g/cm³, namely lead, copper, nickel, cadmium, platinum, zinc, mercury, antimony, arsenic, beryllium, chromium, cobalt, molybdenum, selenium, silver, tellurium, thallium, tin, titanium, uranium, and vanadium (1). It should be pointed out that arsenic, a nonmetal, is often discussed as though it were a heavy metal. Heavy metals are natural constituents of the Earth's crust. They are stable and cannot be degraded or destroyed, and therefore they tend to accumulate in soils and sediments. However, human activities have drastically altered the biochemical and geochemical cycles and balance of some heavy metals. The principal man-made sources of heavy metals are industrial point sources, for example, mines, foundries, and smelters, and diffuse sources such as combustion byproducts, traffic, etc. Relatively volatile heavy metals and those that become attached to airborne particles can be widely dispersed on very large scales. Heavy metals wastes from such anthropogenic activities are released into the environment: air, soil, and water where they not only affect water quality but also enter the food chain where they accumulate in organisms such as fish, making it unfit for human consumption and reducing the growth of some plants. In high concentrations, heavy metals cause adverse effects on health such as the deterioration of the immune system, nervous system, and metabolic activities. Heavy metals are persistent. Therefore, even small amounts in seawater or sediment may become significant to the top of the food chain. Some heavy metals of little toxicity or problem in themselves, such as tin, form complexes with organic materials to produce highly toxic compounds like tributyltin (TBT) used as antifouling paint. However, many are needed in trace amounts by the body for its proper functioning to the extent that their absence may affect the very existence of man.

15.1.1   Nature and Occurrence of Selenium, Nickel, and Beryllium

15.1.1.1   Selenium

Selenium (Se) was discovered in 1817 by the Swedish chemist, Jons Jacob Berzelius, while analyzing a red deposit on the wall of lead chambers used in the production of sulfuric acid. Selenium is a nonmetallic mineral, essentially classified as a metalloid, which lies between sulfur and tellurium in

Group VIA and between arsenic and bromine in Period 4 of the periodic table (2). It can exist either as a gray crystal, red powder, or in a vitreous black form. Se closely resembles sulfur in chemical properties with respect to atomic size, bond energies, ionization potentials, and electron affinities. The major difference between two of these elements is that Se exists in reduced quadrivalent form whereas sulfur occurs in oxidized quadrivalent form (2).

Selenium is naturally present in the Earth's crust at an average concentration of about 0.05 milligram per kilogram (mg/kg), surface water (with a concentration in seawater of about 0.45 microgram per liter [μg/L]), and groundwater. Moreover, higher concentrations may also be found in soils near volcanoes and in minerals such as clausthalite, naumannite, tiemannite, and senenos-ulfur. Even then, selenium does not exist in concentrated deposits. Environmental Se is believed to come from weathering processes of rocks and soils, and as a byproduct of copper refinery slimes (3). Selenium exists in nature as six stable isotopes of which selenium-80 is the most prevalent, comprising about half of natural selenium. The other five stable isotopes and their relative abundances are selenium-74 (0.9%), selenium-76 (9.4%), selenium-77 (7.6%), selenium-78 (24%), and selenium-82 (8.7%). In addition, nine unstable isotopes of Se ranging in atomic weights from 70 to 73, 75, 79, 81, and 83–85 decay thereby emitting radiation. These represent the radioactive isotopes of selenium (4). Of the nine major radioactive selenium isotopes, only one, that is, selenium-79 has a half-life long enough to warrant concern of the U.S. Department of Energy (DOE) environmental management sites such as one at Hanford. As a matter of fact, selenium-79 decays by emitting a beta particle with a half-life of 650,000 years albeit with no attendant gamma radiation. The low specific activity and relatively low energy of its beta-particle limits the radioactive hazards of this isotope. As for the other radioisotopes, the half-life of selenium-75 is 120 days, while the half-lives of all other isotopes are less than 8 h (4).

Around the globe, selenium-79 is present in trace amounts in soil from radioactive fallout, at certain nuclear facilities, such as reactors and facilities that process spent nuclear fuels. For example, the highest concentrations of selenium-79 at the Hanford site are in areas that contain waste from processing irradiated fuel, such as in tanks in the central portion of the site. Selenium is generally one of the less mobile radioactive metals in soil as it preferentially adheres well to soil particles. Concentrations in sandy soil are estimated to be 150 times higher than in interstitial water (in pore spaces between the soil particles), and it is even less mobile in clay soils where concentration ratios exceed 700. The low fission yield of selenium-79 limits its presence at DOE sites. As a consequence, it is generally not a major groundwater contaminant at these sites. Its concentration in plants is typically 0.025 (or 2.5%) of that in soil, with much higher levels in seleniferous plants. Certain foods are especially high in selenium, such as garlic (4).

15.1.1.2 Nickel

Nickel (symbol: Ni, atomic number: 28, atomic weight: 58.7, and oxidation states: +2 and +3) is a hard, silvery-white, magnetic, malleable, ductile metallic element (5) that is naturally present in various ores and to a lesser extent in soil. It occurs in minerals such as garnierite, millerite, niccolite, pentlandite, and pyrrhotite, with the latter two being the principal ores (6). It is also found in most meteorites and often serves as one of the criteria for distinguishing a meteorite from other minerals. Besides, nickel occurs in nature as five stable isotopes (isotopes are different forms of an element that have the same number of protons in the nucleus but a different number of neutrons), and six radioisotopes. Included are (a) nickel-58, this is the most prevalent form, comprising about two-thirds of natural nickel, (b) nickel-60 (26%), (c) nickel-61 (1.1%), (d) nickel-62 (3.6%), and (e) nickel-64 (0.9%) (6). Of the six major radioactive isotopes, only two: nickel-59 and nickel-63 have half-lives long enough to warrant concern. The half-lives of all other nickel isotopes are less than 6 days. Nickel-59 decays with a half-life of 75,000 years by electron capture, and nickel-63 decays with a half-life of 96 years by emitting a beta particle. Both isotopes are present in wastes resulting from the reprocessing of spent nuclear fuel. Nickel-63 is generally the isotope of most concern at U.S. DOE environmental management sites such as Hanford. The long half-life of nickel-59 (with its subsequent low specific activity) combined with its low decay energy limits the radioactive hazards associated with this isotope. Nevertheless, nickel-59 and nickel-63 remain the radionuclides of concern in spent nuclear fuel (as a component of the fuel hardware) and the radioactive wastes associated with operating nuclear reactors and fuel reprocessing plants, of which nickel-63 is present in much higher concentrations than nickel-59 (6).

Nickel is needed in a number of operations, where it is used in alloy form. As such, a number of industrial nations trade in this precious metal. Currently, most of the world's supply of nickel is from Canada; while other sources include Cuba, the former Soviet Union, China, and Australia. The United States has no large deposits of nickel and accounts for less than 1% of the annual world output. For this reason, most of the nickel used in the United States is imported, and about 30% of the annual consumption is from recycled sources (6).

15.1.1.3 Beryllium

Beryllium metal has an atomic number 4, atomic weight 9, and an oxidation state of +2. It is a brittle and hard grayish metal, which normally forms compounds with a sweet taste, but with no specific smell. While some compounds vary in their solubility in water, most of them are insoluble (Table 15.1), and settle to the bottom as particles (7). Beryllium reacts readily with some strong acids, producing hydrogen. In addition, beryllium has a high affinity for oxygen to the extent that a surface film of beryllium oxide forms when the metal is exposed to air. This provides resistance to corrosion, which accompanied with its low density and high electrical and thermal conductivity, makes beryllium an important constituent of many alloys (8). Beryllium (Be) is found in natural deposits as ores containing other elements, and in some precious gemstones such as emeralds, aquamarine, and beryl, as well as I silicate mineral rocks, phenacite, bertrandite, bromellite, chrysoberyl, coal

TABLE 15.1
Physical and Chemical Properties of Beryllium Compounds

	Beryllium Oxide	Beryllium Sulfate	Beryllium Hydroxide	Beryllium Carbonate		Beryllium Fluoride	Beryllium Chloride	Beryllium Nitrate
Molecular formula	BeO	$BeSO_4$	$Be(OH)_2$	$BeCO_3 + Be(OH)_2$		BeF_2	$BeCl_2$	$Be(NO_3)_2 \cdot 3H_2O$
Molecular weight	25.01	105.07	43.03	112.05		47.01	79.93	187.07
CAS registry number	1304-56-9	13510-49-1	13327-32-7	13106-47-3		7787-49-7	7787-47-5	13597-99-4
Specific gravity (20°)	3.01	2.44	1.92	NR		1.986 (25°)	1.899 (25°)	1.557
Boiling point (°C)	3,900	NR	NR	NR		NR	482.3	142
Melting point (°C)	2,530 + 30	Decomposes 550-600	NR	NR		555	399.2	60
Vapor pressure (mm Hg)	NR	NR	NR	NR		NR	1,291	NR
Water solubility (mg/L)	0.2 (30°C)	Insoluble in cold water; converted to tetrahydrate in hot water	Slightly soluble	Insoluble in cold water; decomposes in hot water		Extremely soluble	Very soluble	Very soluble

Source: USEPA, *Toxicological Review of Beryllium Compounds (CAS No. 7440-41-7)*, US Environmental Protection Agency, Washington, DC, April 1998.

Note: NR, Not reported.

deposits, soil, and volcanic dust. Although not naturally found in surface water, beryllium enters water bodies by erosion from rocks and soil. Besides, beryllium may also be introduced into the environment as a waste of coal and fuel combustion, tobacco smoking, as well as other industrial processes (7,9).

The concentration of beryllium in the Earth's crust generally ranges from 1 to 15 milligrams per kilogram (mg/kg) or parts per million (ppm). For instance in the United States, the average concentration of naturally occurring beryllium in soil is 0.6 ppm, with levels typically ranging from 0.1 to 40 ppm. Beryllium concentrations in sandy soils are estimated to be up to 250 times higher than in the interstitial water, that is, the water in the pore space between the soil particles. Notably, much higher concentrations may be found in loam and clay soils, respectively (9).

15.1.2 GENERAL USES OF SELENIUM, NICKEL, AND BERYLLIUM

15.1.2.1 Selenium

Selenium is an essential element for humans, animals (2), and some bacteria and fungi (10), and is incorporated in 25 known human proteins, which are important in various cellular processes (10). Typically, selenium is an important component of proteins (selenoproteins) such as enzymes (selenoenzymes) comprising, among others, (a) the glutathione (GSH) peroxidase family, (b) iodothyronine deiodinases, (c) thioredoxin reductases, (d) selenoprotein P, which is abundant in the plasma and contains up to 10 Se-cysteine moieties per molecule (2,10,11), and (e) formate dehydrogenase from bacteria (12). These enzymes require, for their catalytic activity, selenium, which is in the form of selenocyctein (SeC). As a matter of fact, GSH peroxidase contains 4 g/atom of Se/mole of enzyme (12). The precise molecular mechanism behind the effects of Se in physiologic and in pathologic conditions remain unknown. The recommended dietary intakes for selenium are presented in Table 15.2.

TABLE 15.2
Recommended Dietary Intakes for Selenium[a]

Age Group	Se Levels (μg/day)
Infants	
0–6 months	10
7–12 months	15
Children	
1–3 years	25
4–7 years	30
Adolescents	
8–11 years	50
12–18 years	85
Adults	
19–64 years	
Males	85
Females	70
Pregnancy	+10
Lactation	+15

Source: Tinggi, U., *Toxicol. Lett.*, 137, 103–110, 2003.

[a] These values were published by the Australian National Health and Medical Research Council (NHMRC) (1991).

To understand the nutritional importance of Se, the various roles of the various selenoenzymes has to be explored. For instance, GSH peroxidase is involved in the reduction of hydrogen peroxide to water at the expense of the oxidation of GSH, the enzyme's cofactor. In doing so, it eliminates the toxic effects of hydrogen peroxide, on the one hand, and those of reactive oxygen and lipid peroxide (12,13). Furthermore, the GSH peroxidase family of selonoenzymes has a strong antioxidant role in the gastrointestinal tract (GPx2), extracellular space and plasma (GPx3), and in cell membrane and sperm (GPx4). Moreover, GPx4, also known as phospholipid hydroperoxide (GPx), is involved in detoxification of lipid peroxides inside the cell membranes. On the other hand, GPx5, also called epididymal GPx, is restricted to the epididymis, while the newly discovered, GPx6, is located in the olfactory epithelium and embryonic tissues. Nevertheless, its function in these tissues has not yet been deciphered (14).

Maintenance of cellular redox state is another important function of the GPx family of selenoenzymes. Moreover, GPx are physiologically involved in mediation of such events as differentiation, signal transduction, and regulation of pro-inflammatory cytokine production, as well as the antioxidant defense during spermiogenesis, maturation of spermatozoa, and embryonic development (14). 5^1-deiodinases (iodothyronine deiodinases) are involved in thyroid metabolism, where they are responsible for the conversion of tetraiodothyronine to triiodothyronine, the active thyroid hormone (11,12).

The thioredoxin reductase (TrxR) family has at least three members (TrxR1, TrxR2, and TrxR3), all of which are involved in the thioredoxin system, operating in conjunction with other non-SeC containing protein: thioredoxin peroxidase. The biological role of the system is to provide antioxidant defense, regulate other antioxidant enzymes, modulate several transcription factors, regulate apoptosis, and modulate protein phosphorylation (14).

Selenium in both inorganic (sodium selenate [SeL]) and organic (selenomethionine [SeM] and Se-methylselenocysteine [SeMC]) forms has been demonstrated to exhibit anticancer properties and various mechanisms for preventing cancer with Se have been postulated. Se plays two fundamental roles in cancer prevention. First, some selenoproteins, such as members of the GSH peroxidase family and possibly others, exhibit antioxidant activities, which may account for the prevention of human prostate cancer (10). Second, Se compounds can produce anticarcinogenic metabolites such as methylselenol (CH_3SeH) and hydrogen selenide (H_2Se), which can induce cell death, that is, apoptosis. This is due to their ability to generate oxidative stress. In cancer cells Se compounds have been shown to inhibit cell growth and induce tumor cell apoptosis. Although the exact mechanism is not known, it is surmised that inorganic SeL would trigger apoptosis more in cancer cells by oxidation of thiols and production of reactive oxygen species (ROS) than in normal cells. The differences in response of tumor and normal cells to Se compounds suggest that Se-induced apoptosis is a consequence of the transformation from normal to cancer cells. Typically, SeL exerts its cancer preventing effect by rapidly oxidizing critical thiol-containing cellular substrates, and may therefore be more an effective anticarcinogen than the more frequently used dietary supplements, SeM and SeMC, which require first to be enzymatically converted to methylselenol (CH_3SeH). Likewise, CH_3SeH would eventually undergo direct oxidation to methylselenide (CH_3Se^-), the effective anticarcinogenic agent. SeMC is currently the most promising anticarcinogen because it is the immediate precursor of CH_3SeH and does not produce large amount of H_2Se (10). However, its conversion requires β-lyase, which is not known to be ubiquitous *in vivo* in humans. On the other hand, although SeM can also be directly converted to CH_3Se^- in the presence of methioninase, there is no evidence that this enzyme is present in most human tissues in significant amounts (10). Although inorganic selenium compound, SeL, has shown high *in vitro* anticancer activity, the organic forms, SeM and SeMC, are still considered safer to use in anticancer therapy. Currently, SeMC is considered as one of the most effective Se compounds identified thus far against mammary cancer in animals and it has received the most recent attention as possibly the most useful natural Se compound for cancer reduction. The preference of the organic form of selenium to the inorganic one seems to hinge upon three factors: (a) nutritional bioavailability, (b) less toxicity, and (c) possession of potent cancer chemopreventive activity. These factors are mainly influenced by chemical form and oxidation state of Se. Therefore, the different chemical forms of Se compounds available for human dietary supplementation emphasize the need to establish both the beneficial and detrimental doses of each Se compound (10).

The biological role of selenium, notwithstanding, Se is also extensively applied in a number of industries, including the photocell and solar cell industries, photography industry, where it is used in exposure meters and as a toner, glass industry where Se is used to decolorize glass and to impart a scarlet red color to clear glass, glazes, and enamels, electronics industries, where Se is used in rectifiers, rubber processing industry, where it is used as a vulcanizing agent, making of alloys (additive in stainless steel), constituent of insecticides, and pharmaceutical industry, where it used in manufacture of antidandruff shampoos (4).

15.1.2.2   Nickel

The usefulness of nickel may be underscored from its properties, which render it applicable in a number of processes. Nickel, often chosen for its anticorrosive properties and heat resistance, is most frequently used as a component of several alloys, especially stainless steel. Alloys of nickel provide good corrosion resistance, toughness, strength at high and low temperature, and a wide range of special magnetic and electronic properties. These properties contribute to food and water safety, enhanced product and building lifetimes, cost and energy efficiency, and reliable end uses. In consideration of sustainability, nickel is a desirable material in that its products have long useful lifetimes, and it is one of the most recycled materials globally (15). For instance, nickel is used in various coins and as a component of several alloys, including nichrome and permalloy. While Alnico, an alloy of aluminum, nickel, cobalt, and other metals, is used to make high-strength, permanent magnets, nickel alloy steels are used in heavy machinery, manufacturing, armaments, tools, and high-temperature equipment, such as gas turbines, and in environmental devices used to control emissions such as scrubbers (6). Besides, nickel–titanium (NiTi) shape memory alloys have recently attracted much attention due to their distinctive shape memory effect (SME) and superelasticity (SE) that may not be found in Ti and stainless steels. Due to their SME and SE, nickel–titanium shape memory alloys (NiTi) have attracted much attention as orthopedic materials (16).

Furthermore, nickel compounds are currently being developed for application in chemical-looping technologies. These technologies, which include a chemical-looping combustion category and two categories of chemical-looping reforming, are widely recognized for their potential for power or hydrogen production. Notably, the three techniques are all based on oxygen carriers that circulate between an air- and a fuel reactor, providing the fuel with undiluted oxygen. Two different oxygen carriers, both nickel compounds; $NiO/NiAl_2O_4$ (40/60 wt/wt) and $NiO/MgAl_2O_4$ (60/40 wt/wt) have shown promise and have been evaluated in both continuous and pulse experiments, performed in a batch laboratory fluidized bed at 950°C, using methane as fuel. Of the two oxygen carriers, $NiO/MgAl_2O_4$ offers several advantages at elevated temperatures, that is, higher methane conversion, higher selectivity to reforming, and lesser tendency for carbon formation (17).

Another technology for which nickel is finding promising application is the polymer electrolyte membrane fuel cell (PEMFC) technology. Originally this technology remained on the cusp of commercial viability, being limited only to use in niche applications. The primary reasons for this are the high cost of manufacture and the steady loss in power output during long-term, continuous operation. In addition, its acceptance in transportation markets has been hindered by its current size and weight. To overcome this impasse, a new low-cost, nickel clad bipolar plate concept is currently being developed for use in polymer electrolyte membrane fuel cells (PEMFCs). The development of the nickel clad bipolar plates relies on a newly developed technique, whereby, a powder-pack boronization process is used to establish a passivation layer on the electrolyte exposed surfaces of the bipolar plate in the final stage of manufacture. Under moderate boronization conditions a homogeneous Ni_3B layer grows on the exposed surfaces of the nickel. The thickness of this layer depends on the time and temperature of boronization. At higher temperatures and longer reaction times, a Ni_2B over-layer forms on top of the Ni_3B during boronization. Preliminary results indicate that boronization dramatically improves the corrosion resistance of nickel (18).

15.1.2.3 Beryllium

The greatest use of beryllium is in the making of alloys. Alloys of beryllium with copper are used in the manufacturer of instruments, aircraft parts, springs, electrical connectors, circuit breakers, bearings, gear parts, cameral shutters, and many other industrial components. Other important beryllium alloys with various properties are beryllium–aluminum, beryllium–copper–cobalt, and beryllium–nickel (8,19,20). In addition, beryllium is a component of nuclear reactors, where it serves as a neutron source, owing to its low neutron-absorbing capacity. Besides, pure beryllium metal is also used in missiles and rocket parts, heat shields, and mirrors, while beryllium oxide is used in the electronic industry in manufacturing of insulators, resistors, spark plugs, and microwave tubes (9,20).

15.1.3 SOURCES OF SELENIUM, NICKEL, AND BERYLLIUM AS ENVIRONMENTAL POLLUTANTS

15.1.3.1 Selenium

Selenium pollution is a worldwide phenomenon, associated with a broad spectrum of human activities, which range from mining, agricultural, petrochemical, as well as industrial manufacturing operations (21). These activities release various amounts of selenium into the atmospheric, soil, and aquatic environments. Industrial operations such as coal mining and combustion for power generation enrich and generate high levels of selenium in the wastes. It should, however, be noted that the enrichment of selenium starts naturally during its formation in coal deposits, and other mineral ores in the Earth's crust. Selenium enrichment factor in coal, which is defined as *the ratio of selenium in coal to selenium in the surrounding soils and mineral layers*, may be used to measure the level of selenium concentration and generation by the different operations during power production from coal. When coal is burnt for electricity production, various categories of solid waste and liquid effluents are highly enriched, with an enrichment factor exceeding 65. This ratio is even higher (about 1250 times) in ash that remains after the complete combustion of coal. These are among the highest for trace elements (21). Solid wastes including fly ash, bottom ash, and scrubber ash, in contact with water, generate contaminated leachate because of their oxidation and alkaline pH, which promotes dissolution of selenium ions: selenite and selenate. These selenium ions concentrate further and by the time the discharge or disposal water is sent to the drains, the concentration of Se has arisen to 1000–2700 µg/L (21). Such waste would eventually be released in terrestrial and aquatic/marine environments, where it impacts negatively on both flora and fauna.

Environmental pollution by selenium may come from wastes from gold, silver, and nickel mining operations. The increasing values for gold, silver, and nickel coupled with the new technologies that make extraction of these metals, most especially gold, a profitable endeavor using ore grades, has made the exploitation of low-grade ores that were of either of little or no interest just decades ago, an economically feasible venture. Likewise, several companies in the western United States have invested in such processes as the *Heap-Leach* process, to replace traditional *Deep Shaft and Open Pit* methods. In the *Heap-Leach* process, cyanide-laden water is allowed to percolate through ore piles to dissolve and eventually leach out gold (21). Selenium, which is an important component of the mineral matrix ore deposits, accumulates in the tailings and other surface residuals and ends up in the terrestrial and aquatic environments. For example, many of the mines in North America have left a legacy of environmental damage to lakes and fish populations because of the contaminants that leach from improperly disposed tailings and surface residuals (21).

It should be noted that Se is not obtained from any natural mineral ore of its own, but rather, naturally found in association with other metals. Indeed many metal ores contain variable amounts of selenium. Consequently, the physical–chemical treatments of such ores to extract the desired metal often release selenium and other contaminants into process water and residual solid waste (21).

For instance, when ores containing copper, nickel, and zinc metals are smelted during production, selenium is volatilized and emitted into the atmosphere as vapor. Accordingly, this selenium vapor cools and may either coalesce or adhere to atmospheric dust particles, and subsequently deposit into terrestrial and aquatic systems by either dry or wet deposition. Furthermore, these processes are thought of as crucial factors in the cycling of selenium near smelting facilities (21). Besides, the levels of selenium in copper ores may sometimes exceed those found in coal. For instance, Lemly's report of a study by Nriagu and Wong (1983) suggests that the concentration of Se in copper ore range from 20 to 82 µg Se/g ore in case of copper and 0.4–24 µg Se/g ore in case of coal (21). This implies that contamination from copper smelting operations may by far exceed that from coal and may also span a large area up to as far as 100–200 km from the source of contamination or pollution. It follows that smelting contributes to selenium inputs in distant aquatic systems due to the mechanism by which it is transported in the atmosphere in vapor/particle phase. In this respect, selenium pollution operates on the same principle as the acid rain phenomenon. This means that the emitted Se vapor, like gas phase pollutants, reach aquatic systems and form deposition corridors downwind from major sources (21). Therefore, large-scale metal smelting operations should be viewed as an important contributor to this phenomenon for selenium.

As noted above, selenium is used in the manufacture of various components of electronic equipment. Many such components and industrial waste are dumped in municipal landfills leading to accumulation of selenium in the landfill leachate. As a matter of fact, municipal landfills may generate leach water that contains elevated concentrations of 5–50 µg Se/L of leach if seleneferous materials have deposited there. Such selenium-laden leachates have been generated in many landfills in the United States, the United Kingdom, Sweden, Hong Kong, and Japan (21). Besides, the global distribution of electronics and computer/copier industries, coupled with the practice of landfill disposal of their solid wastes, makes this source an important localized threat of selenium contamination (21).

Apart from the activities discussed above, oil transport, refining, and utilization produce a variety of selenium-laden wastes. Crude oil contains much higher concentrations of selenium than coal: 500–200 µg Se/L in crude oil compared to 0.4–24 µg Se/g in coal, respectively. As a consequence, the potential for hazardous amounts that may be released in process waters and effluents is relatively high. Once in the aquatic environment, selenium rapidly bioaccumulates, causing reproductive failure in fish and aquatic birds. Table 15.3 summarizes the different worldwide locations where selenium environmental pollution has caused devastating effects on aquatic fish and birds (21).

In arid and semiarid regions agriculture is sustained on irrigation. Agricultural irrigation practices in these regions typically use water applications that far exceed what is needed to support crops. Normally the excess is used to flush away salts that tend to accumulate in crop-root zones as evaporation occurs, thereby inhibiting plant growth. Moreover, subsurface irrigation drainage is produced due to a specific set of soil conditions. For example, shallow subsurface (3–10 cm) layers of clay impede the vertical movement of irrigation water as it percolates downward. If not removed, this would result in water logging of the crop-root zone and subsequent buildup of salts as excess water evaporates from the soil surface. This is exactly the problem irrigation was intended to solve in the first place (21). To solve this problem, shallow groundwater must be removed and both wells and surface canals are used to forcefully pump and drain the water away, or rows of permeable clay tile or perforated plastic pipe (3–7 cm) are installed below the surface of agricultural fields. The latter is the method of choice in the western Unites States, whereby, once these drains are installed, irrigation water is applied liberally thus satisfying the water needs of crops while also flushing away excess salts. The resultant subsurface wastewater is pumped or allowed to drain into ponds for evaporative disposal, or into creeks and sloughs that are tributaries to major wetlands, streams, and rivers. Besides, subsurface irrigation drainage is characterized by alkaline pH, elevated concentration of salts, trace elements, and nitrogenous compounds, and low concentrations of pesticides. Although the natural biological and chemical filter provided by the soil effectively degrades and removes most pesticides as irrigation water percolates downward to form surface

TABLE 15.3
Worldwide Locations Where Selenium Has Contaminated Fish and Wildlife Populations

Location	Cause of Selenium Pollution	Major Aquatic Life Contaminated
North Carolina	Coal combustion waste	Reservoir fish
Pennsylvania	Coal landfill waste	Stream fish
West Virginia	Coal mining waste	Stream and lake fish
Minnesota	Municipal landfill leachate	Stream fish
Texas	Coal combustion waste	Reservoir fish
Louisiana	Oil refinery waste	Aquatic birds
Utah	Irrigation drainage	Fish, aquatic birds
Idaho	Phosphate mining waste	Fish, aquatic birds
California	Irrigation drainage	Aquatic birds, fish
Yukon, Canada	Gold mining waste	Stream fish
British Columbia, Canada	Coal mining waste	Stream fish
Ontario, Canada	Metal smelting waste	Stream and lake fish
Chihuahua, Mexico	Irrigation drainage	Stream and river fish
Quito, Ecuador	Gold and silver mining waste	Stream fish
Tefe, Brazil	Gold mining waste	Stream fish
Buenos Aires, Argentina	Gold mining waste	Stream fish
London	Municipal landfill leachate	Stream fish
Stockholm, Sweden	Municipal landfill leachate	Stream fish
Torun, Poland	Nickel and silver mining waste	Stream fish
Perrier Vittel, France	Gold and nickel mining waste	Stream fish
Cairo, Egypt	Irrigation drainage	Fish, aquatic birds
Niamey, Niger	Gold mining waste	Stream fish
Cape Town, South Africa	Gold mining waste	Fish, aquatic birds
Jerusalem, Israel	Irrigation drainage	Fish, aquatic birds
Gorkiy, Russia	Coal combustion waste	Stream and river fish
New Delhi, India	Oil refinery waste	Fish, aquatic birds
Wan Chai, Hong Kong	Municipal landfill leachate	Fish, aquatic birds
Vladivostok, Russia	Metal smelting waste	Stream, estuarine fish
Tokyo, Japan	Municipal landfill leachate	Fish, aquatic birds
New South Wales, Australia	Coal combustion waste	Lake, estuarine fish

Source: Lemly, A.D., *Ecotoxicol. Environ. Saf.*, 59, 44–56, 2004.

drainage, naturally occurring trace elements in the soil, such as selenium (up to 1400 μg Se/L), are leached out under the alkaline oxidizing conditions prevalent in arid climates and are carried in solutions in the drain water (21).

When the subsurface irrigation drain water is discharged into surface waters, a variety of serious biological effects can take place; namely, (a) surface and ground water quality is degraded through salinization and contamination with toxic or potentially toxic trace elements (e.g., selenium, boron, molybdenum, and chromium and (b) long-term impacts may occur if selenium enters aquatic food chains. A typical example is the Kesterson National Wildlife Refuge, California, where thousands of fish and water fowl were poisoned in 1985 from selenium and other trace elements that leached from soils on the west side of the San Joaquin valley and were carried to the refuge in irrigation return flows that were used for wetland management. In this case, it was established that selenium bioaccumulated in aquatic food chains and contaminated 500 ha of shallow marshes. Elevated levels of selenium were also found in every animal group inhabiting these wetlands, from fish and birds to insects, frogs, snakes, and mammals (21).

As has already been pointed out, the agricultural irrigation drainage discharges contain high levels of selenium and other trace elements. This issue emerged as a serious problem in California in the mid-1980s. Of the several mitigation approaches was the application of phytoremediation approach in constructed wetlands. This approach became prominent in the 1990s and has been used to remove selenium from oil refinery effluents. However, the application of constructed wetlands poses a number of risks: (a) as selenium-laden wastewater is being treated as it flows through, the apparent benefits to downstream water quality would be more than offset by the toxic hazards created within the wetland because of bioaccumulation and (b) since wetlands constitute attractive habitats for fish and wildlife, their application increases the likelihood that the fish and wildlife would be exposed to hazardous levels of selenium. The direct consequence of this approach would be a net loss of benefits and creation of an ecological liability that did not exist previously (21). In other words, treatment wetlands may, instead, create selenium problems rather than solve them. Typical example involves a treatment wetland that was constructed at a Chevron corporation's oil refinery in Richmond, California in the mid-1990s. A 40-ha constructed wetland intended for "water enhancement" of conventional pollutant (biochemical oxygen demand [BOD], total organic carbon [TOC], total suspended solids [TSS], and ammonia, among others) was also effective in removing selenium from the waste stream. This was initially viewed as an unanticipated benefit. Besides, the wetland provided an attractive habitat to a large number of migratory waterfowl and shorebirds. However, owing to bioaccumulation, selenium levels exceeded toxic thresholds for wildlife, and the water birds were poisoned. The major problem with the wetland technology is the focus on the removal without parallel studies on the effects of bioaccumulated target pollutant. Such studies, therefore, tend to underestimate the risk to wildlife of the bioaccumulated form of the pollutant, that is, selenium in this case.

In the course of reducing air pollution by seleniferous fly ash produced by combustion of coal, treatment technologies have been developed that may achieve 99.5% reduction of airborne particulate emissions. After removal from air, most fly ash is disposed in landfills that are generally built on clay soils (to impede downward movement of contaminants or upward movement of groundwater), capped with a layer of clay (to impede infiltration of rain water) and topsoil, and revegetated. The problem with this disposal method is that, overtime, landfills become unstable so that either the surface clay cap or the underlying clay develop cracks, resulting in either rainwater or groundwater infiltrating and in the process, leaching selenium. If this happens, selenium-laden seepage (50–200 mg Se/L) may be transported offsite, and ultimately drained into either streams or other surface water bodies. Consequently, selenium would bioaccumulate in the food web, ending up in the fish and wildlife population (21). As a matter of fact, it has been recognized that the design specifications for fly-ash landfills, even under the best conditions, may still produce contaminated leachate. A typical example representative of this problem occurred in 1991 in eastern Pennsylvania, whereby the planned construction of 65-ha fly-ash landfill was halted based on the results of initial environmental risk assessment, which had showed potential health risks caused by selenium-laden leachate to native brook trout (21).

15.1.3.2 Nickel

There are essentially two processes through which nickel may gain entry into the environment: (a) naturally from the weathering of crustal rocks and (b) from human activities involving the metal. For instance, nickel is present in crustal rock at a concentration of about 90 mg/kg, while its concentration in seawater is about 2 milligrams per liter (mg/L). Radioactive isotopes of nickel are also important environmental pollutants. It is now known that trace amounts of nickel-59 and nickel-63 are present around the globe from radioactive fallout. These radioisotopes may also be present, as contaminants from operating reactors and processing spent fuels, at some nuclear facilities. Even then, the environmental threat from radioactive nickel isotopes remains real but low because nickel is generally one of the less mobile radioactive metals in the environment. Notably, the ratio of the concentration of nickel in plants to that in soil is low, estimated at 0.06 (or 6%). This may be due to the fact that Ni adheres quite well to soil, reaching higher concentration ratios of about 400 (nickel association with sandy soils) and in excess of 600 (nickel association with clay soils), respectively, than in interstitial water (i.e., water in the pore spaces between the soil particles). These ratios further suggest that Ni has poor solubility in aqueous media around the plant root zones. Moreover, nickel is generally not a major contaminant in groundwater at DOE sites (6).

15.1.3.3 Beryllium

There are two processes by which beryllium may be released into the environment namely, (a) the natural process whereby it enters the environment (waterways) as a result of the weathering of rocks and soil that contain this metal; (b) through human activities comprising combustion of coal and fuel oil (9), (c) industrial activities such as copper rolling and drawing, nonferrous metal smelting, nonferrous metal rolling and drawing, aluminum foundries, blast furnaces, steel works, and petroleum refining industries (22), and (d) cigarette smoking (20). Since beryllium is relatively insoluble in water and adsorbs tightly to soil, its contamination of drinking water is minimal; and its bioaccumulation in the food chain insignificant (23,24).

The natural means of beryllium release to the environment appears to insignificantly pollute it, as the levels are so low. This leaves human activities as the major sources of environmental pollution by beryllium. For instance activities such as combustion of coal and fuel oil, coupled with the various industrial processes, release beryllium-containing particulates and fly ash into the atmosphere. In addition, air-bone particles of beryllium alloys, oxides, and ceramics are also released from various processing industries (23,24). As these particles pollute the atmosphere, some settle in both water and on land. However, land appears to be the most heavily contaminated due to the reasons highlighted above, that is, the high adsorption capacity of soil particles to beryllium. Tables 15.4 and 15.5 document various levels of beryllium (in pounds; 1 lb = 0.454 kg) released in 1987 and 1993 from various industrial activities. From Table 15.4, for instance, higher concentrations of beryllium in soil than water are notable.

Owing to the high levels of the metal in soil, some would end up being taken up by some plants. The most efficient plants are those that have a natural ability to accumulate beryllium. Also probable, is the effect of pH on solubility of beryllium and its compounds. Since this metal is soluble in strong acids, it is probable also that its solubility may be improved in acidic pH. If this is true, then plants growing in an acidic environment may be able to accumulate it. Notably beryllium levels in drinking water range from 0.01 to 0.7 parts per billion (ppb), while some plants, some of which constitute major foodstuffs, contain Be with a median concentration of 22.5 µg/kg. This median concentration has been reported across 38 different food types, ranging from less than 0.1 to 2200 µg/kg (in kidney beans). Besides, the environment may be polluted by cigarette smoking.

TABLE 15.4
Anthropogenic Beryllium Releases to Aquatic and Land Environments[a]

Major Industries	Beryllium Levels (in Pounds) Released to the Aquatic Environment	Beryllium Levels (in Pounds) Released to Land Environment
Copper rolling, drawing	405	180,502
Nonferrous metal smelting	481	151,790
Aluminum Foundries	5	1,000
Nonferrous rolling, drawing	4	8,000
Blast furnaces, steelworks	250	250
Petroleum refining	142	174

Source: USEPA, *Technical Fact Sheet on Beryllium.* US Environmental Protection Agency, Washington, DC, USA. 2015.
[a] Data estimated for the period; from 1987 to 1993.

TABLE 15.5
Natural and Anthropogenic Emissions of Beryllium to the Atmosphere

Emission Source	Total U.S. Production in Metric Tons (10^6 tons/year)	Emission Factor (g/ton)	Emissions (tons/year)
Natural			
Windblown dust	8.2	0.6	5.0
Volcanic particles	0.41	0.6	0.2
Total			**5.2**
Anthropogenic			
Coal combustion	640	0.28	180
Fuel oil	148	0.048	7.1
Beryllium ore processing	0.008	37.5[a]	0.3
Total			**187.4**

Source: USEPA, *Toxicological Review of Beryllium Compounds (CAS No. 7440-41-7)*, US Environmental Protection Agency, Washington, DC, April 1998.

[a] The production of beryllium ore is expressed in equivalent tons of beryl; the emission factor of 37.5 is hypothetical. 1 ton = 2000 lb = 907 kg.

Typically, one cigarette contains about 0.5–0.7 μg Be, with about 5%–10% escaping into side stream smoke (9). From the foregoing; it may safely be inferred that beryllium may enter the body by eating Be-rich foods, inhaling contaminated air, or smoking a cigarette.

15.2   TOXICITY OF SELENIUM, NICKEL, AND BERYLLIUM

15.2.1   SELENIUM

The early observation of Se toxicity is believed to have been made by Marco Polo during his travel to some regions in China in the thirteenth century when he described a disease called "hoof rot" in horses. The presence of hoof rot disease was localized in areas with high Se soil. Another disease associated with chronic poisoning in animals is called "blind staggers." The Se poisoning comes from feeding, by animals, on plant species classified as primary indicator plants, which naturally accumulate Se up to 1000 mg/kg. The signs of the disease include weight loss, blindness, ataxia, disorientation, and respiratory distress. It is now recognized that acute toxicity of organic and inorganic resorbed selenium compounds has an LD_{50} in the range of 2–5 mg/kg (2,11). Furthermore, Se toxicity in animals was widely recognized in the 1930s in South Dakota, when it was discovered that livestock that grazed in an area of high Se soil developed the disorders known as "alkali disease" and "blind staggers." The "alkali disease" results from chronic poisoning of horses and cattle from continuous ingestion of low Se accumulating plants (classified as secondary indicator plants) containing over 5 but usually less than 50 mg/kg Se. The disease is characterized by dystrophic changes in hooves and a rough hair coat (2).

15.2.1.1   Biochemical Basis of Selenium Toxicity

The primary manifestations of selenium toxicity are due to a simple but important flaw in the process of protein synthesis. Sulfur is a key component of proteins, and sulfur-to-sulfur linkages (ionic disulfide bonds) between strands of amino acids are necessary for protein molecules to coil into their tertiary (helix) structure which, in turn, is necessary for proper functioning of proteins, either as components of cellular structure (tissue synthesis) or as enzymes in cellular metabolism. Selenium is similar to sulfur with regard to its basic chemical and physical properties (has same valence states and forms analogs of hydrogen sulfide, thiosulfate, sulfite, and sulfate). Mammalian and fish cells studies show that they do not discriminate well between selenium and sulfur which are thereby incorporated equally in proteins during protein biosynthesis. This is further supported by the observation that the pathological manifestations and teratogenic features due to selenium toxicity are also the same in fish and in mammals, suggesting that the mechanistic features underlying toxicity are also essentially the same (21). Selenium substitution for sulfur and its eventual incorporation into proteins is postulated to follow this sequence of events: selenium in excessive amounts is erroneously substituted for sulfur, resulting in the formation of a triselenium linkage (Se–Se–Se) or a selenotrisulfide linkage (S–Se–S). Either a triselenium linkage (Se–Se–Se) or a selenotrisulfide linkage (S–Se–S) prevents the formation of the necessary disulfide chemical bonds (S–S). The end result is distorted, dysfunctional enzymes and protein molecules, which impair normal cellular biochemistry (21).

The toxicity of selenium may also result from its effect on the GSH antioxidant system. Inorganic selenium such as sodium selenate (SeL) and organic selenium such as selenomethionine (SeM) exert their toxicity via formation of superoxide radicals and hydrogen selenide (H_2Se). Separate studies have demonstrated that the toxicity due to inorganic selenium (SeL, where Se is in the +4 oxidation state) results from thiol oxidation, redox cycling, and superoxide generation in cells. On the other hand, organic selenium (SeM) may follow either of these metabolic fates: initially, SeM gets incorporated into general body protein in place of methionine thereby increasing the soluble and tissue levels of Se. Second, SeM gets converted into toxic hydrogen selenide (H_2Se) (10). In a redox cycle between selenide and GSH, hydrogen peroxide and superoxide are produced whereby the electrons come from reduced GSH and are transferred by a selenite radical (21).

Apart from substitution for sulfur and disruption of the proper functioning of the GSH system, Se toxicity may indirectly result from mutational damages to DNA. It should be noted that damage to DNA is one of the most important factors that is responsible for the toxic effects of many compounds. Likewise, in their study to determine whether there was DNA damage induction following SeL, SeM, and SeMC treatments in yeast, *Saccharomyces cerevisiae*, double-stranded breaks (DSB) induction by these Se compounds was examined. In contrast to SeM and SeMC, only SeL-induced DSB in exponentially growing cells, indicating that DSB induction may be the basis of the SeL toxicity. Although DSB after SeL exposure may not be induced directly, it is highly probable that the initial single-stranded DNA damage undergoes conversion to DSB. This single-stranded to double-stranded DNA damage conversion is known to happen most frequently in replicating cells, in which either unrepaired DNA single-strand breaks (SSB) or other DNA damage types processed via SSB intermediates are converted to DSB. These breaks lead to the deletion of 1–4 bp, resulting in frame-shift mutations in the open-reading frame (ORF). Accordingly, this will result into production of either defective or nonfunctional proteins (10). In the following sections, typical well-documented cases of the outcomes of selenium-induced errors in protein biosynthesis as well as those owing to substitution of Se for sulfur (25), are explored in detail.

15.2.1.2 Toxicity in Fish

The most well-documented overt toxic symptom in fish is reproductive teratogenesis. Selenium consumed in the diet of adult fish is deposited in eggs, where larval fish metabolizes it after hatching. Consequently, a variety of lethal and sublethal deformities occur in the developing fish, affecting both hard and soft tissues (26). Substitution of selenium for sulfur can also impair proper formation of proteins in juvenile and adult fish, and many internal organs and tissues can develop pathological alterations that are symptomatic of chronic selenosis (27). Pathological alterations in fish are discussed below.

15.2.1.2.1 Gills

The primary structure of adult teleost gills is the semicircular gill arch, usually four pairs. Each arch contains a double row of filaments, each filament having a row of microscopic lamellae projecting from each side. The lamellae contain the blood sinusoids and capillary beds, and are covered by a thin epithelial cell layer, typically two cells thick, underlain by supporting pillar cells, which maintain potency of vascular lumina. Gill lamellae are normally thin, delicate structures, which are necessary for effective gas exchange in respiration. Gills from green sunfish (*Lepomis cyanellus*) exposed to selenium contamination in Belews Lake exhibited extensively dilated blood sinusoids and swollen lamellae (telangiectasia) packed with erythrocytes. Besides, this condition was also associated with hemorrhage of the gill tissue often. Consequently, the selenium-induced dilation of gill lamellae causes impaired blood flow, ineffective gas exchange (reduced respiratory capacity), and metabolic stress response (increased respiratory demand and oxygen consumption) that can lead to death (25).

15.2.1.2.2   Hematology

Hematocrit values (packed cell volumes) are good indicators of hemoglobin levels in blood. Studies of green sunfish from Belews Lake showed that they exhibited significantly reduced hematocrit values (packed erythrocyte volumes) as compared with fish from an uncontaminated reference lake (33% vs. 39%). In addition, blood from this fish had significantly elevated numbers of lymphocytes, with thrombocytes constituting a higher percentage of total leucocytes, while hemoblasts were less numerous than in reference fish (28). These shifts in hematological parameters reflect important changes in the overall health of the fish. For instance, reductions in hematocrit are associated with anemia and lowered mean corpuscular hemoglobin concentration (MCHC) (29). Likewise, reduced MCHC causes impaired respiratory capacity, because selenium can bind to hemoglobin, rendering it incapable of carrying oxygen. A decrease in respiratory capacity would quickly lead to metabolic stress, because, the fish must expend more energy to meet respiratory demands. On the other hand, lower numbers of hemoblasts reflect reduced erythropoiesis and delayed replacement of aging red blood cells in circulation, which also contributes to the reduced respiratory capacity and metabolic stress (29). Furthermore, the elevated levels of lymphocytes signal a generalized immune response triggered by physiological stress and a reduced state of health (25,29).

15.2.1.2.3   Internal Organs: Liver

In an extensive review, Lemly (25) reports a number of pathological conditions caused by selenium toxicity. One such study involved an investigation of the pathological effects of Se toxicity on the liver cells of green sunfish. The structural features of liver tissue from normal green sunfish consist of bilaminar arrays of hepatocytes (liver plates) separated by small blood sinusoids. When blood enters the liver from the hepatic artery and hepatic portal vein, it moves between the liver plates in the sinusoids, and ultimately collects in central veins, which empty into the hepatic veins. Furthermore, parenchymal hepatocytes typically contain numerous mitochondria, rough endoplasmic reticulum, well-developed nucleoli, and both central and peripheral chromatin islands (27). In contrast, Kupffer cells, (phagocytic tissue histocytes) are rarely present in healthy individuals, while lymphocytes, though present, are not numerous. Green sunfish from Belews Lake exhibited several histopathological changes in liver tissue, including (a) lymphocyte infiltration, which was apparent along with extensive vacuolization of parenchymal hepatocytes around central veins, (b) the numbers of Kupffer cells increased, while the central veins became distended and swollen due to loss of surrounding parenchymal cells, (c) cell nuclei were often deformed and pleomorphic, and (d) numerous perisinusoidal lipid droplets (unmetabolized residues) were present (25). Collectively, these ultrastructural changes reflect a degeneration of tissue structure that is sufficient to significantly alter liver function. This liver pathology syndrome is characteristic of chronic selenosis in fish and other vertebrates (25).

15.2.1.2.4   Internal Organs: Kidney

At the ultrastructure level, the kidney of normal fish is quite similar to that of humans, and is made up of glomeruli, mesangial cells, podocytes, endothelial and tubular cells, and both capillary and central veins (which collect and transport urine). Belews Lake green sunfish, which had accumulated high levels of selenium, showed focal intracapillary proliferative glomerulonephritis (25). In this condition, excessive numbers of mesangial cells are present along with an abnormally abundant matrix and periglomerular fibrosis (which can lead to a hardening of the tissue). Besides, numerous tubular casts were present, while the tubular epithelium was desquamated, and vacuolated. In some instances the tubular epithelium was destroyed, thereby rendering the tubular system of the mesonephros incapable of functioning properly. These renal changes in Belews Lake fish were reminiscent of symptoms caused by chronic selenium poisoning in other vertebrates (25).

15.2.1.2.5   Internal Organs: Heart

In the hearts of fish from Belews Lake, a clear pathological pattern occurred. Observable symptoms include the filling of the pericardial spaces surrounding the heart with inflammatory cells, a condition often diagnosed as severe pericarditis, and inflammatory cells also present within the ventricular myocardial tissue, a condition known as myocarditis. The occurrence of pericarditis and myocarditis is attributed to the direct action of selenium on heart tissue, coupled with indirect effects of selenium on the kidney owing to induced glomerulonephritis and associated uremia (25).

15.2.1.2.6   Internal Organs: Ovary

Selenium toxicity produces profound effects on the reproductive organs of fish. For instance, a number of pathological changes have been observed in the ovaries of fish from Belews Lake, which include, among others, swollen, necrotic, and ruptured mature egg follicles, especially in gravid individuals. Such pathology changes have not observed in fish from aquatic environments with either low levels or no selenium intoxication. As a matter of fact, 19 species of fish were affected and the aquatic ecosystem totally altered for over a decade (25,30,31).

15.2.1.2.7   Internal Organs: Eyes

Another organ affected by selenium poisoning is the eye. In this organs selenium poisoning causes selenium-induced cataracts in fish, affecting both the lens and cornea, and has been induced experimentally in mammals by dietary exposure to selenite (25,32). Once again fish from Belews Lake sometimes had corneal cataracts on their eyes (up to 8.1% of fish), while none were found in fish from other lakes. By 1992, selenium residues had fallen in fish, commensurate with reduced selenium inputs to Belews Lake, and the prevalence of cataracts had also fallen, to about 1% (25,26). Apart from cataracts, another abnormality of the eyes that is associated with selenium poisoning in fish is a condition known as edema-induced exophthalmos, or protruding eyeballs. This condition presents with edema emanating from accumulation of fluid in the body cavity and head. Further still, this condition is secondarily caused by tissue damage, specifically an upset in cell permeability as a consequence of distorted selenoproteins in the membrane structure that, in turn, causes internal organs to become "leaky." The ensuing excess fluid creates pressure, which is sufficient to swell the abdomen and force the eyes to protrude from their sockets. If blood is present in the fluid, noticeable hemorrhage around the eyes may be noticeable. Up to 21% of some fish species in Belews Lake exhibited exophthalmos, with the greatest prevalence occurring in crappie, *Pomoxis* spp. (25,26).

15.2.1.3   Teratogenic Deformities

Developmental malformations are among the most conspicuous and diagnostic symptoms of chronic selenium poisoning in fish. Lemly (25) reports studies done using terata, which are permanent biomarkers of toxicity, to reliably identify and evaluate impacts of selenium on fish populations. These deformities in fish, which affect either feeding or respiration may be lethal shortly after hatching, whereby few individuals bearing such terata would survive to join the juvenile population. Besides, terata that are not directly lethal, but which distort the spine and fins, have been found to reduce the swimming ability of fish, leading to increased susceptibility to predation. This is an important indirect cause of mortality. These two factors generally prevent most deformed individuals from surviving to adulthood. In Belews Lake, the reproductive impacts on piscivorous species eliminated much of the predation pressure and allowed many of the deformed individuals of nonpiscivorous species to persist into the juvenile and adult life stages (30). Several types of teratogenic deformities were evident in Belews Lake fish, and many individuals exhibited multiple malformations. The most overt terata were spinal deformities consisting of kyphosis, lordosis, and scoliosis. Less obvious but no less common were terata involving the mouth and fins. The prevalence of deformities varied among species and between years, reaching a high of 70% in green sunfish during 1982. Notable in this period was a close parallel between levels of selenium in fish tissues and frequency of deformities. Terata became more common as selenium increased from 1975 to 1982, peaked in 1982, and decreased in frequency following the cessation of selenium inputs to the lake in 1986 (25,26). By 1996, as the selenium residues fell by 85%–95% from their 1982 high, the prevalence of deformities also fell to as low as 6% or less. Belews Lake was the first site to provide conclusive evidence that exposure to elevated selenium causes teratogenic deformities in natural populations of freshwater fish (25).

15.2.1.4 Apoptosis and Necrosis

Apoptosis, described as programmed active cell death and necrosis, and described as passive cell death with swelling of cells (11), are known consequences of acute toxicity by selenium compounds. Physiologically, apoptosis is fine-tuned in healthy tissue and the balance between cell death and proliferation is the basis of renewal of tissue and control of growth. In tumor cells, the apoptosis program is suppressed or at least partly disabled, giving rise to uncontrolled growth. It is, therefore, tempting to assume that transformed cells are preferentially prone to apoptosis induced by cytotoxic concentrations of selenium compounds, while normal cells would either die by necrosis or be more resistant than tumor cells. However, studies by Weiller and coworkers (11) have demonstrated that bioavailable selenium compounds induce necrotic cell death in transformed hepatocytes as well as in primary cells; leading to the conclusion that at least in the model system investigated preferential apoptotic toxicity to tumor cells is rather unlikely. Further studies in other systems/models should enable the elucidation of a plausible mechanism.

15.2.2 NICKEL

Nickel(II) ion at toxic levels may be released in wastewater streams of various industries such as battery manufacturing plants, among others, that use nickel in any production process step (s). The nickel would end up in the soil or sediment where it may strongly get attached to particles containing iron or manganese. Under acidic conditions, nickel is more mobile in soil and may seep into groundwater. One may be exposed to nickel by drinking water, eating food, and by skin contact with soil, water, and metals containing nickel. Gastrointestinal absorption from food or water is the principal source of internally deposited nickel in the general population. About 5% of the amount ingested is absorbed into the bloodstream through the intestines, while 20%–35% of inhaled nickel is absorbed through the lungs. Of the nickel that reaches the blood, 68% is rapidly excreted in urine, while 2% remains in the kidneys with a very short biological half-life of 0.2 days (about 5 h). The remaining 30% is evenly distributed to all remaining tissues of the body, including the kidneys, and clears with a biological half-life of more than 3 years (1200 days). Nickel can be absorbed into the skin where it may stay, instead of being absorbed into the blood (6).

The toxic effects of nickel have been demonstrated in a number of animal models. Based on such studies, large amounts of nickel in the food of rats and mice causes lung disease and affects the stomach, blood, liver, kidneys, and immune system. Effects on reproduction and birth defects were also found in rats and mice eating or drinking very high levels of nickel. However, the most common adverse health effect of nickel in humans is an allergic reaction to nickel. The most common reaction is a skin rash at the site of contact. Other adverse effects such as unfavorable osteogenesis process and osteonectin synthesis activity as well as high cell death rate have been reported (16). Effects reported in studies of workers chronically exposed to airborne nickel dust include chronic bronchitis, reduced lung function, and cancer of the lung and nasal sinus. It is worth noting that the USEPA has classified nickel subsulfide (a relatively insoluble form of nickel) as a known human carcinogen (6).

In addition, nickel is a radiogenic health hazard only if it is taken into the body. External gamma exposure, however, is not a concern because nickel-63 and nickel-59 do not emit significant gamma radiation. For instance, nickel-63 decays by emitting a beta particle, while nickel-59 decays by electron capture, in which low-energy gamma radiation is emitted. Once taken into the body, radioactive nickel presents emit both beta particles and gamma radiations, which may induce cancer (6).

15.2.3 BERYLLIUM

Beryllium is recognized as being both toxic and carcinogenic; acute exposure has been shown to cause, among others, chronic lung disease in humans, liver necrosis in rats and rabbits, while long-term exposure elicits pulmonary carcinomas and bone marrow sarcomas in experimental animals (33). Before exerting its toxic effects, beryllium and its compounds (oxides, salts) and alloys should either gain entry into the body by either ingestion or inhalation or absorbed through the skin. Entry by ingestion mainly occurs when either the various food types or water contaminated with beryllium are taken. Children and to a lesser extent adults, may be exposed to a limited extent, by ingesting soil. However, the fate of beryllium depends on the form that enters the body. For instance, most beryllium compounds neither dissolve easily nor are they well absorbed (less than 1%) from the gastrointestinal tract. Consequently, if ingested, they are generally excreted in faces (9). Dermal or skin absorption of superfine or ultrafine particles, although expected to be very low, may sometimes occur. Another means of entry of beryllium is through inhalation of dust and fumes containing the metal. Once this has occurred, the beryllium particles would be deposited in the lungs where any of the following may take place: (a) some deposited particles may clear slowly from the lungs and (b) soluble beryllium compounds may be converted to less soluble compounds in about 2–8 weeks. Inhaled beryllium is excreted mainly in the urine (9).

The toxic effects of beryllium were first recognized in Europe in the 1930s. In the 1940s reports of diseases related to beryllium surfaced among workers exposed to beryllium-containing phosphors in the fluorescent lamp industry and nuclear weapons industry (34). Currently, there are two major health effects associated to beryllium namely, acute beryllium disease (ABD) and chronic beryllium disease (CBD) (also called berylliosis). Other possible health effects associated with beryllium are summarized in Table 15.6. Beryllium sensitization (BeS) occurs in the initial stages

---

TABLE 15.6
Possible Human Health Effects of Beryllium Exposure

Target Organ	Disorder
Respiratory tract	Bronchiolitis
	Acute pneumonitis
	CBD or Berylliosis
	Lung cancer
	Pulmonary hypertension[a]
	Pneumothorax[a]
Skin	Contact dermatitis
	Subcutaneous granulomatous nodules
	Ulceration
	Delayed wound healing
Lymphatic/hematologic	Hilar and mediastinal lymphadenopathy[a]
	BeS

Source: ATSDR. Case *Studies in Environmental Medicine, Beryllium Toxicity*, Agency for Toxic Substances and Disease Registry. Atlanta, GA, 2008.

[a] Occurs in association with CBD.

of exposure to beryllium and is found in 1%–16% of exposed workers. In contrast, sensitization may occur in some individuals albeit without beryllium disease. In such individuals there are no associated symptoms and clinical abnormalities revealed by either the pulmonary function test or chest radiography. However, a future risk of developing CBD has been found in these individuals, occurring at a rate of 6%–8% per year (35–38).

ABD occurs after exposure to a high level (more than 1 milligram per cubic meter [mg/m^3]) of relatively soluble forms of beryllium (9) such as beryllium chloride, beryllium fluoride, beryllium nitrate, and beryllium sulphate tetrahydrate (20). The manifestation of the ABD depends on the site of exposure, the form of beryllium, and the dosage. Depending on the site of exposure, beryllium-containing particles may lodge in the skin, respiratory system, or pulmonary region, and the gastrointestinal tract. It has been observed that in cases of either the skin or dermal exposure, contact dermatitis, BeS, ulcerations, and delayed wound healing may be caused. While soluble beryllium compounds specifically cause contact dermatitis, beryllium ulcers occur where a beryllium crystal penetrated the skin most specifically at a site of previous trauma (20,34,39). Besides, the use of beryllium-containing dental prostheses may cause the equivalent of oral contact dermatitis and hand lesions in individuals making oral prostheses (40).

ABD affecting the respiratory system follows after inhalation of high levels of particulate insoluble dust-containing beryllium as well as soluble compounds of the metal. Clinical features are dose-related and usually occur within days but may be delayed for several weeks. The acute disease manifests as inflammation of the upper or lower respiratory tract or both, resulting in what is described as chemical pneumonitis. The disease appears suddenly after short exposure to high concentrations. As with the case for dermal exposure, the form of beryllium is also important in ABD. For instance, while insoluble forms of beryllium may cause such symptoms as irritation of the nose, nasal discharge, and mild epistaxis, soluble forms may cause pneumonitis or rhinitis, and bronchitis (20). In some cases, complaints have been made relating to symptoms such as metallic tastes, anorexia, fatigue, and diarrhea after acute beryllium inhalation, indicating gastrointestinal toxicity (20).

Chronic exposure to beryllium causes CBD, also known as berylliosis, and continues to occur in industries where beryllium and its alloys are processed, smelted, fabricated, and machined. CBD is a disorder in which a delayed type IV hypersensitivity response to beryllium occurs in susceptible individuals leading to noncaseating granuloma formation (41). The period between exposure and disease onset varies. In some instances there may be a latent period of several weeks or years between exposure and the onset of symptoms. Like ABD, both soluble and insoluble forms of beryllium are involved. For example, inhalation of poorly soluble or insoluble beryllium compounds such as beryllium oxide and beryllium dust is the main cause of chronic pneumonitis associated with infiltration of lymphocytes, histiocytes, and plasma cells (20,42). Although the lungs are the main target organ in CBD there is frequently widespread systemic manifestation, thereby contrasting with the acute illness.

Apart from these diseases, the USEPA describes beryllium as a likely human carcinogen especially when inhaled (34,43–45) and it is classified group B1, a probable human carcinogen. In addition to beryllium metal, beryllium compounds including beryllium–aluminum alloy, beryllium chloride, beryllium fluoride, beryllium hydroxide, beryllium oxide, beryllium phosphate, beryllium sulfate, beryllium zinc silicate, and beryl ore, are reasonably anticipated to be carcinogenic. As a matter of fact, the International Agency for Research on Cancer (1993, 2001) has also classified beryllium and beryllium compounds in group 1B encompassing compounds carcinogenic to humans. Evidence in support of the conclusion that beryllium is a possible carcinogen came from two separate studies. One such study involved analyzing the mortality of 689 patients included in the North American beryllium disease case registry mortality from lung cancer (standardized mortality ratio [SMR] = 2.0) and nonmalignant beryllium disease. This study showed significant increase in the cancers, with deaths from lung cancer occurring more frequently in those with acute rather than CBD (20,46).

TABLE 15.7

Observed and Expected Lung Cancer Cases before and after External Adjustment for Differences in Smoking Habits between Exposed Cohorts and U.S. Population, and Corresponding SMRs with 95% CI, Workers from Lorain, Reading, and All Other Plants

	Lung Cancer Observed Cases	No Adjustment for Smoking			Adjustment for Smoking		
Plant		Expected Cases	SMR (CI)	P-Value	Expected Cases	SMR (CI)	P-Value
Lorain	57	33.8	1.69 (1.28–2.19)	0.0003	38.2	1.49 (1.13–1.93)	0.005
Reading	120	96.9	1.24 (1.03–1.48)	0.026	109.8	1.09 (0.91–1.31)	0.353
All others	103	90.8	1.13 (0.93–1.38)	0.222	102.8	1.00 (0.82–1.22)	0.990
Total	280	221.5	1.26 (1.12–1.42)	0.0002	250.8	1.12 (0.99–1.26)	0.074

Source: USEPA, *Toxicological Review of Beryllium Compounds (CAS No. 7440-41-7)*, US Environmental Protection Agency, Washington, DC, (April, 1998).

In another study by Ward and coworkers (20) a significant increase in SMR for lung cancer in workers at two beryllium plants in operation before 1950 was found. It is from these studies and similar ones that the IARC working group on the carcinogenicity of beryllium concluded that there is "sufficient evidence in human for the carcinogenicity of beryllium and beryllium compounds" (20, p. RTECS-CC4025000). Table 15.7 is a summary of the data from a study of War and coworkers compiled by the USEPA (43). Presented are the observed and expected lung cancer cases before and after external adjustment for differences in smoking habits between exposed cohorts and U.S. population, and corresponding standardized mortality ratios (SMRs) with 95% confidence intervals (CIs), workers from Lorain, Reading, and all other plants. An observed increase in lung cancer cases among workers with beryllium exposures, coupled with the lack of evidence for confounding by cigarette smoking (Sanderson and coworkers) provides further evidence that beryllium is a human lung carcinogen (34).

The mechanism of carcinogenesis is not well understood but some studies have indicated interference, by beryllium, with the hormonal regulation of gene expression (33). Evidence in support of this possible mechanism comes from studies conducted by Perry and coworkers, where the effects of metallo-carcinogen beryllium on the regulation of gene expression were assessed by analyzing the hormonal regulation of the synthesis of tyrosine aminotransferase in beryllium-treated hepatoma cell cultures. The study revealed that induction of enzyme synthesis by glucocorticoids was specifically impaired, reducing it to a low level of 50% of the untreated cells. Since induction by insulin or cyclic adenosine 3′,5′-monophosphate (cAMP) were not influenced by the metal, it shows that beryllium exerts its action by selectively impairing the mechanism involved in the steroid-mediated regulation of gene expression, which occurs at transcription level (33). Such alterations in gene expression, however, are typical of cancer cells. Consequently, the loss of capacity to regulate gene expression, which is thought to constitute a major component of the cascade of molecular dysfunction leading to cancer, is expected to occur early in the cascade. The carcinogenicity of beryllium appears to take this route (33).

Despite such suggestive evidence, studies done by some other investigators have disputed the reported risks to lung cancer by workers in beryllium industries. Besides, mutation and chromosomal aberration assays have yielded somewhat contradictory results. Furthermore, only a limited number of studies have addressed the underlying mechanisms of the carcinogenicity and mutagenicity of beryllium. Therefore, it is likely that the different chemical forms of beryllium probably have different mutagenic and carcinogenic effects, thereby explaining why there is still some confusion as to the mechanisms underlying the carcinogenicity of beryllium and its accompanying cancer risk to humans (34).

15.3 POLLUTION MITIGATION PROCESSES

15.3.1 Safety and Health Requirements for Treatment Processes

Before any safety measure is carried out, risk factors for persons employed in such industries need to be established. For instance, persons working with either Se or Ni risk contracting cancer because of exposure to radionuclides. Owing to this possibility, the USEPA has developed toxicity values to estimate the risk of getting cancer or other adverse health effects associated with the chemical toxicity of selenium and nickel. It should be noted that the toxicity value for estimating cancer risk following inhalation exposure is called a unit risk (UR). The UR is an estimate of the chance that a person will get cancer from continuous exposure to a chemical in air at a concentration of 1 mg/m^3. For example, using the inhalation UR, the USEPA estimates that a person would have a one-in-a million chance of developing cancer if exposed daily over a lifetime to air containing 0.002 microgram per cubic meter (μg/m^3) nickel subsulfide (6). Consequently, personnel involved with demineralization treatment processes should be aware of the chemicals being used (should check or consult the material safety data sheet [MSDS] information), the electrical shock hazards, and the hydraulic pressures required to operate the equipment. General industry safety, health, and self-protection practices should be followed, including proper use of tools (5).

15.3.2 Physicochemical Processes

15.3.2.1 Coagulation and Filtration

Precipitation of metals has long been the primary method of treating metal-laden industrial waste-water. The process involves the conversion of soluble heavy metal salts to insoluble salts that will precipitate. The conversion process usually uses pH adjustment, addition of a chemical precipitant (coagulant), and flocculation. Typically, metals precipitate from the solution as hydroxides, sulfides, or carbonates. However, the precipitation process can generate very fine particles that are held in suspension by electrostatic surface charges. These charges cause clouds of counter ions to form around the particles, giving rise to repulsive forces that prevent aggregation and reduce the effectiveness of subsequent solid–liquid separation processes. Therefore, chemical coagulants are often added to overcome the repulsive forces of the particles (47). Coagulants act by destabilizing the particles (colloids) by neutralizing the forces that keep them apart. This action by coagulants is described as coagulation. Cationic coagulants provide positive electric charges to reduce the negative charge (zeta potential) of the colloids. As a result, the particles collide to form larger particles (flocs). When flocs have been formed, flocculants are added whose action involves the formation of bridges between the flocs, and binding of the particles into large agglomerates or clumps. Bridging occurs when segments of the polymer chain adsorb on different particles and help particles aggregate. This action by flocculants is described as flocculation. An anionic flocculant will react against a positively charged suspension, adsorbing on the particles and causing destabilization either by bridging or charge neutralization. In this process it is essential that the flocculating agent be added by slow and gentle mixing to allow for contact between the small flocs and to agglomerate them into larger particles. The newly formed agglomerated particles are quite fragile and can be broken apart by shear forces during mixing. The precipitate can then be removed from the treated water by physical methods such as clarification (settling) and/or filtration (48). In the treatment of Se laden wastewater, Fe$_2$(SO$_4$)$_3$ has been proven to be the most effective coagulant for Se (+4) removal; while Al$_2$(SO$_4$)$_3$ is the most effective for Se (+6) removal. Filtration provides final removal by dual media filtering of all floc and suspended solids. The same process is applicable on both nickel and beryllium removal (4,5). The use of coagulation and filtration is desirable because of the advantages it offers: (a) it requires the lowest capital costs, (b) lowest overall operating costs, (c) it is proven and reliable, and (d) has low pretreatment requirements. However, the process requires caring for the operator who handles the chemical, it produces high sludge volume and may be prone to interference with removal efficiency if waters are high in sulfate (4).

15.3.2.2 Lime Softening

Lime softening (LS) uses chemical additions followed by an upflow through the solid contact clarifier (SCC) to accomplish coagulation, flocculation, and clarification. Prior to this treatment, a pretreatment test involving the use of a test jar may be performed to establish the optimum pH for coagulation, and what adjustment in pH is required. The chemicals added include (a) $Ca(OH)_2$ to precipitate carbonates and (b) Na_2CO_3 to precipitate noncarbonate hardness.

When $Ca(OH)_2$ is used, followed by the upflow SCC step, coagulation and flocculation (agglomeration of the suspended material into larger particles) and final clarification take place. Thereafter, the clarified water flows up and over the weirs, while the settled particles are removed by either pumping or by other collection mechanisms, for example, filtration. LS using these two types of chemical additions: $Ca(OH)_2$ and Na_2CO_3 is currently being used in Se, while $Ca(OH)_2$ is sufficient in softening soluble Ni and Be. The chemical treatments raise the pH to about 10 required to precipitate carbonate hardness and heavy metals, like Ni. Like the coagulation and filtration process described above, LS offers several benefits: (a) other heavy metals are also precipitated, thereby reducing corrosion of pipes, (b) it is a proven and reliable process, and (c) the process has low pretreatment requirements.

These benefits, notwithstanding, LS may be disadvantageous in the following ways: (a) operator care is required with chemical handling, (b) the process produces high sludge volume, and (c) waters high in sulfate may cause significant interference with removal efficiencies and in case of Ni, insoluble Be and Ni compounds may be formed at low carbonate levels requiring coagulation and flocculation (3,5,7).

15.3.2.3 Activated Alumina

In contrast to the two previously explained processes above, activated alumina (AA) uses an extremely porous media in a physical/chemical separation process known as adsorption, where molecules adhere to a surface with which they come into contact due to forces of attraction at the surface. An AA medium is made by aluminum ore being activated by passing oxidizing gases through the material at extremely high temperatures. In addition, this activation process produces pores with high adsorption properties. Contaminated water is passed through a cartridge or canister of AA. As the wastewater flows through, the medium adsorbs the metal ion contaminants. The adsorption process depends on the following factors: (a) physical properties of the AA, such as method of activation, pore size distribution, and surface area; (b) either the chemical/electrical nature of the alumina source or the method of activation and the amount of oxygen and hydrogen associated with them. These parameters make the alumina surfaces become filled thereby making the actively adsorbed contaminants displace the less actively adsorbed ones; (c) chemical composition and concentration of contaminants; (d) temperature and pH of the water, whereby adsorption usually increases as temperature and pH decreases; and (e) the flow-rate and exposure time to the AA, whereby low contaminant concentration and flow rate with extended contact times increase the media life. Commercial preparations are available in powder, pellet, or granular form. AA devices include pour-through for treating small volumes; faucet-mounted (with or without by-pass) for point of use (POU); in-line (with or without by-pass) for treating large volumes at several faucets; and high-volume commercial units for treating community water supply systems. Careful selection of alumina to be used is based on the contaminants in the water and manufacturer's recommendations. Like the coagulation and filtration process described above, and LS processes, AA offers several benefits: (a) it is a well-established process, (b) it is suitable for some organic chemicals, some pesticides, and trihalomethanes (THMs), (c) it is suitable for home use because the process is typically inexpensive, with simple filter replacement requirements, and (d) it improves taste and smell; removes chlorine. These benefits, notwithstanding, AA may be disadvantageous in the following ways: (a) its effectiveness is based on contaminant type, concentration, and rate of water usage, (b) microbes such as bacteria may grow on the alumina surface, thereby affecting their adsorptive properties, (c) the process requires adequate water flow and pressure for backwashing/flushing, and (d) the process requires careful monitoring (7).

15.3.3   REVERSE OSMOSIS

Reverse osmosis (RO) is a filtration process typically used for water. It works by using pressure to force a solution through a membrane, retaining the solute on one side and allowing the pure solvent to pass to the other side. This is the reverse of the normal osmosis process, which is the natural movement of solvent from an area of low solute concentration, through a membrane, to an area of high solute concentration when no external pressure is applied (49). RO membranes reject ions based on size and electrical charge. The membranes used for RO have a dense barrier layer in the polymer matrix where most separation occurs. In most cases the membrane is designed to allow only water to pass through while preventing the passage of solutes (such as salt ions). Although various membrane types have been developed, the common RO membrane materials in use are the asymmetric cellulose acetate or polyamide thin film composite. Moreover, the common membrane construction is either the spiral wound or the hollow fine fiber. Each material and construction method has specific benefits and limitations, which depends upon the raw water characteristics and pretreatment. A typical large RO installation includes a high-pressure feed pump, parallel first and second stage membrane elements (in pressure vessels); valving; and feed, permeate, and concentrate piping. All materials and construction methods require regular maintenance. Factors influencing membrane selection are cost, recovery, rejection, raw water characteristics, and pretreatment. Factors influencing performance are raw water characteristics, pressure, temperature, and regular monitoring and maintenance. A typical RO process requires that a high pressure be exerted on the high concentration side of the membrane. Notably, a pressure of 2–17 bar (30–250 psi) is used for fresh and brackish water, while 40–70 bar (600–1000 psi) is used for seawater, which has around 24 bar (350 psi) natural osmotic pressure that must be overcome (7,49). In essence, there are two forces influencing the movement of water: the pressure caused by the difference in solute concentration between the two compartments (the osmotic pressure) and the externally applied pressure. The process is currently applied in a number of purification operations namely, drinking water purification, water and wastewater purification, desalination, concentration of liquid foods, maple syrup production, and hydrogen production to prevent the formation of minerals (49). It can therefore also be of use in removing metal ions such as Se, Ni, and Be.

Before using the RO membranes the raw water or any material to be purified by this process ought to be pretreated. Pretreatment when working with RO membranes is important due to the nature of their spiral wound design. The material is engineered in such a fashion to allow only one way flow through the system. As such the spiral wound design does not allow for backpulsing with either water or air agitation to scour its surface and remove solids. Since accumulated material cannot be removed from the membrane surface systems they are highly susceptible to fouling (loss of production capacity). Therefore, pretreatment is a necessity for any RO system. This should involve performing a careful review of raw water characteristics and pretreatment needs to prevent membranes from fouling, scaling, or other membrane degradation. Removal of suspended solids from the raw water is necessary to prevent colloidal and biofouling, while removal of dissolved solids is necessary to prevent scaling and chemical attack. Large installation pretreatment can include media filters to remove suspended particles; ion-exchange softening or antiscalant to remove hardness; temperature and pH adjustment to maintain efficiency; acid to prevent scaling and membrane damage; activated carbon or bisulfite to remove chlorine (postdisinfection may be required); and cartridge (micro) filters to remove some dissolved particles and any remaining suspended particles (7,49).

15.3.4   ELECTRODIALYSIS AND ELECTRODIALYSIS REVERSAL

Electrodialysis (ED), which is a membrane separation process based on the selective migration of aqueous ions through ion-exchange membranes as a result of an electrical driving force, is one of the most important processes applied in desalting solutions (50,51). Moreover, in the ED process, only the dissolved solids move through the membranes, while the solvent does not, meaning that either practical concentrations or depletion of electrolyte solutions is possible. Further still, the transport direction and the transport rate for each ion in the ED process depend, among others, on its charge and mobility, solution conductivity, relative concentrations, applied voltage, as well as the ion separation, which is also closely related to the characteristics of the ion-exchange membranes, especially its permselectivity in the system being used (51). In its practical application, two main streams flow in parallel through the membrane stack. One of the streams, which is progressively desalted, is referred to as the product stream, while the other main, a fraction of which is recirculated to reduce the quantity of waste water, is called the concentrate stream. Since the stream increases in concentration, an addition of acid and conditioning chemicals may be required to prevent membrane stack scaling (51).

In order to improve the efficiency of an ED stack, there is need to periodically reverse the polarity of the electrodes, which, in turn, would reverse the direction of ion movement within the membrane stack. As a result, the dilute stream becomes the concentrate stream and vice versa. This is called electrodialysis reversal (EDR). EDR employs periodic reversal of the DC electric field. Typical field reversal frequencies range from 15 to 30 min. When the field is reversed, the electric driving force is reversed, which tends to remove deposited colloids into the brine stream (52). Typically, in EDR, ions migrate through ion-selective semi-permeable membranes as a result of their attraction to two electrically charged electrodes. A typical EDR system includes a membrane stack with a number of cell pairs, each consisting of a cation transfer membrane, a demineralized flow spacer, an anion transfer membrane, and a concentrate flow spacer. Electrode compartments are at opposite ends of the stack. Typically, the membranes are either cation or anion exchange resins cast in sheet form; the spacers are high-density polyethylene (HDPE); and the electrodes are inert metal. EDR stacks are tank contained and often staged. Membrane selection is based on careful review of raw water characteristics. The influent feed water (chemically treated to prevent precipitation) and the concentrated reject flow, respectively, in parallel across the membranes and through the demineralized flow spacers, and finally through the concentrate flow spacers. In addition, the electrodes are continually flushed to reduce either fouling or scaling. Even then, careful consideration of flush feed water is required. A single-stage EDR system usually removes 50% of the total dissolved solids (TDS); therefore, for water with more than 1000 mg/L TDS, blending with higher quality water or a second stage is required to meet 500 mg/L TDS. Since EDR uses the technique of regularly reversing the polarity of the electrodes, thereby freeing accumulated ions on the membrane surface, additional plumbing and electrical controls may be required. This would increase membrane life, eliminate the requirement to add chemicals, and eases cleaning.

Like in RO, there is need to pretreat the raw water to acceptable limits of pH, organics, turbidity, and other raw water characteristics. Typically, the pretreatment step requires chemical feed to prevent scaling, acid addition for pH adjustment, and a cartridge filter for prefiltration. After use, the membrane, and electrodes require regular maintenance. Sometimes solids are not all removed during the pretreatment process, and these will accumulate on the membrane. Such solids can be washed off by turning the power off and letting water circulate through the stack. In order to restore the electrodes, they should be washed to flush out byproducts of electrode reaction such as hydrogen, formed in the cathode space, as well as oxygen and chlorine gas, formed in the anode spacer. As a matter of fact, if the chlorine is not removed, then toxic chlorine gas may form. Depending on raw water characteristics and metal ions (Se, Ni, and Be) concentration, the membranes will require regular replacement. In addition to reversing the polarity, the EDR system also requires flushing at high volume and low pressure. Flushing is continuously required to clean the electrodes.

Like the other processes described earlier, the EDR process also offers several benefits: (a) EDR can operate with minimal fouling or scaling, requires either no chemical addition or small chemical addition and thus is suitable for higher TDS sources, (b) can be operated at low pressure and therefore is typically quieter than RO, and (c) EDR extends membrane life and as a consequence reduces maintenance costs. Even with the above benefits, EDR may be disadvantageous in the following ways: (a) EDR is not suitable for high levels of Fe and Mn, H_2S, and chlorine, (b) EDR may suffer from a limited current density, current leakage, and back diffusion, and (c) at 50% rejection of TDS per pass, the process is limited to water with 3000 mg/L TDS or less (3,5,7).

15.3.5   Ion Exchange

Ion-exchange materials are insoluble substances containing loosely held ions which are able to be exchanged with other ions in solutions which come in contact with them. These exchanges take place without any physical alteration to the ion-exchange material. Ion exchangers are insoluble acids or bases, which have salts that are also insoluble, and this enables them to exchange either positively charged ions (cation exchangers) or negatively charged ones (anion exchangers). Many natural substances such as proteins, cellulose, living cells, and soil particles exhibit ion-exchange properties which play an important role in the way they function in nature (53). In solution, salts separate into positively charged cations and negatively charged anions. Deionization can reduce the amounts of these ions. Cation IX is a reversible chemical process in which ions from an insoluble, permanent, solid resin bed are exchanged for ions in water. The process relies on the fact that water solutions must be electrically neutral, therefore ions in the resin bed are exchanged with ions of similar charge in the water. As a result of the exchange process, no reduction in ions is obtained. In the case of beryllium reduction, operation begins with a fully recharged cation resin bed, having enough positively charged ions to carry out the cation exchange. Usually a polymer resin bed is composed of millions of medium sand grain size, spherical beads. As water passes through the resin bed, the positively charged ions are released into the water, being substituted or replaced with the metal ions in the water (ion exchange). When the resin becomes exhausted of positively charged ions, the bed must be regenerated by passing a strong, usually NaCl (or KCl), solution over the resin bed, displacing the metal ions such as Be^{2+} with Na or K ions. Many different types of cation resins can be used to reduce dissolved beryllium concentrations. The use of IX to reduce concentrations of metal ions such as Be^{2+} will be dependent on the specific chemical characteristics of the raw water. Cation IX, commonly termed water softening, can be used with low flows, up to 200 gallons per minute (gpm) (1 gpm = 3.785 L/min), and when the ratio of hardness to Be is greater than 1 (7).

Like in RO and EDR processes, a pretreatment step of the raw water is required. The raw water should be pretreated to accepted limits for pH, organics, turbidity, and other raw water characteristics. Pretreatment involving both media and carbon filtration may also be required to reduce excessive amounts of TSS, which could plug the resin bed. Besides, the IX resin requires regular regeneration, the frequency of which depends on the raw water characteristics and metal ions concentration (7).

Like the mitigation processes described above, the ion-exchange process also offers some benefits: (a) the process does not require acid addition, degasification, and repressurization, (b) it is easy to operate and also highly reliable, (c) it involves lower initial cost, since resins will not wear out with regular regeneration, (d) it is effective, and therefore, widely used, (e) it is suitable for small- and large-scale installations, and (f) a variety of specific resins are available for removing specific contaminants. However, as seen for other operations, ion-exchange process may be disadvantageous in the following ways: (a) a pretreatment LS step may be required, (b) the process requires salt storage and regular regeneration, making it cumbersome, (c) the process creates a concentrate disposal problem, (d) it is usually not feasible in treating wastewater with high levels of TDS, and (e) the resins are sensitive to the presence of competing ions (7).

15.4   BIOSORPTION

It has been recognized that the physicochemical methods such as those briefly outlined in the foregoing Section 15.3 are either ineffective or expensive especially when the heavy metal ions are in solutions and at concentrations in the order of 1100 mg of dissolved heavy metal ions/L. For instance, activated carbon is only able to remove around 30–40 mg/g of Cd, Zn, and Cr in water and is nonregenerable, making it quite costly when used in wastewater treatment. In addition, the precipitation method often results in sludge production, while ion exchange, which is considered a better alternative technique, is not economically appealing because of high operational cost. As a result of these, biological methods such as biosorption/bioaccumulation for the removal of heavy metal ions may provide an attractive alternative to physicochemical methods (54).

Biosorption strategies consist of a group of applications involving the detoxification of hazardous substances such as heavy metals by transferring them from one medium to another by means of biosorbents, which may either be microbes or plants. Biosorption options are generally characterized as being less disruptive and may henceforth be carried out on site, thereby eliminating the need to transport the toxic materials to treatment sites (54). Biosorption is a very cost-effective method because biosorbents are prepared from naturally abundant biomass, which include, among others, nonliving plant biomass materials such as maize cob and husk, sunflower stalks, *Medicago sativa* (alfalfa), cassava waste, wild cocoyam, sphagnum peat moss, sawdust, chitosan, sago waste, peanut skins, shea butter seed husks, banana pith, coconut fiber, sugar-beet pulp, wheat bran, sugarcane bagasse (54), and *Cassia fistula* (55). Several studies have shown that these biomass materials are effective in the removal of trace metals from the environment. Typically, *C. fistula* has been studied for possible application as a very promising biosorbent for the removal Ni(II) from synthetic aqueous solutions. The feasibility of using *C. fistula* as a biosorbent lies in its numerous ionizable chemical groups comprising carboxyl, carbonyl, alcoholic, and amino groups. Such groups make it a good option for use as a biosorbent in metal biosorption (56).

In their study to explore the ability of *C. fistula* waste biomass to remove Ni(II) from industrial effluents, Hanif and coworkers found that *C. fistula* biomass is very effective in removing Ni(II) from wastewater produced by various industries. These ranged from the ghee industry (GI), nickel chrome plating industry (Ni–Cr PI), battery manufacturing industry (BMI), tannery industry: lower heat unit (TILHU), tannery industry: higher heat unit (TIHHU), textile industry: dyeing unit (TIDU), and textile industry: finishing unit (TIFU). In these industries, the initial Ni(II) concentration in their industrial effluents was found to be 34.89, 183.56, 21.19, 43.29, 47.26, 31.38, and 31.09 mg/L in GI, Ni–Cr PI, BMI, TILHU, TIHHU, TIDU, and TIFU, respectively. After biosorption the final Ni(II) concentration in industrial effluents was found to be 0.05, 17.26, 0.03, 0.05, 0.1, 0.07 and 0.06 mg/L in GI, Ni–Cr PI, BMI, TILHU, TIHHU, TIDU, and TIFU, respectively. Accordingly, the percentage (%) sorption Ni(II) ability of *C. fistula* from seven industries included in their study was in the order: TILHU (99.88) >GI (99.85) ≈BMI (99.85) > TIFU (99.80) > TIHHU (99.78) > TIDU (99.77)»Ni–Cr PI (90.59). Due to unique high Ni(II) sorption capacity of *C. fistula* waste biomass it can be concluded that it is an excellent biosorbent for Ni(II) uptake from industrial effluents (55).

In a separate study, Igwe and Abia (57) determined the equilibrium adsorption isotherms of Cd(II), Pb(II), and Zn(II) ions detoxification from wastewater using unmodified and ethylenediaminetetraacetic acid (EDTA)-modified maize husks as biosorbent. This study established that maize husks are excellent adsorbents for the removal of these metal ions, with the amount of metal ions adsorbed increasing as the initial concentrations increased. The study further established that EDTA modification of maize husks enhances the adsorption capacity of maize husks, which is attributed to the chelating ability of EDTA. Therefore, this study demonstrates that maize husks, which are generally considered as biomass waste, may be used as adsorbents for heavy metal removal from wastewater streams from various industries and would therefore find application in various parts of the world where development is closely tied to affordable cost as well as environmental cleanliness (57).

15.5 STANDARDS AND REGULATIONS FOR SELENIUM, NICKEL, AND BERYLLIUM

In order to establish and enforce to protect the public from adverse health effects resulting from a drinking water contaminant, the USEPA, under the authority of the Safe Drinking Water Act (SDWA), Public Law 93523, Title XIV of the Public Health Service Act, is mandated to set up the National Primary Drinking Water Regulations (NPDWRs) for contaminants occurring in drinking water. In these regulations are included the drinking water standards which set either (a) treatment techniques to control a contaminant or (b) the maximum contaminant level (MCL) allowable for the contaminant in drinking water. The MCL is set when an appropriate method of detection for such a contaminant exists. A treatment technique approach is used when it is not possible to quantify the contaminant at the level necessary to protect public health. In addition, secondary standards are established based on nonhealth-related esthetic qualities of appearance, taste, and odor. While the primary standards are federally enforceable, the secondary maximum contaminant level (SMCL) standards are not, except at the request of a community, as is the practice in California. On the other hand, in the USEPA Office of Water it is also mandated to develop the maximum contaminant level goals (MCLGs), which are used as the first step toward promulgation of NPDWRs. It is noteworthy that the MCLGs are nonenforceable health goals, which are to be set at levels at which no known or anticipated adverse effects on the health of persons would occur, and which allow for an adequate margin of safety (58). Table 15.8 shows the standards set for selenium, nickel, and beryllium metals, respectively. The discharge standards expressed in terms of MCL and MCLGs are, respectively, 50 ppb for Se, 100 ppb for Ni, and 4 ppb for Be. Besides, for beta-particle emission, they are 50 pCi/L (MCL) for Se-79, Ni-59, and Ni-63 isotopes, while the MCLG has been set at 0 pCi/L (1, 22, 59–63).

TABLE 15.8

Standards and Regulations for Selenium, Nickel, and Beryllium

Metal	MCLs[a]	MCLGs[b]	Units	Regulatory Agency	Focus	Likely Source of Contaminant
Selenium	50	50	ppb	U.S. EPA	Drinking water	Discharge from petroleum, glass, and metal refineries; erosion of natural deposits; discharge from mines, chemical manufacturing and runoff from livestock lots (feed additives)
Selnium-79 Isotope	50	0	pCi/L	U.S. EPA	Drinking water	Decay of natural and man-made deposits
Nickel	100	100	ppb	U.S. EPA	Drinking water	Pollution from mining and refining operations; natural occurrence in soil
Nickel isotopes Nickel-59 and Nickel-63	50	0	pCi/L	U.S. EPA		Decay of natural and man-made deposits
Beryllium	4	4	ppb	U.S. EPA	Drinking water	Copper rolling, drawing, nonferrous metal smelting, rolling and drawing, aluminum foundries, blast furnaces, steelworks and petroleum industries

Source: Shammas, N.K. and Wang, L.K. Water quality characteristics and drinking water standards. In: *Water Engineering: Hydraulics, Distribution and Treatment.* John Wiley & Sons, Inc., Hoboken, NJ, pp. 297–324, 2016; Pasco County. *Annual Water Quality Report for Pasco County Utilities.* Southeast No. 1 Service Area PWS ID No. 6512685. Pasco County, FL, Available at www.dep.state.fl.us/swapp 2007; Dover City, *Drinking Water Quality Report*, City of Dover Department of Public Utilities, Dover City, DE, 2005; Houston. *Drinking Water Quality Report*, Harris County Municipal Utility District, Houston, TX, 2007; US EPA. *Consumer Fact Sheet on Selenium*, US Environmental Protection Agency, Washington, DC, 2015; USEPA. *Technical Fact Sheet on Nickel*, US Environmental Protection Agency, Washington, DC, 2015.

Note: ppb: parts per billion or micrograms per liter (μg/L). pCi/L: picocurie per liter.

[a] The highest level of a contaminant allowed in drinking water. MCLs are set as close to the MCLGs as feasible using the best available treatment technology.

[b] The level of a contaminant in drinking water below which there is no known or expected risk to health. MCLGs allow for a margin of safety.

REFERENCES

1. Shammas, N.K. and Wang, L.K. Water quality characteristics and drinking water standards. In: *Water Engineering: Hydraulics, Distribution and Treatment.* John Wiley & Sons, Inc., Hoboken, NJ, pp. 297–324, 2016.
2. Tinggi, U. Essentiality and toxicity of selenium and its status in Australia: A review. *Toxicology Letters,* 137, pp. 103–110, 2003.
3. DOI. *Selenium Fact Sheet.* US Department of Interior, Bureau of Reclamation, Washington, DC, pp. 1–6, 2001.
4. ANL. *Selenium, Human Health Fact Sheet.* Argonne National Laboratory, EVS, Argonne, IL, August 2005.
5. DOI. *Nickel Fact Sheet.* US Department of Interior, Bureau of Reclamation, Washington, DC, pp. 1–6, 2001.
6. ANL. *Nickel, Human Health Fact Sheet.* Argonne National Laboratory, EVS, Argonne, IL, August 2005.
7. DOI. *Beryllium Fact Sheet.* US Department of Interior, Bureau of Reclamation, Washington, DC, pp. 1–6, 2001.
8. IPCS. *Environmental Health Criteria 106. Beryllium.* World Health Organization, International Programme on Chemical Safety, Geneva, 1990.
9. ANL. *Beryllium, Human Health Fact Sheet.* Argonne National Laboratory, EVS, Argonne, IL, August, 2005.
10. Letavayová, L., Vlasáková, D., Spallholz E.J., Brozmanová, J., and Chovanec, M. Toxicity and mutagenicity of selenium compounds in *Saccharomyces cerevisiae. Mutation Research,* 638, pp. 1–10, 2008.
11. Weiller, M., Latta, M., Kresse, M., Lucas, R., and Wendel, A. Toxicity of nutritionally available selenium compounds in primary and transformed hepatocytes. *Toxicology,* 201, pp. 21–30, 2004.
12. Spallholz, E.J. and Hoffman, J.D. Selenium toxicity: Cause and effect in aquatic birds. *Aquatic Toxicology,* 57, pp. 27–37, 2002.
13. Barschak, G.A., Sitta, A., Deon, M., Barden, T.A., Schmitt, O.G., Dutra-Filho, S.C., Wajner, M., and Vargas, R.C. Erythrocyte glutathione peroxidase activity and plasma concentration are reduced in maple syrup urine disease patients during treatment. *International Journal of Developmental Neuroscience,* 25, pp. 335–338, 2007.
14. Pappas, A.C., Zoidis, E., Surai, P.F., and Zervas, G. Selenoproteins and maternal nutrition. *Comparative Biochemistry and Physiology, Part B: Biochemistry and Molecular Biology,* 151(4), pp. 361–372, 2008.
15. Rostkowski, K., Rauch, J., Drakonakis, K., Reck, B., Gordon, R.B., and Graedel, T.E. "Bottom–up" study of in-use nickel stocks in New Haven, CT. *Resources, Conservation and Recycling,* 50, pp. 58–70, 2007.
16. Yeung, K.W.K., Poon, R.W.Y., Liu, X.M., Chu, P.K., Chung, C.Y., Liu, X.Y., Chan, S. et al. Nitrogen plasma-implanted nickel titanium alloys for orthopedic use. *Surface and Coatings Technology,* 201, pp. 5607–5612, 2007.
17. Johansson, M., Mattisson, T., Lyngfelt, A., and Abad, A. Using continuous and pulse experiments to compare two promising nickel-based oxygen carriers for use in chemical-looping technologies. *Fuel,* 87, pp. 988–1001, 2008.
18. Weil, K.S., Kim, J.Y., Xia, G., Coleman, J., and Yang, Z.G. Boronization of nickel and nickel clad materials for potential use in polymer electrolyte membrane fuel cells. *Surface and Coatings Technology,* 201, pp. 4436–4441, 2006.
19. Williams, W.J. Beryllium disease. In: Parkes W. R. (Ed.), *Occupational Lung Disorders,* 3rd ed. Butterworth-Heinemann Ltd., Oxford, pp. 571–592, 1994.
20. Bradberry, S.M., Beer, S.T., and Vale, J.A. *Ukpid Monograph; Beryllium,* RTECS-CC4025000, 1996.
21. Lemly, A.D. Aquatic selenium pollution is a global environmental safety issue. *Ecotoxicology and Environmental Safety,* 59, pp. 44–56, 2004.
22. US EPA. *Technical Fact Sheet on Beryllium.* US Environmental Protection Agency, Washington, DC, 2015.
23. Taylor, T.P., Ding, M., Ehler, D.S., Foreman, T.M., Kaszuba, J.P., and Sauer. N.N. Beryllium in the environment: A review. *Journal of Environmental Science and Health—Part A, Toxic/Hazardous Substances and Environmental Engineering,* 38(2), pp. 439–469, 2003.
24. Kolanz, M.E., Madl, A.K., Kelsh, M.A., Kent, M.S., Kalmes, R.M., and Paustenbach, D.J.A. Comparison and critique of historical and current exposure assessment methods for beryllium: Implications for evaluating risk of chronic beryllium disease. *Applied Occupational and Environmental Hygiene* 16(5), pp. 593–614, 2001.

25. Lemly, A.D. Symptoms and implications of selenium toxicity in fish: The Belews Lake case example. *Aquatic Toxicology*, 57, pp. 39–49, 2002.

26. Lemly, A.D. Teratogenic effects of selenium in natural populations of freshwater fish. *Ecotoxicology and Environmental Safety*, 26, pp. 181–204, 1993.

27. Sorensen, E.M.B. The effects of selenium on freshwater teleosts. In: Hodgson, E. (Ed.), *Reviews in Environmental Toxicology* 2. Elsevier, New York, pp. 59–116, 1986.

28. Sorensen, E.M.B., Cumbie, P.M., Bauer, T.L., Bell, J.S., and Harlan, C.W. Histopathological, hematological, condition-factor, and organ weight changes associated with selenium accumulation in fish from Belews Lake, North Carolina. *Archives of Environmental Contamination and Toxicology*, 13, pp. 153–162, 1984.

29. Lemly, A.D. Metabolic stress during winter increases the toxicity of selenium to fish. *Aquatic Toxicology*, 27, pp. 133–158, 1993.

30. Lemly, A.D. Toxicology of selenium in a freshwater reservoir: Implications for environmental hazard evaluation and safety. *Ecotoxicology and Environmental Safety*, 10, pp. 314–338, 1985.

31. Lemly, A.D. Ecosystem recovery following selenium contamination in a freshwater reservoir. *Ecotoxicology and Environmental Safety*, 36, pp. 275–281, 1997.

32. Shearer, T.R., David, L.L., Anderson, R.S. Selenite cataract: A review. *Current Eye Research*, 6, pp. 289–300, 1987.

33. Perry, T.S., Kulkarni, B.S., Lee, K., and Kenney, T.F. Selective effect of the metallocarcinogen beryllium on hormonal regulation of gene expression in cultured cells. *Cancer Research*, 42, pp. 473–476, 1982.

34. ATSDR. Case *Studies in Environmental Medicine, Beryllium Toxicity*. Agency for Toxic Substances and Disease Registry, Atlanta, GA, 2008.

35. Saltini, C., Richeldi, L., Losi, M., Amicosante, M., Voorter, C., van den Berg-Loonen, E., Dweik, R.A., Wiedemann, H.P., Deubner, D.C., and Tinelli, C. Major histocompatibility locus genetic markers of beryllium sensitization and disease. *European Respiratory Journal*, 18, pp. 677–684, 2001.

36. Henneberger, P.K., Cumro, D., Deubner, D.D., Kent, M.S., McCawley, M., and Kreiss, K. Beryllium sensitization and disease among long-term and short-term workers in a beryllium ceramics plant. *International Archives of Occupational and Environmental Health*, 74 (3), pp. 167–76, 2001.

37. Newman, L.S., Maier, L.A., Martyny, J.W., Mroz, M.M., VanDyke, M.V., and Sackett, H.S. Letter to the editor: Beryllium workers' health risks. *Journal of Occupational and Environmental Hygiene*, 2 (6), pp. D48–D50, 2005.

38. Newman, L.S., Mroz, M.M., Balkissoon, R., and Maier, L.A. Beryllium sensitization progresses to chronic beryllium disease: A longitudinal study of disease risk. *American Journal of Respiratory and Critical Care Medicine*, 171, pp. 54–60, 2005.

39. Berlin, J.M., Taylor, J.S., Sigel, J.E., Bergfeld, W.F., and Dweik, R.A. Beryllium dermatitis. *Journal of American Academy of Dermatology*, 49 (5), pp. 939–41, 2003.

40. Grimaudo, N.J. Biocompatibility of nickel and cobalt dental alloys. *General Dentistry*, 49 (5), pp. 498–503, 2001.

41. Tinkle, S.S., Kittle, L.A., and Newman, L.S. Partial IL-10 inhibition of the cell-mediated immune response in chronic beryllium disease. *Journal of Immunology*, 163 (5), pp. 2747–2753, 1999.

42. Saltini, C. and Amicosante, M. Beryllium disease. *American Journal of the Medical Sciences,* 321 (1), pp. 89–98, 2001.

43. US EPA. *Toxicological Review of Beryllium Compounds (CAS No. 7440-41-7),* US Environmental Protection Agency, Washington, DC, April 1998.

44. Sanderson, W.T., Ward, E.M., Steenland, K., and Petersen, M.R. National Institute for Occupational Safety and Health. Lung cancer case–control study of beryllium workers. *American Journal of Industrial Medicine*, 39 (2), pp. 133–44, 2001.

45. ATSDR. *Toxicological Profile for Beryllium*. US Department of Health and Human Services, Agency for Toxic Substances and Disease Registry, Atlanta, GA, 2002.

46. Steenland, K. and Ward, E. Lung cancer incidence among patients with beryllium disease: A cohort mortality study. *Journal of the National Cancer Institute*, 83 (19), pp. 1380–1385, 1991.

47. Wang, L.K., Hung, Y.T., and Shammas, N.K. (Eds.). *Physicochemical Treatment Processes,* Humana Press, Totowa, NJ, pp. 47–228, 2005.

48. Shammas, N.K. and Wang, L.K. Coagulation. In: *Water Engineering: Hydraulics, Distribution and Treatment*. John Wiley & Sons, Inc., Hoboken, NJ, pp. 417–438, 2016.

49. Shammas, N.K. and Wang, L.K. Alternative and membrane filtration technologies. In: *Water Engineering: Hydraulics, Distribution and Treatment*. John Wiley & Sons, Inc., Hoboken, NJ, pp. 513–544, 2016.

50. Shaposhnik, V.A. and Kesore, K. An early history of electrodialysis with permselective membranes. *Journal of Membrane Science*, 136, p. 35, 1997.

51. Valerdi-Pérez, R., López-Rodríguez, M., and Ibáñez-Mengual, J.A. Characterizing an electrodialysis reversal pilot plant. *Desalination*, 137, pp. 199–206, 2001.

52. Allison, P.R. *Surface and Wastewater Desalination by Electrodialysis Reversal*. Technical Paper, Water and Process Technologies, March 2008.

53. Chen, J.P., Wang, L.K., Yang, L., and Lim, S.F. Emerging biosorption, adsorption, ion exchange and membrane technologies. In: Wang, L.K., Hung, Y.T., and Shammas, N.K. (Eds.), *Advanced Physicochemical Treatment Technologies*, Humana Press, Totowa, NJ, pp. 367–390, 2007.

54. Igwe, J.C. and Abia, A.A. A bioseparation process for removing heavy metals from wastewater using biosorbents. *African Journal of Biotechnology*, 5 (12), pp. 1167–1179, 2006.

55. Hanif, A.M., Nadeema, R., Zafar, N.M., Akhtar, K., and Haq Nawaz Bhatti, N.H. Kinetic studies for Ni(II) biosorption from industrial wastewater by *Cassia fistula* (golden shower) biomass. *Journal of Hazardous Materials*, 145, pp. 501–505, 2007.

56. Crist, R.H., Oberholser, K., Shank, N., and Nguyen, M. Nature of bonding between metallic ions and algal cell walls. *Environmental Science and Technology*, 15, pp. 1212–1217, 1981.

57. Igwe, C.J. and Abia, A.A. Equilibrium sorption isotherm studies of Cd(II), Pb(II) and Zn(II) ions detoxification from wastewater using unmodified and EDTA-modified maize husk. *Electronic Journal of Biotechnology*, 10 (4), pp. 536–548, 2007.

58. US EPA. *Drinking Water Standards and Health Advisories Table*. US Environmental Protection Agency, Washington, DC, December 2006.

59. Pasco County. *Annual Water Quality Report for Pasco County Utilities*. Southeast No. 1 Service Area PWS ID No. 6512685. Pasco County, FL. Available at www.dep.state.fl.us/swapp, 2007.

60. Dover City. *Drinking Water Quality Report*. City of Dover Department of Public Utilities, Dover City, DE, 2005.

61. Houston. *Drinking Water Quality Report*. Harris County Municipal Utility District, Houston, TX, 2007.

62. US EPA. *Consumer Fact Sheet on Selenium*. US Environmental Protection Agency, Washington, DC, 2015.

63. US EPA. *Technical Fact Sheet on Nickel*. US Environmental Protection Agency, Washington, DC, 2015.

Index